STUDENT'S SOLUTIONS MANUAL

ELKA BLOCK
FRANK PURCELL

CALCULUS WITH APPLICATIONS

and

CALCULUS WITH APPLICATIONS, BRIEF VERSION

ELEVENTH EDITIONS

Margaret L. Lial

American River College

Raymond N. Greenwell

Hofstra University

Nathan P. Ritchey

Edinboro University

Boston Columbus Indianapolis New York San Francisco
Amsterdam Cape Town Dubai London Madrid Milan Munich Paris Montreal Toronto
Delhi Mexico City Sao Paulo Sydney Hong Kong Seoul Singapore Taipei Tokyo

www.pearsonhighered.com

PEARSON

CONTENTS

CHAPTER 3   THE DERIVATIVE

CHAPTER 4   CALCULATING THE DERIVATIVE

CHAPTER 5   GRAPHS AND THE DERIVATIVE

CHAPTER 6   APPLICATIONS OF THE DERIVATIVE

CHAPTER 7 INTEGRATION

CHAPTER 8 FURTHER TECHNIQUES AND APPLICATIONS OF INTEGRATION

CHAPTER 9 MULTIVARIABLE CALCULUS

CHAPTER 10 DIFFERENTIAL EQUATIONS

ALGEBRA REFERENCE

R.1 Polynomials

Your Turn 1

$3(x^2 - 4x - 5) - 4(3x^2 - 5x - 7)$

$= 3x^2 - 12x - 15 - 12x^2 + 20x + 28$

$= -9x^2 + 8x + 13$

Your Turn 2

$(3y + 2)(4y^2 - 2y - 5)$

$= (3y)(4y^2 - 2y - 5) + (2)(4y^2 - 2y - 5)$

$= 12y^3 - 6y^2 - 15y + 8y^2 - 4y - 10$

$= 12y^3 + 2y^2 - 19y - 10$

Your Turn 3

$(2x + 7)(3x - 1)$

$= (2x)(3x) + (2x)(-1) + (7)(3x) + (7)(-1)$

$= 6x^2 - 2x + 21x - 7$

$= 6x^2 + 19x - 7$

Your Turn 4

$(3x + 2y)^3$

$= (3x + 2y)(3x + 2y)(3x + 2y)$

$= (9x^2 + 6xy + 6xy + 4y^2)(3x + 2y)$

$= (9x^2 + 12xy + 4y^2)(3x + 2y)$

$= 9x^2(3x + 2y) + 12xy(3x + 2y) + 4y^2(3x + 2y)$

$= 27x^3 + 18x^2y + 36x^2y + 24xy^2 + 12xy^2 + 8y^3$

$= 27x^3 + 54x^2y + 36xy^2 + 8y^3$

R.1 Exercises

1. $(2x^2 - 6x + 11) + (-3x^2 + 7x - 2)$

$= 2x^2 - 6x + 11 - 3x^2 + 7x - 2$

$= (2 - 3)x^2 + (7 - 6)x + (11 - 2)$

$= -x^2 + x + 9$

3. $-6(2q^2 + 4q - 3) + 4(-q^2 + 7q - 3)$

$= (-12q^2 - 24q + 18)$

$\qquad + (-4q^2 + 28q - 12)$

$= (-12q^2 - 4q^2)$

$\qquad + (-24q + 28q) + (18 - 12)$

$= -16q^2 + 4q + 6$

5. $(0.613x^2 - 4.215x + 0.892) - 0.47(2x^2 - 3x + 5)$

$= 0.613x^2 - 4.215x + 0.892 - 0.94x^2$

$\qquad\qquad\qquad\qquad + 1.41x - 2.35$

$= -0.327x^2 - 2.805x - 1.458$

7. $-9m(2m^2 + 3m - 1)$

$= -9m(2m^2) - 9m(3m) - 9m(-1)$

$= -18m^3 - 27m^2 + 9m$

9. $(3t - 2y)(3t + 5y)$

$= (3t)(3t) + (3t)(5y) + (-2y)(3t) + (-2y)(5y)$

$= 9t^2 + 15ty - 6ty - 10y^2$

$= 9t^2 + 9ty - 10y^2$

11. $(2 - 3x)(2 + 3x)$

$= (2)(2) + (2)(3x) + (-3x)(2) + (-3x)(3x)$

$= 4 + 6x - 6x - 9x^2$

$= 4 - 9x^2$

13. $\left(\dfrac{2}{5}y + \dfrac{1}{8}z \right)\left(\dfrac{3}{5}y + \dfrac{1}{2}z \right)$

$= \left(\dfrac{2}{5}y \right)\left(\dfrac{3}{5}y \right) + \left(\dfrac{2}{5}y \right)\left(\dfrac{1}{2}z \right) + \left(\dfrac{1}{8}z \right)\left(\dfrac{3}{5}y \right)$

$\qquad\qquad\qquad\qquad + \left(\dfrac{1}{8}z \right)\left(\dfrac{1}{2}z \right)$

$= \dfrac{6}{25}y^2 + \dfrac{1}{5}yz + \dfrac{3}{40}yz + \dfrac{1}{16}z^2$

$= \dfrac{6}{25}y^2 + \left(\dfrac{8}{40} + \dfrac{3}{40} \right)yz + \dfrac{1}{16}z^2$

$= \dfrac{6}{25}y^2 + \dfrac{11}{40}yz + \dfrac{1}{16}z^2$

15. $(3p - 1)(9p^2 + 3p + 1)$

$= (3p - 1)(9p^2) + (3p - 1)(3p) + (3p - 1)(1)$

$= 3p(9p^2) - 1(9p^2) + 3p(3p)$
$\qquad\qquad\qquad - 1(3p) + 3p(1) - 1(1)$

$= 27p^3 - 9p^2 + 9p^2 - 3p + 3p - 1$

$= 27p^3 - 1$

17. $(2m + 1)(4m^2 - 2m + 1)$

$= 2m(4m^2 - 2m + 1) + 1(4m^2 - 2m + 1)$

$= 8m^3 - 4m^2 + 2m + 4m^2 - 2m + 1$

$= 8m^3 + 1$

19. $(x + y + z)(3x - 2y - z)$

$= x(3x) + x(-2y) + x(-z) + y(3x) + y(-2y)$
$\qquad\qquad + y(-z) + z(3x) + z(-2y) + z(-z)$

$= 3x^2 - 2xy - xz + 3xy - 2y^2 - yz + 3xz$
$\qquad\qquad\qquad\qquad\qquad - 2yz - z^2$

$= 3x^2 + xy + 2xz - 2y^2 - 3yz - z^2$

21. $(x + 1)(x + 2)(x + 3)$

$= [x(x + 2) + 1(x + 2)](x + 3)$

$= [x^2 + 2x + x + 2](x + 3)$

$= [x^2 + 3x + 2](x + 3)$

$= x^2(x + 3) + 3x(x + 3) + 2(x + 3)$

$= x^3 + 3x^2 + 3x^2 + 9x + 2x + 6$

$= x^3 + 6x^2 + 11x + 6$

23. $(x + 2)^2 = (x + 2)(x + 2)$

$= x(x + 2) + 2(x + 2)$

$= x^2 + 2x + 2x + 4$

$= x^2 + 4x + 4$

25. $(x - 2y)^3$

$= [(x - 2y)(x - 2y)](x - 2y)$

$= (x^2 - 2xy - 2xy + 4y^2)(x - 2y)$

$= (x^2 - 4xy + 4y^2)(x - 2y)$

$= (x^2 - 4xy + 4y^2)x$
$\qquad\qquad + (x^2 - 4xy + 4y^2)(-2y)$

$= x^3 - 4x^2y + 4xy^2 - 2x^2y + 8xy^2 - 8y^3$

$= x^3 - 6x^2y + 12xy^2 - 8y^3$

R.2 Factoring

Your Turn 1

Factor $4z^4 + 4z^3 + 18z^2$.

$4z^4 + 4z^3 + 18z^2$

$= (2z^2) \cdot 2z^2 + (2z^2) \cdot 2z + (2z^2) \cdot 9$

$= (2z^2)(2z^2 + 2z + 9)$

Your Turn 2

$x^2 - 3x - 10 = (x + 2)(x - 5)$

since $(2)(-5) = -10$ and $-5 + 2 = -3$.

Your Turn 3

Factor $6a^2 + 5ab - 4b^2$.

$6a^2 + 5ab - 4b^2 = (2a - b)(3a + 4b)$

R.2　Exercises

1. $7a^3 + 14a^2 = 7a^2 \cdot a + 7a^2 \cdot 2$

$\qquad\qquad = 7a^2(a + 2)$

3. $13p^4q^2 - 39p^3q + 26p^2q^2$

$= 13p^2q \cdot p^2q - 13p^2q \cdot 3p + 13p^2q \cdot 2q$

$= 13p^2q(p^2q - 3p + 2q)$

5. $m^2 - 5m - 14 = (m - 7)(m + 2)$

since $(-7)(2) = -14$ and $-7 + 2 = -5$.

7. $z^2 + 9z + 20 = (z + 4)(z + 5)$

since $4 \cdot 5 = 20$ and $4 + 5 = 9$.

9. $a^2 - 6ab + 5b^2 = (a - b)(a - 5b)$

since $(-b)(-5b) = 5b^2$ and $-b + (-5b) = -6b$.

11. $y^2 - 4yz - 21z^2 = (y + 3z)(y - 7z)$

since $(3z)(-7z) = -21z^2$ and $3z + (-7z) = -4z$.

13. $3a^2 + 10a + 7$

The possible factors of $3a^2$ are $3a$ and a and the possible factors of 7 are 7 and 1. Try various combinations until one works.

$3a^2 + 10a + 7 = (a + 1)(3a + 7)$

15. $21m^2 + 13mn + 2n^2 = (7m + 2n)(3m + n)$

17. $3m^3 + 12m^2 + 9m = 3m(m^2 + 4m + 3)$

$\qquad\qquad\qquad\qquad = 3m(m + 1)(m + 3)$

19. $24a^4 + 10a^3b - 4a^2b^2$

$\quad = 2a^2(12a^2 + 5ab - 2b^2)$

$\quad = 2a^2(4a - b)(3a + 2b)$

21. $x^2 - 64 = x^2 - 8^2$

$\qquad\quad = (x + 8)(x - 8)$

23. $10x^2 - 160 = 10(x^2 - 16)$

$\qquad\qquad\quad = 10(x^2 - 4^2)$

$\qquad\qquad\quad = 10(x + 4)(x - 4)$

25. $z^2 + 14zy + 49y^2 = z^2 + 2 \cdot 7zy + 7^2y^2$

$\qquad\qquad\qquad\quad = (z + 7y)^2$

27. $9p^2 - 24p + 16 = (3p)^2 - 2 \cdot 3p \cdot 4 + 4^2$

$\qquad\qquad\qquad\quad = (3p - 4)^2$

29. $27r^3 - 64s^3 = (3r)^3 - (4s)^3$

$\qquad\qquad\qquad = (3r - 4s)(9r^2 + 12rs + 16s^2)$

31. $x^4 - y^4 = (x^2)^2 - (y^2)^2$

$\qquad\quad = (x^2 + y^2)(x^2 - y^2)$

$\qquad\quad = (x^2 + y^2)(x + y)(x - y)$

R.3 Rational Expressions

Your Turn 1

Write in lowest terms $\dfrac{z^2 + 5z + 6}{2z^2 + 7z + 3}$.

$\dfrac{z^2 + 5z + 6}{2z^2 + 7z + 3} = \dfrac{(z + 3)(z + 2)}{(z + 3)(2z + 1)}$

$\qquad\qquad\qquad = \dfrac{z + 2}{2z + 1}$

Your Turn 2

Perform each of the following operations.

(a) $\dfrac{z^2 + 5z + 6}{2z^2 - 5z - 3} \cdot \dfrac{2z^2 - z - 1}{z^2 + 2z - 3}$

$\quad = \dfrac{(z + 2)(z + 3)}{(2z + 1)(z - 3)} \cdot \dfrac{(2z + 1)(z - 1)}{(z + 3)(z - 1)}$

$\quad = \dfrac{(z + 2)\cancel{(z + 3)}\;\cancel{(2z + 1)}\;\cancel{(z - 1)}}{\cancel{(2z + 1)}(z - 3)\cancel{(z + 3)}\cancel{(z - 1)}}$

$\quad = \dfrac{z + 2}{z - 3}$

(b) $\dfrac{a - 3}{a^2 + 3a + 2} + \dfrac{5a}{a^2 - 4}$

$\quad = \dfrac{a - 3}{(a + 2)(a + 1)} + \dfrac{5a}{(a - 2)(a + 2)}$

$\quad = \dfrac{a - 3}{(a + 2)(a + 1)} \cdot \dfrac{(a - 2)}{(a - 2)}$

$\qquad\quad + \dfrac{5a}{(a - 2)(a + 2)} \cdot \dfrac{(a + 1)}{(a + 1)}$

$\quad = \dfrac{(a^2 - 5a + 6) + (5a^2 + 5a)}{(a - 2)(a + 2)(a + 1)}$

$\quad = \dfrac{6a^2 + 6}{(a - 2)(a + 2)(a + 1)}$

$\quad = \dfrac{6(a^2 + 1)}{(a - 2)(a + 2)(a + 1)}$

R.3 Exercises

1. $\dfrac{5v^2}{35v} = \dfrac{5 \cdot v \cdot v}{5 \cdot 7 \cdot v} = \dfrac{v}{7}$

3. $\dfrac{8k + 16}{9k + 18} = \dfrac{8(k + 2)}{9(k + 2)} = \dfrac{8}{9}$

5. $\dfrac{4x^3 - 8x^2}{4x^2} = \dfrac{4x^2(x - 2)}{4x^2} = x - 2$

7. $\dfrac{m^2 - 4m + 4}{m^2 + m - 6} = \dfrac{(m - 2)(m - 2)}{(m - 2)(m + 3)}$

$\qquad\qquad\qquad = \dfrac{m - 2}{m + 3}$

9. $\dfrac{3x^2 + 3x - 6}{x^2 - 4} = \dfrac{3(x + 2)(x - 1)}{(x + 2)(x - 2)} = \dfrac{3(x - 1)}{x - 2}$

11. $\dfrac{m^4 - 16}{4m^2 - 16} = \dfrac{(m^2 + 4)(m + 2)(m - 2)}{4(m + 2)(m - 2)}$

$\qquad\qquad = \dfrac{m^2 + 4}{4}$

13. $\dfrac{9k^2}{25} \cdot \dfrac{5}{3k} = \dfrac{3 \cdot 3 \cdot 5k^2}{5 \cdot 5 \cdot 3k} = \dfrac{3k^2}{5k} = \dfrac{3k}{5}$

15. $\dfrac{3a + 3b}{4c} \cdot \dfrac{12}{5(a + b)} = \dfrac{3(a + b)}{4c} \cdot \dfrac{3 \cdot 4}{5(a + b)}$

$\qquad\qquad\qquad = \dfrac{3 \cdot 3}{c \cdot 5}$

$\qquad\qquad\qquad = \dfrac{9}{5c}$

17. $\dfrac{2k - 16}{6} \div \dfrac{4k - 32}{3} = \dfrac{2k - 16}{6} \cdot \dfrac{3}{4k - 32}$

$\qquad\qquad = \dfrac{2(k - 8)}{6} \cdot \dfrac{3}{4(k - 8)}$

$\qquad\qquad = \dfrac{1}{4}$

19. $\dfrac{4a + 12}{2a - 10} \div \dfrac{a^2 - 9}{a^2 - a - 20}$

$\qquad = \dfrac{4(a + 3)}{2(a - 5)} \cdot \dfrac{(a - 5)(a + 4)}{(a - 3)(a + 3)}$

$\qquad = \dfrac{2(a + 4)}{a - 3}$

21. $\dfrac{k^2 + 4k - 12}{k^2 + 10k + 24} \cdot \dfrac{k^2 + k - 12}{k^2 - 9}$

$\qquad = \dfrac{(k + 6)(k - 2)}{(k + 6)(k + 4)} \cdot \dfrac{(k + 4)(k - 3)}{(k + 3)(k - 3)}$

$\qquad = \dfrac{k - 2}{k + 3}$

23. $\dfrac{2m^2 - 5m - 12}{m^2 - 10m + 24} \div \dfrac{4m^2 - 9}{m^2 - 9m + 18}$

$\qquad = \dfrac{2m^2 - 5m - 12}{m^2 - 10m + 24} \cdot \dfrac{m^2 - 9m + 18}{4m^2 - 9}$

$\qquad = \dfrac{(2m + 3)(m - 4)(m - 6)(m - 3)}{(m - 6)(m - 4)(2m - 3)(2m + 3)}$

$\qquad = \dfrac{m - 3}{2m - 3}$

25. $\dfrac{a + 1}{2} - \dfrac{a - 1}{2} = \dfrac{(a + 1) - (a - 1)}{2}$

$\qquad\qquad = \dfrac{a + 1 - a + 1}{2}$

$\qquad\qquad = \dfrac{2}{2} = 1$

27. $\dfrac{6}{5y} - \dfrac{3}{2} = \dfrac{6 \cdot 2}{5y \cdot 2} - \dfrac{3 \cdot 5y}{2 \cdot 5y} = \dfrac{12 - 15y}{10y}$

29. $\dfrac{1}{m - 1} + \dfrac{2}{m} = \dfrac{m}{m}\left(\dfrac{1}{m - 1}\right) + \dfrac{m - 1}{m - 1}\left(\dfrac{2}{m}\right)$

$\qquad\qquad = \dfrac{m + 2m - 2}{m(m - 1)}$

$\qquad\qquad = \dfrac{3m - 2}{m(m - 1)}$

31. $\dfrac{8}{3(a - 1)} + \dfrac{2}{a - 1} = \dfrac{8}{3(a - 1)} + \dfrac{3}{3}\left(\dfrac{2}{a - 1}\right)$

$\qquad\qquad = \dfrac{8 + 6}{3(a - 1)}$

$\qquad\qquad = \dfrac{14}{3(a - 1)}$

33. $\dfrac{4}{x^2 + 4x + 3} + \dfrac{3}{x^2 - x - 2}$

$\qquad = \dfrac{4}{(x + 3)(x + 1)} + \dfrac{3}{(x - 2)(x + 1)}$

$\qquad = \dfrac{4(x - 2)}{(x - 2)(x + 3)(x + 1)}$

$\qquad\qquad + \dfrac{3(x + 3)}{(x - 2)(x + 3)(x + 1)}$

$\qquad = \dfrac{4(x - 2) + 3(x + 3)}{(x - 2)(x + 3)(x + 1)}$

$\qquad = \dfrac{4x - 8 + 3x + 9}{(x - 2)(x + 3)(x + 1)}$

$\qquad = \dfrac{7x + 1}{(x - 2)(x + 3)(x + 1)}$

35. $\dfrac{3k}{2k^2 + 3k - 2} - \dfrac{2k}{2k^2 - 7k + 3}$

$\qquad = \dfrac{3k}{(2k - 1)(k + 2)} - \dfrac{2k}{(2k - 1)(k - 3)}$

$\qquad = \left(\dfrac{k - 3}{k - 3}\right)\dfrac{3k}{(2k - 1)(k + 2)}$

$\qquad\qquad - \left(\dfrac{k + 2}{k + 2}\right)\dfrac{2k}{(2k - 1)(k - 3)}$

$\qquad = \dfrac{(3k^2 - 9k) - (2k^2 + 4k)}{(2k - 1)(k + 2)(k - 3)}$

$\qquad = \dfrac{k^2 - 13k}{(2k - 1)(k + 2)(k - 3)}$

$\qquad = \dfrac{k(k - 13)}{(2k - 1)(k + 2)(k - 3)}$

37. $\dfrac{2}{a + 2} + \dfrac{1}{a} + \dfrac{a - 1}{a^2 + 2a}$

$\qquad = \dfrac{2}{a + 2} + \dfrac{1}{a} + \dfrac{a - 1}{a(a + 2)}$

$\qquad = \left(\dfrac{a}{a}\right)\dfrac{2}{a + 2} + \left(\dfrac{a + 2}{a + 2}\right)\dfrac{1}{a} + \dfrac{a - 1}{a(a + 2)}$

$\qquad = \dfrac{2a + a + 2 + a - 1}{a(a + 2)}$

$\qquad = \dfrac{4a + 1}{a(a + 2)}$

R.4 Equations

Your Turn 1

Solve $3x - 7 = 4(5x + 2) - 7x$.

$3x - 7 = 20x + 8 - 7x$

$3x - 7 = 13x + 8$

$-10x = 15$

$x = -\dfrac{15}{10}$

$x = -\dfrac{3}{2}$

Your Turn 2

Solve $2m^2 + 7m = 15$.

$2m^2 + 7m - 15 = 0$

$(2m - 3)(m + 5) = 0$

$2m - 3 = 0 \quad \text{or} \quad m + 5 = 0$

$m = \dfrac{3}{2} \quad \text{or} \quad m = -5$

Your Turn 3

Solve $z^2 + 6 = 8z$.

$z^2 - 8z + 6 = 0$

Use the quadratic formula with
$a = 1$, $b = -8$, and $c = 6$.

$z = \dfrac{-(-8) \pm \sqrt{(-8)^2 - 4(1)(6)}}{2(1)}$

$= \dfrac{8 \pm \sqrt{64 - 24}}{2}$

$= \dfrac{8 \pm \sqrt{40}}{2}$

$= \dfrac{8 \pm \sqrt{4 \cdot 10}}{2}$

$= \dfrac{8 \pm 2\sqrt{10}}{2}$

$= 4 \pm \sqrt{10}$

Your Turn 4

Solve $\dfrac{1}{x^2 - 4} + \dfrac{2}{x - 2} = \dfrac{1}{x}$.

$\dfrac{1}{(x - 2)(x + 2)} + \dfrac{2}{x - 2} = \dfrac{1}{x}$

$(x - 2)(x + 2)(x) \cdot \dfrac{1}{(x - 2)(x + 2)}$

$+ (x - 2)(x + 2)(x) \cdot \dfrac{2}{x - 2}$

$= (x - 2)(x + 2)(x) \cdot \dfrac{1}{x}$

$x + 2x^2 + 4x = x^2 - 4$

$x^2 + 5x + 4 = 0$

$(x + 1)(x + 4) = 0$

$x = -1 \quad \text{or} \quad x = -4$

Neither of these values makes a denominator equal to zero, so both are solutions.

R.4 Exercises

1. $2x + 8 = x - 4$

    $x + 8 = -4$

    $x = -12$

    The solution is -12.

    The solution is $\frac{3}{4}$.

3. $0.2m - 0.5 = 0.1m + 0.7$

    $10(0.2m - 0.5) = 10(0.1m + 0.7)$

    $2m - 5 = m + 7$

    $m - 5 = 7$

    $m = 12$

    The solution is 12.

5. $3r + 2 - 5(r + 1) = 6r + 4$

    $3r + 2 - 5r - 5 = 6r + 4$

    $-3 - 2r = 6r + 4$

    $-3 = 8r + 4$

    $-7 = 8r$

    $-\dfrac{7}{8} = r$

    The solution is $-\frac{7}{8}$.

7. $2[3m - 2(3 - m) - 4] = 6m - 4$

    $2[3m - 6 + 2m - 4] = 6m - 4$

    $2[5m - 10] = 6m - 4$

    $10m - 20 = 6m - 4$

    $4m - 20 = -4$

    $4m = 16$

    $m = 4$

    The solution is 4.

9. $x^2 + 5x + 6 = 0$

$(x + 3)(x + 2) = 0$

$x + 3 = 0$ or $x + 2 = 0$

$x = -3$ or $x = -2$

The solutions are -3 and -2.

11. $m^2 = 14m - 49$

$m^2 - 14m + 49 = 0$

$(m)^2 - 2(7m) + (7)^2 = 0$

$(m - 7)^2 = 0$

$m - 7 = 0$

$m = 7$

The solution is 7.

13. $12x^2 - 5x = 2$

$12x^2 - 5x - 2 = 0$

$(4x + 1)(3x - 2) = 0$

$4x + 1 = 0$ or $3x - 2 = 0$

$4x = -1$ or $3x = 2$

$x = -\dfrac{1}{4}$ or $x = \dfrac{2}{3}$

The solutions are $-\frac{1}{4}$ and $\frac{2}{3}$.

15. $4x^2 - 36 = 0$

Divide both sides of the equation by 4.

$x^2 - 9 = 0$

$(x + 3)(x - 3) = 0$

$x + 3 = 0$ or $x - 3 = 0$

$x = -3$ or $x = 3$

The solutions are -3 and 3.

17. $12y^2 - 48y = 0$

$12y(y) - 12y(4) = 0$

$12y(y - 4) = 0$

$12y = 0$ or $y - 4 = 0$

$y = 0$ or $y = 4$

The solutions are 0 and 4.

19. $2m^2 - 4m = 3$

$2m^2 - 4m - 3 = 0$

$m = \dfrac{-(-4) \pm \sqrt{(-4)^2 - 4(2)(-3)}}{2(2)}$

$= \dfrac{4 \pm \sqrt{40}}{4} = \dfrac{4 \pm \sqrt{4 \cdot 10}}{4}$

$= \dfrac{4 \pm \sqrt{4}\sqrt{10}}{4}$

$= \dfrac{4 \pm 2\sqrt{10}}{4} = \dfrac{2 \pm \sqrt{10}}{2}$

The solutions are $\frac{2+\sqrt{10}}{2} \approx 2.5811$ and $\frac{2-\sqrt{10}}{2} \approx -0.5811$.

21. $k^2 - 10k = -20$

$k^2 - 10k + 20 = 0$

$k = \dfrac{-(-10) \pm \sqrt{(-10)^2 - 4(1)(20)}}{2(1)}$

$k = \dfrac{10 \pm \sqrt{100 - 80}}{2}$

$k = \dfrac{10 \pm \sqrt{20}}{2}$

$k = \dfrac{10 \pm \sqrt{4}\sqrt{5}}{2}$

$k = \dfrac{10 \pm 2\sqrt{5}}{2}$

$k = \dfrac{2(5 \pm \sqrt{5})}{2}$

$k = 5 \pm \sqrt{5}$

The solutions are $5 + \sqrt{5} \approx 7.2361$ and $5 - \sqrt{5} \approx 2.7639$.

23. $2r^2 - 7r + 5 = 0$

$(2r - 5)(r - 1) = 0$

$2r - 5 = 0$ or $r - 1 = 0$

$2r = 5$

$r = \dfrac{5}{2}$ or $r = 1$

The solutions are $\frac{5}{2}$ and 1.

25. $3k^2 + k = 6$

$3k^2 + k - 6 = 0$

$$k = \frac{-1 \pm \sqrt{1 - 4(3)(-6)}}{2(3)}$$

$$= \frac{-1 \pm \sqrt{73}}{6}$$

The solutions are $\frac{-1+\sqrt{73}}{6} \approx 1.2573$ and

$\frac{-1-\sqrt{73}}{6} \approx -1.5907$.

27. $\dfrac{3x - 2}{7} = \dfrac{x + 2}{5}$

$$35\left(\frac{3x - 2}{7}\right) = 35\left(\frac{x + 2}{2}\right)$$

$$5(3x - 2) = 7(x + 2)$$

$$15x - 10 = 7x + 14$$

$$8x = 24$$

$$x = 3$$

The solution is $x = 3$.

29. $\dfrac{4}{x - 3} - \dfrac{8}{2x + 5} + \dfrac{3}{x - 3} = 0$

$$\frac{4}{x - 3} + \frac{3}{x - 3} - \frac{8}{2x + 5} = 0$$

$$\frac{7}{x - 3} - \frac{8}{2x + 5} = 0$$

Multiply both sides by $(x - 3)(2x + 5)$. Note that $x \neq 3$ and $x \neq \frac{5}{2}$.

$$(x - 3)(2x + 5)\left(\frac{7}{x - 3} - \frac{8}{2x + 5}\right)$$
$$= (x - 3)(2x + 5)(0)$$

$$7(2x + 5) - 8(x - 3) = 0$$

$$14x + 35 - 8x + 24 = 0$$

$$6x + 59 = 0$$

$$6x = -59$$

$$x = -\frac{59}{6}$$

Note: It is especially important to check solutions of equations that involve rational expressions. Here, a check shows that $-\frac{59}{6}$ is a solution.

31. $\dfrac{2m}{m - 2} - \dfrac{6}{m} = \dfrac{12}{m^2 - 2m}$

$$\frac{2m}{m - 2} - \frac{6}{m} = \frac{12}{m(m - 2)}$$

Multiply both sides by $m(m - 2)$. Note that $m \neq 0$ and $m \neq 2$.

$$m(m - 2)\left(\frac{2m}{m - 2} - \frac{6}{m}\right)$$
$$= m(m - 2)\left(\frac{12}{m(m - 2)}\right)$$

$$m(2m) - 6(m - 2) = 12$$

$$2m^2 - 6m + 12 = 12$$

$$2m^2 - 6m = 0$$

$$2m(m - 3) = 0$$

$$2m = 0 \quad \text{or} \quad m - 3 = 0$$

$$m = 0 \quad \text{or} \quad m = 3$$

Since $m \neq 0$, 0 is not a solution. The solution is 3.

33. $\dfrac{1}{x - 2} - \dfrac{3x}{x - 1} = \dfrac{2x + 1}{x^2 - 3x + 2}$

$$\frac{1}{x - 2} - \frac{3x}{x - 1} = \frac{2x + 1}{(x - 2)(x - 1)}$$

Multiply both sides by $(x - 2)(x - 1)$.
Note that $x \neq 2$ and $x \neq 1$.

$$(x - 2)(x - 1)\left(\frac{1}{x - 2} - \frac{3x}{x - 1}\right)$$
$$= (x - 2)(x - 1) \cdot \left[\frac{2x + 1}{(x - 2)(x - 1)}\right]$$

$$(x - 2)(x - 1)\left(\frac{1}{x - 2}\right) - (x - 2)(x - 1) \cdot \left(\frac{3x}{x - 1}\right)$$
$$= \frac{(x - 2)(x - 1)(2x + 1)}{(x - 2)(x - 1)}$$

$$(x - 1) - (x - 2)(3x) = 2x + 1$$

$$x - 1 - 3x^2 + 6x = 2x + 1$$

$$-3x^2 + 7x - 1 = 2x + 1$$

$$-3x^2 + 5x - 2 = 0$$

$$3x^2 - 5x + 2 = 0$$

$$(3x - 2)(x - 1) = 0$$

$$3x - 2 = 0 \quad \text{or} \quad x - 1 = 0$$

$$x = \frac{2}{3} \quad \text{or} \quad x = 1$$

1 is not a solution since $x \neq 1$. The solution is $\frac{2}{3}$.

35. $\dfrac{5}{b + 5} - \dfrac{4}{b^2 + 2b} = \dfrac{6}{b^2 + 7b + 10}$

$$\frac{5}{b + 5} - \frac{4}{b(b + 2)} = \frac{6}{(b + 5)(b + 2)}$$

Multiply both sides by $b(b + 5)(b + 2)$. Note that $b \neq 0, b \neq -5$, and $b \neq -2$.

$$b(b+5)(b+2)\left(\frac{5}{b+5} - \frac{4}{b(b+2)}\right)$$

$$= b(b+5)(b+2)\left(\frac{6}{(b+5)(b+2)}\right)$$

$$5b(b+2) - 4(b+5) = 6b$$

$$5b^2 + 10b - 4b - 20 = 6b$$

$$5b^2 - 20 = 0$$

$$b^2 - 4 = 0$$

$$(b+2)(b-2) = 0$$

$$b+2 = 0 \quad \text{or} \quad b-2 = 0$$

$$b = -2 \quad \text{or} \quad b = 2$$

Since $b \neq -2, -2$ is not a solution. The solution is 2.

37. $\dfrac{4}{2x^2 + 3x - 9} + \dfrac{2}{2x^2 - x - 3}$

$$= \frac{3}{x^2 + 4x + 3}$$

$$\frac{4}{(2x-3)(x+3)} + \frac{2}{(2x-3)(x+1)}$$

$$= \frac{3}{(x+3)(x+1)}$$

Multiply both sides by $(2x-3)(x+3)(x+1)$. Note that $x \neq \frac{3}{2}$, $x \neq -3$, and $x \neq -1$.

$$(2x-3)(x+3)(x+1)$$

$$\cdot \left(\frac{4}{(2x-3)(x+3)} + \frac{2}{(2x-3)(x+1)}\right)$$

$$= (2x-3)(x+3)(x+1)\left(\frac{3}{(x+3)(x+1)}\right)$$

$$4(x+1) + 2(x+3) = 3(2x-3)$$

$$4x + 4 + 2x + 6 = 6x - 9$$

$$6x + 10 = 6x - 9$$

$$10 = -9$$

This is a false statement. Therefore, there is no solution.

R.5 Inequalities

Your Turn 1

Solve $3z - 2 > 5z + 7$.

$$3z - 2 > 5z + 7$$

$$3z - 2 + 2 > 5z + 7 + 2$$

$$3z > 5z + 9$$

$$3z - 5z > 5z - 5z + 9$$

$$-2z > 9$$

$$\frac{-2z}{-2} < \frac{9}{-2}$$

$$z < -\frac{9}{2}$$

Your Turn 2

Solve $3y^2 \leq 16y + 12$.

$$3y^2 \leq 16y + 12$$

$$3y^2 - 16y - 12 \leq 0$$

First solve the equation $3y^2 - 16y - 12 = 0$.

$$3y^2 - 16y - 12 = 0$$

$$(3y + 2)(y - 6) = 0$$

$$3y + 2 = 0 \quad \text{or} \quad y - 6 = 0$$

$$y = -\frac{2}{3} \quad \text{or} \quad y = 6$$

Determine three intervals on the number line and choose a test point in each interval.

Choose -1 from interval A: $3(-1)^2 - 16(-1) - 12 > 0$

Choose 0 from interval B: $3(0)^2 - 16(0) - 12 < 0$

Choose 7 from interval C: $3(7)^2 - 16(7) - 12 > 0$

The numbers in interval B satisfy the inequality, and since the sign was less than or equal to, the boundary points of interval B are also part of the solution. The solution is $[-2/3, 6]$.

Your Turn 3

Solve $\dfrac{k^2 - 35}{k} \geq 2$.

First solve the corresponding equation $\dfrac{k^2 - 35}{k} = 2$.

$$\frac{k^2 - 35}{k} = 2$$
$$k^2 - 35 = 2k$$
$$k^2 - 2k - 35 = 0$$
$$(k - 7)(k + 5) = 0$$
$$k = 7 \text{ or } k = -5$$

The denominator is 0 when $k = 0$, so there are four intervals to consider:

$$(-\infty, -5), (-5, 0), (0, 7), \text{ and } (7, \infty).$$

Choose a test point in each interval.

$$k = -8; \frac{(-8)^2 - 35}{-8} < 2$$

$$k = -1; \frac{(-1)^2 - 35}{-1} > 2$$

$$k = 5; \frac{(-5)^2 - 35}{5} < 2$$

$$k = 10; \frac{(10)^2 - 35}{10} > 2$$

The second and fourth intervals are part of the solution. Since the inequality is greater than or equal to, we can include the endpoints -5 and 7 but not the endpoint 0, which makes the denominator 0. The solution is $[-5, 0) \cup [7, \infty)$.

R.5 Exercises

1. $x < 4$

Because the inequality symbol means "less than," the endpoint at 4 is not included. This inequality is written in interval notation as $(-\infty, 4)$. To graph this interval on a number line, place an open circle at 4 and draw a heavy arrow pointing to the left.

3. $1 \le x < 2$

The endpoint at 1 is included, but the endpoint at 2 is not. This inequality is written in interval notation as $[1, 2)$. To graph this interval, place a closed circle at 1 and an open circle at 2; then draw a heavy line segment between them.

5. $-9 > x$

This inequality may be rewritten as $x < -9$, and is written in interval notation as $(-\infty, -9)$. Note that the endpoint at -9 is not included. To graph this interval, place an open circle at -9 and draw a heavy arrow pointing to the left.

7. $[-7, -3]$

This represents all the numbers between -7 and -3, including both endpoints. This interval can be written as the inequality $-7 \le x \le -3$.

9. $(-\infty, -1]$

This represents all the numbers to the left of -1 on the number line and includes the endpoint. This interval can be written as the inequality $x \le -1$.

11. Notice that the endpoint -2 is included, but 6 is not. The interval shown in the graph can be written as the inequality $-2 \le x < 6$.

13. Notice that both endpoints are included. The interval shown in the graph can be written as $x \le -4$ or $x \ge 4$.

15.
$$6p + 7 \le 19$$
$$6p \le 12$$
$$\left(\frac{1}{6}\right)(6p) \le \left(\frac{1}{6}\right)(12)$$
$$p \le 2$$

The solution in interval notation is $(-\infty, 2]$.

17.
$$m - (3m - 2) + 6 < 7m - 19$$
$$m - 3m + 2 + 6 < 7m - 19$$
$$-2m + 8 < 7m - 19$$
$$-9m + 8 < -19$$
$$-9m < -27$$
$$-\frac{1}{9}(-9m) > -\frac{1}{9}(-27)$$
$$m > 3$$

The solution is $(3, \infty)$.

19.
$$3p - 1 < 6p + 2(p - 1)$$
$$3p - 1 < 6p + 2p - 2$$
$$3p - 1 < 8p - 2$$
$$-5p - 1 < -2$$
$$-5p < -1$$
$$-\frac{1}{5}(-5p) > -\frac{1}{5}(-1)$$
$$p > \frac{1}{5}$$

The solution is $\left(\frac{1}{5}, \infty\right)$.

21. $-11 < y - 7 < -1$

$-11 + 7 < y - 7 + 7 < -1 + 7$

$-4 < y < 6$

The solution is $(-4, 6)$.

23. $-2 < \dfrac{1 - 3k}{4} \le 4$

$4(-2) < 4\left(\dfrac{1 - 3k}{4}\right) \le 4(4)$

$-8 < 1 - 3k \le 16$

$-9 < -3k \le 15$

$-\dfrac{1}{3}(-9) > -\dfrac{1}{3}(-3k) \ge -\dfrac{1}{3}(15)$

Rewrite the inequalities in the proper order.

$-5 \le k < 3$

The solution is $[-5, 3)$.

25. $\dfrac{3}{5}(2p + 3) \ge \dfrac{1}{10}(5p + 1)$

$10\left(\dfrac{3}{5}\right)(2p + 3) \ge 10\left(\dfrac{1}{10}\right)(5p + 1)$

$6(2p + 3) \ge 5p + 1$

$12p + 18 \ge 5p + 1$

$7p \ge -17$

$p \ge -\dfrac{17}{7}$

The solution is $\left[-\frac{17}{7}, \infty\right)$.

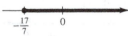

27. $(m - 3)(m + 5) < 0$

Solve $(m - 3)(m + 5) = 0$.

$(m - 3)(m + 5) = 0$

$m = 3$ or $m = -5$

Intervals: $(-\infty, -5), (-5, 3), (3, \infty)$

For $(-\infty, -5)$, choose -6 to test for m.

$(-6 - 3)(-6 + 5) = -9(-1) = 9 \not< 0$

For $(-5, 3)$, choose 0.

$(0 - 3)(0 + 5) = -3(5) = -15 < 0$

For $(3, \infty)$, choose 4.

$(4 - 3)(4 + 5) = 1(9) = 9 \not< 0$

The solution is $(-5, 3)$.

29. $y^2 - 3y + 2 < 0$

$(y - 2)(y - 1) < 0$

Solve $(y - 2)(y - 1) = 0$.

$y = 2$ or $y = 1$

Intervals: $(-\infty, 1), (1, 2), (2, \infty)$

For $(-\infty, 1)$, choose $y = 0$.

$0^2 - 3(0) + 2 = 2 \not< 0$

For $(1, 2)$, choose $y = \dfrac{3}{2}$.

$$\left(\dfrac{3}{2}\right)^2 - 3\left(\dfrac{3}{2}\right) + 2 = \dfrac{9}{4} - \dfrac{9}{2} + 2$$

$$= \dfrac{9 - 18 + 8}{4}$$

$$= -\dfrac{1}{4} < 0$$

For $(2, \infty)$, choose 3.

$3^2 - 3(3) + 2 = 2 \not< 0$

The solution is $(1, 2)$.

31. $x^2 - 16 > 0$

Solve $x^2 - 16 = 0$.

$x^2 - 16 = 0$

$(x + 4)(x - 4) = 0$

$x = -4$ or $x = 4$

Intervals: $(-\infty, -4), (-4, 4), (4, \infty)$

For $(-\infty, -4)$, choose -5.

$(-5)^2 - 16 = 9 > 0$

For $(-4, 4)$, choose 0.

$0^2 - 16 = -16 \not> 0$

For $(4, \infty)$, choose 5.

$5^2 - 16 = 9 > 0$

The solution is $(-\infty, -4) \cup (4, \infty)$.

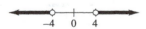

33. $x^2 - 4x \ge 5$

Solve $x^2 - 4x = 5$.

$x^2 - 4x = 5$

$x^2 - 4x - 5 = 0$

$(x + 1)(x - 5) = 0$

$$x + 1 = 0 \quad \text{or} \quad x - 5 = 0$$
$$x = -1 \quad \text{or} \quad x = 5$$

Intervals: $(-\infty, -1), (-1, 5), (5, \infty)$

For $(-\infty, -1)$, choose -2.

$$(-2)^2 - 4(-2) = 12 \geq 5$$

For $(-1, 5)$, choose 0.

$$0^2 - 4(0) = 0 \not\geq 5$$

For $(5, \infty)$, choose 6.

$$(6)^2 - 4(6) = 12 \geq 5$$

The solution is $(-\infty, -1] \cup [5, \infty)$.

35. $3x^2 + 2x > 1$

Solve $3x^2 + 2x = 1$.

$$3x^2 + 2x = 1$$
$$3x^2 + 2x - 1 = 0$$
$$(3x - 1)(x + 1) = 0$$
$$x = \frac{1}{3} \quad \text{or} \quad x = -1$$

Intervals: $(-\infty, -1), \left(-1, \frac{1}{3}\right), \left(\frac{1}{3}, \infty\right)$

For $(-\infty, -1)$, choose -2.

$$3(-2)^2 + 2(-2) = 8 > 1$$

For $\left(-1, \frac{1}{3}\right)$, choose 0.

$$3(0)^2 + 2(0) = 0 \not> 1$$

For $\left(\frac{1}{3}, \infty\right)$, choose 1.

$$3(1)^2 + 2(1) = 5 > 1$$

The solution is $(-\infty, -1) \cup \left(\frac{1}{3}, \infty\right)$.

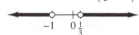

37. $9 - x^2 \leq 0$

Solve $9 - x^2 = 0$.

$$9 - x^2 = 0$$
$$(3 + x)(3 - x) = 0$$
$$x = -3 \quad \text{or} \quad x = 3$$

Intervals: $(-\infty, -3), (-3, 3), (3, \infty)$

For $(-\infty, -3)$, choose -4.

$$9 - (-4)^2 = -7 \leq 0$$

For $(-3, 3)$, choose 0.

$$9 - (0)^2 = 9 \not\leq 0$$

For $(3, \infty)$, choose 4.

$$9 - (4)^2 = -7 \leq 0$$

The solution is $(-\infty, -3] \cup [3, \infty)$.

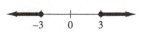

39. $x^3 - 4x \geq 0$

Solve $x^3 - 4x = 0$.

$$x^3 - 4x = 0$$
$$x(x^2 - 4) = 0$$
$$x(x + 2)(x - 2) = 0$$
$$x = 0, \quad \text{or} \quad x = -2, \quad \text{or} \quad x = 2$$

Intervals: $(-\infty, -2), (-2, 0), (0, 2), (2, \infty)$

For $(-\infty, -2)$, choose -3.

$$(-3)^3 - 4(-3) = -15 \not\geq 0$$

For $(-2, 0)$, choose -1.

$$(-1)^3 - 4(-1) = 3 \geq 0$$

For $(0, 2)$, choose 1.

$$(1)^3 - 4(1) = -3 \not\geq 0$$

For $(2, \infty)$, choose 3.

$$(3)^3 - 4(3) = 15 \geq 0$$

The solution is $[-2, 0] \cup [2, \infty)$.

41. $2x^3 - 14x^2 + 12x < 0$

Solve $2x^3 - 14x^2 + 12x = 0$.

$$2x^3 - 14x^2 + 12x = 0$$
$$2x(x^2 - 7x + 6) = 0$$
$$2x(x - 1)(x - 6) = 0$$
$$x = 0, \quad \text{or} \quad x = 1, \quad \text{or} \quad x = 6$$

Intervals: $(-\infty, 0), (0, 1), (1, 6), (6, \infty)$

For $(-\infty, 0)$, choose -1.

$$2(-1)^3 - 14(-1)^2 + 12(-1) = -28 < 0$$

For $(0, 1)$, choose $\frac{1}{2}$.

$$2\left(\frac{1}{2}\right)^3 - 14\left(\frac{1}{2}\right)^2 + 12\left(\frac{1}{2}\right) = \frac{11}{4} \not< 0$$

For $(1, 6)$, choose 2.

$$2(2)^3 - 14(2)^2 + 12(2) = -16 < 0$$

For $(6, \infty)$, choose 7.

$$2(7)^3 - 14(7)^2 + 12(7) = 84 \not< 0$$

The solution is $(-\infty, 0) \cup (1, 6)$.

43. $\dfrac{m-3}{m+5} \leq 0$

Solve $\dfrac{m-3}{m+5} = 0.$

$(m+5)\dfrac{m-3}{m+5} = (m+5)(0)$

$m - 3 = 0$

$m = 3$

Set the denominator equal to 0 and solve.

$m + 5 = 0$

$m = -5$

Intervals: $(-\infty, -5), (-5, 3), (3, \infty)$

For $(-\infty, -5)$, choose -6.

$\dfrac{-6-3}{-6+5} = 9 \not\leq 0$

For $(-5, 3)$, choose 0.

$\dfrac{0-3}{0+5} = -\dfrac{3}{5} \leq 0$

For $(3, \infty)$, choose 4.

$\dfrac{4-3}{4+5} = \dfrac{1}{9} \not\leq 0$

Although the $\leq$ symbol is used, including -5 in the solution would cause the denominator to be zero.

The solution is $(-5, 3]$.

45. $\dfrac{k-1}{k+2} > 1$

Solve $\dfrac{k-1}{k+2} = 1.$

$k - 1 = k + 2$

$-1 \neq 2$

The equation has no solution. Solve $k + 2 = 0$.

$k = -2$

Intervals: $(-\infty, -2), (-2, \infty)$

For $(-\infty, -2)$, choose -3.

$\dfrac{-3-1}{-3+2} = 4 > 1$

For $(-2, \infty)$, choose 0.

$\dfrac{0-1}{0+2} = -\dfrac{1}{2} \not> 1$

The solution is $(-\infty, -2)$.

47. $\dfrac{2y+3}{y-5} \leq 1$

Solve $\dfrac{2y+3}{y-5} = 1.$

$2y + 3 = y - 5$

$y = -8$

Solve $y - 5 = 0$.

$y = 5$

Intervals: $(-\infty, -8), (-8, 5), (5, \infty)$

For $(-\infty, -8)$, choose $y = -10$.

$\dfrac{2(-10)+3}{-10-5} = \dfrac{17}{15} \not\leq 1$

For $(-8, 5)$, choose $y = 0$.

$\dfrac{2(0)+3}{0-5} = -\dfrac{3}{5} \leq 1$

For $(5, \infty)$, choose $y = 6$.

$\dfrac{2(6)+3}{6-5} = \dfrac{15}{1} \not\leq 1$

The solution is $[-8, 5)$.

49. $\dfrac{2k}{k-3} \leq \dfrac{4}{k-3}$

Solve $\dfrac{2k}{k-3} = \dfrac{4}{k-3}.$

$\dfrac{2k}{k-3} = \dfrac{4}{k-3}$

$\dfrac{2k}{k-3} - \dfrac{4}{k-3} = 0$

$\dfrac{2k-4}{k-3} = 0$

$2k - 4 = 0$

$k = 2$

Set the denominator equal to 0 and solve for k.

$k - 3 = 0$

$k = 3$

Intervals: $(-\infty, 2), (2, 3), (3, \infty)$

For $(-\infty, 2)$, choose 0.

$\dfrac{2(0)}{0-3} = 0$ and $\dfrac{4}{0-3} = -\dfrac{4}{3}$, so

$\dfrac{2(0)}{0-3} \not\leq \dfrac{4}{0-3}.$

For $(2, 3)$, choose $\dfrac{5}{2}$.

$\dfrac{2\left(\frac{5}{2}\right)}{\frac{5}{2}-3} = \dfrac{5}{-\frac{1}{2}} = -10$

and $\dfrac{4}{\frac{5}{2}-3} = \dfrac{4}{-\frac{1}{2}} = -8$, so

$$\frac{2\left(\frac{5}{2}\right)}{\frac{5}{2} - 3} \le \frac{4}{\frac{5}{2} - 3}.$$

For $(3, \infty)$, choose 4.

$$\frac{2(4)}{4 - 3} = 8 \quad \text{and} \quad \frac{4}{4 - 3} = 4, \text{so}$$

$$\frac{2(4)}{4 - 3} \not\le \frac{4}{4 - 3}.$$

The solution is $[2, 3)$.

51. $\dfrac{2x}{x^2 - x - 6} \ge 0$

Solve $\dfrac{2x}{x^2 - x - 6} = 0.$

$$\frac{2x}{x^2 - x - 6} = 0$$

$$2x = 0$$

$$x = 0$$

Set the denominator equal to 0 and solve for x.

$$x^2 - x - 6 = 0$$

$$(x + 2)(x - 3) = 0$$

$$x + 2 = 0 \quad \text{or} \quad x - 3 = 0$$

$$x = -2 \quad \text{or} \quad x = 3$$

Intervals: $(-\infty, -2), (-2, 0), (0, 3), (3, \infty)$

For $(-\infty, -2)$, choose -3.

$$\frac{2(-3)}{(-3)^2 - (-3) - 6} = -1 \not\ge 0$$

For $(-2, 0)$, choose -1.

$$\frac{2(-1)}{(-1)^2 - (-1) - 6} = \frac{1}{2} \ge 0$$

For $(0, 3)$, choose 2.

$$\frac{2(2)}{2^2 - 2 - 6} = -1 \not\ge 0$$

For $(3, \infty)$, choose 4.

$$\frac{2(4)}{4^2 - 4 - 6} = \frac{4}{3} \ge 0$$

The solution is $(-2, 0] \cup (3, \infty)$.

53. $\dfrac{z^2 + z}{z^2 - 1} \ge 3$

Solve

$$\frac{z^2 + z}{z^2 - 1} = 3.$$

$$z^2 + z = 3z^2 - 3$$

$$-2z^2 + z + 3 = 0$$

$$-1(2z^2 - z - 3) = 0$$

$$-1(z + 1)(2z - 3) = 0$$

$$z = -1 \quad \text{or} \quad z = \frac{3}{2}$$

Set $z^2 - 1 = 0$.

$$z^2 = 1$$

$$z = -1 \quad \text{or} \quad z = 1$$

Intervals: $(-\infty, -1), (-1, 1), \left(1, \frac{3}{2}\right), \left(\frac{3}{2}, \infty\right)$

For $(-\infty, -1)$, choose $x = -2$.

$$\frac{(-2)^2 + 3}{(-2)^2 - 1} = \frac{7}{3} \not\ge 3$$

For $(-1, 1)$, choose $x = 0$.

$$\frac{0^2 + 3}{0^2 - 1} = -3 \not\ge 3$$

For $\left(1, \frac{3}{2}\right)$, choose $x = \frac{3}{2}$.

$$\frac{\left(\frac{3}{2}\right)^2 + 3}{\left(\frac{3}{2}\right)^2 - 1} = \frac{21}{5} \ge 3$$

For $\left(\frac{3}{2}, \infty\right)$, choose $x = 2$.

$$\frac{2^2 + 3}{2^2 - 1} = \frac{7}{3} \not\ge 3$$

The solution is $\left(1, \frac{3}{2}\right]$.

R.6 Exponents

Your Turn 1

$$\left(\frac{2}{3}\right)^{-3} = \frac{1}{\left(\frac{2}{3}\right)^3} = \frac{1}{\frac{8}{27}} = \frac{27}{8}$$

Your Turn 2

Simplify $\left(\dfrac{y^2 z^{-4}}{y^{-3} z^4}\right)^{-2}$.

$$\left(\frac{y^2z^{-4}}{y^{-3}z^4}\right)^{-2} = \frac{\left(y^2z^{-4}\right)^{-2}}{\left(y^{-3}z^4\right)^{-2}} = \frac{y^{(2)(-2)}z^{(-4)(-2)}}{y^{(-3)(-2)}z^{(4)(-2)}}$$

$$= \frac{y^{-4}z^8}{y^6z^{-8}} = \frac{z^{8-(-8)}}{y^{6-(-4)}} = \frac{z^{16}}{y^{10}}$$

Your Turn 3

$125^{1/3} = 5$, since $5^3 = 125$.

Your Turn 4

$$16^{-3/4} = (16^{1/4})^{-3} = 2^{-3} = \frac{1}{2^3} = \frac{1}{8}$$

Your Turn 5

$$\left(\frac{x^{1/2}x^4}{x^{3/2}}\right)^{1/3} = \left(\frac{x^{(1/2)+4}}{x^{3/2}}\right)^{1/3} = \left(\frac{x^{9/2}}{x^{3/2}}\right)^{1/3}$$

$$= \left(x^{(9/2)-(3/2)}\right)^{1/3} = \left(x^3\right)^{1/3}$$

$$= x^{3(1/3)} = x^1 = x$$

Your Turn 6

Factor $5z^{1/3} + 4z^{-2/3}$.

$$5z^{1/3} + 4z^{-2/3} = z^{-2/3}\left(5z^{(1/3)+(2/3)} + 4z^{(-2/3)+(2/3)}\right)$$

$$= z^{-2/3}(5z + 4)$$

R.6 Exercises

1. $8^{-2} = \dfrac{1}{8^2} = \dfrac{1}{64}$

3. $5^0 = 1$, by definition.

5. $-(-3)^{-2} = -\dfrac{1}{(-3)^2} = -\dfrac{1}{9}$

7. $\left(\dfrac{1}{6}\right)^{-2} = \dfrac{1}{\left(\frac{1}{6}\right)^2} = \dfrac{1}{\frac{1}{36}} = 36$

9. $\dfrac{4^{-2}}{4} = 4^{-2-1} = 4^{-3} = \dfrac{1}{4^3} = \dfrac{1}{64}$

11. $\dfrac{10^8 \cdot 10^{-10}}{10^4 \cdot 10^2}$

$$= \frac{10^{8+(-10)}}{10^{4+2}} = \frac{10^{-2}}{10^6}$$

$$= 10^{-2-6} = 10^{-8}$$

$$= \frac{1}{10^8}$$

13. $\dfrac{x^4 \cdot x^3}{x^5} = \dfrac{x^{4+3}}{x^5} = \dfrac{x^7}{x^5} = x^{7-5} = x^2$

15. $\dfrac{(4k^{-1})^2}{2k^{-5}} = \dfrac{4^2 k^{-2}}{2k^{-5}} = \dfrac{16k^{-2-(-5)}}{2}$

$$= 8k^{-2+5} = 8k^3$$

17. $\dfrac{3^{-1} \cdot x \cdot y^2}{x^{-4} \cdot y^5} = 3^{-1} \cdot x^{1-(-4)} \cdot y^{2-5}$

$$= 3^{-1} \cdot x^{1+4} \cdot y^{-3}$$

$$= \frac{1}{3} \cdot x^5 \cdot \frac{1}{y^3}$$

$$= \frac{x^5}{3y^3}$$

19. $\left(\dfrac{a^{-1}}{b^2}\right)^{-3} = \dfrac{(a^{-1})^{-3}}{(b^2)^{-3}} = \dfrac{a^{(-1)(-3)}}{b^{2(-3)}}$

$$= \frac{a^3}{b^{-6}} = a^3 b^6$$

21. $a^{-1} + b^{-1} = \dfrac{1}{a} + \dfrac{1}{b}$

$$= \left(\frac{b}{b}\right)\left(\frac{1}{a}\right) + \left(\frac{a}{a}\right)\left(\frac{1}{b}\right)$$

$$= \frac{b}{ab} + \frac{a}{ab}$$

$$= \frac{b+a}{ab}$$

$$= \frac{a+b}{ab}$$

23. $\dfrac{2n^{-1} - 2m^{-1}}{m + n^2} = \dfrac{\frac{2}{n} - \frac{2}{m}}{m + n^2}$

$$= \frac{\frac{2}{n} \cdot \frac{m}{m} - \frac{2}{m} \cdot \frac{n}{n}}{(m + n^2)}$$

$$= \frac{2m - 2n}{mn(m + n^2)}$$

$$\text{or} \quad \frac{2(m - n)}{mn(m + n^2)}$$

25. $(x^{-1} - y^{-1})^{-1} = \dfrac{1}{\frac{1}{x} - \frac{1}{y}}$

$$= \frac{1}{\frac{1}{x} \cdot \frac{y}{y} - \frac{1}{y} \cdot \frac{x}{x}}$$

$$= \frac{1}{\frac{y}{xy} - \frac{x}{xy}}$$

$$= \frac{1}{\frac{y-x}{xy}}$$

$$= \frac{xy}{y-x}$$

27. $121^{1/2} = (11^2)^{1/2} = 11^{2(1/2)} = 11^1 = 11$

29. $32^{2/5} = (32^{1/5})^2 = 2^2 = 4$

31. $\left(\frac{36}{144}\right)^{1/2} = \frac{36^{1/2}}{144^{1/2}} = \frac{6}{12} = \frac{1}{2}$

This can also be solved by reducing the fraction first.

$$\left(\frac{36}{144}\right)^{1/2} = \left(\frac{1}{4}\right)^{1/2} = \frac{1^{1/2}}{4^{1/2}} = \frac{1}{2}$$

33. $8^{-4/3} = (8^{1/3})^{-4} = 2^{-4} = \frac{1}{2^4} = \frac{1}{16}$

35. $\left(\frac{27}{64}\right)^{-1/3} = \frac{27^{-1/3}}{64^{-1/3}} = \frac{64^{1/3}}{27^{1/3}} = \frac{4}{3}$

37. $3^{2/3} \cdot 3^{4/3} = 3^{(2/3)+(4/3)} = 3^{6/3} = 3^2 = 9$

39. $\frac{4^{9/4} \cdot 4^{-7/4}}{4^{-10/4}} = 4^{9/4-7/4-(-10/4)}$

$$= 4^{12/4} = 4^3 = 64$$

41. $\left(\frac{x^6 y^{-3}}{x^{-2} y^5}\right)^{1/2} = (x^{6-(-2)} y^{-3-5})^{1/2}$

$$= (x^8 y^{-8})^{1/2}$$

$$= (x^8)^{1/2} (y^{-8})^{1/2}$$

$$= x^4 y^{-4}$$

$$= \frac{x^4}{y^4}$$

43. $\frac{7^{-1/3} \cdot 7 r^{-3}}{7^{2/3} \cdot (r^{-2})^2} = \frac{7^{-1/3+1} r^{-3}}{7^{2/3} \cdot r^{-4}}$

$$= 7^{-1/3+3/3-2/3} r^{-3-(-4)}$$

$$= 7^0 r^{-3+4} = 1 \cdot r^1 = r$$

45. $\frac{3k^2 \cdot (4k^{-3})^{-1}}{4^{1/2} \cdot k^{7/2}}$

$$= \frac{3k^2 \cdot 4^{-1} k^3}{2 \cdot k^{7/2}}$$

$$= 3 \cdot 2^{-1} \cdot 4^{-1} k^{2+3-(7/2)}$$

$$= \frac{3}{8} \cdot k^{3/2}$$

$$= \frac{3k^{3/2}}{8}$$

47. $\frac{a^{4/3}}{a^{2/3}} \cdot \frac{b^{1/2}}{b^{-3/2}} = a^{4/3-2/3} b^{1/2-(-3/2)}$

$$= a^{2/3} b^2$$

49. $\frac{k^{-3/5} \cdot h^{-1/3} \cdot t^{2/5}}{k^{-1/5} \cdot h^{-2/3} \cdot t^{1/5}}$

$$= k^{-3/5-(-1/5)} h^{-1/3-(-2/3)} t^{2/5-1/5}$$

$$= k^{-3/5+1/5} h^{-1/3+2/3} t^{2/5-1/5}$$

$$= k^{-2/5} h^{1/3} t^{1/5} = \frac{h^{1/3} t^{1/5}}{k^{2/5}}$$

51. $3x^3(x^2 + 3x)^2 - 15x(x^2 + 3x)^2$

$$= 3x \cdot x^2(x^2 + 3x)^2 - 3x \cdot 5(x^2 + 3x)^2$$

$$= 3x(x^2 + 3x)^2(x^2 - 5)$$

53. $10x^3(x^2 - 1)^{-1/2} - 5x(x^2 - 1)^{1/2}$

$$= 5x \cdot 2x^2(x^2 - 1)^{-1/2} - 5x(x^2 - 1)^{-1/2}(x^2 - 1)^1$$

$$= 5x(x^2 - 1)^{-1/2}[2x^2 - (x^2 - 1)]$$

$$= 5x(x^2 - 1)^{-1/2}(x^2 + 1)$$

55. $x(2x + 5)^2(x^2 - 4)^{-1/2} + 2(x^2 - 4)^{1/2}(2x + 5)$

$$= (2x + 5)^2(x^2 - 4)^{-1/2}(x)$$
$$+ (x^2 - 4)^1(x^2 - 4)^{-1/2}(2)(2x + 5)$$

$$= (2x + 5)(x^2 - 4)^{-1/2}$$
$$\cdot [(2x + 5)(x) + (x^2 - 4)(2)]$$

$$= (2x + 5)(x^2 - 4)^{-1/2} \cdot (2x^2 + 5x + 2x^2 - 8)$$

$$= (2x + 5)(x^2 - 4)^{-1/2}(4x^2 + 5x - 8)$$

R.7 Radicals

Your Turn 1

Simplify $\sqrt{28x^9y^5}$.

$$\sqrt{28x^9y^5} = \sqrt{4 \cdot x^8 \cdot y^4 \cdot 7xy}$$
$$= 2x^4y^2\sqrt{7xy}$$

Your Turn 2

Rationalize the denominator in $\dfrac{5}{\sqrt{x} - \sqrt{y}}$.

$$\frac{5}{\sqrt{x} - \sqrt{y}} = \frac{5}{\sqrt{x} - \sqrt{y}} \cdot \frac{\sqrt{x} + \sqrt{y}}{\sqrt{x} + \sqrt{y}}$$
$$= \frac{5\left(\sqrt{x} + \sqrt{y}\right)}{x - y}$$

Your Turn 3

$$\frac{4 + \sqrt{x}}{16 - x} = \frac{4 + \sqrt{x}}{16 - x} \cdot \frac{4 - \sqrt{x}}{4 - \sqrt{x}}$$
$$= \frac{16 - x}{(16 - x)(4 - \sqrt{x})} = \frac{1}{4 - \sqrt{x}}$$

R.7   Exercises

1. $\sqrt[3]{125} = 5$ because $5^3 = 125$.

3. $\sqrt[5]{-3125} = -5$ because $(-5)^5 = -3125$.

5. $\sqrt{2000} = \sqrt{4 \cdot 100 \cdot 5}$
    $$= 2 \cdot 10\sqrt{5}$$
    $$= 20\sqrt{5}$$

7. $\sqrt{27} \cdot \sqrt{3} = \sqrt{27 \cdot 3} = \sqrt{81} = 9$

9. $7\sqrt{2} - 8\sqrt{18} + 4\sqrt{72}$
    $$= 7\sqrt{2} - 8\sqrt{9 \cdot 2} + 4\sqrt{36 \cdot 2}$$
    $$= 7\sqrt{2} - 8(3)\sqrt{2} + 4(6)\sqrt{2}$$
    $$= 7\sqrt{2} - 24\sqrt{2} + 24\sqrt{2}$$
    $$= 7\sqrt{2}$$

11. $4\sqrt{7} - \sqrt{28} + \sqrt{343}$
     $$= 4\sqrt{7} - \sqrt{4}\sqrt{7} + \sqrt{49}\sqrt{7}$$
     $$= 4\sqrt{7} - 2\sqrt{7} + 7\sqrt{7}$$
     $$= (4 - 2 + 7)\sqrt{7}$$
     $$= 9\sqrt{7}$$

13. $\sqrt[3]{2} - \sqrt[3]{16} + 2\sqrt[3]{54}$
     $$= \sqrt[3]{2} - (\sqrt[3]{8 \cdot 2}) + 2(\sqrt[3]{27 \cdot 2})$$
     $$= \sqrt[3]{2} - \sqrt[3]{8}\sqrt[3]{2} + 2(\sqrt[3]{27}\sqrt[3]{2})$$
     $$= \sqrt[3]{2} - 2\sqrt[3]{2} + 2(3\sqrt[3]{2})$$
     $$= \sqrt[3]{2} - 2\sqrt[3]{2} + 6\sqrt[3]{2}$$
     $$= 5\sqrt[3]{2}$$

15. $\sqrt{2x^3y^2z^4} = \sqrt{x^2y^2z^4 \cdot 2x} = xyz^2\sqrt{2x}$

17. $\sqrt[3]{128x^3y^8z^9} = \sqrt[3]{64x^3y^6z^9 \cdot 2y^2}$
     $$= \sqrt[3]{64x^3y^6z^9}\sqrt[3]{2y^2}$$
     $$= 4xy^2z^3\sqrt[3]{2y^2}$$

19. $\sqrt{a^3b^5} - 2\sqrt{a^7b^3} + \sqrt{a^3b^9}$
     $$= \sqrt{a^2b^4ab} - 2\sqrt{a^6b^2ab} + \sqrt{a^2b^8ab}$$
     $$= ab^2\sqrt{ab} - 2a^3b\sqrt{ab} + ab^4\sqrt{ab}$$
     $$= (ab^2 - 2a^3b + ab^4)\sqrt{ab}$$
     $$= ab\sqrt{ab}(b - 2a^2 + b^3)$$

21. $\sqrt{a} \cdot \sqrt[3]{a} = a^{1/2} \cdot a^{1/3}$
     $$= a^{1/2 + (1/3)}$$
     $$= a^{5/6} = \sqrt[6]{a^5}$$

23. $\sqrt{16 - 8x + x^2}$
     $$= \sqrt{(4 - x)^2}$$
     $$= |4 - x|$$

25. $\sqrt{4 - 25z^2} = \sqrt{(2 + 5z)(2 - 5z)}$

     This factorization does not produce a perfect square, so the expression $\sqrt{4 - 25z^2}$ cannot be simplified.

27. $\dfrac{5}{\sqrt{7}} = \dfrac{5}{\sqrt{7}} \cdot \dfrac{\sqrt{7}}{\sqrt{7}} = \dfrac{5\sqrt{7}}{7}$

29. $\dfrac{-3}{\sqrt{12}} = \dfrac{-3}{\sqrt{4 \cdot 3}}$
     $$= \frac{-3}{2\sqrt{3}} \cdot \frac{\sqrt{3}}{\sqrt{3}} = \frac{-3\sqrt{3}}{6} = -\frac{\sqrt{3}}{2}$$

31. $\dfrac{3}{1 - \sqrt{2}} = \dfrac{3}{1 - \sqrt{2}} \cdot \dfrac{1 + \sqrt{2}}{1 + \sqrt{2}}$
     $$= \frac{3(1 + \sqrt{2})}{1 - 2}$$
     $$= -3(1 + \sqrt{2})$$

33. $\dfrac{6}{2+\sqrt{2}} = \dfrac{6}{2+\sqrt{2}} \cdot \dfrac{2-\sqrt{2}}{2-\sqrt{2}}$

$\quad = \dfrac{6(2-\sqrt{2})}{4 - 2\sqrt{2} + 2\sqrt{2} - \sqrt{4}}$

$\quad = \dfrac{6(2-\sqrt{2})}{4-2} = \dfrac{6(2-\sqrt{2})}{2}$

$\quad = 3(2-\sqrt{2})$

35. $\dfrac{1}{\sqrt{r}-\sqrt{3}} = \dfrac{1}{\sqrt{r}-\sqrt{3}} \cdot \dfrac{\sqrt{r}+\sqrt{3}}{\sqrt{r}+\sqrt{3}}$

$\quad = \dfrac{\sqrt{r}+\sqrt{3}}{r-3}$

37. $\dfrac{y-5}{\sqrt{y}-\sqrt{5}} = \dfrac{y-5}{\sqrt{y}-\sqrt{5}} \cdot \dfrac{\sqrt{y}+\sqrt{5}}{\sqrt{y}+\sqrt{5}}$

$\quad = \dfrac{(y-5)(\sqrt{y}+\sqrt{5})}{y-5}$

$\quad = \sqrt{y}+\sqrt{5}$

39. $\dfrac{\sqrt{x}+\sqrt{x+1}}{\sqrt{x}-\sqrt{x+1}} = \dfrac{\sqrt{x}+\sqrt{x+1}}{\sqrt{x}-\sqrt{x+1}} \cdot \dfrac{\sqrt{x}+\sqrt{x+1}}{\sqrt{x}+\sqrt{x+1}}$

$\quad = \dfrac{x + 2\sqrt{x(x+1)} + (x+1)}{x-(x+1)}$

$\quad = \dfrac{2x + 2\sqrt{x(x+1)} + 1}{-1}$

$\quad = -2x - 2\sqrt{x(x+1)} - 1$

41. $\dfrac{1+\sqrt{2}}{2} = \dfrac{\left(1+\sqrt{2}\right)\left(1-\sqrt{2}\right)}{2\left(1-\sqrt{2}\right)}$

$\quad = \dfrac{1-2}{2\left(1-\sqrt{2}\right)}$

$\quad = -\dfrac{1}{2\left(1-\sqrt{2}\right)}$

43. $\dfrac{\sqrt{x}+\sqrt{x+1}}{\sqrt{x}-\sqrt{x+1}}$

$\quad = \dfrac{\sqrt{x}+\sqrt{x+1}}{\sqrt{x}-\sqrt{x+1}} \cdot \dfrac{\sqrt{x}-\sqrt{x+1}}{\sqrt{x}-\sqrt{x+1}}$

$\quad = \dfrac{x-(x+1)}{x - 2\sqrt{x} \cdot \sqrt{x+1} + (x+1)}$

$\quad = \dfrac{-1}{2x - 2\sqrt{x(x+1)} + 1}$

LINEAR FUNCTIONS

1.1 Slopes and Equations of Lines

Your Turn 1

Find the slope of the line through $(1, 5)$ and $(4, 6)$.

Let $(x_1, y_1) = (1, 5)$ and $(x_2, y_2) = (4, 6)$.

$$m = \frac{6 - 5}{4 - 1} = \frac{1}{3}$$

Your Turn 2

Find the equation of the line with x-intercept -4 and y-intercept 6.

We know that $b = 6$ and that the line crosses the axes at $(-4, 0)$ and $(0, 6)$. Use these two intercepts to find the slope m.

$$m = \frac{6 - 0}{0 - (-4)} = \frac{6}{4} = \frac{3}{2}$$

Thus the equation for the line in slope-intercept form is $y = \frac{3}{2}x + 6$.

Your Turn 3

Find the slope of the line whose equation is $8x + 3y = 5$.

Solve the equation for y.

$$8x + 3y = 5$$
$$3y = -8x + 5$$
$$y = -\frac{8}{3}x + \frac{5}{3}$$

The slope is $-8/3$.

Your Turn 4

Find the equation (in slope-intercept form) of the line through $(2, 9)$ and $(5, 3)$.

First find the slope.

$$m = \frac{3 - 9}{5 - 2} = \frac{-6}{3} = -2$$

Now use the point-slope form, with $(x_1, y_1) = (5, 3)$.

$$y - y_1 = m(x - x_1)$$
$$y - 3 = -2(x - 5)$$
$$y - 3 = -2x + 10$$
$$y = -2x + 13$$

Your Turn 5

Find (in slope-intercept form) the equation of the line that passes through the point $(4, 5)$ and is parallel to the line $3x - 6y = 7$.

First find the slope of the line $3x - 6y = 7$ by solving this equation for y.

$$3x - 6y = 7$$
$$6y = 3x - 7$$
$$y = \frac{3}{6}x - \frac{7}{6}$$
$$y = \frac{1}{2}x - \frac{7}{6}$$

Since the line we are to find is parallel to this line, it will also have slope 1/2. Use the point-slope form with $(x_1, y_1) = (4, 5)$.

$$y - y_1 = m(x - x_1)$$
$$y - 5 = \frac{1}{2}(x - 4)$$
$$y - 5 = \frac{1}{2}x - 2$$
$$y = \frac{1}{2}x + 3$$

Your Turn 6

Find (in slope-intercept form) the equation of the line that passes through the point $(3, 2)$ and is perpendicular to the line $2x + 3y = 4$.

First find the slope of the line $2x + 3y = 4$ by solving this equation for y.

$$2x + 3y = 4$$
$$3y = -2x + 4$$
$$y = -\frac{2}{3}x + \frac{4}{3}$$

Since the line we are to find is perpendicular to a line with slope $-2/3$, , it will have slope 3/2. (Note that $(-2/3)(3/2) = -1$.)

Use the point-slope form with $(x_1, y_1) = (3, 2)$.

$$y - y_1 = m(x - x_1)$$
$$y - 2 = \frac{3}{2}(x - 3)$$
$$y - 2 = \frac{3}{2}x - \frac{9}{2}$$
$$y = \frac{3}{2}x - \frac{5}{2}$$

1.1 Warmup Exercises

W1. $\dfrac{15 - (-3)}{-2 - 4} = \dfrac{18}{-6} = -3$

W2. $y - (-3) = -2(x + 5)$
$$y + 3 = -2x - 10$$
$$y = -2x - 13$$

W3. $y - \dfrac{1}{2} = \dfrac{2}{5}\left(x + \dfrac{1}{3}\right)$
$$y - \frac{1}{2} = \frac{2}{5}x + \frac{2}{15}$$
$$y = \frac{2}{5}x + \frac{4}{30} + \frac{15}{30}$$
$$y = \frac{2}{5}x + \frac{19}{30}$$

W4. $2x - 3y = 7$
$$-3y = -2x + 7$$
$$y = \frac{2}{3}x - \frac{7}{3}$$

1.1 Exercises

1. Find the slope of the line through $(4, 5)$ and $(-1, 2)$.
$$m = \frac{5 - 2}{4 - (-1)}$$
$$= \frac{3}{5}$$

3. Find the slope of the line through $(8, 4)$ and $(8, -7)$.
$$m = \frac{4 - (-7)}{8 - 8} = \frac{11}{0}$$
The slope is undefined; the line is vertical.

5. $y = x$

Using the slope-intercept form, $y = mx + b$, we see that the slope is 1.

7. $5x - 9y = 11$

Rewrite the equation in slope-intercept form.

$$9y = 5x - 11$$
$$y = \frac{5}{9}x - \frac{11}{9}$$
The slope is $\frac{5}{9}$.

9. $x = 5$

This is a vertical line. The slope is undefined.

11. $y = 8$

This is a horizontal line, which has a slope of 0.

13. Find the slope of a line parallel to $6x - 3y = 12$.

Rewrite the equation in slope-intercept form.
$$-3y = -6x + 12$$
$$y = 2x - 4$$

The slope is 2, so a parallel line will also have slope 2.

15. The line goes through $(1, 3)$, with slope $m = -2$. Use point-slope form.
$$y - 3 = -2(x - 1)$$
$$y = -2x + 2 + 3$$
$$y = -2x + 5$$

17. The line goes through $(-5, -7)$ with slope $m = 0$. Use point-slope form.
$$y - (-7) = 0[x - (-5)]$$
$$y + 7 = 0$$
$$y = -7$$

19. The line goes through $(4, 2)$ and $(1, 3)$. Find the slope, then use point-slope form with either of the two given points.
$$m = \frac{3 - 2}{1 - 4} = -\frac{1}{3}$$
$$y - 3 = -\frac{1}{3}(x - 1)$$
$$y = -\frac{1}{3}x + \frac{1}{3} + 3$$
$$y = -\frac{1}{3}x + \frac{10}{3}$$

21. The line goes through $\left(\frac{2}{3}, \frac{1}{2}\right)$ and $\left(\frac{1}{4}, -2\right)$.
$$m = \frac{-2 - \frac{1}{2}}{\frac{1}{4} - \frac{2}{3}} = \frac{-\frac{4}{2} - \frac{1}{2}}{\frac{3}{12} - \frac{8}{12}}$$
$$m = \frac{\frac{-5}{2}}{-\frac{5}{12}} = \frac{60}{10} = 6$$

$$y - (-2) = 6\left(x - \frac{1}{4}\right)$$

$$y + 2 = 6x - \frac{3}{2}$$

$$y = 6x - \frac{3}{2} - 2$$

$$y = 6x - \frac{3}{2} - \frac{4}{2}$$

$$y = 6x - \frac{7}{2}$$

23. The line goes through $(-8, 4)$ and $(-8, 6)$.

$$m = \frac{4 - 6}{-8 - (-8)} = \frac{-2}{0};$$

which is undefined.

This is a vertical line; the value of x is always -8.
The equation of this line is $x = -8$.

25. The line has x-intercept -6 and y-intercept -3.
Two points on the line are $(-6, 0)$ and $(0, -3)$.
Find the slope; then use slope-intercept form.

$$m = \frac{-3 - 0}{0 - (-6)} = \frac{-3}{6} = -\frac{1}{2}$$
$$b = -3$$

$$y = -\frac{1}{2}x - 3$$

$$2y = -x - 6$$

$$x + 2y = -6$$

27. The vertical line through $(-6, 5)$ goes through the point $(-6, 0)$, so the equation is $x = -6$.

29. Write an equation of the line through $(-4, 6)$, parallel to $3x + 2y = 13$.

Rewrite the equation of the given line in slope-intercept form.

$$3x + 2y = 13$$
$$2y = -3x + 13$$
$$y = -\frac{3}{2}x + \frac{13}{2}$$

The slope is $-\frac{3}{2}$.

Use $m = -\frac{3}{2}$ and the point $(-4, 6)$ in the point-slope form.

$$y - 6 = -\frac{3}{2}[x - (-4)]$$

$$y = -\frac{3}{2}(x + 4) + 6$$

$$y = -\frac{3}{2}x - 6 + 6$$

$$y = -\frac{3}{2}x$$

$$2y = -3x$$

$$3x + 2y = 0$$

31. Write an equation of the line through $(3, -4)$, perpendicular to $x + y = 4$.

Rewrite the equation of the given line as

$$y = -x + 4.$$

The slope of this line is -1. To find the slope of a perpendicular line, solve

$$-1m = -1.$$

$$m = 1$$

Use $m = 1$ and $(3, -4)$ in the point-slope form.

$$y - (-4) = 1(x - 3)$$
$$y = x - 3 - 4$$
$$y = x - 7$$
$$x - y = 7$$

33. Write an equation of the line with y-intercept 4, perpendicular to $x + 5y = 7$.

Find the slope of the given line.

$$x + 5y = 7$$
$$5y = -x + 7$$
$$y = -\frac{1}{5}x + \frac{7}{5}$$

The slope is $-\frac{1}{5}$, so the slope of the perpendicular line will be 5. If the y-intercept is 4, then using the slope-intercept form we have

$$y = mx + b$$
$$y = 5x + 4, \quad \text{or} \quad 5x - y = -4$$

35. Do the points $(4, 3)$, $(2, 0)$, and $(-18, -12)$ lie on the same line?

Find the slope between $(4, 3)$ and $(2, 0)$.

$$m = \frac{0 - 3}{2 - 4} = \frac{-3}{-2} = \frac{3}{2}$$

Find the slope between $(4, 3)$ and $(-18, -12)$.

$$m = \frac{-12 - 3}{-18 - 4} = \frac{-15}{-22} = \frac{15}{22}$$

Since these slopes are not the same, the points do not lie on the same line.

37. A parallelogram has 4 sides, with opposite sides parallel. The slope of the line through $(1, 3)$ and $(2, 1)$ is

$$m = \frac{3-1}{1-2}$$
$$= \frac{2}{-1}$$
$$= -2.$$

The slope of the line through $\left(-\frac{5}{2}, 2\right)$ and $\left(-\frac{7}{2}, 4\right)$ is

$$m = \frac{2-4}{-\frac{5}{2}-\left(-\frac{7}{2}\right)} = \frac{-2}{1} = -2.$$

Since these slopes are equal, these two sides are parallel.

The slope of the line through $\left(-\frac{7}{2}, 4\right)$ and $(1, 3)$ is

$$m = \frac{4-3}{-\frac{7}{2}-1} = \frac{1}{-\frac{9}{2}} = -\frac{2}{9}.$$

Slope of the line through $\left(-\frac{5}{2}, 2\right)$ and $(2, 1)$ is

$$m = \frac{2-1}{-\frac{5}{2}-2} = \frac{1}{-\frac{9}{2}} = -\frac{2}{9}.$$

Since these slopes are equal, these two sides are parallel.

Since both pairs of opposite sides are parallel, the quadrilateral is a parallelogram.

39. The line goes through $(0, 2)$ and $(-2, 0)$

$$m = \frac{2-0}{0-(-2)} = \frac{2}{2} = 1$$

The correct choice is (a).

41. The line appears to go through $(0, 0)$ and $(-1, 4)$.

$$m = \frac{4-0}{-1-0} = \frac{4}{-1} = -4$$

43. (a) See the figure in the textbook.

Segment MN is drawn perpendicular to segment PQ. Recall that MQ is the length of segment MQ.

$$m_1 = \frac{\Delta y}{\Delta x} = \frac{MQ}{PQ}$$

From the diagram, we know that $PQ = 1$.

Thus, $m_1 = \frac{MQ}{1}$, so MQ has length m_1.

(b) $$m_2 = \frac{\Delta y}{\Delta x} = \frac{-QN}{PQ} = \frac{-QN}{1}$$
$$QN = -m_2$$

(c) Triangles MPQ, PNQ, and MNP are right triangles by construction. In triangles MPQ and MNP,

$$\text{angle } M = \text{angle } M,$$

and in the right triangles PNQ and MNP,

$$\text{angle } N = \text{angle } N.$$

Since all right angles are equal, and since triangles with two equal angles are similar, triangle MPQ is similar to triangle MNP and triangle PNQ is similar to triangle MNP.

Therefore, triangles MNQ and PNQ are similar to each other.

(d) Since corresponding sides in similar triangles are proportional,

$$MQ = k \cdot PQ \quad \text{and} \quad PQ = k \cdot QN.$$
$$\frac{MQ}{PQ} = \frac{k \cdot PQ}{k \cdot QN}$$
$$\frac{MQ}{PQ} = \frac{PQ}{QN}$$

From the diagram, we know that $PQ = 1$.

$$MQ = \frac{1}{QN}$$

From (a) and (b), $m_1 = MQ$ and $-m_2 = QN$.

Substituting, we get

$$m_1 = \frac{1}{-m_2}.$$

Multiplying both sides by m_2, we have

$$m_1 m_2 = -1.$$

45. $y = x - 1$

Three ordered pairs that satisfy this equation are $(0, -1)$, $(1, 0)$, and $(4, 3)$. Plot these points and draw a line through them.

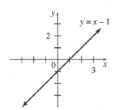

47. $y = -4x + 9$

Three ordered pairs that satisfy this equation are $(0, 9)$, $(1, 5)$, and $(2, 1)$. Plot these points and draw a line through them.

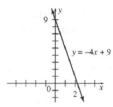

49. $2x - 3y = 12$

Find the intercepts.

If $y = 0$, then

$$2x - 3(0) = 12$$
$$2x = 12$$
$$x = 6$$

so the x-intercept is 6.

If $x = 0$, then

$$2(0) - 3y = 12$$
$$-3y = 12$$
$$y = -4$$

so the y-intercept is -4.

Plot the ordered pairs $(6, 0)$ and $(0, -4)$ and draw a line through these points. (A third point may be used as a check.)

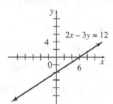

51. $3y - 7x = -21$

Find the intercepts.

If $y = 0$, then

$$3(0) + 7x = -21$$
$$-7x = -21$$
$$x = 3$$

so the x-intercept is 3.

If $x = 0$, then

$$3y - 7(0) = -21$$
$$3y = -21$$
$$y = -7$$

So the y-intercepts is -7.

Plot the ordered pairs $(3, 0)$ and $(0, -7)$ and draw a line through these points. (A third point may be used as a check.)

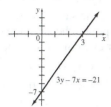

53. $y = -2$

The equation $y = -2$, or, equivalently, $y = 0x - 2$, always gives the same y-value, -2, for any value of x. The graph of this equation is the horizontal line with y-intercept -2.

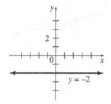

55. $x + 5 = 0$

This equation may be rewritten as $x = -5$. For any value of y, the x-value is -5. Because all ordered pairs that satisfy this equation have the same first number, this equation does not represent a function. The graph is the vertical line with x-intercept -5.

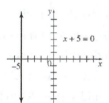

57. $y = 2x$

Three ordered pairs that satisfy this equation are $(0, 0)$, $(-2, -4)$, and $(2, 4)$. Use these points to draw the graph.

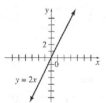

59. $x + 4y = 0$

If $y = 0$, then $x = 0$, so the x-intercept is 0. If $x = 0$, then $y = 0$, so the y-intercept is 0. Both intercepts give the same ordered pair, $(0, 0)$. To get a second point, choose some other value of x (or y). For example if $x = 4$, then

$$x + 4y = 0$$
$$4 + 4y = 0$$
$$4y = -4$$
$$y = -1,$$

giving the ordered pair $(4, -1)$. Graph the line through $(0, 0)$ and $(4, -1)$.

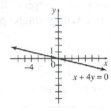

61. (a) The line goes through $(2, 27{,}000)$ and $(5, 63{,}000)$.

$$m = \frac{63{,}000 - 27{,}000}{5 - 2}$$
$$= 12{,}000$$
$$y - 27{,}000 = 12{,}000(x - 2)$$
$$y - 27{,}000 = 12{,}000x - 24{,}000$$
$$y = 12{,}000x + 3000$$

(b) Let $y = 100{,}000$; find x.

$$100{,}000 = 12{,}000x + 3000$$
$$97{,}000 = 12{,}000x$$
$$8.08 = x$$

Sales would surpass \$100,000 after 8 years, 1 month.

63. (a)

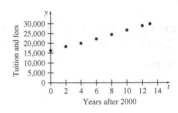

Yes, the data appear to lie roughly along a straight line.

(b) The line goes through $(0, 16{,}072)$ and $(13, 30{,}094)$.

$$m = \frac{30{,}094 - 16{,}072}{13 - 0} \approx 1078.6$$
$$b = 16{,}072$$
$$y = 1078.6t + 16{,}072$$

The slope 1078.6 indicates that tuition and fees have increased approximately \$1079 per year.

(c) The year 2025 is too far in the future to rely on this equation to predict costs; too many other factors may influence these costs by then.

65. (a) The line goes through $(3, 100)$ and $(32, 229.6)$.

$$m = \frac{229.6 - 100}{32 - 3} \approx 4.469$$

Use the point $(3, 100)$ and the point-slope form.

$$y - 100 = 4.469(t - 3)$$
$$y = 4.469t - 13.407 + 100$$
$$y = 4.469t + 86.593$$

(b) The year 2000 corresponds to $t = 2000 - 1980 = 20$.

$$y = 4.469(20) + 86.593$$
$$y \approx 176.0$$

The predicted value is slightly more than the actual CPI of 172.2.

(c) The annual CPI is increasing at a rate of approximately 4.5 units per year.

67. (a) Let x = age.

$$u = 0.85(220 - x) = 187 - 0.85x$$
$$l = 0.7(200 - x) = 154 - 0.7x$$

(b) $u = 187 - 0.85(20) = 170$
$l = 154 - 0.7(20) = 140$

The target heart rate zone is 140 to 170 beats per minute.

(c) $u = 187 - 0.85(40) = 153$
$l = 154 - 0.7(40) = 126$

The target heart rate zone is 126 to 153 beats per minute.

(d) $154 - 0.7x = 187 - 0.85(x + 36)$
$154 - 0.7x = 187 - 0.85x - 30.6$
$154 - 0.7x = 156.4 - 0.85x$
$0.15x = 2.4$
$x = 16$

The younger woman is 16; the older woman is $16 + 36 = 52$. $l = 0.7(220 - 16) \approx 143$ beats per minute.

69. Let $x = 0$ correspond to 1900. Then the "life expectancy from birth" line contains the points $(0, 46)$ and $(110, 78.7)$.

$$m = \frac{78.7 - 46}{110 - 0} = \frac{32.7}{110} \approx 0.297$$

Since $(0, 46)$ is one of the points, the line is given by the equation.

$$y = 0.297x + 46.$$

The "life expectancy from age 65" line contains the points $(0, 76)$ and $(110, 84.1)$.

$$m = \frac{84.1 - 76}{110 - 0} = \frac{8.1}{110} \approx 0.074$$

Since $(0, 76)$ is one of the points, the line is given by the equation

$$y = 0.074x + 76.$$

Set the two expressions for y equal to determine where the lines intersect. At this point, life expectancy should increase no further.

$$0.297x + 46 = 0.074x + 76$$
$$0.223x = 30$$
$$x \approx 135$$

Determine the y-value when $x = 129$. Use the first equation.

$$y = 0.297(135) + 46$$
$$= 40.095 + 46$$
$$= 86.095$$

Thus, the maximum life expectancy for humans is about 86 years.

71. (a) The line goes through $(50, 249{,}187)$ and $(112, 1{,}031{,}631)$.

$$m = \frac{1{,}031{,}631 - 249{,}187}{112 - 50} \approx 12{,}620.06$$

Use the point $(50, 249{,}187)$ and the point-slope form.

$$y - 249{,}187 = 12{,}620.06(t - 50)$$
$$y = 12{,}620.06t - 631{,}003 + 249{,}187$$
$$y = 12{,}620.06t - 381{,}816$$

(b) The year 2015 corresponds to $t = 115$.

$$y = 12{,}620.06(115) - 381{,}816$$
$$y \approx 1{,}069{,}491$$

The number of immigrants admitted to the United States in 2015 will be about 1,069,491.

(c) The equation $y = 12{,}620.16t - 381{,}816$ has $-381{,}816$ for the y-intercept, indicating that the number of immigrants admitted in

the year 1900 was $-381{,}816$. Realistically, the number of immigrants cannot be a negative value, so the equation cannot be used for valid predicted values.

73. (a) Plot the points $(15, 1600)$, $(200, 15{,}000)$, $(290, 24{,}000)$, and $(520, 40{,}000)$.

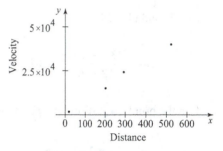

The points lie approximately on a line, so there appears to be a linear relationship between distance and time.

(b) The graph of any equation of the form $y = mx$ goes through the origin, so the line goes through $(520, 40{,}000)$ and $(0, 0)$.

$$m = \frac{40{,}000 - 0}{520 - 0} \approx 76.9$$
$$b = 0$$
$$y = 76.9x + 0$$
$$y = 76.9x$$

(c) Let $y = 60{,}000$; solve for x.

$$60{,}000 = 76.9x$$
$$780.23 \approx x$$

Hydra is about 780 megaparsecs from earth.

(d) $A = \dfrac{9.5 \times 10^{11}}{m}$, $m = 76.9$

$$A = \frac{9.5 \times 10^{11}}{76.9}$$
$$= 12.4 \text{ billion years}$$

75. (a) Let $t = 0$ correspond to 2000. Then the line representing the percent of respondents who got their news from the newspaper contains the points $(6, 40)$ and $(12, 29)$.

$$m = \frac{29 - 40}{12 - 6} \approx -1.83$$

Use the point $(6, 40)$ and the point-slope form.

$$y_n - 40 = -1.83(t - 6)$$
$$y_n = -1.83t + 10.98 + 40$$
$$y_n = -1.83t + 50.98$$

(b) The line representing the percent of respondents who got their news online contains the points $(6, 23)$ and $(12, 39)$.

$$m = \frac{39 - 23}{12 - 6} \approx 2.67$$

Use the point $(6, 23)$ and the point-slope form.

$$y_o - 23 = 2.67(t - 6)$$
$$y_o = 2.67t - 16.02 + 23$$
$$y_o = 2.67t + 6.98$$

(c) The number of respondents who got their news from newspapers is decreasing by about 1.83% per year, while the number of respondents who got their news online is increasing by about 2.67% per year.

1.2 Linear Functions and Applications

Your Turn 1

For $g(x) = -4x + 5$, calculate $g(-5)$.

$$g(x) = -4x + 5$$
$$g(-5) = -4(-5) + 5$$
$$= 20 + 5$$
$$= 25$$

Your Turn 2

For the demand and supply functions given in Example 2, find the quantity of watermelon demanded and supplied at a price of $3.30 per watermelon.

$$p = D(q) = 9 - 0.75q$$
$$3.30 = 9 - 0.75q$$
$$0.75q = 5.7$$
$$q = \frac{5.7}{0.75} = 7.6$$

Since the quantity is in thousands, 7600 watermelon are demanded at a price of $3.30.

$$p = S(q) = 0.75q$$

$$3.30 = 0.75q$$
$$q = \frac{3.3}{0.75} = 4.4$$

Since the quantity is in thousands, 4400 watermelon are supplied at a price of $3.30.

Your Turn 3

Set the two price expressions equal and solve for the equilibrium quantity q.

$$10 - 0.85q = 0.4q$$
$$10 = 1.25q$$
$$q = \frac{10}{1.25} = 8$$

The equilibrium quantity is 8000 watermelon. Use either price expression to find the equilibrium price p.

$$p = 0.4q$$
$$p = 0.4(8) = 3.2$$

The equilibrium price is $3.20 per watermelon.

Your Turn 4

The marginal cost is the slope of the cost function $C(x)$, so this function has the form $C(x) = 15x + b$. To find b, use the fact that producing 80 batches costs $1930.

$$C(x) = 15x + b$$
$$C(80) = 15(80) + b$$
$$1930 = 1200 + b$$
$$b = 730$$

Thus the cost function is $C(x) = 15x + 730$.

Your Turn 5

The fixed cost is b, this function has the form $C(x) = mx + 7145$. To find m, use the fact that producing 100 items costs $7965.

$$C(x) = mx + 7145$$
$$C(100) = 100m + 7145$$
$$7965 = 100m + 7145$$
$$820 = 100m$$
$$m = 8.2$$

Thus the cost function is $C(x) = 8.2x + 7145$.

Your Turn 6

The cost function is $C(x) = 35x + 250$ and the revenue function is $R(x) = 58x$. Thus the profit function is

$$P(x) = R(x) - C(x)$$
$$= 58x - (35x + 250)$$
$$= 23x - 250$$

The profit is to be $8030.

$$P(x) = 23x - 250$$
$$8030 = 23x - 250$$
$$23x = 8280$$
$$x = \frac{8280}{23} = 360$$

Sale of 360 units will produce $8030 profit.

1.2   Warmup Exercises

W1. $3(x - 2)^2 + 6(x + 4) - 5x + 4$

$3(5 - 2)^2 + 6(5 + 4) - 5(5) + 4$

$= 3(3)^2 + 6(9) - 5(5) + 4$

$= 27 + 54 - 25 + 4$

$= 60$

W2.

1.2   Exercises

1. $f(2) = 7 - 5(2) = 7 - 10 = -3$

3. $f(-3) = 7 - 5(-3) = 7 + 15 = 22$

5. $g(1.5) = 2(1.5) - 3 = 3 - 3 = 0$

7. $g\left(-\frac{1}{2}\right) = 2\left(-\frac{1}{2}\right) - 3 = -1 - 3 = -4$

9. $f(t) = 7 - 5(t) = 7 - 5t$

11. This statement is true.

When we solve $y = f(x) = 0$, we are finding the value of x when $y = 0$, which is the x-intercept. When we evaluate $f(0)$, we are finding the value of y when $x = 0$, which is the y-intercept.

13. This statement is true.

Only a vertical line has an undefined slope, but a vertical line is not the graph of a function. Therefore, the slope of a linear function cannot be undefined.

15. The fixed cost is constant for a particular product and does not change as more items are made. The marginal cost is the rate of change of cost at a specific level of production and is equal to the slope of the cost function at that specific value; it approximates the cost of producing one additional item.

19. $10 is the fixed cost and $2.25 is the cost per hour.

Let $x = $ number of hours;

$R(x) = $ cost of renting a snowboard for x hours.

Thus,

$R(x) = $ fixed cost + (cost per hour) · (number of hours)

$R(x) = 10 + (2.25)(x)$

$= 2.25x + 10$

21. $2 is the fixed cost and $0.75 is the cost per half-hour.

Let $x = $ the number of half-hours;

$C(x) = $ the cost of parking a car for x half-hours.

Thus,
$$C(x) = 2 + 0.75x$$
$$= 0.75x + 2$$

23. Fixed cost, $100; 50 items cost $1600 to produce.

Let $C(x) = $ cost of producing x items.

$C(x) = mx + b$, where b is the fixed cost.

$$C(x) = mx + 100$$

Now,

$C(x) = 1600$ when $x = 50$, so

$$1600 = m(50) + 100$$
$$1500 = 50m$$
$$30 = m.$$

Thus, $C(x) = 30x + 100$.

25. Marginal cost: $75; 50 items cost $4300.

$$C(x) = 75x + b$$

Now, $C(x) = 4300$ when $x = 50$.

$$4300 = 75(50) + b$$
$$4300 = 3750 + b$$
$$550 = b$$

Thus, $C(x) = 75x + 550$.

27. $D(q) = 16 - 1.25q$

(a) $D(0) = 16 - 1.25(0) = 16 - 0 = 16$

When 0 watches are demanded, the price is $16.

(b) $D(4) = 16 - 1.25(4) = 16 - 5 = 11$

When 400 watches are demanded, the price is $11.

(c) $D(8) = 16 - 1.25(8) = 16 - 10 = 6$

When 800 watches are demanded, the price is $6.

(d) Let $D(q) = 8$. Find q.

$$8 = 16 - 1.25q$$

$$\frac{5}{4}q = 8$$

$$q = 6.4$$

When the price is \$8, the number of watches demanded is 640.

(e) Let $D(q) = 10$. Find q.

$$10 = 16 - 1.25q$$

$$\frac{5}{4}q = 6$$

$$q = 4.8$$

When the price is \$10, the number of watches demanded is 480.

(f) Let $D(q) = 12$. Find q.

$$12 = 16 - 1.25q$$

$$\frac{5}{4}q = 4$$

$$q = 3.2$$

When the price is \$12, the number of watches demanded is 320.

(g)

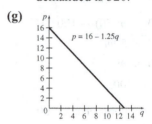

(h) $S(q) = 0.75q$

Let $S(q) = 0$. Find q.

$$0 = 0.75q$$

$$0 = q$$

When the price is \$0, the number of watches supplied is 0.

(i) Let $S(q) = 10$. Find q.

$$10 = 0.75q$$

$$\frac{40}{3} = q$$

$$q = 13.\overline{3}$$

When the price is \$10, The number of watches supplied is about 1333.

(j) Let $S(q) = 20$. Find q.

$$20 = 0.75q$$

$$\frac{80}{3} = q$$

$$q = 26.\overline{6}$$

When the price is \$20, the number of watches demanded is about 2667.

(k)

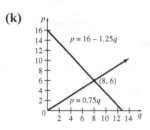

(l)
$$D(q) = S(q)$$
$$16 - 1.25q = 0.75q$$
$$16 = 2q$$
$$8 = q$$

$$S(8) = 0.75(8) = 6$$

The equilibrium quantity is 800 watches, and the equilibrium price is \$6.

29. $p = S(q) = \dfrac{2}{5}q; \ p = D(q) = 100 - \dfrac{2}{5}q$

(a)

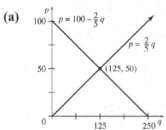

(b) $S(q) = D(q)$

$$\frac{2}{5}q = 100 - \frac{2}{5}q$$

$$\frac{4}{5}q = 100$$

$$q = 125$$

$$S(125) = \frac{2}{5}(125) = 50$$

The equilibrium quantity is 125, the equilibrium price is \$50.

31. Use the supply function to find the equilibrium quantity that corresponds to the given equilibrium price of \$4.50.

$$S(q) = p = 0.3q + 2.7$$
$$4.50 = 0.3q + 2.7$$
$$1.8 = 0.3q$$
$$6 = q$$

The line that represents the demand function goes through the given point $(2, 6.10)$ and the equilibrium point $(6, 4.50)$.

$$m = \frac{4.50 - 6.10}{6 - 2} = -0.4$$

Use point-slope form and the point $(2, 6.10)$.

$$D(q) - 6.10 = -0.4(q - 2)$$
$$D(q) = -0.4q + 0.8 + 6.10$$
$$D(q) = -0.4q + 6.9$$

33. $C(x) = 5x + 20; R(x) = 15x$

(a) $\qquad C(x) = R(x)$
$$5x + 20 = 15x$$
$$20 = 10x$$
$$2 = x$$

The break-even quantity is 2 units.

(b) $\qquad P(x) = R(x) - C(x)$
$$P(x) = 15x - (5x + 20)$$
$$P(100) = 15(100) - (5 \cdot 100 + 20)$$
$$= 1500 - 520$$
$$= 980$$

The profit from 100 units is $980.

(c) $\qquad P(x) = 500$
$$15x - (5x + 20) = 500$$
$$10x - 20 = 500$$
$$10x = 520$$
$$x = 52$$

For a profit of $500, 52 units must be produced.

35. (a) $C(x) = mx + b; \; m = 3.50; \; C(60) = 300$
$$C(x) = 3.50x + b$$

Find b.
$$300 = 3.50(60) + b$$
$$300 = 210 + b$$
$$90 = b$$
$$C(x) = 3.50x + 90$$

(b) $R(x) = 9x$
$$C(x) = R(x)$$
$$3.50x + 90 = 9x$$
$$90 = 5.5x$$
$$16.36 = x$$

Joanne must produce and sell 17 shirts.

(c) $P(x) = R(x) - C(x); P(x) = 500$
$$500 = 9x - (3.50x + 90)$$
$$500 = 5.5x - 90$$
$$590 = 5.5x$$
$$107.27 = x$$

To make a profit of $500, Joanne must produce and sell 108 shirts.

37. (a) Using the points $(100, 11.02)$ and $(400, 40.12)$,

$$m = \frac{40.12 - 11.02}{400 - 100}$$
$$= \frac{29.1}{300} = 0.097.$$

$$y - 11.02 = 0.097(x - 100)$$
$$y - 11.02 = 0.097x - 9.7$$
$$y = 0.097x + 1.32$$
$$C(x) = 0.097x + 1.32$$

(b) The fixed cost is given by the constant in $C(x)$. It is $1.32.

(c) $C(1000) = 0.097(1000) + 1.32$
$$= 97 + 1.32$$
$$= 98.32$$

The total cost of producing 1000 cups is $98.32.

(d) $C(1001) = 0.097(1001) + 1.32$
$$= 97.097 + 1.32$$
$$= 98.417$$

The total cost of producing 10001 cups is $98.42.

(e) Marginal cost $= 98.417 - 98.32$
$$= \$0.097 \quad \text{or} \quad 9.7¢$$

(f) The marginal cost for *any* cup is the slope, $0.097 or 9.7¢. This means the cost of producing one additional cup of coffee would be 9.7¢.

39. $C(x) = 85x + 900$
$R(x) = 105x$

Set $C(x) = R(x)$ to find the break-even quantity.
$$85x + 900 = 105x$$
$$900 = 20x$$
$$45 = x$$

The break-even quantity is 45 units. You should decide not to produce since no more than 38 units can be sold.

$$P(x) = R(x) - C(x)$$
$$= 105x - (85x + 900)$$
$$= 20x - 900$$

The profit function is $P(x) = 20x - 900$.

41. $C(x) = 70x + 500$

$R(x) = 60x$

$$70x + 500 = 60x$$
$$10x = -500$$
$$x = -50$$

This represents a break-even quantity of -50 units. It is impossible to make a profit when the break-even quantity is negative. Cost will always be greater than revenue.

$$P(x) = R(x) - C(x)$$
$$= 60x - (70x + 500)$$
$$= -10x - 500$$

The profit function is $P(x) = -10x - 500$.

43. Since the fixed cost is $400, the cost function is $C(x) = mx + 100$, where m is the cost per unit.

The revenue function is $R(x) = px$, where p is the price per unit.

The profit $P(x) = R(x) - C(x)$ is 0 at the given break-even quantity of 80.

$$P(x) = px - (mx + 400)$$
$$P(x) = px - mx - 400$$
$$P(x) = Mx - 400 \qquad (\text{Let } M = p - m.)$$
$$P(80) = M \cdot 80 - 400$$
$$0 = 80M - 400$$
$$400 = 80M$$
$$5 = M$$

So, the linear profit function is $P(x) = 5x - 400$, and the marginal profit is 5.

45. (a) $f(x) = 34x + 230$

$$1000 = 34x + 230$$
$$770 = 34x$$
$$x = 22.647$$

Approximately 23 acorns per square meter would produce 1000 deer tick larvae per 400 square meters.

(b) The slope is 34, which indicates that the number of deer tick larvae per 400 square meters in the spring will increase by 34 for each additional acorn per square meter in the fall.

47. Use the formula derived in Example 8 in this section of the textbook.

$$F = \frac{9}{5}C + 32 \quad \text{or} \quad C = \frac{5}{9}(F - 32)$$

(a) $F = 58$; find C.

$$C = \frac{5}{9}(58 - 32)$$
$$C = \frac{5}{9}(26) = 14.4$$

The temperature is $14.4°$C.

(b) $F = -20$; find C.

$$C = \frac{5}{9}(F - 32)$$
$$C = \frac{5}{9}(-20 - 32)$$
$$C = \frac{5}{9}(-52) = -28.9$$

The temperature is $-28.9°$C.

(c) $C = 50$; find F.

$$F = \frac{9}{5}C + 32$$
$$F = \frac{9}{5}(50) + 32$$
$$F = 90 + 32 = 122$$

The temperature is $122°$F.

49. If the temperatures are numerically equal, then $F = C$.

$$F = \frac{9}{5}C + 32$$
$$C = \frac{9}{5}C + 32$$
$$-\frac{4}{5}C = 32$$
$$C = -40$$

The Celsius and Fahrenheit temperatures are numerically equal at $-40°$.

1.3   The Least Squares Line

Your Turn 1

x	y	xy	x^2	y^2
1	3	3	1	9
2	4	8	4	16
3	6	18	9	36
4	5	20	16	25
5	7	35	25	49
6	8	48	36	64
$\Sigma x =$ 21	$\Sigma y =$ 33	$\Sigma xy =$ 132	$\Sigma x^2 =$ 91	$\Sigma y^2 =$ 199

The number of data points is $n = 6$. Putting the column totals into the formula for the slope m, we get

$$m = \frac{n(\Sigma xy) - (\Sigma x)(\Sigma y)}{n(\Sigma x^2) - (\Sigma x)^2}$$

$$m = \frac{6(132) - (21)(33)}{6(91) - (21)^2}$$

$$m \approx 0.9429$$

$$b = \frac{\Sigma y - m(\Sigma x)}{n}$$

$$= \frac{33 - (0.9429)(21)}{6} \approx 2.2$$

The least square line is $Y = 0.9429x + 2.2$.

Your Turn 2
Put the column totals computed in Your Turn 1 into the formula for the correlation r.

$$r = \frac{n(\Sigma xy) - (\Sigma x)(\Sigma y)}{\sqrt{n(\Sigma x^2) - (\Sigma x)^2} \cdot \sqrt{n(\Sigma y^2) - (\Sigma y)^2}}$$

$$= \frac{6(132) - (21)(33)}{\sqrt{6(91) - (21)^2} \cdot \sqrt{6(199) - (33)^2}}$$

$$\approx 0.9429$$

1.3   Exercises

3. (a)

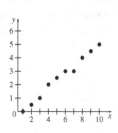

b)

x	y	xy	x^2	y^2
1	0	0	1	0
2	0.5	1	4	0.25
3	1	3	9	1
4	2	8	16	4
5	2.5	12.5	25	6.25
6	3	18	36	9
7	3	21	49	9
8	4	32	64	16
9	4.5	40.5	81	20.25
10	5	50	100	25
55	25.5	186	385	90.75

$$r = \frac{n(\Sigma xy) - (\Sigma x)(\Sigma y)}{\sqrt{n(\Sigma x^2) - (\Sigma x)^2} \cdot \sqrt{n(\Sigma y^2) - (\Sigma y)^2}}$$

$$= \frac{10(186) - (55)(25.5)}{\sqrt{10(385) - (55)^2} \cdot \sqrt{10(90.75) - (25.5)^2}}$$

$$\approx 0.993$$

(c)   The least squares line is of the form
$Y = mx + b$. First solve for m.

$$m = \frac{n(\Sigma xy) - (\Sigma x)(\Sigma y)}{n(\Sigma x^2) - (\Sigma x)^2}$$

$$= \frac{10(186) - (55)(25.5)}{10(385) - (55)^2}$$

$$= 0.5545454545 \approx 0.555$$

Now find b.

$$b = \frac{\Sigma y - m(\Sigma x)}{n}$$

$$= \frac{25.5 - 0.5545454545(55)}{10}$$

$$= -0.5$$

Thus, $Y = 0.555x - 0.5$.

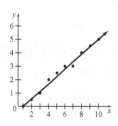

(d)   Let $x = 11$. Find Y.

$$Y = 0.55(11) - 0.5 = 5.6$$

5.

x	y	xy	x^2	y^2
1	1	1	1	1
1	2	2	1	4
2	1	2	4	1
2	2	4	4	4
9	9	81	81	81
15	15	90	91	91

(a) $n = 5$

$$m = \frac{n(\sum xy) - (\sum x)(\sum y)}{n(\sum x^2) - (\sum x)^2}$$

$$= \frac{5(90) - (15)(15)}{5(91) - (15)^2}$$

$$= 0.9782608 \approx 0.9783$$

$$b = \frac{\sum y - m(\sum x)}{n}$$

$$= \frac{15 - (0.9782608)(15)}{5} \approx 0.0652$$

Thus, $Y = 0.9783x + 0.0652$.

$$r = \frac{n(\sum xy) - (\sum x)(\sum y)}{\sqrt{n(\sum x^2) - (\sum x)^2} \cdot \sqrt{n(\sum y^2) - (\sum y)^2}}$$

$$= \frac{5(90) - (15)(15)}{\sqrt{5(91) - (15)^2} \cdot \sqrt{5(91) - (15)^2}} \approx 0.9783$$

(b)

x	y	xy	x^2	y^2
1	1	1	1	1
1	2	2	1	4
2	1	2	4	1
2	2	4	4	4
6	6	9	10	10

$n = 4$

$$m = \frac{n(\sum xy) - (\sum x)(\sum y)}{n(\sum x^2) - (\sum x)^2}$$

$$= \frac{4(9) - (6)(6)}{4(10) - (6)^2} = 0$$

$$b = \frac{\sum y - m(\sum x)}{n} = \frac{6 - (0)(6)}{4} = 1.5$$

Thus, $Y = 0x + 1.5$, or $Y = 1.5$.

$$r = \frac{n(\sum xy) - (\sum x)(\sum y)}{\sqrt{n(\sum x^2) - (\sum x)^2} \cdot \sqrt{n(\sum y^2) - (\sum y)^2}}$$

$$= \frac{4(9) - (6)(6)}{\sqrt{4(10) - (6)^2} \cdot \sqrt{4(10) - (6)^2}}$$

$$= 0$$

(c)

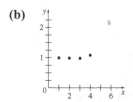

The point $(9, 9)$ is an outlier that has a strong effect on the least squares line and the correlation coefficient.

7.

x	y	xy	x^2	y^2
1	1	1	1	1
2	1	2	4	1
3	1	3	9	1
4	1.1	4.4	16	1.21
10	4.1	10.4	30	4.21

(a) $n = 4$

$$r = \frac{n(\sum xy) - (\sum x)(\sum y)}{\sqrt{n(\sum x^2) - (\sum x)^2} \cdot \sqrt{n(\sum y^2) - (\sum y)^2}}$$

$$= \frac{4(10.4) - (10)(4.1)}{\sqrt{4(30) - (10)^2} \cdot \sqrt{4(4.21) - (4.1)^2}}$$

$$= 0.7745966 \approx 0.7746$$

(b)

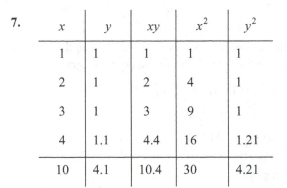

(c)   Yes; because the data points are either on or very close to the horizontal line $y = 1$, it seems that the data should have a strong linear relationship. The correlation coefficient does not describe well a linear relationship if the data points fit a horizontal line.

9. $\quad nb + (\sum x)m = \sum y$

$(\sum x)b + (\sum x^2)m = \sum xy$

$nb + (\sum x)m = \sum y$

$nb = (\sum y) - (\sum x)m$

$b = \dfrac{\sum y - m(\sum x)}{n}$

$(\sum x)\left(\dfrac{\sum y - m(\sum x)}{n}\right) + (\sum x^2)m = \sum xy$

$(\sum x)[(\sum y) - m(\sum x)] + nm(\sum x^2) = n(\sum xy)$

$(\sum x)(\sum y) - m(\sum x)^2 + nm(\sum x^2) = n(\sum xy)$

$nm(\sum x^2) - m(\sum x)^2 = n(\sum xy) - (\sum x)(\sum y)$

$m[n(\sum x^2) - (\sum x)^2] = n(\sum xy) - (\sum x)(\sum y)$

$m = \dfrac{n(\sum xy) - (\sum x)(\sum y)}{n(\sum x^2) - (\sum x)^2}$

11. (a) $\quad m = \dfrac{n(\sum xy) - (\sum x)(\sum y)}{n(\sum x^2) - (\sum x)^2}$

$= \dfrac{10(512.775) - (75)(70.457)}{10(645) - (75)^2}$

$= -0.189727 \approx 0.1897$

$b = \dfrac{\sum y - m(\sum x)}{n}$

$= \dfrac{70.457 - (-0.1897)(75)}{10}$

$= 8.46845 \approx 8.469$

Thus, $Y = 00.1897x + 8.469$.

(b) The year 2020 corresponds to $x = 20$.

$Y = -0.1897(20) + 8.469 = 4.675$

If the trend continues linearly, there will be about 4675 banks in 2020.

(c) Let $Y = 4$ (since Y is the number of banks in thousands) and find x.

$4 = -0.1897x + 8.469$

$-4.469 = -0.1897x$

$23.558 = x$

$24 \approx x$

The number of U.S. banks will drop below 4000 in the year $2000 + 24 = 2024$.

(d)

$r = \dfrac{n(\sum xy) - (\sum x)(\sum y)}{\sqrt{n(\sum x^2) - (\sum x)^2} \cdot \sqrt{n(\sum y^2) - (\sum y)^2}}$

$= \dfrac{10(512.775) - (75)(70.457)}{\sqrt{10(645) - (75)^2} \cdot \sqrt{10(499.481335) - (70.457)^2}}$

≈ -0.9847

Since r is very close to -1, the data has a strong linear relationship, and the least squares line fits the data very well.

13.

x	y	xy	x^2	y^2
5	89.7	448.5	25	8046.09
6	84.1	504.6	36	7072.81
7	81.9	573.3	49	6707.61
8	77.9	623.2	64	6068.41
9	73.5	661.5	81	5402.25
10	68.2	682.0	100	4651.24
11	63.8	701.8	121	4070.44
12	59.6	715.2	144	3552.16
68	598.7	4910.1	620	45,571.01

(a) $\quad m = \dfrac{n(\sum xy) - (\sum x)(\sum y)}{n(\sum x^2) - (\sum x)^2}$

$= \dfrac{8(4910.1) - (68)(598.7)}{8(620) - (68)^2}$

≈ -4.2583

$b = \dfrac{\sum y - m(\sum x)}{n}$

$= \dfrac{598.7 - (-4.2583)(68)}{8}$

≈ 111.033

Thus, $Y = -4.26x + 111.0$.

(b) The percent of households with landlines is decreasing at a rate of about 4.26% per year.

(c) The year 2015 corresponds to $x = 15$.

$Y = -4.26(15) + 111.0 = 47.1$

If the trend continues linearly, the percent of households will be about 47.1% in 2012.

The mean earning for workers with a bachelor's degree will exceed \$75,00 in the year $1900 + 121 = 2021$.

(d)   Let $Y = 40$ and find x.

$$40 = -4.26x + 111.0$$
$$-71 = -4.26x$$
$$16.67 = x$$
$$17 \approx x$$

The percent of households with landlines will dip below 40% in the year $2000 + 17 = 2017$.

(e)

$$r = \frac{n(\sum xy) - (\sum x)(\sum y)}{\sqrt{n(\sum x^2) - (\sum x)^2} \cdot \sqrt{n(\sum y^2) - (\sum y)^2}}$$

$$= \frac{8(4910.1) - (68)(598.7)}{\sqrt{8(620) - (68)^2} \cdot \sqrt{8(45,571.01) - (598.7)^2}}$$

$$\approx -0.9973$$

This means that the least squares line fits the data points extremely well.

15. (a)

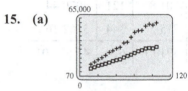

The data points lie in a linear pattern.

(b) $Y = 703.91x - 45,220$.

     $r = 0.9964$; there is a strong positive correlation among the data.

(c)   The mean earnings of high school graduates are growing by about \$704 per year.

(d) $Y = 1404.76x - 94,521$.

     $r = 0.9943$; there is a strong positive correlation among the data.

(e)   The mean earning of workers with a bachelor's degree are growing by about \$1405 per year.

(f)
$$75,000 = 703.91x - 45,220$$
$$120,220 = 703.91x$$
$$x = 170.79$$

The mean earnings for high school graduates will exceed \$75,000 in the year $1900 + 171 = 2071$.

$$75,000 = 1404.76x - 94,521$$
$$1169,521 = 1404.76x$$
$$x = 120.68$$

17. (a)

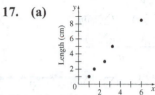

Yes, the data appear to be linear.

(b)

x	y	xy	x^2	y^2
5.8	8.6	49.88	33.64	73.96
1.5	1.9	2.85	2.25	3.61
2.3	3.1	7.13	5.29	9.61
1.0	1.0	1.0	1.0	1.0
3.3	5.0	16.5	10.89	25.0
13.9	19.6	77.36	53.07	113.18

$$m = \frac{n(\sum xy) - (\sum x)(\sum y)}{n(\sum x^2) - (\sum x)^2}$$

$$= \frac{5(77.36) - (13.9)(19.6)}{5(53.07) - 13.9^2}$$

$$= 1.585250901 \approx 1.585$$

$$b = \frac{\sum y - m(\sum x)}{n}$$

$$= \frac{19.6 - 1.585250901(13.9)}{5} \approx -0.487$$

$$Y = 1.585x - 0.487$$

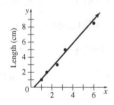

(c)   No, it gives negative values for small widths.

(d) $r = \dfrac{5(77.36) - (13.9)(19.6)}{\sqrt{5(53.07) - 13.9^2} \cdot \sqrt{5(113.18) - 19.6^2}}$

     ≈ 0.999

19. (a)

x	y	xy	x^2	y^2
88.6	20.0	1772	7849.96	400.0
71.6	16.0	1145.6	5126.56	256.0
93.3	19.8	1847.34	8704.89	392.04
84.3	18.4	1551.12	7106.49	338.56

80.6	17.1	1378.26	6496.36	292.41
75.2	15.5	1165.6	5655.04	240.25
69.7	14.7	1024.59	4858.09	216.09
82.0	17.1	1402.2	6724	292.41
69.4	15.4	1068.76	4816.36	237.16
83.3	16.2	1349.46	3938.89	262.44
79.6	15.0	1194	6336.16	225
82.6	17.2	1420.72	6822.76	295.84
80.6	16.0	1289.6	6496.36	256.0
83.5	17.0	1419.5	6972.25	289.0
76.3	14.4	1098.72	5821.69	207.36
1200.6	249.8	20,127.47	96,725.86	4200.56

$$m = \frac{n(\sum xy) - (\sum x)(\sum y)}{n(\sum x^2) - (\sum x)^2}$$

$$= \frac{15(20,127.47) - (1200.6)(249.8)}{15(96,725.86) - 1200.6^2}$$

$$= 0.211925009 \approx 0.212$$

$$b = \frac{\sum y - m(\sum x)}{n}$$

$$= \frac{249.8 - 0.211925(1200.6)}{15} \approx -0.309$$

$$Y = 0.212x - 0.309$$

(b) Let $x = 73$; find Y.

$$Y = 0.212(73) - 0.309 \approx 15.2$$

If the temperature were 73° F, you would expect to hear 15.2 chirps per second.

(c) Let $Y = 18$; find x.

$$18 = 0.212x - 0.309$$

$$18.309 = 0.212x$$

$$86.4 \approx x$$

When the crickets are chirping 18 times per second, the temperature is 86.4°F.

(d)

$$r = \frac{15(20,127) - (1200.6)(249.8)}{\sqrt{15(96,725.86) - (1200.6)^2} \cdot \sqrt{15(4200.56) - (249.8)^2}}$$

$$= 0.835$$

21.

x	y	xy	x^2	y^2
0	8.414	0.000	0	70.7954
5	10.989	54.945	25	120.7581
10	13.359	133.590	100	178.4629
15	15.569	233.535	225	242.3938
20	17.604	352.080	400	309.9008

25	19.971	499.275	625	398.8408
30	22.315	669.45	900	497.9592
105	108.221	1942.875	2275	1819.111

(a)

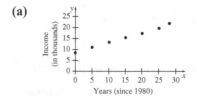

Yes, the data appear to lie along a straight line.

(b)

$$r = \frac{n(\sum xy) - (\sum x)(\sum y)}{\sqrt{n(\sum x^2) - (\sum x)^2} \cdot \sqrt{n(\sum y^2) - (\sum y)^2}}$$

$$= \frac{7(1942.875) - (105)(108.221)}{\sqrt{7(2275) - (105)^2} \cdot \sqrt{7(1819.111) - (108.221)^2}}$$

$$= 0.9996$$

The value indicates a strong linear correlation.

(c)

$$m = \frac{n(\sum xy) - (\sum x)(\sum y)}{n(\sum x^2) - (\sum x)^2}$$

$$= \frac{7(1942.85) - (105)(108.221)}{7(2275) - (105)^2} \approx 0.45651 \approx 0.4565$$

$$b = \frac{\sum y - m(\sum x)}{n}$$

$$= \frac{108.221 - (0.45651)(105)}{7} \approx 8.612$$

Thus, $Y = 0.4565x + 8.612$.

(d) The year 2018 corresponds to $x = 2018 - 1980 = 38$.

$$Y = 0.45651(38) + 8.612 = 25.959$$

The predicted poverty level in the year 2018 is $25,959.

23. (a)

x	y	xy	x^2	y^2
540	20	10,800	291,600	400
510	16	8160	260,100	256
490	10	4900	240,100	100
560	8	4480	313,600	64
470	12	5640	220,900	144
600	11	6600	360,000	121

540	10	5400	291,600	100
580	8	4640	336,400	64
680	15	10,200	462,400	225
560	8	4480	313,600	64
560	13	7280	313,600	169
500	14	7000	250,000	196
470	10	4700	220,900	100
440	10	4400	193,600	100
520	11	5720	270,400	121
620	11	6820	384,400	121
680	8	5440	462,400	64
550	8	4400	302,500	64
620	7	4340	384,400	49
10,490	210	115,400	5,872,500	2522

$$m = \frac{n(\sum xy) - (\sum x)(\sum y)}{n(\sum x^2) - (\sum x)^2}$$

$$m = \frac{19(115,400) - (10,490)(210)}{19(5,872,500) - 10,490^2}$$

$$m = -0.0066996227 \approx -0.0067$$

$$b = \frac{\sum y - m(\sum x)}{n}$$

$$b = \frac{210 - (-0.0066996227)(10,490)}{19} \approx 14.75$$

$$Y = -0.0067x + 14.75$$

(b) Let $x = 420$; find Y.
$$Y = -0.0067(420) + 14.75$$
$$= 11.936 \approx 12$$

(c) Let $x = 620$; find Y.
$$Y = -0.0067(620) + 14.75$$
$$= 10.596 \approx 11$$

(d)
$$r = \frac{19(115,400) - (10,490)(210)}{\sqrt{19(5,872,500) - (10,490)^2} \cdot \sqrt{19(2522) - 210^2}}$$
$$\approx -0.13$$

(e) There is no linear relationship between a student's math SAT and mathematics placement test scores.

25. (a)

x	y	xy	x^2	y^2
150	5000	750,000	22,500	25,000,000
175	5500	962,500	30,625	30,250,000
215	6000	1,290,000	46,225	36,000,000

250	6500	1,625,000	62,500	42,250,000
280	7000	1,960,000	78,400	49,000,000
310	7500	2,325,000	96,100	56,250,000
350	8000	2,800,000	122,500	64,000,000
370	8500	3,145,000	136,900	72,250,000
420	9000	3,780,000	176,400	81,000,000
450	9500	4,275,000	202,500	90,250,000
2970	72,500	22,912,500	974,650	546,250,000

$$m = \frac{n(\sum xy) - (\sum x)(\sum y)}{n(\sum x^2) - (\sum x)^2}$$

$$m = \frac{10(22,912,500) - (2970)(72,500)}{10(974,650) - 2970^2}$$

$$m = 14.90924806$$
$$\approx 14.9$$

$$b = \frac{\sum y - m(\sum x)}{n}$$

$$b = \frac{72,500 - 14.9(2970)}{10}$$

$$b \approx 2820$$

$$Y = 14.9x + 2820$$

(b) Let $x = 150$; find Y.
$$Y = 14.9(150) + 2820$$
$$Y \approx 5060, \text{ compared to actual } 5000$$

Let $x = 280$; Find Y.
$$Y = 14.9(280) + 2820$$
$$\approx 6990, \text{ compared to actual } 7000$$

Let $x = 420$; find Y.
$$Y = 14.9(420) + 2820$$
$$\approx 9080, \text{ compared to actual } 9000$$

(c) Let $x = 230$; find Y.
$$Y = 14.9(230) + 2820$$
$$\approx 6250$$

Adam would need to buy a 6500 BTU air conditioner.

27. (a) Use a calculator's statistical features to obtain the least squares line.
$$Y = -0.1271x + 113.61$$

(b) $Y = -0.3458x + 146.65$

(c) Set the two expressions for Y equal and solve for x.

$$-0.1271x + 113.61 = -0.3458x + 146.65$$
$$0.2187x = 33.04$$
$$x \approx 151$$

The women's record will catch up with the men's record in $1900 + 151$, or in the year 2051.

(d) $r_{men} \approx -0.9762$

$r_{women} \approx -0.9300$

Both sets of data points closely fit a line with negative slope.

(e)
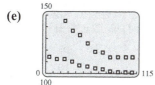

29.

x	y
0	0.0
2.317	11.5
3.72	18.9
5.6	27.8
7.08	32.8
7.5	36.0
8.5	43.9
10.6	51.5
11.93	58.4
15.23	71.8
17.82	80.9
18.97	85.2
20.83	91.3
23.38	100.5

(a) Skaggs' average speed was $100.5/23.38 \approx 4.299$ miles per hour.

(b)
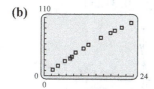

The data appear to lie approximately on a straight line.

(c) Using a graphing calculator,

$$Y = 4.317x + 3.419.$$

(d) Using a graphing calculator,

$$r \approx 0.9971$$

Yes, the least squares line is a very good fit to the data.

(e) A good value for Skaggs' average speed would be the slope of the least squares line, or

$$m = 4.317 \text{ miles per hour.}$$

This value is faster than the average speed found in part (a). The value 4.317 miles per hour is most likely the better value because it takes into account all 14 data pairs.

Chapter 1 Review Exercises

1. False; a line can have only one slant, so its slope is unique.

2. False; the equation $y = 3x + 4$ has slope 3.

3. True; the point $(3, -1)$ is on the line because $-1 = -2(3) + 5$ is a true statement.

4. False; the points $(2, 3)$ and $(2, 5)$ do not have the same y-coordinate.

5. True; the points $(4, 6)$ and $(5, 6)$ do have the same y-coordinate.

6. False; the x-intercept of the line $y = 8x + 9$ is $-\frac{9}{8}$.

7. True; $f(x) = \pi x + 4$ is a linear function because it is in the form $y = mx + b$, where m and b are real numbers.

8. False; $f(x) = 2x^2 + 3$ is not linear function because it isn't in the form $y = mx + b$, and it is a second-degree equation.

9. False; the line $y = 3x + 17$ has slope 3, and the line $y = -3x + 8$ has slope -3. Since $3 \cdot -3 \neq -1$, the lines cannot be perpendicular.

10. False; the line $4x + 3y = 8$ has slope $-\frac{4}{3}$, and the line $4x + y = 5$ has slope -4. Since the slopes are not equal, the lines cannot be parallel.

11. False; a correlation coefficient of zero indicates that there is no linear relationship among the data.

12. True; a correlation coefficient always will be a value between -1 and 1.

13. Marginal cost is the rate of change of the cost function; the fixed cost is the initial expenses before production begins.

15. Through $(-3, 7)$ and $(2, 12)$

$$m = \frac{12 - 7}{2 - (-3)} = \frac{5}{5} = 1$$

17. Through the origin and $(11, -2)$

$$m = \frac{-2 - 0}{11 - 0} = -\frac{2}{11}$$

19. $4x + 3y = 6$

$$3y = -4x + 6$$

$$y = -\frac{4}{3}x + 2$$

Therefore, the slope is $m = -\frac{4}{3}$.

21. $y + 4 = 9$

$$y = 5$$

$$y = 0x + 5$$

$$m = 0$$

23. $y = 5x + 4$

$$m = 5$$

25. Through $(5, -1)$; slope $\frac{2}{3}$

Use point-slope form.

$$y - (-1) = \frac{2}{3}(x - 5)$$

$$y + 1 = \frac{2}{3}(x - 5)$$

$$3(y + 1) = 2(x - 5)$$

$$3y + 3 = 2x - 10$$

$$3y = 2x - 13$$

$$y = \frac{2}{3}x - \frac{13}{3}$$

27. Through $(-6, 3)$ and $(2, -5)$

$$m = \frac{-5 - 3}{2 - (-6)} = \frac{-8}{8} = -1$$

Use point-slope form.

$$y - 3 = -1[x - (-6)]$$

$$y - 3 = -x - 6$$

$$y = -x - 3$$

29. Through $(2, -10)$, perpendicular to a line with undefined slope

A line with undefined slope is a vertical line. A line perpendicular to a vertical line is a horizontal line with equation of the form $y = k$. The desired

line passed through $(2, -10)$, so $k = -10$. Thus, an equation of the desired line is $y = -10$.

31. Through $(3, -4)$ parallel to $4x - 2y = 9$

Solve $4x - 2y = 9$ for y.

$$-2y = -4x + 9$$

$$y = 2x - \frac{9}{2}$$

$$m = 2$$

The desired line has the same slope. Use the point-slope form.

$$y - (-4) = 2(x - 3)$$

$$y + 4 = 2x - 6$$

$$y = 2x - 10$$

Rearrange.

$$2x - y = 10$$

33. Through $(-1, 4)$; undefined slope

Undefined slope means the line is vertical.

The equation of the vertical line through $(-1, 4)$ is $x = -1$.

35. Through $(3, -5)$, parallel to $y = 4$

Find the slope of the given line.

$y = 0x + 4$, so $m = 0$, and the required line will also have slope 0.

Use the point-slope from.

$$y - (-5) = 0(x - 3)$$

$$y + 5 = 0$$

$$y = -5$$

37. $y = 4x + 3$

Let $x = 0$: $y = 4(0) + 3$
                 $y = 3$

Let $y = 0$: $0 = 4x + 3$
                 $-3 = 4x$
                 $-\frac{3}{4} = x$

Draw the line through $(0, 3)$ and $\left(-\frac{3}{4}, 0\right)$.

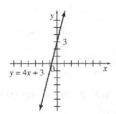

39. $3x - 5y = 15$

$$-5y = -3x + 15$$

$$y = \frac{3}{5}x - 3$$

When $x = 0$, $y = -3$.

When $y = 0$, $x = 5$.

Draw the line through $(0, -3)$ and $(5, 0)$.

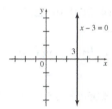

41. $x - 3 = 0$

$$x = 3$$

This is the vertical line through $(3, 0)$.

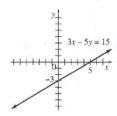

43. $y = 2x$

When $x = 0$, $y = 0$.

When $x = 1$, $y = 2$.

Draw the line through $(0, 0)$ and $(1, 2)$.

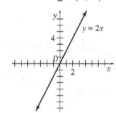

45. (a) Let $t = 0$ represent the year 2000. The line goes through $(0, 100)$ and $(13, 440)$.

$$m = \frac{440 - 100}{13 - 0}$$

$$= \frac{340}{13} \approx 26.2$$

$$y - 100 = 26.2(t - 0)$$

$$y = 26.2t + 100$$

(b) The imports from China are increasing by about $26.2 billion per year.

(c) Let $t = 15$.

$$y = 26.2t + 100$$

$$y = 26.2(15) + 100$$

$$y = 493$$

The amount of imports from China in 2015 will be about $493 billion.

(d) Let $y = 600$; solve for t.

$$600 = 26.2t + 100$$

$$500 = 26.2t$$

$$t \approx 19.084$$

The imports from China would be at least $600 billion in the year $2000 + 20 = 2020$.

47. (a) Let $t = 0$ represent the year 2000. The line goes through $(7, 55{,}627)$ and $(12, 51{,}017)$.

$$m = \frac{51{,}017 - 55{,}627}{12 - 7}$$

$$= \frac{-4610}{5} \approx -922$$

$$y - 55{,}627 = -922(t - 7)$$

$$y = -922t + 6454 + 55{,}627$$

$$y = -922t + 62{,}081$$

(b) The median income for all U.S. households is decreasing by about $922 per year.

(c) Let $t = 15$.

$$y = -922t + 62{,}081$$

$$y = -922(15) + 62{,}081$$

$$y = 48{,}251$$

The median income for all U.S. households in 2015 will be about $48,251.

(d) Let $y = 40{,}000$; solve for x.

$$40{,}000 = -922t + 62{,}081$$

$$-22{,}081 = -922t$$

$$t \approx 23.949$$

The median income would drop below $40,000 in the year $2000 + 24 = 2024$.

49. (a) The line that represents the supply function goes through the points $(60, 40)$ and $(100, 60)$.

$$m = \frac{60 - 40}{100 - 60} \approx 0.5$$

Use $(60, 40)$ and the point-slope form.

$$p - 40 = 0.5(q - 60)$$

$$p = 0.5q - 30 + 40$$

$$p = S(q) = 0.5q + 10$$

(b) The line that represents the demand function goes through the points (50, 47.50) and (80, 32.50).

$$m = \frac{32.50 - 47.50}{80 - 50} \approx -0.5$$

Use $(50, 47.50)$ and the point-slope form.

$$p - 47.50 = -0.5(q - 50)$$
$$p = 0.5q + 25 + 47.50$$
$$p = D(q) = -0.5q + 72.50$$

(c) Set supply equal to demand and solve for q.

$$0.5q + 10 = -0.5q + 72.50$$
$$1q = 62.50$$
$$q = 62.5$$

$$S(62.5) = 0.5(62.5) + 10 = 41.25$$

The equilibrium quantity is about 62.5 dietary supplement pills, and the equilibrium price is about \$41.25.

51. Fixed cost is \$2000; 36 units cost \$8480.

Two points on the line are (0, 2000) and (36, 8480), so

$$m = \frac{8480 - 2000}{36 - 0} = \frac{6480}{36} = 180.$$

Use point-slope form.

$$y = 180x + 2000$$
$$C(x) = 180x + 2000$$

53. Thirty units cost \$1500; 120 units cost \$5640. Two points on the line are (30, 1500), (120, 5640), so

$$m = \frac{5640 - 1500}{120 - 30} = \frac{4140}{90} = 46.$$

Use point-slope form.

$$y - 1500 = 46(x - 30)$$
$$y = 46x - 1380 + 1500$$
$$y = 46x + 120$$
$$C(x) = 46x + 120$$

55. (a) $C(x) = 3x + 160;\ R(x) = 7x$

$$C(x) = R(x)$$
$$3x + 160 = 7x$$
$$160 = 4x$$
$$40x = x$$

The break-even quantity is 40 pounds.

(b) $R(40) = 7 \cdot 40 = \$280$

The revenue for 40 pounds is \$280.

57.

x	y	x	y
2	26,150	8	28,350
3	27,550	9	28,966
4	28,050	10	29,793
5	28,400	11	30,659
6	28,450	12	30,910
7	28,200		

(a) Use the points (2, 26,150) and (12, 30,910) to find the slope.

$$m = \frac{30,910 - 26,150}{12 - 2} = 476$$

$$y - 26,150 = 476(t - 2)$$
$$y - 26,150 = 476t - 952$$
$$y = 476t + 25,198$$

(b) Use the points (4, 28,050) and (12, 30,910) to find the slope.

$$m = \frac{30,910 - 28,050}{12 - 4} = 357.5$$

$$y - 28,050 = 357.5(t - 4)$$
$$y - 28,050 = 357.5t - 1430$$
$$y = 357.5t + 26,620$$

(c) Using a graphing calculator, the least squares line is $Y = 386t + 25,975$.

(d)

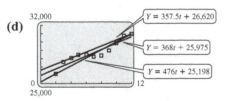

(e) $r = 0.9356$

59.

t	Y (Beef)	y (Pork)	y (Chicken)
0	64.5	47.8	54.2
12	54.5	42.6	56.6

(a) For $b(t)$: Use the points (0, 64.5) and (12, 54.4) to find the slope.

$$m = \frac{54.5 - 64.5}{12 - 0} = -0.833$$

$$y - 64.5 = -0.833(t - 0)$$
$$y = -0.833t + 64.5$$

For $p(t)$: Use the points $(0, 47.8)$ and $(12, 42.6)$ to find the slope.

$$m = \frac{42.6 - 47.8}{12 - 0} = -0.433$$
$$y - 47.8 = -0.433(t - 0)$$
$$y = -0.433t + 47.8$$

For $c(t)$: Use the points $(0, 54.2)$ and $(12, 56.6)$ o find the slope.

$$m = \frac{56.6 - 54.2}{12 - 0} = 0.2$$
$$y - 54.2 = 0.2(t - 0)$$
$$y = 0.2t + 54.2$$

(b) Beef is decreasing by about 0.833 lb/yr; pork is decreasing by about 0.433 lb/yr; chicken is increasing by about 0.2 lb/yr.

(c) $-0.833t + 64.5 = 0.2t + 54.2$

$$-1.033t = -10.3$$
$$t \approx 9.97$$

The consumption of chicken surpassed the consumption of beef in the year $2000 + 10 = 2010$.

(d) $y = -0.833t + 64.5$

$$y = -0.833(15) + 64.5 \approx 52.005$$

$$y = -0.433t + 47.8$$
$$y = -0.433(15) + 47.8 \approx 41.305$$

$$y = 0.2t + 54.2$$
$$y = 0.2(15) + 54.2 = 57.2$$

The consumption of beef will be about 52.0 lb, the consumption of pork will be about 41.3 lb, and the consumption of chicken will be about 57.2 lb in 2015.

61. (a)

x	y	xy	x^2	y^2
130	170	22,100	16,900	28,900
138	160	22,080	19,044	25,600
142	173	24,566	20,164	29,929
159	181	28,779	25,281	32,761
165	201	33,165	27,225	40,401
200	192	38,400	40,000	36,864
210	240	50,400	44,100	57,600
250	290	72,500	62,500	84,100
1394	1607	291,990	255,214	336,155

$$m = \frac{n(\sum xy) - (\sum x)(\sum y)}{n(\sum x^2) - (\sum x)^2}$$

$$m = \frac{8(291,990) - (1394)(1607)}{8(225,214) - 1394^2}$$

$$m = 0.9724399854 \approx 0.9724$$

$$b = \frac{\sum y - m(\sum x)}{n}$$

$$b = \frac{1607 - 0.9724(1394)}{8} \approx 31.43$$

$$Y = 0.9724x + 31.43$$

(b) Let $x = 190$; find Y.

$$Y = 0.9724(190) + 31.43$$
$$Y = 216.19 \approx 216$$

The cholesterol level for a person whose blood sugar level is 190 would be about 216.

(c) $r = \dfrac{8(291,990) - (1394)(1607)}{\sqrt{8(255,214) - 1394^2} \cdot \sqrt{8(336,155) - 1607^2}}$

$$= 0.933814 \approx 0.93$$

63. (a) Use the points $(0, 6400)$ and $(12, 9520)$ to find the slope.

$$m = \frac{9520 - 6400}{12 - 0} = 260$$

$$b = 6400$$

The linear equation for the number of families below the poverty level since 2000 is $y = 260t + 6400$.

(b) Use the points $(4, 7835)$ and $(12, 9520)$ to find the slope.

$$m = \frac{9520 - 7835}{12 - 4} = 210.6$$

$$y - 7835 = 210.6(t - 4)$$
$$y - 7835 = 210.6t - 842.4$$
$$y = 210.6t + 6992.6$$

The linear equation for the number of families below the poverty level since 2000 is $y = 210.6t + 6992.6$.

(c) Using a graphing calculator, the least squares line is $Y = 247.1t + 6532.0$.

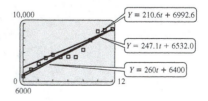

    (d) The least squares line best describes the data.
Since the data seems to fit a straight line, a
linear model describes the data well.

    (e) Using a graphing calculator, $r \approx 0.9515$.

65. Use a graphing calculator to find these
correlations.

    (a) Correlation between years since 2000 and
length: $r = 0.4529$

    (b) Correlation between length and rating:
$r = 0.3955$

    (c) Correlation between years since 2000 and
rating: $r = -0.4768$

    (d)

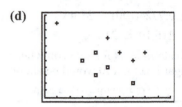

    This calculator graph plots year (on the
horizontal axis) versus rating (on the vertical
axis). Squares represent movies with lengths
no more than 110 minutes, and plus signs
represent movies with lengths 115 minutes or
more.

Chapter 2

NONLINEAR FUNCTIONS

2.1 Properties of Functions

Your Turn 1

$$y = \frac{1}{\sqrt{x^2 - 4}}$$

Since the denominator cannot be zero and the radicand cannot be negative, the domain includes only those values of x satisfying $x^2 - 4 > 0$. Using the methods for solving a quadratic inequality, we get the domain $(-\infty, -2) \cup (2, \infty)$. Since the denominator can never be negative, y cannot be negative or zero. So, the range is $(0, \infty)$.

Your Turn 2

$$f(x) = 2x^2 - 3x - 4$$

(a)
$$\begin{aligned}
f(x + h) &= 2(x + h)^2 - 3(x + h) - 4 \\
&= 2(x^2 + 2xh + h^2) - 3(x + h) - 4 \\
&= 2x^2 + 4xh + 2h^2 - 3x - 3h - 4
\end{aligned}$$

(b)
$$\begin{aligned}
f(x) &= -5 \\
2x^2 - 3x - 4 &= -5 \\
2x^2 - 3x + 1 &= 0 \\
(2x - 1)(x - 1) &= 0
\end{aligned}$$

$$x = \frac{1}{2} \text{ or } x = 1$$

2.1 Warmup Exercises

W1. $4x^2 - 9 = 0$

$(2x - 3)(2x + 3) = 0$

$x = \dfrac{3}{2} \text{ or } x = -\dfrac{3}{2}$

W2. $2x^2 + 9x + 5 = 0$

$(2x - 1)(x + 5) = 0$

$x = \dfrac{1}{2} \text{ or } x = -5$

W3. $16 - x^2 \geq 0$

$x^2 \leq 16$

$-4 \leq x \leq 4$

$16 - x^2 \geq 0$ on the interval $[-4, 4]$.

W4. Solve $21 + 4x - x^2 = 0$.

$(7 - x)(3 + x) = 0$

$x = 7 \text{ or } x = -3$

These roots divide the real line into three intervals:

$$(-\infty, -3) \quad (-3, 7) \quad (7, \infty)$$

Test a point in each interval.

At $x = -5$, $21 + 4x - x^2 = -24 < 0$

At $x = 0$, $21 + 4x - x^2 = 21 > 0$

At $x = 10$, $21 + 4x - x^2 = -39 < 0$

So, $21 + 4x - x^2 \geq 0$ on the interval $[-3, 7]$.

2.1 Exercises

1. The x-value of 82 corresponds to two y-values, 93 and 14. In a function, each value of x must correspond to exactly one value of y.

    The rule is not a function.

3. Each x-value corresponds to exactly one y-value.

    The rule is a function.

5. $$y = x^3 + 2$$

    Each x-value corresponds to exactly one y-value.

    The rule is a function.

7. $x = |y|$

    Each value of x (except 0) corresponds to two y-values.

    The rule is not a function.

9. $y = 2x + 3$

x	-2	-1	0	1	2	3
y	-1	1	3	5	7	9

Pairs: $(-2, -1)$, $(-1, 1)$, $(0, 3)$, $(1, 5)$, $(2, 7)$, $(3, 9)$

Range: $\{-1, 1, 3, 5, 7, 9\}$

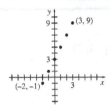

11. $2y - x = 5$

$$2y = 5 + x$$

$$y = \frac{1}{2}x + \frac{5}{2}$$

x	-2	-1	0	1	2	3
y	$\frac{3}{2}$	2	$\frac{5}{2}$	3	$\frac{7}{2}$	4

Pairs: $\left(-2, \frac{3}{2}\right)$, $(-1, 2)$, $\left(0, \frac{5}{2}\right)$, $(1, 3)$, $\left(2, \frac{7}{2}\right)$, $(3, 4)$.

Range: $\left\{\frac{3}{2}, 2, \frac{5}{2}, 3, \frac{7}{2}, 4\right\}$

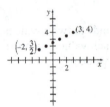

13. $y = x(x + 2)$

x	-2	-1	0	1	2	3
y	0	-1	0	3	8	15

Pairs: $(-2, 0), (-1, -1), (0, 0), (1, 3), (2, 8), (3, 15)$

Range: $\{-1, 0, 3, 8, 15\}$

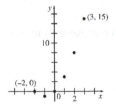

15. $y = x^2$

x	-2	-1	0	1	2	3
y	4	1	0	1	4	9

Pairs: $(-2, 4)$, $(-1, 1)$, $(0, 0)$, $(1, 1)$, $(2, 4)$, $(3, 9)$

Range: $\{0, 1, 4, 9\}$

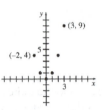

17. $f(x) = 2x$

x can take on any value, so the domain is the set of real numbers, $(-\infty, \infty)$.

19. $f(x) = x^4$

x can take on any value, so the domain is the set of real numbers, $(-\infty, \infty)$.

21. $f(x) = \sqrt{4 - x^2}$

For $f(x)$ to be a real number, $4 - x^2 \geq 0$. Solve $4 - x^2 = 0$.

$$(2 - x)(2 + x) = 0$$
$$x = 2 \quad \text{or} \quad x = -2$$

The numbers form the intervals $(-\infty, -2)$, $(-2, 2)$, and $(2, \infty)$.

Values in the interval $(-2, 2)$ satisfy the inequality; $x = 2$ and $x = -2$ also satisfy the inequality. The domain is $[-2, 2]$.

23. $f(x) = (x - 3)^{1/2} = \sqrt{x - 3}$

For $f(x)$ to be a real number,

$$x - 3 \geq 0$$
$$x \geq 3.$$

The domain is $[3, \infty)$.

25. $f(x) = \dfrac{2}{1 - x^2} = \dfrac{2}{(1 - x)(1 + x)}$

Since division by zero is not defined, $(1 - x) \cdot (1 + x) \neq 0$.

When $(1 - x)(1 + x) = 0$,

$1 - x = 0$ or $1 + x = 0$

$x = 1$ or $x = -1$

Thus, x can be any real number except ± 1.
The domain is

$$(-\infty, -1) \cup (-1, 1) \cup (1, \infty).$$

27. $f(x) = -\sqrt{\dfrac{2}{x^2 - 16}} = -\sqrt{\dfrac{2}{(x - 4)(x + 4)}}.$

$(x - 4) \cdot (x + 4) > 0$, since $(x - 4) \cdot (x + 4)$
< 0 would produce a negative radicand and
$(x - 4) \cdot (x + 4) = 0$ would lead to division
by zero.

Solve $(x - 4) \cdot (x + 4) = 0$.

$x - 4 = 0$ or $x + 4 = 0$

$x = 4$ or $x = -4$

Use the values -4 and 4 to divide the number line
into 3 intervals, $(-\infty, -4)$, $(-4, 4)$ and $(4, \infty)$.
Only the values in the intervals $(-\infty, -4)$ and
$(4, \infty)$ satisfy the inequality.
The domain is $(-\infty, -4) \cup (4, \infty)$.

29. $f(x) = \sqrt{x^2 - 4x - 5} = \sqrt{(x - 5)(x + 1)}$

See the method used in Exercise 21.

$$(x - 5)(x + 1) \geq 0$$

when $x \geq 5$ and when $x \leq -1$. The domain is
$(-\infty, -1] \cup [5, \infty)$.

31. $f(x) = \dfrac{1}{\sqrt{3x^2 + 2x - 1}} = \dfrac{1}{\sqrt{(3x - 1)(x + 1)}}$

$(3 - 2)(x + 1) > 0$, since the radicand cannot be
negative and the denominator of the function
cannot be zero.

Solve $(3 - 1)(x + 1) = 0$.

$3 - 1 = 0$ or $x + 1 = 0$

$x = \frac{1}{3}$ or $x = -1$

Use the values -1 and $\frac{1}{3}$ to divide the number line

into 3 intervals, $(-\infty, -1)$, $(-1, 4)$ and $\left(\frac{1}{3}, \infty\right)$.

Only the values in the intervals $\left(-\infty, -1\right)$ and

$\left(\frac{1}{3}, \infty\right)$ satisfy the inequality.

The domain is $(-\infty, -1) \cup \left(\frac{1}{3}, \infty\right)$.

33. By reading the graph, the domain is all numbers
greater than or equal to -5 and less than 4. The
range is all numbers greater than or equal to -2
and less than or equal to 6.

Domain: $[-5, 4)$; range: $[-2, 6]$

35. By reading the graph, x can take on any value,
but y is less than or equal to 12.

Domain: $(-\infty, \infty)$; range: $(-\infty, 12]$

37. The domain is all real numbers between the end
points of the curve, or $[-2; 4]$.

The range is all real numbers between the minimum
and maximum values of the function or $[0, 4]$.

(a) $f(-2) = 0$

(b) $f(0) = 4$

(c) $f\left(\dfrac{1}{2}\right) = 3$

(d) From the graph, $f(x) = 1$ when $x = -1.5, 1.5$,
or 2.5.

39. The domain is all real numbers between
the endpoints of the curve, or $[-2, 4]$.

The range is all real numbers between the minimum
and maximum values of the function or $[-3, 2]$.

(a) $f(-2) = -3$

(b) $f(0) = -2$

(c) $f\left(\dfrac{1}{2}\right) = -1$

(d) From the graph, $f(x) = 1$ when $x = 2.5$.

41. $f(x) = 3x^2 - 4x + 1$

(a) $f(4) = 3(4)^2 - 4(4) + 1$

$= 48 - 16 + 1$

$= 33$

(b) $f\left(-\dfrac{1}{2}\right) = 3\left(-\dfrac{1}{2}\right)^2 - 4\left(-\dfrac{1}{2}\right) + 1$

$= \dfrac{3}{4} + 2 + 1$

$= \dfrac{15}{4}$

(c) $f(a) = 3(a)^2 - 4(a) + 1$

$= 3a^2 - 4a + 1$

(d) $f\left(\dfrac{2}{m}\right) = 3\left(\dfrac{2}{m}\right)^2 - 4\left(\dfrac{2}{m}\right) + 1$

$\qquad = \dfrac{12}{m^2} - \dfrac{8}{m} + 1$

$\qquad$ or $\dfrac{12 - 8m + m^2}{m^2}$

(e) $\qquad\quad f(x) = 1$

$\qquad 3x^2 - 4x + 1 = 1$

$\qquad\quad 3x^2 - 4x = 0$

$\qquad\quad x(3x - 4) = 0$

$\qquad x = 0 \ \text{ or } \ x = \dfrac{4}{3}$

43. $f(x) = \begin{cases} \dfrac{2x+1}{x-4} & \text{if } x \neq 4 \\ 7 & \text{if } x = 4 \end{cases}$

(a) $f(4) = 7$

(b) $f\left(-\dfrac{1}{2}\right) = \dfrac{2\left(-\frac{1}{2}\right) + 1}{\left(-\frac{1}{2}\right) - 4} = \dfrac{0}{-\frac{9}{2}} = 0$

(c) $f(a) = \dfrac{2a+1}{a-4} \ \text{ if } a \neq 4$

$\qquad f(a) = 7 \ \text{ if } \ a = 4$

(d) $f\left(\dfrac{2}{m}\right) = \dfrac{2\left(\frac{2}{m}\right) + 1}{\frac{2}{m} - 4} = \dfrac{\frac{4}{m} + 1}{\frac{2}{m} - 4}$

$\qquad = \dfrac{\frac{4+m}{m}}{\frac{2-4m}{m}} = \dfrac{4+m}{2-4m} \ \text{ if } m \neq \dfrac{1}{2}$

$\qquad f\left(\dfrac{2}{m}\right) = 7 \ \text{ if } \ m = \dfrac{1}{2}$

(e) $\dfrac{2x+1}{x-4} = 1$

$\qquad 2x + 1 = x - 4$

$\qquad\qquad x = -5$

45. $f(x) = 6x^2 - 2$

$\qquad f(t + 1) = 6(t + 1)^2 - 2$

$\qquad\qquad = 6(t^2 + 2t + 1) - 2$

$\qquad\qquad = 6t^2 + 12t + 6 - 2$

$\qquad\qquad = 6t^2 + 12t + 4$

47. $g(r + h)$

$\qquad = (r + h)^2 - 2(r + h) + 5$

$\qquad = r^2 + 2hr + h^2 - 2r - 2h + 5$

49. $g\left(\dfrac{3}{q}\right) = \left(\dfrac{3}{q}\right)^2 - 2\left(\dfrac{3}{q}\right) + 5$

$\qquad = \dfrac{9}{q^2} - \dfrac{6}{q} + 5$

$\qquad$ or $\dfrac{9 - 6q + 5q^2}{q^2}$

51. $f(x) = 2x + 1$

(a) $f(x + h) = 2(x + h) + 1$

$\qquad\qquad = 2x + 2h + 1$

(b) $f(x + h) - f(x)$

$\qquad = 2x + 2h + 1 - 2x - 1$

$\qquad = 2h$

(c) $\dfrac{f(x + h) - f(x)}{h}$

$\qquad = \dfrac{2h}{h}$

$\qquad = 2$

53. $f(x) = 2x^2 - 4x - 5$

(a) $f(x + h)$

$\qquad = 2(x + h)^2 - 4(x + h) - 5$

$\qquad = 2(x^2 + 2hx + h^2) - 4x - 4h - 5$

$\qquad = 2x^2 + 4hx + 2h^2 - 4x - 4h - 5$

(b) $f(x + h) - f(x)$

$\qquad = 2x^2 + 4hx + 2h^2 - 4x - 4h - 5$

$\qquad\quad - (2x^2 - 4x - 5)$

$\qquad = 2x^2 + 4hx + 2h^2 - 4x - 4h - 5$

$\qquad\quad - 2x^2 + 4x + 5$

$\qquad = 4hx + 2h^2 - 4h$

(c) $\dfrac{f(x + h) - f(x)}{h}$

$\qquad = \dfrac{4hx + 2h^2 - 4h}{h}$

$\qquad = \dfrac{h(4x + 2h - 4)}{h}$

$\qquad = 4x + 2h - 4$

55. $f(x) = \dfrac{1}{x}$

    (a) $f(x + h) = \dfrac{1}{x + h}$

    (b) $f(x + h) - f(x)$

$$= \dfrac{1}{x + h} - \dfrac{1}{x}$$

$$= \left(\dfrac{x}{x}\right)\dfrac{1}{x + h} - \dfrac{1}{x}\left(\dfrac{x + h}{x + h}\right)$$

$$= \dfrac{x - (x + h)}{x(x + h)}$$

$$= \dfrac{-h}{x(x + h)}$$

    (c) $\dfrac{f(x + h) - f(x)}{h}$

$$= \dfrac{1}{h}\left[\dfrac{-h}{x(x + h)}\right]$$

$$= \dfrac{-1}{x(x + h)}$$

57. A vertical line drawn anywhere through the graph will intersect the graph in only one place. The graph represents a function.

59. A vertical line drawn through the graph may intersect the graph in two places. The graph does not represent a function.

61. A vertical line drawn anywhere through the graph will intersect the graph in only one place. The graph represents a function.

63. $f(x) = 3x$
$$f(-x) = 3(-x)$$
$$= -(3x)$$
$$= -f(x)$$

The function is odd.

65. $f(x) = 2x^2$
$$f(-x) = 2(-x)^2$$
$$= 2x^2$$
$$= f(x)$$

The function is even.

67. $f(x) = \dfrac{1}{x^2 + 4}$

$$f(-x) = \dfrac{1}{(-x)^2 + 4}$$

$$= \dfrac{1}{x^2 + 4}$$

$$= f(x)$$

The function is even.

69. $f(x) = \dfrac{x}{x^2 - 9}$

$$f(-x) = \dfrac{-x}{(-x)^2 - 9}$$

$$= -\dfrac{x}{x^2 - 9}$$

$$= -f(x)$$

The function is odd.

71. **(a)** No

    (b) Year, t

    (c) Price of silver in dollars per ounce, $S(t)$

    (d) [2000, 2013]

    (e) [4, 35]

    (f) $15

    (g) 2011

73. If x is a whole number of days, the cost of renting a saw in dollars is $S(x) = 28x + 8$. For x in whole days and a fraction of a day, substitute the next whole number for x in $28x + 8$, because a fraction of a day is charged as a whole day.

    (a) $S\left(\dfrac{1}{2}\right) = S(1) = 28(1) + 8 = 36$

    The cost is $36.

    (b) $S(1) = 28(1) + 8 = 36$

    The cost is $36.

    (c) $S\left(1\dfrac{1}{4}\right) = S(2) = 28(2) + 8$

$$= 56 + 8 = 64$$

    The cost is $64.

    (d) $S\left(3\dfrac{1}{2}\right) = S(4) = 28(4) + 8$

$$= 112 + 8 = 120$$

    The cost is $120.

    (e) $S(4) = 28(4) + 8 = 112 + 8 = 120$

    The cost is $120.

(f) $S\left(4\dfrac{1}{10}\right) = S(5) = 28(5) + 8$

$$= 140 + 8 = 148$$

The cost is $148.

(g) $S\left(4\dfrac{9}{10}\right) = S(5) = 28(5) + 8$

$$= 140 + 8 = 148$$

The cost is $148.

(h) To continue the graph, continue the horizontal bars up and to the right.

(i) The independent variable is x, the number of full and partial days.

(j) The dependent variable is S, the cost of renting a saw.

(k) S is not a linear function. Its graph is not a continuous straight line. S is a step function.

75. **(a)**

$$f(250,000) = 0.40(150,000) + 0.333(100,000)$$
$$= 93,300$$

The maximum amount that an attorney can receive for a $250,000 jury award is $93,300.

(b) $f(350,000) = 0.40(150,000)$
$$+ 0.333(150,000)$$
$$+ 0.30(50,000)$$
$$= 124,950$$

The maximum amount that an attorney can receive for a $350,000 jury award is $124,500.

(c)

$$f(550,000) = 0.40(150,000) + 0.333(150,000)$$
$$+ 0.30(200,000) + 0.24(50,000)$$
$$= 181,950$$

The maximum amount that an attorney can receive for a $550,000 jury award is $181,950.

(d)

```
f(x)
160,000 ┤                    ·(500,000, 169,950)
120,000 ┤
        ┤           ·(300,000, 109,950)
80,000  ┤
        ┤    ·(150,000, 60,000)
40,000  ┤
      0 ┼──┬──┬──┬──┬──┬──┬──→ x
          200,000    600,000
```

77. **(a)** The curve in the graph crosses the point with x-coordinate 17:37 and y-coordinate of approximately 140. So, at time 17 hours, 37 minutes the whale reaches a depth of about 140 m.

(b) The curve in the graph crosses the point with x-coordinate 17:39 and y-coordinate of approximately 240. So, at time 17 hours, 39 minutes the whale reaches a depth of about 250 m.

79. **(a)** **(i)** By the given function f, a muskrat weighing 800 g expends

$$f(800) = 0.01(800)^{0.88}$$
$$\approx 3.6, \text{ or approximately}$$

3.6 kcal/km when swimming at the surface of the water.

(ii) A sea otter weighing 20,000 g expends

$$f(20,000) = 0.01(20,000)^{0.88}$$
$$\approx 61, \text{ or approximately}$$

61 kcal/km when swimming at the surface of the water.

(b) If z is the number of kilograms of an animal's weight, then $x = g(z) = 1000z$ is the number of grams since 1 kilogram equals 1000 grams.

(c) $f(g(z)) = f(1000z)$
$$= 0.01(1000z)^{0.88}$$
$$= 0.01(1000^{0.88})z^{0.88}$$
$$\approx 4.4z^{0.88}$$

81. **(a)** $P = 2l + 2w$

However, $lw = 500$, so $l = \dfrac{500}{w}$.

$$P(w) = 2\left(\dfrac{500}{w}\right) + 2w$$

$$P(w) = \dfrac{1000}{w} + 2w$$

(b) Since $l = \dfrac{500}{w}$, $w \neq 0$ but w could be any positive value. Therefore, the domain of P is $0 < w < \infty$, or $(0, \infty)$.

(c)

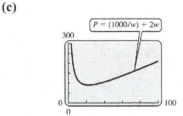

2.2 Quadratic Functions; Translation and Reflection

Your Turn 1

$f(x) = 2x^2 - 6x - 1$

(a)
$$\begin{aligned}
y &= 2x^2 - 6x - 1 \\
&= 2(x^2 - 3x) - 1 \\
&= 2\left(x^2 - 3x + \frac{9}{4}\right) - 1 - 2\left(\frac{9}{4}\right) \\
&= 2\left(x - \frac{3}{2}\right)^2 - \frac{11}{2}
\end{aligned}$$

(b) $y = 2(0)^2 - 6(0) - 1 = -1$

The y-intercept is $(0, -1)$.

(c)
$$0 = 2\left(x - \frac{3}{2}\right)^2 - \frac{11}{2}$$

$$\frac{11}{2} = 2\left(x - \frac{3}{2}\right)^2$$

$$\frac{11}{4} = \left(x - \frac{3}{2}\right)^2$$

$$\pm\frac{\sqrt{11}}{2} = x - \frac{3}{2}$$

$$\frac{3 \pm \sqrt{11}}{2} = x$$

The x-intercepts are
$\left((3 - \sqrt{11})/2, 0\right)$ and $\left((3 + \sqrt{11})/2, 0\right)$.

(d) Since $y = 2\left(x - \frac{3}{2}\right)^2 - \frac{11}{2}$ is in the form

$y = a(x - h)^2 + k$, we get $h = \frac{3}{2}$ and

$k = -\frac{11}{2}$. So the vertex is $\left(\frac{3}{2}, -\frac{11}{2}\right)$.

(e)

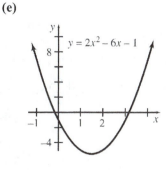

Your Turn 2

$f(x) = x^2 - 4x - 5$

(a) The y-intercept is at $(0, -5)$.

(b) To find the x-intercepts, solve $f(x) = 0$.

$$\begin{aligned}
x^2 - 4x - 5 &= 0 \\
(x - 5)(x + 1) &= 0 \\
x - 5 = 0 \quad &\text{or} \quad x + 1 = 0 \\
x = 5 \quad &\text{or} \quad x = -1
\end{aligned}$$

The x-intercepts are $(-1, 0)$ and $(5, 0)$.

(c) The x-coordinate of the vertex is $\dfrac{-b}{2a} = \dfrac{4}{2} = 2$.

The corresponding y-coordinate is $f(2) = -9$.

The vertex is at $(2, -9)$.

(d) The axis of symmetry is the vertical line through the vertex, which has the equation $x = 2$.

(e)

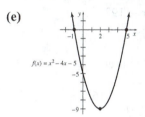

Your Turn 3

Let x be the number of \$40 decreases in the price.

Price per person $= 1650 - 40x$

Number of people $= 900 + 80x$

$$\begin{aligned}
R(x) &= (1650 - 40x)(900 + 80x). \\
&= 1{,}485{,}000 + 96{,}000x - 3200x^2
\end{aligned}$$

$$x = \frac{-b}{2a} = \frac{-96{,}000}{2(-3200)} = \frac{-96{,}000}{-6400} = 15$$

The y-coordinate is

$$\begin{aligned}
y &= 1{,}485{,}000 + 96{,}000(15) - 3200(15)^2 \\
&= 2{,}205{,}000
\end{aligned}$$

The maximum revenue is \$2,205,000 when $1650 - 40(15) = \$1050$ per person is charged.

Your Turn 4

$R(x) = -x^2 + 40x$ and $C(x) = 8x + 192$

(a)
$$R(x) = C(x)$$
$$-x^2 + 40x = 8x + 192$$
$$0 = x^2 - 32x + 192$$
$$0 = (x - 24)(x - 8)$$

The two graphs cross when $x = 8$ or $x = 20$. The minimum break-even quantity is $x = 8$, so the deli owner must sell 8 lb of cream cheese to break even.

(b) The maximum revenue for $R(x) = -x^2 + 40x$ is at
$$x = \frac{-b}{2a} = \frac{-40}{2(-1)} = 20.$$

The maximum revenue is
$$R(20) = -20^2 + 40(20) = 400, \text{ or } \$400.$$

(c) $P(x) = R(x) - C(x)$
$$= -x^2 + 40x - 8x - 192$$
$$= -x^2 + 32x - 192$$

The maximum point of $P(x)$ is at
$$x = \frac{-b}{2a} = \frac{-32}{2(-1)} = 16.$$

$$P(16) = -(16)^2 + 32(16) - 192 = 64$$

So, the maximum profit is $64.

Your Turn 5

(a)

(b)

2.2 Warmup Exercises

W1. The expression is a perfect square.
$$4x^2 - 36x + 81 = (2x)^2 + 2(2x)(-9) + (-9)^2 = (2x - 9)^2$$

W2. The expression is a perfect square.
$$x^2 + \frac{5}{4}x + \frac{25}{64} = (x)^2 + 2(x)\left(\frac{5}{8}\right) + \left(\frac{5}{8}\right)^2 = \left(x + \frac{5}{8}\right)^2$$

W3. $3x^2 - x - 10 = 0$
$$(3x + 5)(x - 2) = 0$$
$$3x + 5 = 0 \quad \text{or} \quad x - 2 = 0$$
$$x = -\frac{5}{3} \quad \text{or} \quad x = 2$$

W4. $2x^2 + 5x - 4 = 0$

Use the quadratic formula with
$a = 2, b = 5,$ and $c = -4.$

$$x = \frac{-5 \pm \sqrt{25 - 4(2)(-4)}}{2(2)} = \frac{-5 \pm \sqrt{57}}{4}$$

2.2   Exercises

3. The graph of $y = x^2 - 3$ is the graph of $y = x^2$ translated 3 units downward.
This is graph **D**.

5. The graph of $y = (x - 3)^2 + 2$ is the graph of $y = x^2$ translated 3 units to the right and 2 units upward.
This is graph **A**.

7. The graph of $y = -(3 - x)^2 + 2$ is the same as the graph of $y = -(x - 3)^2 + 2$. This is the graph of $y = x^2$ reflected in the x-axis, translated 3 units to the right, and translated 2 units upward.
This is graph **C**.

9. $y = 3x^2 + 9x + 5$
$$= 3(x^2 + 3x) + 5$$
$$= 3\left(x^2 + 3x + \frac{9}{4}\right) + 5 - 3\left(\frac{9}{4}\right)$$
$$= 3\left(x + \frac{3}{2}\right)^2 - \frac{7}{4}$$

The vertex is $\left(-\frac{3}{2}, -\frac{7}{4}\right)$.

11. $y = -2x^2 + 8x - 9$

$= -2(x^2 - 4x) - 9$

$= -2(x^2 - 4x + 4) - 9 - (-2)(4)$

$= -2(x - 2)^2 - 1$

The vertex is $(2, -1)$.

13. $y = x^2 + 5x + 6$

$y = (x + 3)(x + 2)$

Set $y = 0$ to find the x-intercepts.

$$0 = (x + 3)(x + 2)$$
$$x = -3, x = -2$$

The x-intercepts are -3 and -2. Set $x = 0$ to find the y-intercept.

$$y = 0^2 + 5(0) + 6$$
$$y = 6$$

The y-intercept is 6.

The x-coordinate of the vertex is

$$x = \frac{-b}{2a} = \frac{-5}{2} = -\frac{5}{2}.$$

Substitute to find the y-coordinate.

$$y = \left(-\frac{5}{2}\right)^2 + 5\left(-\frac{5}{2}\right) + 6 = \frac{25}{4} - \frac{25}{2} + 6 = -\frac{1}{4}$$

The vertex is $\left(-\frac{5}{2}, -\frac{1}{4}\right)$.

The axis is $x = -\frac{5}{2}$, the vertical line through the vertex.

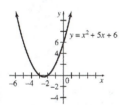

15. $y = -2x^2 - 12x - 16$

$= -2(x^2 + 6x + 8)$

$= -2(x + 4)(x + 2)$

Let $y = 0$.

$$0 = -2(x + 4)(x + 2)$$
$$x = -4, x = -2$$

-4 and -2 are the x-intercepts. Let $x = 0$.

$$y = -2(0)^2 + 12(0) - 16$$

-16 is the y-intercept.

Vertex: $x = \dfrac{-b}{2a} = \dfrac{12}{-4} = -3$

$$y = -2(-3)^2 - 12(-3) - 16$$
$$= -18 + 36 - 16 = 2$$

The vertex is $(-3, 2)$.

The axis is $x = -3$, the vertical line through the vertex.

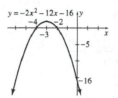

17. $y = 2x^2 + 8x - 8$

Let $y = 0$.

$$2x^2 + 8x - 8 = 0$$
$$x^2 + 4x - 4 = 0$$

$$x = \frac{-4 \pm \sqrt{4^2 - 4(1)(-4)}}{2(1)}$$
$$= \frac{-4 \pm \sqrt{32}}{2} = \frac{-4 \pm 4\sqrt{2}}{2}$$
$$= -2 \pm 2\sqrt{2}$$

The x-intercepts are $-2 \pm 2\sqrt{2} \approx 0.83$ and -4.83.

Let $x = 0$.

$$y = 2(0)^2 + 8(0) - 8 = -8$$

The y-intercept is -8.

The x-coordinate of the vertex is

$$x = \frac{-b}{2a} = -\frac{8}{4} = -2.$$

If $x = -2$,

$$y = 2(-2)^2 + 8(-2) - 8$$
$$= 8 - 16 - 8 = -16.$$

The vertex is $(-2, -16)$.

The axis is $x = -2$.

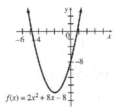

$f(x) = 2x^2 + 8x - 8$

19. $f(x) = 2x^2 - 4x + 5$

Let $f(x) = 0$.

$$0 = 2x^2 - 4x + 5$$

$$x = \frac{-(-4) \pm \sqrt{(-4)^2 - 4(2)(5)}}{2(2)}$$

$$= \frac{4 \pm \sqrt{16 - 40}}{4}$$

$$= \frac{4 \pm \sqrt{-24}}{4}$$

Since the radicand is negative, there are no x-intercepts.

Let $x = 0$.

$$y = 2(0)^2 - 4(0) + 5$$
$$y = 5$$

5 is the y-intercept.

Vertex: $x = \dfrac{-b}{2a} = \dfrac{-(-4)}{2(2)} = \dfrac{4}{4} = 1$

$$y = 2(1)^2 - 4(1) + 5 = 2 - 4 + 5 = 3$$

The vertex is $(1, 3)$.

The axis is $x = 1$.

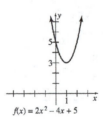

$f(x) = 2x^2 - 4x + 5$

21. $f(x) = -2x^2 + 16x - 21$

Let $f(x) = 0$

Use the quadratic formula.

$$x = \frac{-16 \pm \sqrt{16^2 - 4(-2)(-21)}}{2(-2)}$$

$$= \frac{-16 \pm \sqrt{88}}{-4}$$

$$= \frac{-16 \pm 2\sqrt{22}}{-4}$$

$$= 4 \pm \frac{\sqrt{22}}{2}$$

The x-intercepts are $4 + \dfrac{\sqrt{22}}{2} \approx 6.35$

and $4 - \dfrac{\sqrt{22}}{2} \approx 1.65$.

Let $x = 0$.

$$y = -2(0)^2 + 16(0) - 21$$
$$y = -21$$

-21 is the y-intercept.

Vertex: $x = \dfrac{-b}{2a} = \dfrac{-16}{2(-2)} = \dfrac{-16}{-4} = 4$

$$y = -2(4)^2 + 16(4) - 21$$
$$= -32 + 64 - 21 = 11$$

The vertex is $(4, 11)$.

The axis is $x = 4$.

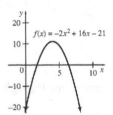

$f(x) = -2x^2 + 16x - 21$

23. $y = \dfrac{1}{3}x^2 - \dfrac{8}{3}x + \dfrac{1}{3}$

Let $y = 0$.

$$0 = \frac{1}{3}x^2 - \frac{8}{3}x + \frac{1}{3}$$

Multiply by 3.

$$0 = x^2 - 8x + 1$$

$$x = \frac{-(-8) \pm \sqrt{(-8)^2 - 4(1)(1)}}{2(1)}$$

$$= \frac{8 \pm \sqrt{64 - 4}}{2} = \frac{8 \pm \sqrt{60}}{2}$$

$$= \frac{8 \pm 2\sqrt{15}}{2} = 4 \pm \sqrt{15}$$

The x-intercepts are $4 + \sqrt{15} \approx 7.87$
and $4 - \sqrt{15} \approx 0.13$.

Let $x = 0$.

$$y = \frac{1}{3}(0)^2 - \frac{8}{3}(0) + \frac{1}{3}$$

$\frac{1}{3}$ is the y-intercept.

Vertex: $x = \dfrac{-b}{2a} = \dfrac{-\left(-\frac{8}{3}\right)}{2\left(\frac{1}{3}\right)} = \dfrac{\frac{8}{3}}{\frac{2}{3}} = 4$

$$y = \frac{1}{3}(4)^2 - \frac{8}{3}(4) + \frac{1}{3}$$

$$= \frac{16}{3} - \frac{32}{3} + \frac{1}{3} = -\frac{15}{3} = -5$$

The vertex is $(4, -5)$.

The axis is $x = 4$.

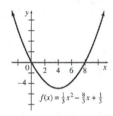

25. The graph of $y = \sqrt{x + 2} - 4$ is the graph
    of $y = \sqrt{x}$ translated 2 units to the left and
    4 units downward.
    This is graph **D**.

27. The graph of $y = \sqrt{-x + 2} - 4$ is the graph
    of $y = \sqrt{-(x - 2)} - 4$, which is the graph
    of $y = \sqrt{x}$ reflected in the y-axis, translated
    2 units to the right, and translated 4 units downward.

    This is graph **C**.

29. The graph of $y = -\sqrt{x + 2} - 4$ is the graph of
    $y = \sqrt{x}$ reflected in the x-axis, translated 2 units
    to the left, and translated 4 units downward.
    This is graph **E**.

31. The graph of $y = -f(x)$ is the graph of $y = f(x)$
    reflected in the x-axis.

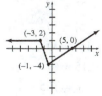

33. The graph of $y = f(-x)$ is the graph
    of $y = f(x)$ reflected in the y-axis.

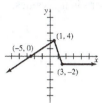

35. $f(x) = \sqrt{x - 2} + 2$

    Translate the graph of $f(x) = \sqrt{x}$ 2 units right
    and 2 units up.

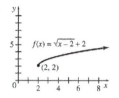

37. $f(x) = -\sqrt{2 - x} - 2$
    $\qquad = -\sqrt{-(x - 2)} - 2$

    Reflect the graph of $f(x)$ vertically and horizontally.
    Translate the graph 2 units right and 2 units down.

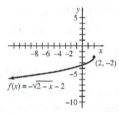

39. If $0 < a < 1$, the graph of $f(ax)$ will be the graph
    of $f(x)$ stretched horizontally.

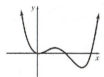

41. If $-1 < a < 0$, the graph of $f(ax)$ will be
    reflected horizontally, since a is negative. It will
    be stretched horizontally.

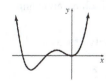

43. If $0 < a < 1$, the graph of $a f(x)$ will be flatter than the graph of $f(x)$. Each y-value is only a fraction of the height of the original y-values.

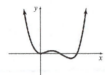

45. If $-1 < a < 0$, the graph will be reflected vertically, since a will be negative. Also, because a is a fraction, the graph will be flatter because each y-value will only be a fraction of its original height.

47. **(a)** Since the graph of $y = f(x)$ is reflected vertically to obtain the graph of $y = -f(x)$, the x-intercept is unchanged. The x-intercept of the graph of $y = f(x)$ is r.

(b) Since the graph of $y = f(x)$ is reflected horizontally to obtain the graph of $y = f(-x)$, the x-intercept of the graph of $y = f(-x)$ is $-r$.

(c) Since the graph of $y = f(x)$ is reflected both horizontally and vertically to obtain the graph of $y = -f(-x)$, the x-intercept of the graph of $y = -f(-x)$ is $-r$.

49. **(a)**

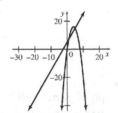

(b) Break-even quantities are values of $x =$ batches of widgets for which revenue and cost are equal.

Set $R(x) = C(x)$ and solve for x.

$$-x^2 + 8x = 2x + 5$$
$$x^2 - 6x + 5 = 0$$
$$(x - 5)(x - 1) = 0$$
$$x - 5 = 0 \text{ or } x - 1 = 0$$
$$x = 5 \text{ or } \quad x = 1$$

So, the break-even quantities are 1 and 5. The minimum break-even quantity is 1 batch of widgets.

(c) The maximum revenue occurs at the vertex of R. Since $R(x) = -x^2 + 8x$, then the x-coordinate of the vertex is

$$x = -\frac{b}{2a} = -\frac{8}{2(-1)} = 4.$$

So, the maximum revenue is

$$R(4) = -4^2 + 8(4) = 16, \text{ or } \$16,000.$$

(d) The maximum profit is the maximum difference $R(x) - C(x)$. Since

$$P(x) = R(x) - C(x)$$
$$= -x^2 + 8x - (2x + 5)$$
$$= -x^2 + 6x - 5$$

is a quadratic function, we can find the maximum profit by finding the vertex of P. This occurs at

$$x = -\frac{b}{2a} = \frac{-6}{2(-1)} = 3.$$

Therefore, the maximum profit is

$$P(3) = -(3)^2 + 6(3) - 5 = 4, \text{ or } \$4000.$$

51. **(a)**

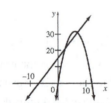

(b) Break-even quantities are values of $x =$ batches of widgets for which revenue equals cost.

Set $R(x) = C(x)$ and solve for x.

$$-\frac{4}{5}x^2 + 10x = 2x + 15$$
$$\frac{4}{5}x^2 - 8x + 15 = 0$$
$$4x^2 - 40x + 75 = 0$$
$$4x^2 - 10x - 30x + 75 = 0$$
$$2x(2x - 5) - 15(2x - 5) = 0$$
$$(2x - 5)(2x - 15) = 0$$
$$2x - 5 = 0 \quad \text{or } 2x - 15 = 0$$
$$x = 2.5 \quad \text{or} \quad x = 7.5$$

So, the break-even quantities are 2.5 and 7.5 with the minimum break-even quantity being 2.5 batches of widgets.

(c) The maximum revenue occurs at the vertex of R. Since $R(x) = -\frac{4}{5}x^2 + 10x,$ then the x-coordinate of the vertex is

$$x = -\frac{b}{2a} = -\frac{10}{2\left(-\frac{4}{5}\right)} = 6.25.$$

So, the maximum revenue is

$$R(6.25) = 31.25, \text{ or } \$31,250.$$

(d) The maximum profit is the maximum difference $R(x) - C(x)$. Since

$$P(x) = R(x) - C(x)$$
$$= -\frac{4}{5}x^2 + 10x - (2x + 15)$$
$$= -\frac{4}{5}x^2 + 8x - 15$$

is a quadratic function, we can find the maximum profit by finding the vertex of P. This occurs at

$$x = -\frac{b}{2a} = -\frac{8}{2\left(\frac{4}{5}\right)} = 5.$$

Therefore, the maximum profit is

$$P(5) = -\frac{4}{5}5^2 + 8(5) - 15 = 5, \text{ or } \$5000.$$

53. $R(x) = 8000 + 70x - x^2$
$$= -x^2 + 70x + 8000$$

The maximum revenue occurs at the vertex.

$$x = \frac{-b}{2a} = \frac{-70}{2(-1)} = 35$$

$$y = 8000 + 70(35) - (35)^2$$
$$= 8000 + 2450 - 1225$$
$$= 9225$$

The vertex is $(35, 9225)$.

The maximum revenue of \$9225 is realized when 35 seats are left unsold.

55. Let $x =$ the number of \$25 increases.

(a) Rent per apartment: $800 + 25x$

(b) Number of apartments rented: $80 - x$

(c) Revenue:

$$R(x) = (\text{number of apartments rented})$$
$$\times (\text{rent per apartment})$$

$$= (80 - x)(800 + 25x)$$
$$= -25x^2 + 1200x + 64,000$$

(d) Find the vertex:

$$x = \frac{-b}{2a} = \frac{-1200}{2(-25)} = 24$$

$$y = -25(24)^2 + 1200(24) + 64,000$$
$$= 78,400$$

The vertex is $(24, 78,400)$. The maximum revenue occurs when $x = 24$.

(e) The maximum revenue is the y-coordinate of the vertex, or \$78,400.

57. $p = 500 - x$

(a) The revenue is

$$R(x) = px$$
$$= (500 - x)(x)$$
$$= 500x - x^2.$$

(b)

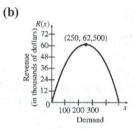

(c) From the graph, the vertex is halfway between $x = 0$ and $x = 500,$ so $x = 250$ units corresponds to maximum revenue. Then the price is

$$p = 500 - x$$
$$= 500 - 250 = \$250.$$

Note that price, p, cannot be read directly from the graph of

$$R(x) = 500x - x^2.$$

(d) $R(x) = 500x - x^2$
$$= -x^2 + 500x$$

Find the vertex.

$$x = \frac{-b}{2a} = \frac{-500}{2(-1)} = 250$$

$$y = -(250)^2 + 500(250)$$
$$= 62,500$$

The vertex is $(250, 62,500)$.

The maximum revenue is \$62,500.

59. **(a)**
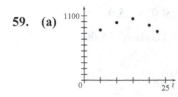

(b) The vertex is at $(15, 1037)$, so f has the form
$$a(t - 15)^2 + 1037.$$

Using the point $(10, 986)$ we have
$$a(10 - 15)^2 + 1037 = 986$$
$$a = \frac{986 - 1037}{25} = -2.04$$
Thus $f(t) = -2.04(t - 15)^2 + 1037.$

(c) The calculator regression produces the function $f(t) = -2.69t^2 + 72.4t + 548$. The graph shows this function together with the data points in the window $[0, 25]$ by $[0, 1100]$.

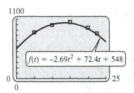

(d) The graph shows the functions found in (b) and (c) in the window $[0, 25]$ by $[0, 1100]$.

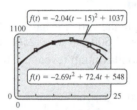

No, the calculagtor fit is better. The fit from (b) gives too large an estimate at the last two time values.

61. $S(x) = 1 - 0.058x - 0.076x^2$

(a) $0.50 = 1 - 0.058x - 0.076x^2$

$$0.076x^2 + 0.058x - 0.50 = 0$$
$$76x^2 + 58x - 500 = 0$$
$$38x^2 + 29x - 250 = 0$$
$$x = \frac{-29 \pm \sqrt{(29)^2 - 4(38)(-250)}}{2(38)}$$
$$= \frac{-29 \pm \sqrt{38,841}}{76}$$

$$\frac{-29 - \sqrt{38,841}}{76} \approx -2.97$$
and $\dfrac{-29 + \sqrt{38,841}}{76} \approx 2.21$

We ignore the negative value.
The value $x = 2.2$ represents 2.2 decades or 22 years, and 22 years after 65 is 87.
The median length of life is 87 years.

(b) If nobody lives, $S(x) = 0$.

$$1 - 0.058x - 0.076x^2 = 0$$
$$76x^2 + 58x - 1000 = 0$$
$$38x^2 + 29x - 500 = 0$$
$$x = \frac{-29 \pm \sqrt{(29)^2 - 4(38)(-500)}}{2(38)}$$
$$= \frac{-29 \pm \sqrt{76,841}}{76}$$

$$\frac{-29 - \sqrt{76,841}}{76} \approx -4.03$$
and $\dfrac{-29 + \sqrt{76,841}}{76} \approx 3.27$

We ignore the negative value.
The value $x = 3.3$ represents 3.3 decades or 33 years, and 33 years after 65 is 98.
Virtually nobody lives beyond 98 years.

63. **(a)** The vertex of the quadratic function
$$y = 0.057x - 0.001x^2 \text{ is at}$$
$$x = -\frac{b}{2a} = -\frac{0.057}{2(-0.001)} = 28.5.$$

Since the coefficient of the leading term, -0.001, is negative, then the graph of the function opens downward, so a maximum is reached at 28.5 weeks of gestation.

(b) The maximum splenic artery resistance reached at the vertex is
$$y = 0.057(28.5) - 0.001(28.5)^2 \approx 0.81.$$

(c) The splenic artery resistance equals 0, when $y = 0$.

$$0.057x - 0.001x^2 = 0 \quad \text{Substitute in the expression in } x \text{ for } y$$

$$x(0.057 - 0.001x) = 0 \quad \text{Factor}$$

$$x = 0 \text{ or } 0.057 - 0.001x = 0 \quad \text{Set each factor equal to 0.}$$

$$x = \frac{0.057}{0.001} = 57$$

So, the splenic artery resistance equals 0 at 0 weeks or 57 weeks of gestation.

No, this is not reasonable because at $x = 0$ or 57 weeks, the fetus does not exist.

65. (a)

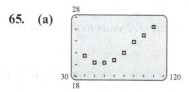

(b) Quadratic

(c) $f(t) = 0.00207t^2 - 0.226t + 26.8$

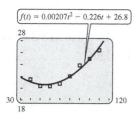

(d) Given that $(h, k) = (60, 20.3)$, the equation has the form

$$f(t) = a(t - 60)^2 + 20.3.$$

Since $(110, 26.5)$ is also on the curve,

$$26.5 = a(110 - 60)^2 + 20.3$$
$$6.2 = 2500a$$
$$a = 0.00248$$

A quadratic function that models the data is

$$f(t) = 0.00248(t - 60)^2 + 20.3.$$

(e)

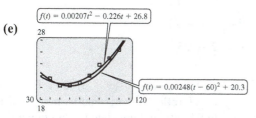

The two graphs are very close.

67. $h(t) = 32t - 16t^2$
$$= -16t^2 + 32t$$

(a) Find the vertex.

$$x = \frac{-b}{2a} = \frac{-32}{-32} = 1$$
$$y = -16(1)^2 + 32(1)$$
$$= 16$$

The vertex is $(1, 16)$, so the maximum height is 16 ft.

(b) When the object hits the ground, $h = 0$, so

$$32t - 16t^2 = 0$$
$$16t(2 - t) = 0$$
$$t = 0 \text{ or } t = 2.$$

When $t = 0$, the object is thrown upward. When $t = 2$, the object hits the ground; that is, after 2 sec.

69. Let $x = $ the width.

Then $380 - 2x = $ the length.

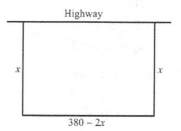

$$\text{Area} = x(380 - 2x) = -2x^2 + 380x$$

Find the vertex:

$$x = \frac{-b}{2a} = \frac{-380}{-4} = 95$$
$$y = -2(95)^2 + 380(95) = 18,050$$

The graph of the area function is a parabola with vertex $(95, 18,050)$.

The maximum area of 18,050 sq ft occurs when the width is 95 ft and the length is

$$380 - 2x = 380 - 2(95) = 190 \text{ ft.}$$

71. Draw a sketch of the arch with the vertex at the origin.

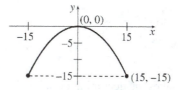

Since the arch is a parabola that opens downward, the equation of the parabola is the form $y = a(x - h)^2 + k$, where the vertex $(h, k) = (0, 0)$ and $a < 0$. That is, the equation is of the form $y = ax^2$.

Since the arch is 30 meters wide at the base and 15 meters high, the points $(15, -15)$ and $(-15, -15)$

are on the parabola. Use $(15, -15)$ as one point on the parabola.

$$-15 = a(15)^2$$

$$a = \frac{-15}{15^2} = -\frac{1}{15}$$

So, the equation is

$$y = -\frac{1}{15}x^2.$$

Ten feet from the ground (the base) is at $y = -5$. Substitute -5 for y and solve for x.

$$-5 = -\frac{1}{15}x^2$$

$$x^2 = -5(-15) = 75$$

$$x = \pm\sqrt{75} = \pm 5\sqrt{3}$$

The width of the arch ten feet from the ground is then

$$5\sqrt{3} - (-5\sqrt{3}) = 10\sqrt{3} \text{ meters}$$

$$\approx 17.32 \text{ meters.}$$

2.3 Polynomial and Rational Functions

Your Turn 1

Graph $f(x) = 64 - x^6$.

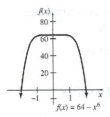

Your Turn 2

Graph $y = \dfrac{4x - 6}{x - 3}$.

The value $x = 3$ makes the denominator 0, so 3 is not in the domain and the line $x = 3$ is a vertical asymptote. Let x get larger and larger. Then $\frac{4x-6}{x-3} \approx \frac{4x}{x} = 4$ as x gets larger and larger. So, the line $y = 4$ is a horizontal asymptote.

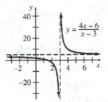

2.3 Warmup Exercises

W1. The graph of $f(x)$ shifted 2 units to the left and 3 units down.

W2. The graph of $f(x)$ reflected across the x-axis and shifted 3 units to the right.

W3. The graph of $f(x)$ reflected across the y-axis and shifted 2 units up.

W4. The graph of $f(x)$ reflected across the y-axis and shifted 2 units to the right.

2.3 Exercises

3. The graph of $f(x) = (x - 2)^3 + 3$ is the graph of $y = x^3$ translated 2 units to the right and 3 units upward.

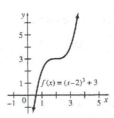

$$0 = (x - 2)^3 + 3$$

$$x - 2 = -\sqrt[3]{3}$$

$$x = 2 - \sqrt[3]{3}$$

5. The graph of $f(x) = -(x + 3)^4 + 1$ is the graph of $y = x^4$ reflected horizontally, translated 3 units to the left, and translated 1 unit upward.

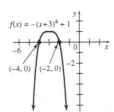

$$0 = -(x + 3)^4 + 1$$

$$x + 3 = \pm 1$$

$$x = -3 \pm 1$$

$$x = -4, \ x = -2$$

7. The graph of $y = x^3 - 7x - 9$ has the right end up, the left end down, at most two turning points, and a y-intercept of -9.
This is graph **D**.

9. The graph of $y = -x^3 - 4x^2 + x + 6$ has the right end down, the left end up, at most two turning points, and a y-intercept of 6.
This is graph **E**.

11. The graph of $y = x^4 - 5x^2 + 7$ has both ends up, at most three turning points, and a y-intercept of 7.
This is graph **I**.

13. The graph of $y = -x^4 + 2x^3 + 10x + 15$ has both ends down, at most three turning points, and a y-intercept of 15.
This is graph **G**

15. The graph of $y = -x^5 + 4x^4 + x^3 - 16x^2 + 12x + 5$ has the right end down, the left end up, at most four turning points, and a y-intercept of 5.
This is graph **A**.

17. The graph of $y = \dfrac{2x^2 + 3}{x^2 + 1}$ has no vertical asymptote, the line with equation $y = 2$ as a horizontal asymptote, and a y-intercept of 3.
This is graph **D**.

19. The graph $y = \dfrac{-2x^2 - 3}{x^2 + 1}$ has no vertical asymptote, the line with equation $y = -2$ as a horizontal asymptote, and a y-intercept of -3.
This is graph **E**.

21. The right end is up and the left end is up. There are three turning points.
The degree is an even integer equal to 4 or more.
The x^n term has a $+$ sign.

23. The right end is up and the left end is down. There are four turning points. The degree is an odd integer equal to 5 or more. The x^n term has a $+$ sign.

25. The right end is down and the left end is up. There are six turning points. The degree is an odd integer equal to 7 or more. The x^n term has a $-$ sign.

27. $y = \dfrac{-4}{x + 2}$

The function is undefined for $x = -2$, so the line $x = -2$ is a vertical asymptote.

x	-102	-12	-7	-5	-3	-1	8	98
$x + 2$	-100	-10	-5	-3	-1	1	10	100
y	0.04	0.4	0.8	1.3	4	-4	-0.4	-0.04

The graph approaches $y = 0$, so the line $y = 0$ (the x-axis) is a horizontal asymptote.

Asymptotes: $y = 0$, $x = -2$

x-intercept:

y-intercept:

-2, the value when $x = 0$

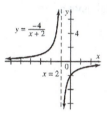

29. $y = \dfrac{2}{3 + 2x}$

$3 + 2x = 0$ when $2x = -3$ or $x = -\frac{3}{2}$, so the line $x = -\frac{3}{2}$ is a vertical asymptote.

x	-51.5	-6.5	-2	-1	3.5	48.5
$3 + 2x$	-100	-10	-1	1	10	100
y	-0.02	-0.2	-2	2	0.2	0.02

The graph approaches $y = 0$, so the line $y = 0$ (the x-axis) is a horizontal asymptote.

Asymptote: $y = 0$, $x = -\frac{3}{2}$

x-intercept:

y-intercept:

$\frac{2}{3}$, the value when $x = 0$

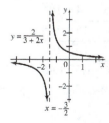

31. $y = \dfrac{2x}{x-3}$

$x - 3 = 0$ when $x = 3$, so the line $x = 3$ is a vertical asymptote.

x	-97	-7	-1	1	2	2.5
$2x$	-194	-14	-2	2	4	5
$x-3$	-100	-10	-4	-2	-1	-0.5
y	1.94	1.4	0.5	-1	-4	-10

x	3.5	4	5	7	11	103
$2x$	7	8	10	14	22	206
$x-3$	0.5	1	2	4	8	100
y	14	8	5	3.5	2.75	2.06

As x gets larger,

$$\frac{2x}{x-3} \approx \frac{2x}{x} = 2.$$

Thus, $y = 2$ is a horizontal asymptote.

Asymptotes: $y = 2$, $x = 3$

x-intercept:

0, the value when $y = 0$

y-intercept:

0, the value when $x = 0$

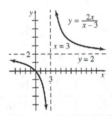

33. $y = \dfrac{x+1}{x-4}$

$x - 4 = 0$ when $x = 4$, so $x = 4$ is a vertical asymptote.

x	-96	-6	-1	0	3
$x+1$	-95	-5	0	1	4
$x-4$	-100	-10	-5	-4	-1
y	0.95	0.5	0	-0.25	-4

x	3.5	4.5	5	14	104
$x+1$	4.5	5.5	6	15	105
$x-4$	-0.5	0.5	1	10	100
y	-9	11	6	1.5	1.05

As x gets larger,

$$\frac{x+1}{x-4} \approx \frac{x}{x} = 1.$$

Thus, $y = 1$ is a horizontal asymptote.

Asymptotes: $y = 1$, $x = 4$

x-intercept: -1, the value when $y = 0$

y-intercept: $-\frac{1}{4}$, the value when $x = 0$

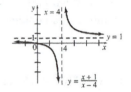

35. $y = \dfrac{3-2x}{4x+20}$

$4x + 20 = 0$ when $4x = -20$ or $x = -5$, so the line $x = -5$ is a vertical asymptote.

x	-8	-7	-6	-4	-3	-2
$3-2x$	-26	-23	-20	-14	-11	-8
$4x+20$	-12	-8	-4	4	8	12
y	2.17	2.88	5	-3.5	-1.38	-0.67

As x gets larger,

$$\frac{3-2x}{4x+20} \approx \frac{-2x}{4x} = -\frac{1}{2}.$$

Thus, the line $y = -\frac{1}{2}$ is a horizontal asymptote.

Asymptotes: $x = -5$, $y = -\frac{1}{2}$

x-intercept:

$\frac{3}{2}$, the value when $y = 0$

y-intercept:

$\frac{3}{20}$, the value when $x = 0$

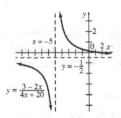

37. $y = \dfrac{-x-4}{3x+6}$

$3x + 6 = 0$ when $3x = -6$ or $x = -2$, so the line $x = -2$ is a vertical asymptote.

x	-5	-4	-3	-1	0	1
$-x-4$	1	0	-1	-3	-4	-5
$3x+6$	-9	-6	-3	3	6	9
y	-0.11	0	0.33	-1	-0.67	-0.56

As x gets larger,

$$\frac{-x-4}{3x+6} \approx \frac{-x}{3x} = -\frac{1}{3}.$$

The line $y = -\frac{1}{3}$ is a horizontal asymptote.

Asymptotes: $y = -\frac{1}{3}, x = -2$

x-intercept:

-4, the value when $y = 0$

y-intercept:

$-\frac{2}{3}$, the value when $x = 0$

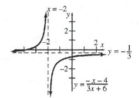

39. $y = \dfrac{x^2 + 7x + 12}{x + 4}$

$= \dfrac{(x+3)(x+4)}{x+4}$

$= x + 3, x \neq -4$

There are no asymptotes, but there is a hole at $x = -4$.

x-intercept: -3, the value when $y = 0$.

y-intercept: 3, the value when $x = 0$.

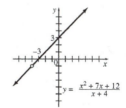

41. For a vertical asymptote at $x = 1$, put $x - 1$ in the denominator. For a horizontal asymptote at $y = 2$, the degree of the numerator must equal the degree of the denominator and the quotient of their leading terms must equal 2. So, $2x$ in the numerator would cause y to approach 2 as x gets larger.

So, one possible answer is $y = \dfrac{2x}{x-1}$.

43. $f(x) = (x-1)(x-2)(x+3)$,

$g(x) = x^3 + 2x^2 - x - 2$,

$h(x) = 3x^3 + 6x^2 - 3x - 6$

(a) $f(1) = (0)(-1)(4) = 0$

(b) $f(x)$ is zero when $x = 2$ and when $x = -3$.

(c) $g(-1) = (-1)^3 + 2(-1)^2 - (-1) - 2$
$= -1 + 2 + 1 - 2 = 0$

$g(1) = (1)^3 + 2(1)^2 - (1) - 2$
$= 1 + 2 - 1 - 2$
$= 0$

$g(-2) = (-2)^3 + 2(-2)^2 - (-2) - 2$
$= -8 + 8 + 2 - 2$
$= 0$

(d) $g(x) = [x - (-1)](x - 1)[x - (-2)]$
$g(x) = (x + 1)(x - 1)(x + 2)$

(e) $h(x) = 3g(x)$
$= 3(x + 1)(x - 1)(x + 2)$

(f) If f is a polynomial and $f(a) = 0$ for some number a, then one factor of the polynomial is $x - a$.

45. $f(x) = \dfrac{1}{x^5 - 2x^3 - 3x^2 + 6}$

(a) Two vertical asymptotes appear, one at $x = -1.4$ and one at $x = 1.4$.

(b) Three vertical asymptotes appear, one at $x = -1.414$, one at $x = 1.414$, and one at $x = 1.442$.

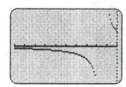

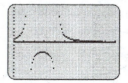

47. $\overline{C}(x) = \dfrac{220,000}{x + 475}$

(a) If $x = 25$,

$$\overline{C}(25) = \frac{220,000}{25 + 475} = \frac{220,000}{500} = \$440.$$

If $x = 50$,

$$\overline{C}(50) = \frac{220,000}{50 + 475} = \frac{220,000}{525} \approx \$419.$$

If $x = 100$,

$$\overline{C}(100) = \frac{220,000}{100 + 475} = \frac{220,000}{575} \approx \$383.$$

If $x = 200$,

$$\overline{C}(200) = \frac{220,000}{200 + 475} = \frac{220,000}{675} \approx \$326.$$

If $x = 300$,

$$\overline{C}(300) = \frac{220,000}{300 + 475} = \frac{220,000}{775} \approx \$284.$$

If $x = 400$,

$$\overline{C}(400) = \frac{220,000}{400 + 475} = \frac{220,000}{875} \approx \$251.$$

(b) A vertical asymptote occurs when the denominator is 0.

$$x + 475 = 0$$
$$x = -475$$

A horizontal asymptote occurs when $\overline{C}(x)$ approaches a value as x gets larger. In this case, $\overline{C}(x)$ approaches 0.

The asymptotes are $x = -475$ and $y = 0$.

(c) x-intercepts:

$$0 = \frac{220,000}{x + 475}; \text{ no such } x, \text{ so no } x\text{-intercepts}$$

y-intercepts:

$$\overline{C}(0) = \frac{220,000}{0 + 475} \approx 463.2$$

(d) Use the following ordered pairs: (25,440), (50,419), (100,383), (200,326), (300,284), (400,251).

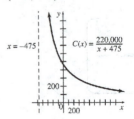

49. Quadratic functions with roots at $x = 0$ and $x = 100$ are of the form $f(x) = ax(100 - x)$.

$f_1(x)$ has a maximum of 100, which occurs at the vertex. The x-coordinate of the vertex lies between the two roots.

The vertex is (50, 100).

$$100 = a(50)(100 - 50)$$
$$100 = a(50)(50)$$
$$\frac{100}{2500} = a$$
$$\frac{1}{25} = a$$
$$f_1(x) = \frac{1}{25}x(100 - x) \text{ or } \frac{x(100 - x)}{25}$$

$f_2(x)$ has a maximum of 250, occurring at (50, 250).

$$250 = a(50)(100 - 50)$$
$$250 = a(50)(50)$$
$$\frac{250}{2500} = a$$
$$\frac{1}{10} = a$$
$$f_2(x) = \frac{1}{10}x(100 - x) \text{ or } \frac{x(100 - x)}{10}$$

$$f_1(x) \cdot f_2(x) = \left[\frac{x(100 - x)}{25}\right] \cdot \left[\frac{x(100 - x)}{10}\right]$$
$$= \frac{x^2(100 - x)^2}{250}$$

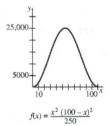

$$f(x) = \frac{x^2(100 - x)^2}{250}$$

51. $y = \dfrac{6.7x}{100 - x}$,

Let $x =$ percent of pollutant; $y =$ cost in thousands.

(a) $x = 50$

$$y = \frac{6.7(50)}{100 - 50} = 6.7$$

The cost is $6700.

$$x = 70$$

$$y = \frac{6.7(70)}{100 - 70} \approx 15.6$$

The cost is $15,600.

$x = 80$

$$y = \frac{6.7(80)}{100 - 80} = 26.8$$

The cost is $26,800.

$x = 90$

$$y = \frac{6.7(90)}{100 - 90} = 60.3$$

The cost is $60,300.

$x = 95$

$$y = \frac{6.7(95)}{100 - 95}$$

The cost is $127,300.

$x = 98$

$$y = \frac{6.7(98)}{100 - 98} = 328.3$$

The cost is $328,300.

$x = 99$

$$y = \frac{6.7(99)}{100 - 99} = 668.3$$

The cost is $663,300.

(b) No, because $x = 100$ makes the denominator zero, so $x = 100$ is a vertical asymptote.

(c)

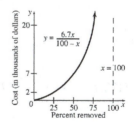

53. (a)

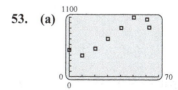

(b) $y = -0.03132t^2 + 12.46t + 383.1$

(c)

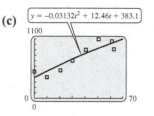
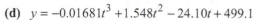

(d) $y = -0.01681t^3 + 1.548t^2 - 24.10t + 499.1$

(e)

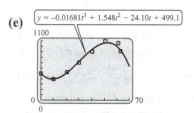

55. $A(x) = 0.003631x^3 - 0.03746x^2 + 0.1012x + 0.009$

(a)

x	0	1	2	3	4	5
$A(x)$	0.009	0.076	0.091	0.073	0.047	0.032

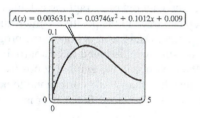

(b) The peak of the curve comes at about $x = 2$ hours.

(c) The curve rises to a y-value of 0.08 at about $x = 1.1$ hours and stays at or above that level until about $x = 2.7$ hours.

57. $f(x) = \dfrac{\lambda x}{1 + (ax)^b}$

(a) A reasonable domain for the function is $[0, \infty)$. Populations are not measured using negative numbers and they may get extremely large.

(b) If $\lambda = a = b = 1$, the function becomes

$$f(x) = \frac{x}{1 + x^2}.$$

(c) If $\lambda = a = 1$ and $b = 2$, the function becomes

$$f(x) = \frac{x}{1 + x}.$$

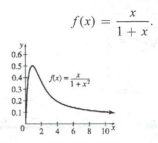

(d) As seen from the graphs, when b increases, the population of the next generation, $f(x)$, gets smaller when the current generation, x, is larger.

59. (a) When $c = 30, w = \frac{30^3}{100} - \frac{1500}{30} = 220$, so the brain weights 220 g when its circumference measures 30 cm. When

$c = 40, w = \frac{40^3}{100} - \frac{1500}{40} = 602.5$, so the brain weighs 602.5 g when its circumference is 40 cm. When $c = 50$, $w = \frac{50^3}{100} - \frac{1500}{50}$ $= 1220$, so the brain weighs 1220 g when its circumference is 50 cm.

(b) Set the window of a graphing calculator so you can trace to the positive x-intercept of the function. Using a "root" or "zero" program, this x-intercept is found to be approximately 19.68. Notice in the graph that positive c values less than 19.68 correspond to negative w values. Therefore, the answer is $c < 19.68$.

(c)

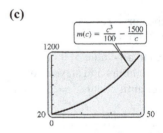

(d) One method is to graph the line $y = 700$ on the graph found in part (c) and use an "intercept" program to find the point of intersection of the two graphs. This point has the approximate coordinates (41.9, 700). Therefore, an infant has a brain weighing 700 g when the circumference measures 41.9 cm.

61. (a)

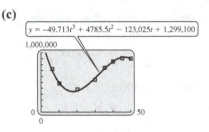

(b)

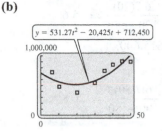

(c)

2.4 Exponential Functions

Your Turn 1

$$25^{x/2} = 125^{x+3}$$
$$(5^2)^{x/2} = (5^3)^{x+3}$$
$$5^x = 5^{3x+9}$$
$$x = 3x + 9$$
$$2x = -9$$
$$x = -\frac{9}{2}$$

Your Turn 2

$$A = P\left(1 + \frac{r}{m}\right)^{tm}$$
$$= 4400\left(1 + \frac{0.0325}{4}\right)^{5(4)}$$
$$= 5172.97$$

The interest amounts to $5172.97 - 4400 = \$772.97$.

Your Turn 3

$$A = Pe^{rt}$$
$$= 800e^{4(0.0315)}$$
$$= 907.43$$

The amount after 4 years will be $907.43.

2.4 Warmup Exercises

W1. $\left(2^3\right)^{-2x} = 2^{[(3)(-2x)]} = 2^{-6x}$

W2. $\left(3^5\right)^{7x-3} = 3^{[(5)(7x-3)]} = 3^{35x-15}$

W3. $5^x \cdot 5^{2x+1} = 5^{x+2x+1} = 5^{3x+1}$

W4. $x^a \cdot \dfrac{1}{x^b} = x^a \cdot x^{-b} = x^{a+(-b)} = x^{a-b}$

2.4 Exercises

1.

number of folds	1	2	3	4	5 ...	10 ...	50
layers of paper	2	4	8	16	32 ...	1024 ...	2^{50}

$2^{50} = 1.125899907 \times 10^{15}$

3. The graph of $y = 3^x$ is the graph of an exponential function $y = a^x$ with $a > 1$.

This is graph **E**.

5. The graph of $y = \left(\frac{1}{3}\right)^{1-x}$ is the graph

of $y = (3^{-1})^{1-x}$ or $y = 3^{x-1}$. This is

the graph of $y = 3^x$ translated 1 unit
to the right.
This is graph **C**.

7. The graph of $y = 3(3)^x$ is the same as the graph

of $y = 3^{x+1}$. This is the graph of $y = 3^x$

translated 1 unit to the left.

This is graph **F**.

9. The graph of $y = 2 - 3^{-x}$ is the same as the

graph of $y = -3^{-x} + 2$. This is the graph

of $y = 3^x$ reflected in the x-axis, reflected
in the y-axis, and translated up 2 units.

This is graph **A**.

11. The graph of $y = 3^{x-1}$ is the graph of $y = 3^x$
translated 1 unit to the right.

This is graph **C**.

13. $2^x = 32$

$2^x = 2^5$

$x = 5$

15. $3^x = \dfrac{1}{81}$

$3^x = \dfrac{1}{3^4}$

$3^x = 3^{-4}$

$x = -4$

17. $4^x = 8^{x+1}$

$(2^2)^x = (2^3)^{x+1}$

$2^{2x} = 2^{3x+3}$

$2x = 3x + 3$

$-x = 3$

$x = -3$

19. $16^{x+3} = 64^{2x-5}$

$(2^4)^{x+3} = (2^6)^{2x-5}$

$2^{4x+12} = 2^{12x-30}$

$4x + 12 = 12x - 30$

$42 = 8x$

$\dfrac{21}{4} = x$

21. $e^{-x} = (e^4)^{x+3}$

$e^{-x} = e^{4x+12}$

$-x = 4x + 12$

$-5x = 12$

$x = -\dfrac{12}{5}$

23. $5^{-|x|} = \dfrac{1}{25}$

$5^{-|x|} = 5^{-2}$

$|x| = 2$

$x = 2 \text{ or } x = -2$

25. $5^{x^2+x} = 1$

$5^{x^2+x} = 5^0$

$x^2 + x = 0$

$x(x + 1) = 0$

$$x = 0 \quad \text{or} \quad x + 1 = 0$$
$$x = 0 \quad \text{or} \qquad x = -1$$

27. $\quad 27^x = 9^{x^2+x}$

$$(3^3)^x = (3^2)^{x^2+x}$$

$$3^{3x} = 3^{2x^2+2x}$$

$$3x = 2x^2 + 2x$$

$$0 = 2x^2 - x$$

$$0 = x(2x - 1)$$

$$x = 0 \quad \text{or} \quad 2x - 1 = 0$$

$$x = 0 \quad \text{or} \qquad x = \frac{1}{2}$$

29. Graph of $y = 5e^x + 2$

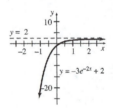

31. Graph of $y = -3e^{-2x} + 2$

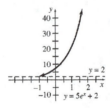

37. $A = P\left(1 + \dfrac{r}{m}\right)^{tm}$, $P = 10{,}000$, $r = 0.04$, $t = 5$

(a) annually, $m = 1$

$$A = 10{,}000\left(1 + \frac{0.04}{1}\right)^{5(1)}$$

$$= 10{,}000(1.04)^5$$

$$= \$12{,}166.53$$

$$\text{Interest} = \$12{,}166.53 - \$10.000$$

$$= \$21{,}66.53$$

(b) semiannually, $m = 2$

$$A = 10{,}000\left(1 + \frac{0.04}{2}\right)^{5(2)}$$

$$= 10{,}000(1.02)^{10}$$

$$= \$12{,}189.94$$

$$\text{Interest} = \$12{,}189.94 - \$10{,}000$$

$$= \$2189.94$$

(c) quarterly, $m = 4$

$$A = 10{,}000\left(1 + \frac{0.04}{4}\right)^{5(4)}$$

$$= 10{,}000(1.01)^{20}$$

$$= \$12{,}201.90$$

$$\text{Interest} = \$12{,}201.90 - \$10{,}000$$

$$= \$2201.90$$

(d) monthly, $m = 12$

$$A = 10{,}000\left(1 + \frac{0.04}{12}\right)^{5(12)}$$

$$= 10{,}000(1.00\overline{3})^{60}$$

$$= \$12{,}209.97$$

$$\text{Interest} = \$12{,}209.97 - \$10{,}000$$

$$= \$2209.97$$

(e) $\qquad A = 10{,}000e^{(0.04)(5)} = \$12{,}214.03$

$$\text{Interest} = \$12{,}214.03 - \$10{,}000$$

$$= \$2214.03$$

39. For 6% compounded annually for 2 years,

$$A = 18{,}000(1 + 0.06)^2$$

$$= 18{,}000(1.06)^2$$

$$= 20{,}224.80$$

For 5.9% compounded monthly for 2 years,

$$A = 18{,}000\left(1 + \frac{0.059}{12}\right)^{12(2)}$$

$$= 18{,}000\left(\frac{12.059}{12}\right)^{24}$$

$$= 20{,}248.54$$

The 5.9% investment is better. The additional interest is

$$\$20{,}248.54 - \$20{,}224.80 = \$23.74.$$

41. $A = Pe^{rt}$

   (a) $r = 3\%$

   $A = 10e^{0.03(3)} = \$10.94$

   (b) $r = 4\%$

   $A = 10e^{0.04(3)} = \$11.27$

   (c) $r = 5\%$

   $A = 10e^{0.05(3)} = \$11.62$

43.

$$1240 = 600\left(1 + \frac{r}{4}\right)^{(14)(4)}$$

$$\frac{1240}{600} = \left(1 + \frac{r}{4}\right)^{56}$$

$$\frac{31}{15} = \left(1 + \frac{r}{4}\right)^{56}$$

$$1 + \frac{r}{4} = \left(\frac{31}{15}\right)^{1/56}$$

$$4 + r = 4\left(\frac{31}{15}\right)^{1/56}$$

$$r = 4\left(\frac{31}{15}\right)^{1/56} - 4$$

$$r \approx 0.0522$$

The required interest rate is 5.22%.

45. (a) The cost in 10 years will be

$(165,000)(1.024)^{10} \approx 209,162.35$ or \$209,162.

   (b) The cost in 5 years will be

$(50)(1.024)^{5} \approx 56.294995$ or \$56.29.

47. (a)

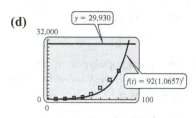

The GDP appears to grow exponentially.

   (b) $f(t) = f_0 a^t$

$f(0) = f_0 a^0 = 92$, so $f_0 = 92$

The year 2010 corresponds to $t = 80$.

$f(80) = 92a^{80} = 14,965$

$$a = \frac{14,965}{92}^{1/80} = 1.0657$$

Thus $f(t) = 92(1.0657)^t$.

(c) The average annual percentage increse in GDP is $1.0657 - 1 = 0.0657 = 6.57\%$.

(d)

The graph shows that the GDP will be approximately equal to $(2)(14,965) = 29,930$ when t is about 91, that is, in 2021.

49. $A(t) = 3100e^{0.0166t}$

   (a) 1970: $t = 10$

$$A(20) = 3100e^{(0.0166)(10)}$$
$$= 3100e^{0.166}$$
$$\approx 3659.78$$

The function gives a population of about 3660 million in 1970.

This is very close to the actual population of about 3686 million.

   (b) 2000: $t = 40$

$$A(50) = 3100e^{0.0166(40)}$$
$$= 3100e^{0.664}$$
$$\approx 6021.90$$

The function gives a population of 6022 million in 2000.

   (c) 2015: $t = 55$

$$= 3100e^{0.0166(55)}$$
$$= 3100e^{0.913}$$
$$= 7724.54$$

From the function, we estimate that the world population in 2015 will be 7725 million.

51. (a) Hispanic population:

$$h(t) = 37.79(1.021)^t$$
$$h(5) = 37.79(1.021)^5$$
$$\approx 41.93$$

The projected Hispanic population in 2005 is 41.93 million, which is slightly less than the actual value of 42.69 million.

(b) Asian population:

$$h(t) = 11.14(1.023)^t$$

$$h(5) = 11.14(1.023)^t$$

$$\approx 12.48$$

The projected Asian population in 2005 is 12.48 million, which is very close to the actual value of 12.69 million.

(c) Annual Hispanic percent increase:

$$1.021 - 1 = 0.021 = 2.1\%$$

Annual Asian percent increase:

$$1.023 - 1 = 0.023 = 2.3\%$$

The Asian population is growing at a slightly faster rate.

(d) Black population:

$$b(t) = 0.5116t + 35.43$$

$$b(5) = 0.5116(5) + 35.43$$

$$\approx 37.99$$

The projected Black population in 2005 is 37.99 million, which is extremely close to the actual value of 37.91 million.

(e) Hispanic population:

Double the actual 2005 value is

$$2(42.69) = 85.38 \text{ million.}$$

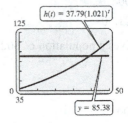

The doubling point is reached when $t \approx 39$, or in the year 2039.

Asian population:

Double the actual 2005 value is

$$2(12.69) = 25.38 \text{ million.}$$

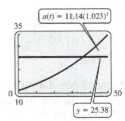

The doubling point is reached when $t \approx 36$, or in the 2036.

Black population:

Double the actual 2005 value is

$$2(37.91) = 7582 \text{ million.}$$

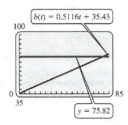

The doubling point is reached when $t \approx 79$, or in the year 2079.

53. **(a)** When $x = 0$, $P = 1013$.

When $x = 10,000$, $P = 265$.

First we fit $P = ae^{kx}$.

$$1013 = ae^0$$

$$a = 1013$$

$$P = 1013e^{kx}$$

$$265 = 1013e^{k(10,000)}$$

$$\frac{265}{1013} = e^{10,000k}$$

$$10,000k = \ln\left(\frac{265}{1013}\right)$$

$$k = \frac{\ln\left(\frac{265}{1013}\right)}{10,000} \approx -1.34 \times 10^{-4}$$

Therefore $P = 1013e^{(-1.34 \times 10^{-4})x}$.

Next we fit $P = mx + b$.

We use the points $(0, 1013)$ and $(10,000, 265)$.

$$m = \frac{265 - 1013}{10,000 - 0} = -0.0748$$

$$b = 1013$$

Therefore $P = -0.0748x + 1013$.

Finally, we fit $P = \frac{1}{ax+b}$.

$$1013 = \frac{1}{a(0) + b}$$

$$b = \frac{1}{1013} \approx 9.87 \times 10^{-4}$$

$$P = \frac{1}{ax + \frac{1}{1013}}$$

$$265 = \frac{1}{10,000a + \frac{1}{1013}}$$

$$\frac{1}{265} = 10,000a + \frac{1}{1013}$$

$$10,000a = \frac{1}{265} - \frac{1}{1013}$$

$$a = \frac{\frac{1}{265} - \frac{1}{1013}}{10,000} \approx 2.79 \times 10^{-7}$$

Therefore,

$$P = \frac{1}{(2.79 \times 10^{-7})x + (9.87 \times 10^{-4})}.$$

(b)

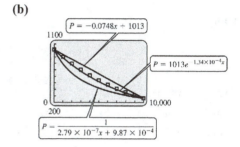

$P = 1013e^{(-1.34 \times 10^{-4})x}$ is the best fit.

(c) $P(1500) = 1013e^{-1.34 \times 10^{-4}(1500)} \approx 829$

$P(11,000) = 1013e^{-1.34 \times 10^{-4}(11,000)} \approx 232$

We predict that the pressure at 1500 meters will be 829 millibars, and at 11,000 meters will be 232 millibars.

(d) Using exponential regression, we obtain $P = 1038(0.99998661)^x$ which differs slightly from the function found in part (b) which can be rewritten as

$$P = 1013(0.99998660)^x.$$

55. (a) $C = mt + b$

Use the points (1, 24,322) and (11, 237,016) to find m.

$$m = \frac{237,016 - 24,322}{11 - 1} = \frac{212,694}{10} = 21,269.4$$

Use the point (1, 24,322) to find b.
$$y = 21,269.4t + b$$
$$24,322 = (21,269.4)(1) + b$$
$$b = 3052.6$$
$$C = 21,269.4t + 3052.6$$

$$C = at^2 + b$$

Use the points (1, 24,322) and (11, 237,016) to find a and b.

$$24,322 = a(1)^2 + b \quad 237,016 = a(11)^2 + b$$
$$24,322 = a + b \quad\quad 237,016 = 121a + b$$

$$121a + b = 237,016$$
$$\underline{a + b = 24,322}$$
$$120a = 212,694$$

$$a = \frac{212,694}{120} \approx 1772.45$$

$$a + b = 24,322$$
$$b = 24,322 - 1772.45$$
$$b = 22,549.55$$

$$C = 1772.45t^2 + 22,549.55$$

$$C = ab^t$$

Use the points (1, 24,322) and (11, 237,016) to find a and b.

$$24,322 = ab^1 = ab \quad\quad 237,016 = ab^{11}$$

$$\frac{237,016}{24,322} = \frac{ab^{11}}{ab} = b^{10}$$

$$b = \sqrt[10]{\frac{237,016}{24,322}} \approx 1.255667 \approx 1.2557$$

$$ab = 24,322$$
$$(1.2557)a = 24,322$$

$$a = \frac{24,322}{1.255667} \approx 19,369.64$$

$$C = 19,369.64(1.2557)^t$$

(b)

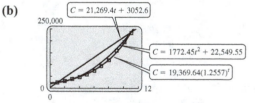

The function $C = 19,369.64(1.2557)^t$ is the best fit.

(c) The regression function is

$C = 19,259.86(1.2585)^t$. This is very close to the function in part (b).

(d) $x = 12$ corresponds to 2012.

$$\begin{aligned} C &= 21{,}269.4t + 3052.6 \\ &= 21{,}269.4(12) + 3052.6 \\ &\approx 258{,}285 \end{aligned}$$

$$\begin{aligned} C &= 1772.45t^2 + 22{,}549.55 \\ &= 1772.45(12)^2 + 22{,}549.55 \\ &\approx 277{,}782 \end{aligned}$$

$$\begin{aligned} C &= 19{,}369.64(1.2557)^t \\ &= 19{,}369.64(1.2557)^{12} \\ &\approx 297{,}682 \end{aligned}$$

$$\begin{aligned} C &= 19{,}259.86(1.2585)^t \\ &= 19{,}259.86(1.2585)^{12} \\ &\approx 304{,}013 \end{aligned}$$

2.5 Logarithmic Functions

Your Turn 1

$5^{-2} = \dfrac{1}{25}$ means $\log_5\left(\dfrac{1}{25}\right) = -2$

Your Turn 2

$\log_3\left(\dfrac{1}{81}\right)$

We seek a number x such that

$$3^x = \frac{1}{81}$$
$$3^x = (3)^{-4}$$
$$x = -4$$

Your Turn 3

$$\begin{aligned} \log_a\left(\frac{x^2}{y^3}\right) &= \log_a x^2 - \log_a y^3 \\ &= 2\log_a x - 3\log_a y \end{aligned}$$

Your Turn 4

$\log_3 50 = \dfrac{\ln 50}{\ln 3} \approx 3.561$

Your Turn 5

$$\begin{aligned} \log_2 x + \log_2(x+2) &= 3 \\ \log_2 x(x+2) &= 3 \\ x(x+2) &= 2^3 \\ x^2 + 2x &= 8 \\ x^2 + 2x - 8 &= 0 \\ x = -4 \text{ or } x &= 2 \end{aligned}$$

Since the logarithm of a negative number is undefined, the only solution is $x = 2$.

Your Turn 6

$$\begin{aligned} 2^{x+1} &= 3^x \\ \ln 2^{x+1} &= \ln 3^x \\ (x+1)\ln 2 &= x\ln 3 \\ x\ln 2 + \ln 2 &= x\ln 3 \\ x\ln 3 - x\ln 2 &= \ln 2 \\ x(\ln 3 - \ln 2) &= \ln 2 \\ x\ln\left(\tfrac{3}{2}\right) &= \ln 2 \\ x &= \frac{\ln 2}{\ln 1.5} \approx 1.7095 \end{aligned}$$

Your Turn 7

$e^{0.025x} = (e^{0.025})^x \approx 1.0253^x$

2.5 Warmup Exercises

W1.
$$\begin{aligned} x^2 - x + 3 &= 9 \\ x^2 - x - 6 &= 0 \\ (x-3)(x+2) &= 0 \\ x - 3 = 0 \text{ or } x + 2 &= 0 \\ x = 3 \qquad \text{or} \quad x &= -2 \end{aligned}$$

W2. Assume that x is not equal to 5. $\dfrac{x}{x-5} = 4$ implies

$$\begin{aligned} x &= 4x - 20 \\ 3x &= 20 \\ x &= \frac{20}{3} \end{aligned}$$

W3. $2 = 8^{x+1} = \left(2^3\right)^{x+1} = 2^{3x+3}$

Therefore,

$$2^1 = 2^{3x+3}$$
$$1 = 3x + 3$$
$$-2 = 3x$$
$$x = -\frac{2}{3}$$

W4. $e^{x-4} = 1 = e^0$

Therefore,
$$x - 4 = 0$$
$$x = 4$$

2.5 Exercises

1. $5^3 = 125$

Since $a^y = x$ means $y = \log_a x$, the equation in logarithmic form is
$$\log_5 125 = 3.$$

3. $3^4 = 81$

The equation in logarithmic form is
$$\log_3 81 = 4.$$

5. $3^{-2} = \frac{1}{9}$

The equation in logarithmic form is
$$\log_3 \frac{1}{9} = -2.$$

7. $\log_2 32 = 5$

Since $y = \log_a x$ means $a^y = x$, the equation in exponential form is
$$2^5 = 32.$$

9. $\ln \frac{1}{e} = -1$

The equation in exponential form is
$$e^{-1} = \frac{1}{e}.$$

11. $\log 100{,}000 = 5$
$$\log_{10} 100{,}000 = 5$$
$$10^5 = 100{,}000$$

When no base is written, $\log_{10}$ is understood.

13. Let $\log_8 64 = x.$
$$\text{Then, } 8^x = 64$$
$$8^x = 8^2$$
$$x = 2.$$
Thus, $\log_8 64 = 2.$

15. $\log_4 64 = x$
$$4^x = 64$$
$$4^x = 4^3$$
$$x = 3$$

17. $\log_2 \frac{1}{16} = x$
$$2^x = \frac{1}{16}$$
$$2^x = 2^{-4}$$
$$x = -4$$

19. $\log_2 \sqrt[3]{\frac{1}{4}} = x$
$$2^x = \left(\frac{1}{4}\right)^{1/3}$$
$$2^x = \left(\frac{1}{2^2}\right)^{1/3}$$
$$2^x = 2^{-2/3}$$
$$x = -\frac{2}{3}$$

21. $\ln e = x$

Recall that ln means $\log_e$.
$$e^x = e$$
$$x = 1$$

23. $\ln e^{5/3} = x$
$$e^x = e^{5/3}$$
$$x = \frac{5}{3}$$

25. The logarithm to the base 3 of 4 is written $\log_3 4$. The subscript denotes the base.

27. $\log_5 (3k) = \log_5 3 + \log_5 k$

29. $\log_3 \dfrac{3p}{5k}$

$= \log_3 3p - \log_3 5k$

$= (\log_3 3 + \log_3 p) - (\log_3 5 + \log_3 k)$

$= 1 + \log_3 p - \log_3 5 - \log_3 k$

31. $\ln \dfrac{3\sqrt{5}}{\sqrt[3]{6}}$

$= \ln 3\sqrt{5} - \ln \sqrt[3]{6}$

$= \ln 3 \cdot 5^{1/2} - \ln 6^{1/3}$

$= \ln 3 + \ln 5^{1/2} - \ln 6^{1/3}$

$= \ln 3 + \dfrac{1}{2}\ln 5 - \dfrac{1}{3}\ln 6$

33. $\log_b 32 = \log_b 2^5$

$\qquad\quad = 5\log_b 2$

$\qquad\quad = 5a$

35. $\log_b 72b = \log_b 72 + \log_b b$

$\qquad\qquad = \log_b 72 + 1$

$\qquad\qquad = \log_b 2^3 \cdot 3^3 + 1$

$\qquad\qquad = \log_b 2^3 + \log_b 3^2 + 1$

$\qquad\qquad = 3\,\log_b 2 + 2\log_b 3 + 1$

$\qquad\qquad = 3a + 2c + 1$

37. $\log_5 30 = \dfrac{\ln 30}{\ln 5}$

$\qquad\qquad \approx \dfrac{3.4012}{1.6094}$

$\qquad\qquad \approx 2.113$

39. $\log_{1.2} 0.95 = \dfrac{\ln 0.95}{\ln 1.2}$

$\qquad\qquad\quad \approx -0.281$

41. $\quad\log_x 36 = -2$

$\qquad\quad x^{-2} = 36$

$\quad (x^{-2})^{-1/2} = 36^{-1/2}$

$\qquad\qquad x = \dfrac{1}{6}$

43. $\log_8 16 = z$

$\qquad 8^z = 16$

$\quad (2^3)^z = 2^4$

$\qquad 2^{3z} = 2^4$

$\qquad 3z = 4$

$\qquad z = \dfrac{4}{3}$

45. $\log_r 5 = \dfrac{1}{2}$

$\qquad r^{1/2} = 5$

$\quad (r^{1/2})^2 = 5^2$

$\qquad r = 25$

47. $\log_5 (9x - 4) = 1$

$\qquad 5^1 = 9x - 4$

$\qquad 9 = 9x$

$\qquad 1 = x$

49. $\log_9 m - \log_9 (m - 4) = -2$

$\qquad\quad \log_9 \dfrac{m}{m - 4} = -2$

$\qquad\qquad 9^{-2} = \dfrac{m}{m - 4}$

$\qquad\qquad \dfrac{1}{81} = \dfrac{m}{m - 4}$

$\qquad\quad m - 4 = 81m$

$\qquad\qquad -4 = 80m$

$\qquad\quad -0.05 = m$

This value is not possible since $\log_9 (-0.05)$ does not exist.

Thus, there is no solution to the original equation.

51. $\log_3 (x - 2) + \log_3 (x + 6) = 2$

$\qquad \log_3 [(x - 2)(x + 6)] = 2$

$\qquad\quad (x - 2)(x + 6) = 3^2$

$\qquad\qquad x^2 + 4x - 12 = 9$

$\qquad\qquad x^2 + 4x - 21 = 0$

$\qquad\qquad (x + 7)(x - 3) = 0$

$\qquad\qquad x = -7 \ \text{ or } \ x = 3$

$x = -7$ does not check in the original equation.
The only solution is 3.

53. $\log_2 (x^2 - 1) - \log_2 (x + 1) = 2$

$$\log_2 \frac{x^2 - 1}{x + 1} = 2$$

$$2^2 = \frac{x^2 - 1}{x + 1}$$

$$4 = \frac{(x - 1)(x + 1)}{x + 1}$$

$$4 = x - 1$$

$$x = 5$$

55. $\ln x + \ln 3x = -1$

$$\ln 3x^2 = -1$$

$$3x^2 = e^{-1}$$

$$x^2 = \frac{e^{-1}}{3}$$

$$x = \sqrt{\frac{e^{-1}}{3}} = \frac{1}{\sqrt{3e}} \approx 0.3502$$

57. $2^x = 6$

$$\ln 2^x = \ln 6$$

$$x \ln 2 = \ln 6$$

$$x = \frac{\ln 6}{\ln 2} \approx 2.5850$$

59. $e^{k-1} = 6$

$$\ln e^{k-1} = \ln 6$$

$$(k - 1)\ln e = \ln 6$$

$$k - 1 = \frac{\ln 6}{\ln e}$$

$$k - 1 = \frac{\ln 6}{1}$$

$$k = 1 + \ln 6$$

$$\approx 2.7918$$

61. $3^{x+1} = 5^x$

$$\ln 3^{x+1} = \ln 5^x$$

$$(x + 1)\ln 3 = x \ln 5$$

$$x \ln 3 + \ln 3 = x \ln 5$$

$$x \ln 5 - x \ln 3 = \ln 3$$

$$x(\ln 5 - \ln 3) = \ln 3$$

$$x = \frac{\ln 3}{\ln(5/3)} \approx 2.1507$$

63. $5(0.10)^x = 4(0.12)^x$

$$\ln[5(0.10)^x] = \ln[4(0.12)^x]$$

$$\ln 5 + x \ln 0.10 = \ln 4 + x \ln 0.12$$

$$x(\ln 0.12 - \ln 0.10) = \ln 5 - \ln 4$$

$$x = \frac{\ln 5 - \ln 4}{\ln 0.12 - \ln 0.10}$$

$$= \frac{\ln 1.25}{\ln 1,2}$$

$$\approx 1.2239$$

65. $10^{x+1} = e^{(\ln 10)(x+1)}$

67. $e^{3x} = (e^3)^x \approx 20.09^x$

69. $f(x) = \log(5 - x)$

$$5 - x > 0$$

$$-x > -5$$

$$x < 5$$

The domain of f is $x < 5$.

71. $\log A - \log B = 0$

$$\log \frac{A}{B} = 0$$

$$\frac{A}{B} = 10^0 = 1$$

$$A = B$$

$$A - B = 0$$

Thus, solving $\log A - \log B = 0$ is equivalent to solving $A - B = 0$

73. Let $m = \log_a \frac{x}{y}$, $n = \log_a x$, and $p = \log_a y$.

Then $a^m = \frac{x}{y}$, $a^n = x$, and $a^p = y$.

Substituting gives

$$a^m = \frac{x}{y} = \frac{a^n}{a^p} = a^{n-p}.$$

So $m = n - p$.

Therefore,

$$\log_a \frac{x}{y} = \log_a x - \log_a y.$$

75. From Example 8, the doubling time t in years when $m = 1$ is given by

$$t = \frac{\ln 2}{\ln(1 + r)}.$$

(a) Let $r = 0.03$.

$$t = \frac{\ln 2}{\ln(1.03)}$$

$$= 23.4 \text{ years}$$

(b) Let $r = 0.06$.

$$t = \frac{\ln 2}{\ln 1.06}$$

$$= 11.9 \text{ years}$$

(c) Let $r = 0.08$.

$$t = \frac{\ln 2}{\ln 1.08}$$

$$= 9.0 \text{ years}$$

(d) Since $0.001 \le 0.03 \le 0.05$, for $r = 0.03$, we use the rule of 70.

$$\frac{70}{100r} = \frac{70}{100(0.03)} = 23.3 \text{ years}$$

Since $0.05 \le 0.06 \le 0.12$, for $r = 0.06$, we use the rule of 72.

$$\frac{72}{100r} = \frac{72}{100(0.06)} = 12 \text{ years}$$

For $r = 0.08$, we use the rule of 72.

$$\frac{72}{100(0.08)} = 9 \text{ years}$$

77.

$$A = Pe^{rt}$$

$$1240 = 600e^{r \cdot 14}$$

$$\frac{31}{15} = e^{14r}$$

$$\ln\left(\frac{31}{15}\right) = \ln e^{14r}$$

$$\ln\left(\frac{31}{15}\right) = 14r$$

$$\frac{\ln\left(\frac{31}{15}\right)}{14} = r$$

$$0.0519 \approx r$$

The interest rate should be 5.19%.

79. $f(t) = 92(1.0657)^t$ where $t = 0$ corresponds to 1930.

$$92(1.0657)^t = 29,930$$

$$e^{(t)\ln(1.0657)} = \frac{29,930}{92}$$

$$t\ln(1.0657) = \ln\left(\frac{29,930}{92}\right)$$

$$t = \frac{1}{\ln(1.0657)}\ln\left(\frac{29,930}{92}\right) = 90.911$$

$t = 91$ corresponds to 2021.

81. (a) $y(t) = y_0 e^{kt}$ implies that $e^{kt} = \dfrac{y(t)}{y_0}$

$$k = \frac{1}{t}\ln\left(\frac{y(t)}{y_0}\right).$$

(b) Using the data given in Exercise 50 in Section 2.4, and with 2006 corresponding to $t = 0$,

$y_0 = 680.5$. The years 2015, 2020, and 2025 correspond to the t values 9, 14 and 19, and the corresponding $y(t)$ are 758.6, 805.8, and 859.3

The three values of k are:

$$\frac{1}{9}\ln\left(\frac{758.6}{680.5}\right) \approx 0.01207$$

$$\frac{1}{14}\ln\left(\frac{808.8}{680.5}\right) \approx 0.01207$$

$$\frac{1}{19}\ln\left(\frac{859.3}{680.5}\right) \approx 0.01228$$

83. (a) The total number of individuals in the community is $50 + 50$, or 100.

Let $P_1 = \dfrac{50}{100} = 0.5$, $P_2 = 0.5$.

$$H = -1[P_1 \ln P_1 + P_2 \ln P_2]$$

$$= -1[0.5\ln 0.5 + 0.5\ln 0.5]$$

$$\approx 0.693$$

(b) For 2 species, the maximum diversity is $\ln 2$.

(c) Yes, $\ln 2 \approx 0.693$.

85. (a) 3 species, $\frac{1}{3}$ each:

$$P_1 = P_2 = P_3 = \frac{1}{3}$$

$$H = -(P_1 \ln P_1 + P_2 \ln P_2 + P_3 \ln P_3)$$

$$= -3\left(\frac{1}{3}\ln\frac{1}{3}\right)$$

$$= -\ln\frac{1}{3}$$

$$\approx 1.099$$

(b) 4 species, $\frac{1}{4}$ each:

$$P_1 = P_2 = P_3 = P_4 = \frac{1}{4}$$

$$H = -(P_1 \ln P_1 + P_2 \ln P_2 + P_3 \ln P_3 + P_4 \ln P_4)$$

$$= -4\left(\frac{1}{4} \ln \frac{1}{4}\right)$$

$$= -\ln \frac{1}{4}$$

$$\approx 1.386$$

(c) Notice that

$$-\ln \frac{1}{3} = \ln (3^{-1})^{-1} = \ln 3 \approx 1.099$$

and

$$-\ln \frac{1}{4} = \ln (4^{-1})^{-1} = \ln 4 \approx 1.386$$

by Property c of logarithms, so the populations are at a maximum index of diversity.

87. $C(t) = C_0 e^{-kt}$

When $t = 0$, $C(t) = 2$, and when $t = 3$, $C(t) = 1$.

$$2 = C_0 e^{-k(0)}$$

$$C_0 = 2$$

$$1 = 2e^{-3k}$$

$$\frac{1}{2} = e^{-3k}$$

$$-3k = \ln \frac{1}{2} = \ln 2^{-1} = -\ln 2$$

$$k = \frac{\ln 2}{3}$$

$$T = \frac{1}{k} \ln \frac{C_2}{C_1}$$

$$T = \frac{1}{\frac{\ln 2}{3}} \ln \frac{5C_1}{C_1}$$

$$T = \frac{3 \ln 5}{\ln 2}$$

$$T \approx 7.0$$

The drug should be given about every 7 hours.

89. (a) $h(t) = 37.79(1.021)^t$

Double the 2005 population is $2(42.69)$
$= 85.38$ million

$$85.38 = 37.79(1.021)^t$$

$$\frac{85.38}{37.79} = (1.021)^t$$

$$\log_{1.021}\left(\frac{85.38}{37.79}\right) = t$$

$$t = \frac{\ln\left(\frac{85.38}{37.79}\right)}{\ln 1.021}$$

$$\approx 39.22$$

The Hispanic population is estimated to double their 2005 population in 2039.

(b) $a(t) = 11.14(1.023)^t$

Double the 2005 population is $2(12.69) = 25.38$ million

$$25.38 = 11.14(1.023)^t$$

$$\frac{25.38}{11.14} = (1.023)^t$$

$$\log_{1.023}\left(\frac{25.38}{11.14}\right) = t$$

$$t = \frac{\ln\left(\frac{25.38}{11.14}\right)}{\ln 1.023}$$

$$\approx 36.21$$

The Asian population is estimated to double their 2005 population in 2036.

91. $C = B \log_2\left(\frac{s}{n} + 1\right)$

$$\frac{C}{B} = \log_2\left(\frac{s}{n} + 1\right)$$

$$2^{C/B} = \frac{s}{n} + 1$$

$$\frac{s}{n} = 2^{C/B} - 1$$

93. Let I_1 be the intensity of the sound whose decibel rating is 85.

(a)

$$10 \log \frac{I_1}{I_0} = 85$$

$$\log \frac{I_1}{I_0} = 8.5$$

$$\log I_1 - \log I_0 = 8.5$$

$$\log I_1 = 8.5 + \log I_0$$

Let I_2 be the intensity of the sound whose decibel rating is 75.

$$10 \log \frac{I_2}{I_0} = 75$$

$$\log \frac{I_2}{I_0} = 7.5$$

$$\log I_2 - \log I_0 = 7.5$$

$$\log I_0 = \log I_2 - 7.5$$

Substitute for I_0 in the equation for $\log I_1$.

$$\log I_1 = 8.5 + \log I_0$$
$$= 8.5 + \log I_2 - 7.5$$
$$= 1 + \log I_2$$
$$\log I_1 - \log I_2 = 1$$
$$\log \frac{I_1}{I_2} = 1$$

Then $\frac{I_1}{I_2} = 10$, so $I_2 = \frac{1}{10} I_1$. This means the intensity of the sound that had a rating of 75 decibels is $\frac{1}{10}$ as intense as the sound that had a rating of 85 decibels.

95. $\text{pH} = -\log[H^+]$

(a) For pure water:

$$7 = -\log[H^+]$$
$$-7 = \log[H^+]$$
$$10^{-7} = [H^+]$$

For acid rain:

$$4 = -\log[H^+]$$
$$-4 = \log[H^+]$$
$$10^{-4} = [H^+]$$
$$\frac{10^{-4}}{10^{-7}} = 10^3 = 1000$$

The acid rain has a hydrogen ion concentration 1000 times greater than pure water.

(b) For laundry solution:

$$11 = -\log[H^+]$$
$$10^{-11} = [H^+]$$

For black coffee:

$$5 = -\log[H^+]$$
$$10^{-5} = [H^+]$$
$$\frac{10^{-5}}{10^{-11}} = 10^6 = 1,000,000$$

The coffee has a hydrogen ion concentration 1,000,000 times greater than the laundry mixture.

2.6 Applications: Growth and Decay; Mathematics of Finance

Your Turn 1

$$y = y_0 e^{kt}$$
$$18 = 5 e^{k(16)}$$
$$e^{16k} = \ln(18/5)$$
$$16k = \ln(18/5)$$
$$k = \frac{\ln(18/5)}{16} \approx 0.08$$
$$y = 5 e^{0.08t}$$

Your Turn 2

Let $A(t) = \frac{1}{10} A_0$ and $k = -(\ln 2/5600)$

$$A(t) = A_0 e^{kt}$$
$$\frac{1}{10} A_0 = A_0 e^{-(\ln 2/5600)t}$$
$$\frac{1}{10} = e^{-(\ln 2/5600)t}$$
$$\ln\left(\frac{1}{10}\right) = \ln e^{-(\ln 2/5600)t}$$
$$\ln\left(\frac{1}{10}\right) = -(\ln 2/5600)t$$
$$t = -\frac{5600}{\ln 2} \ln\left(\frac{1}{10}\right)$$
$$\approx 18602.80$$

The age of the sample is about 18,600 years.

Your Turn 3

(a) 4.25% compounded monthly

$$\left(1 + \frac{0.0425}{12}\right)^{12} - 1 = (1.0035417)^{12} - 1$$

$$\approx 0.0433$$

The effective rate is 4.33%.

(b) 3.75% compounded continuously

$$e^{0.0375} - 1 \approx 0.0382$$

The effective rate is 3.82%.

Your Turn 4

$A = 50,000$, $r = 0.315$, and $m = 4$.

$$50,000 = 30,000\left(1 + \frac{0.0315}{4}\right)^{4t}$$

$$\frac{5}{3} = (1.007875)^{4t}$$

$$t = \frac{\ln(5/3)}{4\ln(1.007875)} \approx 16.28$$

We need to round up to the nearest quarter, so $30,000 will grow to $50,000 in 16.5 years.

Your Turn 5

$$A = Pe^{rt}$$

$$4500 = 3200e^{7r}$$

$$1.40625 = e^{7r}$$

$$\ln 1.40625 = \ln e^{7r}$$

$$7r = \ln 1.40625$$

$$r = \frac{\ln 1.40625}{7} \approx 0.0487$$

The interest rate needed is 4.87%.

2.6 Warmup Exercises

W1. $e^{2t} = 5$

$$2t = \ln(5)$$

$$t = \frac{\ln(5)}{2} \approx 0.8047$$

W2. $e^{k+2} = 3$

$$k + 2 = \ln(3)$$

$$k = \ln(3) - 2 \approx -0.9014$$

W3. $10e^{3t} = 45$

$$e^{3t} = 4.5$$

$$3t = \ln(4.5)$$

$$t = \frac{\ln(4.5)}{3} \approx 0.5014$$

2.6 Exercises

5. Assume that $y = y_0 e^{kt}$ represents the amount remaining of a radioactive substance decaying with a half-life of T. Since $y = y_0$ is the amount of the substance at time $t = 0$, then $y = \frac{y_0}{2}$ is the amount at time $t = T$. Therefore, $\frac{y_0}{2} = y_0 e^{kT}$, and solving for k yields

$$\frac{1}{2} = e^{kT}$$

$$\ln\left(\frac{1}{2}\right) = kT$$

$$k = \frac{\ln\left(\frac{1}{2}\right)}{T}$$

$$= -\frac{\ln 2}{T}.$$

7. $r = 4\%$ compounded quarterly, $m = 4$

$$r_E = \left(1 + \frac{r}{m}\right)^m - 1$$

$$= \left(1 + \frac{0.04}{4}\right)^4 - 1$$

$$\approx 0.0406$$

$$\approx 4.06\%$$

9. $r = 8\%$ compounded continuously

$$r_E = e^r - 1$$

$$= e^{0.08} - 1$$

$$\approx 0.0833 = 8.33\%$$

11. $A = \$10,000$, $r = 6\%$, $m = 4$, $t = 8$

$$P = A\left(1 + \frac{r}{m}\right)^{-tm}$$

$$= 10,000\left(1 + \frac{0.06}{4}\right)^{-8(4)}$$

$$\approx \$6209.93$$

13. $A = \$7300$, $r = 5\%$ compounded continuously, $t = 3$

$$A = Pe^{rt}$$

$$P = \frac{A}{e^{rt}}$$

$$= \frac{7300}{e^{0.05(3)}} \approx \$6283.17$$

15. $r = 9\%$ compounded semiannually

$$r_E = \left(1 + \frac{0.09}{2}\right)^2 - 1$$

$$\approx 0.0920 = 9.20\%$$

17. $r = 6\%$ compounded monthly

$$r_E = \left(1 + \frac{0.06}{12}\right)^{12} - 1$$

$$\approx 0.0617$$

$$\approx 6.17\%$$

19. (a) $A = \$307,000, t = 3, r = 6\%, m = 2$

$$A = P\left(1 + \frac{r}{m}\right)^{mt}$$

$$307,000 = P\left(1 + \frac{0.06}{2}\right)^{3(2)}$$

$$307,000 = P(1.03)^6$$

$$\frac{307,000}{(1.03)^6} = P$$

$$\$257,107.67 = P$$

(b) Interest $= 307,000 - 257,107.67$

$$= \$49,892.33$$

(c) $\qquad P = \$200,000$

$$A = 200,000(1.03)^6$$

$$= 238,810.46$$

The additional amount needed is

$$307,000 - 238,810.46$$

$$= \$68,189.54.$$

(d) $\qquad A = Pe^{rt}$

$$307,000 = 200,000e^{3r}$$

$$1.535 = e^{3r}$$

$$\ln 1.535 = \ln e^{3r}$$

$$3r = \ln 1.535$$

$$r = \frac{\ln 1.535}{3} \approx 0.1428$$

The interest rate needed is 14.28%.

21. $P = \$60,000$

(a) $r = 8\%$ compounded quarterly:

$$A = P\left(1 + \frac{r}{m}\right)^{tm}$$

$$= 60,000\left(1 + \frac{0.08}{47}\right)^{5(4)}$$

$$\approx \$89,156.84$$

$r = 7.75\%$ compounded continuously

$$A = Pe^{rt}$$

$$= 60,000e^{0.775(5)}$$

$$\approx \$88,397.58$$

Christine will earn more money at 8% compound-ded quarterly.

(b) She will earn $759.26 more.

(c) $r = 8\%, m = 4$:

$$r_E = \left(1 + \frac{r}{m}\right)^m - 1$$

$$= \left(1 + \frac{0.08}{4}\right)^4 - 1$$

$$\approx 0.0824$$

$$= 8.24\%$$

$r = 7.75\%$ compounded continuously:

$$r_E = e^r - 1$$

$$= e^{0.0775} - 1$$

$$\approx 0.0806$$

$$= 8.06\%$$

(d) $A = \$80,000$

$$A = Pe^{rt}$$

$$80,000 = 60,000e^{0.0775t}$$

$$\frac{4}{3} = e^{0.0775t}$$

$$\ln \frac{4}{3} = \ln e^{0.0775t}$$

$$\ln 4 - \ln 3 = 0.0775t$$

$$\frac{\ln 4 - \ln 3}{0.0775} = t$$

$$3.71 = t$$

$60,000 will grow to $80,000 in about 3.71 years.

(e) $60,000\left(1 + \dfrac{0.08}{4}\right)^{4x} \geq 80,000$

$$(1.02)^{4x} \geq \frac{80{,}000}{60{,}000}$$

$$(1.02)^{4x} \geq \frac{4}{3}$$

$$\log (1.02)^{4x} \geq \log \left(\frac{4}{3} \right)$$

$$4x \log (1.02) \geq \log \left(\frac{4}{3} \right)$$

$$x \geq \frac{\log \left(\frac{4}{3} \right)}{4 \log (1.02)} \approx 3.63$$

We need to round up to the nearest quarter, so it will take 3.75 years.

23. $S(x) = 1000 - 800e^{-x}$

(a) $S(0) = 1000 - 800e^0$
$$= 1000 - 800$$
$$= 200$$

(b) $S(x) = 500$
$$500 = 1000 - 800e^{-x}$$
$$-500 = -800e^{-x}$$
$$\frac{5}{8} = e^{-x}$$
$$\ln \frac{5}{8} = \ln e^{-x}$$
$$-\ln \frac{5}{8} = x$$
$$0.47 \approx x$$

Sales reach 500 in about $\frac{1}{2}$ year.

(c) Since $800e^{-x}$ will never actually be zero, $S(x) = 1000 - 800e^{-x}$ will never be 1000.

(d) Graphing the function $y = S(x)$ on a graphing calculator will show that there is a horizontal asymptote at $y = 1000$. This indicates that the limit on sales is 1000 units.

25. (a) $P = P_0 e^{kt}$
When $t = 1650$, $P = 500$.
When $t = 2010$, $P = 6756$.

$$500 = P_0 e^{1650k}$$
$$6756 = P_0 e^{2010k}$$
$$\frac{6756}{500} = \frac{P_0 e^{2010k}}{P_0 e^{1650k}}$$
$$\frac{6756}{500} = e^{360k}$$
$$360k = \ln \left(\frac{6765}{500} \right)$$
$$k = \frac{\ln \left(\frac{6765}{500} \right)}{360}$$
$$k = 0.007232$$

Substitute this value into $500 = P_0 e^{1650k}$ to find P_0.

$$500 = P_0 e^{1650(0.007232)}$$
$$P_0 = \frac{500}{e^{1650(0.007232)}}$$
$$P_0 = 0.003286$$

Therefore, $P(t) = 0.003286 e^{0.007232t}$.

(b) $P(1) = 0.003286 e^{0.007232}$
$$\approx 0.0033 \text{ million, or } 3300.$$

The exponential equation gives a world population of only 3300 in the year 1.

(c) No, the answer in part (b) is too small. Exponential growth does not accurately describe population growth for the world over a long period of time.

27. $y = y_0 e^{kt}$
$y = 40{,}000$, $y_0 = 25{,}000$, $t = 10$

(a) $40{,}000 = 25{,}000 e^{k(10)}$
$$1.6 = e^{10k}$$
$$\ln 1.6 = 10k$$
$$0.047 = k$$
The equation is
$$y = 25{,}000 e^{0.047t}.$$

(b) $y = 25{,}000 e^{0.047t}$
$$= 25{,}000 (e^{0.047})^t$$
$$= 25{,}000 (1.048)^t$$

(c)
$$y = 60{,}000$$
$$60{,}000 = 25{,}000e^{0.047t}$$
$$2.4 = e^{0.047t}$$
$$\ln 2.4 = 0.047t$$
$$18.6 = t$$

There will be 60,000 bacteria in about 18.6 hours.

29. $f(t) = 500\,e^{0.1t}$

(a) $f(t) = 3000$
$$3000 = 500e^{0.1t}$$
$$6 = e^{0.1t}$$
$$\ln 6 = 0.1t$$
$$17.9 \approx t$$

It will take 17.9 days.

(b) If $t = 0$ corresponds to January 1, the date January 17 should be placed on the product. January 18 would be more than 17.9 days.

31. (a) From the graph, the risks of chromosomal abnormality per 1000 at ages 20, 35, 42, and 49 are 2, 5, 24, and 125, respectively.

(Note: It is difficult to read the graph accurately. If you read different values from the graph, your answers to parts (b)-(e) may differ from those given here.)

(b) $y = Ce^{kt}$

When $t = 20$, $y = 2$, and when $t = 35$, $y = 5$.
$$2 = Ce^{20k}$$
$$5 = Ce^{35k}$$
$$\frac{5}{2} = \frac{Ce^{35k}}{Ce^{20k}}$$
$$2.5 = e^{15k}$$
$$15k = \ln 2.5$$
$$k = \frac{\ln 2.5}{15} \quad k = 0.061$$

(c) $y = Ce^{kt}$

When $t = 42$, $y = 29$, and when $t = 49$, $y = 125$.

$$24 = Ce^{42k}$$
$$125 = Ce^{49k}$$
$$\frac{125}{24} = \frac{Ce^{49k}}{Ce^{42k}}$$
$$\frac{125}{24} = e^{7k}$$
$$7k = \ln\left(\frac{125}{24}\right)$$
$$k = \frac{\ln\left(\frac{125}{24}\right)}{7} \quad k \approx 0.24$$

(d) Since the values of k are different, we cannot assume the graph is of the form $y = Ce^{kt}$.

(e) The results are summarized in the following table.

n	Value of k for [20, 35]	Value of k for [42, 49]
2	0.00093	0.0017
3	2.3×10^{-5}	2.5×10^{-5}
4	6.3×10^{-7}	4.1×10^{-7}

The value of n should be somewhere between 3 and 4.

33.
$$\frac{1}{2}A_0 = A_0 e^{-0.053t}$$
$$\frac{1}{2} = e^{-0.053t}$$
$$\ln\frac{1}{2} = -0.053t$$
$$-\ln 2 = -0.053t$$
$$t = \frac{\ln 2}{0.52}$$
$$t \approx 13$$

The half-life of plutonium 241 is about 13 years.

35. (a)
$$A(t) = A_0\left(\frac{1}{2}\right)^{t/13}$$
$$A(100) = 4.0\left(\frac{1}{2}\right)^{100/13}$$
$$A(100) \approx 0.0193$$

After 100 years, about 0.0193 gram will remain.

(b)

$$0.1 = 4.0\left(\frac{1}{2}\right)^{t/13}$$

$$\frac{0.1}{4.0} = \left(\frac{1}{2}\right)^{t/13}$$

$$\ln 0.025 = \frac{t}{13}\ln\left(\frac{1}{2}\right)$$

$$t = \frac{13\ln 0.025}{\ln\left(\frac{1}{2}\right)}$$

$$t \approx 69.19$$

It will take 69 years.

37. (a) $y = y_0 e^{kt}$

When $t = 0$, $y = 500$, so $y_0 = 500$.

When $t = 3$, $y = 386$.

$$386 = 500e^{3k}$$

$$\frac{386}{500} = e^{3k}$$

$$e^{3k} = 0.772$$

$$3k = \ln 0.772$$

$$k = \frac{\ln 0.772}{3}$$

$$k \approx -0.0863$$

$$y = 500e^{-0.0863t}$$

(b) From part (a), we have

$$k = \frac{\ln\left(\frac{386}{500}\right)}{3}.$$

$$y = 500e^{kt}$$

$$= 500e^{[\ln(386/500)/3]t}$$

$$= 500e^{\ln(386/500)\cdot(t/3)}$$

$$= 500\left[e^{\ln(386/500)}\right]^{t/3}$$

$$= 500(386/500)^{t/3}$$

$$= 500(0.722)^{t/3}$$

(c) $\frac{1}{2}y_0 = y_0 e^{-0.0863t}$

$$\ln\frac{1}{2} = -0.0863t$$

$$t = \frac{\ln\left(\frac{1}{2}\right)}{-0.0863}$$

$$t \approx 8.0$$

The half-life is about 8.0 days.

39. $y = 40e^{-0.004t}$

(a) $t = 180$

$$y = 40e^{-0.004(180)} = 40e^{-0.72}$$

$$\approx 19.5 \text{ watts}$$

(b)

$$20 = 20e^{-0.0004t}$$

$$\frac{1}{2} = e^{-0.004t}$$

$$\ln\frac{1}{2} = -0.004t$$

$$\frac{\ln 2}{0.004} = t$$

$$173 \approx t$$

It will take about 173 days.

(c) The power will never be completely gone. The power will approach 0 watts but will never be exactly 0.

41. $P(t) = 100e^{-0.1t}$

(a) $P(4) = 100e^{-0.1(4)} \approx 67\%$

(b) $P(10) = 100e^{-0.1(10)} \approx 37\%$

(c)

$$10 = 100e^{-0.1t}$$

$$0.1 = e^{-0.1t}$$

$$\ln(0.1) = -0.1t$$

$$\frac{-\ln(0.1)}{0.1} = t$$

$$23 \approx t$$

It would take about 23 days.

(d)

$$1 = 100e^{-0.1t}$$

$$0.01 = e^{-0.1t}$$

$$\ln(0.01) = -0.1t$$

$$\frac{-\ln(0.01)}{0.1} = t$$

$$46 \approx t$$

It would take about 46 days.

43. $t = 9, T_0 = 18, C = 5, k = .6$

$$f(t) = T_0 + Ce^{-kt}$$

$$f(t) = 18 + 5e^{-0.6(9)}$$

$$= 18 + 5e^{-5.4}$$

$$\approx 18.02$$

The temperature is about 18.02°.

45. $C = -14.6, k = 0.6, T_0 = 18°,$

$f(t) = 10°$

$f(t) = T_0 + Ce^{-kt}$

$f(t) = 18 + (-14.6)e^{-0.6t}$

$-8 = -14.6e^{-0.6t}$

$0.5479 = e^{-0.6t}$

$\ln 0.5479 = -0.6t$

$\dfrac{-\ln 0.5479}{0.6} = t$

$1 \approx t$

It would take about 1 hour for the pizza to thaw.

Chapter 2 Review Exercises

1. True

2. False; for example $f(x) = \frac{x}{x+1}$ is a rational function but not an exponential function.

3. True

4. True

5. False; an exponential function has the form $f(x) = a^x$.

6. False; the vertical asymptote is at $x = 6$.

7. True

8. False; the domain includes all numbers except $x = 2$ and $x = -2$.

9. False; the amount is $A = 2000\left(1 + \frac{0.04}{12}\right)^{24}$.

10. False; the logarithmic function $f(x) = \log_a x$ is not defined for $a = 1$.

11. False; $\ln(5 + 7) = \ln 12 \neq \ln 5 + \ln 7$

12. False; $(\ln 3)^4 \neq 4\ln 3$ since $(\ln 3)^4$ means $(\ln 3)(\ln 3)(\ln 3)(\ln 3)$.

13. False; $\log_{10} 0$ is undefined since $10^x = 0$ has no solution.

14. True

15. False; $\ln(-2)$ is undefined.

16. False; $\frac{\ln 4}{\ln 8} = 0.6667$ and

$\ln 4 - \ln 8 = \ln(1/2) \approx -0.6931$.

17. True

18. True

23. $y = (2x - 1)(x + 1)$

$= 2x^2 + x - 1$

x	-3	-2	-1	0	1	2	3
y	14	5	0	-1	1	9	20

Pairs: $(-3, 14), (-2, 5), (-1, 0), (0, -1), (1, 2),$
$(2, 9), (3, 20)$

Range: $\{-1, 0, 2, 5, 9, 14, 20\}$

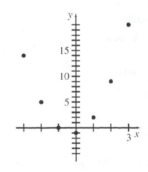

25. $f(x) = 5x^2 - 3$ and $g(x) = -x^2 + 4x + 1$

(a) $f(-2) = 5(-2)^2 - 3 = 17$

(b) $g(3) = -(3)^2 + 4(3) + 1 = 4$

(c) $f(-k) = 5(-k)^2 - 3 = 5k^2 - 3$

(d) $g(3m) = -(3m)^2 + 4(3m) + 1$

$= -9m^2 + 12m + 1$

(e) $f(x + h) = 5(x + h)^2 - 3$

$= 5(x^2 + 2xh + h^2) - 3$

$= 5x^2 + 10xh + 5h^2 - 3$

(f)

$$g(x + h) = -(x + h)^2 + 4(x + h) + 1$$
$$= -(x^2 + 2xh + h^2) + 4x + 4h + 1$$
$$= -x^2 - 2xh - h^2 + 4x + 4h + 1$$

(g) $\dfrac{f(x + h) - f(x)}{h}$

$$= \frac{5(x + h)^2 - 3 - (5x^2 - 3)}{h}$$

$$= \frac{5(x^2 + 2hx + h^2) - 3 - 5x^2 + 3}{h}$$

$$= \frac{5x^2 + 10hx + 5h^2 - 5x^2}{h}$$

$$= \frac{10hx + 5h^2}{h} = 10x + 5h$$

(h)

$$\frac{g(x + h) - g(x)}{h}$$

$$= \frac{-(x + h)^2 + 4(x + h) + 1 - (-x^2 + 4x + 1)}{h}$$

$$= \frac{-(x^2 + 2xh + h^2) + 4x + 4h + 1 + x^2 - 4x - 1}{h}$$

$$= \frac{-x^2 - 2xh - h^2 + 4h + x^2}{h}$$

$$= \frac{-2xh - h^2 + 4h}{h}$$

$$= -2x - h + 4$$

27. $y = \dfrac{3x - 4}{x}$

$x \neq 0$

Domain: $(-\infty, 0) \cup (0, \infty)$

29. $y = \ln(x + 7)$

$$x + 7 > 0$$
$$x > -7$$

Domain: $(-7, \infty)$.

31. $y = 2x^2 + 3x - 1$

The graph is a parabola.

Let $y = 0$.

$$0 = 2x^2 + 3x - 1$$

$$x = \frac{-3 \pm \sqrt{3^2 - 4(2)(-1)}}{2(2)}$$

$$= \frac{-3 \pm \sqrt{9 + 8}}{4}$$

$$= \frac{-3 \pm 17}{4}$$

The x-intercepts are $\dfrac{-3 + \sqrt{17}}{4} \approx 0.28$ and

$\dfrac{-3 - \sqrt{17}}{4} \approx -1.48$.

Let $x = 0$.

$$y = 2(0)^2 + 3(0) - 1$$

-1 is the y-intercept.

Vertex: $x = \dfrac{-b}{2a} = \dfrac{-3}{2(2)} = -\dfrac{3}{4}$

$$y = 2\left(-\frac{3}{4}\right)^2 + 3\left(-\frac{3}{4}\right) - 1$$

$$= \frac{9}{8} - \frac{9}{4} - 1$$

$$= -\frac{17}{8}$$

The vertex is $\left(-\frac{3}{4}, -\frac{17}{8}\right)$.

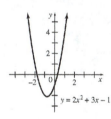

33. $y = -x^2 + 4x + 2$

Let $y = 0$.

$$0 = -x^2 + 4x + 2$$

$$x = \frac{-4 \pm \sqrt{4^2 - 4(-1)(2)}}{2(-1)}$$

$$= \frac{-4 \pm \sqrt{24}}{-2}$$

$$= 2 \pm \sqrt{6}$$

The x-intercepts are $2 + \sqrt{6} \approx 4.45$ and

$2 - \sqrt{6} \approx -0.45$.

Let $x = 0$.

$$y = -0^2 + 4(0) + 2$$

2 is the *y*-intercept.

Vertex: $x = \dfrac{-b}{2a} = \dfrac{-4}{2(-1)} = \dfrac{-4}{-2} = 2$

$$y = -2^2 + 4(2) + 2 = 6$$

The vertex is (2, 6).

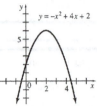

35. $f(x) = x^3 - 3$

Translate the graph of $f(x) = x^3$ 3 units down.

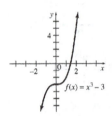

37. $y = -(x - 1)^4 + 4$

Translate the graph of $y = x^4$ 1 unit to the right and reflect vertically. Translate 4 units upward.

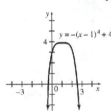

39. $f(x) = \dfrac{8}{x}$

Vertical asymptote: $x = 0$

Horizontal asymptote:

$\dfrac{8}{x}$ approaches zero as *x* gets larger.

$y = 0$ is an asymptote.

x	−4	−3	−2	−1	1	2	3	4
y	−2	−2.7	−4	−8	8	4	2.7	2

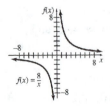

41. $f(x) = \dfrac{4x - 2}{3x + 1}$

Vertical asymptote:

$$3x + 1 = 0$$

$$x = -\dfrac{1}{3}$$

Horizontal asymptote:

As *x* gets larger,

$$\dfrac{4x - 2}{3x - 1} \approx \dfrac{4x}{3x} = \dfrac{4}{3}.$$

$y = \dfrac{4}{3}$ is an asymptote.

x	− 3	−2	−1	0	1	2	3
y	1.75	2	3	−2	0.5	0.86	1

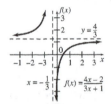

43. $y = 4^x$

x	−2	−1	0	1	2
y	$\dfrac{1}{16}$	$\dfrac{1}{4}$	1	4	16

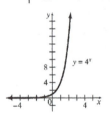

45. $y = \left(\dfrac{1}{5}\right)^{2x-3}$

x	0	1	2
y	125	5	$\dfrac{1}{5}$

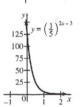

47. $y = \log_2(x - 1)$

$2^y = x - 1$

$x = 1 + 2^y$

x	2	3	5	9
y	0	1	2	3

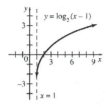

49. $y = -\ln(x + 3)$

$-y = \ln(x + 3)$

$e^{-y} = x + 3$

$e^{-y} - 3 = x$

x	-2.63	-2	-0.28	4.39
y	1	0	-1	-2

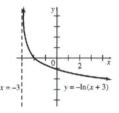

51. $2^{x+2} = \dfrac{1}{8}$

$2^{x+2} = \dfrac{1}{2^3}$

$2^{x+2} = 2^{-3}$

$x + 2 = -3$

$x = -5$

53. $9^{2y+3} = 27^y$

$(3^2)^{2y+3} = (3^3)^y$

$3^{4y+6} = 3^{3y}$

$4y + 6 = 3y$

$y = -6$

55. $3^5 = 243$

The equation in logarithmic form is

$$\log_3 243 = 5.$$

57. $e^{0.8} = 2.22554$

The equation in logarithmic form is

$$\ln 2.22554 \approx 0.8.$$

59. $\log_2 32 = 5$

The equation in exponential form is

$$2^5 = 32.$$

61. $\ln 82.9 = 4.41763$

The equation in exponential form is

$$e^{4.41763} \approx 82.9.$$

63. $\log_3 81 = x$

$3^x = 81$

$3^x = 3^4$

$x = 4$

65. $\log_4 8 = x$

$4^x = 8$

$(2^2)^x = 2^3$

$2x = 3$

$x = \dfrac{3}{2}$

67. $\log_5 3k + \log_5 7k^3$

$= \log_5 3k(7k^3)$

$= \log_5(21k^4)$

69. $4\log_3 y - 2\log_3 x$

$= \log_3 y^4 - \log_3 x^2$

$= \log_3\left(\dfrac{y^4}{x^2}\right)$

71. $6^p = 17$

$\ln 6^p = \ln 17$

$p\ln 6 = \ln 17$

$p = \dfrac{\ln 17}{\ln 6}$

≈ 1.581

73.
$$2^{1-m} = 7$$
$$\ln 2^{1-m} = \ln 7$$
$$(1-m)\ln 2 = \ln 7$$
$$1 - m = \frac{\ln 7}{\ln 2}$$
$$-m = \frac{\ln 7}{\ln 2} - 1$$
$$m = 1 - \frac{\ln 7}{\ln 2}$$
$$\approx -1.807$$

75.
$$e^{-5-2x} = 5$$
$$\ln e^{-5-2x} = \ln 5$$
$$-5 - 2x = \ln 5$$
$$-2x = \ln 5 + 5$$
$$x = \frac{\ln 5 + 5}{-2}$$
$$\approx -3.305$$

77.
$$\left(1 + \frac{m}{3}\right)^5 = 15$$
$$\left[\left(1 + \frac{m}{3}\right)^5\right]^{1/5} = 15^{1/5}$$
$$1 + \frac{m}{3} = 15^{1/5}$$
$$\frac{m}{3} = 15^{1/5} - 1$$
$$m = 3(15^{1/5} - 1)$$
$$\approx 2.156$$

79.
$$\log_k 64 = 6$$
$$k^6 = 64$$
$$k^6 = 2^6$$
$$k = 2$$

81.
$$\log(4p + 1) + \log p = \log 3$$
$$\log[p(4p + 1)] = \log 3$$
$$\log(4p^2 + p) = \log 3$$

$$4p^2 + p = 3$$
$$4p^2 + p - 3 = 0$$
$$(4p - 3)(p + 1) = 0$$
$$4p - 3 = 0 \quad \text{or} \quad p + 1 = 0$$
$$p = \frac{3}{4} \qquad\qquad p = -1$$

p cannot be negative, so $p = \frac{3}{4}$.

83. $f(x) = a^x; a > 0, a \neq 1$

(a) The domain is $(-\infty, \infty)$.

(b) The range is $(0, \infty)$.

(c) The y-intercept is 1.

(d) The x-axis, $y = 0$, is a horizontal asymptote.

(e) The function is increasing if $a > 1$.

(f) The function is decreasing if $0 < a < 1$.

87. $y = \dfrac{7x}{100 - x}$

(a) $y = \dfrac{7(80)}{100 - 80} = \dfrac{560}{20} = 28$

The cost is \$28,000.

(b) $y = \dfrac{7(50)}{100 - 50} = \dfrac{350}{50} = 7$

The cost is \$7000.

(c) $\dfrac{7(90)}{100 - 90} = \dfrac{630}{10} = 63$

The cost is \$63,000.

(d) Plot the points (80, 28), (50, 7), and (90, 63).

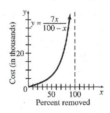

(e) No, because all of the pollutant would be removed when $x = 100$, at which point the denominator of the function would be zero.

89. $P = \$2781.36, r = 4.8\%, t = 6, m = 4$

$$A = P\left(1 + \frac{r}{m}\right)^{tm}$$

$$A = 2781.36\left(1 + \frac{0.048}{4}\right)^{(6)(4)}$$

$$= 2781.36(1.012)^{24}$$

$$= \$3703.31$$

$$\text{Interest} = \$3703.31 - \$2781.36$$

$$= \$921.95$$

91. $P = \$12,104, r = 6.2\%, t = 4$

$$A = Pe^{rt}$$

$$A = 12,104e^{0.062(4)}$$

$$= 12,104e^{0.248}$$

$$= \$15,510.79$$

93. $P = \$12,000, r = 0.05, t = 8$

$$A = 12,000e^{0.05(8)}$$

$$= 12,000e^{0.40}$$

$$= \$17,901.90$$

95. $2100 deposited at 4% compounded quarterly.

$$A = P\left(1 + \frac{r}{m}\right)^{tm}$$

To double:

$$2(2100) = 2100\left(1 + \frac{0.04}{4}\right)^{t \cdot 4}$$

$$2 = 1.01^{4t}$$

$$\ln 2 = 4t \ln 1.01$$

$$t = \frac{\ln 2}{4 \ln 1.01}$$

$$\approx 17.4$$

Because interest is compounded quarterly, round the result up to the nearest quarter, which is 17.5 years or 70 quarters.

To triple:

$$3(2100) = 2100\left(1 + \frac{0.04}{4}\right)^{t \cdot 4}$$

$$3 = 1.01^{4t}$$

$$\ln 3 = 4t \ln 1.01$$

$$t = \frac{\ln 3}{4 \ln 1.01}$$

$$\approx 27.6$$

Because interest is compounded quarterly, round the result up to the nearest quarter, which is 27.75 years or 111 quarters.

97. $r = 6\%, m = 12$

$$r_E = \left(1 + \frac{r}{m}\right)^{m} - 1$$

$$= \left(1 + \frac{0.06}{12}\right)^{12} - 1$$

$$= 0.0617 = 6.17\%$$

99. $A = \$2000, r = 6\%, t = 5, m = 1$

$$P = A\left(1 + \frac{r}{m}\right)^{-tm}$$

$$= 2000\left(1 + \frac{0.06}{1}\right)^{-5(1)}$$

$$= 2000(1.06)^{-5}$$

$$= \$1494.52$$

101. $r = 7\%, t = 8, m = 2, P = 10,000$

$$A = P\left(1 + \frac{r}{m}\right)^{tm}$$

$$= 10,000\left(1 + \frac{0.07}{2}\right)^{8(2)}$$

$$= 10,000(1.035)^{16}$$

$$= \$17,339.86$$

103. $P = \$6000, A = \$8000, t = 3$

$$A = Pe^{rt}$$

$$8000 = 6000e^{3r}$$

$$\frac{4}{3} = e^{3r}$$

$$\ln 4 - \ln 3 = 3r$$

$$r = \frac{\ln 4 - \ln 3}{3}$$

$$r \approx 0.0959 \text{ or about } 9.59\%$$

105. (a) $n = 1000 - (p - 50)(10), \ p \geq 50$

$$= 1000 - 10p + 500$$

$$= 1500 - 10p$$

(b) $R = pn$

$$R = p(1500 - 10p)$$

(c) $p \geq 50$

Since n cannot be negative,

$$1500 - 10p \geq 0$$

$$-10p \geq -1500$$

$$p \leq 150.$$

Therefore, $50 \leq p \leq 150$.

(d) Since $n = 1500 - 10p$,

$$10p = 1500 - n$$

$$p = 150 - \frac{n}{10}.$$

$$R = pn$$

$$R = \left(150 - \frac{n}{10}\right)n$$

(e) Since she can sell at most 1000 tickets, $0 \leq n \leq 1000$.

(f) $R = -10p^2 + 1500p$

$$\frac{-b}{2a} = \frac{-1500}{2(-10)} = 75$$

The price producing maximum revenue is $75.

(g) $R = -\frac{1}{10}n^2 + 150n$

$$\frac{-b}{2a} = \frac{-150}{2\left(-\frac{1}{10}\right)} = 750$$

The number of tickets producing maximum revenue is 750.

(h) $R(p) = -10p^2 + 1500p$

$$R(75) = -10(75)^2 + 1500(75)$$

$$= -56,250 + 112,500$$

$$= 56,250$$

The maximum revenue is $56,250.

(i)

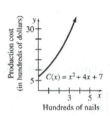

(j) The revenue starts at \$50,000 when the price is \$50, rises to a maximum of \$56,250 when the price is \$75, and falls to 0 when the price is \$150.

107. $C(x) = x^2 + 4x + 7$

(a)

Production cost (in hundreds of dollars)

$C(x) = x^2 + 4x + 7$

Hundreds of nails

(b)
$$C(x + 1) - C(x)$$

$$= (x + 1)^2 + 4(x + 1) + 7 - (x^2 + 4x + 7)$$

$$= x^2 + 2x + 1 + 4x + 4 + 7 - x^2 - 4x - 7$$

$$= 2x + 5$$

(c) $A(x) = \dfrac{C(x)}{x} = \dfrac{x^2 + 4x + 7}{x}$

$$= x + 4 + \frac{7}{x}$$

(d) $A(x + 1) - A(x)$

$$= (x + 1) + 4 + \frac{7}{x + 1} - \left(x + 4 + \frac{7}{x}\right)$$

$$= x + 1 + 4 + \frac{7}{x + 1} - x - 4 - \frac{7}{x}$$

$$= 1 + \frac{7}{x + 1} - \frac{7}{x}$$

$$= 1 + \frac{7x - 7(x + 1)}{x(x + 1)}$$

$$= 1 + \frac{7x - 7x - 7}{x(x + 1)}$$

$$= 1 - \frac{7}{x(x + 1)}$$

109. $F(x) = -\frac{2}{3}x^2 + \frac{14}{3}x + 96$

The maximum fever occurs at the vertex of the parabola.

$$x = \frac{-b}{2a} = \frac{-\frac{14}{3}}{-\frac{4}{3}} = \frac{7}{2}$$

$$y = -\frac{2}{3}\left(\frac{7}{2}\right)^2 + \frac{14}{3}\left(\frac{7}{2}\right) + 96$$

$$= -\frac{2}{3}\left(\frac{49}{4}\right) + \frac{49}{3} + 96$$

$$= -\frac{49}{6} + \frac{49}{3} + 96$$

$$= -\frac{49}{6} + \frac{98}{6} + \frac{576}{6} = \frac{625}{6} \approx 104.2$$

The maximum fever occurs on the third day. It is about 104.2°F.

111. (a)

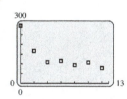

(b) $y = 2.384t^2 - 42.55t + 269.2$

$y = -0.4931t^3 + 11.26t^2$
$\qquad - 82.00t + 292.9$

$y = 213.8(0.9149)^t$

(c)

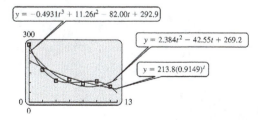

The cubic function seems to best capture the behavior of the data.

(d) $x = 20$ corresponds to 2015.

$$y = 2.384(20)^2 - 42.55(20)$$
$$+ 269.2 \approx 372$$

$$y = -0.4931(20)^3 + 11.26(20)^2$$
$$- 82.00(20) + 292.9$$
$$\approx -788$$

$$y = 213.8(0.9149)^{20} \approx 36$$

The only realistic value is given by the exponential function because the pattern of the data suggests that the number of cases decrease over time.

113. This function has a maximum value at $x \approx 187.9$. At $x \approx 187.9$, $y \approx 345$. The largest girth for which this formula gives a reasonable answer is 187.9 cm. The predicted mass of a polar bear with this girth is 345 kg.

115. $p(t) = \dfrac{1.79 \cdot 10^{11}}{(2026.87 - t)^{0.99}}$

(a) $p(2010) \approx 10.915$ billion

This is about 4.006 billion more than the estimate of 6.909 billion.

(b) $p(2020) \approx 26.56$ billion

$p(2025) \approx 96.32$ billion

117. Graph

$$y = c(t) = e^{-t} - e^{-2t}$$

on a graphing calculator and locate the maximum point. A calculator shows that the x-coordinate of the maximum point is about 0.69, and the y-coordinate is exactly 0.25. Thus, the maximum concentration of 0.25 occurs at about 0.69 minutes.

119. (a) $S = 21.35 + 104.6 \ln A$

(b) $S = 85.49A^{0.3040}$

(c)

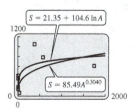

(d) $S \approx 742.2$

$S \approx 694.7$

Neither number is close to the actual number of 421.

121. $t = (1.26 \times 10^9) \dfrac{\ln\left[1 + 8.33\left(\frac{A}{K}\right)\right]}{\ln 2}$

(a) $A = 0, K > 0$

$t = (1.26 \times 10^9) \dfrac{\ln[1 + 8.33(0)]}{\ln 2}$

$= (1.26 \times 10^9)(0) = 0$ years

(b) $t = (1.26 \times 10^9) \dfrac{\ln[1 + 8.33(0.212)]}{\ln 2}$

$= (1.26 \times 10^9) \dfrac{\ln 2.76596}{\ln 2}$

$= 1,849,403,169$

or about 1.85×10^9 years

(c) As r increases, t increases, but at a slower and slower rate. As r decreases, t decreases at a faster and faster rate

THE DERIVATIVE

3.1 Limits

Your Turn 1

$f(x) = x^2 + 2$

x	0.9	0.99	0.999	0.9999	1	1.0001	1.001	1.01	1.1
$f(x)$	2.81	2.9801	2.998001	2.99980001	3	3.00020001	3.002001	3.0201	3.21

The table suggests that, as x get closer and closer to 1 from either side, $f(x)$ gets closer and closer to 3.

So, $\lim\limits_{x \to 1}(x^2 + 2) = 3$.

Your Turn 2

$$f(x) = \frac{x^2 - 4}{x - 2} = \frac{(x + 2)(x - 2)}{(x - 2)}$$

$$= x + 2, \text{provided } x \neq 2$$

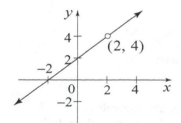

The graph of $y = \frac{x^2 - 4}{x - 2}$ is the graph of $y = x + 2$, except there is a hole at (2, 4).

Looking at the graph, we see that as x is close to, but not equal to 2, $f(x)$ approaches 4.

$$\lim_{x \to 2} \frac{x^2 - 4}{x - 2} = 4$$

Your Turn 3

Find $\lim\limits_{x \to 3} f(x)$ if $f(x) = \begin{cases} 2x - 1 & \text{if } x \neq 3 \\ 1 & \text{if } x = 3 \end{cases}$

The graph of f is shown in the next column.

$$\lim_{x \to 3} f(x) = 5$$

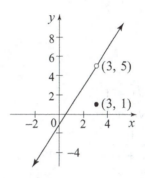

Your Turn 4

Find $\lim\limits_{x \to 0} \dfrac{2x - 1}{x}$.

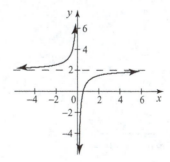

$$\lim_{x \to 0^-} f(x) = \infty$$

$$\lim_{x \to 0^+} f(x) = -\infty$$

Since there is no real number that $f(x)$ approaches as x approaches 0 from either side, nor does $f(x)$ approach either ∞ or $-\infty$, $\lim\limits_{x \to 0} \frac{2x-1}{x}$ does not exist.

Your Turn 5

Let $\lim\limits_{x \to 2} f(x) = 3$ and $\lim\limits_{x \to 2} g(x) = 4$.

$$\lim_{x \to 2}\left[f(x) + g(x)\right]^2 = \left[\lim_{x \to 2}\left[f(x) + g(x)\right]\right]^2$$

$$= \left[\lim_{x \to 2} f(x) + \lim_{x \to 2} g(x)\right]^2$$

$$= [3 + 4]^2$$

$$= 7^2$$

$$= 49$$

Your Turn 6

$$\lim_{x \to -3} \frac{x^2 - x - 12}{x + 3} = \lim_{x \to -3} \frac{(x - 4)(x + 3)}{x + 3} \quad (x \neq 3)$$

$$= \lim_{x \to -3} x - 4$$

$$= (-3) - 4$$

$$= -7$$

Your Turn 7

$$\lim_{x \to 1} \frac{\sqrt{x} - 1}{x - 1} = \lim_{x \to 1} \frac{\sqrt{x} - 1}{x - 1} \cdot \frac{\sqrt{x} + 1}{\sqrt{x} + 1}$$

$$= \lim_{x \to 1} \frac{(\sqrt{x})^2 - 1}{(x - 1)(\sqrt{x} + 1)}$$

$$= \lim_{x \to 1} \frac{x - 1}{(x - 1)(\sqrt{x} + 1)}$$

$$= \lim_{x \to 1} \frac{1}{\sqrt{x} + 1} = \frac{1}{1 + 1} = \frac{1}{2}$$

Your Turn 8

$$\lim_{x \to \infty} \frac{2x^2 + 3x - 4}{6x^2 - 5x + 7} = \lim_{x \to \infty} \frac{\dfrac{2x^2}{x^2} + \dfrac{3x}{x^2} - \dfrac{4}{x^2}}{\dfrac{6x^2}{x^2} - \dfrac{5x}{x^2} + \dfrac{7}{x^2}}$$

$$= \lim_{x \to \infty} \frac{2 + \dfrac{3}{x} - \dfrac{4}{x^2}}{6 - \dfrac{5}{x} + \dfrac{7}{x^2}}$$

$$= \frac{2 + 0 - 0}{6 - 0 + 0} = \frac{2}{6} = \frac{1}{3}$$

3.1 Warmup Exercises

W1. Some trial with the factors of 8 and 15 shows that
$$8x^2 + 22x + 15 = (2x + 3)(4x + 5).$$

W2. Some trial with the factors of 12 shows that
$$12x^2 - 7x - 12 = (3x - 4)(4x + 3).$$

W3. $\dfrac{3x^2 + x - 14}{x^2 - 4} = \dfrac{(3x + 7)(x - 2)}{(x + 2)(x - 2)} = \dfrac{3x + 7}{x + 2}$,
provided that x is not equal to 2 or -2.

W4. $\dfrac{2x^2 + x - 15}{x^2 - 9} = \dfrac{(2x - 5)(x + 3)}{(x + 3)(x - 3)} = \dfrac{2x - 5}{x - 3}$,
provided that x is not equal to 3 or -3.

3.1 Exercises

1. Since $\lim\limits_{x \to 2^-} f(x)$ does not equal $\lim\limits_{x \to 2^+} f(x)$, $\lim\limits_{x \to 2} f(x)$ does not exist. The answer is c.

3. Since $\lim\limits_{x \to 4^-} f(x) = \lim\limits_{x \to 4^+} f(x) = 6$, $\lim\limits_{x \to 4} f(x) = 6$. The answer is b.

5. **(a)** By reading the graph, as x gets closer to 3 from the left or right, $f(x)$ gets closer to 3.

   $$\lim_{x \to 3} f(x) = 3$$

   (b) By reading the graph, as x gets closer to 0 from the left or right, $f(x)$ gets closer to 1.

   $$\lim_{x \to 0} f(x) = 1.$$

7. **(a)** By reading the graph, as x gets closer to 0 from the left or right, $f(x)$ gets closer to 0.

   $$\lim_{x \to 0} f(x) = 0$$

   (b) By reading the graph, as x gets closer to 2 from the left, $f(x)$ gets closer to -2, but as x gets closer to 2 from the right, $f(x)$ gets closer to 1.

   $$\lim_{x \to 2} f(x) \text{ does not exist.}$$

9. (a) (i) By reading the graph, as x gets closer to -2 from the left, $f(x)$ gets closer to -1.

$$\lim_{x \to -2^-} f(x) = -1$$

(ii) By reading the graph, as x gets closer to -2 from the right, $f(x)$ gets closer to $-\frac{1}{2}$.

$$\lim_{x \to -2^+} f(x) = -\frac{1}{2}$$

(iii) Since $\lim_{x \to -2^-} f(x) = -1$ and $\lim_{x \to -2^+}$ $f(x) = -\frac{1}{2}$, $\lim_{x \to -2} f(x)$ does not exist.

(iv) $f(-2)$ does not exist since there is no point on the graph with an x-coordinate of -2.

(b) (i) By reading the graph, as x gets closer to -1 from the left, $f(x)$ gets closer to $-\frac{1}{2}$.

$$\lim_{x \to -1^-} f(x) = -\frac{1}{2}$$

(ii) By reading the graph, as x gets closer to -1 from the right, $f(x)$ gets closer to $-\frac{1}{2}$.

$$\lim_{x \to -1^+} f(x) = -\frac{1}{2}$$

(iii) Since $\lim_{x \to -1^-} f(x) = -\frac{1}{2}$ and $\lim_{x \to -1^+} f(x) = -\frac{1}{2}$, $\lim_{x \to -1} f(x) = -\frac{1}{2}$.

(iv) $f(-1) = -\frac{1}{2}$ since $\left(-1, -\frac{1}{2}\right)$ is a point of the graph.

11. By reading the graph, as x moves further to the right, $f(x)$ gets closer to 3.

Therefore, $\lim_{x \to \infty} f(x) = 3$.

13. $\lim_{x \to 2} F(x)$ in Exercise 6 exists because

$\lim_{x \to 2^-} F(x) = 4$ and $\lim_{x \to 2^+} F(x) = 4$.

$\lim_{x \to -2} f(x)$ in Exercise 9 does not exist since

$\lim_{x \to -2^-} f(x) = -1$, but $\lim_{x \to -2^+} f(x) = -\frac{1}{2}$.

15. From the table, as x approaches 1 from the left or the right, $f(x)$ approaches 4.

$$\lim_{x \to 1} f(x) = 4$$

17. $k(x) = \dfrac{x^3 - 2x - 4}{x - 2}$; find $\lim_{x \to 2} k(x)$.

x	1.9	1.99	1.999
$k(x)$	9.41	9.9401	9.9941

x	2.001	2.01	2.1
$k(x)$	10.006	10.0601	10.61

As x approaches 2 from the left or the right, $k(x)$ approaches 10.

$$\lim_{x \to 2} k(x) = 10$$

19. $h(x) = \dfrac{\sqrt{x} - 2}{x - 1}$; find $\lim_{x \to 1} h(x)$.

x	0.9	0.99	0.999
$h(x)$	10.51317	100.50126	1000.50013

x	1.001	1.01	1.1
$h(x)$	-999.50012	-99.50124	-9.51191

$$\lim_{x \to 1^-} = \infty$$

$$\lim_{x \to 1^+} = -\infty$$

Thus, $\lim_{x \to 1} h(x)$ does not exist.

21. $\lim_{x \to 4} [f(x) - g(x)] = \lim_{x \to 4} f(x) - \lim_{x \to 4} g(x)$
$$= 9 - 27 = -18$$

23. $\lim_{x \to 4} \dfrac{f(x)}{g(x)} = \dfrac{\lim_{x \to 4} f(x)}{\lim_{x \to 4} g(x)} = \dfrac{9}{27} = \dfrac{1}{3}$

25. $\lim_{x \to 4} \sqrt{f(x)} = \lim_{x \to 4} [f(x)^{1/2}]$
$$= \left[\lim_{x \to 4} f(x) \right]^{1/2}$$
$$= 9^{1/2} = 3$$

27. $\lim_{x \to 4} 2^{f(x)} = 2^{\lim_{x \to 4} f(x)}$
$$= 2^9$$
$$= 512$$

29. $\lim\limits_{x\to 4} \dfrac{f(x) + g(x)}{2g(x)}$

$= \dfrac{\lim\limits_{x\to 4}[f(x) + g(x)]}{\lim\limits_{x\to 4} 2g(x)}$

$= \dfrac{\lim\limits_{x\to 4} f(x) + \lim\limits_{x\to 4} g(x)}{2 \lim\limits_{x\to 4} g(x)}$

$= \dfrac{9 + 27}{2(27)} = \dfrac{36}{54} = \dfrac{2}{3}$

31. $\lim\limits_{x\to 3} \dfrac{x^2 - 9}{x - 3} = \lim\limits_{x\to 3} \dfrac{(x - 3)(x + 3)}{x - 3}$

$= \lim\limits_{x\to 3} (x + 3)$

$= \lim\limits_{x\to 3} x + \lim\limits_{x\to 3} 3$

$= 3 + 3$

$= 6$

33. $\lim\limits_{x\to 1} \dfrac{5x^2 - 7x + 2}{x^2 - 1} = \lim\limits_{x\to 1} \dfrac{(5x - 2)(x - 1)}{(x + 1)(x - 1)}$

$= \lim\limits_{x\to 1} \dfrac{5x - 2}{x + 1}$

$= \dfrac{5 - 2}{2}$

$= \dfrac{3}{2}$

35. $\lim\limits_{x\to -2} \dfrac{x^2 - x - 6}{x + 2} = \lim\limits_{x\to -2} \dfrac{(x - 3)(x + 2)}{x + 2}$

$= \lim\limits_{x\to -2} (x - 3)$

$= \lim\limits_{x\to -2} x + \lim\limits_{x\to -2} (-3)$

$= -2 - 3$

$= -5$

37. $\lim\limits_{x\to 0} \dfrac{\frac{1}{x+3} - \frac{1}{3}}{x}$

$= \lim\limits_{x\to 0} \left(\dfrac{1}{x + 3} - \dfrac{1}{3} \right)\left(\dfrac{1}{x} \right)$

$= \lim\limits_{x\to 0} \left[\dfrac{3}{3(x + 3)} - \dfrac{x + 3}{3(x + 3)} \right]\left(\dfrac{1}{x} \right)$

$= \lim\limits_{x\to 0} \dfrac{3 - x - 3}{3(x + 3)(x)}$

$= \lim\limits_{x\to 0} \dfrac{-x}{3(x + 3)x}$

$= \lim\limits_{x\to 0} \dfrac{-1}{3(x + 3)}$

$= \dfrac{-1}{3(0 + 3)}$

$= -\dfrac{1}{9}$

39. $\lim\limits_{x\to 25} \dfrac{\sqrt{x} - 5}{x - 25}$

$= \lim\limits_{x\to 25} \dfrac{\sqrt{x} - 5}{x - 25} \cdot \dfrac{\sqrt{x} + 5}{\sqrt{x} + 5}$

$= \lim\limits_{x\to 25} \dfrac{x - 25}{(x - 25)(\sqrt{x} + 5)}$

$= \lim\limits_{x\to 25} \dfrac{1}{\sqrt{x} + 5}$

$= \dfrac{1}{\sqrt{25} + 5}$

$= \dfrac{1}{10}$

41. $\lim\limits_{h\to 0} \dfrac{(x + h)^2 - x^2}{h}$

$= \lim\limits_{h\to 0} \dfrac{x^2 + 2hx + h^2 - x^2}{h}$

$= \lim\limits_{h\to 0} \dfrac{2hx + h^2}{h}$

$= \lim\limits_{h\to 0} \dfrac{h(2x + h)}{h}$

$= \lim\limits_{h\to 0} (2x + h)$

$= 2x + 0 = 2x$

43. $\lim_{x \to \infty} \dfrac{3x}{7x - 1} = \lim_{x \to \infty} \dfrac{\frac{3x}{x}}{\frac{7x}{x} - \frac{1}{x}}$

$$= \lim_{x \to \infty} \dfrac{3}{7 - \frac{1}{x}}$$

$$= \dfrac{3}{7 - 0} = \dfrac{3}{7}$$

45. $\lim_{x \to -\infty} \dfrac{-3x^2 + 2x}{2x^2 - 2x + 1}$

$$= \lim_{x \to -\infty} \dfrac{\frac{3x^2}{x^2} + \frac{2x}{x^2}}{\frac{2x^2}{x^2} - \frac{2x}{x^2} + \frac{1}{x^2}}$$

$$= \lim_{x \to -\infty} \dfrac{3 + \frac{2}{x}}{2 - \frac{2}{x} + \frac{1}{x^2}}$$

$$= \dfrac{3 - 0}{2 + 0 + 0} = \dfrac{3}{2}$$

47. $\lim_{x \to \infty} \dfrac{3x^3 + 2x - 1}{2x^4 - 3x^3 - 2}$

$$= \lim_{x \to \infty} \dfrac{\frac{3x^3}{x^4} + \frac{2x}{x^4} - \frac{1}{x^4}}{\frac{2x^4}{x^4} - \frac{3x^3}{x^4} - \frac{2}{x^4}}$$

$$= \lim_{x \to \infty} \dfrac{\frac{3}{x} + \frac{2}{x^3} - \frac{1}{x^4}}{2 - \frac{3}{x} - \frac{2}{x^4}}$$

$$= \dfrac{0 + 0 - 0}{2 - 0 - 0} = 0$$

49. $\lim_{x \to \infty} \dfrac{2x^3 - x - 3}{6x^2 - x - 1}$

$$= \lim_{x \to \infty} \dfrac{\frac{2x^3}{x^2} - \frac{x}{x^2} - \frac{3}{x^2}}{\frac{6x^2}{x^2} - \frac{x}{x^2} - \frac{1}{x^2}}$$

$$= \lim_{x \to \infty} \dfrac{2x - \frac{1}{x} - \frac{3}{x^2}}{6 - \frac{1}{x} - \frac{1}{x^2}} = \infty$$

The limit does not exist.

51. $\lim_{x \to \infty} \dfrac{2x^2 - 7x^4}{9x^2 + 5x - 6} = \lim_{x \to \infty} \dfrac{\frac{2x^2}{x^2} - \frac{7x^4}{x^2}}{\frac{9x^2}{x^2} + \frac{5x}{x^2} - \frac{6}{x^2}}$

$$= \lim_{x \to \infty} \dfrac{2 - 7x^2}{9 + \frac{5}{x} - \frac{6}{x^2}}$$

The denominator approaches 9, while the numerator becomes a negative number that is larger and larger in magnitude, so

$$\lim_{x \to \infty} \dfrac{2x^2 - 7x^4}{9x^2 + 5x - 6} = -\infty \text{ (does not exist).}$$

53. $\lim_{x \to -1^-} f(x) = 1$ and $\lim_{x \to -1^+} f(x) = 1$.

Therefore $\lim_{x \to -1} f(x) = 1$.

55. (a) $\lim_{x \to 3} f(x) = 2$.

(b) $\lim_{x \to 5} f(x)$ does not exist since $\lim_{x \to 5^-} f(x) = 2$

and $\lim_{x \to 5^+} f(x) = 8$.

57. The denominator

$$x^2 - 3x + 2 = (x - 1)(x - 2). \text{ Thus}$$

$\lim_{x \to 2} \dfrac{3x^2 + kx - 2}{x^2 - 3x + 2}$ will exist when the numerator

contains a factor of $x - 2$. This occurs when

$3(2)^2 + k(2) - 2 = 0$, which requires $k = -5$.

The fraction is then

$$\dfrac{3x^2 - 5x - 2}{(x - 1)(x - 2)} = \dfrac{(3x + 1)(x - 2)}{(x - 1)(x - 2)} = \dfrac{3x + 1}{x - 1} \text{ pro}$$

vided $x \neq 2$. The limit of the fraction as $x \to 2$

is $\dfrac{3(2) + 1}{2 - 1} = 7$.

59. Find $\lim_{x \to 3} f(x)$, where $f(x) = \dfrac{x^2 - 9}{x - 3}$.

x	2.9	2.99	2.999	3.001	3.01	3.1
$f(x)$	5.9	5.99	5.999	6.001	6.01	6.1

$$\lim_{x \to 3} f(x) = \lim_{x \to 3} \dfrac{x^2 - 9}{x - 3} = 6.$$

61. Find $\lim_{x \to 1} f(x)$, where $f(x) = \dfrac{5x^2 - 7x + 2}{x^2 - 1}$.

x	0.9	0.99	0.999	1.001	1.01	1.1
$f(x)$	1.316	1.482	1.498	1.502	1.517	1.667

$$\lim_{x \to 1} f(x) = \lim_{x \to 1} \dfrac{5x^2 - 7x + 2}{x^2 - 1} = 1.5 = \dfrac{3}{2}.$$

63. (a) $\lim\limits_{x \to -2} \dfrac{3x}{(x+2)^3}$ does not exist since

$$\lim\limits_{x \to -2^+} \dfrac{3x}{(x+2)^3} = -\infty$$

and $\lim\limits_{x \to -2^-} \dfrac{3x}{(x+2)^3} = \infty.$

(b) Since $(x+2)^3 = 0$ when $x = -2$, $x = -2$ is the vertical asymptote of the graph of $F(x)$.

(c) The two answers are related. Since $x = -2$ is a vertical asymptote, we know that $\lim\limits_{x \to -2} F(x)$ does not exist.

67. (a) $\lim\limits_{x \to -\infty} e^x = 0$ since, as the graph goes further to the left, e^x gets closer to 0.

(b) The graph of e^x has a horizontal asymptote at $y = 0$ since $\lim\limits_{x \to -\infty} e^x = 0.$

69. (a) $\lim\limits_{x \to 0^+} \ln x = -\infty$ (does not exist) since, as the graph gets closer to $x = 0$, the value of $\ln x$ get smaller.

(b) The graph of $y = \ln x$ has a vertical asymptote at $x = 0$ since $\lim\limits_{x \to 0^+} \ln x = -\infty.$

73. $\lim\limits_{x \to 1} \dfrac{x^4 + 4x^3 - 9x^2 + 7x - 3}{x - 1}$

(a)

x	1.01	1.001	1.0001	0.99	0.999	0.9999
$f(x)$	5.0908	5.009	5.0009	4.9108	4.991	4.9991

As $x \to 1^-$ and as $x \to 1^+$, we see that $f(x) \to 5.$

(b) Graph

$$y = \dfrac{x^4 + 4x^3 - 9x^2 + 7x - 3}{x - 1}$$

on a graphing calculator. One suitable choice for the viewing window is $[-6, 6]$ by $[-10, 40]$ with Xscl $= 1$, Yscl $= 10.$

Because $x - 1 = 0$ when $x = 1$, we know that the function is undefined at this x-value. The graph does not show an asymptote at $x = 1$. This indicates that the rational expression that defines this function is not written in lowest terms, and that the graph

should have an open circle to show a "hole" in the graph at $x = 1$. The graphing calculator doesn't show the hole, but if we try to find the value of the function at $x = 1$, we see that it is undefined. (Using the TABLE feature on a TI-84 Plus, we see that for $x = 1$, the y-value is listed as "ERROR.")

By viewing the function near $x = 1$, and using the ZOOM feature, we see that as x gets close to 1 from the left or the right, y gets close to 5, suggesting that

$$\lim\limits_{x \to 1} \dfrac{x^4 + 4x^3 - 9x^2 + 7x - 3}{x - 1} = 5.$$

75. $\lim\limits_{x \to -1} \dfrac{x^{1/3} + 1}{x + 1}$

(a)

x	-1.01	-1.001	-1.0001
$f(x)$	0.33223	0.33322	0.33332

x	-0.99	-0.999	-0.9999
$f(x)$	0.33445	0.33344	0.33334

We see that as $x \to -1^-$ and as $x \to -1^+$, $f(x) \to 0.3333$ or $\frac{1}{3}.$

(b) Graph $\quad y = \dfrac{x^{1/3} + 1}{x + 1}.$

One suitable choice for the viewing window is $[-5, 5]$ by $[-2, 2]$

Because $x + 1 = 0$ when $x = -1$, we know that the function is undefined at this x-value. The graph does not show an asymptote at $x = -1$. This indicates that the rational expression that defined this function is not written lowest terms, and that the graph should have an open circle to show a "hole" in the graph at $x = -1$. The graphing calculator doesn't show the hole, but if we try to find the value of the function at $x = -1$, we see that it is undefined. (Using the TABLE feature on a TI-83, we see that for $x = -1$, the y-value is listed as "ERROR.")

By viewing the function near $x = -1$ and using the ZOOM feature, we see that as x gets close to -1 from the left or right, y gets close to 0.3333, suggesting that

$$\lim\limits_{x \to -1} \dfrac{x^{1/3} + 1}{x + 1} = 0.3333 \text{ or } \frac{1}{3}.$$

77. $\lim\limits_{x\to\infty} \dfrac{\sqrt{9x^2+5}}{2x}$

Graph the functions on a graphing calculator. A good choice for the viewing window is $[-10, 10]$ by $[-5, 5]$.

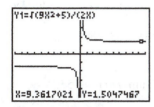

(a) The graph appears to have horizontal asymptotes at $y = \pm 1.5$. We see that as $x \to \infty, y \to 1.5$, so we determine that

$$\lim\limits_{x\to\infty} \dfrac{\sqrt{9x^2+5}}{2x} = 1.5.$$

(b) As $x \to \infty$,

$$\dfrac{\sqrt{9x^2+5}}{2x} \to \dfrac{3|x|}{2x}.$$

Since $x > 0, |x| = x$, so

$$\dfrac{3|x|}{2x} = \dfrac{3x}{2x} = \dfrac{3}{2}.$$

Thus,

$$\lim\limits_{x\to\infty} \dfrac{\sqrt{9x^2+5}}{2x} = \dfrac{3}{2} \text{ or } 1.5.$$

79. $\lim\limits_{x\to-\infty} \dfrac{\sqrt{36x^2+2x+7}}{3x}$

Graph this function on a graphing calculator. A good choice for the viewing window is $[-10, 10]$ by $[-5, 5]$.

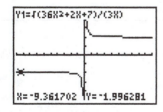

(a) The graph appears to have horizontal asymptotes at $y = \pm 2$. We see that as $x \to -\infty, y \to -2$, so we determine that

$$\lim\limits_{x\to-\infty} \dfrac{\sqrt{36x^2+2x+7}}{3x} = -2.$$

(b) As $x \to -\infty$,

$$\dfrac{\sqrt{36x^2+2x+7}}{3x} \to \dfrac{6|x|}{3x}.$$

Since $x < 0, |x| = -x$, so

$$\dfrac{6|x|}{3x} = \dfrac{6(-x)}{3x} = -2.$$

Thus,

$$\lim\limits_{x\to-\infty} \dfrac{\sqrt{36x^2+2x+7}}{3x} = -2.$$

81. $\lim\limits_{x\to\infty} \dfrac{(1+5x^{1/3}+2x^{5/3})^3}{x^5}$

Graph this function on a graphing calculator. A good choice for the viewing window is $[-20, 20]$ by $[0, 20]$ with $\text{Xscl} = 5, \text{Yscl} = 5$.

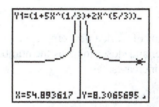

(a) The graph appears to have a horizontal asymptote at $y = 8$. We see that as $x \to \infty, y \to 8$, so we determine that

$$\lim\limits_{x\to\infty} \dfrac{(1+5x^{1/3}+2x^{5/3})^3}{x^5} = 8.$$

(b) As $x \to \infty$,

$$\dfrac{(1+5x^{1/3}+2x^{5/3})^3}{x^5} \to \dfrac{8x^5}{x^5} = 8.$$

Thus, $\lim\limits_{x\to\infty} \dfrac{(1+5x^{1/3}+2x^{5/3})^3}{x^5} = 8.$

85. **(a)** $\lim\limits_{x\to 94} T(x) = 7.25$ cents

(b) $\lim\limits_{x\to 13^-} T(x) = 7.25$ cents

(c) $\lim\limits_{x\to 13^+} T(x) = 7.5$ cents

(d) $\lim\limits_{x\to 13} T(x) = $ does not exist

(e) $T(13) = 7.5$ cents

87. $C(x) = 15{,}000 + 6x$

$$\overline{C}(x) = \frac{C(x)}{x} = \frac{15{,}000 + 6x}{x} = \frac{15{,}000}{x} + 6$$

$$\lim_{x \to \infty} \overline{C}(x) = \lim_{x \to \infty} \frac{15{,}000}{x} + 6 = 0 + 6 = 6$$

This means that the average cost approaches \$6 as the number of DVDs produced becomes very large.

89. $P(s) = \dfrac{63s}{s + 8}$

$$\lim_{s \to \infty} \frac{63s}{s + 8} = \lim_{s \to \infty} \frac{\frac{63s}{s}}{\frac{s}{s} + \frac{8}{s}}$$

$$= \lim_{s \to \infty} \frac{63}{1 + \frac{8}{s}}$$

$$= \frac{63}{1 + 0}$$

$$= 63$$

The number of items of work a new employee produces gets closer and closer to 63 as the number of days of training increases.

91. $\displaystyle\lim_{n \to \infty} \left[\frac{R}{i - g} \left[1 - \left(\frac{1 + g}{1 + i} \right)^n \right] \right]$

$$= \frac{R}{i - g} \lim_{n \to \infty} \left[1 - \left(\frac{1 + g}{1 + i} \right)^n \right]$$

$$= \frac{R}{i - g} \left[\lim_{n \to \infty} 1 - \lim_{n \to \infty} \left(\frac{1 + g}{1 + i} \right)^n \right]$$

assuming $i > g$,

$$= \frac{R}{i - g} [1 - 0]$$

$$= \frac{R}{i - g}$$

93. (a) $D(t) = 155(1 - e^{-0.0133t})$

$$D(20) = 155\left(1 - e^{-0.0133(20)}\right)$$

$$= 155(1 - e^{-0.266})$$

$$\approx 36.2$$

The depth of the sediment layer deposited below the bottom of the lake in 1970 was 36.2 cm.

(b) $\displaystyle\lim_{t \to \infty} D(t) = \lim_{t \to \infty} 155(1 - e^{-0.0133t})$

$$= 155 \lim_{t \to \infty} (1 - e^{-0.0133t})$$

$$= 155\left(\lim_{t \to \infty} 1 - \lim_{t \to \infty} e^{-0.0133t} \right)$$

$$= 155(1) - 155 \lim_{t \to \infty} e^{-0.0133t}$$

$$= 155 - 155(0) = 155$$

Thus,

$$\lim_{t \to \infty} D(t) = 155.$$

Going back in time (t is years before 1990), the depth of the sediment approaches 155 cm.

95. (a) $p_2 = \dfrac{1}{2} + \left(0.7 - \dfrac{1}{2} \right)[1 - 2(0.2)]^2 = 0.572$

(b) $p_4 = \dfrac{1}{2} + \left(0.7 - \dfrac{1}{2} \right)[1 - 2(0.2)]^4 = 0.526$

(c) $p_8 = \dfrac{1}{2} + \left(0.7 - \dfrac{1}{2} \right)[1 - 2(0.2)]^8 = 0.503$

(d) $\displaystyle\lim_{n \to \infty} P_n = \lim_{n \to \infty} \left[\frac{1}{2} + \left(p_0 - \frac{1}{2} \right)(1 - 2p)^n \right]$

$$= \frac{1}{2} + \lim_{n \to \infty} \left(p_0 - \frac{1}{2} \right)(1 - 2p)^n$$

$$= \frac{1}{2} + \left(p_0 - \frac{1}{2} \right) \lim_{n \to \infty} (1 - 2p)^n$$

$$= \frac{1}{2} + \left(p_0 - \frac{1}{2} \right) \cdot 0 = \frac{1}{2}$$

The number in parts (a), (b), and (c) represent the probability that the legislator will vote yes on the second, fourth, and eighth votes. In (d), as the number of roll calls increases, the probability gets close to 0.5, but is never less than 0.5.

3.2 Continuity

Your Turn 1

$$f(x) = \sqrt{5x + 3}$$

The square root function is discontinuous wherever $5x + 3 < 0$. There is a discontinuity when

$$5a + 3 < 0, \text{ or } a < -\frac{3}{5}.$$

Your Turn 2

$$f(x) = \begin{cases} 5x - 4 & \text{if } x < 0 \\ x^2 & \text{if } 0 \le x \le 3 \\ x + 6 & \text{if } x > 3 \end{cases}$$

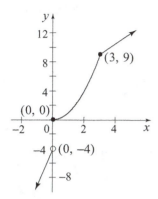

$$\lim_{x \to 0^-} f(x) = -4$$

$$\lim_{x \to 0^+} f(x) = 0$$

Because $\lim\limits_{x \to 0^-} f(x) \ne \lim\limits_{x \to 0^+} f(x)$, the limit doesn't exist, so f is discontinuous at $x = 0$.

3.2 Warmup Exericses

W1.

If $x \ne 2$, $\dfrac{2x^2 - 11x + 14}{x^2 - 5x + 6} = \dfrac{(x - 2)(2x - 7)}{(x - 2)(x - 3)}$

$= \dfrac{2x - 7}{x - 3}$.

Thus $\lim\limits_{x \to 2} \dfrac{2x^2 - 11x + 14}{x^2 - 5x + 6} = \lim\limits_{x \to 2} \dfrac{2x - 7}{x - 3}$

$= \dfrac{2(2) - 7}{2 - 3} = 3$.

W2.

If $x \ne 4$, $\dfrac{3x^2 - 4x - 32}{x^2 - 6x + 8} = \dfrac{(x - 4)(3x + 8)}{(x - 4)(x - 2)}$

$= \dfrac{3x + 8}{x - 2}$.

Thus $\lim\limits_{x \to 4} \dfrac{3x^2 - 4x - 32}{x^2 - 6x + 8} = \lim\limits_{x \to 4} \dfrac{3x + 8}{x - 2}$

$= \dfrac{3(4) + 8}{4 - 2} = 10$.

W3. $\lim\limits_{x \to 3^-} f(x) = \lim\limits_{x \to 3^-} (x + 1) = 4$

$\lim\limits_{x \to 3^+} f(x) = \lim\limits_{x \to 3^+} (2x - 2) = 4$

Thus $\lim\limits_{x \to 3} f(x) = 4$.

W4. $\lim\limits_{x \to 5^-} f(x) = \lim\limits_{x \to 5^-} 2x - 2 = 8$

$\lim\limits_{x \to 5^+} f(x) = \lim\limits_{x \to 5^+} 4x + 1 = 21$

Thus $\lim\limits_{x \to 5} f(x)$ does not exist.

W5. Since in an open interval containing 6, $f(x)$ is defined as $4x + 1$,

$\lim\limits_{x \to 6} f(x) = \lim\limits_{x \to 6} (4x + 1) = 25$.

3.2 Exercises

1. Discontinuous at $x = -1$

  (a) $f(-1)$ does not exist.

  (b) $\lim\limits_{x \to -1^-} f(x) = \dfrac{1}{2}$

  (c) $\lim\limits_{x \to -1^+} f(x) = \dfrac{1}{2}$

  (d) $\lim\limits_{x \to -1} f(x) = \dfrac{1}{2}$ (since (a) and (b) have the same answers)

  (e) $f(-1)$ does not exist.

3. Discontinuous at $x = 1$

  (a) $f(1) = 2$

  (b) $\lim\limits_{x \to 1^-} f(x) = -2$

  (c) $\lim\limits_{x \to 1^+} f(x) = -2$

  (d) $\lim\limits_{x \to 1} f(x) = -2$ (since (a) and (b) have the same answers)

  (e) $\lim\limits_{x \to 1} f(x) \ne f(1)$

5. Discontinuous at $x = -5$ and $x = 0$

  (a) $f(-5)$ does not exist. $f(0)$ does not exist.

  (b) $\lim\limits_{x \to -5^-} f(x) = \infty$ (limit does not exist)

  $\lim\limits_{x \to 0^-} f(x) = 0$

  (c) $\lim\limits_{x \to -5^+} f(x) = -\infty$ (limit does not exist)

  $\lim\limits_{x \to 0^+} f(x) = 0$

(d) $\lim\limits_{x \to -5} f(x)$ does not exist, since the answers to (a) and (b) are different. $\lim\limits_{x \to 0} f(x) = 0$, since the answers to (a) and (b) are the same.

(e) $f(-5)$ does not exist and $\lim\limits_{x \to -5} f(x)$ does not exist. $f(0)$ does not exist.

7. $f(x) = \dfrac{5 + x}{x(x - 2)}$

$f(x)$ is discontinuous at $x = 0$ and $x = 2$ since the denominator equals 0 at these two values.

$\lim\limits_{x \to 0} f(x)$ does not exist since $\lim\limits_{x \to 0^-} f(x) = \infty$ and $\lim\limits_{x \to 0^+} f(x) = -\infty$.

$\lim\limits_{x \to 2} f(x)$ does not exist since

$\lim\limits_{x \to 2^-} f(x) = -\infty$ and $\lim\limits_{x \to 2^+} f(x) = \infty$.

9. $f(x) = \dfrac{x^2 - 4}{x - 2}$

$f(x)$ is discontinuous at $x = 2$ since the denominator equals zero at that value.

Since for $x \neq 2$

$$\frac{x^2 - 4}{x - 2} = \frac{(x + 2)(x - 2)}{x - 2} = x + 2,$$

$\lim\limits_{x \to 2} f(x) = 2 + 2 = 4.$

11. $p(x) = x^2 - 4x + 11$

Since $p(x)$ is a polynomial function, it is continuous everywhere and thus discontinuous nowhere.

13. $p(x) = \dfrac{|x + 2|}{x + 2}$

$p(x)$ is discontinuous at $x = -2$ since the denominator is zero at that value.

since $\lim\limits_{x \to -2^-} p(x) = -1$ and $\lim\limits_{x \to -2^+} p(x) = 1$,

$\lim\limits_{x \to -2} p(x)$ does not exist.

15. $k(x) = e^{\sqrt{x-1}}$

The function is undefined for $x < 1$, so the function is discontinuous for $a < 1$. The limit as x approaches any $a < 1$ does not exist because the function is undefined for $x < 1$.

17. As x approaches 0 from the left or the right, $\left|\dfrac{x}{x-1}\right|$ approaches 0 and $r(x) = \ln\left|\dfrac{x}{x-1}\right|$ goes to $-\infty$. So $\lim\limits_{x \to 0} r(x)$ does not exist. As x approaches 1 from the left or the right, $\left|\dfrac{x}{x-1}\right|$ goes to ∞ and so does $r(x) = \ln\left|\dfrac{x}{x-1}\right|$. So $\lim\limits_{x \to 1} r(x)$ does not exist.

19. $f(x) = \begin{cases} 1 & \text{if } x < 2 \\ x + 3 & \text{if } 2 \leq x \leq 4 \\ 7 & \text{if } x > 4 \end{cases}$

(a)

(b) $f(x)$ is discontinuous at $x = 2$.

(c) $\lim\limits_{x \to 2^-} f(x) = 1$

$\lim\limits_{x \to 2^+} f(x) = 5$

21. $g(x) = \begin{cases} 11 & \text{if } x < -1 \\ x^2 + 2 & \text{if } -1 \leq x \leq 3 \\ 11 & \text{if } x > 3 \end{cases}$

(a)

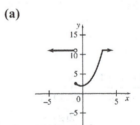

(b) $g(x)$ is discontinuous at $x = -1$.

(c) $\lim\limits_{x \to -1^-} g(x) = 11$

$\lim\limits_{x \to -1^+} g(x) = (-1)^2 + 2 = 3$

23. $h(x) = \begin{cases} 4x + 4 & \text{if } x \leq 0 \\ x^2 - 4x + 4 & \text{if } x > 0 \end{cases}$

(a)

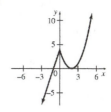

(b) There are no points of discontinuity.

25. Find k so that $kx^2 = x + k$ for $x = 2$.

$$k(2)^2 = 2 + k$$
$$4k = 2 + k$$
$$3k = 2$$
$$k = \frac{2}{3}$$

27. $\dfrac{2x^2 - x - 15}{x - 3} = \dfrac{(2x + 5)(x - 3)}{x - 3} = 2x + 5$

Find k so that $2x + 5 = kx - 1$ for $x = 3$.

$$2(3) + 5 = k(3) - 1$$
$$6 + 5 = 3k - 1$$
$$11 = 3k - 1$$
$$12 = 3k$$
$$4 = k$$

31. $f(x) = \dfrac{x^2 + x + 2}{x^3 - 0.9x^2 + 4.14x - 5.4} = \dfrac{P(x)}{Q(x)}$

(a) Graph

$$Y_1 = \frac{P(x)}{Q(x)} = \frac{x^2 + x + 2}{x^3 - 0.9x^2 + 4.14x - 5.4}$$

on a graphing calculator. A good choice for the viewing window is $[-3,3]$ by $[-10,10]$.

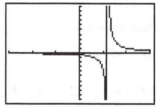

The graph has a vertical asymptote at $x = 1.2$, which indicates that f is discontinuous at $x = 1.2$.

(b) Graph

$$Y_2 = Q(x) = x^3 - .09x^2 + 4.14x - 5.4$$

using the same viewing window.

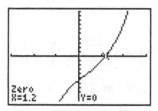

We see that this graph has one x-intercept, 1.2. This indicates that 1.2 is the only real solution of the equation $Q(x) = 0$.

This result verifies our answer from part (a) because a rational function of the form

$$f(x) = \frac{P(x)}{Q(x)}$$

will be discontinuous wherever $Q(x) = 0$.

33. $g(x) = \dfrac{x + 4}{x^2 + 2x - 8}$

$$= \frac{x + 4}{(x - 2)(x + 4)}$$

$$= \frac{1}{x - 2}, x \neq -4$$

If $g(x)$ is defined so that $g(-4) = \frac{1}{-4-2} = -\frac{1}{6}$, then the function becomes continuous at -4. It cannot be made continuous at 2. The correct answer is (a).

35. (a) $\lim\limits_{x \to 6} P(x)$

As x approaches 6 from the left or the right, the value of $P(x)$ for the corresponding point on the graph approaches 500.

Thus, $\lim\limits_{x \to 6} P(x) = \$500$.

(b) $\lim\limits_{x \to 10^-} P(x) = \$1500$

because, as x approaches 10 from the left, $P(x)$ approaches $\$1500$.

(c) $\lim\limits_{x \to 10^+} P(x) = \$1000$ because, as x approaches 10 from the right, $P(x)$ approaches $\$1000$.

(d) Since $\lim\limits_{x \to 10^+} P(x) \neq \lim\limits_{x \to 10^-} P(x)$, $\lim\limits_{x \to 10} P(x)$ does not exist.

(e) From the graph, the function is discontinuous at $x = 10$. This may be the result of a change of shifts.

(f) From the graph, the second shift will be as profitable as the first shift when 15 units are produced.

37. In dollars,

$$F(x) = \begin{cases} 1.25x & \text{if } 0 < x \le 100 \\ 1.00x & \text{if } x > 100. \end{cases}$$

(a) $F(80) = 1.25(80) = \$100$

(b) $F(150) = 1.00(150) = \$150$

(c) $F(100) = 1.25(100) = \$125$

(d) F is discontinuous at $x = 100$.

39. $C(x)$ is a step function.

(a) $\lim\limits_{x \to 3^-} C(x) = \$1.40$

(b) $\lim\limits_{x \to 3^+} C(x) = \$1.61$

(c) $\lim\limits_{x \to 3} C(x)$ does not exist.

(d) $C(3) = \$1.40$

(e) $\lim\limits_{x \to 8.5^-} C(x) = \$2.66$

(f) $\lim\limits_{x \to 8.5^+} C(x) = \$2.66$

(g) $\lim\limits_{x \to 8.5} C(x) = \$2.66$

(h) $C(8.5) = \$2.66$

(i) $C(x)$ is discontinuous at $1, 2, 3, \ldots, 11, 12$

41.

$$W(t) = \begin{cases} 48 + 3.64t + 0.6363t^2 + 0.00963t^3, & 1 \le t \le 28 \\ -1{,}004 + 65.8t, & 28 < t \le 56 \end{cases}$$

(a) $W(25) = 48 + 3.64(25) + 0.6363(25)^2$

$$+ 0.00963(25)^3$$

$$\approx 687.156$$

A male broiler at 25 days weighs about 687 grams.

(b) $W(t)$ is not a continuous function.

At $t = 28$

$$\lim\limits_{t \to 28^-} W(t)$$

$$= \lim\limits_{t \to 28^-} 48 + 3.64t + 0.6363t^2 + 0.00963t^3$$

$$= 48 + 3.64(28) + 0.6363(28)^2 + 0.00963(28)^3$$

$$\approx 860.18$$

and $\lim\limits_{t \to 28^-} W(t) \neq \lim\limits_{t \to 28^+} (-1004 + 65.8t)$

$$= -1004 + 65.8(28)$$

$$= 838.4$$

so $\lim\limits_{t \to 28^-} W(t) \neq \lim\limits_{t \to 28^+} W(t)$

Thus $W(t)$ is discontinuous.

(c)

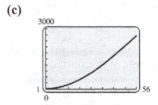

3.3 Rates of Change

Your Turn 1

$$A(t) = 11.14(1.023)^t$$

Average rate of change from $t = 0$ (2000) to $t = 10$ (2010) is

$$\frac{A(10) - A(0)}{10 - 0} = \frac{11.14(1.023)^{10} - 11.14(1.023)^0}{10}$$

$$= \frac{13.98 - 11.14}{10} = \frac{2.84}{10} \approx 0.0284,$$

or 0.0284 million.

The U.S. Asian population increased, on average, by 284,000 people per year.

Your Turn 2

$$\frac{A(2013) - A(2011)}{2103 - 2011} = \frac{65.80 - 68.35}{2} = -1.325$$

The average change per year is a decrease of $1.325 billion.

Your Turn 3

For $t = 2$, the instantaneous velocity is

$$\lim\limits_{h \to 0} \frac{s(2+h) - s(2)}{h} \text{ feet per second.}$$

$$s(2 + h) = 2(2 + h)^2 - 5(2 + h) + 40$$

$$= 2(4 + 4h + h^2) - 10 - 5h + 40$$

$$= 8 + 8h + h^2 - 10 - 5h + 40$$

$$= h^2 + 3h + 38$$

$$s(2) = 2(2)^2 - 5(2) + 40$$

$$= 2(4) - 10 + 40$$

$$= 38$$

$$\lim_{h \to 0} \frac{s(2+h)-s(2)}{h} = \lim_{h \to 0} \frac{h^2 + 3h + 38 - 38}{h}$$

$$= \lim_{h \to 0} \frac{h^2 + 3h}{h}$$

$$= \lim_{h \to 0} \frac{h(h+3)}{h}$$

$$= \lim_{h \to 0} h + 3 = 3$$

or 3 feet per second.

Your Turn 4

$$C(x) = x^2 - 2x + 12$$

The instantaneous rate of change of cost when $x = 4$ is

$$\lim_{h \to 0} \frac{C(h+4) - C(4)}{h}$$

$$= \lim_{h \to 0} \frac{[(h+4)^2 - 2(h+4) + 12] - [4^2 - 2(4) + 12]}{h}$$

$$= \lim_{h \to 0} \frac{h^2 + 8h + 16 - 2h - 8 + 12 - 16 + 8 - 12}{h}$$

$$= \lim_{h \to 0} \frac{h^2 + 6h}{h} = \lim_{h \to 0} \frac{h(h+6)}{h}$$

$$= \lim_{h \to 0} h + 6 = 0 + 6 = 6$$

When $x = 4$, the cost increases at a rate of \$6 per unit.

Your Turn 5

$$A(t) = 11.14(1.023)^t, \quad t = 10 \text{ corresponds to 2010}$$

$$\lim_{h \to 0} \frac{11.14(1.023)^{10+h} - 11.14(1.023)^{10}}{h}$$

Use the TABLE feature on a TI-84 Plus calculator.

h	$\dfrac{11.14(1.023)^{10+h} - 11.14(1.023)^{10}}{h}$
1	0.32164
0.1	0.31836
0.01	0.31803
0.001	0.318
0.0001	0.318
0.00001	0.318

The limit seems to be approaching 0.318 million. The instantaneous rate of change in the U.S. Asian population is about 318,000 people per year in 2010.

3.3 Warmup Exercises

W1. $f(x) = 2x^2 + 3x + 4$

$$f(5+h) = 2(5+h)^2 + 3(5+h) + 4$$

$$= 2(25 + 10h + h^2) + 15 + 3h + 4$$

$$= 50 + 20h + 2h^2 + 19 + 3h$$

$$= 2h^2 + 23h + 69$$

W2. $f(x) = 2x^2 + 3x + 4$

$$\frac{f(2+h) - f(2)}{h}$$

$$= \frac{2(2+h)^2 + 3(2+h) + 4 - (2(2)^2 + 3(2) + 4)}{h}$$

$$= \frac{2h^2 + 8h + 8 + 6 + 3h + 4 - 18}{h}$$

$$= \frac{2h^2 + 11h}{h} = 2h + 11$$

W3. $f(x) = \dfrac{2}{x+1}$

$$f(3+h) = \frac{2}{(3+h)+1} = \frac{2}{4+h}$$

W4. $f(x) = \dfrac{2}{x+1}$

$$\frac{f(4+h) - f(4)}{h}$$

$$= \frac{1}{h} \cdot \left[\frac{2}{(4+h)+1} - \frac{2}{5} \right]$$

$$= \frac{1}{h} \cdot \left[\frac{(2)(5) - (2)(5+h)}{(5)(5+h)} \right]$$

$$= \frac{1}{h} \cdot \left(\frac{-2h}{(5)(5+h)} \right) = -\frac{2}{(5)(5+h)}$$

3.3 Exercises

1. $y = x^2 + 2x = f(x)$ between $x = 1$ and $x = 3$

Average rate of change

$$= \frac{f(3) - f(1)}{3 - 1}$$

$$= \frac{15 - 3}{2}$$

$$= 6$$

3. $y = -3x^3 + 2x^2 - 4x + 1 = f(x)$ between $x = -2$ and $x = 1$

Average rate of change

$$= \frac{f(1) - f(-2)}{1 - (-2)}$$

$$= \frac{(-4) - (-41)}{1 - (-2)} = \frac{-45}{3} = -15$$

5. $y = \sqrt{x} = f(x)$ between $x = 1$ and $x = 4$
Average rate of change

$$= \frac{f(4) - f(1)}{4 - 1}$$

$$= \frac{2 - 1}{3}$$

$$= \frac{1}{3}$$

7. $y = e^x = f(x)$ between $x = -2$ and $x = 0$

Average rate of change

$$= \frac{f(0) - f(-2)}{0 - (-2)}$$

$$= \frac{1 - e^{-2}}{2}$$

$$\approx 0.4323$$

9.

$$\lim_{h \to 0} \frac{s(6 + h) - s(6)}{h}$$

$$= \lim_{h \to 0} \frac{(6 + h)^2 + 5(6 + h) + 2 - [6^2 + 5(6) + 2]}{h}$$

$$= \lim_{h \to 0} \frac{h^2 + 17h + 68 - 68}{h} = \lim_{h \to 0} \frac{h^2 + 17h}{h}$$

$$= \lim_{h \to 0} \frac{h(h + 17)}{h} = \lim_{h \to 0} (h + 17) = 17$$

The instantaneous velocity at $t = 6$ is 17.

11. $s(t) = 5t^2 - 2t - 7$

$$\lim_{h \to 0} \frac{s(2 + h) - s(2)}{h}$$

$$= \lim_{h \to 0} \frac{[5(2 + h)^2 - 2(2 + h) - 7] - [5(2)^2 - 2(2) - 7]}{h}$$

$$= \lim_{h \to 0} \frac{[20 + 20h + 5h^2 - 4 - 2h - 7] - [20 - 4 - 7]}{h}$$

$$= \lim_{h \to 0} \frac{9 + 18h + 5h^2 - 9}{h} = \lim_{h \to 0} \frac{18h + 5h^2}{h}$$

$$= \lim_{h \to 0} \frac{h(18 + 5h)}{h} = \lim_{h \to 0} (18 + 5h) = 18$$

The instantaneous velocity at $t = 2$ is 18.

13. $s(t) = t^3 + 2t + 9$

$$\lim_{x \to 0} \frac{s(1 + h) - s(1)}{h}$$

$$= \lim_{h \to 0} \frac{[(1 + h)^3 + 2(1 + h) + 9] - [(1)^3 + 2(1) + 9]}{h}$$

$$= \lim_{h \to 0} \frac{[1 + 3h + 3h^2 + h^3 + 2 + 2h + 9] - [1 + 2 + 9]}{h}$$

$$= \lim_{h \to 0} \frac{h^3 + 3h^2 + 5h + 12 - 12}{h}$$

$$= \lim_{h \to 0} \frac{h^3 + 3h^2 + 5h}{h} = \lim_{h \to 0} \frac{h(h^2 + 3h + 5)}{h}$$

$$= \lim_{h \to 0} (h^2 + 3h + 5) = 5$$

The instantaneous velocity at $t = 1$ is 5.

15. $f(x) = x^2 + 2x$ at $x = 0$

$$\lim_{h \to 0} \frac{f(0 + h) - f(0)}{h}$$

$$= \lim_{h \to 0} \frac{(0 + h)^2 + 2(0 + h) - [0^2 + 2(0)]}{h}$$

$$= \lim_{h \to 0} \frac{h^2 + 2h}{h}$$

$$= \lim_{h \to 0} \frac{h(h + 2)}{h}$$

$$= \lim_{h \to 0} h + 2 = 2$$

The instantaneous rate of change at $x = 0$ is 2.

17. $g(t) = 1 - t^2$ at $t = -1$

$$\lim_{h \to 0} \frac{g(-1 + h) - g(-1)}{h}$$

$$= \lim_{h \to 0} \frac{1 - (-1 + h)^2 - [1 - (-1)^2]}{h}$$

$$= \lim_{h \to 0} \frac{1 - (1 - 2h + h^2) - 1 + 1}{h}$$

$$= \lim_{h \to 0} \frac{2h - h^2}{h}$$

$$= \lim_{h \to 0} \frac{h(2 - h)}{h}$$

$$= \lim_{h \to 0} (2 - h) = 2$$

The instantaneous rate of change at $t = -1$ is 2.

19. $f(x) = x^x$ at $x = 2$

h	
0.01	$\dfrac{f(2 + 0.01) - f(2)}{0.01}$
	$= \dfrac{2.01^{2.01} - 2^2}{0.01}$
	$= 6.84$
0.001	$\dfrac{f(2 + 0.001) - f(2)}{0.001}$
	$= \dfrac{2.001^{2.001} - 2^2}{0.001}$
	$= 6.779$
0.0001	$\dfrac{f(2 + 0.0001) - f(2)}{0.00001}$
	$= \dfrac{2.0001^{2.0001} - 2^2}{0.0001}$
	$= 6.773$
0.00001	$\dfrac{f(2 + 0.00001) - f(2)}{0.00001}$
	$= \dfrac{2.00001^{2.00001} - 2^2}{0.00001}$
	$= 6.7727$
0.000001	$\dfrac{f(2 + 0.000001) - f(2)}{0.000001}$
	$= \dfrac{2.000001^{2.000001} - 2^2}{0.000001}$
	$= 6.7726$

The instantaneous rate of change at $x = 2$ is 6.773.

21. $f(x) = x^{\ln x}$ at $x = 2$

h	
0.01	$\dfrac{f(2 + 0.01) - f(2)}{0.01}$
	$= \dfrac{2.01^{\ln 2.01} - 2^{\ln 2}}{0.01}$
	$= 1.1258$
0.001	$\dfrac{f(2 + 0.001) - f(2)}{0.001}$
	$= \dfrac{2.001^{\ln 2.001} - 2^{\ln 2}}{0.001}$
	$= 1.1212$

h	
0.0001	$\dfrac{f(2 + 0.0001) - f(2)}{0.0001}$
	$= \dfrac{2.0001^{\ln 2.0001} - 2^{\ln 2}}{0.0001}$
	$= 1.1207$
0.00001	$\dfrac{f(2 + 0.00001) - f(2)}{0.00001}$
	$= \dfrac{2.00001^{\ln 2.00001} - 2^{\ln 2}}{0.00001}$
	$= 1.1207$

The instantaneous rate of change at $x = 2$ is 1.121.

25. $P(x) = 2x^2 - 5x + 6$

(a) $P(4) = 18$

$P(2) = 4$

Average rate of change of profit

$$= \frac{P(4) - P(2)}{4 - 2}$$

$$= \frac{18 - 4}{2} = \frac{14}{2} = 7,$$

which is \$700 per item.

(b) $P(3) = 9$

$P(2) = 4$

Average rate of change of profit

$$= \frac{P(3) - P(2)}{3 - 2} = \frac{9 - 4}{1} = 5$$

which is \$500 per item.

(c) $\lim\limits_{h\to 0} \dfrac{P(2+h)-P(2)}{h}$

$= \lim\limits_{h\to 0} \dfrac{2(2+h)^2 - 5(2+h) + 6 - 4}{h}$

$= \lim\limits_{h\to 0} \dfrac{8 + 8h + 2h^2 - 10 - 5h + 2}{h}$

$= \lim\limits_{h\to 0} \dfrac{2h^2 + 3h}{h} = \lim\limits_{h\to 0} \dfrac{h(2h+3)}{h}$

$= \lim\limits_{h\to 0}(2h+3) = 3,$

which is $300 per item.

(d) $\lim\limits_{h\to 0} \dfrac{P(4+h)-P(4)}{h}$

$= \lim\limits_{h\to 0} \dfrac{2(4+h)^2 - 5(4+h) + 6 - 18}{h}$

$= \lim\limits_{h\to 0} \dfrac{32 + 16h + 2h^2 - 20 - 5h - 12}{h}$

$= \lim\limits_{h\to 0} \dfrac{2h^2 + 11h}{h}$

$= \lim\limits_{h\to 0} \dfrac{h(2h+11)}{h}$

$= \lim\limits_{h\to 0} 2h + 11 = 11,$

which is $1100 per item.

27. $N(p) = 80 - 5p^2, 1 \le p \le 4$

(a) Average rate of change of demand is

$\dfrac{N(3) - N(2)}{3 - 2} = \dfrac{35 - 60}{1}$

$= -25$ boxes per dollar.

(b) Instantaneous rate of change when p is 2 is

$\lim\limits_{h\to 0} \dfrac{N(2+h) - N(2)}{h}$

$= \lim\limits_{h\to 0} \dfrac{80 - 5(2+h)^2 - [80 - 5(2)^2]}{h}$

$= \lim\limits_{h\to 0} \dfrac{80 - 20 - 20h - 5h^2 - (80 - 20)}{h}$

$= \lim\limits_{h\to 0} \dfrac{-5h^2 - 20h}{h} = -20$ boxes per dollar.

Around the $2 point, a $1 price increase (say, from $1.50 to $2.50) causes a drop in demand of about 20 boxes.

(c) Instantaneous rate of change when p is 3 is

$\lim\limits_{h\to 0} \dfrac{80 - 5(3+h)^2 - [80 - 5(3)^2]}{h}$

$= \lim\limits_{h\to 0} \dfrac{80 - 45 - 30h - 5h^2 - 80 + 45}{h}$

$= \lim\limits_{h\to 0} \dfrac{-30h - 5h^2}{h}$

$= -30$ boxes per dollar.

(d) As the price increases, the demand decreases; this is an expected change.

29. $A(t) = 1000e^{0.03t}$

(a) Average rate of change in the total amount from $t = 0$ to $t = 5$:

$\dfrac{A(5) - A(0)}{5 - 0} = \dfrac{1000e^{0.03(5)} - 1000e^{0.03(0)}}{5}$

$= 32.3668,$

which is $32.37 per year.

(b) Average rate of change in the total amount from $t = 5$ to $t = 10$:

$\dfrac{A(10) - A(5)}{10 - 5} = \dfrac{1000e^{0.03(10)} - 1000e^{0.03(5)}}{5}$

$= 37.6049,$

which is $37.60 per year.

(c) Instantaneous rate of change for $t = 5$:

$\lim\limits_{h\to 0} \dfrac{1000e^{0.03(5+h)} - 1000e^{0.03(5)}}{h}$

Use the TABLE feature on a TI-84 Plus calculator to estimate the limit.

h	$\dfrac{1000e^{0.03(5+h)} - 1000e^{0.03(5)}}{h}$
1	35.3831
0.1	34.9074
0.01	34.8603
0.001	34.8556
0.0001	34.8551
0.00001	34.8550

The limit seems to be approaching 34.855. So, the instantaneous rate of change for $t = 5$ is about $34.86 per year.

31. Let $P(t) =$ the price per gallon of gasoline for the month t, where $t = 1$ represents January, $t = 2$ represents February, and so on.

(a) $P(1) = 339$ (cents)

$P(3) = 378$ (cents)

Average change in price from January to March:

$$\frac{P(3) - P(1)}{3 - 1} = \frac{378 - 339}{2} = 19.5$$

On average, the price of gasoline increased about 19.5 cents per gallon per month.

(b) $P(3) = 378$ (cents), $P(12) = 336$ (cents)

Average change in price from March to December:

$$\frac{P(12) - P(3)}{12 - 3} = \frac{336 - 378}{9} \approx -4.7$$

On average, the price of gasoline decreased about 4.7 cents per gallon per month.

(c) $P(1) = 339$ (cents), $P(12) = 336$ (cents)

Average change in price from January to December:

$$\frac{P(12) - P(1)}{12 - 1} = \frac{336 - 339}{12} \approx -0.3$$

On average, the price of gasoline decreased about 0.3 cents per gallon per month.

33. (a) $\dfrac{A(15) - A(0)}{15} = \dfrac{11.14\left(1.023^{15} - 1.023^{0}\right)}{15}$

≈ 0.302

A gives the population in millions so this is an average rate of change of 302,000 people per year.

(b) Use the TABLE feature on a TI-84 Plus calculator to estimate the limit.

h	$\dfrac{11.14(1.023)^{15+h} - 11.14(1.023)^{15}}{h}$
1	0.3604
0.1	0.3567
0.01	0.3563
0.001	0.3563

The limit seems to be approaching 0.356. So, the instantaneous rate of change of the Asian population in 2015 is about 356,000 people per year.

35. (a) 2006 to 2008:

$$\frac{47,800 - 48,600}{2} = -400 \text{ per year}$$

(b) 2008 to 2010:

$$\frac{47,500 - 47,800}{2} = -150 \text{ per year}$$

(c) 2007 to 1011:

$$\frac{49,273 - 56,000}{4} \approx -1682 \text{ per year}$$

37. (a)

$F(t) = -10.28 + 175.9te^{-t/1.3}$

(b) The average rate of change during the first hour is

$$\frac{F(1) - F(0)}{1 - 0} \approx 81.51$$

kilojoules per hour per hour.

(c) Store $F(t)$ in a function menus of a graphing calculator. Store $\dfrac{Y_1(1+X) - Y_1(1)}{X}$ as Y_2 in the function menu, where Y_1 represents $F(t)$. Substitute small values for X in Y_2 perhaps with use of a table feature of the graphing calculator. As X is allowed to get smaller, Y_2 approaches 18.81 kilojoules per hour per hour.

(d) Through use of a MAX/feature program of a graphing calculator, the maximum point seen in part (a) is estimated to occur at approximately $t = 1.3$ hours.

39. Let $I(t)$ represent immigration (in thousands) in year t.

(a) $\dfrac{I(1960) - I(1910)}{1960 - 1910} = \dfrac{265 - 1042}{50}$

$= -15.54$

The average rate of change is $-15,540$ immigrants per year.

(b) $\dfrac{I(2010) - I(1960)}{2010 - 1960} = \dfrac{1043 - 265}{50}$

$= 15.56$

The average rate of change is 15,560 immigrants per year.

(c) $\dfrac{I(2010) - I(1910)}{2010 - 1910} = \dfrac{1043 - 1042}{100}$

$= 0.01$

The average rate of change is 10 immigrants per year.

(d) $\dfrac{-15,540 + 15,560}{2} = \dfrac{20}{2} = 10$

They are equal. This will not be true for all time periods. (It is true only for time periods of equal length.)

(e) 2013 is 53 years after 1960.

$1,043,000 + 53(15,560/\text{year}) \approx 1,090,000$

The predicted number of immigrants in 2013 is about 1,090,000 immigrants. The predicted value is about 99,000 more than the actual number of 990,553.

41. (a) $\dfrac{T(3000) - T(1000)}{3000 - 1000} = \dfrac{23 - 15}{2000}$

$= \dfrac{8}{2000}$

$= \dfrac{4}{1000}$

From 1000 to 3000 ft, the temperature changes about $4°$ per 1000 ft; the temperature rises (on the average).

(b) $\dfrac{T(5000) - T(1000)}{5000 - 1000} = \dfrac{22 - 15}{4000}$

$= \dfrac{7}{4000}$

$= \dfrac{1.75}{1000}$

From 1000 to 5000 ft, the temperature changes about $1.75°$ per 1000 ft; the temperature rises (on the average).

(c) $\dfrac{T(9000) - T(3000)}{9000 - 3000} = \dfrac{15 - 23}{6000}$

$= \dfrac{-8}{6000}$

$= \dfrac{-\frac{4}{3}}{1000}$

From 3000 to 9000 ft, the temperature changes about $-\frac{4}{3}°$ per 1000 ft; the temperature falls (on the average).

(d) $\dfrac{T(9000) - T(1000)}{9000 - 1000} = \dfrac{15 - 15}{8000} = 0$

From 1000 to 9000 ft, the temperature changes about $0°$ per 1000 ft; the temperature stays constant (on the average).

(e) The temperature is highest at 3000 ft and lowest at 1000 ft. If 7000 ft is changed to 10,000 ft, the lowest temperature would be at 10,000 ft.

(f) The temperature at 9000 ft is the same as 1000 ft.

43. (a) Average rate of change from 0.5 to 1:

$\dfrac{f(1) - f(0.5)}{1 - 0.5} = \dfrac{55 - 30}{0.5} = 50$ mph

Average rate of change from 1 to 1.5:

$\dfrac{f(1.5) - f(1)}{1.5 - 1} = \dfrac{80 - 55}{0.5} = 50$ mph

Estimate of instantaneous velocity is

$\dfrac{50 + 50}{2} = 50$ mph.

(b) Average rate of change from 1.5 to 2:

$\dfrac{f(2) - f(1.5)}{2 - 1.5} = \dfrac{104 - 80}{0.5} = 48$ mph

Average rate of change from 2 to 2.5

$\dfrac{f(2.5) - f(2)}{2.5 - 2} = \dfrac{124 - 104}{0.5} = 40$ mph

Estimate of instantaneous velocity is

$\dfrac{48 + 40}{2} = 44$ mph.

3.4 Definition of the Derivative

Your Turn 1

(a) $f(x) = x^2 - x, \ x = -2$ and $x = 1.$

Slope of secant line

$= \dfrac{f(1) - f(-2)}{1 - (-2)} = \dfrac{0 - 6}{3} = -2$

Use the point-slope form and the point $(1, f(1))$, or $(1, 0)$.

$$y - y_1 = m(x - x_1)$$
$$y - 0 = -2(x - 1)$$
$$y = -2x + 2$$

(b)   slope of tangent at $(-2, 6)$

$= \lim_{h \to 0} \dfrac{f(-2 + h) - f(x)}{h}$

$$= \lim_{h \to 0} \frac{[(-2+h)^2 - (-2+h)] - [(-2)^2 - (-2)]}{h}$$

$$= \lim_{h \to 0} \frac{\left[4 - 4h + h^2 + 2 - h\right] - \left[4 + 2\right]}{h}$$

$$= \lim_{h \to 0} \frac{-5h + h^2}{h} = \lim_{h \to 0} (-5 + h)$$

$$= -5$$

The equation of the tangent line is
$$y - 6 = (-5)(x - (-2))$$
$$y = 6 - 5x - 10$$
$$y = -5x - 4.$$

Your Turn 2

$$f(x) = x^2 - x$$

$$f'(x) = \lim_{h \to 0} \frac{f(x+h) - f(x)}{h}$$

$$= \lim_{h \to 0} \frac{[(x+h)^2 - (x+h)] - [x^2 - x]}{h}$$

$$= \lim_{h \to 0} \frac{x^2 + 2xh + h^2 - x - h - x^2 + x}{h}$$

$$= \lim_{h \to 0} \frac{2xh - h + h^2}{h} = \lim_{h \to 0} \frac{h(2x - 1 + h)}{h}$$

$$= \lim_{h \to 0} (2x - 1 + h) = 2x - 1 + 0$$

$$= 2x - 1$$

$$f'(-2) = 2(-2) - 1 = -5$$

Your Turn 3

$$f(x) = x^3 - 1$$

$$f'(x) = \lim_{h \to 0} \frac{f(x+h) - f(x)}{h}$$

$$= \lim_{h \to 0} \frac{[(x+h)^3 - 1] - [x^3 - 1]}{h}$$

$$= \lim_{h \to 0} \frac{x^3 + 3x^2h + 3xh^2 + h^3 - 1 - x^3 + 1}{h}$$

$$= \lim_{h \to 0} \frac{3x^2h + 3xh^2 + h^3}{h}$$

$$= \lim_{h \to 0} \frac{h(3x^2 + 3xh + h^2)}{h}$$

$$= \lim_{h \to 0} (3x^2 + 3xh + h^2)$$

$$= 3x^2 + 0 + 0$$

$$= 3x^2$$

$$f'(-1) = 3(-1)^2 = 3$$

Your Turn 4

$$f(x) = -\frac{2}{x}$$

$$f'(x) = \lim_{h \to 0} \frac{f(x+h) - f(x)}{h}$$

$$= \lim_{h \to 0} \frac{\left(-\frac{2}{x+h}\right) - \left(-\frac{2}{x}\right)}{h}$$

$$= \lim_{h \to 0} \left[-\frac{2}{x+h} + \frac{2}{x} \right] \cdot \frac{1}{h}$$

$$= \lim_{h \to 0} \frac{-2x + 2x + 2h}{x(x+h)} \cdot \frac{1}{h}$$

$$= \lim_{h \to 0} \frac{2h}{x(x+h)} \cdot \frac{1}{h} = \lim_{h \to 0} \frac{2}{x(x+h)}$$

$$= \frac{2}{x(x+0)}$$

$$= \frac{2}{x^2}$$

Your Turn 5

$f(x) = 2\sqrt{x}$

$$f'(x) = \lim_{h \to 0} \frac{f(x+h) - f(x)}{h}$$

$$= \lim_{h \to 0} \frac{2\sqrt{x+h} - 2\sqrt{x}}{h}$$

$$= \lim_{h \to 0} \frac{2\sqrt{x+h} - 2\sqrt{x}}{h} \cdot \frac{2\sqrt{x+h} + 2\sqrt{x}}{2\sqrt{x+h} + 2\sqrt{x}}$$

$$= \lim_{h \to 0} \frac{4(x+h) - 4x}{h\left(2\sqrt{x+h} + 2\sqrt{x}\right)}$$

$$= \lim_{h \to 0} \frac{4x + 4h - 4x}{h\left(2\sqrt{x+h} + 2\sqrt{x}\right)}$$

$$= \lim_{h \to 0} \frac{4h}{h\left(2\sqrt{x+h} + 2\sqrt{x}\right)}$$

$$= \lim_{h \to 0} \frac{4}{2\sqrt{x+h} + 2\sqrt{x}}$$

$$= \frac{4}{2\sqrt{x} + 2\sqrt{x}} = \frac{4}{4\sqrt{x}}$$

$$= \frac{1}{\sqrt{x}}$$

Your Turn 6

$C(x) = 10x - 0.002x^2$

$$C'(x) = \lim_{h \to 0} \frac{C(x+h) - C(x)}{h}$$

$$= \lim_{h \to 0} \frac{10(x+h) - 0.002(x+h)^2 - 10x + 0.002x^2}{h}$$

$$= \lim_{h \to 0} \frac{10x + 10h - 0.002x^2 - 0.004xh - 0.002h^2 - 10x + 0.002x^2}{h}$$

$$= \lim_{h \to 0} \frac{10h - 0.004xh - 0.002h^2}{h}$$

$$= \lim_{h \to 0} \frac{h(10 - 0.004x - 0.002h)}{h}$$

$$= \lim_{h \to 0} (10 - 0.004x - 0.002h)$$

$$= 10 - 0.004x + 0$$

$$= 10 - 0.004x$$

$$C'(100) = 10 - 0.004(100) = 10 - 0.4 = 9.60$$

The rate of change of the cost when $x = 100$ is $9.60.

Your Turn 7

$f(x) = 2\sqrt{x}$ at $x = 4$

From Your Turn 5, we have

$$f'(x) = \frac{1}{\sqrt{x}}.$$

At $x = 4$: $f'(4) = \frac{1}{\sqrt{4}} = \frac{1}{2}$ and $f(4) = 2\sqrt{4} = 2(2) = 4$

Slope of the tangent line at $(4, f(4))$, or $(4, 4)$ is $\frac{1}{2}$.

$$y - y_1 = m(x - x_1)$$

$$y - 4 = \frac{1}{2}(x - 4)$$

$$y - 4 = \frac{1}{2}x - 2$$

$$y = \frac{1}{2}x + 2$$

3.4 Warmup Exercises

W1.

$f(x) = 3x^2 - 2x - 5$

$\dfrac{f(x + h) - f(x)}{h}$

$= \dfrac{3(x + h)^2 - 2(x + h) - 5 - (3x^2 - 2x - 5)}{h}$

$= \dfrac{3x^2 + 6xh + 3h^2 - 2x - 2h - (3x^2 - 2x - 5)}{h}$

$= \dfrac{6xh + 3h^2 - 2h}{h} = 6x + 3h - 2$

W2.

$f(x) = \dfrac{3}{x - 2}$

$\dfrac{f(x + h) - f(x)}{h} = \dfrac{1}{h} \cdot \left[\dfrac{3}{x + h - 2} - \dfrac{3}{x - 2} \right]$

$= \dfrac{1}{h} \cdot \left[\dfrac{(3)(x - 2) - (3)(x + h - 2)}{(x + h - 2)(x - 2)} \right]$

$= \dfrac{1}{h} \cdot \left[\dfrac{3x - 6 - 3x - 3h + 6}{(x + h - 2)(x - 2)} \right]$

$= \dfrac{1}{h} \cdot \left[\dfrac{-3h}{(x + h - 2)(x - 2)} \right]$

$= -\dfrac{3}{(x + h - 2)(x - 2)}$

W3. The line $5x + 6y = 7$ has slope $-5/6$ so any line parallel to this line has equation $y = -(5/6)x + b$. If the line passes through $(4, -1)$, then $-1 = -(5/6)4 + b$

$$b = 1 + 5/3 = 7/3$$

and the equation of the line is
$y = -(5/6)x + 7/3$.

W4. The line through $(6, 2)$ and $(-2, 5)$ has slope equal to $\dfrac{2 - 5}{6 - (-2)} = \dfrac{-3}{8} = -\dfrac{3}{8}$. This line has equation $y = -(3/8)x + b$. Since it passes through $(6, 2)$, $2 = -(3/8)(6) + b$

$$b = 2 + 9/4 = 17/4$$

and the equation of the line is
$y = -(3/8)x + 17/4$.

3.4 Exercises

1. **(a)** $f(x) = 5$ is a horizontal line and has slope 0; the derivative is 0.

   (b) $f(x) = x$ has slope 1; the derivative is 1.

   (c) $f(x) = -x$ has slope of -1, the derivative is -1.

   (d) $x = 3$ is vertical and has undefined slope; the derivative does not exist.

   (e) $y = mx + b$ has slope m; the derivative is m.

3. $f(x) = \dfrac{x^2 - 1}{x + 2}$ is not differentiable when $x + 2 = 0$ or $x = -2$ because the function is undefined and a vertical asymptote occurs there.

5. Using the points $(5, 3)$ and $(6, 5)$, we have

$$m = \frac{5 - 3}{6 - 5} = \frac{2}{1}$$
$$= 2.$$

7. Using the points $(-2, 2)$ and $(2, 3)$, we have

$$m = \frac{3 - 2}{2 - (-2)} = \frac{1}{4}.$$

9. Using the points $(-3, -3)$ and $(0, -3)$, we have

$$m = \frac{-3 - (-3)}{0 - 3} = \frac{0}{-3} = 0.$$

11. $f(x) = 3x - 7$

Step 1 $f(x + h)$
$$= 3(x + h) - 7$$
$$= 3x + 3h - 7$$

Step 2 $f(x + h) - f(x)$
$$= 3x + 3h - 7 - (3x - 7)$$
$$= 3x + 3h - 7 - 3x + 7$$
$$= 3h$$

Step 3 $\dfrac{f(x + h) - f(x)}{h} = \dfrac{3h}{h} = 3$

Step 4 $f'(x) = \lim\limits_{h \to 0} \dfrac{f(x + h) - f(x)}{h}$
$$= \lim_{h \to 0} 3 = 3$$

$f'(-2) = 3, \ f'(0) = 3 \ f'(3) = 3$

13. $f(x) = -4x^2 + 9x + 2$

Step 1 $f(x + h)$
$$= -4(x + h)^2 + 9(x + h) + 2$$
$$= -4(x^2 + 2xh + h^2) + 9x + 9h + 2$$
$$= -4x^2 - 8xh - 4h^2 + 9x + 9h + 2$$

Step 2 $f(x + h) - f(x)$
$$= -4x^2 - 8xh - 4h^2 + 9x + 9h + 2$$
$$-(-4x^2 + 9x + 2)$$
$$= -8xh - 4h^2 + 9h$$
$$= h(-8x - 4h + 9)$$

Step 3 $\dfrac{f(x + h) - f(x)}{h}$
$$= \dfrac{h(-8x - 4h + 9)}{h}$$
$$= -8x - 4h + 9$$

Step 4 $f'(x) = \lim\limits_{h \to 0} \dfrac{f(x + h) - f(x)}{h}$
$$= \lim_{h \to 0} (-8x - 4h + 9)$$
$$= -8x + 9$$
$$f'(-2) = -8(-2) + 9 = 25$$
$$f'(0) = -8(0) + 9 = 9$$
$$f'(3) = -8(3) + 9 = -15$$

15. $f(x) = \dfrac{12}{x}$

$$f(x + h) = \dfrac{12}{x + h}$$

$f(x + h) - f(x) = \dfrac{12}{x + h} - \dfrac{12}{x}$
$$= \dfrac{12x - 12(x + h)}{x(x + h)}$$
$$= \dfrac{12x - 12x - 12h}{x(x + h)}$$
$$= \dfrac{-12h}{x(x + h)}$$

$\dfrac{f(x + h) - f(x)}{h} = \dfrac{-12h}{hx(x + h)}$
$$= \dfrac{-12}{x(x + h)}$$
$$= \dfrac{-12}{x^2 + xh}$$

$f'(x) = \lim\limits_{h \to 0} \dfrac{f(x + h) - f(x)}{h}$
$$= \lim_{h \to 0} \dfrac{-12}{x^2 + xh}$$
$$= \dfrac{-12}{x^2}$$

$f'(-2) = \dfrac{-12}{(-2)^2} = \dfrac{-12}{4} = -3$

$f'(0) = \dfrac{-12}{0^2}$ which is undefined so $f'(0)$ does not exist.

$$f'(3) = \dfrac{-12}{3^2}$$
$$= \dfrac{-12}{9} = -\dfrac{4}{3}$$

17. $f(x) = \sqrt{x}$

Steps 1-3 are combined.

$\dfrac{f(x + h) - f(x)}{h}$
$$= \dfrac{\sqrt{x + h} - \sqrt{x}}{h}$$
$$= \dfrac{\sqrt{x + h} - \sqrt{x}}{h} \cdot \dfrac{\sqrt{x + h} + \sqrt{x}}{\sqrt{x + h} + \sqrt{x}}$$
$$= \dfrac{x + h - x}{h(\sqrt{x + h} + \sqrt{x})}$$
$$= \dfrac{1}{\sqrt{x + h} + \sqrt{x}}$$

$f'(x) = \lim\limits_{h \to 0} \dfrac{f(x + h) - f(x)}{h}$
$$= \lim_{h \to 0} \dfrac{1}{\sqrt{x + h} + \sqrt{x}} = \dfrac{1}{2\sqrt{x}}$$

$f'(-2) = \frac{1}{2\sqrt{-2}}$ which is undefined so $f'(-2)$

does not exist.

$f'(0) = \frac{1}{2\sqrt{0}} = \frac{1}{0}$ which is undefined so $f'(0)$

does not exist.

$f'(3) = \frac{1}{2\sqrt{3}}$

19. $f(x) = 2x^3 + 5$

Steps 1-3 are combined.

$$\frac{f(x + h) - f(x)}{h}$$

$$= \frac{2(x + h)^3 + 5 - (2x^3 + 5)}{h}$$

$$= \frac{2(x^3 + 3x^2h + 3xh^2 + h^3) + 5 - 2x^3 - 5}{h}$$

$$= \frac{2x^3 + 6x^2h + 6xh^2 + 2h^3 + 5 - 2x^3 - 5}{h}$$

$$= \frac{6x^2h + 6xh^2 + 2h^3}{h}$$

$$= \frac{h(6x^2 + 6xh + 2h^2)}{h}$$

$$= 6x^2 + 6xh + 2h^2$$

$$f'(x) = \lim_{h \to 0}(6x^2 + 6xh + 2h^2) = 6x^2$$

$$f'(-2) = 6(-2)^2 = 24$$

$$f'(0) = 6(0)^2 = 0$$

$$f'(3) = 6(3)^2 = 54$$

21. (a) $f(x) = x^2 + 2x;\ x = 3, x = 5$

Slope of secant line $= \dfrac{f(5) - f(3)}{5 - 3}$

$$= \frac{(5)^2 + 2(5) - [(3)^2 + 2(3)]}{2}$$

$$= \frac{35 - 15}{2}$$

$$= 10$$

Now use $m = 10$ and $(3, f(3)) = (3,15)$ in the point-slope form.

$$y - 15 = 10(x - 3)$$
$$y - 15 = 10x - 30$$
$$y = 10x - 30 + 15$$
$$y = 10x - 15$$

(b) $f(x) = x^2 + 2x;\ x = 3$

$$\frac{f(x + h) - f(x)}{h}$$

$$= \frac{[(x + h)^2 + 2(x + h)] - (x^2 + 2x)}{h}$$

$$= \frac{(x^2 + 2hx + h^2 + 2x + 2h) - (x^2 + 2x)}{h}$$

$$= \frac{2hx + h^2 + 2h}{h} = 2x + h + 2$$

$$f'(x) = \lim_{h \to 0}(2x + h + 2) = 2x + 2$$

$f'(3) = 2(3) + 2 = 8$ is the slope of the tangent line at $x = 3$.

Use $m = 8$ and $(3,15)$ in the point-slope form.

$$y - 15 = 8(x - 3)$$
$$y = 8x - 9$$

23. (a) $f(x) = \dfrac{5}{x};\ x = 2, x = 5$

Slope of secant line $= \dfrac{f(5) - f(2)}{5 - 2}$

$$= \frac{\frac{5}{5} - \frac{5}{2}}{3} = \frac{1 - \frac{5}{2}}{3}$$

$$= -\frac{1}{2}$$

Now use $m = -\dfrac{1}{2}$ and $(5, f(5)) = (5, 1)$ in the point-slope form.

$$y - 1 = -\frac{1}{2}[x - 5]$$

$$y - 1 = -\frac{1}{2}x + \frac{5}{2}$$

$$y = -\frac{1}{2}x + \frac{5}{2} + 1$$

$$y = -\frac{1}{2}x + \frac{7}{2}$$

(b) $f(x) = \dfrac{5}{x}; x = 2$

$$\frac{f(x+h) - f(x)}{h} = \frac{\frac{5}{x+h} - \frac{5}{x}}{h}$$

$$= \frac{\frac{5x - 5(x+h)}{(x+h)x}}{h}$$

$$= \frac{5x - 5x - 5h}{h(x+h)(x)}$$

$$= \frac{-5h}{h(x+h)x}$$

$$= \frac{-5}{(x+h)x}$$

$$f'(x) = \lim_{h \to 0} \frac{-5}{(x+h)(x)} = -\frac{5}{x^2}$$

$f'(2) = \dfrac{-5}{2^2} = -\dfrac{5}{4}$ is the slope of the tangent line at $x = 2$.

Now use $m = -\dfrac{5}{4}$ and $\left(2, \dfrac{5}{2}\right)$ in the point-slope form.

$$y - \frac{5}{2} = -\frac{5}{4}(x - 2)$$

$$y - \frac{5}{2} = -\frac{5}{4}x + \frac{10}{4}$$

$$y = -\frac{5}{4}x + 5$$

25. (a) $f(x) = 4\sqrt{x}; x = 9, x = 16$

Slope of secant line $= \dfrac{f(16) - f(9)}{16 - 9}$

$$= \frac{4\sqrt{16} - 4\sqrt{9}}{7}$$

$$= \frac{16 - 12}{7} = \frac{4}{7}$$

Now use $m = \dfrac{4}{7}$ and $(9, f(9)) = (9, 12)$ in the point-slope form.

$$y - 12 = \frac{4}{7}(x - 9)$$

$$y - 12 = \frac{4}{7}x - \frac{36}{7}$$

$$y = \frac{4}{7}x - \frac{36}{7} + 12$$

$$y = \frac{4}{7}x + \frac{48}{7}$$

(b) $f(x) = 4\sqrt{x}; x = 9$

$$\frac{f(x+h) - f(x)}{h}$$

$$= \frac{4\sqrt{x+h} - 4\sqrt{x}}{h} \cdot \frac{4\sqrt{x+h} + 4\sqrt{x}}{4\sqrt{x+h} + 4\sqrt{x}}$$

$$= \frac{16(x+h) - 16x}{h(4\sqrt{x+h} + 4\sqrt{x})}$$

$$f'(x) = \lim_{h \to 0} \frac{16(x+h) - 16x}{h(4\sqrt{x+h} + 4\sqrt{x})}$$

$$= \lim_{h \to 0} \frac{16h}{h(4\sqrt{x+h} + 4\sqrt{x})}$$

$$= \lim_{h \to 0} \frac{4}{(\sqrt{x+h} + \sqrt{x})} = \frac{4}{2\sqrt{x}}$$

$$= \frac{2}{\sqrt{x}}$$

$f'(9) = \dfrac{2}{\sqrt{9}} = \dfrac{2}{3}$ is the slope of the tangent line at $x = 9$.

Use $m = \dfrac{2}{3}$ and $(9, 12)$ in the point-slope form.

$$y - 12 = \frac{2}{3}(x - 9)$$

$$y = \frac{2}{3}x + 6$$

27. $f(x) = -4x^2 + 11x$

$$\frac{f(x+h) - f(x)}{h}$$

$$= \frac{-4(x+h)^2 + 11(x+h) - (-4x^2 + 11x)}{h}$$

$$= \frac{-8xh - 4h^2 + 11h}{h}$$

$$f'(x) = \lim_{h \to 0} (-8x - 4h + 11) = -8x + 11$$

$$f'(2) = -8(2) + 11 = -5$$

$$f'(16) = -8(16) + 11 = -117$$

$$f'(-3) = -8(-3) + 11 = 35$$

29. $f(x) = e^x$

$$\frac{f(x+h) - f(x)}{h} = \frac{e^{x+h} - e^x}{h}$$

$$f'(x) = \lim_{h \to 0} \frac{e^{x+h} - e^x}{h}$$

$f'(2) \approx 7.3891; f'(16) \approx 8{,}886{,}111; f'(-3) \approx 0.0498$

31. $f(x) = -\dfrac{2}{x}$

$$\dfrac{f(x+h) - f(x)}{h} = \dfrac{\frac{-2}{x+h} - \left(\frac{-2}{x}\right)}{h}$$

$$= \dfrac{\frac{-2x + 2(x+h)}{(x+h)x}}{h}$$

$$= \dfrac{2h}{h(x+h)x} = \dfrac{2}{(x+h)x}$$

$$f'(x) = \lim_{h \to 0} \dfrac{2}{(x+h)x} = \dfrac{2}{x^2}$$

$$f'(2) = \dfrac{2}{2^2} = \dfrac{1}{2}$$

$$f'(16) = \dfrac{2}{16^2} = \dfrac{2}{256} = \dfrac{1}{128}.$$

$$f'(-3) = \dfrac{2}{(-3)^2} = \dfrac{2}{9}$$

33. $f(x) = \sqrt{x}$

$$\dfrac{f(x+h) - f(x)}{h}$$

$$= \dfrac{\sqrt{x+h} - \sqrt{x}}{h} \cdot \dfrac{\sqrt{x+h} + \sqrt{x}}{\sqrt{x+h} + \sqrt{x}}$$

$$= \dfrac{(x+h) - x}{h(\sqrt{x+h} + \sqrt{x})}$$

$$= \dfrac{h}{h(\sqrt{x+h} + \sqrt{x})} = \dfrac{1}{\sqrt{x+h} + \sqrt{x}}$$

$$f'(x) = \lim_{h \to 0} \dfrac{1}{\sqrt{x+h} + \sqrt{x}} = \dfrac{1}{2\sqrt{x}}$$

$$f'(2) = \dfrac{1}{2\sqrt{2}}$$

$$f'(16) = \dfrac{1}{2\sqrt{16}} = \dfrac{1}{8}$$

$$f'(-3) = \dfrac{1}{2\sqrt{-3}} \text{ is not a real number, so}$$

$f'(-3)$ does not exist.

35. At $x = 0$, the graph of $f(x)$ has a sharp point. Therefore, there is no derivative for $x = 0$.

37. For $x = -3$ and $x = 0$, the tangent to the graph of $f(x)$ is vertical. For $x = -1$, there is a gap in the graph of $f(x)$. For $x = 2$, the function $f(x)$ does not exist. For $x = 3$ and $x = 5$, the graph of $f(x)$ has sharp points. Therefore, no derivative exists for $x = -3$, $x = -1$, $x = 0$, $x = 2$, $x = 3$, and $x = 5$.

39. (a) The rate of change of $f(x)$ is positive when $f(x)$ is increasing, that is, on $(a, 0)$ and (b, c).

  (b) The rate of change of $f(x)$ is negative when $f(x)$ is decreasing, that is, on $(0, b,)$.

  (c) The rate of change is zero when the tangent to the graph is horizontal, that is, at $x = 0$ and $x = b$.

41. The zeros of graph (b) correspond to the turning points of graph (a), the points where the derivative is zero. Graph (a) gives the distance, while graph (b) gives the velocity.

43. $f(x) = x^x$, $a = 3$

(a)

h	
0.01	$\dfrac{f(3 + 0.01) - f(3)}{0.01}$ $= \dfrac{3.01^{3.01} - 3^3}{0.01}$ $= 57.3072$
0.001	$\dfrac{f(3 + 0.001) - f(3)}{0.001}$ $= \dfrac{3.001^{3.001} - 3^3}{0.001}$ $= 56.7265$
0.00001	$\dfrac{f(3 + 0.00001) - f(3)}{0.00001}$ $= \dfrac{3.00001^{3.00001} - 3^3}{0.00001}$ $= 56.6632$
0.000001	$\dfrac{f(3 + 0.000001) - f(3)}{0.000001}$ $= \dfrac{3.000001^{3.000001} - 3^3}{0.000001}$ $= 56.6626$
0.0000001	$\dfrac{f(3 + 0.0000001) - f(3)}{0.0000001}$ $= \dfrac{3.0000001^{3.0000001} - 3^3}{0.0000001}$ $= 56.6625$

It appears that $f'(3) \approx 56.66$.

(b) Graph the function on a graphing calculator and move the cursor to an x-value near $x = 3$. A good choice for the initial viewing window is $[0, 4]$ by $[0, 60]$ with Xscl $= 1$, Yscl $= 10$.

Now zoom in on the function several times. Each time you zoom in, the graph will look less like a curve and more like a straight line. Use the TRACE feature to select two points on the graph, and record their coordinates. Use these two points to compute the slope. The result will be close to the most accurate value found in part (a), which is 56.66.

Note: In this exercise, the method used in part (a) gives more accurate results than the method used in part (b).

45. $f(x) = x^{1/x}$, $a = 3$

(a)

h	
0.01	$\dfrac{f(3 + 0.01) - f(3)}{0.01}$
	$= \dfrac{3.01^{1/3.01} - 3^{1/3}}{0.01}$
	$= -0.0160$
0.001	$\dfrac{f(3 + 0.001) - f(3)}{0.001}$
	$= \dfrac{3.001^{1/3.001} - 3^{1/3}}{0.001}$
	$= -0.0158$
0.0001	$\dfrac{f(3 + 0.0001) - f(3)}{0.0001}$
	$= \dfrac{3.0001^{1/3.0001} - 3^{1/3}}{0.0001}$
	$= -0.0158$

It appears that $f'(3) = -0.0158$.

(b) Graph the function on a graphing calculator and move the cursor to an x-value near $x = 3$. A good choice for the initial viewing window is $[0, 5]$ by $[0, 3]$.

Follow the procedure outlined in the solution for Exercise 43, part (b). Note that near $x = 3$, the graph is very close to a horizontal line, so we expect that it slope will be close to 0. The final result will be close to the value found in part (a) of this exercise, which is -0.0158.

49. $D(p) = -2p^2 - 4p + 300$

D is demand; p is price.

(a) Given that $D'(p) = -4p - 4$, the rate of change of demand with respect to price is $-4p - 4$, the derivative of the function $D(p)$.

(b) $D'(10) = -4(10) - 4$

$\quad\quad\quad = -44$

The demand is decreasing at the rate of about 44 items for each increase in price of $1.

51. $R(x) = 20x - \dfrac{x^2}{500}$

(a) $R'(x) = 20 - \dfrac{1}{250}x$

At $y = 1000$,

$$R'(1000) = 20 - \frac{1}{250}(1000)$$

$$= \$16 \text{ per table}.$$

(b) The marginal revenue for the l00lst table is approximately $R'(1000)$. From (a), this is about $16.

(c) The actual revenue is

$$R(1001) - R(1000) = 20(1001) - \frac{1001^2}{500}$$

$$-\left[20(1000) - \frac{1000^2}{500}\right]$$

$$= 18,015.998 - 18,000$$

$$= \$15.998 \text{ or } \$16.$$

(d) The marginal revenue gives a good approximation of the actual revenue from the sale of the l00lst table.

53. (a) $f(x) = 0.0000329x^3 - 0.00450x^2$

$\quad\quad\quad\quad + 0.0613x + 2.34$

$f(10) = 2.54$

$f(20) = 2.03$

$f(30) = 1.02$

(b)

$$Y_1 = 0.0000329x^3 - 0.00450x^2 + 0.0613x + 2.34$$

$$nDeriv(Y_1, x, 0) \approx 0.061$$

$$nDeriv(Y_1, x, 10) \approx -0.019$$

$$nDeriv(Y_1, x, 20) \approx -0.079$$

$$nDeriv(Y_1, x, 30) \approx -0.120$$

$$nDeriv(Y_1, x, 35) \approx -0.133$$

55. The derivative at $(2, 4000)$ can be approximated by the slope of the line through $(0, 2000)$ and $(2, 4000)$. The derivative is approximately

$$\frac{4000 - 2000}{2 - 0} = \frac{2000}{2} = 1000.$$

Thus the shellfish population is increasing at a rate of 1000 shellfish per unit time.

The derivative at about (10,10,300) can be approximated by the slope of the line through (10,10,300) and (13,12,000). The derivative is approximately

$$\frac{12,000 - 10,300}{13 - 10} = \frac{1700}{3} \approx 570.$$

The shellfish population is increasing at a rate of about 570 shellfish per unit time. The derivative at about (13, 11,250) can be approximated by the slope of the line through (13, 11,250) and (16, 12,000). The derivative is approximately

$$\frac{12,000 - 11,000}{16 - 11} = \frac{1000}{5} \approx 200.$$

The shellfish population is increasing at a rate of 200 shellfish per unit time.

57. (a) Set $M(v) = 150$ and solve for v.

$$0.0312443v^2 - 101.39v + 82,264 = 150$$
$$0.0312443v^2 - 101.39v + 82,114 = 0$$

Solve using the quadratic formula.
Let D equal the discriminant.

$$\begin{aligned} D &= b^2 - 4ac \\ &= (-101.39)^2 - 4(0.0312443)(82,114) \\ &\approx 17.55 \end{aligned}$$

$$v = \frac{101.39 \pm \sqrt{D}}{2(0.0312443)}$$

$v \approx 1690$ meter per second or

$v \approx 1560$ meters per second.

Since the functions is defined only for $v \geq 1620$, the only solution is 1690 meters per second.

(b) Calculate $\displaystyle\lim_{h \to 0} \frac{M(1700 + h) - M(1700)}{h}$

$$\begin{aligned} M(1700 + h) \\ &= 0.0312443(1700 + h)^2 - 101.39(1700 + h) + 82,264 \\ &= 90,296.027 + 106.23062h + 0.0312443h^2 - 172,363 - 101.39h + 82,264 \\ &= 0.01312443h^2 + 4.84062h + 197.027 \end{aligned}$$

$M(1700) = 197.027$, so the derivative of $M(v)$ at $v = 1700$ is

$$\lim_{h \to 0} \left[\frac{0.0312443h^2 + 4.84062h + 197.027 - 197.027}{h} \right]$$

$$= \lim_{h \to 0} \left[\frac{0.0312443h^2 + 4.84062h}{h} \right]$$

$$= \lim_{h \to 0} \left(0.0312443h + 4.84062 \right)$$

$$= 4.84062$$

$$\approx 4.84 \text{ days per meter per second}$$

The increase in velocity for this cheese from 1700 m/s to 1701 m/s indicates that the approximate age of the cheese has increased by 4.84 days.

59. (a) From the graph, $T(0.5) \approx 185$. A tangent drawn at this point appears to intersect the $T = 1$ vertical line at about 320, so the tangent has a slope of about

$$\frac{T(1) - T(0.5)}{1 - 0.5} = \frac{320 - 185}{0.5} = 270.$$

$T'(0.5) \approx 270$; at 9:00 AM the temperature is increasing at about $270°$ per hour.

(b) A tangent drawn to the curve at $T = 3$ appears to intersect the $T = 2$ line at 480 and the $T = 4$ line at 180, so the tangent has a slope of about

$$\frac{T(4) - T(2)}{4 - 2} = \frac{180 - 480}{2} = -150.$$

$T'(3) \approx -150$; at 11:30 AM the temperature is decreasing at about $150°$ per hour.

(c) At $T = 4$ the graph appears to have a horizontal tangent, so $T'(4) \approx 0$; the temperature is staying constant at 12:30 PM.

(d) At about 11:15 AM.

61. (a) At 40 oz the tangent looks horizontal; thus the derivative for a 40-ounce bat is about 0 mph per oz.
The slope of the graph at $x = 30$ can be estimated using the points $(30, 79)$ and $(32, 80)$.

$$\text{slope} = \frac{80 - 79}{32 - 30} = 0.5$$

Thus, the derivative for a 30 ounce bat is about 0.5 mph per oz .

(b) The optimal bat is 40 oz.

3.5 Graphical Differentiation

Your Turn 1

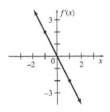

Your Turn 2

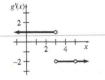

W1. The tangent at $(2, 13)$ goes through the points $(0, 9)$ and $(4, 17)$, so the tangent has slope

$$\frac{17 - 9}{4 - 0} = 2.$$

W2. The tangent at $(3, 4)$ goes through the point $(1, 14)$, so the tangent has slope

$$\frac{4 - 14}{3 - 1} = -5.$$

3.5 Exercises

3. Since the x-intercepts of the graph of f' occur whenever the graph of f has a horizontal tangent line, Y_1 is the derivative of Y_2. Notice that Y_1 has 2 x-intercepts; each occurs at an x-value where the tangent line to Y_2 is horizontal.

Note also that Y_1 is positive whenever Y_2 is increasing, and that Y_1 is negative whenever Y_2 is decreasing.

5. Since the x-intercepts of the graph of f' occur whenever the graph of f has a horizontal tangent line, Y_2 is the derivative of Y_1. Notice that Y_2 has 1 x-intercept which occurs at the x-value where the tangent line to Y_1 is horizontal. Also notice that the range on which Y_1 is increasing, Y_2 is positive and the range on which it is decreasing, Y_2 is negative.

6. Since the x-intercepts of the graph f' occur whenever the graph of f has a horizontal tangent line, Y_1 is the derivative of Y_2. Notice that Y_1 has 4 x-intercepts; each occurs at an x-value where the tangent line to Y_2 is horizontal.

Note also that Y_1 is negative whenever Y_2 is decreasing and Y_1 is positive whenever Y_2 is increasing.

7. To graph f', observe the intervals where the slopes of tangent lines are positive and where they are negative to determine where the derivative is positive and where it is negative. Also, whenever f has a horizontal tangent, f' will be 0, so the graph of f' will have an x-intercept. The x-values of the three turning point on the graph of f become the three x-intercepts of the graph of f.

Estimate the magnitude of the slope at several points by drawing tangents to the graph of f.

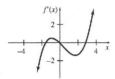

9. On the interval $(-\infty, -2)$, the graph of f is a horizontal line, so its slope is 0. Thus, on this interval, the graph of f' is $y = 0$ on $(-\infty, -2)$. On the interval $(-2, 0)$, the graph of f is a straight line, so its slope is constant. To find this slope, use the points $(-2, 2)$ and $(0, 0)$.

$$m = \frac{2 - 0}{-2 - 0} = \frac{2}{-2} = -1$$

On the interval $(0, 1)$, the slope is also constant. To find this slope, use the points $(0, 0)$ and $(1, 1)$.

$$m = \frac{1 - 0}{1 - 0} = 1$$

On the interval $(1, \infty)$, the graph is again a horizontal line, so $m = 0$. The graph of f' will be made up of portions of the y-axis and the lines $y = -1$ and $y = 1$.

Because the graph of f has "sharp points" or "corners" at $x = -2$, $x = 0$, and $x = 1$, we know that $f'(-2)$, $f'(0)$, and $f'(1)$ do not exist. We show this on the graph of f' by using open circles at the endpoints of the portions of the graph.

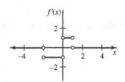

11. On the interval $(-\infty, -2)$, the graph of f is a straight line, so its slope is constant. To find this slope, use the points $(-4, 2)$ and $(-2, 0)$.

$$m = \frac{0 - 2}{-2 - (-4)} = \frac{-2}{2} = -1$$

On the interval $(2, \infty)$, the slope of f is also constant. To find this slope, use the points $(2, 0)$ and $(3, 2)$.

$$m = \frac{2 - 0}{3 - 2} = \frac{2}{1} = 2$$

Thus, we have $f'(x) = -1$ on $(-\infty, -2)$ and $f'(x) = 2$ on $(2, \infty)$.

Because f is discontinuous at $x = -2$ and $x = 2$, we know that $f'(-2)$ and $f'(2)$ do not exist, which we indicate with open circles at $(-2, -1)$ and $(2, 2)$ on the graph of f'.

On the interval $(-2, 2)$ all tangent lines have positive slopes, so the graph of f' will be above the y-axis. Notice that the slope of f (and thus the y-value of f') decreases on $(-2, 0)$ and increases on $(0, 2)$ with a minimum value on this interval of about 1 at $x = 0$.

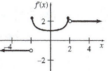

13. We observe that the slopes of tangent lines are positive on the interval $(-\infty, 0)$ and negative on the interval $(0, \infty)$, so the value of f' will be positive on $(-\infty, 0)$ and negative on $(0, \infty)$. Since f is undefined at $x = 0$, $f'(0)$ does not exist.

Notice that the graph of f becomes very flat when $|x| \to \infty$. The *value* of f approaches 0 and also the *slope* approaches 0. Thus, $y = 0$ (the x-axis) is a horizontal asymptote for both the graph of f and the graph of f'.

As $x \to 0^-$ and $x \to 0^+$, the graph of f gets very steep, so $|f'(x)| \to \infty$. Thus, $x = 0$ (the y-axis) is a vertical asymptote for both the graph of f and the graph of f'.

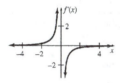

15. The slope of $f(x)$ is undefined at $x = -2, -1, 0, 1,$ and 2, and the graph approaches vertical (unbounded slope) as x approaches those values. Accordingly, the graph of $f'(x)$ has vertical asymptotes at $x = -2, -1, 0, 1,$ and 2. $f(x)$ has turning points (zero slope) at $x = -1.5, -0.5, 0.5,$ and 1.5, so the graph of $f'(x)$ crosses the x-axis at those values. Elsewhere, the graph of $f'(x)$ is negative where $f(x)$ is decreasing and positive where $f(x)$ is increasing.

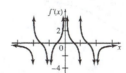

17. The graph of G decreases steadily with varying degrees of steepness. The steepness increases (that is, the slopes of the tangent lines becomes more negative) between $t = 12$ and $t = 16$. Since G is discontinuous at $t = 16$, $G'(16)$ doesn't exist.

The graph continues to decrease after $t = 16$, but the slopes of the tangent lines become less negative as the curve gets flatter. So, the derivative values are increasing toward 0.

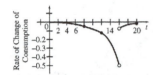

19. The growth rate of the function $y = f(t)$ is given by the derivative of this function $y' = f(t)$. We use the graph of f to sketch the graph of f'. First, notice as x increases, y increases throughout the domain of f, but at a slower and slower rate.
The slope of f is positive but always decreasing, and approaches 0 as t gets large. Thus, y' will always be positive and decreasing. It will approach but never reach 0.

To plot point on the graph of f', we need to estimate the slope of f at several points. From the graph of f, we obtain the values given in the following table.

t	y'
2	1000
10	700
13	250

Use these points to sketch the graph.

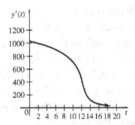

21.

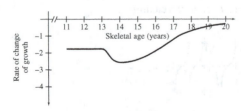

About 9 cm; about 2.6 cm less per year

23.

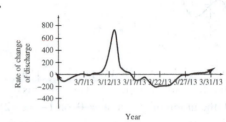

Chapter 3   Review Exercises

1. True

2. True

3. True

4. False; for example, if $f(x) = \frac{x^2-4}{x+2}$,

$\lim\limits_{x \to -2} f(x) = -4$, but the graph of $f(x) = \frac{x^2-4}{x+2}$

has a hole at the point $(-2, -4)$.

5. True

6. False; for example, the rational function $f(x) = \frac{5}{x+1}$ is discontinuous at $x = -1$.

7. False; the derivative gives the instantaneous rate of change of a function.

8. True

9. True

10. True

11. False; the slope of the tangent line gives the instantaneous rate of change.

12. False; for example, the function $f(x) = |x|$ is continuous at $x = 0$, but $f'(0)$ does not exist. The graph of $f(x) = |x|$ has a "corner" at $x = 0$.

17. (a) $\lim\limits_{x \to -3^-} = 4$

(b) $\lim\limits_{x \to -3^+} = 4$

(c) $\lim\limits_{x \to -3} = 4$ (since parts (a) and (b) have the same answer)

(d) $f(-3) = 4$, since $(-3, 4)$ is a point of the graph.

19. (a) $\lim\limits_{x \to 4^-} f(x) = \infty$

(b) $\lim\limits_{x \to 4^+} f(x) = -\infty$

(c) $\lim\limits_{x \to 4} f(x)$ does not exist since limits in (a) and (b) do not exist.

(d) $f(4)$ does not exist since the graph has no point with an x-value of 4.

21. $\lim\limits_{x \to -\infty} g(x) = \infty$ since the y-value gets very large as the x-value gets very small.

23. $\lim\limits_{x \to 6} \dfrac{2x + 7}{x + 3} = \dfrac{2(6) + 7}{6 + 3} = \dfrac{19}{9}$

25. $\lim\limits_{x \to 4} \dfrac{x^2 - 16}{x - 4} = \lim\limits_{x \to 4} \dfrac{(x - 4)(x + 4)}{x - 4}$

$\qquad = \lim\limits_{x \to 4} (x + 4)$

$\qquad = 4 + 4$

$\qquad = 8$

27. $\lim\limits_{x \to -4} \dfrac{2x^2 + 3x - 20}{x + 4} = \lim\limits_{x \to -4} \dfrac{(2x - 5)(x + 4)}{x + 4}$

$\qquad = \lim\limits_{x \to -4} (2x - 5)$

$\qquad = 2(-4) - 5$

$\qquad = -13$

29. $\lim\limits_{x \to 9} \dfrac{\sqrt{x} - 3}{x - 9} = \lim\limits_{x \to 9} \dfrac{\sqrt{x} - 3}{x - 9} \cdot \dfrac{\sqrt{x} + 3}{\sqrt{x} + 3}$

$\qquad = \lim\limits_{x \to 9} \dfrac{x - 9}{(x - 9)(\sqrt{x} + 3)}$

$\qquad = \lim\limits_{x \to 9} \dfrac{1}{\sqrt{x} + 3}$

$\qquad = \dfrac{1}{\sqrt{9} + 3}$

$\qquad = \dfrac{1}{6}$

31. $\lim\limits_{x \to \infty} \dfrac{2x^2 + 5}{5x^2 - 1} = \lim\limits_{x \to \infty} \dfrac{\dfrac{2x^2}{x^2} + \dfrac{5}{x^2}}{\dfrac{5x^2}{x^2} - \dfrac{1}{x^2}}$

$\qquad = \lim\limits_{x \to \infty} \dfrac{2 + \dfrac{5}{x^2}}{5 - \dfrac{1}{x^2}}$

$\qquad = \dfrac{2 + 0}{5 - 0}$

$\qquad = \dfrac{2}{5}$

33. $\lim\limits_{x \to -\infty} \left(\dfrac{3}{8} + \dfrac{3}{x} - \dfrac{6}{x^2} \right)$

$\qquad = \lim\limits_{x \to -\infty} \dfrac{3}{8} + \lim\limits_{x \to -\infty} \dfrac{3}{x} - \lim\limits_{x \to -\infty} \dfrac{6}{x^2}$

$\qquad = \dfrac{3}{8} + 0 - 0$

$\qquad = \dfrac{3}{8}$

35. As shown on the graph, $f(x)$ is discontinuous at x_2 and x_4.

37. $f(x)$ is discontinuous at $x = 0$ and $x = -\frac{1}{3}$ since that is where the denominator of $f(x)$ equals 0. $f(0)$ and $f\left(-\frac{1}{3}\right)$ do not exist.

$\lim\limits_{x \to 0} f(x)$ does not exist since $\lim\limits_{x \to 0^+} f(x) = -\infty$, but $\lim\limits_{x \to 0^-} f(x) = \infty$. $\lim\limits_{x \to -\frac{1}{3}} f(x)$ does not exist since $\lim\limits_{x \to -\frac{1}{3}^-} = -\infty$, but $\lim\limits_{x \to -\frac{1}{3}^+} f(x) = \infty$.

39. $f(x)$ is discontinuous at $x = -5$ since that is where the denominator of $f(x)$ equals 0. $f(-5)$ does not exist.

$\lim\limits_{x \to -5} f(x)$ does not exist since $\lim\limits_{x \to -5^-} f(x) = \infty$, but $\lim\limits_{x \to -5^+} f(x) = -\infty$.

41. $f(x) = x^2 + 3x - 4$ is continuous everywhere since f is a polynomial function.

43. (a)

(b) The graph is discontinuous at $x = 1$.

(c) $\lim\limits_{x \to 1^-} f(x) = 0$; $\lim\limits_{x \to 1^+} f(x) = 2$

45. $f(x) = \dfrac{x^4 + 2x^3 + 2x^2 - 10x + 5}{x^2 - 1}$

(a) Find the values of $f(x)$ when x is close to 1.

x	y
1.1	2.6005
1.01	2.06
1.001	2.006
1.0001	2.0006
0.99	1.94
0.999	1.994
0.9999	1.9994

It appears that $\lim\limits_{x \to 1} f(x) = 2$.

(b) Graph

$$y = \frac{x^4 + 2x^3 + 2x^2 - 10x + 5}{x^2 - 1}$$

on a graphing calculator. One suitable choice for the viewing window is $[-2, 6]$ by $[-10, 10]$. Because $x^2 - 1 = 0$ when $x = -1$ or $x = 1$, this function is discontinuous at these two x-values. The graph shows a vertical asymptote at $x = -1$ but not at $x = 1$. The graph should have an open circle to show a "hole" in the graph at $x = 1$. The graphing calculator doesn't show the hole, but trying to find the value of the function of $x = 1$ will show that this value is undefined.

By viewing the function near $x = 1$ and using the ZOOM feature, we see that as x gets close to 1 from the left or the right, y gets close to 2, suggesting that

$$\lim\limits_{x \to 1} \frac{x^4 + 2x^3 + 2x^2 - 10x + 5}{x^2 - 1} = 2.$$

47. $y = 6x^3 + 2 = f(x)$; from $x = 1$ to $x = 4$

$$f(4) = 6(4)^3 + 2 = 386$$
$$f(1) = 6(1)^3 + 2 = 8$$

Average rate of change:

$$= \frac{386 - 8}{4 - 1} = \frac{378}{3} = 126$$

$$y' = 18x$$

Instantaneous rate of change at $x = 1$:

$$f'(1) = 18(1) = 18$$

49. $y = \dfrac{-6}{3x - 5} = f(x)$; from $x = 4$ to $x = 9$

$$f(9) = \frac{-6}{3(9) - 5} = \frac{-6}{22} = -\frac{3}{11}$$

$$f(4) = \frac{-6}{3(4) - 5} = -\frac{6}{7}$$

Average rate of change:

$$= \frac{\frac{-3}{11} - \left(-\frac{6}{7}\right)}{9 - 4} = \frac{\frac{-21 + 66}{77}}{5} = \frac{45}{5(77)} = \frac{9}{77}$$

$$y' = \frac{(3x - 5)(0) - (-6)(3)}{(3x - 5)^2} = \frac{18}{(3x - 5)^2}$$

Instantaneous rate of change at $x = 4$:

$$f'(4) = \frac{18}{(3 \cdot 4 - 5)^2} = \frac{18}{7^2} = \frac{18}{49}$$

51. (a) $f(x) = 3x^2 - 5x + 7$; $x = 2$, $x = 4$

Slope of secant line

$$= \frac{f(4) - f(2)}{4 - 2}$$

$$= \frac{[3(4)^2 - 5(4) + 7] - [3(2)^2 - 5(2) + 7]}{2}$$

$$= \frac{35 - 9}{2}$$

$$= 13$$

Now use $m = 13$ and $2, f(2) = (2, 9)$ in the point-slope form.

$$y - 9 = 13(x - 2)$$
$$y - 9 = 13x - 26$$
$$y = 13x - 26 + 9$$
$$y = 13x - 17$$

(b) $f(x) = 3x^2 - 5x + 7;\ x = 2$

$$\frac{f(x + h) - f(x)}{h}$$

$$= \frac{[3(x + h)^2 - 5(x + h) + 7] - [3x^2 - 5x + 7]}{h}$$

$$= \frac{3x^2 + 6xh + 3h^2 - 5x - 5h + 7 - 3x^2 + 5x - 7}{h}$$

$$= \frac{6xh + 3h^2 - 5h}{h}$$

$$= 6x + 3h - 5$$

$$f'(x) = \lim_{h \to 0} 6x + 3h - 5$$

$$= 6x - 5$$

$$f'(2) = 6(2) - 5$$

$$= 7$$

Now use $m = 7$ and $(2, f(2)) = (2, 9)$ in the point-slope form.

$$y - 9 = 7(x - 2)$$
$$y - 9 = 7x - 14$$
$$y = 7x - 14 + 9$$
$$y = 7x - 5$$

53. (a) $f(x) = \dfrac{12}{x - 1};\ x = 3,\ x = 7$

Slope of secant line $= \dfrac{f(7) - f(3)}{7 - 3}$

$$= \frac{\frac{12}{7-1} - \frac{12}{3-1}}{4}$$

$$= \frac{2 - 6}{4}$$

$$= -1$$

Now use $m = -1$ and $(3, f(x)) = (3, 6)$ in the point-slope form.

$$y - 6 = -1(x - 3)$$
$$y - 6 = -x + 3$$
$$y = -x + 3 + 6$$
$$y = -x + 9$$

(b) $f(x) = \dfrac{12}{x - 1};\ x = 3$

$$\frac{f(x + h) - f(x)}{h} = \frac{\frac{12}{x+h-1} - \frac{12}{x-1}}{h}$$

$$= \frac{12(x - 1) - 12(x + h - 1)}{h(x - 1)(x + h - 1)}$$

$$= \frac{-12h}{h(x - 1)(x + h - 1)}$$

$$= -\frac{12}{(x - 1)(x + h - 1)}$$

$$f'(x) = \lim_{h \to 0} -\frac{12}{(x - 1)(x + h - 1)}$$

$$= -\frac{12}{(x - 1)^2}$$

$$f'(3) = -\frac{12}{(3 - 1)^2}$$

$$= -3$$

Now use $m = -3$ and $(3, f(x)) = (3, 6)$ in the point-slope form.

$$y - 6 = -3(x - 3)$$
$$y - 6 = -3x + 9$$
$$y = -3x + 9 + 6$$
$$y = -3x + 15$$

55. $y = 4x^2 + 3x - 2 = f(x)$

$$y' = \lim_{h \to 0} \frac{f(x + h) - f(x)}{h}$$

$$= \lim_{h \to 0} \frac{[4(x + h)^2 + 3(x + h) - 2] - [4x^2 + 3x - 2]}{h}$$

$$= \lim_{h \to 0} \frac{4(x^2 + 2xh + h^2) + 3x + 3h - 2 - 4x^2 - 3x + 2}{h}$$

$$= \lim_{h \to 0} \frac{4x^2 + 8xh + 4h^2 + 3x + 3h - 2 - 4x^2 - 3x + 2}{h}$$

$$= \lim_{h \to 0} \frac{8xh + 4h^2 + 3h}{h}$$

$$= \lim_{h \to 0} \frac{h(8x + 4h + 3)}{h}$$

$$= \lim_{h \to 0} (8x + 4h + 3)$$

$$= 8x + 3$$

57. $f(x) = (\ln x)^x$, $x_0 = 3$

(a)

h	
0.01	$\dfrac{f(3 + 0.01) - f(3)}{0.01}$
	$= \dfrac{(\ln 3.01)^{3.01} - (\ln 3)^3}{0.01} = 1.3385$
0.001	$\dfrac{f(3 + 0.001) - f(3)}{0.001}$
	$= \dfrac{(\ln 3.001)^{3.001} - (\ln 3)^3}{0.001} = 1.3323$
0.0001	$\dfrac{f(3 + 0.0001) - f(3)}{0.0001}$
	$= \dfrac{(\ln 3.0001)^{3.0001} - (\ln 3)^3}{0.0001} = 1.3317$
0.00001	$\dfrac{f(3 + 0.00001) - f(3)}{0.00001}$
	$= \dfrac{(\ln 3.00001)^{3.00001} - (\ln 3)^3}{0.00001} = 1.3317$

It appears that $f'(3) \approx 1.332$.

(b) Using a graphing calculator will confirm this result.

59. On the interval $(-\infty, 0)$, the graph of f is a straight line, so its slope is constant. To find this slope, use the points $(-2, 2)$ and $(0, 0)$.

$$m = \frac{0 - 2}{0 - (-2)} = \frac{-2}{2} = -1$$

Thus, the value of f' will be -1 on this interval.

The graph of f has a sharp point at 0, so $f'(0)$ does not exist. To show this, we use an open circle on the graph of f' at $(0, -1)$.

We also observe that the slope of f is positive but decreasing from $x = 0$ to about $x = 1$, and then negative from there on. As $x \to \infty$, $f(x) \to 0$ and also $f'(x) = 0$.

Use this information to complete the graph of f'.

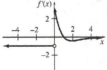

61. $\displaystyle\lim_{x \to \infty} \frac{cf(x) - dg(x)}{f(x) - g(x)}$

$$= \frac{\displaystyle\lim_{x \to \infty}[cf(x) - dg(x)]}{\displaystyle\lim_{x \to \infty}[f(x) - g(x)]}$$

$$= \frac{\displaystyle\lim_{x \to \infty}[cf(x)] - \lim_{x \to \infty}[dg(x)]}{\displaystyle\lim_{x \to \infty}[f(x)] - \lim_{x \to \infty}[g(x)]}$$

$$= \frac{c \displaystyle\lim_{x \to \infty}[f(x)] - d \lim_{x \to \infty}[g(x)]}{\displaystyle\lim_{x \to \infty}[f(x)] - \lim_{x \to \infty}[g(x)]}$$

$$= \frac{c \cdot c - d \cdot d}{c - d} = \frac{(c + d)(c - d)}{c - d}$$

$$= c + d$$

The answer is (e).

63. $C(x) = \begin{cases} 1.50x & \text{for } 0 < x \le 125 \\ 1.35x & \text{for } x > 125 \end{cases}$

(a) $C(100) = 1.50(100) = \$150$

(b) $C(125) = 1.50(125) = \$187.50$

(c) $C(140) = 1.35(140) = \$189$

(d)

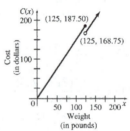

(e) By reading the graph, $C(x)$ is discontinuous at $x = \$125$.

The average cost per pound is given by $\overline{C}(x) = \dfrac{C(x)}{x}$.

$$\overline{C}(x) = \begin{cases} 1.50 & \text{for } 0 < x \le 125 \\ 1.35 & \text{for } x > 125 \end{cases}$$

(f) $\overline{C}(100) = \$1.50$

(g) $\overline{C}(125) = \$1.50$

(h) $\overline{C}(140) = \$1.35$

The marginal cost is given by

$$C(x) = \begin{cases} 1.50 & \text{for } 0 < x \le 125 \\ 1.35 & \text{for } x > 125. \end{cases}$$

(i) $C'(100) = 1.50$; the 101st pound will cost $\$1.50$.

(j) $C'(140) = 1.35$; the 141st pound will cost $\$1.35$.

65. (b) The value of x for which the average cost is smallest is $x = 7.5$. This can be found by drawing a line from the origin to any point of $C(x)$. At $x = 7.5$, you will get a line with the smallest slope.

(c) The marginal cost equals the average cost at the point where the average cost is smallest.

67.

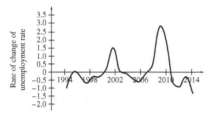

The annual unemployment rate in 2012 was about 8.2%. Our sketch of the rate of change of the unemployment rate indicates that the rate of change in 2012 was approximately -0.5% per year.

69. $V(t) = -t^2 + 6t - 4$

(a)

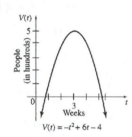

$V(t) = -t^2 + 6t - 4$

(b) The x-intercepts of the parabola are 0.8 and 5.2, so a reasonable domain would be $[0.8, 5.2]$, which represents the time period from 0.8 to 5.2 weeks.

(c) The number of cases reaches a maximum at the vertex;

$$x = \frac{-b}{2a} = \frac{-6}{-2} = 3$$

$$V(3) = -3^2 + 6(3) - 4 = 5$$

The vertex of the parabola is $(3, 5)$. This represents a maximum at 3 weeks of 500 cases.

(d) The rate of change function is

$$V'(t) = -2t + 6.$$

(e) The rate of change in the number of cases at the maximum is

$$V'(3) = -2(3) + 6 = 0.$$

(f) The sign of the rate of change up to the maximum is $+$ because the function is increasing. The sign of the rate of change after the maximum is $-$ because the function is decreasing.

71. (a)

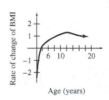

(b)

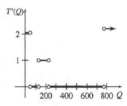

73. (a) The graph is discontinuous nowhere.

(b) The graph is not differentiable where the graph makes a sudden change, namely at $x = 50$, $x = 130$, $x = 230$, and $x = 770$.

(c)

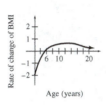

CALCULATING THE DERIVATIVE

4.1 Techniques for Finding Derivatives

Your Turn 1

$$f(t) = \frac{1}{\sqrt{t}} = t^{-1/2}$$

$$f'(t) = -\frac{1}{2}t^{-1/2-1}$$

$$= -\frac{1}{2}t^{-3/2}$$

$$= -\frac{1}{2t^{3/2}} \text{ or } -\frac{1}{2\sqrt{t^3}}$$

Your Turn 2

$$y = 3\sqrt{x} = 3x^{1/2}$$

$$\frac{dy}{dx} = 3 \cdot \frac{1}{2}x^{-1/2}$$

$$= \frac{3}{2}x^{-1/2}$$

$$= \frac{3}{2\sqrt{x}}$$

Your Turn 3

$$h(t) = -3t^2 + 2\sqrt{t} + \frac{5}{t^4} - 7$$

$$= 3t^2 + 2t^{1/2} + 5t^{-4} - 7$$

$$h'(t) = -6t + t^{-1/2} - 20t^{-5}$$

$$= -6t + \frac{1}{\sqrt{t}} - \frac{20}{t^5}$$

Your Turn 4

Find the marginal cost of the cost function
$C(x) = 5x^3 - 10x^2 + 75$ when $x = 100$.

$$C'(x) = 15x^2 - 20x$$

$$C'(100) = 15(100)^2 - 20(100)$$

$$= 150,000 - 2000$$

$$= 148,000$$

The marginal cost when $x = 100$ is $148,000.

Your Turn 5

Find the marginal revenue of the demand function
$p = 16 - 1.25q$ when $q = 5$.

$$R(q) = qp = q(16 - 1.25q) = 16q - 1.25q^2$$

$$R'(q) = 16 - 2.5q$$

$$R'(5) = 16 - 2.5(5) = 3.5$$

The marginal revenue for 5 units is $3.50 per unit.

4.1 Warmup Exercises

W1. A line parallel to $4x + 2y = 9$ will have slope $(-4)/2 = -2$ and equation
$y = -2x + b.$
If the line passes through $(2, 7)$,
$7 = -2(2) + b$
$b = 11$
and the equation is $y = -2x + 11.$

W2. A line parallel to $2x - 5y = 3$ will have slope $2/5$ and equation $y = (2/5)x + b.$
If the line passes through $(-3, 4)$,
$4 = (2/5)(-3) + b$
$b = 4 + 6/5 = 26/5$
and the equation is $y = (2/5)x + 26/5.$

4.1 Exercises

1. $y = 12x^3 - 8x^2 + 7x + 5$

$$\frac{dy}{dx} = 12(3x^{3-1}) - 8(2x^{2-1}) + 7x^{1-1} + 0$$

$$= 36x^2 - 16x + 7$$

3. $y = 3x^4 - 6x^3 + \frac{x^2}{8} + 5$

$$\frac{dy}{dx} = 3(4x^{4-1}) - 6(3x^{3-1}) + \frac{1}{8}(2x^{2-1}) + 0$$

$$= 12x^3 - 18x^2 + \frac{1}{4}x$$

5. $y = 6x^{3.5} - 10x^{0.5}$

$\dfrac{dy}{dx} = 6(3.5x^{3.5-1}) - 10(0.5x^{0.5-1})$

$= 21x^{2.5} - 5x^{-0.5}$ or $21x^{2.5} - \dfrac{5}{x^{0.5}}$

7. $y = 8\sqrt{x} + 6x^{3/4} = 8x^{1/2} + 6x^{3/4}$

$\dfrac{dy}{dx} = 8\left(\dfrac{1}{2}x^{1/2-1}\right) + 6\left(\dfrac{3}{4}x^{3/4-4}\right)$

$= 4x^{-1/2} + \dfrac{9}{2}x^{-1/4}$ or $\dfrac{4}{x^{1/2}} + \dfrac{9}{2x^{1/4}}$

9. $y = 10x^{-3} + 5x^{-4} - 8x$

$\dfrac{dy}{dx} = 10(-3x^{-3-1}) + 5(-4x^{-4-1}) - 8x^{1-1}$

$= -30x^{-4} - 20x^{-5} - 8$ or $\dfrac{-30}{x^4} - \dfrac{20}{x^5} - 8$

11. $f(t) = \dfrac{7}{t} - \dfrac{5}{t^3}$

$= 7t^{-1} - 5t^{-3}$

$f'(t) = 7(-1t^{-1-1}) - 5(-3t^{-3-1})$

$= -7t^{-2} + 15t^{-4}$ or $\dfrac{-7}{t^2} + \dfrac{15}{t^4}$

13. $y = \dfrac{6}{x^4} - \dfrac{7}{x^3} + \dfrac{3}{x} + \sqrt{5}$

$= 6x^{-4} - 7x^{-3} + 3x^{-1} + \sqrt{5}$

$\dfrac{dy}{dx} = 6(-4x^{-4-1}) - 7(-3x^{-3-1})$

$+ 3(-1x^{-1-1}) + 0$

$= -24x^{-5} + 21x^{-4} - 3x^{-2}$

or $\dfrac{-24}{x^5} + \dfrac{21}{x^4} - \dfrac{3}{x^2}$

15. $p(x) = -10x^{-1/2} + 8x^{-3/2}$

$p'(x) = -10\left(-\dfrac{1}{2}x^{-3/2}\right) + 8\left(-\dfrac{3}{2}x^{-5/2}\right)$

$= 5x^{-3/2} - 12x^{-5/2}$ or $\dfrac{5}{x^{3/2}} - \dfrac{12}{x^{5/2}}$

17. $y = \dfrac{6}{4\sqrt{x}} = 6x^{-1/4}$

$\dfrac{dy}{dx} = 6\left(-\dfrac{1}{4}\right)x^{-5/4}$

$= -\dfrac{3}{2}x^{-5/4}$ or $\dfrac{-3}{2x^{5/4}}$

19. $f(x) = \dfrac{x^3 + 5}{x} = x^2 + 5x^{-1}$

$f'(x) = 2x^{2-1} + 5(-1x^{-1-1})$

$= 2x - 5x^{-2}$ or $2x - \dfrac{5}{x^2}$

21. $g(x) = (8x^2 - 4x)^2$

$= 64x^4 - 64x^3 + 16x^2$

$g'(x) = 64(4x^{4-1}) - 64(3x^{3-1}) + 16(2x^{2-1})$

$= 256x^3 - 192x^2 + 32x$

23. A quadratic function has degree 2. When the derivative is taken, the power will decrease by 1 and the derivative function will be linear, so the correct choice is (b).

27. $D_x\left[9x^{-1/2} + \dfrac{2}{x^{3/2}}\right]$

$= D_x[9x^{-1/2} + 2x^{-3/2}]$

$= 9\left(-\dfrac{1}{2}x^{-3/2}\right) + 2\left(-\dfrac{3}{2}x^{-5/2}\right)$

$= -\dfrac{9}{2}x^{-3/2} - 3x^{-5/2}$ or $\dfrac{-9}{2x^{3/2}} - \dfrac{3}{x^{5/2}}$

29. $f(x) = \dfrac{x^4}{6} - 3x$

$= \dfrac{1}{6}x^4 - 3x$

$f'(x) = \dfrac{1}{6}(4x^3) - 3$

$= \dfrac{2}{3}x^3 - 3$

$f'(-2) = \dfrac{2}{3}(-2)^3 - 3$

$= -\dfrac{16}{3} - 3$

$= -\dfrac{25}{3}$

31. $y = x^4 - 5x^3 + 2; \ x = 2$

$y' = 4x^3 - 15x^2$

$y'(2) = 4(2)^3 - 15(2)^2$

$\qquad = -28$

The slope of tangent line at $x = 2$ is -28.

Use $m = -28$ and $(x_1, y_1) = (2, -22)$ to obtain the equation.

$$y - (-22) = -28(x - 2)$$
$$y = -28x + 34$$

33. $y = -2x^{1/2} + x^{3/2}$

$y' = -2\left(\dfrac{1}{2}x^{-1/2}\right) + \dfrac{3}{2}x^{1/2}$

$\quad = -x^{-1/2} + \dfrac{3}{2}x^{1/2}$

$\quad = -\dfrac{1}{x^{1/2}} + \dfrac{3x^{1/2}}{2}$

$y'(9) = -\dfrac{1}{(9)^{1/2}} + \dfrac{3(9)^{1/2}}{2}$

$\qquad = -\dfrac{1}{3} + \dfrac{9}{2}$

$\qquad = \dfrac{25}{6}$

The slope of the tangent line at $x = 9$ is $\frac{25}{6}$.

35. $f(x) = 9x^2 - 8x + 4$

$f'(x) = 18x - 8$

Let $f'(x) = 0$ to find the point where the slope of the tangent line is zero.

$$18x - 8 = 0$$
$$18x = 8$$
$$x = \dfrac{8}{18} = \dfrac{4}{9}$$

Find the y-coordinate.

$$f(x) = 9x^2 - 8x + 4$$

$f\left(\dfrac{4}{9}\right) = 9\left(\dfrac{4}{9}\right)^2 - 8\left(\dfrac{4}{9}\right) + 4$

$\qquad = 9\left(\dfrac{16}{81}\right) - \dfrac{32}{9} + 4$

$\qquad = \dfrac{16}{9} - \dfrac{32}{9} + \dfrac{36}{9} = \dfrac{20}{9}$

The slope of the tangent line is zero at one point, $\left(\dfrac{4}{9}, \dfrac{20}{9}\right)$.

37. $f(x) = 2x^3 + 9x^2 - 60x + 4$

$f'(x) = 6x^2 + 18x - 60$

If the tangent line is horizontal, then its slope is zero and $f'(x) = 0$.

$$6x^2 + 18x - 60 = 0$$
$$6(x^2 + 3x - 10) = 0$$
$$6(x + 5)(x - 2) = 0$$
$$x = -5 \ \text{or} \ x = 2$$

Thus, the tangent line is horizontal at $x = -5$ and $x = 2$.

39. $f(x) = x^3 - 4x^2 - 7x + 8$

$f'(x) = 3x^2 - 8x - 7$

If the tangent line is horizontal, then its slope is zero and $f'(x) = 0$.

$$3x^2 - 8x - 7 = 0$$
$$x = \dfrac{8 \pm \sqrt{64 + 84}}{6}$$
$$x = \dfrac{8 \pm \sqrt{148}}{6}$$
$$x = \dfrac{8 \pm 2\sqrt{37}}{6}$$
$$x = \dfrac{2(4 \pm \sqrt{37})}{6}$$
$$x = \dfrac{4 \pm \sqrt{37}}{3}$$

Thus, the tangent line is horizontal at $x = \dfrac{4 \pm \sqrt{37}}{3}$.

41. $f(x) = 6x^2 + 4x - 9$

$f'(x) = 12x + 4$

If the slope of the tangent line is -2, $f'(x) = -2$.

$$12x + 4 = -2$$
$$12x = -6$$
$$x = -\dfrac{1}{2}$$
$$f\left(-\dfrac{1}{2}\right) = -\dfrac{19}{2}$$

The slope of the tangent line is -2 at $\left(-\dfrac{1}{2}, -\dfrac{19}{2}\right)$.

43. $f(x) = x^3 + 6x^2 + 21x + 2$

$f'(x) = 3x^2 + 12x + 21$

If the slope of the tangent line is 9, $f'(x) = 9$.

$$3x^2 + 12x + 21 = 9$$
$$3x^2 + 12x + 12 = 0$$
$$3(x^2 + 4x + 4) = 0$$
$$3(x + 2)^2 = 0$$
$$x = -2$$
$$f(-2) = -24$$

The slope of the tangent line is 9 at $(-2, -24)$.

45. $f(x) = \dfrac{1}{2}g(x) + \dfrac{1}{4}h(x)$

$$f'(x) = \dfrac{1}{2}g'(x) + \dfrac{1}{4}h'(x)$$

$$f'(2) = \dfrac{1}{2}g'(2) + \dfrac{1}{4}h'(2)$$

$$= \dfrac{1}{2}(7) + \dfrac{1}{4}(14) = 7$$

49. $\dfrac{f(x)}{k} = \dfrac{1}{k} \cdot f(x)$

Use the rule for the derivative of a constant times a function.

$$\dfrac{d}{dx}\left[\dfrac{f(x)}{k}\right] = \dfrac{d}{dx}\left[\dfrac{1}{k} \cdot f(x)\right]$$

$$= \dfrac{1}{k}f'(x)$$

$$= \dfrac{f'(x)}{k}$$

51. The demand is given by $q = 5000 - 100p$. Solve for p.

$$p = \dfrac{5000 - q}{100}$$

$$R(q) = q\left(\dfrac{5000 - q}{100}\right)$$

$$= \dfrac{5000q - q^2}{100}$$

$$R'(q) = \dfrac{5000 - 2q}{100}$$

(a) $R'(1000) = \dfrac{5000 - 2(1000)}{100}$

$$= 30$$

(b) $R'(2500) = \dfrac{5000 - 2(2500)}{100}$

$$= 0$$

(c) $R'(3000) = \dfrac{5000 - 2(3000)}{100}$

$$= -10$$

53. $p(q) = \dfrac{1000}{q^2} + 1000$

If R is the revenue function, $R(q) = qp(q)$.

$$R(q) = q\left(\dfrac{1000}{q^2} + 1000\right)$$

$$R(q) = 1000q^{-1} + 1000q$$

$$R'(q) = -1000q^{-2} + 1000$$

$$R'(q) = 1000 - \dfrac{1000}{q^2}$$

$$R'(q) = 1000\left(1 - \dfrac{1}{q^2}\right)$$

$$R'(10) = 1000\left(1 - \dfrac{1}{10^2}\right)$$

$$R'(10) = 1000\left(\dfrac{99}{100}\right)$$

$$= 990$$

The marginal revenue is \$990.

55. (a) $y = kn^p$

By the constant rule and the power rule, the marginal product of labor is

$$\dfrac{dy}{dn} = k\left(p \cdot n^{p-1}\right) = \dfrac{kp}{n^{1-p}}.$$

(b) Since $1 - p > 0$, an increase in n increases the denominator of the marginal product of labor, while the numerator remains constant.

57. (a) In 1982 when $t = 50$:

$$C(50) = 0.007714(50)^2 - 0.03969(50) + 0.726$$

$$= 18.026$$

$$\approx 18 \text{ cents}$$

In 2002 when $t = 70$:

$$C(70) = 0.007714(70)^2 - 0.03969(70) + 0.726$$

$$= 35.746$$

$$\approx 36 \text{ cents}$$

(b) $C'(t) = 0.015428t - 0.03969$

In 1982 when $t = 50$:

$C'(50) = 0.015428(50) - 0.03969$
$= 0.73171$
≈ 0.732 cent per year

In 2002 when $t = 70$:

$C'(70) = 0.015428(70) - 0.03969$
$= 1.04027$
≈ 1.04 cents per year

(c) Using a graphing calculator, a cubic function that models the postage cost data is

$C(t) = -0.0001331t^3 + 0.02458t^2$
$- 0.5841t + 3.101$

$C'(t) = -0.0003993t^2$
$+ 0.04916t - 0.5841$

$C'(50) \approx 0.876$ cent per year
$C'(70) \approx 0.901$ cent per year

59. $N(t) = 0.00437t^{3.2}$

$N'(t) = 0.013984t^{2.2}$

(a) $N'(5) \approx 0.4824$

(b) $N'(10) \approx 2.216$

61. $V(t) = -2159 + 1313t - 60.82t^2$

(a) $V(3) = -2159 + 1313(3) - 60.82(3)^2$
$= 1232.62$ cm^3

(b) $V'(t) = 1313 - 121.64t$
$V'(3) = 1313 - 121.64(3)$
$= 948.08$ cm^3/yr

63. $v = 2.69l^{1.86}$

$\dfrac{dv}{dl} = (1.86)2.69l^{1.86-1} \approx 5.00l^{0.86}$

65. $t = 0.0588s^{1.125}$

(a) When $s = 1609$, $t \approx 238.1$ seconds, or 3 minutes, 58.1 seconds.

(b) $\dfrac{dt}{ds} = 0.0588(1.125s^{1.125-1})$
$= 0.06615s^{0.125}$

When $s = 100$, $\dfrac{dt}{ds} \approx 0.118$ sec/m.

At 100 meters, the fastest possible time increases by 0.118 seconds for each additional meter.

(c) Yes, they have been surpassed. In 2000, the world record in the mile stood at 3:43.13. (Ref:www.runnersworld.com)

67. $\text{BMI} = \dfrac{703w}{h^2}$

(a) $6'8'' = 80$ in.

$\text{BMI} = \dfrac{703(250)}{80^2} \approx 27.5$

(b) $\text{BMI} = \dfrac{703w}{80^2} = 24.9$ implies

$w = \dfrac{24.9(80)^2}{703} \approx 227.$

A 250-lb person needs to lose 23 pounds to get down to 227 lbs.

(c) If $f(h) = \dfrac{703(125)}{h^2} = 87,875h^{-2}$, then

$f'(h) = 87,875(-2h^{-2-1})$
$= -175,750h^{-3} = -\dfrac{175,750}{h^3}$

(d) $f'(65) = -\dfrac{175,750}{65^3} \approx -0.64$

For a 125-lb female with a height of 65 in. (5′5″), the BMI decreases by 0.64 for each additional inch of height.

(e) Sample Chart

ht/wt	140	160	180	200
60	27	31	35	39
65	23	27	30	33
70	20	23	26	29
75	17	20	22	25

(f) Substituting the metric equivalents into the formula

$$BMI = \frac{703w}{h^2} \text{ we have}$$

$$B_m = \frac{703\left(\dfrac{w_m}{0.4536}\right)}{\left(\dfrac{h_m}{0.0254}\right)^2} = \frac{w_m}{h_m^2}.$$

69. $s(t) = 18t^2 - 13t + 8$

   (a) $v(t) = s'(t) = 18(2t) - 13 + 0$
$$= 36t - 13$$

   (b) $v(0) = 36(0) - 13 = -13$
$$v(5) = 36(5) - 13 = 167$$
$$v(10) = 36(10) - 13 = 347$$

71. $s(t) = -3t^3 + 4t^2 - 10t + 5$

   (a) $v(t) = s'(t) = -3(3t^2) + 4(2t) - 10 + 0$
$$= -9t^2 + 8t - 10$$

   (b) $v(0) = -9(0)^2 + 8(0) - 10 = -10$
$$v(5) = -9(5)^2 + 8(5) - 10$$
$$= -225 + 40 - 10 = -195$$
$$v(10) = -9(10)^2 + 8(10) - 10$$
$$= -900 + 80 - 10 = -830$$

73. $s(t) = -16t^2 + 64t$

   (a) $v(t) = s'(t) = -16(2t) + 64 = -32t + 64$
$$v(2) = -32(2) + 64 = -64 + 64 = 0$$
$$v(3) = -32(3) + 64 = -96 + 64 = -32$$

The ball's velocity is 0 ft/sec after 2 seconds and -32 ft/sec after 3 seconds.

   (b) As the ball travels upward, its speed decreases because of the force of gravity until, at maximum height, its speed is 0 ft/sec.

In part (a), we found that $v(2) = 0$.

It takes 2 seconds for the ball to reach its maximum height.

   (c) $s(2) = -16(2)^2 + 64(2)$
$$= -16(4) + 128$$
$$= -64 + 128$$
$$= 64$$

It will go 64 ft high.

75. **(a)** When the tread length is 0.5 m,

$$f = \frac{c}{2T} = \frac{340\,\dfrac{m}{sec}}{(2)(0.5\text{ m})} = 340\,\frac{1}{sec} \text{ or } 340$$

cycles per second.

   (b) $\dfrac{df}{dT} = -\dfrac{c}{2T^2}$

At $T = 0.5$ m,

$$\frac{df}{dT} = -\frac{c}{2T^2} = -\frac{340\,\dfrac{m}{sec}}{(2)(0.5\text{ m})^2} = -680\,\frac{1}{m}\cdot\frac{1}{sec}$$

or -680 cycles/sec/m.

77. $y_1 = 4.13x + 14.63$

$$y_2 = -0.033x^2 + 4.647x + 13.347$$

   (a) When $x = 5$, $y_1 \approx 35$ and $y_2 \approx 36$.

   (b) $\dfrac{dy_1}{dx} = 4.13$

$$\frac{dy_2}{dx} = 0.033(2x) + 4.647$$
$$= -0.066x + 4.647$$

When $x = 5$, $\dfrac{dy_1}{dx} = 4.13$ and $\dfrac{dy_2}{dx} \approx 4.32$.

These values are fairly close and represent the rate of change of four years for a dog for one year of a human, for a dog that is actually 5 years old.

   (c) With the first two points eliminated, the dog age increases in 2-year steps and the human age increases in 8-year steps, for a slope of 4. The equation has the form $y = 4x + b$.

A value of 16 for b makes the numbers come out right. $y = 4x + b$. For a dog of age $x = 5$ years or more, the equivalent human age is given by $y = 4x + 16$.

4.2 Derivatives of Products and Quotients

Your Turn 1

$$y = (x^3 + 7)(4 - x^2)$$
$$\frac{dy}{dx} = (x^3 + 7)(-2x) + (4 - x^2)(3x^2)$$
$$= -2x^4 - 14x + 12x^2 - 3x^4$$
$$= -5x^4 + 12x^2 - 14x$$

Your Turn 2

$$f(x) = \frac{3x + 2}{5 - 2x}$$

$$f'(x) = \frac{(5 - 2x)(3) - (3x + 2)(-2)}{(5 - 2x)^2}$$

$$= \frac{19}{(5 - 2x)^2}$$

Your Turn 3

$$D_x\left[\frac{(5x - 3)(2x + 7)}{3x + 7}\right]$$

$$= \frac{\left[\begin{array}{l}(3x + 7)D_x[(5x - 3)(2x + 7)] \\ - [(5x - 3)(2x + 7)]D_x(3x + 7)\end{array}\right]}{(3x + 7)^2}$$

$$= \frac{\left[\begin{array}{l}(3x + 7)[(5x - 3)(2) + (2x + 7)(5)] \\ - (10x^2 + 29x - 21)(3)\end{array}\right]}{(3x + 7)^2}$$

$$= \frac{\left[\begin{array}{l}(3x + 7)(10x - 6 + 10x + 35) \\ - (30x^2 + 87x - 63)\end{array}\right]}{(3x + 7)^2}$$

$$= \frac{60x^2 + 227x + 203 - 30x^2 - 87x + 63}{(3x + 7)^2}$$

$$= \frac{30x^2 + 140x + 266}{(3x + 7)^2}$$

Your Turn 4

$$C(x) = \frac{4x + 50}{x + 2}$$

The marginal average cost is

$$\frac{d}{dx}\left[\frac{C(x)}{x}\right] = \frac{d}{dx}\left(\frac{4x + 50}{x^2 + 2x}\right)$$

$$= \frac{(x^2 + 2x)(4) - (4x + 50)(2x + 2)}{(x^2 + 2x)^2}$$

$$= \frac{4x^2 + 8x - (8x^2 + 108x + 100)}{(x^2 + 2x)^2}$$

$$= \frac{4x^2 + 8x - 8x^2 - 108x - 100}{(x^2 + 2x)^2}$$

$$= \frac{-4x^2 - 100x - 100}{(x^2 + 2x)^2}$$

4.2 Warmup Exercises

W1.

$$f(x) = 3x^4 + 4x^3 - 5$$

$$f'(x) = (3)(4)x^{4-1} + 4(3)x^{3-1} - 0$$

$$= 12x^3 + 12x^2$$

W2.

$$f(x) = \frac{2}{x^3} + 6\sqrt{x} = 2x^{-3} + 6x^{1/2}$$

$$f'(x) = (2)(-3)x^{-3-1} + 6(1/2)x^{1/2-1}$$

$$= -6x^{-4} + 3x^{-1/2}$$

$$= -\frac{6}{x^4} + \frac{3}{\sqrt{x}}$$

W3.

$$f(x) = 9x^{2/3} + \frac{4}{\sqrt{x}} = 9x^{2/3} + 4x^{-1/2}$$

$$f'(x) = (9)(2/3)x^{(2/3)-1} + (4)(-1/2)x^{(-1/2)-1}$$

$$= 6x^{-1/3} - 2x^{-3/2}$$

4.2 Exercises

1. $y = (3x^2 + 2)(2x - 1)$

$$\frac{dy}{dx} = (3x^2 + 2)(2) + (2x - 1)(6x)$$

$$= 6x^2 + 4 + 12x^2 - 6x$$

$$= 18x^2 - 6x + 4$$

3. $y = (2x - 5)^2$

$$= (2x - 5)(2x - 5)$$

$$\frac{dy}{dx} = (2x - 5)(2) + (2x - 5)(2)$$

$$= 4x - 10 + 4x - 10$$

$$= 8x - 20$$

5. $k(t) = (t^2 - 1)^2 = (t^2 - 1)(t^2 - 1)$

$$k'(t) = (t^2 - 1)(2t) + (t^2 - 1)(2t)$$

$$= 2t^3 - 2t + 2t^3 - 2t$$

$$= 4t^3 - 4t$$

7. $y = (x + 1)(\sqrt{x} + 2)$

$\quad\quad = (x + 1)(x^{1/2} + 2)$

$\quad\dfrac{dy}{dx} = (x + 1)\left(\dfrac{1}{2}x^{-1/2}\right) + (x^{1/2} + 2)(1)$

$\quad\quad = \dfrac{1}{2}x^{1/2} + \dfrac{1}{2}x^{-1/2} + x^{1/2} + 2$

$\quad\quad = \dfrac{3}{2}x^{1/2} + \dfrac{1}{2}x^{-1/2} + 2$

$\quad\quad \text{or } \dfrac{3x^{1/2}}{2} + \dfrac{1}{2x^{1/2}} + 2$

9. $p(y) = (y^{-1} + y^{-2})(2y^{-3} - 5y^{-4})$

$\quad p'(y) = (y^{-1} + y^{-2})(-6y^{-4} + 20y^{-5})$

$\quad\quad\quad + (-y^{-2} - 2y^{-3})(2y^{-3} - 5y^{-4})$

$\quad\quad = -6y^{-5} + 20y^{-6} - 6y^{-6} + 20y^{-7}$

$\quad\quad\quad - 2y^{-5} + 5y^{-6} - 4y^{-6} + 10y^{-7}$

$\quad\quad = -8y^{-5} + 15y^{-6} + 30y^{-7}$

11. $f(x) = \dfrac{6x + 1}{3x + 10}$

$\quad f'(x) = \dfrac{(3x + 10)(6) - (6x + 1)(3)}{(3x + 10)^2}$

$\quad\quad = \dfrac{18x + 60 - 18x - 3}{(3x + 10)^2}$

$\quad\quad = \dfrac{57}{(3x + 10)^2}$

13. $y = \dfrac{5 - 3t}{4 + t}$

$\quad \dfrac{dy}{dx} = \dfrac{(4 + t)(-3) - (5 - 3t)(1)}{(4 + t)^2}$

$\quad\quad = \dfrac{-12 - 3t - 5 + 3t}{(4 + t)^2}$

$\quad\quad = \dfrac{-17}{(4 + t)^2}$

15. $y = \dfrac{x^2 + x}{x - 1}$

$\quad \dfrac{dy}{dx} = \dfrac{(x - 1)(2x + 1) - (x^2 + x)(1)}{(x - 1)^2}$

$\quad\quad = \dfrac{2x^2 + x - 2x - 1 - x^2 - x}{(x - 1)^2}$

$\quad\quad = \dfrac{x^2 - 2x - 1}{(x - 1)^2}$

17. $f(t) = \dfrac{4t^2 + 11}{t^2 + 3}$

$\quad f'(t) = \dfrac{(t^2 + 3)(8t) - (4t^2 + 11)(2t)}{(t^2 + 3)^2}$

$\quad\quad = \dfrac{8t^3 + 24t - 8t^3 - 22t}{(t^2 + 3)^2}$

$\quad\quad = \dfrac{2t}{(t^2 + 3)^2}$

19.

$\quad g(x) = \dfrac{x^2 - 4x + 2}{x^2 + 3}$

$\quad g'(x) = \dfrac{(x^2 + 3)(2x - 4) - (x^2 - 4x + 2)(2x)}{(x^2 + 3)^2}$

$\quad\quad = \dfrac{2x^3 - 4x^2 + 6x - 12 - 2x^3 + 8x^2 - 4x}{(x^2 + 3)^2}$

$\quad\quad = \dfrac{4x^2 + 2x - 12}{(x^2 + 3)^2}$

21. $p(t) = \dfrac{\sqrt{t}}{t - 1}$

$\quad\quad = \dfrac{t^{1/2}}{t - 1}$

$\quad p'(t) = \dfrac{(t - 1)\left(\dfrac{1}{2}t^{-1/2}\right) - t^{1/2}(1)}{(t - 1)^2}$

$\quad\quad = \dfrac{\dfrac{1}{2}t^{1/2} - \dfrac{1}{2}t^{-1/2} - t^{1/2}}{(t - 1)^2}$

$\quad\quad = \dfrac{-\dfrac{1}{2}t^{1/2} - \dfrac{1}{2}t^{-1/2}}{(t - 1)^2}$

$\quad\quad = \dfrac{-\dfrac{\sqrt{t}}{2} - \dfrac{1}{2\sqrt{t}}}{(t - 1)^2} \text{ or } \dfrac{-t - 1}{2\sqrt{t}\,(t - 1)^2}$

23. $y = \dfrac{5x + 6}{\sqrt{x}} = \dfrac{5x + 6}{x^{1/2}} = 5x^{1/2} + 6x^{-1/2}$

$\quad \dfrac{dy}{dx} = \dfrac{5}{2}x^{-1/2} - 3x^{-3/2} \text{ or } \dfrac{5x - 6}{2x\sqrt{x}}$

25. $h(z) = \dfrac{z^{2.2}}{z^{3.2} + 5}$

$h'(z) = \dfrac{(z^{3.2} + 5)(2.2z^{1.2}) - z^{2.2}(3.2z^{2.2})}{(z^{3.2} + 5)^2} = \dfrac{2.2z^{4.4} + 11z^{1.2} - 3.2z^{4.4}}{(z^{3.2} + 5)^2} = \dfrac{-z^{4.4} + 11z^{1.2}}{(z^{3.2} + 5)^2}$

27. $f(x) = \dfrac{(3x^2 + 1)(2x - 1)}{5x + 4}$

$f'(x) = \dfrac{(5x + 4)[(3x^2 + 1)(2) + (6x)(2x - 1)] - (3x^2 + 1)(2x - 1)(5)}{(5x + 4)^2}$

$= \dfrac{(5x + 4)(18x^2 - 6x + 2) - (3x^2 + 1)(10x - 5)}{(5x + 4)^2}$

$= \dfrac{90x^3 - 30x^2 + 10x + 72x^2 - 24x + 8 - 30x^3 + 15x^2 - 10x + 5}{(5x + 4)^2}$

$= \dfrac{60x^3 + 57x^2 - 24x + 13}{(5x + 4)^2}$

29. $h(x) = f(x)g(x)$

$h'(x) = f(x)g'(x) + g(x)f'(x)$

$h'(3) = f(3)g'(3) + g(3)f'(3) = 9(5) + 4(8) = 77$

31.   In the first step, the two terms in the numerator are reversed. The correct work follows.

$$D_x\left(\dfrac{2x + 5}{x^2 - 1}\right)$$

$$= \dfrac{(x^2 - 1)(2) - (2x + 5)(2x)}{(x^2 - 1)^2}$$

$$= \dfrac{2x^2 - 2 - 4x^2 - 10x}{(x^2 - 1)^2}$$

$$= \dfrac{-2x^2 - 10x - 2}{(x^2 - 1)^2}$$

33. $f(x) = \dfrac{x}{x - 2}$, at $(3, 3)$

$m = f'(x) = \dfrac{(x - 2)(1) - x(1)}{(x - 2)^2} = -\dfrac{2}{(x - 2)^2}$

At $(3,3)$,

$$m = -\dfrac{2}{(3 - 2)^2} = -2,$$

Use the point-slope from.

$$y - 3 = -2(x - 3)$$
$$y = -2x + 9$$

35. (a) $f(x) = \dfrac{3x^3 + 6}{x^{2/3}}$

$$f'(x) = \frac{(x^{2/3})(9x^2) - (3x^3 + 6)(\frac{2}{3}x^{-1/3})}{(x^{2/3})^2} = \frac{9x^{8/3} - 2x^{8/3} - 4x^{-1/3}}{x^{4/3}} = \frac{7x^{8/3} - \frac{4}{x^{1/3}}}{x^{4/3}} = \frac{7x^3 - 4}{x^{5/3}}$$

(b) $f(x) = 3x^{7/3} + 6x^{-2/3}$

$$f'(x) = 3\left(\frac{7}{3}x^{4/3}\right) + 6\left(-\frac{2}{3}x^{-5/3}\right) = 7x^{4/3} - 4x^{-5/3}$$

(c) The derivatives are equivalent.

37. $f(x) = \dfrac{u(x)}{v(x)}$

$$f'(x) = \lim_{h \to 0} \frac{f(x + h) - f(x)}{h} = \lim_{h \to 0} \frac{\frac{u(x+h)}{v(x+h)} - \frac{u(x)}{v(x)}}{h} = \lim_{h \to 0} \frac{u(x + h)v(x) - u(x)v(x + h)}{hv(x + h)v(x)}$$

$$= \lim_{h \to 0} \frac{u(x + h)v(x) - u(x)v(x) + u(x)v(x) - u(x)v(x + h)}{hv(x + h)v(x)}$$

$$= \lim_{h \to 0} \frac{v(x)[u(x + h) - u(x)] - u(x)[v(x + h) - v(x)]}{hv(x + h)v(x)}$$

$$= \lim_{h \to 0} \frac{v(x)\frac{u(x+h)-u(x)}{h} - u(x)\frac{v(x+h)-v(x)}{h}}{v(x + h)v(x)} = \frac{v(x) \cdot u'(x) - u(x)v'(x)}{[v(x)]^2}$$

39. Graph the numerical derivative of $f(x) = (x^2 - 2)(x^2 - \sqrt{2})$ for x ranging from -2 to 2. The derivative crosses the x-axis at 0 and at approximately -1.307 and 1.307.

41. (a) $f(x) = x^2 + bx$ and $g(x) = cx + d$

$$f'(x)g'(x) = (2x + b)(c) = 2cx + bc$$

(b) $\big[f(x)g(x)\big]' = f(x)g'(x) + f(x)g'(x)$

$$= (x^2 + bx)(c) + (2x + b)(cx + d)$$

$$= 3cx^2 + 2x(d + bc) + bd$$

43. $C(x) = \dfrac{3x + 2}{x + 4}$

$$\overline{C}(x) = \frac{C(x)}{x} = \frac{3x + 2}{x^2 + 4x}$$

(a) $\overline{C}(10) = \dfrac{3(10) + 2}{10^2 + 4(10)} = \dfrac{32}{140} \approx 0.2286$ hundreds of dollars or $22.86 per unit

(b) $\overline{C}(20) = \dfrac{3(20) + 2}{(20)^2 + 4(20)} = \dfrac{62}{480} \approx 0.1292$ hundreds of dollars or $12.92 per unit

(c) $\overline{C}(x) = \dfrac{3x + 2}{x^2 + 4x}$ per unit

(d) $\overline{C}'(x) = \dfrac{(x^2 + 4x)(3) - (3x + 2)(2x + 4)}{(x^2 + 4x)^2} = \dfrac{3x^2 + 12x - 6x^2 - 12x - 4x - 8}{(x^2 + 4x)^2} = \dfrac{-3x^2 - 4x - 8}{(x^2 + 4x)^2}$

45. $M(d) = \dfrac{100d^2}{3d^2 + 10}$

(a) $M'(d) = \dfrac{(3d^2 + 10)(200d) - (100d^2)(6d)}{(3d^2 + 10)^2} = \dfrac{600d^3 + 2000d - 600d^3}{(3d^2 + 10)^2} = \dfrac{2000d}{(3d^2 + 10)^2}$

(b) $M'(2) = \dfrac{2000(2)}{[3(2)^2 + 10]^2} = \dfrac{4000}{484} \approx 8.3$

This means the new employee can assemble about 8.3 additional bicycles per day after 2 days of training.

$$M'(5) = \dfrac{2000(5)}{[3(5)^2 + 10]^2} = \dfrac{10,000}{7225} \approx 1.4$$

This means the new employee can assemble about 1.4 additional bicycles per day after 5 days of training.

47. $\overline{C}(x) = \dfrac{C(x)}{x}$

Let $u(x) = C(x)$, with $u'(x) = C'(x)$

Let $v(x) = x$ with $v'(x) = 1$. Then, by the quotient rule,

$$\overline{C}(x) = \dfrac{v(x) \cdot u'(x) - u(x) \cdot v'(x)}{[v(x)]^2} = \dfrac{x \cdot C'(x) - C(x) \cdot 1}{x^2} = \dfrac{xC'(x) - C(x)}{x^2}$$

49. Let $C(t)$ be the cost as a function of time and $q(t)$ be the quantity as a function of time.

Then $\overline{C}(t) = \dfrac{C(t)}{q(t)}$ is the revenue as a function of time. Let $t = t_1$ represent last month.

$$\overline{C}'(t) = \dfrac{q(t)C'(t) - C(t)q'(t)}{[g(t)]^2}$$

$$\overline{C}'(t_1) = \dfrac{q(t_1)C'(t_1) - C(t_1)q'(t_1)}{[g(t_1)]^2}$$

$$= \dfrac{(12,500)(1200) - (27,000)(350)}{(12,500)^2}$$

$$= 0.03552$$

The average cost is increasing at a rate of $0.03552 per gallon per month.

51. $f(x) = \dfrac{Kx}{A + x}$

(a) $f'(x) = \dfrac{(A + x)K - Kx(1)}{(A + x)^2}$

$f'(x) = \dfrac{AK}{(A + x)^2}$

(b) $f'(A) = \dfrac{AK}{(A + A)^2}$

$= \dfrac{AK}{4A^2} = \dfrac{K}{4A}$

53. $R(w) = \dfrac{30(w - 4)}{w - 1.5}$

(a) $R(5) = \dfrac{30(5 - 4)}{5 - 1.5}$

≈ 8.57 min

(b) $R(7) = \dfrac{30(7 - 4)}{7 - 1.5}$

≈ 16.36 min

(c)

$$R'(w) = \frac{(w - 1.5)(30) - 30(w - 4)(1)}{(w - 1.5)^2}$$

$$= \frac{30w - 45 - 30w + 120}{(w - 1.5)^2}$$

$$= \frac{75}{(w - 1.5)^2}$$

$$R'(5) = \frac{75}{(5 - 1.5)^2} \approx 6.12 \frac{\min^2}{\text{kcal}}$$

$$R'(7) = \frac{75}{(7 - 1.5)^2} \approx 2.48 \frac{\min^2}{\text{kcal}}$$

55. $f(x) = \dfrac{x^2}{2(1 - x)}$

$$f'(x) = \frac{2(1 - x)(2x) - x^2(-2)}{[2(1 - x)]^2}$$

$$= \frac{4x - 4x^2 + 2x^2}{4(1 - x)^2}$$

$$= \frac{4x - 2x^2}{4(1 - x)^2}$$

$$= \frac{2x(2 - x)}{4(1 - x)^2}$$

$$= \frac{x(2 - x)}{2(1 - x)^2}$$

(a) $f'(0.1) = \dfrac{0.1(2 - 0.1)}{2(1 - 0.1)^2} \approx 0.1173$

(b) $f'(0.6) = \dfrac{0.6(2 - 0.6)}{2(1 - 0.6)^2} = 2.625$

4.3 The Chain Rule

Your Turn 1

Let $f(x) = 2x - 1$ and $g(x) = \sqrt{3x + 5}$.

$$g(0) = \sqrt{3 \cdot 0 + 5} = \sqrt{5}$$
$$f[g(0)] = 2\sqrt{5} - 1$$

$$f(0) = 2 \cdot 0 - 1 = -1$$
$$g[f(0)] - \sqrt{3(-1) + 5} = \sqrt{2}$$

Your Turn 2

Let $f(x) = 2x - 3$ and $g(x) = x^2 + 1$.

$$g[f(x)] = g(2x - 3)$$
$$= (2x - 3)^2 + 1$$
$$= 4x^2 - 12x + 9 + 1$$
$$= 4x^2 - 12x + 10$$

Your Turn 3

Write $h(x) = (2x - 3)^3$ in the form $h(x) = f[g(x)]$.

One possible answer is $h(x) = f[g(x)]$

where $g(x) = 2x - 3$ and $f(x) = x^3$.

Your Turn 4

$$y = (5x^2 - 6x)^{-2}$$

$$\frac{dy}{dx} = -2(5x^2 - 6x)^{-3} \cdot (10x - 6)$$

$$= \frac{-2(10x - 6)}{(5x^2 - 6x)^3} = \frac{-20x + 12}{(5x^2 - 6x)^3}$$

Your Turn 5

$$D_x[(x^2 - 7)^{10}] = 10(x^2 - 7)^9(2x) = 20x(x^2 - 7)^9$$

Your Turn 6

$$y = x^2(5x - 1)^3$$

$$\frac{dy}{dx} = x^2 \cdot 3(5x - 1)^2(5) + (5x - 1)^3 \cdot 2x$$

$$= 15x^2(5x - 1)^2 + 2x(5x - 1)^3$$

$$= x(5x - 1)^2[15x + 2(5x - 1)]$$

$$= x(5x - 1)^2(25x - 2)$$

Your Turn 7

$$D_x\left[\frac{(4x - 1)^3}{x + 3}\right] = \frac{(x + 3)[3(4x - 1)^2(4)] - (4x - 1)^3(1)}{(x + 3)^2}$$

$$= \frac{12(x + 3)(4x - 1)^2 - (4x - 1)^3}{(x + 3)^2}$$

$$= \frac{(4x - 1)^2[12(x + 3) - (4x - 1)]}{(x + 3)^2}$$

$$= \frac{(4x - 1)^2(8x + 37)}{(x + 3)^2}$$

4.3   Warmup Exercises

W1.

$$f(x) = \frac{x^2}{x^3 + 1}$$

$$f'(x) = \frac{(2x)\left(x^3 + 1\right) - \left(x^2\right)\left(3x^2\right)}{\left(x^3 + 1\right)^2}$$

$$= \frac{2x - x^4}{\left(x^3 + 1\right)^2}$$

W2.

$$f(x) = \frac{6\sqrt{x}}{x^4 + 1} = \frac{6x^{1/2}}{x^4 + 1}$$

$$f'(x) = \frac{\left(3x^{-1/2}\right)\left(x^4 + 1\right) - \left(6x^{1/2}\right)\left(4x^3\right)}{\left(x^4 + 1\right)^2}$$

$$= \frac{3\left(x^4 + 1\right) - 6x\left(4x^3\right)}{\left(x^{1/2}\right)\left(x^4 + 1\right)^2}$$

$$= \frac{3 - 21x^4}{\sqrt{x}\left(x^4 + 1\right)^2}$$

W3. $f(x) = \dfrac{x^{2/3}}{x^{1/3} + 2}$

$$f'(x) = \frac{\frac{2}{3}x^{-1/3}\left(x^{1/3} + 2\right) - \left(x^{2/3}\right)\left(\frac{1}{3}x^{-2/3}\right)}{\left(x^{1/3} + 2\right)^2}$$

$$= \frac{2 + 4x^{-1/3} - 1}{3\left(x^{1/3} + 2\right)^2} = \frac{1 + 4x^{-1/3}}{3\left(x^{1/3} + 2\right)^2}$$

4.3   Exercises

In Exercises 1 through 6, $f(x) = 5x^2 - 2x$
and $g(x) = 8x + 3$.

1. $g(2) = 8(2) + 3 = 19$
$f[g(2)] = f[19]$
$$= 5(19)^2 - 2(19)$$
$$= 1805 - 38 = 1767$$

3. $f(2) = 5(2)^2 - 2(2)$
$$= 20 - 4 = 16$$
$g[f(2)] = g[16]$
$$= 8(16) + 3$$
$$= 128 + 3 = 131$$

5. $g(k) = 8k + 3$
$f[g(k)] = f[8k + 3]$
$$= 5(8k + 3)^2 - 2(8k + 3)$$
$$= 5(64k^2 + 48k + 9) - 16k - 6$$
$$= 320k^2 + 224k + 39$$

7. $f(x) = \dfrac{x}{8} + 7; g(x) = 6x - 1$

$$f[g(x)] = \frac{6x - 1}{8} + 7$$

$$= \frac{6x - 1}{8} + \frac{56}{8}$$

$$= \frac{6x + 55}{8}$$

$$g[f(x)] = 6\left[\frac{x}{8} + 7\right] - 1$$

$$= \frac{6x}{8} + 42 - 1$$

$$= \frac{3x}{4} + 41$$

$$= \frac{3x}{4} + \frac{164}{4}$$

$$= \frac{3x + 164}{4}$$

9. $f(x) = \dfrac{1}{x}; g(x) = x^2$

$$f[g(x)] = \frac{1}{x^2}$$

$$g[f(x)] = \left(\frac{1}{x}\right)^2$$

$$= \frac{1}{x^2}$$

11. $f(x) = \sqrt{x + 2}; g(x) = 8x^2 - 6$

$f[g(x)] = \sqrt{(8x^2 - 6) + 2}$

$= \sqrt{8x^2 - 4}$

$g[f(x)] = 8(\sqrt{x + 2})^2 - 6$

$= 8x + 16 - 6$

$= 8x + 10$

13. $f(x) = \sqrt{x + 1}; g(x) = \dfrac{-1}{x}$

$f[g(x)] = \sqrt{\dfrac{-1}{x} + 1}$

$= \sqrt{\dfrac{x - 1}{x}}$

$g[f(x)] = \dfrac{-1}{\sqrt{x + 1}}$

15. $y = (5 - x^2)^{3/5}$

If $f(x) = x^{3/5}$ and $g(x) = 5 - x^2$, then

$$y = f[g(x)] = (5 - x^2)^{3/5}.$$

17. $y = -\sqrt{13 + 7x}$

If $f(x) = -\sqrt{x}$ and

$g(x) = 13 + 7x,$

then $y = f[g(x)] = -\sqrt{13 + 7x}.$

19. $y = (x^2 + 5x)^{1/3} - 2(x^2 + 5x)^{2/3} + 7$

If $f(x) = x^{1/3} - 2x^{2/3} + 7$ and

$g(x) = x^2 + 5x,$

then

$$y = f[g(x)] = (x^2 + 5x)^{1/3} - 2(x^2 + 5x)^{2/3} + 7.$$

21. $y = (8x^4 - 5x^2 + 1)^4$

Let $f(x) = x^4$ and $g(x) = 8x^4 - 5x^2 + 1.$

Then $(8x^4 - 5x^2 + 1)^4 = f[g(x)].$

Use the alternate form of the chain rule.

$\dfrac{dy}{dx} = f'[g(x)] \cdot g'(x)$

$f'(x) = 4x^3$

$f'[g(x)] = 4[g(x)]^3 = 4(8x^4 - 5x^2 + 1)^3$

$g'(x) = 32x^3 - 10x$

$\dfrac{dy}{dx} = 4(8x^4 - 5x^2 + 1)^3(32x^3 - 10x)$

23. $k(x) = -2(12x^2 + 5)^{-6}$

Use the generalized power rule with

$u = 12x^2 + 5, n = -6,$ and $u' = 24x.$

$k'(x) = -2[-6(12x^2 + 5)^{-6-1} \cdot 24x]$

$= -2[-144x(12x^2 + 5)^{-7}]$

$= 288x(12x^2 + 5)^{-7}$

25. $s(t) = 45(3t^3 - 8)^{3/2}$

Use the generalized power rule with

$u = 3t^3 - 8, n = \dfrac{3}{2},$ and $u' = 9t^2.$

$s'(t) = 45\left[\dfrac{3}{2}(3t^3 - 8)^{1/2} \cdot 9t^2\right]$

$= 45\left[\dfrac{27}{2}t^2(3t^3 - 8)^{1/2}\right]$

$= \dfrac{1215}{2}t^2(3t^3 - 8)^{1/2}$

27. $g(t) = -3\sqrt{7t^3 - 1}$

$= -3\sqrt{(7t^3 - 1)^{1/2}}$

Use generalized power rule with

$u = 7t^3 - 1, n = \frac{1}{2},$ and $u' = 21t^2.$

$g'(t) = -3\left[\dfrac{1}{2}(7t^3 - 1)^{-1/2} \cdot 21t^2\right]$

$= -3\left[\dfrac{21}{2}t^2(7t^3 - 1)^{-1/2}\right]$

$= \dfrac{-63}{2}t^2 \cdot \dfrac{1}{(7t^3 - 1)^{1/2}}$

$= \dfrac{-63t^2}{2\sqrt{7t^3 - 1}}$

29. $m(t) = -6t(5t^4 - 1)^4$

Use the product rule and the power rule.

$m'(t) = -6t[4(5t^4 - 1)^3 \cdot 20t^3] + (5t^4 - 1)^4(-6)$

$= -480t^4(5t^4 - 1)^3 - 6(5t^4 - 1)^4$

$= -6(5t^4 - 1)^3[80t^4 + (5t^4 - 1)]$

$= -6(5t^4 - 1)^3(85t^4 - 1)$

31. $y = (3x^4 + 1)^4(x^3 + 4)$

Use the product rule and the power rule.

$\dfrac{dy}{dx} = (3x^4 + 1)^4(3x^2) + (x^3 + 4)[4(3x^4 + 1)^3 \cdot 12x^3]$

$= 3x^2(3x^4 + 1)^4 + 48x^3(x^3 + 4)(3x^4 + 1)^3$

$= 3x^2(3x^4 + 1)^3[3x^4 + 1 + 16x(x^3 + 4)]$

$= 3x^2(3x^4 + 1)^3(3x^4 + 1 + 16x^4 + 64)$

$= 3x^2(3x^4 + 1)^3(19x^4 + 64x + 1)$

33. $q(y) = 4y^2(y^2 + 1)^{5/4}$

Use the product rule and the power rule.

$q'(y) = 4y^2 \cdot \dfrac{5}{4}(y^2 + 1)^{1/4}(2y) + 8y(y^2 + 1)^{5/4}$

$= 10y^3(y^2 + 1)^{1/4} + 8y(y^2 + 1)^{5/4}$

$= 2y(y^2 + 1)^{1/4}[5y^2 + 4(y^2 + 1)^{4/4}]$

$= 2y(y^2 + 1)^{1/4}(9y^2 + 4)$

35. $y = \dfrac{-5}{(2x^3 + 1)^2} = -5(2x^3 + 1)^{-2}$

$\dfrac{dy}{dx} = -5[-2(2x^3 + 1)^{-3} \cdot 6x^2]$

$= -5[-12x^2(2x^3 + 1)^{-3}]$

$= 60x^2(2x^3 + 1)^{-3}$

$= \dfrac{60x^2}{(2x^3 + 1)^3}$

37. $r(t) = \dfrac{(5t - 6)^4}{3t^2 + 4}$

$r'(t)$

$= \dfrac{(3t^2 + 4)[4(5t - 6)^3 \cdot 5] - (5t - 6)^4(6t)}{(3t^2 + 4)^2}$

$= \dfrac{20(3t^2 + 4)(5t - 6)^3 - 6t(5t - 6)^4}{(3t^2 + 4)^2}$

$= \dfrac{2(5t - 6)^3[10(3t^2 + 4) - 3t(5t - 6)]}{(3t^2 + 4)^2}$

$= \dfrac{2(5t - 6)^3(30t^2 + 40 - 15t^2 + 18t)}{(3t^2 + 4)^2}$

$= \dfrac{2(5t - 6)^3(15t^2 + 18t + 40)}{(3t^2 + 4)^2}$

39. $y = \dfrac{3x^2 - x}{(2x - 1)^5}$

$\dfrac{dy}{dx} = \dfrac{(2x - 1)^5(6x - 1) - (3x^2 - x)[5(2x - 1)^4 \cdot 2]}{[(2x - 1)^5]^2}$

$= \dfrac{(2x - 1)^5(6x - 1) - 10(3x^2 - x)(2x - 1)^4}{(2x - 1)^{10}}$

$= \dfrac{(2x - 1)^4[(2x - 1)(6x - 1) - 10(3x^2 - x)]}{(2x - 1)^{10}}$

$= \dfrac{12x^2 - 2x - 6x + 1 - 30x^2 + 10x}{(2x - 1)^6}$

$= \dfrac{-18x^2 + 2x + 1}{(2x - 1)^6}$

43. (a) $D_x(f[g(x)])$ at $x = 1$

$= f'[g(1)] \cdot g'(1)$

$= f'(2) \cdot \left(\dfrac{2}{7}\right)$

$= -7\left(\dfrac{2}{7}\right) = -2$

(b) $D_x(f[g(x)])$ at $x = 2$

$= f'[g(2)] \cdot g'(2)$

$= f'(3) \cdot \left(\dfrac{3}{7}\right)$

$= -8\left(\dfrac{3}{7}\right) = -\dfrac{24}{7}$

45. $f(x) = \sqrt{x^2 + 16}; \; x = 3$

$f(x) = (x^2 + 16)^{1/2}$

$f'(x) = \dfrac{1}{2}(x^2 + 16)^{-1/2}(2x)$

$f'(x) = \dfrac{x}{\sqrt{x^2 + 16}}$

$f'(3) = \dfrac{3}{\sqrt{3^2 + 16}} = \dfrac{3}{5}$

$f(3) = \sqrt{3^2 + 16} = 5$

We use $m = \frac{3}{5}$ and the point $P(3, 5)$ in the point-slope form

$$y - 5 = \frac{3}{5}(x - 3)$$

$$y - 5 = \frac{3}{5}x - \frac{9}{5}$$

$$y = \frac{3}{5}x + \frac{16}{5}$$

47. $f(x) = x(x^2 - 4x + 5)^4; \; x = 2$

$f'(x) = x \cdot 4(x^2 - 4x + 5)^3 \cdot (2x - 4)$

$\qquad + 1 \cdot (x^2 - 4x + 5)^4$

$\quad = (x^2 - 4x + 5)^3$

$\qquad \cdot [4x(2x - 4) + (x^2 - 4x + 5)]$

$\quad = (x^2 - 4x + 5)^3(9x^2 - 20x + 5)$

$f'(2) = (1)^3(1) = 1$

$f(2) = 2(1)^4 = 2$

We use $m = 1$ and the point $P(2, 2)$.

$$y - 2 = 1(x - 2)$$

$$y - 2 = x - 2$$

$$y = x$$

49. $f(x) = \sqrt{x^3 - 6x^2 + 9x + 1}$

$f(x) = (x^3 - 6x^2 + 9x + 1)^{1/2}$

$f'(x) = \dfrac{1}{2}(x^3 - 6x^2 + 9x + 1)^{-1/2}$

$\qquad \cdot (3x^2 - 12x + 9)$

$f'(x) = \dfrac{3(x^2 - 4x + 3)}{2\sqrt{x^3 - 6x^2 + 9x + 1}}$

If the tangent line is horizontal, its slope is zero and $f'(x) = 0$.

$$\frac{3(x^2 - 4x + 3)}{2\sqrt{x^3 - 6x^2 + 9x + 1}} = 0$$

$$3(x^2 - 4x + 3) = 0$$

$$3(x - 1)(x - 3) = 0$$

$$x = 1 \text{ or } x = 3$$

The tangent line is horizontal $x = 1$ and $x = 3$.

53. $D(p) = \dfrac{-p^2}{100} + 500; \; p(c) = 2c - 10$

The demand in terms of the cost is

$D(c) = D[p(c)]$

$\qquad = \dfrac{-(2c - 10)^2}{100} + 500$

$\qquad = \dfrac{-4(c - 5)^2}{100} + 500$

$\qquad = \dfrac{-c^2 + 10c - 25}{25} + 500$

$\qquad = \dfrac{-c^2 + 10c - 25 + 12{,}500}{25}$

$\qquad = \dfrac{-c^2 + 10c + 12{,}475}{25}.$

55. $A = 1500\left(1 + \dfrac{r}{36{,}500}\right)^{1825}$

$\frac{dA}{dr}$ is the rate of change of A with respect to r.

$$\frac{dA}{dr} = 1500(1825)\left(1 + \frac{r}{36{,}500}\right)^{1824}\left(\frac{1}{36{,}500}\right)$$

$$\qquad = 75\left(1 + \frac{r}{36{,}500}\right)^{1824}$$

(a) For $r = 6\%$,

$$\frac{dA}{dr} = 75\left(1 + \frac{6}{36{,}500}\right)^{1824} = \$101.22.$$

(b) For $r = 8\%$,

$$\frac{dA}{dr} = 75\left(1 + \frac{8}{36{,}500}\right)^{1824} = \$111.86.$$

(c) For $r = 9\%$,

$$\frac{dA}{dr} = 75\left(1 + \frac{9}{36{,}500}\right)^{1824} = \$117.59.$$

57. $V = \dfrac{60{,}000}{1 + 0.3t + 0.1t^2}$

The rate of change of the value is

$$V'(t) = \dfrac{(1 + 0.3t + 0.1t^2)(0) - 60{,}000(0.3 + 0.2t)}{(1 + 0.3t + 0.1t^2)^2}$$

$$= \dfrac{-60{,}000(0.3 + 0.2t)}{(1 + 0.3t + 0.1t^2)^2}.$$

(a) 2 years after purchase, the rate of change in the value is

$$V'(2) = \dfrac{-60{,}000[0.3 + 0.2(2)]}{[1 + 0.3(2) + 0.1(2)^2]^2}$$

$$= \dfrac{-60{,}000(0.3 + 0.4)}{(1 + 0.6 + 0.4)^2}$$

$$= \dfrac{-42{,}000}{4}$$

$$= -\$10{,}500.$$

(b) 4 years after purchase, the rate of change in the value is

$$V'(4) = \dfrac{-60{,}000[0.3 + 0.2(4)]}{[1 + 0.3(4) + 0.1(4)^2]^2}$$

$$= \dfrac{-66{,}000}{14.44}$$

$$= -\$4570.64.$$

59. $P(x) = 2x^2 + 1;\ x = f(a) = 3a + 2$

$$P[f(a)] = 2(3a + 2)^2 + 1$$

$$= 2(9a^2 + 12a + 4) + 1$$

$$= 18a^2 + 24a + 9$$

61. (a) $r(t) = 2t;\ A(r) = \pi r^2$

$$A[r(t)] = \pi(2t)^2$$

$$= 4\pi t^2$$

$A = 4\pi t^2$ gives the area of the pollution in terms of the time since the pollutants were first emitted.

(b) $D_t A[r(t)] = 8\pi t$

$$D_t A[r(4)] = 8\pi(4) = 32\pi$$

At 12 P.M., the area of pollution is changing at the rate of 32π mi²/hr.

63. $C(t) = \dfrac{1}{2}(2t + 1)^{-1/2}$

$$C'(t) = \dfrac{1}{2}\left(-\dfrac{1}{2}\right)(2t + 1)^{-3/2}(2)$$

$$= -\dfrac{1}{2}(2t + 1)^{-3/2}$$

(a) $C'(0) = -\dfrac{1}{2}[2(0) + 1]^{-3/2}$

$$= -\dfrac{1}{2}$$

$$= -0.5$$

(b) $C'(4) = -\dfrac{1}{2}[2(4) + 1]^{-3/2}$

$$= -\dfrac{1}{2}(9)^{-3/2}$$

$$= \dfrac{-1}{2}\cdot\dfrac{1}{(\sqrt{9})^3}$$

$$= -\dfrac{1}{54}$$

$$\approx -0.02$$

(c) $C'(7.5) = -\dfrac{1}{2}[2(7.5) + 1]^{-3/2}$

$$= -\dfrac{1}{2}(16)^{-3/2}$$

$$= -\dfrac{1}{2}\left(\dfrac{1}{(\sqrt{16})^3}\right)$$

$$= -\dfrac{1}{128}$$

$$\approx -0.008$$

(d) C is always decreasing because

$$C' = -\tfrac{1}{2}(2t + 1)^{-3/2}$$

is always negative for $t \geq 0$.

(The amount of calcium in the bloodstream will continue to decrease over time.)

65. $V(r) = \dfrac{4}{3}\pi r^3,\ S(r) = 4\pi r^2,\ r(t) = 6 - \dfrac{3}{17}t$

(a) $r(t) = 0$ when $6 - \dfrac{3}{17}t = 0$;

$$t = \dfrac{17(6)}{3} = 34 \text{ min}.$$

(b) $\dfrac{dV}{dr} = 4\pi r^2, \dfrac{dS}{dr} = 8\pi r, \dfrac{dr}{dt} = -\dfrac{3}{17}$

$$\frac{dV}{dt} = \frac{dV}{dr} \cdot \frac{dr}{dt} = -\frac{12}{17}\pi r^2$$

$$= -\frac{12}{17}\pi\left(6 - \frac{3}{17}t\right)^2$$

$$\frac{dS}{dt} = \frac{dS}{dr} \cdot \frac{dr}{dt} = -\frac{24}{17}\pi r$$

$$= -\frac{24}{17}\pi\left(6 - \frac{3}{17}t\right)$$

When $t = 17$,

$$\frac{dV}{dt} = -\frac{12}{17}\pi\left[6 - \frac{3}{17}(17)\right]^2$$

$$= -\frac{108}{17}\pi \ \text{mm}^3/\text{min}$$

$$\frac{dS}{dt} = -\frac{24}{17}\pi\left[6 - \frac{3}{17}(17)\right]$$

$$= -\frac{72}{17}\pi \ \text{mm}^2/\text{min}$$

At $t = 17$ minutes, the volume is decreasing by $\frac{108}{17}\pi$ mm^3 per minute and the surface area is decreasing by $\frac{72}{17}\pi$ mm^2 per minute.

67. (a) $y = ((x^3)^2)^2$

$$\frac{dy}{dx} = 2((x^3)^2) \cdot \frac{d}{dx}((x^3)^2)$$

$$= 2x^6 \cdot 2(x^3) \cdot \frac{d}{dx}(x^3)$$

$$= 2x^6 \cdot 2x^3 \cdot 3x^2$$

$$= 12x^{11}$$

(b) $y = ((x^3)^2)^2 = x^{12}$

$$\frac{dy}{dx} = 12x^{12-1} = 12x^{11}$$

4.4   Derivatives of Exponential Functions

Your Turn 1

(a) $y = 4^{3x}$

$$\frac{dy}{dx} = (\ln 4) \cdot 4^{3x} \cdot 3 = 3(\ln 4)4^{3x}$$

(b) $y = e^{7x^3 + 5}$

$$\frac{dy}{dx} = e^{7x^3 + 5} \cdot 21x^2 = 21x^2 e^{7x^3 + 5}$$

Your Turn 2

$$y = (x^2 + 1)^2 e^{2x}$$

$$\frac{dy}{dx} = (x^2 + 1) \cdot e^{2x} \cdot 2 + e^{2x} \cdot 2(x^2 + 1) \cdot 2x$$

$$= 2(x^2 + 1)e^{2x}(x^2 + 2x + 1)$$

$$= 2e^{2x}(x^2 + 1)(x + 1)^2$$

Your Turn 3

$$f(x) = \frac{100}{5 + 2e^{-0.01x}}$$

$$f'(x) = \frac{(5 + 2e^{-0.01x}) \cdot 0 - 100 \cdot 2e^{-0.01x} \cdot (-0.01)}{(5 + 2e^{-0.01x})^2}$$

$$= \frac{2e^{-0.01x}}{(5 + 2e^{-0.01x})^2}$$

Your Turn 4

$$Q(t) = 100e^{-0.421t}$$

$$\frac{dQ}{dt} = 100 \cdot e^{-0.421t}(-0.421)$$

$$= -42.1e^{-0.421t}$$

After 2 years ($t = 2$), the rate of change of the quantity present is

$$\frac{dQ}{dt} = -42.1e^{-0.421(2)}$$

$$= -42.1e^{-0.842}$$

$$\approx -18.1 \text{ grams per year}$$

4.4   Warmup Exericses

W1.

$$f(x) = \sqrt{x^6 + 5}$$

Let $h(x) = x^{1/2}$ and $g(x) = x^6 + 5$.

Then $f(x) = h(g(x))$

and $f'(x) = h'(g(x))g'(x)$.

$$h'(g(x)) = \frac{1}{2\big(g(x)\big)^{1/2}} = \frac{1}{2\big(x^6 + 5\big)^{1/2}}$$

$$g'(x) = 6x^5$$

$$f'(x) = \frac{6x^5}{2\sqrt{x^6 + 5}} = \frac{3x^5}{\sqrt{x^6 + 5}}$$

1.
$$y = e^{4x}$$

Let $g(x) = 4x$,

with $g'(x) = 4$.

$$\frac{dy}{dx} = 4e^{4x}$$

W2.

$$f(x) = \big(4x^2 + 3x + 2\big)^{10}$$

Let $h(x) = x^{10}$ and $g(x) = 4x^2 + 3x + 2$.

Then $f(x) = h(g(x))$

and $f'(x) = h'(g(x))g'(x)$.

$$h'(g(x)) = 10(g(x))^9 = 10\big(4x^2 + 3x + 2\big)^9$$

$$g'(x) = 8x + 3$$

$$f'(x) = 10\big(4x^2 + 3x + 2\big)^9 (8x + 3)$$

3. $y = -8e^{3x}$

$$\frac{dy}{dx} = -8(3e^{3x}) = -24e^{3x}$$

5. $y = -16e^{2x+1}$

$$g(x) = 2x + 1$$

$$g'(x) = 2$$

$$\frac{dy}{dx} = -16(2e^{2x+1}) = -32e^{2x+1}$$

7. $y = e^{x^2}$

$$g(x) = x^2$$

$$g'(x) = 2x$$

$$\frac{dy}{dx} = 2xe^{x^2}$$

W3. $f(x) = \left(\dfrac{x^2 - 1}{x^2 + 1}\right)^{5/2}$

Let $h(x) = x^{5/2}$ and $g(x) = \dfrac{x^2 - 1}{x^2 + 1}$.

Then $f(x) = h(g(x))$

and $f'(x) = h'(g(x))g'(x)$.

$$h'(g(x)) = \frac{5}{2}\left(\frac{x^2 - 1}{x^2 + 1}\right)^{3/2}$$

$$g'(x) = \frac{(2x)\big(x^2 + 1\big) - \big(x^2 - 1\big)(2x)}{\big(x^2 + 1\big)^2}$$

$$= \frac{4x}{\big(x^2 + 1\big)^2}$$

$$f'(x) = \left[\frac{5}{2}\left(\frac{x^2 - 1}{x^2 + 1}\right)^{3/2}\right]\left[\frac{4x}{\big(x^2 + 1\big)^2}\right]$$

$$= \frac{10x\big(x^2 - 1\big)^{3/2}}{\big(x^2 + 1\big)^{7/2}}$$

9. $y = 3e^{2x^2}$

$$g(x) = 2x^2$$

$$g'(x) = 4x$$

$$\frac{dy}{dx} = 3\left(4xe^{2x^2}\right)$$

$$= 12xe^{2x^2}$$

11. $y = 4e^{2x^2 - 4}$

$$g(x) = 2x^2 - 4$$

$$g'(x) = 4x$$

$$\frac{dy}{dx} = 4\left[(4x)e^{2x^2 - 4}\right]$$

$$= 16xe^{2x^2 - 4}$$

13. $y = xe^x$

Use the product rule.

$$\frac{dy}{dx} = xe^x + e^x \cdot 1 = e^x(x + 1)$$

15. $y = (x + 3)^2 e^{4x}$

Use the product rule.

$$\frac{dy}{dx} = (x + 3)^2(4)e^{4x} + e^{4x} \cdot 2(x + 3)$$

$$= 4(x + 3)^2 e^{4x} + 2(x + 3)e^{4x}$$

$$= 2(x + 3)e^{4x}\big[2(x + 3) + 1\big]$$

$$= 2(x + 3)(2x + 7)e^{4x}$$

17. $y = \dfrac{x^2}{e^x}$

Use the quotient rule.

$$\frac{dy}{dx} = \frac{e^x(2x) - x^2 e^x}{(e^x)^2}$$

$$= \frac{xe^x(2 - x)}{e^{2x}}$$

$$= \frac{x(2 - x)}{e^x}$$

19. $y = \dfrac{e^x + e^{-x}}{x}$

$$\frac{dy}{dx} = \frac{x(e^x - e^{-x}) - (e^x + e^{-x})}{x^2}$$

21. $p = \dfrac{10,000}{9 + 4e^{-0.2t}}$

$$\frac{dp}{dt} = \frac{(9 + 4e^{-0.2t}) \cdot 0 - 10,000[0 + 4(-0.2)e^{-0.2t}]}{(9 + 4e^{-0.2t})^2}$$

$$= \frac{8000e^{-0.2t}}{(9 + 4e^{-0.2t})^2}$$

23. $f(z) = \left(2z + e^{-z^2}\right)^2$

$$f'(z) = 2\left(2z + e^{-z^2}\right)^1\left(2 - 2ze^{-z^2}\right)$$

$$= 4\left(2z + e^{-z^2}\right)\left(1 - ze^{-z^2}\right)$$

25. $y = 7^{3x+1}$

Let g $(x) = 3x + 1$, with $g'(x) = 3$. Then

$$\frac{dy}{dx} = (\ln 7)(7^{3x+1}) \cdot 3 = 3(\ln 7)7^{3x+1}$$

27. $y = 3 \cdot 4^{x^2+2}$

Let $g(x) = x^2 + 2$, with $g'(x) = 2x$. Then

$$\frac{dy}{dx} = 3(\ln 4)4^{x^2+2} \cdot 2x = 6x(\ln 4)4^{x^2+2}$$

29. $s = 2 \cdot 3^{\sqrt{t}}$

Let $g(t) = \sqrt{t}$, with $g'(t) = \frac{1}{2\sqrt{t}}$. Then

$$\frac{ds}{dt} = 2(\ln 3)3^{\sqrt{t}} \cdot \frac{1}{2\sqrt{t}}$$

$$= \frac{(\ln 3)3^{\sqrt{t}}}{\sqrt{t}}$$

31. $y = \dfrac{te^t + 2}{e^{2t} + 1}$

Use the quotient rule and product rule.

$$\frac{dy}{dt} = \frac{(e^{2t} + 1)(te^t + e^t \cdot 1) - (te^t + 2)(2e^{2t})}{(e^{2t} + 1)^2}$$

$$= \frac{(e^{2t} + 1)(te^t + e^t) - (te^t + 2)(2e^{2t})}{(e^{2t} + 1)^2}$$

$$= \frac{te^{3t} + e^{3t} + te^t + e^t - 2te^{3t} - 4e^{2t}}{(e^{2t} + 1)^2}$$

$$= \frac{-te^{3t} + e^{3t} + te^t + e^t - 4e^{2t}}{(e^{2t} + 1)^2}$$

$$= \frac{(1 - t)e^{3t} - 4e^{2t} + (1 + t)e^t}{(e^{2t} + 1)^2}$$

33. $f(x) = e^{x\sqrt{3x+2}}$

Let $g(x) = x\sqrt{3x + 2}$.

$$g'(x) = 1 \cdot \sqrt{3x + 2} + x\left(\frac{3}{2\sqrt{3x + 2}}\right)$$

$$= \sqrt{3x + 2} + \frac{3x}{2\sqrt{3x + 2}}$$

$$= \frac{2(3x + 2)}{2\sqrt{3x + 2}} + \frac{3x}{2\sqrt{3x + 2}}$$

$$= \frac{9x + 4}{2\sqrt{3x + 2}}$$

$$f'(x) = e^{x\sqrt{3x+2}} \cdot \left(\frac{9x + 4}{2\sqrt{3x + 2}}\right)$$

35. $y = y_o e^{kt}$

$\dfrac{dy}{dx} = \dfrac{d}{dt}\left[y_o e^{kt}\right] = y_o k e^{kt}$

$\quad = k(y_o e^{kt})$

$\quad = ky$

37.   Graph the function $y = e^x$.

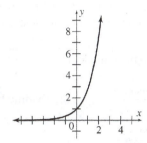

Sketch the lines tangent to the graph at $x = -1$, $0, 1, 2$.

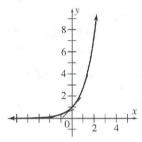

Estimate the slopes of the tangent lines at these points.

At $x = -1$ the slope is a little steeper than $\frac{1}{3}$ or approximately $0.\overline{3}$.

At $x = 0$ the slope is 1.

At $x = 1$ the slope is a little steeper than $\frac{5}{2}$ or 2.5.

At $x = 2$ the slope is a little steeper than $7\frac{1}{3}$ or $7.\overline{3}$.

Note that $e^{-1} \approx 0.36787944$, $e^0 = 1$,

$e^1 = e \approx 2.7182812$, and $e^2 \approx 7.3890561$. The values are close enough to the slopes of the tangent lines to convince us that $\frac{de^x}{dx} = e^x$.

39. $C(x) = \sqrt{900 - 800 \cdot 1.1^{-x}}$

$C(x) = [900 - 800(1.1^{-x})]^{1/2}$

$\quad = \dfrac{1}{2}[900 - 800(1.1^{-x})]^{-1/2}$

$\quad \cdot [-800(\ln 1.1)(1.1^{-x})(-1)]$

$C'(x) = \dfrac{(400 \ln 1.1)(1.1^{-x})}{\sqrt{900 - 800(1.1^{-x})}}$

(a) $C'(0) = \dfrac{400 \ln 1.1}{\sqrt{100}} \approx 3.81$

The marginal cost is $3.81.

(b) $C'(20) = \dfrac{(400 \ln 1.1)(1.1^{-20})}{\sqrt{900 - 800(1.1^{-20})}} \approx 0.20$

The marginal cost is $.20.

(c)   As x becomes larger and larger, $C'(x)$ approaches zero.

41.   (a)

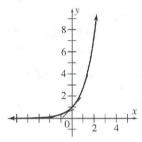

(b)

$S(A) = c - de^{-A\theta}$

$S'(A) = \theta d e^{-A\theta}$

With $c = 60,000$, $d = 50,000$, and $\theta = 0.0006$,

$S(2500) = 60,000 - (50,000)e^{-(2500)(0.0006)}$

$\quad = 48,843.49 = \$48,843.$

$S'(2500) = (0.0006)(50,000)e^{-(2500)(0.0006)}$

$\quad = 6.694$

(The units are dollars of sales per dollar of advertising.)

(c) $S(4000) = 60,000 - (50,000)e^{-(4000)(0.0006)}$

$\quad = 55,464.10 = \$55,464.$

$S'(4000) = (0.0006)(50,000)e^{-(4000)(0.0006)}$

$\quad = 2.722$

43. **(a)** $G_0 = 2$, $m = 250$, $k = 0.0018$

$$G(t) = \frac{250}{1 + \left(\frac{250}{2} - 1\right)e^{-0.0018(250)t}}$$

$$= \frac{250}{1 + 124e^{-0.45t}}$$

(b)

$$G'(t) = \frac{(1 + 124e^{-0.45t})(0) - 250(124e^{-0.45t})(-0.45)}{(1 + 124e^{-0.45t})^2}$$

$$= \frac{13,950e^{-0.45t}}{(1 + 124e^{-0.45t})^2}$$

1995 when $t = 5$:

$$G(5) = \frac{250}{1 + 124e^{-0.45(5)}} \approx 17.8$$

$$G'(5) = \frac{13,950e^{-0.45(5)}}{\left(1 + 124e^{-0.45(5)}\right)^2} \approx 7.4$$

The number of Internet users in 1995 is about 17.8 million, and the growth rate is about 7.4 million users per year .

(c) 2000 when $t = 10$:

$$G(10) = \frac{250}{1 + 124e^{-0.45(10)}} \approx 105.2$$

$$G'(10) = \frac{13,950e^{-0.45(10)}}{\left(1 + 124e^{-0.45(10)}\right)^2} \approx 27.4$$

The number of Internet users in 2000 is about 105.2 million, and the growth rate is about 27.4 million users per year.

(d) 2010 when $t = 20$:

$$G(20) = \frac{250}{1 + 124e^{-0.45(20)}} \approx 246.2$$

$$G'(20) = \frac{13,950e^{-0.45(20)}}{(1 + 124e^{-0.45(20)})^2} \approx 1.7$$

The number is Internet users in 2010 is about 246.2 million, and the growth rate is about 1.7 million users per year.

(e) The rate of growth increases for a while and then gradually decreases to 0.

45. $h(t) = 37.79(1.021)^t$

$h'(t) = 37.79(\ln 1.021)(1.021)^t$

$\quad = 0.789(1.021)^t$

(a) For 2015, $t = 15$:

$$h(15) = 37.79(1.021)^{15} \approx 51.6$$

The Hispanic population in the United States will be about 51,600,000 in 2015.

(b) $h'(15) = 0.789(\ln 1.021)(1.021)^{15} \approx 1.07$

The Hispanic population in the United States will be increasing at the rate of 1,070,000 people per year at the end of the year 2015.

47. **(a)** The general logistic model is

$$G(t) = \frac{mG_0}{G_0 + (m - G_0)e^{-kmt}}, \text{ or, dividing by } G_0,$$

$$G(t) = \frac{m}{1 + \frac{m - G_0}{G_0}e^{-kmt}}.$$

For the cactus wrens,
$k = 0.0125$, $G_0 = 2.33$, and $m = 31.4$.

$$\frac{m - G_0}{G_0} = \frac{31.4 - 2.33}{2.33} = 12.4764 \approx 12.48$$

$$mt = (31.4)(0.0125) = 0.3925 \approx 0.393$$

The model is

$$G(t) = \frac{31.4}{1 + 12.5e^{-0.393t}}.$$

$$G'(t) = \frac{-31.4\left((12.5)(-0.393)e^{-0.393t}\right)}{\left(1 + 12.5e^{-0.393t}\right)^2}$$

$$= \frac{154.25e^{-0.393t}}{\left(1 + 12.5e^{-0.393t}\right)^2}$$

(b) $G(1) = 3.3270 = 3.33$ g

$G'(1) = 1.1690 = 1.17$ g/day

(c) $G(5) = 11.4104 = 11.4$ g

$G'(5) = 2.8547 = 2.85$ g/day

(d) $G(10) = 25.2098 = 25.2$ g

$G'(10) = 1.9531 = 1.95$ g/day

(e) The growth rate of cactus wrens increases for a while and then gradually decreases.

49. $V(t) = 1100[1023e^{-0.02415t} + 1]^{-4}$

(a) $V(240) = 1100[1023e^{-0.02415(240)} + 1]^{-4}$

$\qquad \approx 3.857 \, \text{cm}^3$

(b) $V = \frac{4}{3}\pi r^3$, so $r(V) = \sqrt[3]{\frac{3V}{4\pi}}$

$r(3.857) = \sqrt[3]{\frac{3(3.857)}{4\pi}} \approx 0.973\,\text{cm}$

(c)

$V(t) = 1100[1023e^{-0.02415t} + 1]^{-4} = 0.5$

$[1023e^{-0.02415t} + 1]^{-4} = \frac{1}{2200}$

$(1023e^{-0.02415t} + 1)^4 = 2200$

$1023e^{-0.02415t} + 1 = 2200^{1/4}$

$1023e^{-0.02415t} = 2200^{1/4} - 1$

$e^{-0.02415t} = \frac{2200^{1/4} - 1}{1023}$

$-0.02415t = \ln\left(\frac{2200^{1/4} - 1}{1023}\right)$

$t = \frac{1}{-0.02415}\ln\left(\frac{2200^{1/4} - 1}{1023}\right) \approx 214\,\text{months}$

The tumor has been growing for almost 18 years.

(d) As t goes to infinity, $e^{-0.02415t}$ goes to zero, and $V(t) = 1100[1023e^{-0.02415t} + 1]^{-4}$ goes to 1100 cm³, which corresponds to a sphere with a radius of $\sqrt[3]{\frac{3(1100)}{4\pi}} \approx 6.4$ cm. It makes sense that a tumor growing in a person's body reaches a maximum volume of this size.

(e) By the chain rule,

$\frac{dV}{dt} = 1100(-4)[1023e^{-0.2415t} + 1]^{-5}$

$\cdot (1023)(e^{-0.2415t})(-0.02415)$

$= 108,703.98[1023e^{-0.02415t} + 1]^{-5}e^{-0.2415t}$

At $t = 240$, $\frac{dV}{dt} \approx 0.282$.

At 240 months old, the tumor is increasing in volume at the instantaneous rate of 0.282 cm³/month.

51. $p(x) = 0.001131e^{0.1268x}$

(a) $p(40) = 0.001131e^{0.1268(40)} \approx 0.180$

(b) When $p(x) = 1$,

$0.001131e^{0.1268x} = 1$

$e^{0.1268x} = \frac{1}{0.001131}$

$0.1268x = \ln\frac{1}{0.001131}$

$x = \frac{1}{0.1268}\ln\frac{1}{0.001131}$

≈ 54

This represents the year 2024.

(c) $p'(x) = 0.001131e^{0.1268x}(0.1268)$

$= 0.0001434108e^{0.1268x}$

$p'(44) = 0.0001434108e^{0.1268(44)}$

≈ 0.038

The marginal increase in the proportion per year in 2014 is approximately 0.038.

53. $W_1(t) = 509.7(1 - 0.941e^{-0.00181t})$

$W_2(t) = 498.4(1 - 0.889e^{-0.00219t})^{1.25}$

(a) Both W_1 and W_2 are strictly increasing functions, so they approach their maximum values as t approaches ∞.

$\lim_{t\to\infty} W_1(t) = \lim_{t\to\infty} 509.7(1 - 0.941e^{-0.00181t})$

$= 509.7(1 - 0) = 509.7$

$\lim_{t\to\infty} W_2(t) = \lim_{t\to\infty} 498.4(1 - 0.889e^{-0.00219t})^{1.25}$

$= 498.4(1 - 0)^{1.25} = 498.4$

So, the maximum values of W_1 and W_2 are 509.7 kg and 498.4 kg respectively.

(b) $0.9(509.7) = 509.7(1 - 0.941e^{-0.00181t})$

$0.9 = 1 - 0.941e^{-0.00181t}$

$\frac{0.1}{0.941} \approx e^{-0.00181t}$

$1239 \approx t$

$0.9(498.4) = 498.4(1 - 0.889e^{-0.00219t})^{1.25}$

$0.9 = (1 - 0.889e^{-0.00219t})^{1.25}$

$\frac{1 - 0.9^{0.8}}{0.889} = e^{-0.00219t}$

$1095 \approx t$

Respectively, it will take the average beef cow about 1239 days or 1095 days to reach 90% of its maximum.

(c)

$$W_1'(t) = (509.7)(-0.941)(-0.00181)e^{-0.00181t}$$

$$\approx 0.868126e^{-0.00181t}$$

$$W_1'(750) \approx 0.868126e^{-0.00181(750)}$$

$$\approx 0.22 \text{ kg/day}$$

$$W_2'(t) = (498.4)(1.25)(1 - 0.889e^{-0.00219t})^{0.25}$$

$$\cdot (-0.889)(-0.00219)e^{-0.00219t}$$

$$\approx 1.21292e^{-0.00219t}(1 - 0.889e^{-0.00219t})^{0.25}$$

$$W_2'(750) \approx 1.12192e^{-0.00219(750)}$$

$$\cdot \left(1 - 0.889e^{-0.00219(750)}\right)^{0.25}$$

$$\approx 0.22 \text{ kg/day}$$

Both functions yield a rate of change of about 0.22 kg per day.

(d) Looking at the graph, the growth patterns of the two functions are very similar.

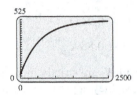

(e) The graphs of the rate of change of the two functions are also very similar.

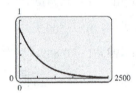

55. (a) $G_0 = 0.00369, m = 1, k = 3.5$

$$G(t) = \cfrac{1}{1 + \left(\cfrac{1}{0.00369} - 1\right)e^{-3.5(1)t}}$$

$$= \frac{1}{1 + 270e^{-3.5t}}$$

(b)

$$G'(t) = -(1 + 270e^{-3.5t})^{-2} \cdot 270e^{-3.5t}(-3.5)$$

$$= \frac{945e^{-3.5t}}{(1 + 270e^{-3.5t})^2}$$

$$G(1) = \frac{1}{1 + 270e^{-3.5(1)}} \approx 0.109$$

$$G'(1) = \frac{945e^{-3.5(1)}}{\left[1 + 270e^{-3.5(1)}\right]^2} \approx 0.341$$

The proportion is 0.109 and the rate of growth is 0.341 per century.

(c) $$G(2) = \frac{1}{1 + 270e^{-3.5(2)}}$$

$$\approx 0.802$$

$$G'(2) = \frac{945e^{-3.5(2)}}{[1 + 270e^{-3.5(2)}]^2}$$

$$\approx 0.555$$

The proportion is 0.802 and the rate of growth is 0.555 per century.

(d) $$G(3) = \frac{1}{1 + 270e^{-3.5(3)}}$$

$$\approx 0.993$$

$$G'(3) = \frac{945e^{-3.5(2)}}{[1 + 270e^{-3.5(2)}]^2}$$

$$\approx 0.0256$$

The proportion is 0.993 and the rate of growth is 0.0256 per century.

(e) The rate of growth increases for a while and then gradually decreases to 0.

57. $G_0 = 1.603$, $m = 6.8$, $k = 0.0440$

(a) $G(t) = \dfrac{6.8}{1 + \left(\frac{6.8}{1.603} - 1\right)e^{-0.0440(6.8)t}}$

$\qquad = \dfrac{6.8}{1 + 3.242e^{-0.2992t}}$

(b) For 2006, $t = 4$:

$G(4) = \dfrac{6.8}{1 + 3.242e^{-0.2992(4)}} \approx 3.435$

$G'(4) = \dfrac{6.59604e^{-0.2992(4)}}{\left(1 + 3.242e^{-0.2992(4)}\right)^2} \approx 0.509$

In 2006, about 3.435 million students enrolled in at least one online course and the growth rate is about 0.509 million students per year.

(c) For 2010, $t = 8$:

$G(8) = \dfrac{6.8}{1 + 3.242e^{-0.2992(8)}} \approx 5.247$

$G'(8) = \dfrac{6.59604e^{-0.2992(8)}}{\left(1 + 3.242e^{-0.2992(8)}\right)^2} \approx 0.359$

In 2010, about 5.247 million students enrolled in at least one online course and the growth rate is about 0.359 million students per year.

(d) For 2014, $t = 12$:

$G(12) = \dfrac{6.8}{1 + 3.242e^{-0.2992(12)}} \approx 6.242$

$G'(12) = \dfrac{6.59604e^{-0.2992(12)}}{\left(1 + 3.242e^{-0.2992(12)}\right)^2} \approx 0.153$

In 2014, about 6.242 million students enrolled in at least one online course and the growth rate is about 0.153 million students per year.

(e) The growth rate increases at first and then decreases.

59. $Q(t) = CV(1 - e^{-t/RC})$

(a) $I_c = \dfrac{dQ}{dt} = CV\left[0 - e^{-t/RC}\left(-\dfrac{1}{RC}\right)\right]$

$\qquad = CV\left(\dfrac{1}{RC}\right)e^{-t/RC}$

$\qquad = \dfrac{V}{R}e^{-t/RC}$

(b) When $C = 10^{-5}$ farads, $R = 10^7$ ohms, and $V = 10$ volts, after 200 seconds

$$I_c = \dfrac{10}{10^7}e^{-200/(10^7 \cdot 10^{-5})}$$

$$\approx 1.35 \times 10^{-7}\,\text{amps}$$

61. Using the given value of $T = 30$,

$$H(D) = 30 + \frac{5}{9}\left[6.11e^{5417.7530 \cdot \left(\frac{1}{273.16} - \frac{1}{D}\right)} - 10\right]$$

$$H'(D) = \frac{18{,}390.2616}{D^2} \cdot e^{5417.7530 \cdot \left(\frac{1}{273.16} - \frac{1}{D}\right)}$$

where $18{,}390.2616$ is approximately equal to $(5/9)(6.11)(5417.7530)$

(a) $H(278) = 29.239 \approx 29.24°$

$H'(278) = 0.3361$

(b) $H(290) = 35.183 \approx 35.18°$

$H'(290) = 0.6918$

(c) $H(300) = 44.462 \approx 44.46°$

$H'(300) = 1.205$

63. **(a)** $s(0) = 693.9 - 34.38\left(e^{0.01003(0)} + e^{-0.01003(0)}\right)$

$= 693.9 - 34.38(2)$

$= 625.14$

The height of the Gateway Arch at its center is 625.14ft.

(b) $m = s'(t) = -34.38(e^{0.01003x}(0.01003) + e^{-0.01003x}(-0.01003))$

$= -34.38(0.1003e^{0.01003x} - 0.01003e^{-0.01003x})$

$m = s'(0) = -34.38(0.01003(1) - 0.01003(1)) = 0$

The slope of the line tangent to the curve when $x = 0$ is 0, which makes sense because the tangent line has to be a horizontal line at the center of the arch (the highest point on the curve)

(c) $s'(150) = -34.38\left(0.01003e^{0.01003(150)} - 0.01003e^{-0.01003(150)}\right)$

$= -1.476$

The rate of change of the height of the arch at $x = 150$ ft is -1.476 ft per foot from the center.

65. $t(r) = 218 + 31(0.933)^n$

(a) $t(64) = 218 + 31(0.933)^{64}$

≈ 218.4 sec

(b) $t'(n) = (31 \ln 0.933)(0.933)^n$

$t'(64) = (31 \ln 0.933)(0.933)^{64}$

≈ -0.025

The record is decreasing by 0.025 seconds per year at the end of 2014.

(c) As $n \to \infty$, $(0.933)^n \to 0$ and $t(n) \to 218$.

If the estimate is correct, then this is the least amount of time that it will ever take a human to run a mile.

4.5   Derivatives of Logarithmic Functions

Your Turn 1

$f(x) = \log_3 x$

$f'(x) = \dfrac{1}{(\ln 3)x}$

Your Turn 2

(a) $y = \ln(2x^3 - 3)$

$\dfrac{dy}{dx} = \dfrac{1}{2x^3 - 3} \cdot \dfrac{d}{dx}(2x^3 - 3)$

$= \dfrac{1}{2x^3 - 3}(6x^2)$

$= \dfrac{6x^2}{2x^3 - 3}$

(b) $f(x) = \log_4(5x + 3x^3)$

$f'(x) = \dfrac{1}{(\ln 4)(5x + 3x^3)} \cdot \dfrac{d}{dx}(5x + 3x^3)$

$= \dfrac{5 + 9x^2}{(\ln 4)(5x + 3x^3)}$

Your Turn 3

(a) $y = \ln|2x + 6|$ **(b)**

$\dfrac{dy}{dx} = \dfrac{1}{2x + 6} \cdot \dfrac{d}{dx}(2x + 6)$

$= \dfrac{2}{2x + 6} = \dfrac{2}{2(x + 3)} = \dfrac{1}{x + 3}$

$f(x) = x^2 \ln 3x$

$f'(x) = x^2 \cdot \dfrac{1}{3x}(3) + \ln 3x(2x)$

$= x + 2x \ln 3x$

(c) $s(t) = \dfrac{\ln(t^2 - 1)}{t + 1}$

$s'(t) = \dfrac{(t + 1)\left(\frac{1}{t^2 - 1}\right)(2t) - \ln(t^2 - 1)(1)}{(t + 1)^2}$

$= \dfrac{\frac{2t}{t - 1} - \ln(t^2 - 1)}{(t + 1)^2}$

$= \dfrac{2t - (t - 1)\ln(t^2 - 1)}{(t - 1)(t + 1)^2}$

4.5   Warmup Exercises

W1.

$f(x) = e^{x^2}$

$f'(x) = \left(\dfrac{d}{dx}x^2\right)e^{x^2} = 2xe^{x^2}$

W2.

$f(x) = x^2 e^{4x}$

$f'(x) = x^2\left(\dfrac{d}{dx}e^{4x}\right) + \left(\dfrac{d}{dx}x^2\right)e^{4x}$

$= x^2\left(4e^{4x}\right) + (2x)e^{4x}$

$= 2x(2x + 1)e^{4x}$

W3.

$f(x) = \dfrac{e^{2x}}{e^{3x} + 1}$

$f'(x) = \dfrac{\left(2e^{ex}\right)\left(e^{3x} + 1\right) - \left(e^{2x}\right)\left(3x^{3x}\right)}{\left(e^{3x} + 1\right)^2}$

$= \dfrac{e^{2x}\left(2e^{3x} + 2 - 3e^{3x}\right)}{\left(e^{3x} + 1\right)^2}$

$= \dfrac{e^{2x}\left(2 - e^{3x}\right)}{\left(e^{3x} + 1\right)^2}$

4.5   Exercises

1. $y = \ln(8x)$

$\dfrac{dy}{dx} = \dfrac{d}{dx}(\ln 8x)$

$= \dfrac{d}{dx}(\ln 8 + \ln x)$

$= \dfrac{d}{dx}(\ln 8) + \dfrac{d}{dx}(\ln x)$

$= 0 + \dfrac{1}{x} = \dfrac{1}{x}$

3. $y = \ln(8 - 3x)$

$g(x) = 8 - 3x$

$g'(x) = -3$

$\dfrac{dy}{dx} = \dfrac{g'(x)}{g(x)} = \dfrac{-3}{8 - 3x}$ or $\dfrac{3}{3x - 8}$

5. $y = \ln |4x^2 - 9x|$

$g(x) = 4x^2 - 9x$

$g'(x) = 8x - 9$

$\dfrac{dy}{dx} = \dfrac{g'(x)}{g(x)} = \dfrac{8x - 9}{4x^2 - 9x}$

7. $y = \ln \sqrt{x + 5}$

$g(x) = \sqrt{x + 5} = (x + 5)^{1/2}$

$g'(x) = \dfrac{1}{2}(x + 5)^{-1/2}$

$\dfrac{dy}{dx} = \dfrac{\frac{1}{2}(x + 5)^{-1/2}}{(x + 5)^{1/2}} = \dfrac{1}{2(x + 5)}$

9. $y = \ln(x^4 + 5x^2)^{3/2}$

$= \dfrac{3}{2} \ln(x^4 + 5x^2)$

$\dfrac{dy}{dx} = \dfrac{3}{2} D_x[\ln(x^4 + 5x^2)]$

$g(x) = x^4 + 5x^2$

$g'(x) = 4x^3 + 10x$

$\dfrac{dy}{dx} = \dfrac{3}{2}\left[\dfrac{4x^3 + 10x}{x^4 + 5x^2}\right]$

$= \dfrac{3}{2}\left[\dfrac{2x(2x^2 + 5)}{x^2(x^2 + 5)}\right]$

$= \dfrac{3(2x^2 + 5)}{x(x^2 + 5)}$

11. $y = -5x \ln(3x + 2)$

Use the product rule.

$\dfrac{dy}{dx} = -5x\left[\dfrac{d}{dx}\ln(3x + 2)\right] + \ln(3x + 2)\left[\dfrac{d}{dx}(-5x)\right]$

$= -5x\left(\dfrac{3}{3x + 2}\right) + [\ln(3x + 2)](-5)$

$= -\dfrac{15x}{3x + 2} - 5\ln(3x + 2)$

13. $s = t^2 \ln|t|$

$\dfrac{ds}{dt} = t^2 \cdot \dfrac{1}{t} + 2t \ln|t|$

$= t + 2t \ln|t|$

$= t(1 + 2\ln|t|)$

15. $y = \dfrac{2 \ln(x + 3)}{x^2}$

Use the quotient rule.

$\dfrac{dy}{dx} = \dfrac{x^2\left(\frac{2}{x+3}\right) - 2\ln(x + 3) \cdot 2x}{(x^2)^2}$

$= \dfrac{\frac{2x^2}{x+3} - 4x\ln(x + 3)}{x^4}$

$= \dfrac{2x^2 - 4x(x + 3)\ln(x + 3)}{x^4(x + 3)}$

$= \dfrac{x[2x - 4(x + 3)\ln(x + 3)]}{x^4(x + 3)}$

$= \dfrac{2x - 4(x + 3)\ln(x + 3)}{x^3(x + 3)}$

17. $y = \dfrac{\ln x}{4x + 7}$

Use the quotient rule.

$\dfrac{dy}{dx} = \dfrac{(4x + 7)\left(\frac{1}{x}\right) - (\ln x)(4)}{(4x + 7)^2}$

$= \dfrac{\frac{4x+7}{x} - 4\ln x}{(4x + 7)^2}$

$= \dfrac{4x + 7 - 4x\ln x}{x(4x + 7)^2}$

19. $y = \dfrac{3x^2}{\ln x}$

$\dfrac{dy}{dx} = \dfrac{(\ln x)(6x) - 3x^2\left(\frac{1}{x}\right)}{(\ln x)^2}$

$= \dfrac{6x \ln x - 3x}{(\ln x)^2}$

21. $y = (\ln|x + 1|)^4$

$\dfrac{dy}{dx} = 4(\ln|x + 1|)^3\left(\dfrac{1}{x + 1}\right)$

$= \dfrac{4(\ln|x + 1|)^3}{x + 1}$

23. $y = \ln|\ln x|$

$g(x) = \ln x$

$g'(x) = \dfrac{1}{x}$

$$\frac{dy}{dx} = \frac{g'(x)}{g(x)}$$

$$= \frac{\frac{1}{x}}{\ln x}$$

$$= \frac{1}{x \ln x}$$

25. $y = e^{x^2} \ln x, \; x > 0$

$$\frac{dy}{dx} = e^{x^2}\left(\frac{1}{x}\right) + (\ln x)(2x)e^{x^2}$$

$$= \frac{e^{x^2}}{x} + 2xe^{x^2} \ln x$$

27. $y = \dfrac{e^x}{\ln x}, \; x > 0$

Use the quotient rule.

$$\frac{dy}{dx} = \frac{(\ln x)e^x - e^x\left(\frac{1}{x}\right)}{(\ln x)^2} \cdot \frac{x}{x}$$

$$= \frac{xe^x \ln x - e^x}{x(\ln x)^2}$$

29. $g(z) = (e^{2z} + \ln z)^3$

$$g'(z) = 3(e^{2z} + \ln z)^2\left(e^{2z} \cdot 2 + \frac{1}{z}\right)$$

$$= 3(e^{2z} + \ln z)^2\left(\frac{2ze^{2z} + 1}{z}\right)$$

31. $y = \log(6x)$

$g(x) = 6x$ and $g'(x) = 6.$

$$\frac{dy}{dx} = \frac{1}{\ln 10}\left(\frac{6}{6x}\right)$$

$$= \frac{1}{x \ln 10}$$

33. $y = \log|1 - x|$

$g(x) = 1 - x$ and $g'(x) = -1.$

$$\frac{dy}{dx} = \frac{1}{\ln 10} \cdot \frac{-1}{1 - x}$$

$$= -\frac{1}{(\ln 10)(1 - x)}$$

$$\text{or } \frac{1}{(\ln 10)(x - 1)}$$

35. $y = \log_5 \sqrt{5x + 2}$

$g(x) = \sqrt{5x + 2}$ and $g'(x) = \dfrac{5}{2\sqrt{5x + 2}}.$

$$\frac{dy}{dx} = \frac{1}{\ln 5} \cdot \frac{\frac{5}{2\sqrt{5x+2}}}{\sqrt{5x + 2}}$$

$$= \frac{5}{2 \ln 5(5x + 2)}$$

37. $y = \log_3 (x^2 + 2x)^{3/2}$

$g(x) = (x^2 + 2x)^{3/2}$ and

$$g'(x) = \frac{3}{2}(x^2 + 2x)^{1/2} \cdot (2x + 2)$$

$$= 3(x + 1)(x^2 + 2x)^{1/2}$$

$$\frac{dy}{dx} = \frac{1}{\ln 3} \cdot \frac{3(x + 1)(x^2 + 2x)^{1/2}}{(x^2 + 2x)^{3/2}}$$

$$= \frac{3(x + 1)}{(\ln 3)(x^2 + 2x)}$$

39. $w = \log_8(2^p - 1)$

$g(p) = 2^p - 1$

$g'(p) = (\ln 2)2^p$

$$\frac{dw}{dp} = \frac{1}{\ln 8} \cdot \frac{(\ln 2)2^p}{(2^p - 1)}$$

$$= \frac{(\ln 2)2^p}{(\ln 8)(2^p - 1)}$$

41. $f(x) = e^{\sqrt{x}} \ln(\sqrt{x} + 5)$

Use the product rule.

$$f'(x) = e^{\sqrt{x}}\left(\frac{\frac{1}{2\sqrt{x}}}{\sqrt{x} + 5}\right) + [\ln(\sqrt{x} + 5)]e^{\sqrt{x}}\left(\frac{1}{2\sqrt{x}}\right)$$

$$= \frac{e^{\sqrt{x}}}{2}\left[\frac{1}{\sqrt{x}(\sqrt{x} + 5)} + \frac{\ln(\sqrt{x} + 5)}{\sqrt{x}}\right]$$

43. $f(t) = \dfrac{\ln(t^2 + 1) + t}{\ln(t^2 + 1) + 1}$

Use the quotient rule.

$$u(t) = \ln(t^2 + 1) + t,\, u'(t) = \frac{2t}{t^2 + 1} + 1$$

$$v(t) = \ln(t^2 + 1) + 1,\, v'(t) = \frac{2t}{t^2 + 1}$$

$$f'(t) = \frac{[\ln(t^2 + 1) + 1]\left(\frac{2t}{t^2+1} + 1\right) - [\ln(t^2 + 1) + t]\left(\frac{2t}{t^2+1}\right)}{[\ln(t^2 + 1) + 1]^2}$$

$$= \frac{\frac{2t\ln(t^2+1)}{t^2+1} + \frac{2t}{t^2+1} + \ln(t^2 + 1) + 1 - \frac{2t\ln(t^2+1)}{t^2+1} - \frac{2t^2}{t^2+1}}{[\ln(t^2 + 1) + 1]^2}$$

$$= \frac{\frac{2t-2t^2}{t^2+1} + \ln(t^2 + 1) + 1}{[\ln(t^2 + 1) + 1]^2}$$

$$= \frac{2t - 2t^2 + (t^2 + 1)\ln(t^2 + 1) + t^2 + 1}{(t^2 + 1)[\ln(t^2 + 1) + 1]^2}$$

$$= \frac{-t^2 + 2t + 1 + (t^2 + 1)\ln(t^2 + 1)}{(t^2 + 1)[\ln(t^2 + 1) + 1]^2}$$

49. Use the derivative of $\ln x$.

$$\frac{d\ln[u(x)v(x)]}{dx} = \frac{1}{u(x)v(x)} \cdot \frac{d[u(x)v(x)]}{dx}$$

$$\frac{d\ln u(x)}{dx} = \frac{1}{u(x)} \cdot \frac{d[u(x)]}{dx}$$

$$\frac{d\ln v(x)}{dx} = \frac{1}{v(x)} \cdot \frac{d[v(x)]}{dx}$$

Then since $\ln[u(x)v(x)] = \ln u(x) + \ln v(x)$,

$$\frac{1}{u(x)v(x)} \cdot \frac{d[u(x)v(x)]}{dx} = \frac{1}{u(x)} \cdot \frac{d[u(x)]}{dx} + \frac{1}{v(x)} \cdot \frac{d[v(x)]}{dx}.$$

Multiply both sides of this equation by $u(x)v(x)$. Then $\dfrac{d[u(x)v(x)]}{dx} = v(x)\dfrac{d[u(x)]}{dx} + u(x)\dfrac{d[v(x)]}{dx}$.

This is the product rule.

51. Graph the function $y = \ln x$. Sketch lines tagent to the graph at $x = \frac{1}{2}, 1, 2, 3, 4$. Estimate the slopes of the tangent lines at these points.

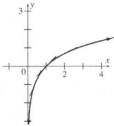

x	slope of tangent
$\frac{1}{2}$	2
1	1
2	$\frac{1}{2}$
3	$\frac{1}{3}$
4	$\frac{1}{4}$

The values of the slopes at x are $\frac{1}{x}$.

Thus we see that $\frac{d \ln x}{dx} = \frac{1}{x}$.

53. (a) $h(x) = u(x)^{v(x)}$

$$\frac{d}{dx} \ln h(x) = \frac{d}{dx} \ln[u(x)^{v(x)}]$$

$$= \frac{d}{dx}[v(x) \ln u(x)]$$

$$= v(x) \frac{d}{dx} \ln u(x) + (\ln u(x)) v'(x)$$

$$= v(x) \frac{u'(x)}{u(x)} + (\ln u(x)) v'(x)$$

$$= \frac{v(x) u'(x)}{u(x)} + (\ln u(x)) v'(x)$$

(b) Since $\frac{d}{dx} \ln h(x) = \frac{h'(x)}{h(x)}$,

$$h'(x) = h(x) \frac{d}{dx} \ln h(x)$$

$$= h(x) \left[\frac{v(x) u'(x)}{u(x)} + (\ln u(x)) v'(x) \right]$$

$$= u(x)^{v(x)} \left[\frac{v(x) u'(x)}{u(x)} + (\ln u(x)) v'(x) \right]$$

55. $h(x) = (x^2 + 1)^{5x}$

$u(x) = x^2 + 1, u'(x) = 2x$

$v(x) = 5x, v'(x) = 5$

$$h'(x) = (x^2 + 1)^{5x} \left[\frac{5x(2x)}{x^2 + 1} + \ln(x^2 + 1) \cdot (5) \right]$$

$$= (x^2 + 1)^{5x} \left[\frac{10x^2}{x^2 + 1} + 5 \ln(x^2 + 1) \right]$$

57. $p = 100 + \dfrac{50}{\ln q}, q > 1$

(a) $R = pq$

$$R = 100q + \frac{50q}{\ln q}$$

The marginal revenue is

$$\frac{dR}{dq} = 100 + \frac{(\ln q)(50) - 50q\left(\frac{1}{q}\right)}{(\ln q)^2}$$

$$= 100 + \frac{50(\ln q - 1)}{(\ln q)^2}.$$

(b) The revenue from one more unit is $\frac{dR}{dq}$ for $q = 8$.

$$100 + \frac{50(\ln 8 - 1)}{(\ln 8)^2} = \$112.48$$

(c) The manager can use the information from (b) to decide if it is reasonable to sell additional items.

59. $C(x) = 5 \log_2 x + 10$

$$\bar{C}(x) = \frac{C(x)}{x} = \frac{5 \log_2 x + 10}{x}$$

$$\bar{C}'(x) = \frac{x \cdot 5 \cdot \frac{1}{\ln 2} \cdot \frac{1}{x} - (5 \log_2 x + 10) \cdot 1}{x^2}$$

$$= \frac{5 - (\ln 2)(5 \log_2 x + 10)}{x^2 \ln 2}$$

(a) $\bar{C}'(10) = \dfrac{5 - (\ln 2)(5 \log_2 10 + 10)}{(10^2) \ln 2}$

$$\approx -\$0.19396$$

(b) $\bar{C}'(20) = \dfrac{5 - (\ln 2)(5 \log_2 20 + 10)}{(20^2) \ln 2}$

$$\approx -\$0.06099$$

61. $\ln\left(\dfrac{N(t)}{N_0}\right) = 9.8901e^{-e^{2.54197-0.2167t}}$

(a) $\dfrac{N(t)}{1000} = e^{9.8901e^{-e^{2.54197-0.2167t}}}$

$N(t) = 1000e^{9.8901e^{-e^{2.54197-0.2167t}}}$

(b) $N'(20) \approx 1{,}307{,}416$ bacteria/hour

Twenty hours into the experiment, the number of bacteria is increasing at a rate of 1,307,416 per hour

(c) $S(t) = \ln\left(\dfrac{N(t)}{N_0}\right)$

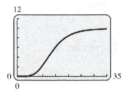

(d)

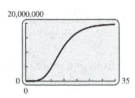

The two graphs have the same general shape, but $N(t)$ is scaled much larger.

(e)

$\displaystyle \lim_{t\to\infty} S(t) = \lim_{t\to\infty} 9.8901e^{-e^{2.54197-0.2167t}}$

$= 9.8901$

$S(t) = \ln\left(\dfrac{N(t)}{N_0}\right)$

$N(t) = N_0 e^{S(t)}$

$\displaystyle \lim_{t\to\infty} N(t) = N_0 e^{\lim_{t\to\infty} S(t)} = 1000e^{9.8901}$

$\approx 19{,}734{,}033$ bacteria

63. $\log y = 1.54 - 0.008x - 0.658\log x$

(a) $y(x) = 10^{(1.54-0.008x-0.658\log x)}$

$= 10^{1.54}(10^{-0.008x})(10^{-0.658\log x})$

$= 10^{1.54}(10^{0.008})^{-x}(10^{\log x})^{-0.658}$

$\approx 34.7(1.0186)^{-x}x^{-0.658}$

(b) **(i)**

$y(20) = 34.7(1.0186)^{-20}20^{-0.658}$

≈ 3.343 imagoes per mated female per day

(ii)

$y(40) = 34.7(1.0186)^{-40}40^{-0.658}$

≈ 1.466 imagoes per mated female per day

(c)

$\dfrac{dy}{dx} = -34.7(1.0186)^{-x}x^{-0.658}\left(\ln(1.0186) + \dfrac{0.658}{x}\right)$

(i)

$\dfrac{dy}{dx}(20) = -0.17160... \approx -0.172$

imagoes per mated female per day per flies per bottle

(ii)

$\dfrac{dy}{dx}(40) = -0.051120... \approx -0.0511$

imagoes per mated female per day per flies per bottle

65. $P(t) = (t + 100)\ln(t + 2)$

$P'(t) = (t + 100)\left(\dfrac{1}{t + 2}\right) + \ln(t + 2)$

$P'(2) = (102)\left(\dfrac{1}{4}\right) + \ln(4) \approx 26.9$ ants/day

$P'(8) = (108)\left(\dfrac{1}{10}\right) + \ln(10) \approx 13.1$ ants/day

67. $M = \dfrac{2}{3}\log\dfrac{E}{0.007}$

(a) $8.9 = \dfrac{2}{3}\log\dfrac{E}{0.007}$

$13.35 = \log\dfrac{E}{0.007}$

$10^{13.35} = \dfrac{E}{0.007}$

$E = 0.007(10^{13.35})$

$\approx 1.567 \times 10^{11}$ kWh

(b) $10,000,000 \times 247$ kWh/month

$\quad = 2,470,000,000$ kWh/month

$\dfrac{1.567 \times 10^{11} \text{ kWh}}{2,470,000,000 \text{ kWh/month}} \approx 63.4$ months

(c) $\quad M = \dfrac{2}{3} \log E - \dfrac{2}{3} \log 0.007$

$\quad \dfrac{dM}{dE} = \dfrac{2}{3}\left(\dfrac{1}{(\ln 10)E}\right) = \dfrac{2}{(3\ln 10)E}$

When $E = 70,000$,

$\quad \dfrac{dM}{dE} = \dfrac{2}{(3\ln 10)70,000}$

$\qquad\qquad \approx 4.14 \times 10^{-6}$

(d) $\dfrac{dM}{dE}$ varies inversely with E, so as E increases, $\dfrac{dM}{dE}$ decreases and approaches zero.

Chapter 4   Review Exercises

1. False; the derivative of π^3 is 0 because π^3 is a constant.

2. True

3. False; the derivative of a product $u(x) \cdot v(x)$ is $u(x) \cdot v'(x) + v(x) \cdot u'(x)$.

4. True

5. False; the chain rule is used to take the derivative of a composition of functions.

6. False; the derivative ce^x is ce^x for any constant c.

7. False; the derivative of 10^x is $(\ln 10)10^x$.

8. True

9. True

10. False; the derivative of $\log x$ is $\dfrac{1}{(\ln 10)x}$, whereas the derivative of $\ln x$ is $\dfrac{1}{x}$.

11. $\quad y = 5x^3 - 7x^2 - 9x + \sqrt{5}$

$\quad \dfrac{dy}{dx} = 5(3x^2) - 7(2x) - 9 + 0$

$\qquad\quad = 15x^2 - 14x - 9$

13. $\quad y = 9x^{8/3}$

$\quad \dfrac{dy}{dx} = 9\left(\dfrac{8}{3}x^{5/3}\right)$

$\qquad\quad = 24x^{5/3}$

15. $\quad f(x) = 3x^{-4} + 6\sqrt{x}$

$\qquad\qquad = 3x^{-4} + 6x^{1/2}$

$\quad f'(x) = 3(-4x^{-5}) + 6\left(\dfrac{1}{2}x^{-1/2}\right)$

$\qquad\qquad = -12x^{-5} + 3x^{-1/2}$ or $-\dfrac{12}{x^5} + \dfrac{3}{x^{1/2}}$

17. $\quad k(x) = \dfrac{3x}{4x + 7}$

$\quad k'(x) = \dfrac{(4x + 7)(3) - (3x)(4)}{(4x + 7)^2}$

$\qquad\quad = \dfrac{12x + 21 - 12x}{(4x + 7)^2}$

$\qquad\quad = \dfrac{21}{(4x + 7)^2}$

19. $\quad y = \dfrac{x^2 - x + 1}{x - 1}$

$\quad \dfrac{dy}{dx} = \dfrac{(x - 1)(2x - 1) - (x^2 - x + 1)(1)}{(x - 1)^2}$

$\qquad\quad = \dfrac{2x^2 - 3x + 1 - x^2 + x - 1}{(x - 1)^2}$

$\qquad\quad = \dfrac{x^2 - 2x}{(x - 1)^2}$

21. $\quad f(x) = (3x^2 - 2)^4$

$\quad f'(x) = 4(3x^2 - 2)^3[3(2x)]$

$\qquad\quad = 24x(3x^2 - 2)^3$

23. $y = \sqrt{2t^7 - 5} = (2t^7 - 5)^{1/2}$

$\dfrac{dy}{dt} = \dfrac{1}{2}(2t^7 - 5)^{1/2}[2(7t^6)]$

$\quad = 7t^6(2t^7 - 5)^{-1/2}$ or $\dfrac{7t^6}{(2t^7 - 5)^{1/2}}$

25. $y = 3x(2x + 1)^3$

$\dfrac{dy}{dx} = 3x(3)(2x + 1)^2(2) + (2x + 1)^3(3)$

$\quad = (18x)(2x + 1)^2 + 3(2x + 1)^3$

$\quad = 3(2x + 1)^2[6x + (2x + 1)]$

$\quad = 3(2x + 1)^2(8x + 1)$

27.

$r(t) = \dfrac{5t^2 - 7t}{(3t + 1)^3}$

$r'(t) = \dfrac{(3t + 1)^3(10t - 7) - (5t^2 - 7t)(3)(3t + 1)^2(3)}{[(3t + 1)^3]^2}$

$\quad = \dfrac{(3t + 1)^3(10t - 7) - 9(5t^2 - 7t)(3t + 1)}{(3t + 1)^6}$

$\quad = \dfrac{(3t + 1)(10t - 7) - 9(5t^2 - 7t)}{(3t + 1)^4}$

$\quad = \dfrac{30t^2 - 11t - 7 - 45t^2 + 63t}{(3t + 1)^4}$

$\quad = \dfrac{-15t^2 + 52t - 7}{(3t + 1)^4}$

29. $p(t) = t^2(t^2 + 1)^{5/2}$

$p'(t) = t^2 \cdot \dfrac{5}{2}(t^2 + 1)^{3/2} \cdot 2t + 2t(t^2 + 1)^{5/2}$

$\quad = 5t^3(t^2 + 1)^{3/2} + 2t(t^2 + 1)^{5/2}$

$\quad = t(t^2 + 1)^{3/2}[5t^2 + 2(t^2 + 1)^1]$

$\quad = t(t^2 + 1)^{3/2}(7t^2 + 2)$

31. $y = -6e^{2x}$

$\dfrac{dy}{dx} = -6(2e^{2x}) = -12e^{2x}$

33. $y = e^{-2x^3}$

$g(x) = -2x^3$

$g'(x) = -6x^2$

$y' = -6x^2 e^{-2x^3}$

35. $y = 5x \cdot e^{2x}$

Use the product rule.

$\dfrac{dy}{dx} = 5x(2e^{2x}) + e^{2x}(5)$

$\quad = 10xe^{2x} + 5e^{2x}$

$\quad = 5e^{2x}(2x + 1)$

37. $y = \ln(2 + x^2)$

$g(x) = 2x + x^2$

$g'(x) = 2x$

$\dfrac{dy}{dx} = \dfrac{2x}{2 + x^2}$

39. $y = \dfrac{\ln|3x|}{x - 3}$

$\dfrac{dy}{dx} = \dfrac{(x - 3)\left(\frac{1}{3x}\right)(3) - (\ln|3x|)(1)}{(x - 3)^2}$

$\quad = \dfrac{x - 3 - x\ln|3x|}{x(x - 3)^2}$

41. $y = \dfrac{xe^x}{\ln(x^2 - 1)}$

$\dfrac{dy}{dx} = \dfrac{\ln(x^2 - 1)[xe^x + e^x] - xe^x\left(\frac{1}{x^2-1}\right)(2x)}{[\ln(x^2 - 1)]^2}$

$\quad = \dfrac{e^x(x + 1)\ln(x^2 - 1) - \frac{2x^2 e^x}{x^2-1}}{[\ln(x^2 - 1)]^2} \cdot \dfrac{x^2 - 1}{x^2 - 1}$

$\quad = \dfrac{e^x(x + 1)(x^2 - 1)\ln(x^2 - 1) - 2x^2 e^x}{(x^2 - 1)[\ln(x^2 - 1)]^2}$

43. $s = (t^2 + e^t)^2$

$s' = 2(t^2 + e^t)(2t + e^t)$

45. $y = 3 \cdot 10^{-x^2}$

$\dfrac{dy}{dx} = 3 \cdot (\ln 10)10^{-x^2}(-2x)$

$\quad = -6x(\ln 10) \cdot 10^{-x^2}$

47. $g(z) = \log_2(z^3 + z + 1)$

$$g'(z) = \frac{1}{\ln 2} \cdot \frac{3z^2 + 1}{z^3 + z + 1}$$

$$= \frac{3z^2 + 1}{(\ln 2)(z^3 + z + 1)}$$

49. $f(x) = e^{2x} \ln(xe^x + 1)$

Use the product rule.

$$f'(x) = e^{2x}\left(\frac{e^x + xe^x}{xe^x + 1}\right) + [\ln(xe^x + 1)](2e^{2x})$$

$$= \frac{(1 + x)e^{3x}}{xe^x + 1} + 2e^{2x}\ln(xe^x + 1)$$

51. **(a)** $D_x(f[g(x)])$ at $x = 2$

$$= f'[g(2)]g'(2)$$

$$= f'(1)\left(\frac{3}{10}\right)$$

$$= -5\left(\frac{3}{10}\right)$$

$$= -\frac{3}{2}$$

(b) $D_x(f[g(x)])$ at $x = 3$

$$= f'[g(3)]g'(3)$$

$$= f'(2)\left(\frac{4}{11}\right)$$

$$= -6\left(\frac{4}{11}\right)$$

$$= -\frac{24}{11}$$

53. $y = x^2 - 6x$; tangent at $x = 2$

$$\frac{dy}{dx} = 2x - 6$$

Slope $= y'(2) = 2(2) - 6 = -2$

Use $(2, -8)$ and point-slope form.

$$y - (-8) = -2(x - 2)$$

$$y + 8 = -2x + 4$$

$$y + 2x = -4$$

$$y = -2x - 4$$

55. $y = \frac{3}{x-1}$; tangent at $x = -1$

$$y = \frac{3}{x - 1} = 3(x - 1)^{-1}$$

$$\frac{dy}{dx} = 3(-1)(x - 1)^{-2}(1)$$

$$= -3(x - 1)^{-2}$$

Slope $= y'(-1) = -3(-1 - 1)^{-2} = -\frac{3}{4}$

Use $\left(-1, -\frac{3}{2}\right)$ and point-slope form.

$$y - \left(-\frac{3}{2}\right) = -\frac{3}{4}[x - (-1)]$$

$$y + \frac{3}{2} = -\frac{3}{4}(x + 1)$$

$$y + \frac{6}{4} = -\frac{3}{4}x - \frac{3}{4}$$

$$y = -\frac{3}{4}x - \frac{9}{4}$$

57. $y = \sqrt{6x - 2}$; tangent at $x = 3$

$$y = \sqrt{6x - 2} = (6x - 2)^{1/2}$$

$$\frac{dy}{dx} = \frac{1}{2}(6x - 2)^{-1/2}(6)$$

$$= 3(6x - 2)^{-1/2}$$

slope $= y'(3) = 3(6 \cdot 3 - 2)^{-1/2}$

$$= 3(16)^{-1/2}$$

$$= \frac{3}{16^{1/2}} = \frac{3}{4}$$

Use $(3, 4)$ and point-slope form.

$$y - 4 = \frac{3}{4}(x - 3)$$

$$y - \frac{16}{4} = \frac{3}{4}x - \frac{9}{4}$$

$$y = \frac{3}{4}x + \frac{7}{4}$$

59. $y = e^x$; $x = 0$

$$\frac{dy}{dx} = e^x$$

The value of $\frac{dy}{dx}$ when $x = 0$ is the slope

$m = e^0 = 1.$

When $x = 0$, $y = e^0 = 1$. Use $m = 1$ with $P(0, 1)$.

$$y - 1 = 1(x - 0)$$

$$y = x + 1$$

61.　$y = \ln x; \; x = 1$

$$\frac{dy}{dx} = \frac{1}{x}$$

The value of $\frac{dy}{dx}$ when $x = 1$ is the

slope $m = \frac{1}{1} = 1$.

When $x = 1$, $y = \ln 1 = 0$.

Use $m = 1$ with $P(1, 0)$.

$$y - 0 = 1(x - 1)$$
$$y = x - 1$$

63.　The slope of the graph of $y = x + k$ is 1.
First, we find the point on the graph of
$f(x) = \sqrt{2x - 1}$ at which the slope is
also 1.

$$f(x) = (2x - 1)^{1/2}$$
$$f'(x) = \frac{1}{2}(2x - 1)^{-1/2}(2)$$
$$f'(x) = \frac{1}{\sqrt{2x - 1}}$$

The slope is 1 when

$$\frac{1}{\sqrt{2x - 1}} = 1$$
$$1 = \sqrt{2x - 1}$$
$$1 = 2x - 1$$
$$2x = 2$$
$$x = 1,$$

and

$$f(1) = 1.$$

Therefore, at $P(1, 1)$ on the graph of
$f(x) = \sqrt{2x - 1}$, the slope is 1.
An equation of the tangent line is

$$y - 1 = 1(x - 1)$$
$$y - 1 = x - 1$$
$$y = x + 0.$$

Any tangent line intersects the curve in exactly
one point.

From this we see that if $k = 0$, there is one point
of intersection.

The graph of f is below the line $y = x + 0$.
Therefore, if $k > 0$, the graph of $y = x + k$
will not intersect the graph.

Consider the point $Q\left(\frac{1}{2}, 0\right)$ on the graph. We find
an equation of the line through Q with slope 1.

$$y - 0 = 1\left(x - \frac{1}{2}\right)$$
$$y = x - \frac{1}{2}$$

The line with a slope of 1 through $Q\left(\frac{1}{2}, 0\right)$ will
intersect the graph in two points. One is Q and
the other is some point on the graph to the right
of Q.

The graph of $y = x + 0$ intersects the graph in
one point, while the graph of $y = x - \frac{1}{2}$
intersects it in two points. If we use a value
of k in $y = x + k$ with $-\frac{1}{2} < k < 0$, we will
have a line with a y-intercept between $-\frac{1}{2}$ and a
0 and a slope of 1 which will intersect the graph
in two points.

If k, the y-intercept, is less than $-\frac{1}{2}$, the graph
of $y = x + k$ will be below point Q and will
intersect the graph of f in exactly one point.

To summarize, the graph of $y = x + k$ will
intersect the graph of $f(x) = \sqrt{2x - 1}$ in

(1) 　no points if $k > 0$;

(2) 　exactly one point if $k = 0$ or if $k < -\frac{1}{2}$;

(3) 　exactly two points if $-\frac{1}{2} \leq k < 0$

65.　Using the result $\widehat{fg} = \hat{f} + \hat{g}$, the total amount
of tuition collected goes up by approximately
$2\% + 3\% = 5\%$.

Let $T =$ tuition per person before the increase
and S the number of students before the increase.
Then the new tuition is $1.03T$ and the new numbers
of students is $1.02S$, so the total amount of tuition
collected is $(1.03T)(1.02S) = 1.0506TS$, which is
an increase of 5.06%.

67.

Let $f(x) = a(n - x)^{-n}$ and $g(x) = bx^n$ for $n \neq 0$.

$$f'(x)g'(x) = a(n - x)^{-n}\left(nbx^{n-1}\right) + (an)(n - x)^{-n-1}\left(bx^n\right)$$

$$= anbx^{n-1}(n - x)^{-n}\left[1 + x(n - x)^{-1}\right]$$

$$= anbx^{n-1}(n - x)^{-n}\left[\frac{n}{x - n}\right] = \frac{abn^2 x^{n-1}}{(n - x)^{n+1}}$$

$$\left[f(x)g(x)\right]' = \left[\left(a(n-x)^{-n}\right)\left(bx^n\right)\right]'$$

$$= \left[ab\left(\frac{x}{n-x}\right)^n\right]' = abn\left(\frac{x}{n-x}\right)^{n-1}\left(\frac{x}{n-x}\right)'$$

$$= abn\left(\frac{x}{n-x}\right)^{n-1}\left(\frac{(1)(n-x)-(x)(-1)}{(n-x)^2}\right)$$

$$= \frac{abn^2x^{n-1}}{(n-x)^{n+1}}$$

69. $C(x) = \sqrt{x+1}$

$$\bar{C}(x) = \frac{C(x)}{x} = \frac{\sqrt{x+1}}{x}$$

$$= \frac{(x+1)^{1/2}}{x}$$

$$\bar{C}'(x) = \frac{x[\frac{1}{2}(x+1)^{-1/2}] - (x+1)^{1/2}(1)}{x^2}$$

$$= \frac{\frac{1}{2}x(x+1)^{-1/2} - (x+1)^{1/2}}{x^2}$$

$$= \frac{x(x+1)^{-1/2} - 2(x+1)^{1/2}}{2x^2}$$

$$= \frac{(x+1)^{-1/2}[x - 2(x+1)]}{2x^2}$$

$$= \frac{(x+1)^{-1/2}(-x-2)}{2x^2}$$

$$= \frac{-x-2}{2x^2(x+1)^{1/2}}$$

71. $C(x) = (x^2 + 3)^3$

$$\bar{C}(x) = \frac{C(x)}{x} = \frac{(x^2+3)^3}{x}$$

$$\bar{C}'(x) = \frac{x[3(x^2+3)^2(2x)] - (x^2+3)^3(1)}{x^2}$$

$$= \frac{6x^2(x^2+3)^2 - (x^2+3)^3}{x^2}$$

$$= \frac{(x^2+3)^2[6x^2 - (x^2+3)]}{x^2}$$

$$= \frac{(x^2+3)^2(5x^2-3)}{x^2}$$

73. $C(x) = 10 - e^{-x}$

$$\bar{C}(x) = \frac{C(x)}{x}$$

$$\bar{C}(x) = \frac{10 - e^{-x}}{x}$$

$$\bar{C}'(x) = \frac{x(e^{-x}) - (10 - e^{-x})\cdot 1}{x^2}$$

$$= \frac{e^{-x}(x+1) - 10}{x^2}$$

75. $S(x) = 1000 + 60\sqrt{x} + 12x$

$$= 1000 + 60x^{1/2} + 12x$$

$$\frac{dS}{dx} = 60\left(\frac{1}{2}x^{-1/2}\right) + 12$$

$$= 30x^{-1/2} + 12 = \frac{30}{\sqrt{x}} + 12$$

(a) $\dfrac{dS}{dx}(9) = \dfrac{30}{\sqrt{9}} + 12 = \dfrac{30}{3} + 12 = 22$

Sales will increase by $22 million when $1000 more is spent on research.

(b) $\dfrac{dS}{dx}(16) = \dfrac{30}{\sqrt{16}} + 12 = \dfrac{30}{4} + 12 = 19.5$

Sales will increase by $19.5 million when $1000 more is spent on research.

(c) $\dfrac{dS}{dx}(25) = \dfrac{30}{\sqrt{25}} + 12 = \dfrac{30}{5} + 12 = 18$

Sales will increase by $18 million when $1000 more is spent on research.

(d) As more money is spent on research, the increase in sales is decreasing.

77. $T(x) = \dfrac{1000 + 60x}{4x + 5}$

$$T'(x) = \frac{(4x+5)(60) - (1000+60x)(4)}{(4x+5)^2}$$

$$= \frac{240x + 300 - 4000 - 240x}{(4x+5)^2}$$

$$= \frac{-3700}{(4x+5)^2}$$

(a) $T'(9) = \dfrac{-3700}{[4(9)+5]^2} = \dfrac{-3700}{1681} \approx -2.201$

Costs will decrease $2201 for the next $100 spent on training.

(b) $T'(19) = \dfrac{-3700}{[4(19) + 5]^2}$

$= \dfrac{-3700}{6561} \approx -0.564$

Costs will decrease $564 for the next $100 spent on training.

(c) Costs will always decrease because

$$T'(x) = \dfrac{-3700}{(4x + 5)^2}$$

will always be negative.

79. $A(r) = 1000e^{12r/100}$

$A'(r) = 1000e^{12r/100} \cdot \dfrac{12}{100}$

$= 120e^{12r/100}$

$A'(5) = 120e^{0.6} \approx 218.65$

The balance increases by approximately $218.65 for every 1% increase in the interest rate when the rate is 5%.

81.

$P(t) = -0.01741t^4 + 0.6790t^3 - 7.141t^2 + 16.95t + 204.5$

$P'(t) = -0.06964t^3 + 2.0370t^2 - 14.282t + 16.95$

(a)

$P'(1) = 4.6354$; the volume of mail is increasing at about 4.64 billion pieces per year.

(b)

$P'(8) = -2.5937$; the volume of mail is decreasing at about 2.59 billion pieces per year.

83. (a) Using the regression feature on a graphing calculator, a cubic function that models the data is

$$y = (1.799 \times 10^{-5})t^3 - (3.177 \times 10^{-4})t^2$$
$$-0.06866t + 2.504.$$

Using the regression feature on a graphing calculator, a quartic function that models the data is

$$y = (-1.112 \times 10^{-6})t^4 + (2.920 \times 10^{-4})t^3$$
$$-0.02238t^2 + 0.6483t - 4.278.$$

(b) Using the cubic function, $\dfrac{dy}{dx}$ at $x = 105$ is about 0.59 dollar per year. Using the quartic function, $\dfrac{dy}{dx}$ at $x = 105$ is about 0.46 dollar per year.

85. $G(t) = \dfrac{m G_0}{G_0 + (m - G_0)e^{-kmt}}$, where

$m = 30,000$, $G_o = 2000$, and $k = 5.10^{-6}$.

(a)

$G(t) = \dfrac{(30,000)(2000)}{2000 + (30,000 - 2000)e^{-5.10^{-6}(30,000)t}}$

$= \dfrac{30,000}{1 + 14e^{-0.15t}}$

(b)

$G(t) = 30,000(1 + 14e^{-0.15t})^{-1}$

$G(6) = 30,000(1 + 14e^{-0.90})^{-1} \approx 4483$

$G'(t) = -30,000(1 + 14e^{-0.15t})^{-2}(-2.1e^{-0.15t})$

$= \dfrac{63,000e^{-0.15t}}{(1 + 14e^{-0.15t})^2}$

$G'(6) = \dfrac{63,000e^{-0.90}}{(1 + 14e^{-0.90})^2} \approx 572$

The population is 4483, and the rate of growth is 572.

87. $M(t) = 3583e^{-e^{-0.020(t-66)}}$

(a) $M(250) = 3583e^{-e^{-0.020(250-66)}}$

≈ 3493.76 grams,

or about 3.5 kilograms

(b) As $t \to \infty$, $-e^{-0.020(t-66)} \to 0$,

$e^{-e^{-0.020(t-66)}} \to 1$, and $M(t) \to 3583$

grams or about 3.6 kilograms.

(c) 50% of 3583 is 1791.5.

$1791.5 = 3583e^{-e^{-0.020(t-66)}}$

$\ln\left(\dfrac{1791.5}{3583}\right) = -e^{-0.020(t-66)}$

$\ln\left(\ln\dfrac{3583}{1791.5}\right) = -0.020(t - 66)$

$t = -\dfrac{1}{0.020}\ln\left(\ln\dfrac{3583}{1791.5}\right) + 66$

≈ 84 days

(d)

$D_t M(t)$

$$= 3583e^{-e^{-0.020(t-66)}} D_t\left(-e^{-0.020(t-66)}\right)$$

$$= 3583e^{-e^{-0.020(t-56)}}\left(-e^{-0.020(t-66)}\right)(-0.020)$$

$$= 71.66e^{-e^{-0.020(t-66)}}\left(e^{-0.020(t-66)}\right)$$

When $t = 250$, $D_t M(t) \approx 1.76$ g/day.

(e)

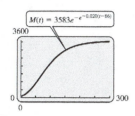

Growth is initially rapid, then tapers off.

(f)

Day	Weight	Rate
50	904	24.90
100	2159	21.87
150	2974	11.08
200	3346	4.59
250	3494	1.76
300	3550	0.66

89. $C(t) = 19{,}259.86(1.2585)^t$

$C'(t) = (19{,}259.86)(\ln 1.2585)(1.2585)^t$

$$= 4428.237(1.2585)^t$$

(a)
$C'(5) = 13{,}979.671 \approx 13{,}980$ megawatts/year

(b)
$C'(10) = 44{,}132.960 \approx 44{,}133$ megawatts/year

(c)
$C'(15) = 139{,}325.033 \approx 139{,}325$ megawatts/year

91. $p(t) = 1.757(1.0248)^t$

$p'(t) = (1.757)(\ln 1.0248)(1.0248)^t$

$$= 0.043042(1.0248)^t$$

$p'(70) = 0.23913 \approx 0.239$

The production of corn is increasing at a rate of 0.239 billion bushels per year in 2000.

93. $f(x) = k(x - 49)^6 + .8$

$f'(x) = k \cdot 6(x - 49)^5$

$$= (3.8 \times 10^{-9})(6)(x - 49)^5$$

$$= (2.28 \times 10^{-8})(x - 49)^5$$

(a) $f'(20) = (2.28 \times 10^{-8})(20 - 49)^5$

$$\approx -0.4677$$

fatalities per 1000 licensed drivers per 100 million miles per year.

At the age of 20, each extra year results in a decrease of 0.4677 fatalities per 1000 licensed drivers per 100 million miles.

(b) $f'(60) = (2.28 \times 10^{-8})(60 - 49)^5$

$$\approx 0.003672$$

fatalities per 1000 licensed drivers per 100 million miles per year.

At the age of 60, each extra year results in an increase of 0.003672 fatalities per 1000 licensed drivers per 100 million miles.

Chapter 5

GRAPHS AND THE DERIVATIVE

5.1   Increasing and Decreasing Functions

Your Turn 1

Scanning across the x-values from left to right we see that the function is increasing on $(-1, 2)$ and $(4, \infty)$ and decreasing on $(-\infty, -1)$ and $(2, 4)$.

Your Turn 2

$$f(x) = -x^3 - 2x^2 + 15x + 10$$
$$f'(x) = -3x^2 - 4x + 15$$

Set $f'(x)$ equal to 0 and solve for x.

$$-3x^2 - 4x + 15 = 0$$
$$3x^2 + 4x - 15 = 0$$
$$(3x - 5)(x + 3) = 0$$
$$x = \frac{5}{3} \quad \text{or} \quad x = -3$$

The only critical numbers are -3 and $5/3$. These points determine three intervals: $(-\infty, -3)$, $(-3, 5/3)$, and $(5/3, \infty)$. Determine the sign of f' in each interval by picking a test point and evaluating $f'(x)$.

$$f'(-4) = -(3(-4) - 5)((-4) + 3) = -(-17)(-1) < 0$$
$$f'(0) = -(3(0) - 5)((0) + 3) = -(-5)(3) > 0$$
$$f'(2) = -(3(2) - 5)((2) + 3) = -(1)(5) < 0$$

The arrows in each interval in the figure below indicate where f is increasing or decreasing.

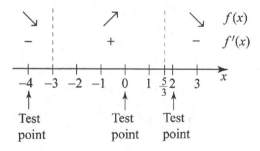

The function $f(x) = -x^3 - 2x^2 + 15x + 10$ is increasing on $(-3, 5/3)$ and decreasing on $(-\infty, -3)$ and $(5/3, \infty)$.

Since $f(-3) = -26$ and $f\left(\frac{5}{3}\right) = \frac{670}{27} \approx 24.8$, the graph goes through $(-3, -26)$ and $\left(\frac{5}{3}, \frac{670}{27}\right)$.

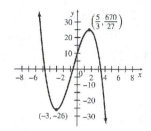

Your Turn 3

$$f(x) = (2x + 4)^{2/5}$$
$$f'(x) = \frac{2}{5}(2x + 4)^{-3/5}(2)$$
$$= \frac{4}{5(2x + 4)^{3/5}}$$

Find the critical numbers. $f'(x)$ is never 0 so we find any values where $f'(x)$ fails to be defined by setting the denominator equal to 0.

$$5(2x + 4)^{3/5} = 0$$
$$(2x + 4)^{3/5} = 0$$
$$2x + 4 = 0^{5/3} = 0$$
$$2x = -4$$
$$x = -2$$

Since $f'(-2)$ does not exist but $f(-2)$ is defined, $x = -2$ is a critical number.

Choosing -3 and -1 as test points, we find that $f'(-3) < 0$ and $f'(-1) > 0$. Thus the function f is decreasing on $(-\infty, -2)$ and increasing on $(-2, \infty)$.

Using the fact that $f(-2) = 0$ and $f(0) = 4^{2/5} \approx 1.74$ we can sketch the following graph. Note that the graph is not smooth at $x = -2$, where the derivative is not defined.

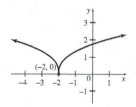

Your Turn 4

$$f(x) = \frac{-2x}{x + 2}; \text{ the domain of } f \text{ excludes } -2$$

$$f'(x) = \frac{(x + 2)(-2) - (-2x)(1)}{(x + 2)^2}$$

$$= \frac{-2x - 4 + 2x}{(x + 2)^2}$$

$$= -\frac{4}{(x + 2)^2}$$

$f'(x)$ is never 0 and fails to exist at $x = -2$; however, f is undefined at $x = -2$ so there are no critical numbers. Apply the first derivative test for numbers on either side of $x = -2$.

$$f'(-3) = -\frac{4}{(-3 + 2)^2} < 0$$

$$f'(-1) = -\frac{4}{(-1 + 2)^2} < 0$$

Since $f'(x)$ is negative wherever it is defined, the function is never increasing and is decreasing on $(-\infty, -2)$ and $(-2, \infty)$.

Next find any asymptotes. Since $x = -2$ makes the denominator of f equal to 0, the line $x = -2$ is a vertical asymptote.

$$\lim_{x \to \infty} \frac{-2x}{x + 2} = \lim_{x \to \infty} \frac{-2}{1 + \frac{2}{x}} = -2$$

The line $y = -2$ is a horizontal asymptote.

The x-intercept is $(0, 0)$. The graph is shown below.

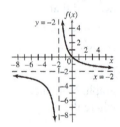

5.1   Warmup Exercises

W1. $f(x) = 8x^2 + 2x - 15$

$$8x^2 + 2x - 15 = 0$$

$$(4x - 5)(2x + 3) = 0$$

$$4x - 5 = 0 \text{ or } 2x + 3 = 0$$

$$x = \frac{5}{4} \text{ or } x = -\frac{3}{2}$$

The roots are 5/4, −3/2.

W2. $f(x) = 12x^3 - 41x^2 - 15x$

$$12x^3 - 41x^2 - 15x = 0$$

$$(x)(3x + 1)(4x - 15) = 0$$

$$x = 0 \text{ or } 3x + 1 = 0 \text{ or } 4x - 15 = 0$$

$$x = 0 \text{ or } x = -\frac{1}{3} \text{ or } x = \frac{15}{4}$$

The roots are 0, −1/3, 15/4.

W3. $f(x) = e^{x^2 - 9} - 1$

$$e^{x^2 - 9} - 1 = 0$$

$$e^{x^2 - 9} = 1 = e^0$$

$$x^2 - 9 = 0$$

$$x = 3 \text{ or } -3$$

The roots are −3, 3.

W4. $f(x) = x^{7/3} - 4x^{1/3}$

$$x^{7/3} - 4x^{1/3} = 0$$

$$x^{1/3}(x^2 - 4) = 0$$

$$x^{1/3}(x + 2)(x - 2) = 0$$

$$x = 0 \text{ or } x = -2 \text{ or } x = 2$$

The roots are 0, −2, 2.

W5. $f(x) = \dfrac{x^2 + 4}{x^3 + 5}$

Use the quotient rule.

$$f'(x) = \frac{\left(x^3 + 5\right)(2x) - \left(x^2 + 4\right)\left(3x^2\right)}{\left(x^3 + 5\right)^2}$$

$$= \frac{2x^4 + 10x - 3x^4 - 12x^2}{\left(x^3 + 5\right)^2}$$

$$= \frac{-x^4 - 12x^2 + 10x}{\left(x^3 + 5\right)^2}$$

$$= \frac{(-x)\left(x^3 + 12x - 10\right)}{\left(x^3 + 5\right)^2}$$

W6. $f(x) = \sqrt{3 - x^2} = \left(3 - x^2\right)^{1/2}$

Use the chain rule with $f(x) = h(g(x))$ where $h(x) = x^{1/2}$ and $g(x) = 3 - x^2$.

$$f'(x) = \frac{1}{2}\left(3 - x^2\right)^{-1/2}(-2x)$$

$$= \frac{-x}{\sqrt{3 - x^2}}$$

W7. $f(x) = x^2 e^{5x^3}$

Use the product and chain rules.

$$f'(x) = \left(x^2\right)\left(15x^2 e^{5x^3}\right) + (2x)\left(e^{5x^3}\right)$$

$$= \left(15x^4 + 2x\right)e^{5x^3}$$

$$= x\left(15x^3 + 2\right)e^{5x^3}$$

W8. $f(x) = 2\ln\left(x^{3/2} + 5\right)$

Use the formula for the derivative of the natural logarithm and the chain rule.

Let $h(x) = 2\ln x$ and $g(x) = x^{3/2} + 5$.

Then $f(x) = h(g(x))$ and $f'(x) = h'(g(x))g'(x)$.

$$h'(x) = \frac{2}{x} \text{ so } h'(g(x)) = \frac{2}{x^{3/2} + 5}$$

$$g'(x) = \frac{3}{2}x^{1/2}$$

$$h'(g(x))g'(x) = \left(\frac{2}{x^{3/2} + 5}\right)\left(\frac{3}{2}x^{1/2}\right)$$

$$= \frac{3x^{1/2}}{x^{3/2} + 5}$$

5.1 Exercises

1. By reading the graph, f is

   (a) increasing on $(1, \infty)$ and

   (b) decreasing on $(-\infty, 1)$.

3. By reading the graph, g is

   (a) increasing on $(-\infty, -2)$ and

   (b) decreasing on $(-2, \infty)$.

5. By reading the graph, h is

   (a) increasing on $(-\infty, -4)$ and $(-2, \infty)$ and

   (b) decreasing on $(-4, -2)$.

7. By reading the graph, f is

   (a) increasing on $(-7, -4)$ and $(-2, \infty)$ and

   (b) decreasing on $(-\infty, -7)$ and $(-4, -2)$.

9. (a) Since the graph of $f'(x)$ is positive for $x < -1$ and $x > 3$, the intervals where

$f(x)$ is increasing are $(-\infty, -1)$ and $(3, \infty)$.

   (b) Since the graph of $f'(x)$ is negative for $-1 < x < 3$, the interval where $f(x)$ is decreasing is $(-1, 3)$.

11. (a) Since the graph of $f'(x)$ is positive for $x < -8, -6 < x < -2.5$ and $x > -1.5$, the intervals where $f(x)$ is increasing are $(-\infty, -8), (-6, -2.5)$, and $(-1.5, \infty)$.

    (b) Since the graph of $f'(x)$ is negative for $-8 < x < -6$ and $-2.5 < x < -1.5$, the intervals where $f(x)$ is decreasing are $(-8, -6)$ and $(-2.5, -1.5)$.

13. $y = 2.3 + 3.4x - 1.2x^2$

    (a) $y' = 3.4 - 2.4x$

        y' is zero when

        $$3.4 - 2.4x = 0$$

        $$x = \frac{3.4}{2.4} = \frac{17}{12}$$

        and there are no values of x where y' does not exist, so the only critical number is $x = \frac{17}{12}$.

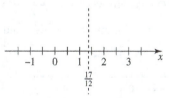

        Test a point in each interval.

        When
        $x = 0, y' = 3.4 - 2.4(0) = 3.4 > 0.$

        When
        $x = 2, y' = 3.4 - 2.4(2) = -1.4 < 0.$

    (b) The function is increasing on $\left(-\infty, \frac{17}{12}\right)$.

    (c) The function is decreasing on $\left(\frac{17}{12}, \infty\right)$.

15. $f(x) = \dfrac{2}{3}x^3 - x^2 - 24x - 4$

   (a) $f'(x) = 2x^2 - 2x - 24$

            $= 2(x^2 - x - 12)$

            $= 2(x + 3)(x - 4)$

     $f'(x)$ is zero when $x = -3$ or $x = 4$, so the critical numbers are -3 and 4.

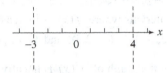

     Test a point in each interval.

$$f'(-4) = 16 > 0$$
$$f'(0) = -24 < 0$$
$$f'(5) = 16 > 0$$

   (b) f is increasing on $(-\infty, -3)$ and $(4, \infty)$.

   (c) f is decreasing on $(-3, 4)$.

17. $f(x) = 4x^3 - 15x^2 - 72x + 5$

   (a) $f'(x) = 12x^2 - 30x - 72$

            $= 6(2x^2 - 5x - 12)$

            $= 6(2x + 3)(x - 4)$

     $f'(x)$ is zero when $x = -\dfrac{3}{2}$ or $x = 4$, so the critical numbers are $-\dfrac{3}{2}$ and 4.

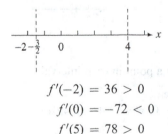

$$f'(-2) = 36 > 0$$
$$f'(0) = -72 < 0$$
$$f'(5) = 78 > 0$$

   (b) f is increasing on $\left(-\infty, -\dfrac{3}{2}\right)$ and $(4, \infty)$.

   (c) f is decreasing on $\left(-\dfrac{3}{2}, 4\right)$.

19. $f(x) = x^4 + 4x^3 + 4x^2 + 1$

   (a) $f'(x) = 4x^3 + 12x^2 + 8x$

            $= 4x(x^2 + 3x + 2)$

            $= 4x(x + 2)(x + 1)$

     $f'(x)$ is zero when $x = 0$, $x = -2$, or $x = -1$, so the critical numbers are 0, -2, and -1.

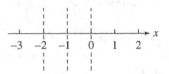

     Test a point in each interval.

$$f'(-3) = -12(-1)(-2) = -24 < 0$$
$$f'(-1.5) = -6(.5)(-.5) = 1.5 > 0$$
$$f'(-.5) = -2(1.5)(.5) = -1.5 < 0$$
$$f'(1) = 4(3)(2) = 24 > 0$$

   (b) f is increasing on $(-2, -1)$ and $(0, \infty)$.

   (c) f is decreasing on $(-\infty, -2)$ and $(-1, 0)$.

21. $y = -3x + 6$

   (a) $y' = -3 < 0$

     There are no critical numbers since y' is never 0 and always exists.

   (b) Since y' is always negative, the function is increasing on no interval.

   (c) y' is always negative, so the function is decreasing everywhere, or on the interval $(-\infty, \infty)$.

23. $f(x) = \dfrac{x + 2}{x + 1}$

   (a) $f'(x) = \dfrac{(x + 1)(1) - (x + 2)(1)}{(x + 1)^2}$

            $= \dfrac{-1}{(x + 1)^2}$

     The derivative is never 0, but it fails to exist at $x = -1$. Since -1 is not in the domain of f, however, -1 is not a critical number.

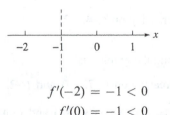

$$f'(-2) = -1 < 0$$
$$f'(0) = -1 < 0$$

   (b) f is increasing on no interval.

   (c) f is decreasing everywhere that it is defined, on $(-\infty, -1)$ and on $(-1, \infty)$.

25. $y = \sqrt{x^2 + 1}$

$\qquad = (x^2 + 1)^{1/2}$

(a) $y' = \dfrac{1}{2}(x^2 + 1)^{-1/2}(2x)$

$\qquad = x(x^2 + 1)^{-1/2}$

$\qquad = \dfrac{x}{\sqrt{x^2 + 1}}$

$y' = 0$ when $x = 0$.

Since y does not fail to exist for any x, and since $y' = 0$ when $x = 0$, 0 is the only critical number.

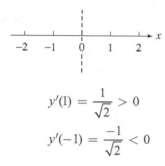

$y'(1) = \dfrac{1}{\sqrt{2}} > 0$

$y'(-1) = \dfrac{-1}{\sqrt{2}} < 0$

(b) y is increasing on $(0, \infty)$.

(c) y is decreasing on $(-\infty, 0)$.

27. $f(x) = x^{2/3}$

(a) $f'(x) = \dfrac{2}{3}x^{-1/3} = \dfrac{2}{3x^{1/3}}$

$f'(x)$ is never zero, but fails to exist when $x = 0$, so 0 is the only critical number.

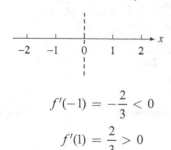

$f'(-1) = -\dfrac{2}{3} < 0$

$f'(1) = \dfrac{2}{3} > 0$

(b) f is increasing on $(0, \infty)$.

(c) f is decreasing on $(-\infty, 0)$.

29. $y = x - 4\ln(3x - 9)$

(a) $y' = 1 - \dfrac{12}{3x - 9} = 1 - \dfrac{4}{x - 3}$

$\qquad = \dfrac{x - 7}{x - 3}$

y' is zero when $x = 7$. The derivative does not exist at $x = 3$, but note that the domain of f is $(3, \infty)$.

Thus, the only critical number is 7.

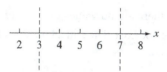

Choose a value in the intervals $(3, 7)$ and $(7, \infty)$.

$f'(4) = -3 < 0$

$f(8) = \dfrac{1}{5} > 0$

(b) The function is increasing on $(7, \infty)$.

(c) The function is decreasing on $(3, 7)$.

31. $f(x) = xe^{-3x}$

(a) $f'(x) = e^{-3x} + x(-3e^{-3x})$

$\qquad = (1 - 3x)e^{-3x}$

$\qquad = \dfrac{1 - 3x}{e^{3x}}$

$f'(x)$ is zero when $x = \dfrac{1}{3}$ and three are no values of x where $f'(x)$ does not exist, so the critical number is $\dfrac{1}{3}$.

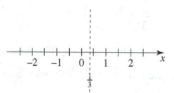

Test a point in each interval.

$f'(0) = \dfrac{1 - 3(0)}{e^{3(0)}} = 1 > 0$

$f'(1) = \dfrac{1 - 3(1)}{e^{3(1)}} = -\dfrac{2}{e^3} < 0$

(b) The function is increasing on $\left(-\infty, \dfrac{1}{3}\right)$.

(c) The function is decreasing on $\left(\frac{1}{3}, \infty\right)$.

33. $f(x) = x^2 2^{-x}$

(a) $f'(x) = x^2[\ln 2(2^{-x})(-1)] + (2^{-x})2x$

$\qquad = 2^{-x}(-x^2 \ln 2 + 2x)$

$\qquad = \dfrac{x(2 - x\ln 2)}{2^x}$

$f'(x)$ is zero when $x = 0$ **or** $x = \dfrac{2}{\ln 2}$

and there are no values of x where $f'(x)$ does not exist. The critical numbers are 0 and $\dfrac{2}{\ln 2}$.

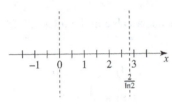

Test a point in each interval.

$f'(-1) = \dfrac{(-1)(2 - (-1)\ln 2)}{2^{-1}} = -2(2 + \ln 2) < 0$

$f'(1) = \dfrac{(1)(2 - (1)\ln 2)}{2^1} = \dfrac{2 - \ln 2}{2} > 0$

$f'(3) = \dfrac{(3)(2 - (3)\ln 2)}{2^3} = \dfrac{3(2 - 3\ln 2)}{8} < 0$

(b) The function is increasing on $\left(0, \dfrac{2}{\ln 2}\right)$.

(c) The function is decreasing on $(-\infty, 0)$ and $\left(\dfrac{2}{\ln 2}, \infty\right)$.

35. $y = x^{2/3} - x^{5/3}$

(a) $y' = \dfrac{2}{3}x^{-1/3} - \dfrac{5}{3}x^{2/3} = \dfrac{2 - 5x}{3x^{1/3}}$

$y' = 0$ when $x = \dfrac{2}{5}$. The derivative does not exist at $x = 0$. So the critical numbers are 0 and $\dfrac{2}{5}$.

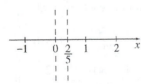

Test a point in each interval.

$y'(-1) = \dfrac{7}{-3} < 0$

$y'\left(\dfrac{1}{5}\right) = \dfrac{1}{3\left(\frac{1}{5}\right)^{1/3}} = \dfrac{5^{1/3}}{2} > 0$

$y'(1) = \dfrac{-3}{3} = -1 < 0$

(b) y is increasing on $\left(0, \dfrac{2}{5}\right)$.

(c) y is decreasing on $(-\infty, 0)$ and $\left(\dfrac{2}{5}, \infty\right)$.

39. $f(x) = ax^2 + bx + c, a < 0$

$f'(x) = 2ax + b$

Let $f'(x) = 0$ to find the critical number.

$$2ax + b = 0$$
$$2ax = -b$$
$$x = \dfrac{-b}{2a}$$

Choose a value in the interval $\left(-\infty, -\dfrac{b}{2a}\right)$.

Since $a < 0$,

$$\dfrac{-b}{2a} - \dfrac{-1}{2a} = \dfrac{-b+1}{2a} < \dfrac{-b}{2a}.$$
$$f'\left(\dfrac{-b+1}{2a}\right) = 2a\left(\dfrac{-b+1}{2a}\right) + b$$
$$= 1 < 0$$

Choose a value in the interval $\left(\dfrac{-b}{2a}, \infty\right)$.

Since $a < 0$,

$$\dfrac{-b}{2a} - \dfrac{-1}{2a} = \dfrac{-b-1}{2a} < \dfrac{-b}{2a}.$$
$$f'\left(\dfrac{-b-1}{2a}\right) = 2a\left(\dfrac{-b-1}{2a}\right) + b$$
$$= -1 > 0$$

f is increasing on $\left(-\infty, \dfrac{-b}{2a}\right)$ and decreasing on $\left(\dfrac{-b}{2a}, \infty\right)$.

This tells us that the curve opens downward and $x = \dfrac{-b}{2a}$ is the x-coordinate of the vertex.

$$f\left(\frac{-b}{2a}\right) = a\left(\frac{-b}{2a}\right)^2 + b\left(\frac{-b}{2a}\right) + c$$

$$= \frac{ab^2}{4a^2} - \frac{b^2}{2a} + c$$

$$= \frac{b^2}{4a} - \frac{2b^2}{4a} + \frac{4bc}{4a}$$

$$= \frac{4ac - b^2}{4a}$$

The vertex is $\left(\frac{-b}{2a}, \frac{4ac-b^2}{4a}\right)$ or $\left(-\frac{b}{2a}, \frac{4ac-b^2}{4a}\right)$.

41. $f(x) = \ln x$

$$f'(x) = \frac{1}{x}$$

$f'(x)$ is undefined at $x = 0$. $f'(x)$ never equals zero. Note that $f(x)$ has a domain of $(0, \infty)$. Pick a value in the interval $(0, \infty)$.

$$f'(2) = \frac{1}{2} > 0$$

$f(x)$ is increasing on $(0, \infty)$.

$f(x)$ is never decreasing.

Since $f(x)$ never equals zero, the tangent line is horizontal nowhere.

43. $f(x) = e^{0.001x} - \ln x$

$$f'(x) = 0.001e^{0.001x} - \frac{1}{x}$$

Note that $f(x)$ is only defined for $x > 0$. Use a graphing calculator to plot $f'(x)$ for $x > 0$.

(a) $f'(x) > 0$ on about $(567, \infty)$, so $f(x)$ is increasing on about $(567, \infty)$.

(b) $f'(x) < 0$ on about $(0, 567)$, so $f(x)$ is decreasing on about $(0, 567)$.

45. $H(r) = \dfrac{300}{1 + 0.03r^2} = 300(1 + 0.03r^2)^{-1}$

$$H'(r) = 300[-1(1 + 0.03r^2)^{-2}(0.06r)]$$

$$= \frac{-18r}{(1 + 0.03r^2)^2}$$

Since r is a mortgage rate (in percent), it is always positive. Thus, $H'(r)$ is always negative.

(a) H is increasing on nowhere.

(b) H is decreasing on $(0, \infty)$.

47. $C(x) = 0.32x^2 - 0.00004x^3$

$$R(x) = 0.848x^2 - 0.0002x^3$$

$$P(x) = R(x) = C(x)$$

$$= (0.848x^2 - 0.0002x^3)$$

$$\quad - (0.32x^2 - 0.00004x^3)$$

$$= 0.528x^2 - 0.00016x^3$$

$$P'(x) = 1.056x - 0.00048x^2$$

$$1.056x - 0.00048x^2 = 0$$

$$x(1.056 - 0.00048x) = 0$$

$$x = 0 \text{ or } x = 2200$$

Choose $x = 1000$ and $x = 3000$ as test points.

$$P'(1000) = 1.056(1000) - 0.00048(1000)^2 = 576$$

$$P'(3000) = 1.056(3000) - 0.00048(3000)^2$$

$$= -1152$$

The function is increasing on $(0, 2200)$.

49. $A(t) = 0.0000329t^3 - 0.00450t^2 + 0.0613t$
$$\quad + 2.34$$

$$A'(t) = 0.0000987t^2 - 0.009t + 0.0613$$

Set $A'(t)$ equal to 0 and solve for t.

$$0.0000987t^2 - 0.009t + 0.0613 = 0$$

$$t = \frac{0.009 \pm \sqrt{(0.009)^2 - 4(0.0000987)(0.0613)}}{(2)(0.0000987)}$$

$$t \approx 83.8 \text{ or } t = 7.4$$

Since $0 \leq t \leq 50, t \approx 7.4$.

Check the sign of A' on either side of this critical number.

$$A'(7) \approx 0.00314 > 0$$

$$A'(8) \approx -0.00438 < 0$$

(a) The projected year-end assets function is increasing on the interval $(0, 7.4)$, or from 2000 to about the middle of 2007.

(b) The projected year-end assets function is decreasing on the interval $(7.4, 50)$, or from about the middle of 2007 to 2050.

51. $A(x) = 0.003631x^3 - 0.03746x^2$
$$\quad + 0.1012x + 0.009$$

$$A'(x) = 0.010893x^2 - 0.07492x + 0.1012$$

Solve for $A'(x) = 0$.

$x \approx 1.85$ or $x \approx 5.03$

Choose $x = 1$ and $x = 4$ as test points.

$A'(1) = 0.010893(1)^2 - 0.07492(1) + 0.1012$

$\qquad = 0.037173$

$A'(4) = 0.010893(4)^2 - 0.07492(4) + 0.1012$

$\qquad = -0.024192$

(a) The function is increasing on $(0, 1.85)$.

(b) The function is decreasing on $(1.85, 5)$.

53. $K(t) = \dfrac{5t}{t^2 + 1}$

$K'(t) = \dfrac{5(t^2 + 1) - 2t(5t)}{(t^2 + 1)^2}$

$\qquad = \dfrac{5t^2 + 5 - 10t^2}{(t^2 + 1)^2}$

$\qquad = \dfrac{5 - 5t^2}{(t^2 + 1)^2}$

$K'(t) = 0$ when

$$\dfrac{5 - 5t^2}{(t^2 + 1)^2} = 0$$

$$5 - 5t^2 = 0$$

$$5t^2 = 5$$

$$t = \pm 1.$$

Since t is the time after a drug is administered, the function applies only for $[0, \infty)$, so we discard $t = -1$. Then 1 divides the domain into two intervals.

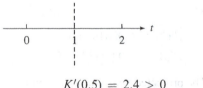

$$K'(0.5) = 2.4 > 0$$
$$K'(2) = -0.6 < 0$$

(a) K is increasing on $(0, 1)$.

(b) K is decreasing on $(1, \infty)$.

55. **(a)** $F(t) = -10.28 + 175.9te^{-t/1.3}$

$F'(t) = (175.9)(e^{-t/1.3})$

$\qquad + (175.9.9t)\left(-\dfrac{1}{1.3}e^{-t/1.3}\right)$

$\qquad = (175.9)(e^{-t/1.3})\left(1 - \dfrac{t}{1.3}\right)$

$\qquad \approx 175.9e^{-t/1.3}(1 - 0.769t)$

(b) $F'(t)$ is equal to 0 at $t = 1.3$. Therefore, 1.3 is a critical number. Since the domain is $(0, \infty)$, test values in the intervals from $(0, 1.3)$ and $(1.3, \infty)$.

$F'(1) \approx 18.83 > 0$ and $F'(2) \approx -20.32 < 0$

$F'(t)$ is increasing on $(0, 1.3)$ and decreasing on $(1.3, \infty)$.

57. **(a)** $R = \sqrt{\dfrac{N_0 - N}{2\pi DN}}$,

Use the chain rule and quotient rule.

$\dfrac{dR}{dN}$

$= \dfrac{1}{2\sqrt{\dfrac{N_0 - N}{2\pi DN}}}\left(\dfrac{(2\pi DN)(-1) - (N_0 - N)(2\pi D)}{(2\pi DN)^2}\right)$

$= \dfrac{1}{2\sqrt{\dfrac{N_0 - N}{2\pi DN}}}\left(\dfrac{-N_0}{(2\pi DN)^2}\right)$

$= \dfrac{-N_0}{2N\sqrt{\dfrac{N_0 - N}{2\pi DN}}(2\pi DN)(2\pi D)}$

$= \dfrac{-N_0}{2N^{3/2}\sqrt{2\pi D(N_0 - N)}}$

(b) As the number of plants increases, the maximum plant radius decreases.

(c) As N approaches 0, dR/dN approaches $-\infty$. As N approaches N_0, dR/dN approaches $-\infty$.

59. $f(x) = \dfrac{1}{\sqrt{2\pi}}e^{-x^2/2}$

$f'(x) = \dfrac{1}{\sqrt{2\pi}}e^{-x^2/2}(-x)$

$\qquad = \dfrac{-x}{\sqrt{2\pi}}e^{-x^2/2}$

$f'(x) = 0$ when $x = 0$.

Choose a value from each of the intervals $(-\infty, 0)$ and $(0, \infty)$.

$$f'(-1) = \frac{1}{\sqrt{2\pi}}e^{-1/2} > 0$$

$$f'(1) = \frac{-1}{\sqrt{2\pi}}e^{-1/2} < 0$$

The function is increasing on $(-\infty, 0)$ and decreasing on $(0, \infty)$.

61. **(a)** $(2500, 5750)$

(b) $(5750, 6000)$

(c) $(2800, 4800)$

(d) $(2500, 2800)$ and $(4800, 6000)$

5.2   Relative Extrema

Your Turn 1

There is an open interval containing x_1 for which $f(x) \geq f(x_1)$.

There is an open interval containing x_2 for which $f(x) \leq f(x_2)$.

There is an open interval containing x_3 for which $f(x) \geq f(x_3)$.

There are relative minima of $f(x_1)$ at $x = x_1$ and $f(x_3)$ at $x = x_3$ and a relative maximum of $f(x_2)$ at $x = x_2$.

Your Turn 2

$$f(x) = -x^3 - 2x^2 + 15x + 10$$
$$f'(x) = -3x^2 - 4x + 15$$

Set $f'(x)$ equal to 0 and solve for x to find the critical numbers.

$$-3x^2 - 4x + 15 = 0$$
$$-(3x^2 + 4x - 15) = 0$$
$$-(3x - 5)(x + 3) = 0$$

$$3x - 5 = 0 \text{ or } x + 3 = 0$$
$$x = 5/3 \quad \text{ or } x = -3$$

The critical numbers are -3 and $5/3$. They divide the domain of f into the three intervals $(-\infty, -3)$, $(-3, 5/3)$, and $(5/3, \infty)$. Pick test points in each interval to determine the sign of f' on that interval. For example, choose $x = -4, x = 0,$ and $x = 2$.

$$f'(-4) = -17$$
$$f'(0) = 15$$
$$f'(2) = -5$$

Thus the derivative is negative on $(-\infty, -3)$, positive on $(-3, 5/3)$, and negative on $(5/3, \infty)$. The graph below shows this information and indicates where f is increasing or decreasing.

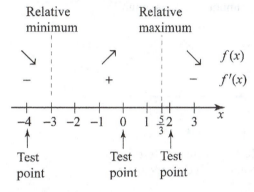

By the first derivative test, we know that the function has a relative minimum of $f(-3) = -26$ at $x = -3$ and a relative maximum of $f(5/3) = 670/27 \approx 24.8$ at $x = 5/3$.

Your Turn 3

$$f(x) = x^{2/3} - x^{5/3}$$
$$f'(x) = \frac{2}{3}x^{-1/3} - \frac{5}{3}x^{2/3}$$
$$= \frac{1}{3}\left(\frac{2 - 5x}{x^{1/3}}\right)$$

$x = 0$ is a critical number because $f(x)$ is defined at $x = 0$ but $f'(x)$ is not. To find any other critical numbers, assume $x \neq 0$ and set $f'(x) = 0$ and solve for x.

$$\frac{1}{3}\left(\frac{2 - 5x}{x^{1/3}}\right) = 0$$
$$2 - 5x = 0$$
$$x = \frac{2}{5}$$

Use the critical numbers to divide the line into three intervals, $(-\infty, 0)$, $(0, 2/5)$, and $(2/5, \infty)$. Choose test points in each interval; for example, choose -1, $-1/5$, and 1.

$$f'(-1) = -2.33$$
$$f'(1/5) = 0.57$$
$$f'(1) = -1$$

Thus the derivative is negative on $(-\infty, -1)$, positive on $(-1, 2/5)$, and negative on $(2/5, \infty)$. The graph below shows this information and indicates where f is increasing or decreasing.

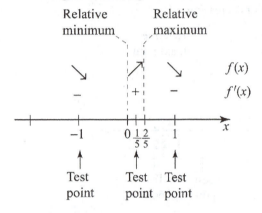

The first derivative test identifies $f(2/5)$ as a relative maximum, and there is a relative minimum at $x = 0$. Thus the function has a relative maximum of

$$f\left(\frac{2}{5}\right) = \frac{3}{5}\left(\frac{2}{5}\right)^{2/3} \approx 0.3257 \text{ at } x = \frac{2}{5}$$

and a relative minimum of $f(0) = 0$ at $x = 0$.

Your Turn 4

$$f(x) = x^2 e^x$$

Using the product rule:

$$f'(x) = x^2 e^x + 2xe^x$$
$$= e^x(x^2 + 2x)$$

Find the critical values:

$$e^x(x^2 + 2x) = 0$$
$$x^2 + 2x = 0$$
$$x(x + 2) = 0$$
$$x + 2 = 0 \quad \text{or} \quad x = 0$$
$$x = -2 \quad \text{or} \quad x = 0$$

There are two critical numbers, dividing the line into the intervals $(-\infty, -2)$, $(-2, 0)$ and $(0, \infty)$. Choose a test point in each interval, for example -3, -1, and 1.

$$f'(-3) = 0.149$$
$$f'(-1) = -0.368$$
$$f'(1) = 8.155$$

Thus the derivative is positive on $(-\infty, -2)$, negative on $(-2, 0)$, and positive on $(0, \infty)$. The graph below shows this information and indicates where f is increasing or decreasing.

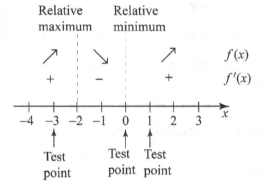

The function has a relative maximum of $4e^{-2} \approx 0.5413$ at $x = -2$ and a relative minimum of $f(0) = 0$ at $x = 0$.

Your Turn 5

$$C(q) = 100 + 10q$$
$$p = D(q) = 50 - 2q$$
$$P(q) = R(q) - C(q)$$
$$= qD(q) - C(q)$$
$$= q(50 - 2q) - (100 + 10q)$$
$$= 50q - 2q^2 - 100 + 10q$$
$$= -2q^2 + 40q - 100$$
$$P'(q) = -4q + 40$$
$$= -4(q - 10)$$

P' is 0 only when $q = 10$, and this is the only critical number. $P'(5) = 20$ and $P(20) = -40$ so P is increasing as q approaches 10 from the left and decreasing to the right of $q = 10$. The value of P at $q = 10$ is $P(10) = -2(100) + 40(1) - 100 = 100$.

Thus the maximum weekly profit is $100 when the demand is 10 items per week. The company should charge $D(10) = 50 - 2(10) = 30$, or $30 per item.

5.2   Warmup Exercises

W1. $f(x) = 2x^3 - 3x^2 - 36x + 4$

$f'(x) = 6x^2 - 6x - 36$

$\qquad = 6(x^2 - x - 6) = 6(x - 3)(x + 2)$

$f'(x) = 0$ when $(x - 3)(x + 2) = 0$,
that is, when $x = 3$ and $x = -2$.
These points divide the real line into three
intervals:
$(-\infty, -2), (-2, 3), (3, \infty)$

Evaluate f' at a test point in each interval, say
$x = -4, x = 0,$ and $x - 4.$

$f'(-4) = 84 > 0$

$f'(0) = -36 < 0$

$f'(4) = 36 > 0$

Thus f is increasing on $(-\infty, -2)$ and $(3, \infty)$
and is decreasing on $(-2, 3)$.

W2. $f(x) = 3x^4 - 10x^3 - 36x^2 - 72$

$f'(x) = 12x^3 - 30x^2 - 72x$

$\qquad = 6x(2x^2 - 5x - 12)$

$\qquad = 6x(2x + 3)(x - 4)$

$f'(x) = 0$ when $6x(2x + 3)(x - 4) = 0,$
that is, when $x = -3/2, x = 0,$ and $x = 4.$
These points divide the real line into four
intervals:
$(-\infty, -3/2), (-3/2, 0), (0, 4), (4, \infty)$

Evaluate f' at a test point in each interval, say
$x = -2, x = -1, x = 1,$ and $x = 5.$

$f'(-2) = -72 < 0$

$f'(-1) = 30 > 0$

$f'(1) = -90 < 0$

$f'(5) = 390 > 0$

Thus f is increasing on
$(-3/2, 0)$ and $(4, \infty)$ and is decreasing on
$(-\infty, -3/2)$ and $(0, 4)$.

5.2   Exercises

1.   As shown on the graph, the relative minimum of
-4 occurs when $x = 1.$

3.   As shown on the graph, the relative maximum of
3 occurs when $x = -2.$

5.   As shown on the graph, the relative maximum of
3 occurs when $x = -4$ and the relative minimum
of 1 occurs when $x = -2.$

7.   As shown on the graph, the relative maximum of
3 occurs when $x = -4;$ the relative minimum of
-2 occurs when $x = -7$ and $x = -2$.

9.   Since the graph of the function is zero at $x = -1$
and $x = 3,$ the critical numbers are -1 and 3.

Since the graph of the derivative is positive on
$(-\infty, -1)$ and negative on $(-1, 3),$ there is a
relative maximum at $-1.$ Since the graph of the
function is negative on $(-1, 3)$ and positive on
$(3, \infty),$ there is a relative minimum at 3.

11.   Since the graph of the derivative is zero at
$x = -8,\ x = -6,\ x = -2.5$ and $x = -1.5,$
the critical numbers are $-8, -6, -2.5,$ and
$-1.5.$

Since the graph of the derivative is positive on
$(-\infty, -8)$ and negative on $(-8, -6),$ there is a
relative maximum at $-8.$ Since the graph of the
derivative is negative on $(-8, -6)$ and positive
on $(-6, -2.5),$ there is a relative minimum at
$-6.$ Since the graph of the derivative is positive
on $(-6, -2.5)$ and negative on $(-2.5, -1.5),$
there is a relative maximum at $-2.5.$ Since the
graph of the derivative is negative on
$(-2.5, -1.5)$ and positive on $(-1.5, \infty),$ there is
a relative minimum at $-1.5.$

13. $f(x) = x^2 - 10x + 33$

$f'(x) = 2x - 10$

$f'(x)$ is zero when $x = 5.$

$f'(0) = -10 < 0$

$f'(6) = 2 > 0$

f is decreasing on $(-\infty, 5)$ and increasing on
$(5, \infty).$ Thus, a relative minimum occurs at
$x = 5.$

$$f(5) = 8$$

Relative minimum of 8 at 5

15. $f(x) = x^3 + 6x^2 + 9x - 8$

$f'(x) = 3x^2 + 12x + 9 = 3(x^2 + 4x + 3)$

$\quad\quad = 3(x + 3)(x + 1)$

$f'(x)$ is zero when $x = -1$ or $x = -3$.

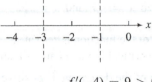

$f'(-4) = 9 > 0$

$f'(-2) = -3 < 0$

$f'(0) = 9 > 0$

Thus, f is increasing on $(-\infty, -3)$, decreasing on $(-3, -1)$, and increasing on $(-1, \infty)$.

f has a relative maximum at -3 and a relative minimum at -1.

$f(-3) = -8$

$f(-1) = -12$

Relative maximum of -8 at -3; relative minimum of -12 at -1

17. $f(x) = -\dfrac{4}{3}x^3 - \dfrac{21}{2}x^2 - 5x + 8$

$f'(x) = -4x^2 - 21x - 5$

$\quad\quad = (-4x - 1)(x + 5)$

$f'(x)$ is zero when $x = -5$, or $x = -\frac{1}{4}$.

$f'(-6) = -23 < 0$

$f'(-4) = 15 > 0$

$f'(0) = -5 < 0$

f is decreasing on $(-\infty, -5)$, increasing on $\left(-5, -\frac{1}{4}\right)$, and decreasing on $\left(-\frac{1}{4}, \infty\right)$. f has a relative minimum at -5 and a relative maximum at $-\frac{1}{4}$.

$f(-5) = -\dfrac{377}{6}$

$f\left(-\dfrac{1}{4}\right) = \dfrac{827}{96}$

Relative maximum of $\frac{827}{96}$ at $-\frac{1}{4}$; relative minimum of $-\frac{377}{6}$ at -5.

19. $f(x) = x^4 - 18x^2 - 4$

$f'(x) = 4x^3 - 36x$

$\quad\quad = 4x(x^2 - 9)$

$\quad\quad = 4x(x + 3)(x - 3)$

$f'(x)$ is zero when $x = 0$ or $x = -3$ or $x = 3$.

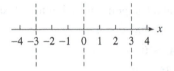

$f'(-4) = 4(-4)^3 - 36(-4) = -112 < 0$

$f'(-1) = -4 + 36 = 32 > 0$

$f'(1) = 4 - 36 = -32 < 0$

$f'(4) = 4(4)^3 - 36(4) = 112 > 0$

f is decreasing on $(-\infty, -3)$ and $(0, 3)$; f is increasing on $(-3, 0)$ and $(3, \infty)$.

$f(-3) = -85$

$f(0) = -4$

$f(3) = -85$

Relative maximum of -4 at 0; relative minimum of -85 at 3 and -3

21. $f(x) = 3 - (8 + 3x)^{2/3}$

$f'(x) = -\dfrac{2}{3}(8 + 3x)^{-1/3}(3)$

$\quad\quad = -\dfrac{2}{(8 + 3x)^{1/3}}$

Critical number:

$8 + 3x = 0$

$x = -\dfrac{8}{3}$

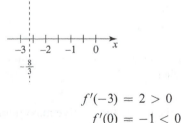

$f'(-3) = 2 > 0$

$f'(0) = -1 < 0$

f is increasing on $\left(-\infty, -\frac{8}{3}\right)$ and decreasing on $\left(-\frac{8}{3}, \infty\right)$.

$$f\left(-\frac{8}{3}\right) = 3$$

Relative maximum of 3 at $-\frac{8}{3}$.

23. $f(x) = 2x + 3x^{2/3}$

$f'(x) = 2 + 2x^{-1/3}$

$\quad = 2 + \dfrac{2}{\sqrt[3]{x}}$

Find the critical numbers.

$f'(x) = 0$ when

$$2 + \frac{2}{\sqrt[3]{x}} = 0$$

$$\frac{2}{\sqrt[3]{x}} = -2$$

$$x = (-1)^3$$

$$x = -1.$$

$f'(x)$ does not exist when

$\sqrt[3]{x} = 0$, that is, when $x = 0$.

$f'(-2) = 2 + \dfrac{2}{\sqrt[3]{-2}} \approx 0.41 > 0$

$f'\left(-\dfrac{1}{2}\right) = 2 + \dfrac{2}{\sqrt[3]{-\frac{1}{2}}}$

$\qquad = 2 + \dfrac{2\sqrt[3]{2}}{-1} \approx -0.52 < 0$

$f'(1) = 2 + \dfrac{2}{\sqrt[3]{1}} = 4 > 0$

f is increasing on $(-\infty, -1)$ and $(0, \infty)$.

f is decreasing on $(-1, 0)$.

$$f(-1) = 2(-1) + 3(-1)^{2/3} = 1$$
$$f(0) = 0$$

Relative maximum of 1 at -1; relative minimum of 0 at 0.

25. $f(x) = x - \dfrac{1}{x}$

$f'(x) = 1 + \dfrac{1}{x^2}$ is never zero, but fails to exist at $x = 0$

Since $f(x)$ also fails to exist at $x = 0$, there are no critical numbers and no relative extrema.

27. $f(x) = \dfrac{x^2 - 2x + 1}{x - 3}$

$f'(x) = \dfrac{(x-3)(2x-2) - (x^2 - 2x + 1)(1)}{(x-3)^2}$

$\quad = \dfrac{x^2 - 6x + 5}{(x-3)^2}$

Find the critical numbers:

$$x^2 - 6x + 5 = 0$$
$$(x - 5)(x - 1) = 0$$
$$x = 5 \quad \text{or} \quad x = 1$$

Note that $f(x)$ and $f'(x)$ do not exist at $x = 3$, so the only critical numbers are 1 and 5.

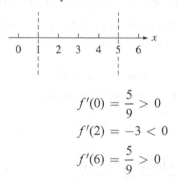

$f'(0) = \dfrac{5}{9} > 0$

$f'(2) = -3 < 0$

$f'(6) = \dfrac{5}{9} > 0$

$f(x)$ is increasing on $(-\infty, 1)$ and $(5, \infty)$.

$f(x)$ is decreasing on $(1, 5)$.

$$f(1) = 0$$
$$f(5) = 8$$

Relative maximum of 0 at 1; relative minimum of 8 at 5

29. $f(x) = x^2 e^x - 3$

$f'(x) = x^2 e^x + 2xe^x$

$\quad = xe^x(x + 2)$

$f'(x)$ is zero at $x = 0$ and $x = -2$.

$$f'(-3) = 3e^{-3} = \frac{3}{e^3} > 0$$

$$f'(-1) = -e^{-1} = \frac{-1}{e} < 0$$

$$f'(1) = 3e^1 > 0$$

f is increasing on $(-\infty, -2)$ and $(0, \infty)$.

f is decreasing on $(-2, 0)$.

$$f(0) = 0 \cdot e^0 - 3 = -3$$
$$f(-2) = (-2)^2 e^{-2} - 3$$
$$= \frac{4}{e^2} - 3$$
$$\approx -2.46$$

Relative minimum of -3 at 0; relative maximum of -2.46 at -2.

31. $f(x) = 2x + \ln x$

$$f'(x) = 2 + \frac{1}{x} = \frac{2x + 1}{x}$$

$f'(x)$ is zero at $x = -\frac{1}{2}$. The domain of $f(x)$ is $(0, \infty)$. Therefore $f'(x)$ is never zero in the domain of $f(x)$.

$f'(1) = 3 > 0$. Since $f(x)$ is always increasing, f has no relative extrema.

33. $f(x) = \dfrac{2^x}{x}$

$$f'(x) = \frac{(x) \ln 2(2^x) - 2^x(1)}{x^2}$$

$$= \frac{2^x(x \ln 2 - 1)}{x^2}$$

Find the critical numbers:

$$x \ln 2 - 1 = 0 \quad \text{or} \quad x^2 = 0$$
$$x = \frac{1}{\ln 2} \qquad\qquad x = 0$$

Since f is defined for $x = 0$, 0 is not a critical number. $x = \frac{1}{\ln 2} \approx 1.44$ is the only critical number.

$$f'(1) \approx -0.6137 < 0$$
$$f'(2) \approx 0.3863 > 0$$

f is decreasing on $\left(0, \frac{1}{\ln 2}\right)$ and increasing on $\left(\frac{1}{\ln 2}, \infty\right)$.

$$f\left(\frac{1}{\ln 2}\right) = \frac{2^{1/\ln 2}}{\frac{1}{\ln 2}}$$

$$= \ln 2 \left(e^{\ln 2}\right)^{1/\ln 2} = e \ln 2$$

Relative minimum of $e \ln 2$ at $\frac{1}{\ln 2}$.

35. $y = -2x^2 + 12x - 5$

$$y' = -4x + 12$$
$$= -4(x - 3)$$

The vertex occurs when $y' = 0$ or when

$$x - 3 = 0$$
$$x = 0$$

When $x = 3$,

$$y = -2(3)^2 + 12(3) - 5 = 13$$

The vertex is $(3, 13)$.

37. $f(x) = x^5 - x^4 + 4x^3 - 30x^2 + 5x + 6$

$$f'(x) = 5x^4 - 4x^3 + 12x^2 - 60x + 5$$

Graph f' on a graphing calculator. A suitable choice for the viewing window is $[-4, 4]$ by $[-50, 50]$, $\text{Yscl} = 10$.

Use the calculator to estimate the x-intercepts of this graph. These numbers are the solutions of the equation $f'(x) = 0$ and thus the critical numbers for f. Rounded to three decimal places, these x-values are 0.085 and 2.161.

Examine the graph of f' near $x = 0.085$ and $x = 2.161$. Observe that $f'(x) > 0$ to the left of $x = 0.085$ and $f'(x) < 0$ to the right of $x = 0.085$. Also observe that $f'(x) < 0$ to the left of $x = 2.161$ and $f'(x) > 0$ to the right of $x = 2.161$. The first derivative test allows us to conclude that f has a relative maximum at $x = 0.085$ and a relative minimum at $x = 2.161$.

$$f(0.085) \approx 6.211$$
$$f(2.161) \approx -57.607$$

Relative maximum of 6.211 at 0.085; relative minimum of -57.607 at 2.161.

39. $f(x) = 2|x + 1| + 4|x - 5| - 20$

Graph this function in the window $[-10, 10]$ by $[-15, 30]$, $\text{Yscl} = 5$.

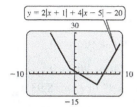

The graph shows that f has no relative maxima, but there is a relative minimum at $x = 5$.

(Note that the graph has a sharp point at $(5, -8)$, indicating that $f'(-5)$ does not exist.)

41. $f' = 0$ when $x = -1$ and when $x = 3$.

f' is positive just to the left of -1 and negative just to the right of -1, so f has a relative maximum at $x = -1$. f' is negative just to the left of 3 and positive just to the right of 3, so f has a relative minimum at $x = 3$.

43. $C(q) = 80 + 18q$; $p = 70 - 2q$

$$P(q) = R(q) - C(q) = pq - C(q)$$
$$= (70 - 2q)q - (80 + 18q)$$
$$= -2q^2 + 52q - 80$$

(a)   Since the graph of P is a parabola that opens downward, we know that its vertex is a maximum point. To find the q-value of this point, we find the critical number.

$$P'(q) = -4q + 52$$
$$P'(q) = 0 \text{ when}$$

$$-4q + 52 = 0$$
$$4q = 52$$
$$q = 13$$

The number of units that produce maximum profit is 13.

(b)   If $q = 13$,

$$p = 70 - 2(13)$$
$$= 44$$

The price that produces maximum profit is $44.

(c) $P(13) = -2(13)^2 + 52(13) - 80 = 258$

The maximum profit is $258.

45. $C(q) = 100 + 20qe^{-0.01q}$; $p = 40e^{-0.01q}$

$$P(q) = R(q) - C(q) = pq - C(q)$$
$$= (40e^{-0.01q})q - (100 + 20qe^{-0.01q})$$
$$= 20qe^{-0.01q} - 100$$

(a) $P'(q) = 20e^{-0.01q} + 20qe^{-0.01q}(-0.01)$
$$= (20 - 0.2q)e^{-0.01q}$$

Solve $P'(q) = 0$.

$$(20 - 0.2q)e^{-0.01q} = 0$$
$$20 - 0.2q = 0$$
$$q = 100$$

Since $e^{-0.01q} > 0$ for all values of q, the sign of $P'(q)$ is the same as the sign of $20 - 0.2q$. For $q < 100$, $P'(q) > 0$; for $q > 100$, $P'(q) < 0$. Therefore, the number of units that produces maximum profit is 100.

(b)   If $q = 100$,

$$p = 40e^{-0.01(100)}$$
$$= 40e^{-1}$$
$$\approx 14.72$$

The price per unit that produces maximum profit is $14.72.

(c) $P(100) = 20(100)e^{-0.01(100)} - 100$
$$= 2000e^{-1} - 100$$
$$\approx 635.76$$

The maximum profit is $635.76.

47. $P(t) = -0.006088t^3 + 0.1904t^2$
$$- 1.076t + 18.39$$

$$P'(t) = -0.018264t^2 + 0.3808t - 1.076$$

Use a graphing calculator to find the critical numbers for $0 \leq t \leq 24$.

The critical numbers are $t \approx 3.370$ and $t \approx 17.479$. These divide the domain of P into three intervals, and we pick a test point in each: Pick $t = 1$, $t = 10$, and $t = 20$.

$$P'(1) \approx -0.713$$
$$P'(10) \approx 0.906$$
$$P'(20) \approx -0.766$$

P is decreasing on $(0, 3.370)$ and $(17.479, 24)$ and increasing on $(3.370, 17.479)$. There will be relative maxima at the left endpoint ($t = 0$) and at $t = 17.479$; there will be relative minima at $t = 3.370$ and at the right endpoint ($t = 24$).

$$P(0) = 18.39$$
$$P(3.370) \approx 16.69$$
$$P(17.479) \approx 25.24$$
$$P(24) \approx 18.08$$

We convert the time values to hours and minutes by multiplying the decimal parts by 60:
$(0.370)(60) \approx 22$ and $(0.479)(60) \approx 29$.

Recalling that the function P gives the power in thousands of megawatts, we have the final result:

Relative maximum of 18,400 megawatts at midnight ($t = 0$); relative minimum of 16,700 megawatts at 3:22 A.M.; relative maximum of 25,200 megawatts at 5:29 P.M.; relative minimum of 18,100 megawatts at midnight ($t = 24$).

49. $p = D(q) = 200e^{-0.1q}$

$$R(q) = pq$$
$$= 200qe^{-0.1q}$$
$$R'(q) = 200qe^{-0.1q}(-0.1) + 200e^{-0.1q}$$
$$= 20e^{-0.1q}(10 - q)$$

$R'(q) = 0$ when $q = 10$, the only critical number. Use the first derivative test to verify that $q = 10$ gives the maximum revenue.

$$R'(9) = 20e^{-0.9} > 0$$
$$R'(11) = -20e^{-1.1} < 0$$

The maximum revenue results when $q = 10$

$p = D(10) = \frac{200}{e} \approx 73.58$, or when telephones are sold at $73.58.

51. $C(x) = 0.002x^3 = 9x + 6912$

$$\bar{C}(x) = \frac{C(x)}{x} = 0.002x^2 + 9 + \frac{6912}{x}$$

$$\bar{C}'(x) = 0.004x - \frac{6912}{x^2}$$

$\bar{C}'(x) = 0$ when

$$0.004x - \frac{6912}{x^2} = 0$$

$$0.004x^3 = 6912$$

$$x^3 = 1,728,000$$

$$x = 120$$

A product level of 120 units will produce the minimum average cost per unit.

53. $a(t) = 0.008t^3 - 0.288t^2 + 2.304t + 7$

$$a'(t) = 0.024t^2 - 0.576t + 2.304$$

Set $a' = 0$ and solve for t.

$$0.024t^2 - 0.576t + 2.304 = 0$$

$$0.024(t^2 - 24t + 96) = 0$$

$$t^2 - 24t + 96 = 0$$

$$t \approx 5.07 \text{ or } t \approx 18.93$$

$t = 5.07 = 5 \text{ hours} + 0.07 \cdot 60 \text{ minutes}$ corresponds to 5:04 P.M.

$t = 18.93 = 18 \text{ hours} + 0.93 \cdot 60 \text{ minutes}$ corresponds to 6:56 A.M.

55. $M(t) = 369(0.93)^t (t)^{0.36}$

$$M'(t) = (369)(0.93)^t \ln(0.93)(t^{0.36})$$
$$+ 369(0.93)^t (0.36)(t)^{-0.64}$$
$$= (369t^{0.36})(0.93^t \ln 0.93)$$
$$+ \frac{132.84(0.93)^t}{t^{0.64}}$$

$M'(t) = 0$ when $t \approx 4.96$.

Verify that $t \approx 4.96$ gives a maximum.

$$M'(4) > 0$$
$$M'(5) < 0$$

$$M(4.96) = 369(0.93)^{4.96}(4.96)^{0.36} \approx 485.22$$

The female moose reaches a maximum weight of about 458.22 kilograms at about 4.96 years.

57. $D(x) = -x^4 + 8x^3 + 80x^2$

$$D'(x) = -4x^3 + 24x^2 + 160x$$
$$= -4x(x + 4)(x - 10)$$

$D'(x) = 0$ when $x = 0$, $x = -4$, or $x = 10$. Disregard the nonpositive values.

Verify that $x = 10$ gives a maximum.

$$D'(9) = 468 > 0$$
$$D'(11) = -660 < 0$$

The speaker should aim for a degree of discrepancy of 10.

59. $s(t) = -16t^2 + 40t + 3$

   $s'(t) = -32t + 40$

   (a)   when $s'(t) = 0$,

   $$-32t + 40 = 0$$
   $$32t = 40$$
   $$t = \frac{40}{32} = \frac{5}{4}$$

   Verify that $t = 5/4$ gives a maximum.

   $$s'(1) = 8$$
   $$s'(2) = -24$$

   Now find the height when $t = 5/4$.

   $$s\left(\frac{5}{4}\right) = -16\left(\frac{5}{4}\right)^2 + 40\left(\frac{5}{4}\right) + 3$$
   $$= 28$$

   The maximum height of the cork is 28 feet.

   (b)   The cork remains in the air as long as $s(t) > 0$. Use the quadratic formula to solve $s(t) = 0$.

   $$-16t^2 + 40t + 3 = 0$$

   $$t = \frac{-40 \pm \sqrt{40^2 - 4(-16)(3)}}{2(-16)}$$

   $$= \frac{5 \pm 2\sqrt{7}}{4}$$

   $$\approx -0.073, 2.573$$

   Only the positive solution is relevant, so the cork stays in the air for about 2.57 seconds.

5.3   Higher Derivatives, Concavity, and the Second Derivative Test

Your Turn 1

$$f(x) = 5x^4 - 4x^3 + 3x$$
$$f'(x) = 20x^3 - 12x^2 + 3$$
$$f''(x) = 60x^2 - 24x$$
$$f''(1) = 60(1^2) - 24(1)$$
$$= 36$$

Your Turn 2

(a)   Use the chain rule.

$$f(x) = (x^3 + 1)^2$$
$$f'(x) = (2)(x^3 + 1)(3x^2)$$
$$= 6x^2(x^3 + 1)$$
$$= 6x^5 + 6x^2$$
$$f''(x) = 30x^4 + 12x$$

(b)   Use the product rule.

$$f(x) = xe^x$$
$$f'(x) = xe^x + (1)e^x$$
$$= xe^x + e^x$$

Now note that we have already found the derivative of the first term of $f'(x)$, which is the original function $f(x)$.

Thus

$$f''(x) = (xe^x + e^x) + e^x$$
$$= 2e^x + xe^x$$

(c)   Use the quotient rule.

$$h(x) = \frac{\ln x}{x}$$

$$h'(x) = \frac{(x)\left(\frac{1}{x}\right) - (\ln x)(1)}{x^2}$$

$$= \frac{1 - \ln x}{x^2}$$

$$h''(x) = \frac{(x^2)\left(-\frac{1}{x}\right) - (1 - \ln x)(2x)}{x^4}$$

$$= \frac{-x - 2x + 2x \ln x}{x^4}$$

$$= \frac{-3 + 2 \ln x}{x^3}$$

Your Turn 3

$$s(t) = t^3 - 3t^2 - 24t + 10$$
$$v(t) = s'(t) = 3t^2 - 6t - 24$$
$$a(t) = v'(t) = s''(t) = 6t - 6$$

First find when v changes sign.

$$v(t) = 3t^2 - 6t - 24 = 0$$
$$3(t^2 - 2t - 8) = 0$$
$$(t - 4)(t + 2) = 0$$
$$t = 4 \text{ or } t = -2$$

Only the positive value of t is relevant. Check the velocity at times before and after 4, say at $t = 2$ and $t = 6$.

$$v(2) = -24 < 0$$
$$v(6) = 48 > 0$$

Thus the car backs up for the first four seconds and then goes forward.

Now find where a changes sign.

$$a(t) = 6t - 6 = 0$$
$$6t - 6 = 0$$
$$t = 1$$

Check the acceleration at times before and after 1, say at $t = 1/2$ and $t = 2$.

$$a(1/2) = -3 < 0$$
$$a(2) = 6 > 0$$

Now we construct the following graph showing the signs of v and a.

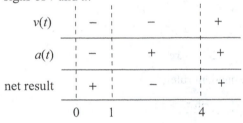

The car speeds up (in the backward direction) for $0 < t < 1$; it slows down (still in the backward direction) for $1 < t < 4$; it speeds up (in the forward direction) for $t > 4$. (All times are in seconds.)

Your Turn 4

$$f(x) = x^5 - 30x^3$$
$$f'(x) = 5x^4 - 90x^2$$
$$f''(x) = 20x^3 - 180x$$

Factor $f''(x)$ and create a number line.

$$f''(x) = 20x^3 - 180x$$
$$= 20(x^3 - 9x)$$
$$= 20(x)(x - 3)(x + 3)$$

The zeros of $f''(x)$ are $-3, 0,$ and 3, and they divide the domain of f into four intervals:

$$(-\infty, -3), (-3, 0), (0, 3), (3, \infty)$$

Choose a test point in each interval and find the sign of $f''(x)$ at the test points. For example, choose $-4, -1, 1,$ and 4.

$$f''(-4) = -560$$
$$f''(-1) = 160$$
$$f''(1) = -160$$
$$f''(4) = 560$$

Here is the corresponding number line.

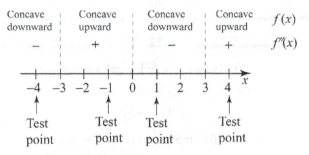

The figure shows that f is concave downward on $(-\infty, -3)$ and $(0, 3)$ and concave upward on $(-3, 0)$ and $(3, \infty)$. Inflection points occur where $f''(x)$ changes sign, so we evaluate f at these points:

$$f(-3) = 567$$
$$f(0) = 0$$
$$f(3) = -567$$

Thus the inflection points are $(-3, 567)$, $(0, 0)$ and $(3, -567)$.

Your Turn 5

$$f(x) = -2x^3 + 3x^2 + 72x$$
$$f'(x) = -6x^2 + 6x + 72$$

Solve the equation $f'(x) = 0$.

$$-6x^2 + 6x + 72 = 0$$
$$x^2 - x - 12 = 0$$
$$(x + 3)(x - 4) = 0$$
$$x = -3 \text{ or } x = 4$$

Now use the second derivative test.

$$f''(x) = -12x + 6$$
$$f''(-3) = 42$$
$$f''(4) = -42$$

Since $f''(-3) > 0$, $x = -3$ leads to a relative minimum; since $f''(4) < 0$, $x = 4$ leads to a relative maximum. Thus we have a relative minimum of $f(-3) = -135$ at $x = -3$ and a relative maximum of $f(4) = 208$ at $x = 4$.

5.3 Warmup Exercises

W1. $f(x) = x^3 - 3x^2 - 72x + 20$

$f'(x) = 3x^2 - 6x - 72$

$= 3(x^2 - 2x - 24)$

$= 3(x - 6)(x + 4)$

The critical numbers are 6 and -4, and they divide the domain into three intervals:
$(-\infty, -4), (-4, 6)$ and $(6, \infty)$

Evaluate f' at a test point in each interval.

$f'(-5) = 33$

$f'(0) = -72$

$f'(8) = 72$

At $x = -4$, f' changes from positive to negative, so f has a relative maximum at $x = -4$; the relative maximum value is $f(-4) = 196$. At

$x = 6$, f' changes from negative to positive, so f as a relative minimum at $x = 6$; the relative minimum value is $f(6) = -304$.

W2. $f(x) = x^4 - 8x^3 - 32x^2 + 10$

$f'(x) = 4x^3 - 24x^2 - 64x$

$= 4x(x^2 - 6x - 16)$

$= 4x(x - 8)(x + 2)$

The critical numbers are -2, 0 and 8, and they divide the domain into four intervals:
$(-\infty, -2), (-2, 0), (0, 8), (8, \infty)$

Evaluate f' at a test point in each interval.

$f'(-3) = -132$

$f'(-1) = 36$

$f'(1) = -84$

$f'(10) = 960$

At $x = -2$ and $x = 8$, f' changes from negative to positive, so f has a relative minimum at these two points. At $x = 0$, f' changes from positive to negative, so f has a relative maximum at this point. The values of the relative minima are $f(-2) = -38$ and $f(8) = -2038$; the value of the relative maximum is $f(0) = 10$.

5.3 Exercises

1. $f(x) = 5x^3 - 7x^2 + 4x + 3$

$f'(x) = 15x^2 - 14x + 4$

$f''(x) = 30x - 14$

$f''(0) = 30(0) - 14 = -14$

$f''(2) = 30(2) - 14 = 46$

3. $f(x) = 4x^4 - 3x^3 - 2x^2 + 6$

$f'(x) = 16x^3 - 9x^2 - 4x$

$f''(x) = 48x^2 - 18x - 4$

$f''(0) = 48(0)^2 - 18(0) - 4 = -4$

$f''(2) = 48(2)^2 - 18(2) - 4 = 152$

5. $f(x) = 3x^2 - 4x + 8$

$f'(x) = 6x - 4$

$f''(x) = 6$

$f''(0) = 6$

$f''(2) = 6$

7. $f(x) = \dfrac{x^2}{1 + x}$

$f'(x) = \dfrac{(1 + x)(2x) - x^2(1)}{(1 + x)^2}$

$= \dfrac{2x + x^2}{(1 + x)^2}$

$f''(x) = \dfrac{(1 + x)^2(2 + 2x) - (2x + x^2)(2)(1 + x)}{(1 + x)^4}$

$= \dfrac{(1 + x)(2 + 2x) - (2x + x^2)(2)}{(1 + x)^3}$

$= \dfrac{2}{(1 + x)^3}$

$f''(0) = 2$

$f''(2) = \dfrac{2}{27}$

9. $f(x) = \sqrt{x^2 + 4} = (x^2 + 4)^{1/2}$

$f'(x) = \dfrac{1}{2}(x^2 + 4)^{-1/2} \cdot 2x$

$= \dfrac{x}{(x^2 + 4)^{1/2}}$

$$f''(x) = \frac{(x^2 + 4)^{1/2}(1) - x\left[\frac{1}{2}(x^2 + 4)^{-1/2}\right]2x}{x^2 + 4}$$

$$= \frac{(x^2 + 4)^{1/2} - \frac{x^2}{(x^2+4)^{1/2}}}{x^2 + 4}$$

$$= \frac{(x^2 + 4) - x^2}{(x^2 + 4)^{3/2}}$$

$$= \frac{4}{(x^2 + 4)^{3/2}}$$

$$f''(0) = \frac{4}{(0^2 + 4)^{3/2}}$$

$$= \frac{4}{4^{3/2}} = \frac{4}{8} = \frac{1}{2}$$

$$f''(2) = \frac{4}{(2^2 + 4)^{3/2}}$$

$$= \frac{4}{8^{3/2}} = \frac{4}{16\sqrt{2}} = \frac{1}{4\sqrt{2}}$$

11. $f(x) = 32x^{3/4}$

$$f'(x) = 24x^{-1/4}$$

$$f''(x) = -6x^{-5/4} = -\frac{6}{x^{5/4}}$$

$f''(0)$ does not exist.

$$f''(2) = -\frac{6}{2^{5/4}}$$

$$= -\frac{3}{2^{1/4}}$$

13. $f(x) = 5e^{-x^2}$

$$f'(x) = 5e^{-x^2}(-2x) = -10xe^{-x^2}$$

$$f''(x) = -10xe^{-x^2}(-2x) + e^{-x^2}(-10)$$

$$= 20x^2e^{-x^2} - 10e^{-x^2}$$

$$f''(0) = 20(0^2)e^{-0^2} - 10e$$

$$= 0 - 10 = -10$$

$$f''(2) = 20(2^2)e^{-(2^2)} - 10e^{-(2^2)}$$

$$= 80e^{-4} - 10e^{-4} = 70e^{-4}$$

$$\approx 1.282$$

15. $f(x) = \dfrac{\ln x}{4x}$

$$f'(x) = \frac{4x\left(\frac{1}{x}\right) - (\ln x)(4)}{(4x)^2}$$

$$= \frac{4 - 4\ln x}{16x^2} = \frac{1 - \ln x}{4x^2}$$

$$f''(x) = \frac{4x^2\left(-\frac{1}{x}\right) - (1 - \ln x)8x}{16x^4}$$

$$= \frac{-4x - 8x + 8x\ln x}{16x^4}$$

$$= \frac{-12x + 8x\ln x}{16x^4}$$

$$= \frac{4x(-3 + 2\ln x)}{16x^4}$$

$$= \frac{-3 + 2\ln x}{4x^3}$$

$f''(0)$ does not exist because $\ln 0$ is undefined.

$$f''(2) = \frac{-3 + 2\ln 2}{4(2)^3} = \frac{-3 + 2\ln 2}{32} \approx 0.050$$

17. $f(x) = 7x^4 + 6x^3 + 5x^2 + 4x + 3$

$$f'(x) = 28x^3 + 18x^2 + 10x + 4$$

$$f''(x) = 84x^2 + 36x + 10$$

$$f'''(x) = 168x + 36$$

$$f^{(4)}(x) = 168$$

19. $f(x) = 5x^5 - 3x^4 + 2x^3 + 7x^2 + 4$

$$f'(x) = 25x^4 - 12x^3 + 6x^2 + 14x$$

$$f''(x) = 100x^3 - 36x^2 + 12x + 14$$

$$f'''(x) = 300x^2 - 72x + 12$$

$$f^{(4)}(x) = 600x - 72$$

21. $f(x) = \dfrac{x-1}{x+2}$

$$f'(x) = \frac{(x+2)-(x-1)}{(x+2)^2} = \frac{3}{(x+2)^2}$$

$$f''(x) = \frac{-3(2)(x+2)}{(x+2)^4} = \frac{-6}{(x+2)^3}$$

$$f'''(x) = \frac{(-6)(-3)(x+2)^2}{(x+2)^6}$$

$$= 18(x+2)^{-4} \quad \text{or} \quad \frac{18}{(x+2)^4}$$

$$f^{(4)}(x) = \frac{-18(4)(x+2)^3}{(x+2)^8}$$

$$= -72(x+2)^{-5} \quad \text{or} \quad \frac{-72}{(x+2)^5}$$

23. $f(x) = \dfrac{3x}{x-2}$

$$f'(x) = \frac{(x-2)(3)-3x(1)}{(x-2)^2} = \frac{-6}{(x-2)^2}$$

$$f''(x) = \frac{-6(-2)(x-2)}{(x-2)^4} = \frac{12}{(x-2)^3}$$

$$f'''(x) = \frac{-12(3)(x-2)^2}{(x-2)^6} = -36(x-2)^{-4}$$

$$\text{or} \quad \frac{-36}{(x-2)^4}$$

$$f^{(4)}(x) = \frac{-36(-4)(x-2)^3}{(x-2)^8} = 144(x-2)^{-5}$$

$$\text{or} \quad \frac{144}{(x-2)^5}$$

25. (a) Any term with power less than n will be 0 after being differentiated n times, so only the term x^n survives, and its derivative is $n!$. Thus $f^{(n)}(x) = n!$.

(b) For $k > n$ each term has derivative 0 and $f^{(k)}(x) = 0$

27. $\quad f(x) = e^x$

$$f'(x) = e^x$$

$$f''(x) = e^x$$

$$f'''(x) = e^x$$

$$f^{(n)}(x) = e^x$$

29. Concave upward on $(2, \infty)$

Concave downward on $(-\infty, 2)$

Inflection point at $(2, 3)$

31. Concave upward on $(-\infty, -1)$ and $(8, \infty)$

Concave downward on $(-1, 8)$

Inflection points at $(-1, 7)$ and $(8, 6)$

33. Concave upward on $(2, \infty)$

Concave downward on $(-\infty, 2)$

No points inflection

35. $\quad f(x) = x^2 + 10x - 9$

$$f'(x) = 2x + 10$$

$$f''(x) = 2 > 0 \text{ for all } x.$$

Always concave upward

No inflection points

37. $\quad f(x) = -2x^3 + 9x^2 + 168x - 3$

$$f'(x) = -6x^2 + 18x + 168$$

$$f''(x) = -12x + 18$$

$$f''(x) = -12x + 18 > 0 \quad \text{when}$$

$$-6(2x - 3) > 0$$

$$2x - 3 < 0$$

$$x < \frac{3}{2}.$$

Concave upward on $\left(-\infty, \frac{3}{2}\right)$

$$f''(x) = -12x + 18 < 0 \text{ when}$$

$$-6(2x - 3) < 0$$

$$2x - 3 > 0$$

$$x > \frac{3}{2}.$$

Concave downward on $\left(\frac{3}{2}, \infty\right)$

$$f''(x) = -12x + 18 = 0 \text{ when}$$

$$-6(2x + 3) = 0$$

$$2x + 3 = 0$$

$$x = \frac{3}{2}.$$

$$f\left(\frac{3}{2}\right) = \frac{525}{2}$$

Inflection point at $\left(\frac{3}{2}, \frac{525}{2}\right)$

39. $f(x) = \dfrac{3}{x-5}$

$f'(x) = \dfrac{-3}{(x-5)^2}$

$f''(x) = \dfrac{-3(-2)(x-5)}{(x-5)^4} = \dfrac{6}{(x-5)^3}$

$f''(x) = \dfrac{6}{(x-5)^3} > 0$ when

$(x-5)^3 > 0$

$x - 5 > 0$

$x > 5.$

Concave upward on $(5, \infty)$

$f''(x) = \dfrac{6}{(x-5)^3} < 0$ when

$(x-5)^3 < 0$

$x - 5 < 0$

$x < 5.$

Concave downward on $(-\infty, 5)$

$f''(x) \neq 0$ for any value for x; it does not exist when $x = 5$. There is a change of concavity there, but no inflection point since $f(5)$ does not exist.

41. $f(x) = x(x+5)^2$

$f'(x) = x(2)(x+5) + (x+5)^2$

$= (x+5)(2x + x + 5)$

$= (x+5)(3x+5)$

$f''(x) = (x+5)(3) + (3x+5)$

$= 3x + 15 + 3x + 5 = 6x + 20$

$f''(x) = 6x + 20 > 0$ when

$2(3x + 10) > 0$

$3x > -10$

$x > -\dfrac{10}{3}.$

Concave upward on $\left(-\dfrac{10}{3}, \infty\right)$

$f''(x) = 6x + 20 < 0$ when

$2(3x + 10) < 0$

$3x < -10$

$x < -\dfrac{10}{3}.$

Concave downward on $\left(-\infty, -\dfrac{10}{3}\right)$

$f\left(-\dfrac{10}{3}\right) = -\dfrac{10}{3}\left(-\dfrac{10}{3} + 5\right)^2$

$= \dfrac{-10}{3}\left(\dfrac{-10+15}{3}\right)^2$

$= -\dfrac{10}{3} \cdot \dfrac{25}{9} = -\dfrac{250}{27}$

Inflection point at $\left(-\dfrac{10}{3}, -\dfrac{250}{27}\right)$

43. $f(x) = 18x - 18e^{-x}$

$f'(x) = 18 - 18e^{-x}(-1) = 18 + 18e^{-x}$

$f''(x) = 18e^{-x}(-1) = -18e^{-x}$

$f''(x) = -18e^{-x} < 0$ for all x

$f(x)$ is never concave upward and always concave downward. There are no points of inflection since $-18e^{-x}$ is never equal to 0.

45. $f(x) = x^{8/3} - 4x^{5/3}$

$f'(x) = \dfrac{8}{3}x^{5/3} - \dfrac{20}{3}x^{2/3}$

$f''(x) = \dfrac{40}{9}x^{2/3} - \dfrac{40}{9}x^{-1/3} = \dfrac{40(x-1)}{9x^{1/3}}$

$f''(x) = 0$ when $x = 1$

$f''(x)$ fails to exist when $x = 0$

Note that both $f(x)$ and $f'(x)$ exist at $x = 0$. Check the sign of $f''(x)$ in the three intervals determined by $x = 0$ and $x = 1$ using test points.

$f''(-1) = \dfrac{40(-2)}{9(-1)} = \dfrac{80}{9} > 0$

$f''\left(\dfrac{1}{8}\right) = \dfrac{40\left(-\dfrac{7}{8}\right)}{9\left(\dfrac{1}{2}\right)} = -\dfrac{70}{9} < 0$

$f''(8) = \dfrac{40(7)}{9(2)} = \dfrac{140}{9} > 0$

Concave upward on $(-\infty, 0)$ and $(1, \infty)$; concave downward on $(0, 1)$

$f(0) = (0)^{8/3} - 4(0)^{5/3} = 0$

$f(1) = (1)^{8/3} - 4(1)^{5/3} = -3$

Inflection points at $(0, 0)$ and $(1, -3)$

47. $f(x) = \ln(x^2 + 1)$

$$f'(x) = \frac{2x}{x^2 + 1}$$

$$f''(x) = \frac{(x^2 + 1)(2) - (2x)(2x)}{(x^2 + 1)^2}$$

$$= \frac{-2x^2 + 2}{(x^2 + 1)^2}$$

$$f''(x) = \frac{-2x^2 + 2}{(x^2 + 1)^2} > 0 \text{ when}$$

$$-2x^2 + 2 > 0$$

$$-2x^2 > -2$$

$$x^2 < 1$$

$$-1 < x < 1$$

Concave upward on $(-1, 1)$

$$f''(x) = \frac{-2x^2 + 2}{(x^2 + 1)^2} < 0 \text{ when}$$

$$-2x^2 + 2 < 0$$

$$-2x^2 < -2$$

$$x^2 > 1$$

$$x > 1 \text{ or } x < -1$$

Concave downward on $(-\infty, -1)$ and $(1, \infty)$

$$f(1) = \ln[(1)^2 + 1] = \ln 2$$

$$f(-1) = \ln[(-1)^2 + 1] = \ln 2$$

Inflection points at $(-1, \ln 2)$ and $(1, \ln 2)$

49. $f(x) = x^2 \log |x|$

$$f'(x) = 2x \log |x| + x^2 \left(\frac{1}{x \ln 10} \right)$$

$$= 2x \log |x| + \frac{x}{\ln 10}$$

$$f''(x) = 2 \log |x| + 2x \left(\frac{1}{x \ln 10} \right) + \frac{1}{\ln 10}$$

$$= 2 \log |x| + \frac{3}{\ln 10}$$

$f''(x) > 0$ when

$$2 \log |x| + \frac{3}{\ln 10} > 0$$

$$2 \log |x| > -\frac{3}{\ln 10}$$

$$\log |x| > -\frac{3}{2 \ln 10}$$

$$\frac{\ln |x|}{\ln 10} > -\frac{3}{2 \ln 10}$$

$$\ln |x| > -\frac{3}{2}$$

$$|x| > e^{-3/2}$$

$$x > e^{-3/2} \text{ or } x < -e^{-3/2}$$

Concave upward on $(-\infty, -e^{-3/2})$ and $(e^{-3/2}, \infty)$

$f''(x) < 0$ when

$$2 \log |x| + \frac{3}{\ln 10} < 0$$

$$2 \log |x| < -\frac{3}{\ln 10}$$

$$\log |x| < -\frac{3}{2 \ln 10}$$

$$\frac{\ln |x|}{\ln 10} < -\frac{3}{2 \ln 10}$$

$$\ln |x| < -\frac{3}{2}$$

$$|x| < e^{-3/2}$$

$$-e^{-3/2} < x < e^{-3/2}$$

Note that $f(x)$ is not defined at $x = 0$.

Concave downward on $(-e^{-3/2}, 0)$ and $(0, e^{-3/2})$.

$$f(-e^{-3/2}) = (-e^{-3/2})^2 \log|-e^{-3/2}|$$

$$= e^{-3} \log e^{-3/2} = -\frac{3e^{-3}}{2 \ln 10}$$

$$f(e^{-3/2}) = (e^{-3/2})^2 \log|e^{-3/2}|$$

$$= e^{-3} \log e^{-3/2} = -\frac{3e^{-3}}{2 \ln 10}$$

Inflection points at $\left(-e^{-3/2}, -\frac{3e^{-3}}{2 \ln 10} \right)$

and $\left(e^{-3/2}, -\frac{3e^{-3}}{2 \ln 10} \right)$

51. Since the graph of $f'(x)$ is increasing on $(-\infty, 0)$ and $(4, \infty)$, the function is concave upward on $(-\infty, 0)$ and $(4, \infty)$. Since the graph of $f'(x)$ is decreasing on $(0, 4)$, the function is concave downward on $(0, 4)$. The inflection points are at 0 and 4.

53. Since the graph of $f'(x)$ is increasing on $(-7, 3)$ and $(12, \infty)$, the function is concave upward on $(-7, 3)$ and $(12, \infty)$. Since the graph of $f'(x)$ is decreasing on $(-\infty, -7)$ and $(3, 12)$, the function is concave downward on $(-\infty, -7)$ and $(3, 12)$. The inflection points are at $-7, 3$, and 12.

55. Choose $f(x) = x^k$, where $1 < k < 2$.

If $k = \frac{4}{3}$, then

$$f'(x) = \frac{4}{3}x^{1/3} \qquad f''(x) = \frac{4}{9}x^{-2/3} = \frac{4}{9x^{2/3}}$$

Critical number: 0

Since $f'(x)$ is negative when $x < 0$ and positive when $x > 0$, $f(x) = x^{4/3}$ has a relative minimum at $x = 0$.

If $k = \frac{5}{3}$, then

$$f(x) = \frac{5}{3}x^{2/3} \qquad f''(x) = \frac{10}{9}x^{-1/3} = \frac{10}{9x^{1/3}}$$

$f''(x)$ is never 0, and does not exist when $x = 0$; so, the only candidate for an inflection point is at $x = 0$.

Since $f''(x)$ is negative when $x < 0$ and positive when $x > 0$, $f(x) = x^{5/3}$ has an inflection point at $x = 0$.

57. **(a)** The slope of the tangent line to $f(x) = e^x$ as $x \to -\infty$ is close to 0 since the tangent line is almost horizontal, and a horizontal line has a slope of 0.

(b) The slope of the tangent line to $f(x) = e^x$ as $x \to 0$ is close to 1 since the first derivative represents the slope of the tangent line, $f'(x) = e^x$, and $e^0 = 1$.

59. $f(x) = -x^2 - 10x - 25$

$$f'(x) = -2x - 10$$
$$= -2(x + 5) = 0$$

Critical number: -5

$f''(x) = -2 < 0$ for all x.

The curve is concave downward, which means a relative maximum occurs at $x = -5$.

61. $f(x) = 3x^3 - 3x^2 + 1$

$$f'(x) = 9x^2 - 6x$$
$$= 3x(3x - 2) = 0$$

Critical numbers: 0 and $\frac{2}{3}$

$$f''(x) = 18x - 6$$

$f''(0) = -6 < 0$, which means that a relative maximum occurs at $x = 0$.

$f''\left(\frac{2}{3}\right) = 6 > 0$, which means that a relative minimum occurs at $x = \frac{2}{3}$.

63. $f(x) = (x + 3)^4$

$$f'(x) = 4(x + 3)^3 = 0$$

Critical number: $x = -3$

$$f''(x) = 12(x + 3)^2$$
$$f''(-3) = 12(-3 + 3)^2 = 0$$

The second derivative test fails.
Use the first derivative test.

$$f'(-4) = 4(-4 + 3)^2$$
$$= 4(-1)^3 = -4 < 0$$

This indicates that f is decreasing on $(-\infty, -3)$.

$$f'(0) = 4(0 + 3)^3$$
$$= 4(3)^3 = 108 > 0$$

This indicates that f is increasing on $(-3, \infty)$. A relative minimum occurs at -3.

65. $f(x) = x^{7/3} + x^{4/3}$

$$f'(x) = \frac{7}{3}x^{4/3} + \frac{4}{3}x^{1/3}$$

$f'(x) = 0$ when

$$\frac{7}{3}x^{4/3} + \frac{4}{3}x^{1/3} = 0$$

$$\frac{x^{1/3}}{3}(7x + 4) = 0$$

$$x = 0 \text{ or } x = -\frac{4}{7}.$$

Critical numbers: $-\frac{4}{7}, 0$

$$f''(x) = \frac{28}{9}x^{1/3} + \frac{4}{9}x^{-2/3}$$

$$f''\left(-\frac{4}{7}\right) = \frac{28}{9}\left(-\frac{4}{7}\right)^{-1/3} + \frac{4}{9}\left(-\frac{4}{7}\right)^{-2/3} \approx -1.9363$$

Relative maximum occurs at $-\frac{4}{7}$.

$f''(0)$ does not exist, so the second derivative test fails.

Use the first derivative test.

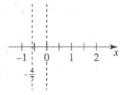

$$f'\left(-\frac{1}{2}\right) = \frac{7}{3}\left(-\frac{1}{2}\right)^{4/3} + \frac{4}{3}\left(-\frac{1}{2}\right)^{1/3} \approx -0.1323$$

This indicates that f is decreasing on $\left(-\frac{4}{7}, 0\right)$.

$$f'(1) = \frac{7}{3}(1)^{4/3} + \frac{4}{3}(1)^{1/3} = \frac{11}{3}$$

This indicates that f is increasing on $(0, \infty)$.
Relative minimum occurs at 0.

67. $f'(x) = x^3 - 6x^2 + 7x + 4$

$f''(x) = 3x^2 - 12x + 7$

Graph f' and f'' in the window $[-5, 5]$ by $[-5, 15]$, Xscl $= 0.5$.

Graph of f':

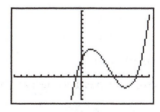

Graph of f'':

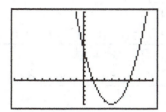

(a) f has relative extrema where $f'(x) = 0$. Use the graph to approximate the x-intercepts of the graph of f'. These numbers are the solutions of the equation $f'(x) = 0$. We find that the critical numbers of f are about $-0.4, 2.4$, and 4.0.

By either looking at the graph of f' and applying the first derivative test or by looking at the graph of f'' and applying the second derivative test, we see that f has relative minima at about -0.4 and 4.0 and a relative maximum at about 2.4.

(b) Examine the graph of f' to determine the intervals where the graph lies above and below the x-axis. We see that $f'(x) > 0$ on about $(-0.4, 2.4)$ and $(4.0, \infty)$, indicating that f is increasing on the same intervals. We also see that $f'(x) < 0$ on about $(-\infty, -0.4)$ and $(2.4, 4.0)$, indicating that f is decreasing on the same intervals.

(c) Examine the graph of f''. We see that this graph has two x-intercepts, so there are two x-values where $f''(x) = 0$. These x-values are about 0.7 and 3.3. Because the sign of f'' changes at these two values, we see that the x-values of the inflection points of the graph of f are about 0.7 and 3.3.

(d) We observe from the graph of f'' that $f''(x) > 0$ on about $(-\infty, 0.7)$ and $(3.3, \infty)$, so f is concave upward on the same intervals.

Likewise, we observe that $f''(x) < 0$ on about $(0.7, 3.3)$, so f is concave downward on the same interval.

69. $f'(x) = \dfrac{1 - x^2}{(x^2 + 1)^2}$

$f''(x) = \dfrac{(x^2 + 1)^2(-2x) - (1 - x^2)(2)(x^2 + 1)2x}{(x^2 + 1)^4}$

$= \dfrac{-2x(x^2 + 1)^2[(x^2 + 1) + 2(1 - x^2)]}{(x^2 + 1)^4}$

$= \dfrac{-2x(3 - x^2)}{(x^2 + 1)^3}$

Graph f' and f'' in the window $[-3, 3]$ by $[-1.5, 1, 5]$, Xscl $= 0.2$.

Graph of f':   Graph of f'':

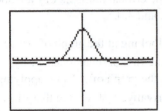   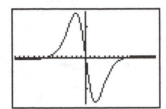

(a)   The critical numbers of f are the x-intercepts of the graph of f'. (Note that there are no values where f' does not exist.) We see from the graph that these x-values are -1 and 1.

By either looking at the graph of f' and applying the first derivative test or by looking at the graph of f'' and applying the second derivative test, we see that f has a relative minimum at -1 and a relative maximum at 1.

(b)   Examine the graph of f' to determine the intervals where the graph lies above and below the x-axis. We see that $f'(x) > 0$ on $(-1, 1)$, indicating that f is increasing on the same interval. We also see that $f'(x) < 0$ on $(-\infty, -1)$ and $(1, \infty)$, indicating that f is decreasing on the same intervals.

(c)   Examine the graph of f''. We see that the graph has three x-intercepts, so there are three values where $f''(x) = 0$. These x-

values are about -1.7, 0, and about 1.7. Because the sign of f'' and thus the concavity of f changes at these three values, we see that the x-values of the inflection points of the graph of f are about -1.7, 0, and about 1.7.

(d)   We observe from the graph of f'' that $f''(x) > 0$ on about $(-1.7, 0)$ and $(1.7, \infty)$, so f is concave upward on the same intervals.

Likewise, we observe that $f''(x) < 0$ on about $(-\infty, -1.7)$ and $(0, 1.7)$, so f is concave downward on the same intervals.

71.   There are many examples. The easiest is $f(x) = \sqrt{x}$. This graph is increasing and concave downward.

$$f'(x) = \frac{1}{2}x^{-1/2} = \frac{1}{2\sqrt{x}}$$

$f'(0)$ does not exist, while $f'(x) > 0$ for all $x > 0$. (Note that the domain of f is $[0, \infty)$.)

As x increases, the value of $f'(x)$ decreases, but remains positive. It approaches zero, but never becomes zero or negative.

73. $R(x) = \dfrac{4}{27}(-x^3 + 66x^2 + 1050x - 400)$

$0 \le x \le 25$

$$R'(x) = \frac{4}{27}(-3x^2 + 132x + 1050)$$

$$R''(x) = \frac{4}{27}(-6x + 132)$$

A point of diminishing returns occurs at a point of inflection, or where $R''(x) = 0$.

$$\frac{4}{27}(-6x + 132) = 0$$

$$-6x + 132 = 0$$

$$6x = 132$$

$$x = 22$$

Test $R''(x)$ to determine whether concavity changes at $x = 22$.

$$R''(20) = \frac{4}{27}(-6 \cdot 20 + 132) = \frac{16}{9} > 0$$

$$R''(24) = \frac{4}{27}(-6 \cdot 24 + 132) = -\frac{16}{9} < 0$$

$R(x)$ is concave upward on $(0, 22)$ and concave downward on $(22, 25)$.

$$R(22) = \frac{4}{27}[-(22)^3 + 66(22)^2 + 1060(22) - 400]$$

$$\approx 6517.9$$

The point of diminishing returns is $(22, 6517.9)$.

75. $R(x) = -0.6x^3 + 3.7x^2 + 5x,\ 0 \le x \le 6$

$R'(x) = -1.8x^2 + 7.4x + 5$

$R''(x) = -3.6x + 7.4$

A point of diminishing returns occurs at a point of inflection or where $R''(x) = 0$.

$$-3.6x + 7.4 = 0$$

$$-3.6x = -7.4$$

$$x = \frac{-7.4}{-3.6x} \approx 2.0556 \approx 2.06$$

Test $R''(x)$ to determine whether concavity changes at $x = 2.06$.

$$R''(2) = -3.6(2) + 7.4$$

$$= -7.2 + 7.4 = 0.2 > 0$$

$$R''(3) = -3.6(3) + 7.4$$

$$= -10.8 + 7.4 = -3.4 < 0$$

$R(x)$ is concave upward on $(0, 2.06)$ and concave downward on $(2.06, 6)$.

$R(2.0556) = -0.6(2.0556)^3 + 3.7(2.0556)^2 + 5(2.0556)$

≈ 20.7

The point of diminishing returns is $(2.06, 20.7)$.

77.   Let $D(q)$ represent the demand function. The revenue function, $R(q)$, is $R(q) = qD(q)$. The marginal revenue is given by

$$R'(q) = qD'(q) + D(q)(1)$$
$$= qD'(q) + D(q).$$
$$R''(q) = qD''(q) + D'(q)(1) + D'(q)$$
$$= qD''(q) + 2D'(q)$$

gives the rate of decline of marginal revenue. $D'(q)$ gives the rate of decline of price. If marginal revenue declines more quickly than price,

$$qD''(q) + 2D'(q) - D'(q) < 0$$
$$\text{or} \quad qD''(q) + D'(q) < 0.$$

79. (a) $R(x) = Cx\left(1 - e^{-kx}\right)$ The derivative will be non-negative when

$$R'(x) = C\left[x \cdot (kx)e^{-kx} + 1 \cdot (1 - e^{-kx})\right]$$

$$= C\left[1 + (kx - 1)e^{-kx}\right]$$

$$(kx - 1)e^{-kx} \geq -1$$

$$kx - 1 \geq -e^{kx}$$

$$1 - kx \leq e^{kx}$$

The left and right sides are equal at

$x = 0$ an $1 - kx$ is a decreasing function and e^{kx} is an increasing function for positive k, so R' will be nonnegative on $[0, 1]$ and R will be increasing on $[0, 1]$

(b) $R'(x) = C\left[1 + (kx - 1)e^{-kx}\right]$

$$R''(x) = C\left[ke^{-kx} + (kx - 1)(-ke^{-kx})\right]$$

$$= C\left[ke^{-kx}\left(2 - kx\right)\right]$$

R'' will be positive when $2 > kx$ or $x < \dfrac{2}{k}$.

So R will be concave up on $[0, 2/k)$

if $k \geq 2$ and on $[0, 1]$ if $k < 2$.

81.   (a) f_0 represents initial population $(t = 0)$.

(b) $(a, f(a))$ is the point where the graph changes concavity or the inflection point.

(c) f_M is the maximum carrying capacity.

83. $K(t) = \dfrac{3t}{t^2 + 4}$

(a) $K'(t) = \dfrac{3(t^2 + 4) - (2t)(3t)}{(t^2 + 4)^2}$

$$= \dfrac{-3t^2 + 12}{(t^2 + 4)^2} = 0$$

$$-3t^2 + 12 = 0$$

$$t^2 = 4$$

$$t = 2 \quad \text{or} \quad t = -2$$

For this application, the domain of K is $[0, \infty)$, so the only critical number is 2.

$$K''(t) = \frac{(t^2 + 4)^2(-6t) - (-3t^2 + 12)(2)(t^2 + 4)(2t)}{(t^2 + 4)^4}$$

$$= \frac{-6t(t^2 + 4) - 4t(-3t^2 + 12)}{(t^2 + 4)^3}$$

$$= \frac{6t^3 - 72t}{(t^2 + 4)^3}$$

$K''(2) = \frac{-96}{512} = -\frac{3}{16} < 0$ implies that $K(t)$ is maximized at $t = 2$.

Thus, the concentration is a maximum after 2 hours.

(b) $K(2) = \dfrac{3(2)}{(2)^2 + 4} = \dfrac{3}{4}$

The maximum concentration

is $\frac{3}{4}\%$.

85. $G(t) = \dfrac{31.4}{1 + 12.5e^{-0.393t}}$

The solution will be easier to follow if we replace the given constants with letters and derive a general solution. Use the quotient rule and the chain rule.

$G(t) = \dfrac{a}{1 + be^{-ct}}$

Use the power rule and the chain rule.

$$G'(t) = \frac{a(-1)\left(-bc \cdot e^{-ct}\right)}{\left(1 + be^{-ct}\right)^2} = \frac{abc \cdot e^{-ct}}{\left(1 + be^{-ct}\right)^2}$$

$$G''(t) = \frac{\left(-abc^2 \cdot e^{-ct}\right)\left[\left(1 + be^{-ct}\right)^2\right] - \left(abc \cdot e^{-ct}\right)\left(2\left(1 + be^{-ct}\right)\left(-bc \cdot e^{-ct}\right)\right)}{\left[\left(1 + be^{-ct}\right)^2\right]^2}$$

$$= \frac{\left(-abc^2 \cdot e^{-ct}\right)\left(1 + be^{-ct}\right) + \left(-abc^2 \cdot e^{-ct}\right)\left(-2be^{-ct}\right)}{\left(1 + be^{-ct}\right)^3}$$

$$= \frac{\left(abc^2 \cdot e^{-ct}\right)\left(be^{-ct} - 1\right)}{\left(1 + be^{-ct}\right)^3}$$

$G''(t)$ will be 0 when

$be^{-ct} = 1$

$b = e^{ct}$

$t = \dfrac{\ln(b)}{c}$

For the cactus wren function, $b = 12.5$ and $c = 0.393$, so

$t = \dfrac{\ln(12.5)}{0.393} = 6.427$. $G(6.427) = 15.7$, so the inflection point is $(6.427, 15.7)$.

87. $L(t) = Be^{-ce^{-kt}}$

$L'(t) = Be^{-ce^{-kt}}(-ce^{-kt})'$

$\qquad = Be^{-ce^{-kt}}[-ce^{-kt}(-kt)']$

$\qquad = Bcke^{-ce^{-kt}-kt}$

$L''(t) = Bcke^{-ce^{-kt}-kt}(-ce^{-kt} - kt)'$

$\qquad = Bcke^{-ce^{-kt}-kt}[-ce^{-kt}(-kt)' - k]$

$\qquad = Bcke^{-ce^{-kt}-kt}(cke^{-kt} - k)$

$\qquad = Bck^2e^{-ce^{-kt}-kt}(ce^{-kt} - 1)$

$L''(t) = 0$ when $ce^{-kt} - 1 = 0$

$\qquad ce^{-kt} - 1 = 0$

$\qquad \dfrac{c}{e^{kt}} = 1$

$\qquad e^{kt} = c$

$\qquad kt = \ln c$

$\qquad t = \dfrac{\ln c}{k}$

Letting $c = 7.267963$ and $k = 0.670840$

$t = \dfrac{\ln 7.267963}{0.670840} \approx 2.96$ years

Verify that there is a point of inflection at $t = \frac{\ln c}{k} \approx 2.96$. For

$L''(t) = Bck^2e^{-ce^{-kt}-kt}(ce^{-kt} - 1),$

we only need to test the factor $ce^{-kt} - 1$ on the intervals determined by $t \approx 2.96$ since the other factors are always positive.

$L''(1)$ has the same sign as

$7.267963e^{-0.670840(1)} - 1 \approx 2.72 > 0.$

$L''(3)$ has the same sign as

$7.267963e^{-0.670840(3)} - 1 \approx -0.029 < 0.$

Therefore L, is concave up on $\left(0, \frac{\ln c}{k} \approx 2.96\right)$

and concave down on $\left(\frac{\ln c}{k}, \infty\right)$, so there is a

point of inflection at $t = \frac{\ln c}{k} \approx 2.96$ years.

This signifies the time when the rate of growth begins to slow down since L changes from concave up to concave down at this inflection point.

89. $v'(x) = -35.98 + 12.09x - 0.4450x^2$

$v'(x) = 12.09 - 0.89x$

$v''(x) = -0.89$

Since $-0.89 < 0$, the function is always concave down.

91.

$P(R) = \dfrac{1}{1 + 2\pi DR^2}$

$P'(R) = \dfrac{-4\pi DR}{\left(1 + 2\pi DR^2\right)^2}$

$P''(R) = \dfrac{(-4\pi D)\left(1 + 2\pi DR^2\right)^2 - (-4\pi DR)(2)\left(1 + 2\pi DR^2\right)(4\pi DR)}{\left(1 + 2\pi DR^2\right)^4}$

$\qquad = \dfrac{(-4\pi D)\left(1 + 2\pi DR^2\right) - (-4\pi DR)(2)(4\pi DR)}{\left(1 + 2\pi DR^2\right)^3}$

$\qquad = \dfrac{(4\pi D)\left(8\pi DR^2 - 1 - 2\pi DR^2\right)}{\left(1 + 2\pi DR^2\right)^3} = \dfrac{(4\pi D)\left(6\pi DR^2 - 1\right)}{\left(1 + 2\pi DR^2\right)^3}$

The inflection point will occur when

$6\pi DR^2 = 1$ or $R = \dfrac{1}{\sqrt{6\pi D}}.$

$\dfrac{1}{\sqrt{6\pi D}} = 0.022$ implies

$D = \dfrac{1}{(0.022)^2 \cdot 6\pi} = 109.61 \approx 110.$

93. $s(t) = -16t^2$

$v(t) = s'(t) = -32t$

(a) $v(3) = -32(3) = -96$ ft/sec

(b) $v(5) = -32(5) = -160$ ft/sec

(c) $v(8) = -32(8) = -256$ ft/sec

(d) $a(t) = v'(t) = s''(t)$

$\qquad = -32$ ft/sec^2

95. $s(t) = 256t - 16t^2$

$v(t) = s'(t) = 256 - 32t$

$a(t) = v'(t) = s''(t) = -32$

To find when the maximum height occurs, set $s'(t) = 0.$

$256 - 32t = 0$

$\qquad t = 8$

Find the maximum height.

$s(8) = 256(8) - 16(8^2)$

$\qquad = 1024$

The maximum height of the ball is 1024 ft. The ball hits the ground when $s = 0$.

$$256t - 16t^2 = 0$$
$$16t(16 - t) = 0$$
$$t = 0 \quad \text{(initial moment)}$$
$$t = 16 \quad \text{(final moment)}$$

The ball hits the ground 16 seconds after being shot.

5.4   Curve Sketching

Your Turn 1

$$f(x) = -x^3 + 3x^2 + 9x - 10$$
$$f'(x) = -3x^2 + 6x + 9$$
$$f''(x) = -6x + 6$$

Use the first derivative to find intervals where the function is increasing or decreasing.

$$-3x^2 + 6x + 9 = 0$$
$$-3(x^2 - 2x - 3) = 0$$
$$(x + 1)(x - 3) = 0$$
$$x = -1 \text{ or } x = 3$$

These critical numbers divide the line into three intervals, $(-\infty, -1)$, $(-1, 3)$ and $(3, \infty)$. Evaluate the derivative at a test point in each region.

$$f'(-2) = -15$$
$$f'(0) = 9$$
$$f'(4) = -4$$

This shows that f is decreasing on $(-\infty, -1)$, increasing on $(-1, 3)$, and decreasing on $(3, \infty)$. By the first derivative test, f has a relative minimum of $f(-1) = -15$ at $x = -1$ and a relative maximum of $f(3) = 17$ at $x = 3$.

Use the second derivative to find the intervals where the function is concave upward or downward.

$$-6x + 6 = 0$$
$$6x = x$$
$$x = 1$$

The value where the second derivative is 0 divides the line into two intervals, $(-\infty, 1)$ and $(1, \infty)$. Evaluate $f''(x)$ at a point in each interval.

$$f''(0) = 6$$
$$f''(2) = -6$$

This shows that f is concave upward on $(-\infty, 1)$ and concave downward on $(1, \infty)$. The graph has an inflection point at $(1, f(1))$, or $(1, 1)$.

This information is summarized in the following table.

Interval	$(-\infty, -1)$	$(-1, 1)$	$(1, 3)$	$(3, \infty)$
Sign of f'	−	+	+	−
Sign of f''	+	+	−	−
f increasing or decreasing	Decreasing	Increasing	Increasing	Decreasing
Concavity of f	Upward	Upward	Downward	Downward
Shape of graph	⌣	⌣	⌢	⌢

Now sketch the graph using this information.

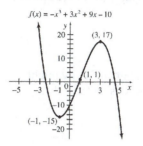

Your Turn 2

$$f(x) = 4x + \frac{1}{x} = \frac{4x^2 + 1}{x}$$

Because $x = 0$ makes the denominator 0 but not the numerator, the line $x = 0$ (that is, the y-axis) is a vertical asymptote. Neither $\lim\limits_{x \to -\infty} f(x)$ nor $\lim\limits_{x \to \infty} f(x)$ exists, so there is no horizontal asymptote. But since the term $1/x$ gets very small as $|x|$ gets large, the graph of f approaches the line $y = 4x$ as $|x|$ becomes larger and larger, so this line is an oblique asymptote.

$f(-x) = -f(x)$, so the graph of the left side can be found by rotating the right side around the origin by 180°.

Find any critical numbers.

$$f'(x) = 4 - \frac{1}{x^2}$$
$$4 - \frac{1}{x^2} = 0$$
$$x^2 = \frac{1}{4}$$
$$x = \frac{1}{2} \text{ or } -\frac{1}{2}$$

The critical numbers are $-1/2$ and $1/2$, which together with the location of the vertical asymptote divide the line into four regions, $(-\infty, -1/2)$, $(-1/2, 0)$, $(0, 1/2)$, and $(1/2, \infty)$. Evaluate the derivative at a test point in each region.

$$f'(-1) = 3$$

$$f'\left(-\frac{1}{4}\right) = -12$$

$$f'\left(\frac{1}{4}\right) = -12$$

$$f'(1) = 3$$

This shows that f has a relative maximum of $f(-1/2) = -4$ at $x = -1/2$ and a relative minimum of $f(1/2) = 4$ at $0\, x = 1/2$.

Now find the intervals where the graph is concave upward or concave downward.

$$f''(x) = \frac{2}{x^3}$$

The second derivative is never 0, but concavity might change where the second derivative is undefined, at $x = 0$. In fact,

$$f''(x) < 0 \text{ for } x < 0,$$

$$f''(x) > 0 \text{ for } x > 0,$$

so the graph is concave downward to the left of the origin and concave upward to the right of the origin. This provides enough information to sketch the graph.

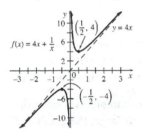

Your Turn 3

$$f(x) = \frac{4x^2}{x^2 + 4}$$

The denominator of f is never 0, so f has no vertical asymptotes. However,

$$\lim_{x \to -\infty} \frac{4x^2}{x^2 + 4} = 4 \text{ and } \lim_{x \to \infty} \frac{4x^2}{x^2 + 4} = 4,$$

so the line $y = 4$ is a horizontal asymptote. Note that $f(-x) = f(x)$, so the graph of f is symmetrical around the y-axis,

Find any critical numbers.

$$f'(x) = \frac{(x^2 + 4)(8x) - (4x^2)(2x)}{(x^2 + 4)^2}$$

$$= \frac{8x^3 + 32x - 8x^3}{(x^2 + 4)^2}$$

$$= \frac{32x}{(x^2 + 4)^2}$$

The derivative is 0 only when $x = 0$, and the denominator of the derivative is never 0. Therefore the critical number divides the line into just two regions, $(-\infty, 0)$ and $(0, \infty)$. Evaluate the derivative at a test point in each region.

$$f'(-1) = -1.28$$

$$f'(1) = 1.28$$

Thus f is decreasing as x approaches 0 from the left, and increasing to the right of $x = 0$, so f has a relative minimum of $f(0) = 0$ at $x = 0$.

Now find the intervals where the graph is concave upward or concave downward.

$$f'(x) = \frac{32x}{(x^2 + 4)^2}$$

$$f''(x) = \frac{(x^2 + 4)^2(32) - (32x)(2)(x^2 + 4)(2x)}{(x^2 + 4)^4}$$

$$= \frac{(x^2 + 4)[(x^2 + 4)(32) - 128x^2]}{(x^2 + 4)^4}$$

$$= \frac{128 - 96x^2}{(x^2 + 4)^3}$$

The denominator is never 0, but the numerator is 0 when

$$128 - 96x^2 = 0$$

$$96x^2 = 128$$

$$3x^2 = 4$$

$$x^2 = \frac{4}{3}$$

$$x = \pm\sqrt{\frac{4}{3}} \approx \pm 1.155$$

The zeros of the second derivative divide the line into three regions,

$$\left(-\infty, -\sqrt{\frac{4}{3}}\right), \left(-\sqrt{\frac{4}{3}}, \sqrt{\frac{4}{3}}\right) \text{ and } \left(\sqrt{\frac{4}{3}}, \infty\right).$$

Evaluate $f''(x)$ at a point in each interval.

$$f''(-2) = -0.5$$

$$f''(0) = 2$$

$$f''(2) = -0.5$$

According to this information, the graph is

concave downward on the interval $\left(-\infty, -\sqrt{\frac{4}{3}}\right)$,

concave upward on the interval $\left(-\sqrt{\frac{4}{3}}, \sqrt{\frac{4}{3}}\right)$,

and concave downward on the interval $\left(\sqrt{\frac{4}{3}}, \infty\right)$.

Thus there will be two inflection points. Find the corresponding values of f.

$$f\left(-\sqrt{\frac{4}{3}}\right) = 1$$

$$f\left(\sqrt{\frac{4}{3}}\right) = 1$$

The inflection points are at $\left(-\sqrt{\frac{4}{3}}, 1\right)$ and $\left(\sqrt{\frac{4}{3}}, 1\right)$.

This provides enough information to sketch the graph.

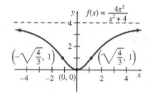

Your Turn 4

$f(x) = (x + 2)e^{-x}$

Since $\lim\limits_{x \to \infty} (x + 2)e^{-x} = 0$, the line $y = 0$ (the x-axis) is a horizontal asymptote for the graph. Neither $f(-x) = -f(x)$ nor $f(-x) = f(x)$ is true, so the graph has no symmetry.

Find any critical numbers.

$$f'(x) = (x + 2)(-e^{-x}) + (1)(e^{-x})$$

$$= -(1 + x)e^{-x}$$

The derivative is 0 at only one point, where $1 + x = 0$ or $x = -1$. This critical number divides the line into two regions, $(-\infty, -1)$ and $(-1, \infty)$. Evaluate the derivative at a test point in each region.

$$f'(-2) \approx 7.389$$

$$f'(0) = -1$$

Thus f is increasing on the interval $(-\infty, -1)$ and decreasing on the interval $(-1, \infty)$, and has a relative maximum of $f(-1) \approx 2.718$ at $x = -1$.

Now find the intervals where the graph is concave upward or concave downward.

$$f'(x) = -(1 + x)e^{-x}$$

$$f''(x) = (1 + x)e^{-x} + (-1)e^{-x}$$

$$= xe^{-x}$$

The second derivative is 0 only at $x = 0$. Evaluate $f''(x)$ at points on either side of $x = 0$.

$$f''(-1) \approx -2.718$$

$$f''(1) \approx 0.368$$

The graph is concave downward to the left of $x = 0$ and concave upward to the right of $x = 0$. There is an inflection point at $(0, f(0))$, or $(0, 2)$.

This provides enough information to sketch the graph.

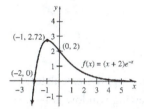

5.4 Warmup Exercises

W1. $f(x) = x^4 - 2x^3 - 12x^2 + 4x + 13$

$$f'(x) = 4x^3 - 6x^2 - 24x + 4$$

$$f''(x) = 12x^2 - 12x - 24$$

$$= 12(x^2 - x - 2)$$

$$= 12(x - 2)(x + 1)$$

$f''(x)$ will be 0 at $x = 2$ and $x = -1$. The factorization shows that each of these values is a single root at which f'' changes sign, so each root locates an inflection point. $f(-1) = 0$ and $f(2) = -27$, so the inflection points for f are $(-1, 0)$ and $(2, -27)$.

W2. $f(x) = x^3 - 21x^2 - 72x + 72x \ln x$

$$f'(x) = 6x^2 - 42x - 71$$

$$+ 72\left(x \cdot \frac{1}{x} + \ln x\right)$$

$$= 6x^2 - 42x + 72 \ln x$$

$$f''(x) = 12x - 42 + \frac{72}{x}$$

Set $f''(x)$ equal to 0, multiply through by x (which will not be 0 since 0 is not in the domain of f), and factor.

$$6x - 42 + \frac{72}{x} = 0$$

$$6x^2 - 42x + 72 = 0$$

$$6(x^2 - 7x + 12) = 0$$

$$6(x - 3)(x - 4) = 0$$

Thus f'' is 0 at $x = 3$ and $x = 4$.

Note that $f''(2) = 6$, $f''(7/2) = -3/7$, and $f''(5) = 12/5$, so the concavity of the graph of f changes at each of the roots.

$f(3) = -140.6997 \approx 140.7$ and

$f(4) = -160.7472 \approx -160.7$, so the inflection points for f are $(3, -140.7)$ and $(4, -160.7)$.

5.4 Exercises

1. Graph $y = x \ln |x|$ on a graphing calculator. A suitable choice for the viewing window is $[-1, 1]$ by $[-1, 1]$, Xscl $= 0.1$, Yscl $= 0.1$.

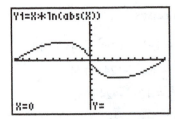

The calculator shows no y-value when $x = 0$ because 0 is not in the domain of this function. However, we see from the graph that

$$\lim_{x \to 0^-} x \ln |x| = 0$$

and

$$\lim_{x \to 0^+} x \ln |x| = 0.$$

Thus,

$$\lim_{x \to 0} x \ln |x| = 0.$$

3. $f(x) = -2x^3 - 9x^2 + 108x - 10$

Domain is $(-\infty, \infty)$.

$f(-x) = -2(-x)^3 - 9(-x)^2 + 108(-x) - 10$

$\qquad = 2x^3 - 9x^2 - 108x - 10$

No symmetry

$f'(x) = -6x^2 - 18x + 108$

$\qquad = -6(x^2 + 3x - 18)$

$\qquad = -6(x + 6)(x - 3)$

$f'(x) = 0$ when $x = -6$ or $x = 3$.

Critical numbers: -6 and 3

Critical points: $(-6, -550)$ and $(3, 179)$

$f''(x) = -12x - 18$

$f''(-6) = 54 > 0$

$f''(3) = -54 < 0$

Relative maximum at 3, relative minimum at -6

Increasing on $(-6, 3)$

Decreasing on $(-\infty, -6)$ and $(3, \infty)$

$f''(x) = -12x - 18 = 0$

$\qquad -6(2x + 3) = 0$

$$x = -\frac{3}{2}$$

Point of inflection at $(-1.5, -185.5)$

Concave upward on $(-\infty, -1.5)$

Concave downward on $(-1.5, \infty)$

y-intercept:

$$y = -2(0)^3 - 9(0)^2 + 108(0) - 10 = -10$$

[Graph showing the curve $f(x) = -2x^3 - 9x^2 + 108x - 10$ with labeled points $(3, 179)$, $(-1.5, -185.5)$, and $(-6, -550)$, with axis markings at -6, 3, and y-values 250, -250, -500.]

$f(x) = -2x^3 - 9x^2 + 108x - 10$

5. $f(x) = -3x^3 + 6x^2 - 4x - 1$

Domain is $(-\infty, \infty)$.

$f(-x) = -3(-x)^3 + 6(-x)^2 - 4(-x) - 1$

$\qquad = 3x^3 + 6x^2 + 4x - 1$

No symmetry

$f'(x) = -9x^2 + 12x - 4$

$\qquad = -(3x - 2)^2$

$(3x - 2)^2 = 0$

$$x = \frac{2}{3}$$

Critical number: $\frac{2}{3}$

$f\left(\frac{2}{3}\right) = -3\left(\frac{2}{3}\right)^3 + 6\left(\frac{2}{3}\right)^2 - 4\left(\frac{2}{3}\right) - 1 = -\frac{17}{9}$

Critical point: $\left(\frac{2}{3}, -\frac{17}{9}\right)$

$f'(0) = -9(0)^2 + 12(0) - 4 = -4 < 0$

$f'(1) = -9(1)^2 + 12(1) - 4 = -1 < 0$

No relative extremum at $\left(\frac{2}{3}, -\frac{17}{9}\right)$

Decreasing on $(-\infty, \infty)$

$f''(x) = -18x + 12$

$\qquad = -6(3x - 2)$

$3x - 2 = 0$

$\qquad x = \dfrac{2}{3}$

Point of inflection at $\left(\frac{2}{3}, -\frac{17}{9}\right)$

$f''(0) = -18(0) + 12 = 12 > 0$

$f''(1) = -18(1) + 12 = -6 < 0$

Concave upward on $\left(-\infty, \frac{2}{3}\right)$

Concave upward on $\left(\frac{2}{3}, \infty\right)$

Point of inflection at $\left(\frac{2}{3}, -\frac{17}{9}\right)$

y-intercept:

$y = -3(0)^3 + 6(0)^2 - 4(0) - 1 = -1$

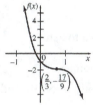

$f(x) = -3x^3 + 6x^2 - 4x - 1$

7. $f(x) = x^4 - 24x^2 + 80$

Domain is $(-\infty, \infty)$.

$f(-x) = (-x)^4 - 24(-x)^2 + 80$

$\qquad = x^4 - 24x^2 + 80 = f(x)$

The graph is symmetric about the y-axis.

$f'(x) = 4x^3 - 48x$

$\qquad\qquad 4x^3 - 48x = 0$

$\qquad\qquad 4x(x^2 - 12) = 0$

$4x(x - 2\sqrt{3})(x + 2\sqrt{3}) = 0$

Critical numbers: $-2\sqrt{3}, 0,$ and $2\sqrt{3}$

Critical points: $(-2\sqrt{3}, -64), (0, 80),$ and $(2\sqrt{3}, -64)$

$f''(x) = 12x^2 - 48$

$f''(-2\sqrt{3}) = 12(-2\sqrt{3})^2 - 48 = 96 > 0$

$f''(0) = 12(0)^2 - 48 = -48 < 0$

$f''(2\sqrt{3}) = 12(2\sqrt{3})^2 - 48 = 96 > 0$

Relative maximum at 0, relative minima at $-2\sqrt{3}$ and $2\sqrt{3}$

Increasing on $(-2\sqrt{3}, 0)$ and $(2\sqrt{3}, \infty)$

Decreasing on $(-\infty, -2\sqrt{3})$ and $(0, 2\sqrt{3})$

$12x^2 - 48 = 0$

$12(x^2 - 4) = 0$

$\qquad x = \pm 2$

Points of inflection at $(-2, 0)$ and $(2, 0)$

Concave upward on $(-\infty, -2)$ and $(2, \infty)$

Concave downward on $(-2, 2)$

x-intercepts: $0 = x^4 - 24x^2 + 80$

Let $u = x^2$.

$u^2 - 24u + 80 = 0$

$(u - 4)(u - 20) = 0$

$u = 4 \quad$ or $\quad u = 20$

$x = \pm 2 \quad$ or $\quad x = \pm 2\sqrt{5}$

y-intercept: $y = (0)^4 - 24(0)^2 + 80 = 80$

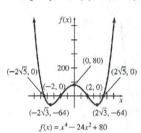

$f(x) = x^4 - 24x^2 + 80$

9. $f(x) = x^4 - 4x^3$

Domain is $(-\infty, \infty)$.

$f(-x) = (-x)^4 - 4(-x)^3$

$\qquad = x^4 + 4x^3 \neq f(x)$ or $-f(x)$

The graph is not symmetric about the y-axis or the origin.

$f'(x) = 4x^3 - 12x^2$

$4x^3 - 12x^2 = 0$

$4x^2(x - 3) = 0$

Critical numbers: 0 and 3

Critical points: $(0, 0)$ and $(3, -27)$

$f''(x) = 12x^2 - 24x$

$f''(0) = 12(0)^2 - 24(0) = 0$

$f''(3) = 12(3)^2 - 24(3) = 36 > 0$

Second derivative test fails for 0. Use first derivative test.

$$f'(-1) = 4(-1)^3 - 12(-1)^2 = -16 < 0$$

$$f'(1) = 4(1)^3 - 12(1)^2 = -8 < 0$$

Neither a relative minimum nor maximum at 0

Relative minimum at 3

Increasing on $(3, \infty)$

Decreasing on $(-\infty, 3)$

$$12x^2 - 24x = 0$$

$$12x(x - 2) = 0$$

$$x = 0 \text{ or } x = 2$$

Points of inflection at $(0, 0)$ and $(2, -16)$

Concave upward on $(-\infty, 0)$ and $(2, \infty)$

Concave downward on $(0, 2)$

x-intercepts: $x^4 - 4x^3 = 0$

$$x^3(x - 4) = 0$$

$$x = 0 \text{ or } x = 4$$

y-intercepts: $y = (0)^4 - 4(0)^3 = 0$

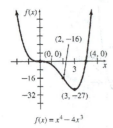

$$f(x) = x^4 - 4x^3$$

11. $f(x) = 2x + \dfrac{10}{x}$

$$= 2x + 10x^{-1}$$

Since $f(x)$ does not exist when $x = 0$, the domain is $(-\infty, 0) \cup (0, \infty)$.

$$f(-x) = 2(-x) + 10(-x)^{-1}$$

$$= -(2x + 10x^{-1})$$

$$= -f(x)$$

The graph is symmetric about the origin.

$$f'(x) = 2 - 10x^{-2}$$

$$2 - \frac{10}{x^2} = 0$$

$$\frac{2(x^2 - 5)}{x^2} = 0$$

$$x = \pm\sqrt{5}$$

Critical numbers: $-\sqrt{5}$ and $\sqrt{5}$

Critical points: $(-\sqrt{5}, -4\sqrt{5})$ and $(\sqrt{5}, 4\sqrt{5})$

Test a point in the intervals $(-\infty, -\sqrt{5})$,

$(-\sqrt{5}, 0)$, $(0, \sqrt{5})$, and $(\sqrt{5}, \infty)$.

$$f'(-3) = 2 - 10(-3)^{-2} = \frac{8}{9} > 0$$

$$f'(-1) = 2 - 10(-1)^{-2} = -8 < 0$$

$$f'(1) = 2 - 10(1)^{-2} = -8 < 0$$

$$f'(3) = 2 - 10(3)^{-2} = \frac{8}{9} > 0$$

Relative maximum at $-\sqrt{5}$

Relative minimum at $\sqrt{5}$

Increasing on $(-\infty, -\sqrt{5})$ and $(\sqrt{5}, \infty)$

Decreasing on $(-\sqrt{5}, 0)$ and $(0, \sqrt{5})$

(Recall that $f(x)$ does not exist at $x = 0$.)

$$f''(x) = 20x^{-3} = \frac{20}{x^3}$$

$f''(x) = \dfrac{20}{x^3}$ is never equal to zero.

There are no inflection points.

Test a point in the intervals $(-\infty, 0)$ and $(0, \infty)$.

$$f''(-1) = \frac{20}{(-1)^3} = -20 < 0$$

$$f''(1) = \frac{20}{(1)^3} = 20 > 0$$

Concave upward on $(0, \infty)$

Concave downward on $(-\infty, 0)$

$f(x)$ is never zero, so there are no x-intercepts.

$f(x)$ does not exist for $x = 0$, so there is no y-intercept.

Vertical asymptote at $x = 0$

$y = 2x$ is an oblique asymptote.

$$f(-x) = 2(-x) + 10(-x)^{-1}$$

$$= -(2x + 10x^{-1})$$

$$= -f(x)$$

$$f(x) = 2x + \frac{10}{x}$$

13. $f(x) = \dfrac{-x + 4}{x + 2}$

Since $f(x)$ does not exist when $x = -2$, the domain is $(-\infty, -2) \cup (-2, \infty)$.

$f(-x) = \dfrac{-(-x) + 4}{(-x) + 2} = \dfrac{x + 4}{-x + 2}$

The graph is not symmetric about the y-axis or the origin.

$f'(-x) = \dfrac{(x + 2)(-1) - (-x + 4)(1)}{(x + 2)^2}$

$= \dfrac{-6}{(x + 2)^2}$

$f'(x) < 0$ and is never zero. $f'(x)$ fails to exist for $x = -2$.

No critical numbers; no relative extrema

Decreasing on $(-\infty, -2)$ and $(-2, \infty)$

$f''(x) = \dfrac{12}{(x + 2)^3}$

$f''(x)$ fails to exist for $x = -2$.

No points of inflection

Test a point in the intervals $(-\infty, -2)$ and $(-2, \infty)$.

$f''(-3) = -12 < 0$

$f''(-1) = 12 > 0$

Concave upward on $(-2, \infty)$

Concave downward on $(-\infty, -2)$

x-intercept: $\dfrac{-x + 4}{x + 2} = 0$

$x = 4$

y-intercept: $y = \dfrac{-0 + 4}{0 + 2} = 2$

Vertical asymptote at $x = -2$

Horizontal asymptote at $y = -1$

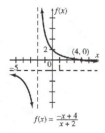

$f(x) = \frac{-x+4}{x+2}$

15. $f(x) = \dfrac{1}{x^2 + 4x + 3}$

$= \dfrac{1}{(x + 3)(x + 1)}$

Since $f(x)$ does not exist when $x = -3$ and $x = -1$, the domain is $(-\infty, -3) \cup (-3, -1) \cup (-1, \infty)$.

$f(-x) = \dfrac{1}{(-x)^2 + 4(-x) + 3} = \dfrac{1}{x^2 - 4x + 3}$

The graph is not symmetric about the y-axis or the origin.

$f'(x) = \dfrac{0 - (2x + 4)}{(x^2 + 4x + 3)^2} = \dfrac{-2(x + 2)}{[(x + 3)(x + 1)]^2}$

Critical number: -2

Test a point in the intervals $(-\infty, -3)$, $(-3, -2)$, $(-2, -1)$, and $(-1, \infty)$.

$f'(-4) = \dfrac{-2(-4 + 2)}{[(-4 + 3)(-4 + 1)]^2} = \dfrac{4}{9} > 0$

$f'\left(-\dfrac{5}{2}\right) = \dfrac{-2\left(-\frac{5}{2} + 2\right)}{\left[\left(-\frac{5}{2} + 3\right)\left(-\frac{5}{2} + 1\right)\right]^2} = \dfrac{16}{9} > 0$

$f'\left(-\dfrac{3}{2}\right) = \dfrac{-2\left(-\frac{3}{2} + 2\right)}{\left[\left(-\frac{3}{2} + 3\right)\left(-\frac{3}{2} + 1\right)\right]^2} = -\dfrac{16}{9} < 0$

$f'(0) = \dfrac{-2(0 + 2)}{[(0 + 3)(0 + 1)]^2} = -\dfrac{4}{9} < 0$

$f(-2) = \dfrac{1}{(-2 + 3)(-2 + 1)} = -1$

Relative maximum at $(-2, -1)$

Increasing on $(-\infty, -3)$ and $(-3, -2)$

Decreasing on $(-2, -1)$ and $(-1, \infty)$

$f''(x) = \dfrac{(x^2 + 4x + 3)^2(-2) - (-2x - 4)(2)(x^2 + 4x + 3)(2x + 4)}{(x^2 + 4x + 3)^4}$

$= \dfrac{-2(x^2 + 4x + 3)[(x^2 + 4x + 3) + (-2x - 4)(2x + 4)]}{(x^2 + 4x + 3)^4}$

$= \dfrac{-2(x^2 + 4x + 3 - 4x^2 - 16x - 16)}{(x^2 + 4x + 3)^3}$

$= \dfrac{-2(-3x^2 - 12x - 13)}{(x^2 + 4x + 3)^3}$

$= \dfrac{2(3x^2 + 12x + 13)}{[(x + 3)(x + 1)]^3}$

Since $3x^2 + 12x + 13 = 0$ has no real solutions, there are no x-values where $f''(x) = 0$. $f''(x)$ does not exist where $x = -3$ and $x = -1$. Since $f(x)$ does not exist at these x-values, there are no points of inflection.

Test a point in the intervals $(-\infty, -3)$, $(-3, -1)$, and $(-1, \infty)$.

$$f''(-4) = \frac{2[3(-4)^2 + 12(-4) + 13]}{[(-4+3)(-4+1)]^3} = \frac{26}{27} > 0$$

$$f''(-2) = \frac{2[3(-2)^2 + 12(-2) + 13]}{[(-2+3)(-2+1)]^3} = -2 < 0$$

$$f''(0) = \frac{2[3(0)^2 + 12(0) + 13]}{[(0+3)(0+1)]^3} = \frac{26}{27} > 0$$

Concave upward on $(-\infty, -3)$ and $(-1, \infty)$

Concave downward on $(-3, -1)$

$f(x)$ is never zero, so there are no x-intercepts.

y-intercept: $y = \dfrac{1}{(0+3)(0+1)} = \dfrac{1}{3}$

Vertical asymptotes where $f(x)$ is undefined at $x = -3$ and $x = -1$.

Horizontal asymptote at $y = 0$

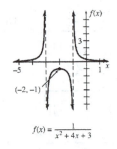

$$f(x) = \frac{1}{x^2 + 4x + 3}$$

17. $f(x) = \dfrac{x}{x^2 + 1}$

Domain is $(-\infty, \infty)$

$$f(-x) = \frac{-x}{(-x)^2 + 1} = -\frac{x}{x^2 + 1} = -f(x)$$

The graph is symmetric about the origin.

$$f'(x) = \frac{(x^2 + 1)(1) - x(2x)}{(x^2 + 1)^2}$$

$$= \frac{1 - x^2}{(x^2 + 1)^2}$$

$$1 - x^2 = 0$$

Critical numbers: 1 and -1

Critical points: $\left(1, \frac{1}{2}\right)$ and $\left(-1, -\frac{1}{2}\right)$

$$f''(x) = \frac{(x^2 + 1)^2(-2x) - (1 - x^2)(2)(x^2 + 1)(2x)}{(x^2 + 1)^4}$$

$$= \frac{-2x^3 - 2x - 4x + 4x^3}{(x^2 + 1)^3} = \frac{2x^3 - 6x}{(x^2 + 1)^3}$$

$$f''(1) = -\frac{1}{2} < 0$$

$$f''(-1) = \frac{1}{2} > 0$$

Relative maximum at 1

Relative minimum at -1

Increasing on $(-1, 1)$

Decreasing on $(-\infty, -1)$ and $(1, \infty)$

$$f''(x) = \frac{2x^3 - 6x}{(x^2 + 1)^3} = 0$$

$$2x^3 - 6x = 0$$

$$2x(x^2 - 3) = 0$$

$$x = 0, x = \pm\sqrt{3}$$

Inflection points at $(0, 0)$, $\left(\sqrt{3}, \frac{\sqrt{3}}{4}\right)$ and $\left(-\sqrt{3}, -\frac{\sqrt{3}}{4}\right)$

Concave upward on $(-\sqrt{3}, 0)$ and $(\sqrt{3}, \infty)$

Concave downward on $(-\infty, -\sqrt{3})$ and $(0, \sqrt{3})$

x-intercept: $0 = \dfrac{x}{x^2 + 1}$

$$0 = x$$

y-intercept: $y = \dfrac{0}{0^2 + 1} = 0$

Horizontal asymptote at $y = 0$

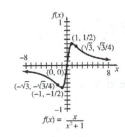

$$f(x) = \frac{x}{x^2 + 1}$$

19. $f(x) = \dfrac{1}{x^2 - 9}$

$$= \dfrac{1}{(x + 3)(x - 3)}$$

Since $f(x)$ does not exist when $x = -3$ and $x = 3$, the domain is $(-\infty, -3) \cup (-3, 3) \cup (3, \infty)$.

$$f(-x) = \frac{1}{(-x)^2 - 9} = \frac{1}{x^2 - 9} = f(x)$$

The graph is symmetric about the y-axis.

$$f'(x) = \frac{-2x}{(x^2 - 9)^2}$$

Critical number: 0

Critical point: $\left(0, -\frac{1}{9}\right)$

Test a point in the intervals $(-\infty, -3)$, $(-3, 0)$, $(0, 3)$, and $(3, \infty)$.

$$f'(-4) = \frac{-2(-4)}{[(-4)^2 - 9]^2} = \frac{8}{49} > 0$$

$$f'(-1) = \frac{-2(-4)}{[(-1)^2 - 9]^2} = \frac{1}{32} > 0$$

$$f'(1) = \frac{-2(1)}{[(1)^2 - 9]^2} = -\frac{1}{32} < 0$$

$$f'(4) = \frac{-2(4)}{[(4)^2 - 9]^2} = -\frac{8}{49} < 0$$

Relative maximum at $\left(0, -\frac{1}{9}\right)$

Increasing on $(-\infty, -3)$ and $(-3, 0)$

Decreasing on $(0, 3)$ and $(3, \infty)$

$$f''(x) = \frac{(x^2 - 9)^2(-2) - (-2x)(2)(x^2 - 9)(2x)}{(x^2 - 9)^4}$$

$$= \frac{-2(x^2 - 9)[(x^2 - 9) + (-2x)(2x)]}{(x^2 + 4)^4}$$

$$= \frac{-2(x^2 - 9 - 4x^2)}{(x^2 - 9)^3}$$

$$= \frac{-2(-3x^2 - 9)}{(x^2 - 9)^3}$$

$$= \frac{6(x^2 + 3)}{[(x + 3)(x - 3)]^3}$$

Since $x^2 + 3 = 0$ has no solutions, there are no x-values where $f''(x) = 0$. $f''(x)$ does not exist where $x = -3$ and $x = 3$. Since $f(x)$ does not exist at these x-values, there are no points of inflection.

Test a point in the intervals $(-\infty, -3)$, $(-3, 3)$, and $(3, \infty)$.

$$f''(-4) = \frac{6[(-4)^2 + 3]}{[(-4 + 3)(-4 - 3)]^3} = \frac{114}{343} > 0$$

$$f''(0) = \frac{6[(0)^2 + 3]}{[(0 + 3)(0 - 3)]^3} = -\frac{2}{81} < 0$$

$$f''(4) = \frac{6[(4)^2 + 3]}{[(4 + 3)(4 - 3)]^3} = \frac{114}{343} > 0$$

Concave upward on $(-\infty, -3)$ and $(3, \infty)$

Concave downward on $(-3, 3)$

$f(x)$ is never zero, so there are no x-intercepts.

y-intercept: $y = \frac{1}{0^2 - 9} = -\frac{1}{9}$

Vertical asymptotes where $f(x)$ is undefined at $x = -3$ and $x = 3$.

Horizontal asymptote at $y = 0$

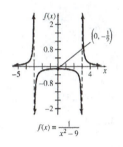

$$f(x) = \frac{1}{x^2 - 9}$$

21. $f(x) = x \ln |x|$

The domain of this function is $(-\infty, 0) \cup (0, \infty)$.

$$f(-x) = -x \ln\left|-x\right|$$
$$= -x \ln |x| = -f(x)$$

The graph is symmetric about the origin.

$$f'(x) = x \cdot \frac{1}{x} + \ln |x|$$
$$= 1 + \ln|x|$$

$f'(x) = 0$ when

$$0 = 1 + \ln |x|$$
$$-1 = \ln |x|$$
$$e^{-1} = |x|$$
$$x = \pm\frac{1}{e} \approx \pm 0.37.$$

Critical numbers: $\pm\frac{1}{e} \approx \pm 0.37$.

$f'(-1) = 1 + \ln |-1| = 1 > 0$

$f'(-0.1) = 1 + \ln |-0.1| \approx -1.3 < 0$

$f'(0.1) = 1 + \ln |0.1| \approx -1.3 < 0$

$f'(1) = 1 + \ln |1| = 1 > 0$

$f\left(\dfrac{1}{e}\right) = \dfrac{1}{e} \ln \left|\dfrac{1}{e}\right| = -\dfrac{1}{e}$

$f\left(-\dfrac{1}{e}\right) = -\dfrac{1}{e} \ln \left|-\dfrac{1}{e}\right| = \dfrac{1}{e}$

Relative maximum of $\left(-\dfrac{1}{e}, \dfrac{1}{e}\right)$; relative

minimum of $\left(\dfrac{1}{e}, -\dfrac{1}{e}\right)$.

Increasing on $\left(-\infty, -\dfrac{1}{e}\right)$ and $\left(\dfrac{1}{e}, \infty\right)$ and

decreasing on $\left(-\dfrac{1}{e}, 0\right)$ and $\left(0, \dfrac{1}{e}\right)$.

$f''(x) = \dfrac{1}{x}$

$f''(-1) = \dfrac{1}{-1} = -1 < 0$

$f''(1) = \dfrac{1}{1} = 1 > 0$

Concave downward on $(-\infty, 0)$;

Concave upward on $(0, \infty)$.

There is no y-intercept.

x-intercept: $0 = x \ln |x|$

$x = 0$ or $\ln|x| = 0$

$|x| = e^0 = 1$

$x = \pm 1$

Since 0 is not in the domain, the only x-intercepts are -1 and 1.

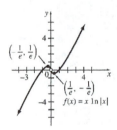

23. $f(x) = \dfrac{\ln x}{x}$

Note that the domain of this function is $(0, \infty)$.

$f(-x) = \dfrac{\ln(-x)}{-x}$ does not exist when $x \geq 0$,

no symmetry.

$f'(x) = \dfrac{x\left(\dfrac{1}{x}\right) - \ln x(1)}{x^2}$

$= \dfrac{1 - \ln x}{x^2}$

Critical numbers:

$1 - \ln x = 0$

$1 = \ln x$

$e^1 = x$

$f(e) = \dfrac{\ln e}{e} = \dfrac{1}{e}$

Critical points: $\left(e, \dfrac{1}{e}\right)$

$f'(1) = \dfrac{1 - \ln 1}{1^2} = \dfrac{1}{1} = 1 > 0$

$f'(3) = \dfrac{1 - \ln 3}{3^2} = -0.01 < 0$

There is a relative maximum at $\left(e, \dfrac{1}{e}\right)$.

The function is increasing on $(0, e)$ and decreasing on (e, ∞).

$f''(x) = \dfrac{x^2\left(-\dfrac{1}{x}\right) - (1 - \ln x)2x}{x^4}$

$= \dfrac{-x - 2x(1 - \ln x)}{x^4}$

$= \dfrac{-x[1 + 2(1 - \ln x)]}{x^4}$

$= \dfrac{-(1 + 2 - 2 \ln x)}{x^3}$

$= \dfrac{-3 + 2 \ln x}{x^3}$

$f''(x) = 0$ when $-3 + 2 \ln x = 0$

$2 \ln x = 3$

$\ln x = \dfrac{3}{2} = 1.5$

$x = e^{1.5} \approx 4.48.$

$f''(1) = \dfrac{-3 + 2 \ln 1}{1^3} = -3 < 0$

$f''(5) = \dfrac{-3 + 2 \ln 5}{5^3} \approx 0.0018 > 0$

Inflection point at $\left(e^{1.5}, \dfrac{1.5}{e^{1.5}}\right) \approx (4.48, 0.33)$

Concave downward on $(0, e^{1.5})$; concave upward on $(e^{1.5}, \infty)$

$$f(e^{1.5}) = \frac{\ln e^{1.5}}{e^{1.5}} = \frac{1.5}{e^{1.5}}$$

$$= \frac{3}{2e^{1.5}} \approx 0.33$$

Since $x \neq 0$, there is no y-intercept.

x-intercept: $f(x) = 0$ when $\ln x = 0$

$$x = e^0 = 1$$

Vertical asymptote at $x = 0$

Horizontal asymptote at $y = 0$

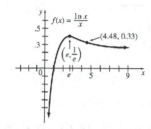

25. $f(x) = xe^{-x}$

Domain is $(-\infty, \infty)$.

$$f(-x) = -xe^x$$

The graph has no symmetry.

$$f'(x) = -xe^{-x} + e^{-x}$$

$$= e^{-x}(1 - x)$$

$f'(x) = 0$ when $e^{-x}(1 - x) = 0$

$$x = 1$$

Critical numbers: 1

Critical points: $\left(1, \frac{1}{e}\right)$

$$f'(0) = e^{-0}(1 - 0) = 1 > 0$$

$$f'(2) = e^{-2}(1 - 2) = \frac{-1}{e^2} < 0$$

Relative maximum at $\left(1, \frac{1}{e}\right)$

Increasing on $(-\infty, 1)$; decreasing on $(1, \infty)$

$$f''(x) = e^{-x}(-1) + (1 - x)(-e^{-x})$$

$$= -e^{-x}(1 + 1 - x)$$

$$= -e^{-x}(2 - x)$$

$f'' = 0$ when $-e^{-x}(2 - x) = 0$

$$x = 2.$$

$$f''(0) = -e^{-0}(2 - 0) = -2 < 0$$

$$f''(3) = -e^{-3}(2 - 3) = \frac{1}{e^3} > 0$$

Inflection point at $\left(2, \frac{2}{e^2}\right)$

Concave downward on $(-\infty, 2)$, concave upward on $(2, \infty)$

x-intercept: $0 = xe^{-x}$

$$x = 0$$

y-intercept: $y = 0 \cdot e^{-0} = 0$

Horizontal asymptote at $y = 0$

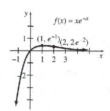

27. $f(x) = (x - 1)e^{-x}$

Domain is $(-\infty, \infty)$

$$f(-x) = (-x - 1)e^x$$

The graph has no symmetry.

$$f'(x) = -(x - 1)e^{-x} + e^{-x}(1)$$

$$= e^{-x}[-(x - 1) + 1]$$

$$= e^{-x}(2 - x)$$

$f'(x) = 0$ when $e^{-x}(2 - x) = 0$

$$x = 2.$$

Critical number: 2

Critical point: $\left(2, \frac{1}{e^2}\right)$

$$f''(x) = -e^{-x} + (2 - x)(-e^{-x})$$

$$= -e^{-x}[1 + (2 - x)]$$

$$= -e^{-x}(3 - x)$$

$$f''(2) = -e^{-2}(3 - 2) = \frac{-1}{e^2} < 0$$

Relative maximum at $\left(2, \frac{1}{e^2}\right)$

$$f'(0) = e^{-0}(2 - 0) = 2 > 0$$

$$f'(3) = e^{-3}(2 - 3) = \frac{-1}{e^3} < 0$$

Increasing on $(-\infty, 2)$; decreasing on $(2, \infty)$.

$f''(x) = 0$ when $-e^{-x}(3 - x) = 0$

$$x = 3.$$

$f''(0) = -e^{-0}(3 - 0) = -3 < 0$

$f''(4) = -e^{-4}(3 - 4) = \dfrac{1}{e^4} > 0$

Inflection point at $\left(3, \dfrac{2}{e^3}\right)$

Concave downward on $(-\infty, 3)$; concave upward on $(3, \infty)$

$f(3) = (3 - 1)e^{-3} = \dfrac{2}{e^3}$

y-intercept: $y = (0 - 1)e^{-0}$

$\qquad\qquad = (-1)(1) = -1$

x-intercept: $0 = (x - 1)e^{-x}$

$\qquad\qquad x - 1 = 0$

$\qquad\qquad\quad x = 1$

Horizontal asymptote at $y = 0$

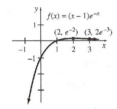

29. $f(x) = x^{2/3} - x^{5/3}$

Domain is $(-\infty, \infty)$.

$f(-x) = x^{2/3} + x^{5/3}$

The graph has no symmetry.

$f'(x) = \dfrac{2}{3}x^{-1/3} - \dfrac{5}{3}x^{2/3}$

$\qquad = \dfrac{2 - 5x}{3x^{1/3}}$

$f'(x) = 0$ when $2 - 5x = 0$

Critical number: $x = \dfrac{2}{5}$

$f\left(\dfrac{2}{5}\right) = \left(\dfrac{2}{5}\right)^{2/3} - \left(\dfrac{2}{5}\right)^{5/3}$

$\qquad = \dfrac{3 \cdot 2^{2/3}}{5^{5/3}} \approx 0.326$

Critical point: $(0.4, 0.326)$

$f''(x) = \dfrac{3x^{1/3}(-5) - (2 - 5x)(3)\left(\frac{1}{3}\right)x^{-2/3}}{(3x^{1/3})^2}$

$\qquad = \dfrac{-15x^{1/3} - (2 - 5x)x^{-2/3}}{9x^{2/3}}$

$\qquad = \dfrac{-15x - (2 - 5x)}{9x^{4/3}}$

$\qquad = \dfrac{-10x - 2}{9x^{4/3}}$

$f''\left(\dfrac{2}{5}\right) = \dfrac{-10\left(\frac{2}{5}\right) - 2}{9\left(\frac{2}{5}\right)^{4/3}}$

$\qquad \approx -2.262 < 0$

Relative maximum at $\left(\dfrac{2}{5}, \dfrac{3 \cdot 2^{2/3}}{5^{5/3}}\right) \approx (0.4, 0.326)$

$f'(x)$ does not exist when $x = 0$

Since $f''(0)$ is undefined, use the first derivative test.

$f'(-1) = \dfrac{2 - 5(-1)}{3(-1)^{1/3}} = \dfrac{7}{-3} < 0$

$f'\left(\dfrac{1}{8}\right) = \dfrac{2 - 5\left(\frac{1}{8}\right)}{3\left(\frac{1}{8}\right)^{1/3}} = \dfrac{11}{12} > 0$

$f'(1) = \dfrac{2 - 5}{3 \cdot 1^{1/3}} = -1 < 0$

Relative minimum at $(0, 0)$

f increases on $\left(0, \dfrac{2}{5}\right)$.

f decreases on $(-\infty, 0)$ and $\left(\dfrac{2}{5}, \infty\right)$.

$f''(x) = 0$ when $-10x - 2 = 0$

$\qquad\qquad\qquad\qquad x = -\dfrac{1}{5}$

$f''(x)$ undefined when $9x^{4/3} = 0$

$\qquad\qquad\qquad\qquad x = 0$

$f''(-1) = \dfrac{-10(-1) - 2}{9(-1)^{4/3}} = \dfrac{8}{9} > 0$

$f''\left(-\dfrac{1}{8}\right) = \dfrac{-10\left(-\frac{1}{8}\right) - 2}{9\left(-\frac{1}{8}\right)^{4/3}} = -\dfrac{4}{3} < 0$

$f''(1) = \dfrac{-10(1) - 2}{9(1)^{4/3}} = -\dfrac{4}{3} < 0$

Concave upward on $\left(-\infty, -\dfrac{1}{5}\right)$

Concave upward on $\left(-\dfrac{1}{5}, \infty\right)$

Inflection point at $\left(-\frac{1}{5}, \frac{6}{5^{5/3}}\right) \approx (-0.2, 0.410)$

y-intercept: $y = 0^{2/3} - 0^{5/3} = 0$

x-intercept: $0 = x^{2/3} - x^{5/3}$

$\qquad = x^{2/3}(1 - x)$

$\qquad\qquad x = 0 \text{ or } x = 1$

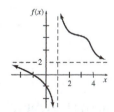

31. For Exercises 3, 7, and 9, the relative maxima or minima are outside the vertical window of $-10 \leq y \leq 10$.

For Exercise 11, the default window shows only a small portion of the graph.

For Exercise 15, the default window does not allow the graph to properly display the vertical asymptotes.

33. For Exercises 17, 19, 23, 25, and 27, the y-coordinate of the relative minimum, relative maximum, or inflection points is so small, it may be hard to distinguish.

For Exercises 35–39 other graphs are possible.

35. (a) indicates a smooth, continuous curve except where there is a vertical asymptote.

 (b) indicates that the function decreases on both sides of the asymptote, so there are no relative extrema.

 (c) gives the horizontal asymptote $y = 2$.

 (d) and (e) indicate that concavity does not change left of the asymptote, but that the right portion of the graph changes concavity at $x = 2$ and $x = 4$.

  There are inflection points at 2 and 4.

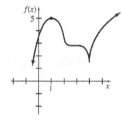

37. (a) indicates that there can be no asymptotes, sharp "corners", holes, or jumps. The graph must be one smooth curve.

(b) and (c) indicate relative maxima at -3 and 4 and a relative minimum at 1.

(d) and (e) are consistent with (g).

(f) indicates turning points at the critical numbers -3 and 4.

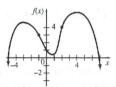

39. (a) indicates that the curve may not contain breaks.

 (b) indicates that there is a sharp "corner" at 4.

 (c) gives a point at $(1, 5)$.

 (d) shows critical numbers.

 (e) and (f) indicate (combined with (c) and (d)) a relative maximum at $(1, 5)$, and (combined with (b)) a relative minimum at 4.

 (g) is consistent with (b).

 (h) indicates the curve is concave upward on $(2, 3)$.

 (i) indicates the curve is concave downward on $(-\infty, 2)$, $(3, 4)$ and $(4, \infty)$.

Chapter 5 Review Exercises

1. True

2. False: The function is increasing on this interval.

3. False: The function could have neither a minimum nor a maximum; consider $f(x) = x^3$ at $c = 0$.

4. True

5. False: Consider $f(x) = x^{3/2}$ at $c = 0$.

6. True

7. False: The function is concave upward.

8. False: Consider $f(x) = x^4$ at $c = 0$.

9. False: Consider $f(x) = x^4$, which has a relative minimum at $c = 0$.

10. False: Polynomials are rational functions, and nonconstant polynomials have neither vertical nor horizontal asymptotes.

11. True

12. False: Consider $f(x) = x^2$ on $(-1, 1)$ with $c = 0$.

17. $f(x) = x^2 + 9x + 8$

$f'(x) = 2x + 9$

$f'(x) = 0$ when $x = -\frac{9}{2}$ and f' exists everywhere.

Critical number: $-\frac{9}{2}$

Test an x-value in the intervals $\left(-\infty, -\frac{9}{2}\right)$ and $\left(-\frac{9}{2}, -\infty\right)$.

$f'(-5) = -1 < 0$

$f'(-4) = 1 > 0$

f is increasing on $\left(-\frac{9}{2}, \infty\right)$ and decreasing on $\left(-\infty, -\frac{9}{2}\right)$.

19. $f(x) = -x^3 + 2x^2 + 15x + 16$

$\begin{aligned} f'(x) &= -3x^2 + 4x + 15 \\ &= -(3x^2 - 4x - 15) \\ &= -(3x + 5)(x - 3) \end{aligned}$

$f'(x) = 0$ when $x = -\frac{5}{3}$ or $x = 3$ and f' exists everywhere.

Critical numbers: $-\frac{5}{3}$ and 3

Test an x-value in the intervals $\left(-\infty, -\frac{5}{3}\right)$, $\left(-\frac{5}{3}, 3\right)$, and $(3, \infty)$.

$f'(-2) = -5 < 0$

$f'(0) = 15 > 0$

$f'(4) = -17 < 0$

f is increasing on $\left(-\frac{5}{3}, 3\right)$ and decreasing on $\left(-\infty, -\frac{5}{3}\right)$ and $(3, \infty)$.

21. $f(x) = \dfrac{16}{9 - 3x}$

$f'(x) = \dfrac{16(-1)(-3)}{(9 - 3x)^2} = \dfrac{48}{(9 - 3x)^2}$

$f'(x) > 0$ for all x $(x \neq 3)$, and f is not defined for $x = 3$.

f is increasing on $(-\infty, 3)$ and $(3, \infty)$ and never decreasing.

23. $f(x) = \ln|x^2 - 1|$

$f'(x) = \dfrac{2x}{x^2 - 1}$

f is not defined for $x = -1$ and $x = 1$.

$f'(x) = 0$ when $x = 0$.

Test an x-value in the intervals $(-\infty, -1)$, $(-1, 0)$, $(0, 1)$, and $(1, \infty)$.

$f'(-2) = -\dfrac{4}{3} < 0$

$f'\left(-\dfrac{1}{2}\right) = \dfrac{4}{3} > 0$

$f'\left(\dfrac{1}{2}\right) = -\dfrac{4}{3} < 0$

$f'(2) = \dfrac{4}{3} > 0$

f is increasing on $(-1, 0)$ and $(1, \infty)$ and decreasing on $(-\infty, -1)$ and $(0, 1)$.

25. $f(x) = -x^2 + 4x - 8$

$f'(x) = -2x + 4 = 0$

Critical number: $x = 2$

$f''(x) = -2 < 0$ for all x, so $f(2)$ is a relative maximum.

$f(2) = -4$

Relative maximum of -4 at 2

27. $f(x) = 2x^2 - 8x + 1$

$f'(x) = 4x - 8 = 0$

Critical number: $x = 2$

$f''(x) = 4 > 0$ for all x, so $f(2)$ is a relative minimum.

$f(2) = -7$

Relative minimum of -7 at 2

29. $f(x) = 2x^3 + 3x^2 - 36x + 20$

$f'(x) = 6x^2 + 6x - 36 = 0$

$6(x^2 + x - 6) = 0$

$(x + 3)(x - 2) = 0$

Critical numbers: -3 and 2

$f''(x) = 12x + 6$

$f''(-3) = -30 < 0$, so a maximum occurs
      at $x = -3$.

$f''(2) = 30 > 0$, so a minimum occurs
      at $x = 2$.

$f(-3) = 101$

$f(2) = -24$

Relative maximum of 101 at -3

Relative minimum of -24 at 2

31. $f(x) = \dfrac{xe^x}{x - 1}$

$f'(x) = \dfrac{(x - 1)(xe^x + e^x) - xe^x(1)}{(x - 1)^2}$

$= \dfrac{x^2e^x - xe^x - xe^x - e^x - xe^x}{(x - 1)^2}$

$= \dfrac{x^2e^x - xe^x - e^x}{(x - 1)^2}$

$= \dfrac{e^x(x^2 - x - 1)}{(x - 1)^2}$

$f'(x)$ is undefined at $x = 1$, but 1 is not in the
domain of $f(x)$.

$f'(x) = 0$ when $x^2 - x - 1 = 0$

$x = \dfrac{1 \pm \sqrt{1 - 4(1)(-1)}}{2}$

$= \dfrac{1 \pm \sqrt{5}}{2}$

$\dfrac{1 + \sqrt{5}}{2} \approx 1.618$ or $\dfrac{1 - \sqrt{5}}{2} = -0.618$

Critical numbers are -0.618 and 1.618.

$f'(1.4) = \dfrac{e^{1.4}(1.4^2 - 1.4 - 1)}{(1.4 - 1)^2} \approx -11.15 < 0$

$f'(2) = \dfrac{e^2(2^2 - 2 - 1)}{(2 - 1)^2} = e^2 \approx 7.39 > 0$

$f'(-1) = \dfrac{e^{-1}[(-1)^2 - (-1) - 1]}{(-1 - 1)^2} \approx 0.09 > 0$

$f'(0) = \dfrac{e^0(0^2 - 0 - 1)}{(0 - 1)^2} = -1 < 0$

There is a relative maximum at $(-0.618, 0.206)$
and a relative minimum at $(1.618, 13.203)$.

33. $f(x) = 3x^4 - 5x^2 - 11x$

$f'(x) = 12x^3 - 10x - 11$

$f''(x) = 36x^2 - 10$

$f''(1) = 36(1)^2 - 10 = 26$

$f''(-3) = 36(-3)^2 - 10 = 314$

35. $f(x) = \dfrac{4x + 2}{3x - 6}$

$f'(x) = \dfrac{(3x - 6)(4) - (4x + 2)(3)}{(3x - 6)^2}$

$= \dfrac{12x - 24 - 12x - 6}{(3x - 6)^2} = \dfrac{-30}{(3x - 6)^2}$

$= -30(3x - 6)^{-2}$

$f''(x) = -30(-2)(3x - 6)^{-3}(3)$

$= 180(3x - 6)^{-3}$ or $\dfrac{180}{(3x - 6)^3}$

$f''(1) = 180[3(1) - 6]^{-3} = -\dfrac{20}{3}$

$f''(-3) = 180[3(-3) - 6]^{-3} = -\dfrac{4}{75}$

37. $f(t) = \sqrt{t^2 + 1} = (t^2 + 1)^{1/2}$

$f'(t) = \dfrac{1}{2}(t^2 + 1)^{-1/2}(2t) = t(t^2 + 1)^{-1/2}$

$f''(t) = (t^2 + 1)^{-1/2}(1)$

$+ t\left[\left(-\dfrac{1}{2}\right)(t^2 + 1)^{-3/2}(2t)\right]$

$$= (t^2 + 1)^{-1/2} - t^2(t^2 + 1)^{-3/2}$$

$$= \frac{1}{(t^2 + 1)^{1/2}} - \frac{t^2}{(t^2 + 1)^{3/2}} = \frac{t^2 + 1 - t^2}{(t^2 + 1)^{3/2}}$$

$$= (t^2 + 1)^{-3/2} \quad \text{or} \quad \frac{1}{(t^2 + 1)^{3/2}}$$

$$f''(1) = \frac{1}{(1 + 1)^{3/2}} = \frac{1}{2^{3/2}} \approx 0.354$$

$$f''(-3) = \frac{1}{(9 + 1)^{3/2}} = \frac{1}{10^{3/2}} \approx 0.032$$

39. $f(x) = -2x^3 - \dfrac{1}{2}x^2 + x - 3$

Domain is $(-\infty, \infty)$

The graph has no symmetry.

$$f'(x) = -6x^2 - x + 1 = 0$$
$$(3x - 1)(2x + 1) = 0$$

Critical numbers: $\dfrac{1}{3}$ and $-\dfrac{1}{2}$

Critical points: $\left(\dfrac{1}{3}, -2.80\right)$ and $\left(-\dfrac{1}{2}, -3.375\right)$

$$f''(x) = -12x - 1$$

$$f''\left(\frac{1}{3}\right) = -5 < 0$$

$$f''\left(-\frac{1}{2}\right) = 5 > 0$$

Relative maximum at $\dfrac{1}{3}$

Relative minimum at $-\dfrac{1}{2}$

Increasing on $\left(-\dfrac{1}{2}, \dfrac{1}{3}\right)$

Decreasing on $\left(-\infty, -\dfrac{1}{2}\right)$ and $\left(\dfrac{1}{3}, \infty\right)$

$$f''(x) = -12x - 1 = 0$$

$$x = -\frac{1}{12}$$

Point of inflection at $\left(-\dfrac{1}{12}, -3.09\right)$

Concave upward on $\left(-\infty, -\dfrac{1}{12}\right)$

Concave downward on $\left(-\dfrac{1}{12}, \infty\right)$

y-intercept:

$$y = -2(0)^3 - \frac{1}{2}(0)^2 + (0) - 3 = -3$$

$$f(x) = -2x^3 - \tfrac{1}{2}x^2 + x - 3$$

41. $f(x) = x^4 - \dfrac{4}{3}x^3 - 4x^2 + 1$

Domain is $(-\infty, \infty)$

The graph has no symmetry.

$$f'(x) = 4x^3 - 4x^2 - 8x = 0$$
$$4x(x^2 - x - 2) = 0$$
$$4x(x - 2)(x + 1) = 0$$

Critical numbers: $0, 2,$ and -1

Critical points: $(0, 1)$, $\left(2, -\dfrac{29}{3}\right)$ and $\left(-1, -\dfrac{2}{3}\right)$

$$f''(x) = 12x^2 - 8x - 8$$
$$= 4(3x^2 - 2x - 2)$$
$$f''(-1) = 12 > 0$$
$$f''(0) = -8 < 0$$
$$f''(2) = 24 > 0$$

Relative maximum at 0

Relative minima at -1 and 2

Increasing on $(-1, 0)$ and $(2, \infty)$

Decreasing on $(-\infty, -1)$ and $(0, 2)$

$$f''(x) = 4(3x^2 - 2x - 2) = 0$$

$$x = \frac{2 \pm \sqrt{4 - (-24)}}{6}$$

$$= \frac{1 \pm \sqrt{7}}{3}$$

Points of inflection at $\left(\dfrac{1 \pm \sqrt{7}}{3}, -5.12\right)$ and

$\left(\dfrac{1 - \sqrt{7}}{3}, 0.11\right)$

Concave upward on $\left(-\infty, \dfrac{1 - \sqrt{7}}{3}\right)$ and $\left(\dfrac{1 + \sqrt{7}}{3}, \infty\right)$

Concave downward on $\left(\dfrac{1 - \sqrt{7}}{3}, \dfrac{1 + \sqrt{7}}{3}\right)$

y-intercept:

$$y = (0)^4 - \frac{4}{3}(0)^3 - 4(0)^2 + 1 = 1$$

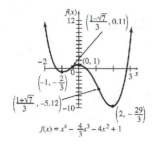

$f(x) = x^4 - \frac{4}{3}x^3 - 4x^2 + 1$

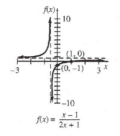

$f(x) = \frac{x-1}{2x+1}$

43. $f(x) = \dfrac{x - 1}{2x + 1}$

Domain is $\left(-\infty, -\frac{1}{2}\right) \cup \left(-\frac{1}{2}, \infty\right)$

The graph has no symmetry.

$$f'(x) = \frac{(2x + 1)(1) - (x - 1)(2)}{(2x + 1)^2}$$

$$= \frac{3}{(2x + 1)^2}$$

f' is never zero.

$f'\left(-\frac{1}{2}\right)$ does not exist, but $-\frac{1}{2}$ is not a critical number because $-\frac{1}{2}$ is not in the domain of f.

Thus, there are no critical numbers, so $f(x)$ has no relative extrema.

Increasing on $\left(-\infty, \frac{1}{2}\right)$ and $\left(\frac{1}{2}, \infty\right)$

$$f''(x) = \frac{-12}{(2x + 1)^3}$$

$f''(0) = -12 < 0$

$f''(-1) = 12 > 0$

No inflection points

Concave upward on $\left(-\infty, -\frac{1}{2}\right)$

Concave downward on $\left(-\frac{1}{2}, \infty\right)$

x-intercept: $\dfrac{x - 1}{2x + 1} = 0$

$x = 1$

y-intercept: $y = \dfrac{0 - 1}{2(0) + 1} = -1$

Vertical asymptote at $x = -\frac{1}{2}$

Horizontal asymptote at $y = \frac{1}{2}$

45. $f(x) = -4x^3 - x^2 + 4x + 5$

Domain is $(-\infty, \infty)$

The graph has no symmetry.

$f'(x) = -12x^2 - 2x + 4$

$\quad = -2(6x^2 + x - 2) = 0$

$\quad (3x + 2)(2x - 1) = 0$

Critical numbers: $-\frac{2}{3}$ and $\frac{1}{2}$

Critical points: $\left(-\frac{2}{3}, 3.07\right)$ and $\left(\frac{1}{2}, 6.25\right)$

$f''(x) = -24x - 2$

$\quad = -2(12x + 1)$

$f''\left(-\frac{2}{3}\right) = 14 > 0$

$f''\left(\frac{1}{2}\right) = -14 < 0$

Relative maximum at $\frac{1}{2}$

Relative minimum at $-\frac{2}{3}$

Increasing on $\left(-\frac{2}{3}, \frac{1}{2}\right)$

Decreasing on $\left(-\infty, -\frac{2}{3}\right)$ and $\left(\frac{1}{2}, \infty\right)$

$f''(x) = -2(12x + 1) = 0$

$$x = -\frac{1}{12}$$

Point of inflection at $\left(-\frac{1}{12}, 4.66\right)$

Concave upward on $\left(-\infty, -\frac{1}{12}\right)$

Concave downward on $\left(-\frac{1}{12}, \infty\right)$

y-intercept:

$y = -4(0)^3 - (0)^2 + 4(0) + 5 = 5$

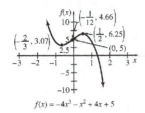

$f(x) = -4x^3 - x^2 + 4x + 5$

47. $f(x) = x^4 + 2x^2$

Domain is $(-\infty, \infty)$

$f(-x) = (-x)^4 - 2(-x)^2$

$\qquad = x^4 + 2x^2 = f(x)$

The graph is symmetric about the y-axis.

$f'(x) = 4x^3 + 4x$

$\qquad = 4x(x^2 + 1) = 0$

Critical number: 0

Critical point: $(0, 0)$

$f''(x) = 12x^2 + 4 = 4(3x^2 + 1)$

$f''(0) = 4 > 0$

Relative minimum at 0

Increasing on $(0, \infty)$

Decreasing on $(-\infty, 0)$

$f''(x) = 4(3x^2 + 1) \neq 0$ for any x

No points of inflection

$f''(-1) = 16 > 0$

$f''(1) = 16 > 0$

Concave upward on $(-\infty, \infty)$

x-intercept: 0; y-intercept: 0

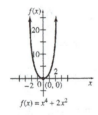

$f(x) = x^4 + 2x^2$

49. $f(x) = \dfrac{x^2 + 4}{x}$

Domain is $(-\infty, 0) \cup (0, \infty)$

$f(-x) = \dfrac{(-x)^2 + 4}{-x}$

$\qquad = \dfrac{x^2 + 4}{-x} = -f(x)$

The graph is symmetric about the origin.

$f'(x) = \dfrac{x(2x) - (x^2 + 4)}{x^2}$

$\qquad = \dfrac{x^2 - 4}{x^2} = 0$

Critical numbers: -2 and 2

Critical points: $(-2, -4)$ and $(2, 4)$

$f''(x) = \dfrac{8}{x^3}$

$f''(-2) = -1 < 0$

$f''(2) = 1 > 0$

Relative maximum at -2

Relative minimum at 2

Increasing on $(-\infty, -2)$ and $(2, \infty)$

Decreasing on $(-2, 0)$ and $(0, 2)$

$f''(x) = \dfrac{8}{x^3} > 0$ for all x.

No inflection points

Concave upward on $(0, \infty)$

Concave downward on $(-\infty, 0)$

No x- or y-intercepts

Vertical asymptote at $x = 0$

Oblique asymptote at $y = x$

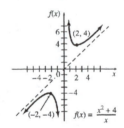

$f(x) = \dfrac{x^2 + 4}{x}$

51. $f(x) = \dfrac{2x}{3 - x}$

Domain is $(-\infty, 3) \cup (3, \infty)$

The graph has no symmetry.

$f'(x) = \dfrac{(3 - x)(2) - (2x)(-1)}{(3 - x)^2}$

$\qquad = \dfrac{6}{(3 - x)^2}$

$f'(x)$ is never zero. $f'(3)$ does not exist, but since 3 is not in the domain of f, it is not a critical number. No critical numbers, so no relative extrema

$f'(0) = \dfrac{2}{3} > 0$

$f'(4) = 6 > 0$

Increasing on $(-\infty, 3)$ and $(3, \infty)$

$$f''(x) = \frac{12}{(3-x)^3}$$

$f''(x)$ is never zero. $f''(3)$ does not exist, but since 3 is not in the domain of f, there is no inflection point at $x = 3$.

$$f''(0) = \frac{12}{27} > 0$$

$f''(4) = -12 < 0$

Concave upward on $(-\infty, 3)$

Concave downward on $(3, \infty)$

x-intercept: 0; y-intercept: 0

Vertical asymptote at $x = 3$

Horizontal asymptote at $y = -2$

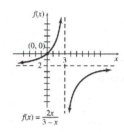

53. $f(x) = xe^{2x}$

Domain is $(-\infty, \infty)$.

$f(-x) = -xe^{-2x}$

The graph has no symmetry.

$$f'(x) = (1)(e^{2x}) + (x)(2e^{2x})$$

$$= e^{2x}(2x + 1)$$

$f'(x) = 0$ when $x = -\frac{1}{2}$.

Critical number: $-\frac{1}{2}$

Critical point: $\left(-\frac{1}{2}, -\frac{1}{2e}\right)$

$f'(-1) = e^{2(-1)}[2(-1) + 1] = -e^{-2} < 0$

$f'(0) = e^{2(0)}[2(0) + 1] = 1 > 0$

No relative maximum

Relative minimum at $\left(-\frac{1}{2}, -\frac{1}{2e}\right)$

Decreasing on $\left(-\infty, -\frac{1}{2}\right)$ and increasing on

$\left(-\frac{1}{2}, \infty\right)$

$f''(x) = 2e^{2x}(2x + 1) + e^{2x}(2)$

$$= 4e^{2x}(x + 1)$$

$f''(x) = 0$ when $x = -1$.

$f''(-2) = 4e^{2(-2)}[(-2) + 1] = -4e^{-4} < 0$

$f''(0) = 4e^{2(0)}[(0) + 1] = 4 > 0$

Inflection point at $(-1, -e^{-2})$

Concave upward on $(-1, \infty)$

Concave downward on $(-\infty, -1)$

x-intercept: $xe^{2x} = 0$

$x = 0$

y-intercept: $y = (0)e^{2(0)} = 0$

Since $\lim\limits_{x \to -\infty} xe^{2x} = 0$, there is a horizontal asymptote at $y = 0$.

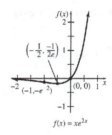

$f(x) = xe^{2x}$

55. $f(x) = \ln(x^2 + 4)$

Domain is $(-\infty, \infty)$.

$f(-x) = \ln[(-x)^2 + 4] = \ln(x^2 + 4) = f(x)$

The graph is symmetric about the y-axis.

$$f'(x) = \frac{2x}{x^2 + 4}$$

$f'(x) = 0$ when $x = 0$.

Critical number: 0

Critical point: $(0, \ln 4)$

$f'(-1) = \frac{2(-1)}{(-1)^2 + 4} = -\frac{2}{5} < 0$

$f'(1) = \frac{2(1)}{(1)^2 + 4} = \frac{2}{5} > 0$

No relative maximum

Relative minimum at $(0, \ln 4)$

Increasing on $(0, \infty)$

Decreasing on $(-\infty, 0)$

$$f''(x) = \frac{(x^2 + 4)(2) - (2x)(2x)}{(x^2 + 4)^2}$$

$$= \frac{-2(x^2 - 4)}{(x^2 + 4)^2}$$

$f''(x) = 0$ when

$x^2 - 4 = 0$

$\quad x = \pm 2$

$f''(-3) = \dfrac{-2[(-3)^2 - 4]}{[(-3)^2 + 4]^2} = -\dfrac{10}{169} < 0$

$f''(0) = \dfrac{-2[(0)^2 - 4]}{[(0)^2 + 4]^2} = \dfrac{1}{2} > 0$

$f''(3) = \dfrac{-2[(3)^2 - 4]}{[(3)^2 + 4]^2} = -\dfrac{10}{169} < 0$

Inflection points at $(-2, \ln 8)$ and $(2, \ln 8)$

Concave upward on $(-2, 2)$

Concave downward on $(-\infty, -2)$ and $(2, \infty)$

Since $f(x)$ never equals zero, there are no x-intercepts.

y-intercept: $y = \ln[(0)^2 + 4] = \ln 4$

No horizontal or vertical asymptotes.

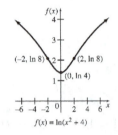

$f(x) = \ln(x^2 + 4)$

57.　$f(x) = 4x^{1/3} + x^{4/3}$

Domain is $(-\infty, \infty)$.

$f(-x) = 4(-x)^{1/3} + (-x)^{4/3} = -4x^{1/3} + x^{4/3}$

The graph is not symmetric about the y-axis or origin.

$f'(x) = \dfrac{4}{3}x^{-2/3} + \dfrac{4}{3}x^{1/3}$

$f'(x) = 0$ when

$\dfrac{4}{3}x^{-2/3} + \dfrac{4}{3}x^{1/3} = 0$

$\dfrac{4}{3}x^{-2/3}(1 + x) = 0$

$\quad x = -1$

$f'(x)$ is not defined when $x = 0$

Critical numbers: -1 and 0

Critical points: $(-1, -3)$ and $(0, 0)$

$f'(-8) = \dfrac{4}{3}(-8)^{-2/3} + \dfrac{4}{3}(-8)^{1/3} = -\dfrac{7}{3} < 0$

$f'\left(-\dfrac{1}{8}\right) = \dfrac{4}{3}\left(-\dfrac{1}{8}\right)^{-2/3} + \dfrac{4}{3}\left(-\dfrac{1}{8}\right)^{1/3} = \dfrac{14}{3} > 0$

$f'(1) = \dfrac{4}{3}(1)^{-2/3} + \dfrac{4}{3}(1)^{1/3} = \dfrac{8}{3} > 0$

No relative maximum

Relative minimum at $(-1, -3)$

Increasing on $(-1, \infty)$

Decreasing on $(-\infty, -1)$

$f''(x) = -\dfrac{8}{9}x^{-5/3} + \dfrac{4}{9}x^{-2/3}$

$f''(x) = 0$ when

$-\dfrac{8}{9}x^{-5/3} + \dfrac{4}{9}x^{-2/3} = 0$

$\dfrac{4}{9}x^{-5/3}(-2 + x) = 0$

$\quad x = 2$

$f''(x)$ is not defined when $x = 0$

$f''(-1) = -\dfrac{8}{9}(-1)^{-5/3} + \dfrac{4}{9}(-1)^{-2/3} = \dfrac{4}{3} > 0$

$f''(1) = -\dfrac{8}{9}(1)^{-5/3} + \dfrac{4}{9}(1)^{-2/3} = -\dfrac{4}{9} < 0$

$f''(8) = -\dfrac{8}{9}(8)^{-5/3} + \dfrac{4}{9}(8)^{-2/3} = \dfrac{1}{12} > 0$

Inflection points at $(0, 0)$ and $(2, 6 \cdot 2^{1/3})$

Concave upward on $(-\infty, 0)$ and $(2, \infty)$

Concave downward on $(0, 2)$

x-intercept: $4x^{1/3} + x^{4/3} = 0$

$\quad x^{1/3}(4 + x) = 0$

$\quad\quad x = 0 \text{ or } x = -4$

y-intercept: $y = 4(0)^{1/3} + (0)^{4/3} = 0$

No horizontal or vertical asymptotes

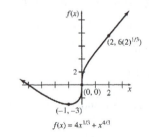

$f(x) = 4x^{1/3} + x^{4/3}$

59.

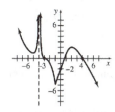

Other graphs are possible.

61. **(a)-(b)** If the price of the stock is falling faster and faster, $P(t)$ would be decreasing, so $P'(t)$ would be negative. $P(t)$ would be concave downward, so $P''(t)$ would also be negative.

63. **(a)** Profit = Income − Cost

$$P(q) = qp - C(q)$$
$$= q(-q^2 - 3q + 299)$$
$$- (-10q^2 + 250q)$$
$$= -q^3 - 3q^2 + 299q$$
$$+ 10q^2 - 250q$$
$$= -q^3 + 7q^2 + 49q$$

(b) $P'(q) = -3q^2 + 14q + 49$
$$= (-3q - 7)(q - 7)$$
$$= -(3q + 7)(q - 7)$$

$q = \frac{7}{3}$ (nonsensical) or $q = 7$

$$P''(q) = -6q + 14$$
$$P''(7) = -6 \cdot 7 + 14 = -28 < 0$$
$$\text{(indicates a maximum)}$$

7 brushes would produce the maximum profit.

(c) $p = -7^2 - 3(7) + 299$
$$= -49 - 21 + 299 = 229$$

$229 is the price that produces the maximum profit.

(d) $P(7) = -7^3 + 7(7^2) + 49(7) = 343$

The maximum profit is $343.

(e) $P''(q) = 0$ when $-6q + 14 = 0$

$$q = \frac{7}{3}$$

$$P''(2) = -6(2) + 14 = 2 > 0$$
$$P''(3) = -6 \cdot 3 + 14 = -4 < 0$$

The point of diminishing returns is $q = \frac{7}{3}$ (between 2 and 3 brushes).

65. $f(t)$ is increasing and concave downward.

$f'(t)$ is positive and decreasing.

$f''(t)$ is negative.

67. **(a)** Since the second derivative has many sign changes, the graph continually changes from concave upward to concave downward. Since there is a nonlinear decline, the graph must be one that declines, levels off, declines, levels off, etc. Therefore, the first derivative has many critical numbers where the first derivative is zero.

(b) The curve is always decreasing except at frequent points of inflection.

69. Sketch the curve for $l_1(v) = 0.08e^{0.33v}$

$$l_1'(v) = 0.0264e^{0.33v}$$

$$e^{0.33v} \neq 0$$

$l_1(v)$ has no critical points.

$$l_1''(v) = 0.008712e^{0.33v}$$

$$e^{0.33v} \neq 0$$

$l_1(v)$ has no inflection points.

Sketch the curve for $l_2 = -0.87v^2 + 28.17v - 211.41$

$$l_2'(v) = -1.74v + 28.17$$
$$-1.74v + 28.17 = 0$$
$$v \approx 16.19$$

Critical point: $(16.19, 16.62)$

$$l_2''(v) = -1.74$$

$l_2(v)$ has no inflection points.

$l_2(v)$ has a relative maximum at $(16.19, 16.62)$.

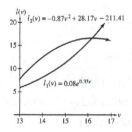

71. $y = 34.7(1.0186)^{-x}(x^{-0.658})$

In chapter 4, the function was originally defined as

$\log y = 1.54 - 0.008x - 0.658 \log x$

so, $0 < x < \infty$, and $0 < y < \infty$.

The function will have a vertical asymptote at $x = 0$ and a horizontal asymptote at $y = 0$.

$$\frac{dy}{dx} = -34.7(1.0186)^{-x}x^{-0.658}\left[\ln(1.0186) + \frac{0.658}{x}\right]$$

For every value in the domain, $\dfrac{dy}{dx} < 0$, so y has no critical points and is decreasing on $(0, \infty)$.

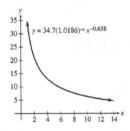

73. $y(t) = A^{c^t}$

$y'(t) = (\ln A)A^{c^t} \cdot \dfrac{d}{dt}c^t$

$\qquad = (\ln A)(\ln c)c^t A^{c^t}$

$y''(t) = (\ln A)(\ln c) \cdot [(\ln c)c^t A^{c^t}$

$\qquad\qquad + c^t(\ln A)(\ln c)c^t A^{c^t}]$

$\qquad = (\ln A)(\ln c)^2 c^t A^{c^t}[1 + (\ln A)c^t]$

$y''(t) = 0$ when $1 + \ln(A)c^t = 0$

$c^t = -\dfrac{1}{\ln A}$

$t \ln c = \ln\left(-\dfrac{1}{\ln A}\right)$

$t = -\dfrac{\ln(-\ln A)}{\ln c}$

$\quad = -\dfrac{\ln[-\ln(0.3982 \cdot 10^{-291})]}{\ln 0.4252}$

By properties of logarithms,

$-\ln(0.3982 \cdot 10^{-291}) = -[\ln(0.3982) + \ln(10^{-291})]$

$\qquad\qquad = -[\ln(0.3982) - 291\ln(10)]$

$\qquad\qquad = -\ln(0.3982) + 291\ln(10)$

So,

$$t = -\frac{\ln[-\ln(0.3982) + 291\ln(10)]}{\ln(0.4252)}$$

≈ 7.6108

At about 7.6108 years the rate of learning to pass the test begins to slow down.

75.   **(a)**   The U.S. total inventory was at a relative maximum in 1965, 1973, 1976, 1983, 1986, and 1988.

(b)   The U.S. total inventory was at its largest relative maximum from 1965 to 1967. During this period, the Soviet total inventory was concave upward. This means that the total inventory was increasing at an increasingly rapid rate.

77.   **(a)**

$$G(x) = \frac{3}{1 + e^{-c(x-50)}} + \frac{1}{1 + e^{-c(x-70)}} + \frac{1}{1 + e^{-c(x-90)}}$$

Each term of the derivative looks like

$$\frac{kce^{-c(x-b)}}{\left(1 + e^{-c(x-b)}\right)^2}$$

for positive k and c, so both numerator and denominator are positive and the sum of three positive terms is positive. We could simply note that each denominator in the formula for $G(x)$ is a decreasing function of x.

(b)   Here is the graph of $G(x)$ with the constant c set to 0.7.

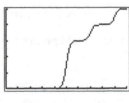

[0, 100] by [0, 5]

The derivative is close to 0 on the intervals [0, 44], [56, 65], [75, 84], and [97, 100].

(c)   From the graph in (b), the inflection points appear to be at the x-values 50, 60, 70, 80 and 90. The corresponding points on the graph are (50, 1.5), (60, 3.0), (70, 3.5), (80, 4.0), and (90, 4.5).

(d)   The first term of the derivative can be written as follows:

$$\frac{3ce^{-c(x-50)}}{\left(1 + e^{-c(x-50)}\right)^2} = \frac{3c}{\left(1 + e^{-c(x-50)}\right)^2 \left(e^{c(x-50)}\right)}$$

$$= \frac{3c}{\left(1 + e^{-c(x-50)}\right)^2 \left(e^{c(x-50)/2}\right)^2}$$

$$= \frac{3c}{\left(e^{c(x-50)/2} + e^{-c(x-50)/2}\right)^2}$$

In whichever direction x moves from 50, one term in the denominator quickly becomes very large, so the fraction only contributes significantly to the derivative when x is close to 50. The other two fractions behave similarly at $x = 70$ and $x = 90$.

Chapter 6

APPLICATIONS OF THE DERIVATIVE

6.1   Absolute Extrema

Your Turn 1

To find the absolute extrema of $f(x) = 3x^{2/3} - 3x^{5/3}$ on $[0, 8]$, first find the critical numbers in the interval $(0, 8)$.

$$f'(x) = 2x^{-1/3} - 5x^{2/3}$$

$$= \frac{2}{x^{1/3}} - 5x^{2/3}$$

$$= \frac{2}{x^{1/3}} - \frac{5x}{x^{1/3}}$$

$$= \frac{1}{x^{1/3}}(2 - 5x)$$

$f'(x)$ is 0 when $2 - 5x = 0$, that is, when $x = \frac{5}{2} = 0.4$. Now evaluate the function f at the critical value 0.4 and at the endpoints 0 and 8.

x	$f(x)$
0	0
0.4	0.977
8	−84

The absolute maximum of approximately 0.977 occurs when $x = 2/5$; the absolute minimum of -84 occurs when $x = 8$.

Your Turn 2

The domain of the function
$$f(x) = -x^4 - 4x^3 + 8x^2 + 20$$
is the open interval $(-\infty, \infty)$. As x approaches $+\infty$ and $-\infty$ the dominant term is $-x^4$, and this approaches $-\infty$. Thus the function has no absolute minimum. To find the absolute maximum, we evaluate f at the critical points.

$$f'(x) = -4x^3 - 12x^2 + 16x$$

$$= -4x(x^2 + 3x - 4)$$

$$= -4x(x + 4)(x - 1)$$

Setting $f'(x)$ equal to 0, we have $x = 0, x = -4$, or $x = 1$.

x	$f(x)$
−4	148
0	20
1	23

The absolute maximum is 148, at $x = -4$. As noted above, there is no absolute minimum.

6.1   Warmup Exercisess

W1. $f(x) = 4x^3 - 7x^2 - 40x + 5$

$$f'(x) = 12x^2 - 14x - 40$$

$$= 2(6x^2 - 7x - 20)$$

f' is defined everywhere, so the only critical numbers will occur where $f'(x) = 0$.

$$2(6x^2 - 7x - 20) = 0$$
$$2(3x + 4)(2x - 5) = 0$$
$$3x + 4 = 0 \quad \text{or} \quad 2x - 5 = 0$$
$$x = -\frac{4}{3} \quad \text{or} \quad x = \frac{5}{2}$$

The critical numbers are $-\dfrac{4}{3}$ and $\dfrac{5}{2}$.

W2. $f(x) = 15x^{7/2} - 7x^{9/2}$

$$f'(x) = \frac{105}{2}x^{5/2} - \frac{63}{2}x^{7/2}$$

$$= \left(\frac{21}{2}x^{5/2}\right)(3x - 5)$$

f' is defined everywhere, so the only critical numbers will occur where $f'(x) = 0$, which happens at $x = 0$ and at $x = 5/3$.

6.1   Exercises

1.   As shown on the graph, the absolute maximum occurs at x_3; there is no absolute minimum. (There is no functional value that is less than all others.)

3. As shown on the graph, there are no absolute extrema.

5. As shown on the graph, the absolute minimum occurs at x_1; there is no absolute maximum.

7. As shown on the graph, the absolute maximum occurs at x_1; the absolute minimum occurs at x_2.

11. $f(x) = x^3 - 6x^2 + 9x - 8; [0,5]$

Find critical numbers:

$$f'(x) = 3x^2 - 12x + 9 = 0$$
$$x^2 - 4x + 3 = 0$$
$$(x - 3)(x - 1) = 0$$
$$x = 1 \text{ or } x = 3$$

x	$f(x)$	
0	-8	Absolute minimum
1	-4	
3	-8	Absolute minimum
5	12	Absolute maximum

13. $f(x) = \frac{1}{3}x^3 + \frac{3}{2}x^2 - 4x + 1; [-5, 2]$

Find critical numbers:

$$f'(x) = x^2 + 3x - 4 = 0$$
$$(x + 4)(x - 1) = 0$$
$$x = -4 \text{ or } x = 1$$

x	$f(x)$	
-4	$\frac{59}{3} \approx 19.67$	Absolute maximum
1	$-\frac{7}{6} \approx -1.17$	Absolute minimum
-5	$\frac{101}{6} \approx 16.83$	
2	$\frac{5}{3} \approx 1.67$	

15. $f(x) = x^4 - 18x^2 + 1; [-4, 4]$

$$f'(x) = 4x^3 - 36x = 0$$
$$4x(x^2 - 9) = 0$$
$$4x(x + 3)(x - 3) = 0$$
$$x = 0 \text{ or } x = -3 \text{ or } x = 3$$

x	$f(x)$	
-4	-31	
-3	-80	Absolute minimum
0	1	Absolute maximum
3	-80	Absolute minimum
4	-31	

17. $f(x) = \frac{1 - x}{3 + x}; [0, 3]$

$$f'(x) = \frac{-4}{(3 + x)^2}$$

No critical numbers

x	$f(x)$	
0	$\frac{1}{3}$	Absolute maximum
3	$-\frac{1}{3}$	Absolute minimum

19. $f(x) = \frac{x - 1}{x^2 + 1}; [1, 5]$

$$f'(x) = \frac{-x^2 + 2x + 1}{(x^2 + 1)^2}$$

$f'(x) = 0$ when

$$-x^2 + 2x + 1 = 0$$
$$x = 1 \pm \sqrt{2},$$

but $1 - \sqrt{2}$ is not in $[1, 5]$.

x	$f(x)$	
1	0	Absolute minimum
5	$\frac{2}{13} \approx 0.15$	
$1 + \sqrt{2}$	$\frac{\sqrt{2} - 1}{2} \approx 0.21$	Absolute maximum

21. $f(x) = (x^2 - 4)^{1/3}; [-2, 3]$

$$f'(x) = \frac{1}{3}(x^2 - 4)^{-2/3}(2x)$$
$$= \frac{2x}{3(x^2 - 4)^{2/3}}$$

$f'(x) = 0$ when $2x = 0$

$$x = 0$$

$f'(x)$ is undefined at $x = -2$ and $x = 2$, but $f(x)$ is defined there, so -2 and 2 are also critical numbers:

x	$f(x)$	
-2	0	
0	$(-4)^{1/3} \approx -1.587$	Absolute minimum
2	0	
3	$5^{1/3} \approx 1.710$	Absolute maximum

23. $f(x) = 5x^{2/3} + 2x^{5/3}; [-2, 1]$

$$f'(x) = \frac{10}{3}x^{-1/3} + \frac{10}{3}x^{2/3}$$

$$= \frac{10}{3x^{1/3}} + \frac{10x^{2/3}}{3}$$

$$= \frac{10x + 10}{3x^{1/3}}$$

$$= \frac{10(x+1)}{3\sqrt[3]{x}}$$

$f'(x) = 0$ when $10(x+1) = 0$

$$x + 1 = 0$$

$$x = -1.$$

$f'(x)$ is undefined at $x = 0$, but $f(x)$ is defined at $x = 0$, so 0 is also a critical number.

x	$f(x)$	
-2	1.587	
-1	3	
0	0	Absolute minimum
1	7	Absolute maximum

25. $f(x) = x^2 - 8 \ln x; [1, 4]$

$$f'(x) = 2x - \frac{8}{x}$$

$f'(x) = 0$ when $2x - \frac{8}{x} = 0$

$$2x = \frac{8}{x}$$

$$2x^2 = 8$$

$$x^2 = 4$$

$$x = -2 \quad \text{or} \quad x = 2$$

but $x = -2$ is not in the given interval.

Although $f'(x)$ fails to exist at $x = 0$, 0 is not in the specified domain for $f(x)$, so 0 is not a critical number.

x	$f(x)$	
1	1	
2	-1.545	Absolute minimum
4	4.910	Absolute maximum

27. $f(x) = x + e^{-3x}; [-1, 3]$

$$f'(x) = 1 - 3e^{-3x}$$

$f'(x) = 0$ when $1 - 3e^{-3x} = 0$

$$-3e^{-3x} = -1$$

$$e^{-3x} = \frac{1}{3}$$

$$-3x = \ln\frac{1}{3}$$

$$x = \frac{\ln 3}{3}$$

x	$f(x)$	
-1	19.09	Absolute maximum
$\dfrac{\ln 3}{3}$	0.6995	Absolute minimum
3	3.000	

29. $f(x) = \dfrac{-5x^4 + 2x^3 + 3x^2 + 9}{x^4 - x^3 + x^2 + 7}; [-1, 1]$

The indicated domain tells us the x-values to use for the viewing window, but we must experiment to find a suitable range for the y-values. In order to show the absolute extrema on $[-1, 1]$, we find that a suitable window is $[-1, 1]$ by $[0, 1.5]$ with Xscl $= 0.1$, Yscl $= 0.1$.

From the graph, we see that on $[-1, 1]$, f has an absolute maximum of 1.356 at about 0.6085 and an absolute minimum of 0.5 at -1.

31. $f(x) = 2x + \dfrac{8}{x^2} + 1, \ x > 0$

$$f'(x) = 2 - \frac{16}{x^3}$$

$$= \frac{2x^3 - 16}{x^3}$$

$$= \frac{2(x-2)(x^2 + 2x + 4)}{x^3}$$

Since the specified domain is $(0, \infty)$, a critical number is $x = 2$.

x	$f(x)$
2	7

There is an absolute minimum of 7 at $x = 2$;
there is no absolute maximum, as can be seen by
looking at the graph of f.

33. $f(x) = -3x^4 + 8x^3 + 18x^2 + 2$

$f'(x) = -12x^3 + 24x^2 + 36x$

$\quad = -12x(x^2 - 2x - 3)$

$\quad = -12x(x - 3)(x + 1)$

Critical numbers are 0, 3, and -1.

x	$f(x)$
-1	9
0	2
3	137

There is an absolute maximum of 137 at $x = 3$;
there is no absolute minimum, as can be seen by
looking at the graph of f.

35. $f(x) = \dfrac{x - 1}{x^2 + 2x + 6}$

$f'(x) = \dfrac{(x^2 + 2x + 6)(1) - (x - 1)(2x + 2)}{(x^2 + 2x + 6)^2}$

$\quad = \dfrac{x^2 + 2x + 6 - 2x^2 + 2}{(x^2 + 2x + 6)^2}$

$\quad = \dfrac{-x^2 + 2x + 8}{(x^2 + 2x + 6)^2}$

$\quad = \dfrac{-(x^2 - 2x - 8)}{(x^2 + 2x + 6)^2}$

$\quad = \dfrac{-(x - 4)(x + 2)}{(x^2 + 2x + 6)^2}$

Critical numbers are 4 and -2.

x	$f(x)$
-2	$-\dfrac{1}{2}$
4	0.1

There is an absolute maximum of 0.1 at $x = 4$
and an absolute minimum of -0.5 at $x = -2$.
This can be verified by looking at the graph of f.

37. $f(x) = \dfrac{\ln x}{x^3}$

$f'(x) = \dfrac{x^3 \cdot \dfrac{1}{x} - 3x^2 \ln x}{x^6}$

$\quad = \dfrac{x^2 - 3x^2 \ln x}{x^6}$

$\quad = \dfrac{x^2(1 - 3\ln x)}{x^6}$

$\quad = \dfrac{1 - 3\ln x}{x^4}$

$f'(x) = 0$ when $x = e^{1/3}$, and $f'(x)$ does not
exist when $x \le 0$. The only critical number is $e^{1/3}$.

x	$f(x)$
$e^{1/3}$	$\dfrac{1}{3}e^{-1} \approx 0.1226$

There is an absolute maximum of 0.1226 at
$x = e^{1/3}$. There is no absolute minimum, as can
be seen by looking at the graph of f.

39. $f(x) = 2x - 3x^{2/3}$

$f'(x) = 2 - 2x^{-1/3} = 2 - \dfrac{2}{\sqrt[3]{x}} = \dfrac{2\sqrt[3]{x} - 2}{\sqrt[3]{x}}$

$f'(x) = 0$ when $2\sqrt[3]{x} - 2 = 0$

$\qquad\qquad\qquad 2\sqrt[3]{x} = 2$

$\qquad\qquad\qquad \sqrt[3]{x} = 1$

$\qquad\qquad\qquad\quad x = 1$

$f'(x)$ is undefined at $x = 0$, but $f(x)$ is defined
at $x = 0$. So the critical numbers are 0 and 1.

(a) On $[-1, 0.5]$

x	$f(x)$
-1	-5
0	0
1	-1
0.5	-0.88988

Absolute minimum of -5 at $x = -1$;
absolute maximum of 0 at $x = 0$

(b) On $[0.5, 2]$

x	$f(x)$
0.5	-0.88988
1	-1
2	-0.7622

Absolute maximum of about -0.76 at $x = 2$; absolute minimum of -1 at $x = 1$.

41. (a) Looking at the graph we see that there are relative maxima of 8496 in 2001, 7556 in 2004, 6985 in 2006, and 6700 in 2008. There are relative minima of 7127 in 2000, 7465 in 2003, 6748 in 2005, 5933 in 2007, and 5014 in 2011.

(b) The absolute maximum is 8496 (in 2001) and the absolute minimum is 5014 (in 2011).

43. $P(x) = -x^3 + 9x^2 + 120x - 400, x \geq 5$

$$P'(x) = -3x^2 + 18x + 120$$
$$= -3(x^2 - 6x - 40)$$
$$= -3(x - 10)(x + 4) = 0$$
$$x = 10 \quad \text{or} \quad x = -4$$

-4 is not relevant since $x \geq 5$, so the only critical number is 10.

The graph of $P'(x)$ is a parabola that opens downward, so $P'(x) > 0$ on the interval $[5, 10)$ and $P'(x) < 0$ on the interval $(10, \infty)$. Thus, $P(x)$ is a maximum at $x = 10$.

Since x is measured in hundreds thousands, 10 hundred thousand or 1,000,000 tires must be sold to maximize profit.

Also,

$$P(10) = -(10)^3 + 9(10)^2 + 120(10) - 400$$
$$= 700.$$

The maximum profit is $700 thousand or $700,000.

45. $C(x) = x^3 + 37x + 250$

(a) $1 \leq x \leq 10$

$$\overline{C}(x) = \frac{C(x)}{x} = \frac{x^3 + 37x + 250}{x}$$
$$= x^2 + 37 + \frac{250}{x}$$

$$\overline{C}'(x) = 2x - \frac{250}{x^2}$$
$$= \frac{2x^3 - 250}{x^2} = 0 \text{ when}$$
$$2x^3 = 250$$
$$x^3 = 125$$
$$x = 5.$$

Test for relative minimum.

$$\overline{C}'(4) = -7.625 < 0$$
$$\overline{C}'(6) \approx 5.0556 > 0$$
$$\overline{C}(5) = 112$$
$$\overline{C}(1) = 1 + 37 + 250 = 288$$
$$\overline{C}(10) = 100 + 37 + 25 = 162$$

The minimum for $1 \leq x \leq 10$ is 112.

(b) $10 \leq x \leq 20$

There are no critical values in this interval. Check the endpoints.

$$\overline{C}(10) = 162$$
$$\overline{C}(20) = 400 + 37 + 12.5 = 449.5$$

The minimum for $10 \leq x \leq 20$ is 162.

47. The value $x = 11.5$ minimizes $\frac{f(x)}{x}$ because this is the point where the line from the origin to the curve is tangent to the curve.

A production level of 11.5 units results in the minimum cost per unit.

49. The value $x = 100$ maximizes $\frac{f(x)}{x}$ because this is the point where the line from the origin to the curve is tangent to the curve.

A production level of 100 units results in the maximum profit per item produced.

53. $f(x) = \dfrac{x^2 + 36}{2x}, 1 \leq x \leq 12$

$$f'(x) = \frac{2x(2x) - (x^2 + 36)(2)}{(2x)^2}$$
$$= \frac{4x^2 - 2x^2 - 72}{4x^2}$$
$$= \frac{2x^2 - 72}{4x^2} = \frac{2(x^2 - 36)}{4x^2}$$
$$= \frac{(x + 6)(x - 6)}{2x^2}$$

$f'(x) = 0$ when $x = 6$ and when $x = -6$. Only 6 is in the interval $1 \leq x \leq 12$.

Test for relative maximum or minimum.

$$f'(5) = \frac{(11)(-1)}{50} < 0$$
$$f'(7) = \frac{(13)(1)}{98} > 0$$

The minimum occurs at $x = 6$, or at 6 months. Since $f(6) = 6$, $f(1) = 18.5$, and $f(12) = 7.5$, the minimum percent is 6%.

55. Since we are only interested in the length during weeks 22 through 28, the domain of the function for this problem is [22, 28]. We now look for any critical numbers in this interval. We find

$$L'(t) = 0.788 - 0.02t$$

There is a critical number at $t = \dfrac{0.788}{0.02} = 39.4$. which is not in the interval. Thus, the maximum value will occur at one of the endpoints.

t	$L(t)$
22	5.4
28	7.2

The maximum length is about 7.2 millimeters.

57. $M(x) = -\dfrac{1}{45}x^2 + 2x - 20,\ 30 \le x \le 65$

$$M'(x) = -\dfrac{1}{45}(2x) + 2 = -\dfrac{2x}{45} + 2$$

When $M'(x) = 0$,

$$-\dfrac{2x}{45} + 2 = 0$$

$$2 = \dfrac{2x}{45}$$

$$45 = x.$$

x	$M(x)$
30	20
45	25
65	$\dfrac{145}{9} \approx 16.1$

The absolute maximum miles per gallon is 25 at 45 mph and the absolute minimum miles per gallon is about 16.1 at 65 mph.

59. Total area $A(x) = \pi\left(\dfrac{x}{2\pi}\right)^2 + \left(\dfrac{12 - x}{4}\right)^2$

$$= \dfrac{x^2}{4\pi} + \dfrac{(12 - x)^2}{16}$$

$$A'(x) = \dfrac{x}{2\pi} - \dfrac{12 - x}{8} = 0$$

$$\dfrac{4x - \pi(12 - x)}{8\pi} = 0$$

$$x = \dfrac{12\pi}{4 + \pi} \approx 5.28$$

x	Area
0	9
5.28	5.04
12	11.46

The total area is minimized when the piece used to form the circle is $\dfrac{12\pi}{4+\pi}$ feet, or about 5.28 feet long.

61. For the solution to Exercise 59, the piece of length x used to form the circle is $\dfrac{12\pi}{4+\pi}$ feet. The circle can be inscribed inside the square if the side of the square equals the diameter of the circle (that is, twice the radius).

$$\text{side of the square} = 2\,(\text{radius})$$

$$\dfrac{12 - x}{4} = 2\left(\dfrac{x}{2\pi}\right)$$

$$\dfrac{12 - x}{4} = \dfrac{x}{\pi}$$

$$4x = 12\pi - \pi x$$

$$x(4 + \pi) = 12\pi$$

$$x = \dfrac{12\pi}{4 + \pi}$$

Therefore, the circle formed by piece of length $x = \dfrac{12\pi}{4+\pi}$ can be inscribed inside the square.

6.2   Applications of Extrema

Your Turn 1

Assign a variable to the quantity to be minimized:

$$M = x^2 y$$

Use the given condition to express M in terms of one variable, say x:

$$x + 3y = 30$$

$$y = \dfrac{30 - x}{3}$$

$$M = x^2\left(\dfrac{30 - x}{3}\right) = 10x^2 - \dfrac{1}{3}x^3$$

Find the domain of M:

Since x and y must be nonnegative, we have $x \ge 0$ and $\dfrac{30-x}{3} \ge 0$ which requires $x \le 30$. Thus the domain of M is [0, 30].

Find the critical numbers:

$$\dfrac{dM}{dx} = 20x - x^2 = x(20 - x)$$

$\frac{dM}{dx}$ is 0 when $x = 0$ (already found as an endpoint) and when $x = 20$.

Evaluate M at the critical numbers and endpoints:

x	M
0	0
20	$\frac{4000}{3}$
30	0

The maximum of M occurs at $x = 20$.

The corresponding value of y is $\frac{30-20}{3}$ or $\frac{10}{3}$.

Your Turn 2

The first steps of the solution follow Example 2. As in Example 2, $300 - x$ will be the distance the professor must run along the trail. The first change in the model occurs when we compute the professor's total time. Since his speed through the woods is now 40 m/min rather than 70 m/min, his total time $T(x)$ is now

$$T(x) = \frac{300 - x}{160} + \frac{\sqrt{800^2 + x^2}}{40}.$$

As before, the domain of T is $[0, 300]$. The derivative of T is

$$T'(x) = -\frac{1}{160} + \frac{1}{40}\left(\frac{1}{2}\right)(800^2 + x^2)^{-1/2}(2x)$$

Now find the critical numbers by setting the derivative of T equal to 0.

$$-\frac{1}{160} + \frac{1}{40}\left(\frac{1}{2}\right)(800^2 + x^2)^{-1/2}(2x) = 0$$

$$\frac{x}{40\sqrt{800^2 + x^2}} = \frac{1}{160}$$

$$4x = \sqrt{800^2 + x^2}$$

$$16x^2 = 800^2 + x^2$$

$$15x^2 = 800^2$$

$$x = \frac{800}{\sqrt{15}}$$

$$x \approx 206.56$$

Calculate the total time at this critical number and at the endpoints of the domain.

x	T
0	21.875
206.56	21.24
300	21.36

The minimum travel time occurs for $x = 206.56$. The professor runs $300 - 206.56$ meters or about 93 meters along the path and then heads into the woods.

Your Turn 3

We follow the procedure of Example 3, but now our volume function is

$$V(x) = x(8 - 2x)^2 = 4(16x - 8x^2 + x^3).$$

For nonnegative side lengths we require $x \geq 0$ and $8 - 2x \geq 0$ or $x \leq 4$; the domain of V is thus $[0, 4]$.

Set the derivative of V equal to 0 and solve for x.

$$V'(x) = 4(16 - 16x + 3x^2) = 4(3x - 4)(x - 4)$$

This derivative has two positive roots, $x = 4$ and $x = 4/3$. We have already identified one of these as an endpoint of the domain and the other lies within the domain. Evaluating V at 0, 4/3 and 4 gives the following table.

x	V
0	0
4/3	$\frac{1024}{27}$
4	0

The maximum volume is $1024/27$ m^3 and occurs when $x = 4/3$ m.

Note that we can solve this problem efficiently using a scaling argument. Since the proportions of the largest-volume box should not depend on the linear scale we adopt (that is, on the units in which we measure length), we can just note that our piece of metal is $8/12$ or the $2/3$ size of the one in Example 3. Our minimizing length x will be $2/3$ of the value we found in Example 3, or $(2/3)(2) = 4/3$ m.

Your Turn 4

Follow the procedure of Example 4 with a volume of 500 cm^3 instead of 1000 cm^3.

$$V = \pi r^2 h = 500$$

so

$$h = \frac{500}{\pi r^2}$$

and

$$S = 2\pi r^2 + 2\pi r \frac{500}{\pi r^2}$$

$$= 2\pi r^2 + \frac{1000}{r}.$$

Excluding a radius of 0 gives a domain for S of $(0, \infty)$. Now find the critical numbers of S.

$$S' = 4\pi r - \frac{1000}{r^2}$$

$$4\pi r = \frac{1000}{r^2}$$

$$r^3 = \frac{250}{\pi}$$

$$r = \left(\frac{250}{\pi}\right)^{1/3} \approx 4.301$$

We can verify that S' is negative to the left of 4.3 and positive to the right (for example, $S'(4) \approx -12.2$ and $S'(5) \approx 22.8$). Thus the function S is decreasing as we move toward 4.3 from left and increasing as we move past 4.3 to the right, so there is a relative minimum at 4.3. Since there is only one critical number, the critical point theorem tells us that it corresponds to an absolute minimum of the area function.

Then $h \approx \frac{500}{\pi(4.301)^2} \approx 8.604$. The minimum surface area is obtained with a radius of 4.3 cm and a height of 8.6 cm.

As with the box in Your Turn 3, we could obtain this answer directly from Example 4 using a scaling argument. Changing the volume by a factor of $1/2$ (from 1000 to 500) scales all linear measures by a factor of $(1/2)^{1/3}$, so we can find the new r and h by dividing the values found in Example 4 by $(1/2)^{1/3}$:

$$r = \frac{5.419}{2^{1/3}} \approx 4.301$$

$$h = \frac{10.84}{2^{1/3}} \approx 8.604$$

6.2 Warmup Exercises

W1. $f(x) = 5x^4 - 18x^3 - 28x^2 + 12$

$f'(x) = 20x^3 - 54x^2 - 56x$

$\qquad = (2x)\left(10x^2 - 27x - 28\right)$

$\qquad = (2x)(5x + 4)(2x - 7)$

$f'(x) = 0$ when

$2x = 0$ or $5x + 4 = 0$ or $2x - 7 = 0$.

Thus the critical numbers are

$$0, -\frac{4}{5}, \frac{7}{2}.$$

W2. $f(x) = \frac{2x + 1}{x^2 + 3}$

$f'(x) = \dfrac{(2)(x^2 + 3) - (2x + 1)(2x)}{(x^2 + 3)^2}$

$\qquad = \dfrac{2x^2 + 6 - 4x^2 - 2x}{(x^2 + 3)^2}$

$\qquad = \dfrac{(-2)(x^2 + x - 3)}{(x^2 + 3)^2}$

$f'(x) = 0$ when $x^2 + x - 3 = 0$.
The roots of this equation are given by the quadratic formula as $x = \dfrac{-1 \pm \sqrt{13}}{2}$,

so the critical numbers include $\dfrac{-1 \pm \sqrt{13}}{2}$.

Since f' is defined everywhere, these will be the only critical numbers.

6.2 Exercises

1. $x + y = 180, P = xy$

(a) $y = 180 - x$

(b) $P = xy = x(180 - x)$

(c) Since $y = 180 - x$ and x and y are nonnegative numbers, $x \geq 0$ and $180 - x \geq 0$ or $x \leq 180$. The domain of P is $[0, 180]$.

(d) $P'(x) = 180 - 2x$

$$180 - 2x = 0$$
$$2(90 - x) = 0$$
$$x = 90$$

(e)

x	P
0	0
90	8100
180	0

(f) From the chart, the maximum value of P is 8100; this occurs when $x = 90$ and $y = 90$.

3. $x + y = 90$

Minimize $x^2 y$.

(a) $y = 90 - x$

(b) Let $P = x^2 y = x^2(90 - x)$
$$= 90x^2 - x^3.$$

(c) Since $y = 90 - x$ and x and y are nonnegative numbers, the domain of P is $[0, 90]$.

(d) $P' = 180x - 3x^2$
$$180x - 3x^2 = 0$$
$$3x(60 - x) = 0$$
$$x = 0 \text{ or } x = 60$$

(e)

x	P
0	0
60	108,000
90	0

(f) The maximum value of $x^2 y$ occurs when $x = 60$ and $y = 30$. The maximum value is 108,000.

5. $C(x) = \dfrac{1}{2}x^3 + 2x^2 - 3x + 35$

The average cost function is

$$A(x) = \bar{C}(x) = \frac{C(x)}{x}$$

$$= \frac{\frac{1}{2}x^3 + 2x^2 - 3x + 35}{x}$$

$$= \frac{1}{2}x^2 + 2x - 3 + \frac{35}{x}$$

$$\text{or } \frac{1}{2}x^2 + 2x - 3 + 35x^{-1}.$$

Then

$$A'(x) = x + 2 - 35x^{-2}$$

$$\text{or } x + 2 - \frac{35}{x^2}.$$

Graph $y = A'(x)$ on a graphing calculator. A suitable choice for the viewing window is $[0,10]$ by $[-10,10]$. (Negative values of x are not meaningful in this application.) Using the calculator, we see that the graph has an x-intercept at $x \approx 2.722$. Thus, 2.722 is a critical number.

Now graph $y = A(x)$ and use this graph to confirm that a minimum occurs at $x \approx 2.722$.

Thus, the average cost is smallest at $x \approx 2.722$. At this value of x, $A \approx 19.007$.

7. $p(x) = 160 - \dfrac{x}{10}$

(a) Revenue from sale of x thousand candy bars:

$$R(x) = 1000xp$$

$$= 1000x\left(160 - \frac{x}{10}\right)$$

$$= 160,000x - 100x^2$$

(b) $R'(x) = 160,000 - 200x$

$$160,000 - 200x = 0$$
$$800 = x$$

The maximum revenue occurs when 800 thousand bars are sold.

(c) $R(800) = 160,000(800) - 100(800)^2$
$$= 64,000,000$$

The maximum revenue is 64,000,000 cents or $640,000.

9. Let $x =$ the width and $y =$ the length.

(a) The perimeter is

$$P = 2x + y$$
$$= 1400,$$

so

$$y = 1400 - 2x.$$

(b) Area $= xy = x(1400 - 2x)$
$$A(x) = 1400x - 2x^2$$

(c) $A' = 1400 - 4x$

$$1400 - 4x = 0$$
$$1400 = 4x$$
$$350 = x$$

$A'' = -4$, which implies that $x = 350$ m leads to the maximum area.

(d) If $x = 350$,

$$y = 1400 - 2(350) = 700.$$

The maximum area is $(350)(700)$

$$= 245,000 \text{ m}^2.$$

11. Let $x = $ the width of the rectangle

$y = $ the total length of the rectangle.

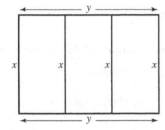

An equation for the fencing is

$$3600 = 4x + 2y$$
$$2y = 3600 - 4x$$
$$y = 1800 - 2x.$$

$$\text{Area} = xy = x(1800 - 2x)$$
$$A(x) = 1800x - 2x^2$$
$$A' = 1800 - 4x$$
$$1800 - 4x = 0$$
$$1800 = 4x$$
$$450 = x$$

$A'' = -4$, which implies that $x = 450$ is the location of a maximum.

If $x = 450$, $y = 1800 - 2(450) = 900$.

The maximum area is

$$(450)(900) = 405,000 \text{ m}^2.$$

13. Let $x = $ length at \$1.50 per meter

$y = $ width at \$3 per meter.

$$xy = 25,600$$
$$y = \frac{25,600}{x}$$

$$\text{Perimeter} = x + 2y = x + \frac{51,200}{x}$$

$$\text{Cost} = C(x) = x(1.5) + \frac{51,200}{x}(3)$$

$$= 1.5x + \frac{153,600}{x}$$

Minimize cost:

$$C'(x) = 1.5 - \frac{153,600}{x^2}$$

$$1.5 - \frac{153,600}{x^2} = 0$$

$$1.5 = \frac{153,600}{x^2}$$

$$1.5x^2 = 153,600$$

$$x^2 = 102,400$$

$$x = 320$$

$$y = \frac{25,600}{320} = 80$$

320 m at \$1.50 per meter will cost \$480. 160 m at \$3 per meter will cost \$480. The total cost will be \$960.

15. Let $x = $ the number of refunds.

Then $90 + x = $ the number of passengers.

(a) Revenue $= R(x) = (90 + x)(1600 - 10x)$

$$= 144,000 + 700x - 10x^2$$

Assume that the number of refunds is nonnegative and that the number of refunds is limited to 160 so that the revenue will be nonnegative. Thus the domain of R is [0,160]. Now set the derivative of R equal to 0 and solve.

$$R' = 700 - 20x = 0$$
$$x = 35$$

Check the value of R at this critical number and at the endpoints of the domain:

x	R
0	144,000
35	156,250
160	0

Thus the maximum revenue is obtained with 35 refunds, which happens when there are 125 passengers.

(b) The maximum revenue is \$156,250.

17. Let x = the number of days to wait.

$$\frac{12,000}{100} = 120 = \text{the number of 100-lb groups collected already.}$$

Then $7.5 - 0.15x$ = the price per 100 lb;

$\quad 4x$ = the number of 100-lb groups collected per day;

$\quad 120 + 4x$ = total number of 100-lb groups collected.

$$\text{Revenue} = R(x)$$
$$= (7.5 - 0.15x)(120 + 4x)$$
$$= 900 + 12x - 0.6x^2$$

$$R'(x) = 12 - 1.2x = 0$$
$$x = 10$$

$R''(x) = -1.2 < 0$ so $R(x)$ is maximized at $x = 10$.

The scouts should wait 10 days at which time their income will be maximized at

$$R(10) = 900 + 12(10) - 0.6(10)^2 = \$960.$$

19. Let x = a side of the base

$\quad h$ = the height of the box.

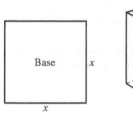 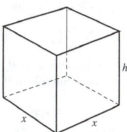

An equation for the volume of the box is

$$V = x^2h,$$
$$\text{so} \quad 32 = x^2h$$
$$h = \frac{32}{x^2}.$$

The box is open at the top so the area of the surface material $m(x)$ in square inches is the area of the base plus the area of the four sides.

$$m(x) = x^2 + 4xh$$
$$= x^2 + 4x\left(\frac{32}{x^2}\right)$$
$$= x^2 + \frac{128}{x}$$
$$m'(x) = 2x - \frac{128}{x^2}$$

$$\frac{2x^3 - 128}{x^2} = 0$$
$$2x^3 - 128 = 0$$
$$2(x^3 - 64) = 0$$
$$x = 4$$

$m'(x) = 2 + \frac{256}{x^3} > 0$ since $x > 0$.

So, $x = 4$ minimizes the surface material. If $x = 4$.

$$h = \frac{32}{x^2} = \frac{32}{16} = 2.$$

The dimensions that will minimize the surface material are 4 in. by 4 in. by 2 in.

21. Let $\quad x$ = the length of the side of the cutout square.

Then $3 - 2x$ = the width of the box and $8 - 2x$ = the length of the box.

$$V(x) = x(3 - 2x)(8 - 2x)$$
$$= 4x^3 - 22x^2 + 24x$$

The domain of V is $\left(0, \frac{3}{2}\right)$.

Maximize the volume.

$$V'(x) = 12x^2 - 44x + 24$$
$$12x^2 - 44x + 24 = 0$$
$$4(3x^2 - 11x + 6) = 0$$
$$4(3x - 2)(x - 3) = 0$$
$$x = \frac{2}{3} \quad \text{or} \quad x = 3$$

3 is not in the domain of V.

$$V''(x) = 24x - 44$$
$$V''\left(\frac{2}{3}\right) = -28 < 0$$

This implies that V is maximized when $x = \frac{2}{3}$.

The box will have maximum volume when $x = \frac{2}{3}$ ft or 8 in.

23. Let $\quad x \quad = \quad$ the length of a side of the top and bottom.

Then $\quad x^2 \quad = \quad$ the area of the top and bottom

and $\quad (3)(2x^2) \quad = \quad$ the cost for the top and bottom

Let $\quad y \quad = \quad$ depth of box.

Then $\quad xy \quad = \quad$ the area of one side,

$\quad 4xy \quad = \quad$ the total area of the sides.

and $\quad (1.50)(4xy) \quad = \quad$ The cost of the sides

The total cost is

$$C(x) = (3)(2x^2) + (1.50)(4xy) = 6x^2 + 6xy.$$

The volume is

$$V = 16{,}000 = x^2 y.$$

$$y = \frac{16{,}000}{x^2}$$

$$C(x) = 6x^2 + 6x\left(\frac{16{,}000}{x^2}\right) = 6x^2 + \frac{96{,}000}{x}$$

$$C'(x) = 12x - \frac{96{,}000}{x^2} = 0$$

$$x^3 = 8000$$

$$x = 20$$

$$C''(x) = 12 + \frac{192{,}000}{x^3} > 0 \text{ at } x = 20, \text{ which}$$

implies that $C(x)$ is minimized when $x = 20$.

$$y = \frac{16{,}000}{(20)^2} = 40$$

So the dimensions of the box are x by x by y, or 20 cm by 20 cm by 40 cm.

$$C(20) = 6(20)^2 + \frac{96{,}000}{20} = 7200$$

The minimum total cost is $7200.

25. (a) $\quad S = 2\pi r^2 + 2\pi rh, \ V = \pi r^2 h$

$$S = 2\pi r^2 + \frac{2V}{r}$$

Treat V as a constant.

$$S' = 4\pi r - \frac{2V}{r^2}$$

$$4\pi r - \frac{2V}{r^2} = 0$$

$$\frac{4\pi r^3 - 2V}{r^2} = 0$$

$$4\pi r^3 - 2V = 0$$

$$2\pi r^3 - V = 0$$

$$2\pi r^3 = V$$

$$2\pi r^3 = \pi r^2 h$$

$$2r = h$$

27. Let x be the number of feet over the minimum 2-foot length and width. The revenue $r(x)$ is then the number sold times the area in square feet times the price per square foot:

$$r(x) = (100 - 6x)(2 + x)^2(5)$$

$$= -30x^3 + 380x^2 + 1880x + 2000$$

$$r'(x) = -90x^2 + 760x + 1880$$

$$= (-10)(x + 2)(9x - 94)$$

The domain of both r and r' is $[0, 16.5]$ since 2-by-2 is the minimum size and the number sold must be nonnegative.

The only positive root of $r'(x)$ is $94/9 \approx 10.444$.

Since $r''(x) = -180x + 760$

and $r''\left(\dfrac{94}{9}\right) = -1120 < 0$,

94/9 represents a local maximum for r. Since x must be a multiple of 0.5, we evaluate r at $x = 10$ and $x = 10.5$.

$$r(10) = 28{,}800$$

$$r(10.5) = 28{,}906.25$$

Check the value of r at the endpoints of the domain of r, which are also possible locations for an absolute maximum.

$$r(0) = 2000 \text{ and } r(16.5) = 1711.25$$

Thus the maximum revenue of $28,906.25 is obtained when the carpets have length and width equal to $2 + 10.5 = 12.5$ feet.

29. From Example 4, we know that the surface area of the can is given by

$$S = 2\pi r^2 + \frac{2000}{r}.$$

Aluminum costs 3¢/cm^2, so the cost of the aluminum to make the can is

$$0.03\left(2\pi r^2 + \frac{2000}{r}\right) = 0.06\pi r^2 + \frac{60}{r}.$$

The perimeter (or circumference) of the circular top is $2\pi r$. Since there is a 2¢/cm charge to seal the top and bottom, the sealing cost is

$$0.02(2)(2\pi r) = 0.08\pi r.$$

Thus, the total cost is given by the function

$$C(r) = 0.06\pi r^2 + \frac{60}{r} + 0.08\pi r$$

$$= 0.06\pi r^2 + 60r^{-1} + 0.08\pi r.$$

Then

$$C'(r) = 0.12\pi r - 60r^{-2} + 0.08\pi$$

$$= 0.12\pi r - \frac{60}{r^2} + 0.08\pi.$$

Graph

$$y = 0.12\pi x - \frac{60}{x^2} + 0.08\pi$$

on a graphing calculator. Since r must be positive in this application, our window should not include negative values of x. A suitable choice for the viewing window is $[0,10]$ by $[-10,10]$. From the graph, we find that $C'(x) = 0$ when $x \approx 5.206$. Thus, the cost is minimized when the radius is about 5.206 cm.

We can find the corresponding height by using the equation

$$h = \frac{1000}{\pi r^2}$$

from Example 4.

If $r = 5.206$.

$$h = \frac{1000}{\pi(5.206)^2} \approx 11.75.$$

To minimize cost, the can should have radius 5.206 cm and height 11.75 cm.

31. In Exercise 29 and 30, we found that the cost of the aluminum to make the can is $0.06\pi r^2 + \frac{60}{r}$, the cost to seal the top and bottom is $0.08\pi r$, and the cost to seal the vertical seam is $\frac{10}{\pi r^2}$. Thus, the total cost is now given by the function

$$C(r) = 0.06\pi r^2 + \frac{60}{r} + 0.08\pi r + \frac{10}{\pi r^2}$$

$$\text{or } 0.06\pi r^2 + 60r^{-1} + 0.08\pi r + \frac{10}{\pi}r^{-2}.$$

Then

$$C'(r) = 0.12\pi r - 60r^{-2} + 0.80\pi - \frac{20}{\pi}r^{-3}$$

$$\text{or } 0.12\pi r - \frac{60}{r^2} + 0.08\pi - \frac{20}{\pi r^3}.$$

Graph

$$y = 0.12\pi r - \frac{60}{r^2} + 0.08\pi - \frac{20}{\pi r^3}$$

on a graphing calculator. A suitable choice for the viewing window is $[0,10]$ by $[-10,10]$. From the graph, we find that $C'(x) = 0$ when $x \approx 5.242$. Thus, the cost is minimized when the radius is about 5.242 cm. To find the corresponding height, use the equation

$$h = \frac{1000}{\pi r^2}$$

from Example 4.

If $r = 5.242$,

$$h = \frac{1000}{\pi(5.242)^2} \approx 11.58.$$

To minimize cost, the can should have radius 5.242 cm and height 11.58 cm.

33. Distance on shore: $9 - x$ miles

Cost on shore: $400 per mile

Distance underwater: $\sqrt{x^2 + 36}$

Cost underwater: $500 per mile

Find the distance from A, that is, $(9 - x)$, to minimize cost, $C(x)$.

$$C(x) = (9 - x)(400) + \left(\sqrt{x^2 + 36}\right)(500)$$

$$= 3600 - 400x + 500(x^2 + 36)^{1/2}$$

$$C'(x) = -400 + 500\left(\frac{1}{2}\right)(x^2 + 36)^{-1/2}(2x)$$

$$= -400 + \frac{500x}{\sqrt{x^2 + 36}}$$

If $C'(x) = 0$,

$$\frac{500x}{\sqrt{x^2 + 36}} = 400$$

$$\frac{5x}{4} = \sqrt{x^2 + 36}$$

$$\frac{25}{16}x^2 = x^2 + 36$$

$$\frac{9}{16}x^2 = 36$$

$$x = \frac{6 \cdot 4}{3} = 8.$$

(Discard the negative solution.)

Then the distance should be

$$9 - x = 9 - 8$$

$$= 1 \text{ mile from point } A.$$

35. $p(t) = \dfrac{20t^3 - t^4}{1000}, \; [0, 20]$

(a) $p'(t) = \dfrac{3}{50}t^2 - \dfrac{1}{250}t^3$

$$= \frac{1}{50}t^2\left[3 - \frac{1}{5}t\right]$$

Critical numbers:

$$\frac{1}{50}t^2 = 0 \quad \text{or} \quad 3 - \frac{1}{5}t = 0$$

$$t = 0 \quad \text{or} \qquad t = 15$$

t	$p(t)$
0	0
15	16.875
20	0

The number of people infected reaches a maximum in 15 days.

(b) $P(15) = 16.875\%$

37. $H(S) = f(S) - S$

$$f(S) = 12S^{0.25}$$

$$H(S) = 12S^{0.25} - S$$

$$H'(S) = 3S^{-0.75} - 1$$

$$H'(S) = 0 \text{ when}$$

$$3S^{-0.75} - 1 = 0$$

$$S^{-0.75} = \frac{1}{3}$$

$$\frac{1}{S^{0.75}} = \frac{1}{3}$$

$$S^{3/4} = 3$$

$$S = 3^{4/3}$$

$$S = 4.327.$$

The number of creatures needed to sustain the population is $S_0 = 4.327$ thousand.

$H''(S) = \dfrac{-2.25}{S^{1.75}} < 0$ when $S = 4.327$, so $H(S)$ is maximized.

$$H(4.327) = 12(4.327)^{0.25} - 4.327$$

$$\approx 12.98$$

The maximum sustainable harvest is 12.98 thousand.

39. $N(t) = 20\left[\dfrac{t}{12} - \ln\left(\dfrac{t}{12}\right)\right] + 30;$

$$1 \le t \le 15$$

$$N'(t) = 20\left[\frac{1}{12} - \frac{12}{t}\left(\frac{1}{12}\right)\right]$$

$$= 20\left(\frac{1}{12} - \frac{1}{t}\right)$$

$$= \frac{20(t - 12)}{12t}$$

$N'(t) = 0$ when $t = 12$.

$N''(t)$ does not exist at $t = 0$, but 0 is not in the domain of N.

Thus, 12 is the only critical number.

To find the absolute extrema on $[1, 15]$, evaluate N at the critical number and at the endpoints.

t	$N(t)$
1	81.365
12	50
15	50.537

Use this table to answer the questions in (a)-(d).

(a) The number of bacteria will be a minimum at $t = 12$, which represents 12 days.

(b) The minimum number of bacteria is given by $N(12) = 50$, which represents 50 bacteria per mL.

(c) The number of bacteria will be a maximum at $t = 1$, which represents 1 day.

(d) The maximum number of bacteria is give by $N(1) = 81.365$, which represents 81.365 bacteria per mL.

41. $r = 0.1, \; P = 100$

$$f(S) = Se^{r(1 - S/P)}$$

$$f'(S) = -\frac{1}{1000} \cdot Se^{0.1(1 - S/100)} + e^{0.1(1 - S/100)}$$

$$f'(S_0) = -0.001S_0e^{0.1(1 - S_0/100)} + e^{0.1(1 - S_0/100)}$$

Graph

$$Y_1 = -0.001xe^{0.1(1 - x/100)} + e^{0.1(1 - x/100)}$$

and

$$Y_2 = 1$$

on the same screen. A suitable choice for the viewing window is $[0, 60]$ by $[0.5, 1.5]$ with $\text{Xscl} = 10$ and $\text{Yscl} = 0.5$. By zooming or using the "intersect" option, we find the graphs intersect when $x \approx 49.37$.

The maximum sustainable harvest is 49.37.

43. Let $x =$ distance from P to A.

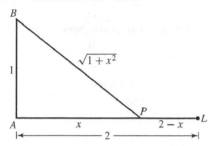

Energy used over land: 1 unit per mile

Energy used over water: $\frac{4}{3}$ units per mile

Distance over land: $(2 - x)$ mi

Distance over water: $\sqrt{1 + x^2}$ mi

Find the location of P to minimize energy used.

$$E(x) = 1(2 - x) + \frac{4}{3}\sqrt{1 + x^2}, \text{ where } 0 \le x \le 2.$$

$$E'(x) = -1 + \frac{4}{3}\left(\frac{1}{2}\right)(1 + x^2)^{-1/2}(2x)$$

If $E'(x) = 0$,

$$\frac{4}{3}x(1 + x^2)^{-1/2} = 1$$

$$\frac{4x}{3(1 + x^2)^{1/2}} = 1$$

$$\frac{4}{3}x = (1 + x^2)^{1/2}$$

$$\frac{16}{9}x^2 = 1 + x^2$$

$$x^2 = \frac{9}{7}$$

$$x = \frac{3}{\sqrt{7}} = \frac{3\sqrt{7}}{7}.$$

x	$E(x)$
0	3.3333
1.134	2.8819
2	2.9814

The absolute minimum occurs at $x \approx 1.134$.

Point P is $\frac{3\sqrt{7}}{7} \approx 1.134$ mi from Point A.

45. (a) $f(S) = aSe^{-bS}$ $\qquad f(S) = Se^{r(1-S/P)}$

$$= e^r Se^{-(r/P)S}$$

Comparing the two terms, replace a with e^r and b with r/P.

(b) Shepherd:

$$f(S) = \frac{aS}{1 + (S/b)^c}$$

$$f'(S) = \frac{[1 + (S/b)^c](a) - (aS)[c(S/b)^{c-1}(1/b)]}{[1 + (S/b)^c]^2}$$

$$= \frac{a + a(S/b)^c - (acS/b)(S/b)^{c-1}}{[1 + (S/b)^c]^2}$$

$$= \frac{a + a(S/b)^c - ac(S/b)^c}{[1 + (S/b)^c]^2}$$

$$= \frac{a[1 + (1 - c)(S/b)^c]}{[1 + (S/b)^c]^2}$$

Ricker:

$$f(S) = aSe^{-bS}$$

$$f'(S) = ae^{-bS} + aSe^{-bS}(-b)$$

$$= ae^{-bS}(1 - bS)$$

Berverton-Holt:

$$f(S) = \frac{aS}{1 + (S/b)}$$

$$f'(S) = \frac{[1 + (S/b)](a) - aS(1/b)}{[1 + (S/b)]^2}$$

$$= \frac{a + a(S/b) - a(S/b)}{[1 + (S/b)]^2}$$

$$= \frac{a}{[1 + (S/b)]^2}$$

(c) Shepherd:

$$f'(0) = \frac{a[1 + (1 - c)(0/b)^c]}{[1 + (0/b)^c]^2} = a$$

Ricker:

$$f'(0) = ae^{-b(0)}[1 - b(0)] = a$$

Beverton-Holt:

$$f'(0) = \frac{a}{[1 + (0/b)]^2} = a$$

The constant a represents the slope of the graph of $f(S)$ at $S = 0$.

(d) First find the critical numbers by solving $f'(S) = 0$.

Shepherd:

$$f'(S) = 0$$
$$a[1 + (1 - c)(S/b)^c] = 0$$
$$(1 - c)(S/b)^c = -1$$
$$(c - 1)(S/b)^c = 1$$

Substitute $b = 248.72$ and $c = 3.24$ and solve for S.

$$(3.24 - 1)(S/248.72)^{3.24} = 1$$

$$\left(\frac{S}{248.72}\right)^{3.24} = \frac{1}{2.24}$$

$$\frac{S}{248.72} = \left(\frac{1}{2.24}\right)^{1/3.24}$$

$$S = 248.72\left(\frac{1}{2.24}\right)^{1/3.24} \approx 193.914$$

Using the Shepherd model, next year's population is maximized when this year's population is about 194,000 tons. This can be verified by examing the graph of $f(S)$.

(e) First find the critical numbers by solving $f'(S) = 0$.

Ricker:

$$f'(S) = 0$$
$$ae^{-bS}(1 - bS) = 0$$
$$1 - bS = 0$$
$$S = \frac{1}{b}$$

Substitute $b = 0.0039$ and solve for S.

$$S = \frac{1}{0.0039}$$
$$S \approx 256.410$$

Using the Ricker model, next year's population is maximized when this year's population is about 256,000 tons. This can be verified by examining the graph of $f(S)$.

47. The goal is to minimize surface area for a fixed volume, that is, to

minimize $2\pi r^2 + 2\pi rh$ when $\pi r^2 h = 65$.

$h = \dfrac{65}{\pi r^2}$ so we can write the surface as a function of r alone:

$$S(r) = 2\pi r^2 + 2\pi r\left(\frac{65}{\pi r^2}\right)$$

$$= 2\pi r^2 + \frac{130}{r}$$

Find the critical numbers.

$$S'(r) = 4\pi r - \frac{130}{r^2}$$

$S'(r) = 0$ implies

$$4\pi r = \frac{130}{r^2}$$

$$r^3 = \frac{130}{4\pi}$$

$$r = \left(\frac{130}{4\pi}\right)^{1/3} = 2.1789 \approx 2.18$$

The formula $S(r) = 2\pi r^2 + 2\pi r\left(\dfrac{65}{\pi r^2}\right)$

shows that letting r go to 0 and increasing r without limit both increase the surface area without limit, so the critical value found above represents a local minimum for the surface area. $r = 2.1789$ implies

$$h = \frac{65}{\pi(2.1789)^2} = 4.3580 \approx 4.36.$$

Thus the radius and height of the cell of minimum surface area are $2.18 \ \mu m$ and $4.36 \ \mu m$.

49. Let $\quad 8 - x =$ the distance the hunter will travel on the river.

Then $\sqrt{9 + x^2} =$ the distance he will travel on land.

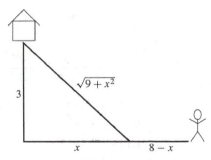

The rate on the river is 5 mph, the rate on land is 2 mph. Using $t = \dfrac{d}{r}$,

$$\frac{8 - x}{5} = \text{the time on the river,}$$

$$\frac{\sqrt{9 - x^2}}{2} = \text{the time on land.}$$

The total time is

$$T(x) = \frac{8 - x}{5} + \frac{\sqrt{9 + x^2}}{2}$$

$$= \frac{8}{5} - \frac{1}{5}x + \frac{1}{2}(9 + x^2)^{1/2}.$$

$$T' = -\frac{1}{5} + \frac{1}{4} \cdot 2x(9 + x^2)^{-1/2}$$

$$-\frac{1}{5} + \frac{x}{2(9 + x^2)^{1/2}} = 0$$

$$\frac{1}{5} = \frac{x}{2(9 + x^2)^{1/2}}$$

$$2(9 + x^2)^{1/2} = 5x$$

$$4(9 + x^2) = 25x^2$$

$$36 + 4x^2 = 25x^2$$

$$36 = 21x^2$$

$$\frac{6}{\sqrt{21}} = x$$

$$\frac{6\sqrt{21}}{21} = \frac{2\sqrt{21}}{7} = x$$

x	$T(x)$
0	3.1
$\frac{2\sqrt{21}}{7}$	2.98
8	4.27

Since the maximum time is 2.98 hr, the hunter should travel $8 - \frac{2\sqrt{21}}{7} = \frac{56 - 2\sqrt{21}}{7}$ or about 6.7 miles along the river.

51. Let $x = \text{width.}$

Then $x = \text{height}$

and $108 - 4x = \text{length.}$

(since length plus girth $= 108$)

$$V(x) = l \cdot w \cdot h$$

$$= (108 - 4x)x \cdot x$$

$$= 108x^2 - 4x^3$$

$$V'(x) = 216x - 12x^2$$

Set $V'(x) = 0,$ and solve for $x.$

$$216x - 12x^2 = 0$$

$$12x(18 - x) = 0$$

$$x = 0 \text{ or } x = 18$$

0 is not in the domain, so the only critical number is 18.

$$\text{Width} = 18$$

$$\text{Height} = 18$$

$$\text{Length} = 108 - 4(18) = 36$$

The dimensions of the box with maximum volume are 36 inches by 18 inches by 18 inches.

6.3 Further Business Applications: Economic Lot Size; Economic Order Quantity; Elasticity of Demand

Your Turn 1

Use Equation (3) with $k = 3,$ $M = 18,000$ and $f = 750.$

$$q = \sqrt{\frac{2fM}{k}}$$

$$= \sqrt{\frac{2(750)(18,000)}{3}}$$

$$= \sqrt{(900)(10,000)}$$

$$= 3000$$

To minimize production costs there should be 3000 cans per batch, requiring $18,000/3,000 = 6$ batches per year.

Your Turn 2

Use Equation (3) with $k = 2, M = 320$ and $f = 30.$

$$q = \sqrt{\frac{2(30)(320)}{2}}$$

$$= \sqrt{9600}$$

$$\approx 97.98$$

Since q is very close to 98 we expect a q of 98 to minimize costs, but we will check both 97 and 98 using $T(q) = \frac{fM}{q} + \frac{kq}{2}.$ $T(97) \approx 195.969$ and $T(98)$ $\approx 195.959,$ so the company should order 98 units in each batch. The time between orders will be about $12\left(\frac{98}{320}\right) = 3.675$ months.

Your Turn 3

$$q = 24{,}000 - 3p^2$$

$$\frac{dq}{dp} = -6p$$

$$E = -\frac{p}{q}\frac{dq}{dp} = \frac{6p^2}{24{,}000 - 3p^2}$$

When $p = 50$,

$$E = \frac{6(50^2)}{24{,}000 - 3(50^2)}$$

$$\approx 0.909,$$

which corresponds to inelastic demand.

Your Turn 4

$$q = 200e^{-0.4p}$$

$$\frac{dq}{dp} = -80e^{-0.4p}$$

$$E = -\frac{p}{q}\frac{dq}{dp}$$

$$= \frac{80pe^{-0.4p}}{200e^{-0.4p}}$$

$$= 0.4p$$

For $p = 100$ we have $E = 0.4(100) = 40$, which corresponds to elastic demand.

Your Turn 5

$$q = 3600 - 3p^2$$

$$\frac{dq}{dp} = -6p$$

$$E = -\frac{p}{q}\frac{dq}{dp}$$

$$= \frac{6p^2}{3600 - 3p^2}$$

$$= \frac{2p^2}{1200 - p^2}$$

The demand has unit elasticity when $E = 1$.

$$\frac{2p^2}{1200 - p^2} = 1$$

$$2p^2 = 1200 - p^2$$

$$3p^2 = 1200$$

$$p^2 = 400$$

$$p = 20$$

Testing value for p smaller and larger than 20 (say 15 and 25) we find

$$E(15) < 1$$
$$E(25) > 1$$

Thus demand is inelastic when $p < 20$ and elastic when $p > 20$.

Revenue is maximized at the price corresponding to unit elasticity, which is $p = \$20$. At this price the revenue is

$$pq = (20)[3600 - 3(20^2)]$$
$$= (20)(2400)$$
$$= 48{,}000$$

or $48,000.

6.3   Warmup Exercises

W1. $f(x) = \dfrac{k}{x} = kx^{-1}$

$$f'(x) = k(-1)x^{-2} = -\frac{k}{x^2}$$

W2. $f(x) = \dfrac{a}{x^2} = ax^{-2}$

$$f'(x) = a(-2)x^{-3} = -\frac{2a}{x^3}$$

6.3   Exercises

1. When $q < \sqrt{\dfrac{2fM}{k}}, T'(q) < -\dfrac{k}{2} + \dfrac{k}{2} = 0$; and when $q > \sqrt{\dfrac{2fM}{k}}, T'(q) > -\dfrac{k}{2} + \dfrac{k}{2} = 0$. Since the function $T(q)$ is decreasing before $q = \sqrt{\dfrac{2fM}{k}}$ and increasing after $q = \sqrt{\dfrac{2fM}{k}}$, there must be a relative minimum at $q = \sqrt{\dfrac{2fM}{k}}$. By the critical point theorem, there is an absolute minimum there.

3. The economic order quantity formula assumes that M, the total units needed per year, is known. Thus, c is the correct answer.

5. The demand function $q(p)$ is positive and increasing, so $\dfrac{dq}{dp}$ is positive. Since p_0 and q_0 are also positive, the elasticity $E = -\dfrac{p_0}{q_0} \cdot \dfrac{dq}{dp}$ is negative.

7. $q = m - np$ for $0 \le p \le \dfrac{m}{n}$

$$\frac{dq}{dp} = -n$$

$$E = -\frac{p}{q} \cdot \frac{dq}{dp}$$

$$E = -\frac{p}{m - np}(-n)$$

$$E = \frac{pn}{m - np} = 1$$

$$pn = m - np$$

$$2np = m$$

$$p = \frac{m}{2n}$$

Thus, $E = 1$ when $p = \frac{m}{2n}$, or at the midpoint of the demand curve on the interval $0 \le p \le \frac{m}{n}$.

9. Use equation (3) with $k = 1$, $M = 100,000$, and $f = 500$.

$$q = \sqrt{\frac{2fM}{k}} = \sqrt{\frac{2(500)(100,000)}{1}}$$

$$= \sqrt{100,000,000} = 10,000$$

10,000 lamps should be made in each batch to minimize production costs.

11. From Exercise 9, $M = 100,000$, and $q = 10,000$. The number of batches per year is

$$\frac{M}{q} = \frac{100,000}{10,000} = 10.$$

13. Here $k = 0.50$, $M = 100,000$, and $f = 60$. We have

$$q = \sqrt{\frac{2fM}{k}} = \sqrt{\frac{2(60)(100,000)}{0.50}}$$

$$= \sqrt{24,000,000} \approx 4898.98$$

$T(4898) = 2449.489792$ and $T(4899) = 2449.489743$, so ordering 4899 copies per order minimizes the annual costs.

15. Using maximum inventory size,

$$T(q) = \frac{fM}{q} + gM + kq; \ (0, \infty)$$

$$T'(q) = \frac{-fM}{q^2} + k$$

Set the derivative equal to 0.

$$\frac{-fM}{q^2} + k = 0$$

$$k = \frac{fM}{q^2}$$

$$q^2 k = fM$$

$$q^2 = \frac{fM}{k}$$

$$q = \sqrt{\frac{fM}{k}}$$

Since $\lim\limits_{q \to 0} T(q) = \infty$,

$\lim\limits_{q \to \infty} T(q) = \infty$, and

$q = \sqrt{\dfrac{fM}{k}}$ is the only critical value in $(0, \infty)$,

$q = \sqrt{\dfrac{fM}{k}}$ is the number of unit that should be ordered or manufactured to minimize total costs.

17. Assuming an annual cost, k_1, for storing a single unit, plus an annual cost per unit, k_2, that must be paid for each unit up to the maximum number of units stored, we have

$$T(q) = \frac{fM}{q} + gM + \frac{k_1 q}{2} + k_2 q; \ (0, \infty)$$

$$T'(q) = \frac{-fM}{q^2} + \frac{k_1}{2} + k_2$$

Set the derivative equal to 0.

$$\frac{-fM}{q^2} + \frac{k_1}{2} + k_2 = 0$$

$$\frac{k_1 + 2k_2}{2} = \frac{fM}{q^2}$$

$$\frac{q^2(k_1 + 2k_2)}{2} = fM$$

$$q^2 = \frac{2fM}{k_1 + 2k_2}$$

$$q = \sqrt{\frac{2fM}{k_1 + 2k_2}}$$

Since $\lim\limits_{q \to 0} T(q) = \infty$, $\lim\limits_{q \to \infty} T(q) = \infty$, and

$q = \sqrt{\dfrac{2fM}{k_1 + 2k_2}}$ is the only critical value in $(0, \infty)$,

$q = \sqrt{\dfrac{2fM}{k_1 + 2k_2}}$ is the number of units that should be ordered or manufactured to minimize the total cost in this case.

19. $q = 50 - \dfrac{p}{4}$

  (a) $\dfrac{dq}{dp} = -\dfrac{1}{4}$

$$E = -\dfrac{p}{q} \cdot \dfrac{dq}{dp}$$

$$= -\dfrac{p}{50 - \frac{p}{4}}\left(-\dfrac{1}{4}\right)$$

$$= -\dfrac{p}{\frac{200-p}{4}}\left(-\dfrac{1}{4}\right)$$

$$= \dfrac{p}{200 - p}$$

  (b) $R = pq$

$$\dfrac{dR}{dp} = q(1 - E)$$

  When R is maximum, $q(1 - E) = 0$.

  Since $q = 0$ means no revenue, set
  $1 - E = 0$, so $E = 1$.

  From (a),

$$\dfrac{p}{200 - p} = 1$$

$$p = 200 - p$$

$$p = 100.$$

$$q = 50 - \dfrac{p}{4}$$

$$= 50 - \dfrac{100}{4}$$

$$= 25$$

  Total revenue is maximized if $q = 25$.

21. (a) $q = 37{,}500 - 5p^2$

$$\dfrac{dq}{dp} = -10p$$

$$E = \dfrac{-p}{q} \cdot \dfrac{dq}{dp}$$

$$= \dfrac{-p}{37{,}500 - 5p^2}(-10p)$$

$$= -\dfrac{10p^2}{37{,}500 - 5p^2}$$

$$= \dfrac{2p^2}{7500 - p^2}$$

  (b) $R = pq$

$$\dfrac{dR}{dp} = q(1 - E)$$

  When R is maximum, $q(1 - E) = 0$. Since
  $q = 0$ means no revenue, set $1 - E = 0$.

$$E = 1$$

  From (a),

$$\dfrac{2p^2}{7500 - p^2} = 1$$

$$2p^2 = 7500 - p^2$$

$$3p^2 = 7500$$

$$p^2 = 2500$$

$$p = \pm 50.$$

  Since p must be positive, $p = 50$.

$$q = 37{,}500 - 5p^2$$

$$= 37{,}500 - 5(50)^2$$

$$= 37{,}500 - 5(2500)$$

$$= 37{,}500 - 12{,}500$$

$$= 25{,}000.$$

23. $p = 400e^{-0.2q}$

  In order to find the derivative $\dfrac{dq}{dp}$, we first need to solve for q in the equation $p = 400e^{-0.2q}$.

  (a) $\dfrac{p}{400} = e^{-0.2q}$

$$\ln\left(\dfrac{p}{400}\right) = \ln\left(e^{-0.2q}\right) = -0.2q$$

$$q = \dfrac{\ln\frac{p}{400}}{-0.2} = -5\ln\left(\dfrac{p}{400}\right)$$

  Now

$$\dfrac{dq}{dp} = -5\dfrac{1}{\frac{p}{400}} \cdot \dfrac{1}{400} = \dfrac{-5}{p}, \text{ and}$$

$$E = -\dfrac{p}{q} \cdot \dfrac{dq}{dp} = -\dfrac{p}{q} \cdot \dfrac{-5}{p} = \dfrac{5}{q}.$$

  (b) $R = pq$

$$\dfrac{dR}{dp} = q(1 - E)$$

  When R is maximum, $q(1 - E) = 0$. Since
  $q = 0$ means no revenue, set $1 - E = 0$.

$$E = 1$$

From part (a),

$$\frac{5}{q} = 1$$

$$5 = q$$

25. $q = 400 - 0.2p^2$

$$\frac{dq}{dp} = 0 - 0.4p$$

$$E = -\frac{p}{q} \cdot \frac{dq}{dp}$$

$$E = -\frac{P}{400 - 0.2p^2}(-0.4p)$$

$$= \frac{0.4p^2}{400 - 0.2p^2}$$

(a) If $p = \$20$,

$$E = \frac{(0.4)(20)^2}{400 - 0.2(20)^2}$$

$$= 0.5.$$

Since $E < 1$, demand is inelastic. This indicates that total revenue increases as price increases.

(b) If $p = \$40$,

$$E = \frac{(0.4)(40)^2}{400 - 0.2(40)^2}$$

$$= 8.$$

Since $E > 1$, demand is elastic. This indicates that total revenue decreases as price increases.

27. $q = 2,431,129p^{-0.06}$

$$\frac{dq}{dp} = (-0.06)[2,431,129p^{-1.06}]$$

$$E = -\frac{p}{q}\frac{dq}{dp}$$

$$= -\frac{p}{2,431,129p^{-0.06}}(-0.06)[2,431,129p^{-1.06}]$$

$$= 0.06$$

At any price (including $100/barrel) the elasticity is 0.06 and the demand is inelastic.

29. $q = 3,751,000p^{-2.826}$

$$\frac{dq}{dp} = (-2.826)[3,751,000p^{-3.826}]$$

$$E = -\frac{p}{q}\frac{dq}{dp}$$

$$= -\frac{p}{3,751,000p^{-2.826}}(-2.826)$$

$$[3,751,000p^{-3.826}]$$

$$= 2.826$$

At any price the elasticity is 2.826 and the demand is elastic.

31. (a) $q = 55.2 - 0.022p$

$$\frac{dq}{dp} = -0.022$$

$$E = -\frac{p}{q} \cdot \frac{dq}{dp}$$

$$= \frac{-p}{55.2 - 0.022p} \cdot (-0.022)$$

$$= \frac{0.022p}{55.2 - 0.022p}$$

When $p = \$166.10$,

$$E = \frac{3.6542}{55.2 - 3.6542} \approx 0.071.$$

(b) Since $E < 1$, the demand for airfare is inelastic at this price.

(c) $R = pq$

$$\frac{dR}{dp} = q(1 - E)$$

When R is maximum, $q(1 - E) = 0$.

Since $q = 0$ means no revenue, set

$$1 - E = 0.$$

$$E = 1$$

From (a),

$$\frac{0.022p}{55.2 - 0.022p} = 1$$

$$0.022p = 55.2 - 0.022p$$

$$0.044p = 55.2$$

$$p \approx 1255$$

Total revenue is maximized if $p \approx \$1255$.

33.

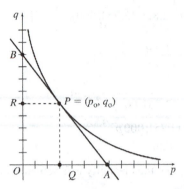

In the figure, we label P_0 as P. The slope of the tangent line is

$$\frac{dq}{dp} = -\frac{BR}{RP} = \frac{OB - OR}{-RP} = \frac{OB - q_0}{-p_0}$$

or

$$-p_0 \frac{dq}{dp} = OB - q_0$$

$$-\frac{p_0}{q_0} \cdot \frac{dq}{dp} = \frac{OB}{q_0} - 1$$

$$= \frac{OB}{OR} - 1$$

Because triangles AOB and PRB are similar,

$$-\frac{p_0}{q_0} \cdot \frac{dq}{dp} = \frac{AB}{AP} - 1$$

$$= \frac{AB - AP}{AP}$$

$$= \frac{PB}{PA}$$

But $E = -\frac{p_0}{q_0} \cdot \frac{dq}{dp}$ so the ratio $\frac{PB}{PA}$ equals the elasticity E.

6.4 Implicit Differentiation

Your Turn 1

$$x^2 + y^2 = xy$$

$$\frac{d}{dx}(x^2 + y^2) = \frac{d}{dx}xy$$

$$2x + 2y\frac{dy}{dx} = x\frac{dy}{dx} + y$$

$$(2y - x)\frac{dy}{dx} = y - 2x$$

$$\frac{dy}{dx} = \frac{y - 2x}{2y - x}$$

Your Turn 2

$$xe^y + x^2 = \ln y$$

$$\frac{d}{dx}(xe^y + x^2) = \frac{d}{dx}\ln y$$

$$\frac{d}{dx}xe^y + \frac{d}{dx}x^2 = \frac{d}{dx}\ln y$$

$$e^y + xe^y\frac{dy}{dx} + 2x = \frac{1}{y}\frac{dy}{dx}$$

$$\left(xe^y - \frac{1}{y}\right)\frac{dy}{dx} = -2x - e^y$$

$$(xye^y - 1)\frac{dy}{dx} = -2xy - ye^y$$

$$\frac{dy}{dx} = \frac{-2xy - ye^y}{xye^y - 1} = \frac{ye^y - 2xy}{1 - xye^y}$$

Your Turn 3

$$y^4 - x^4 - y^2 + x^2 = 0$$

$$\frac{d}{dx}(y^4 - x^4 - y^2 + x^2) = \frac{d}{dx}(0)$$

$$4y^3\frac{dy}{dx} - 4x^3 - 2y\frac{dy}{dx} + 2x = 0$$

$$(4y^3 - 2y)\frac{dy}{dx} = 4x^3 - 2x$$

At the point $(1, 1)$, $4y^3 - 2y \neq 0$ so we can divide both sides of the equation above by this factor.

$$\frac{dy}{dx} = \frac{4x^3 - 2x}{4y^3 - 2y}$$

At $(1, 1)$,

$$\frac{dy}{dx} = \frac{4x^3 - 2x}{4y^3 - 2y} = \frac{4 - 2}{4 - 2} = 1$$

so the slope of the tangent line is $m = 1$. Use the point-slope form of the equation of a line.

$$y - y_1 = m(x - x_1)$$
$$y - 1 = 1(x - 1)$$
$$y = x$$

The equation of the tangent line at the point $(1, 1)$ is $y = x$.

Your Turn 4

$$p = \frac{100,000}{q^2 + 100q} = 100,000(q^2 + 100q)^{-1}$$

$$\frac{d}{dp}p = \frac{d}{dp}[100,000\,(q^2 + 100q)^{-1}]$$

$$1 = -100,000\frac{2q + 100}{(q^2 + 100q)^2}\frac{dq}{dp}$$

$$\frac{dq}{dp} = -\frac{(q^2 + 100q)^2}{100,000\,(2q + 100)}$$

Substitute $q = 200$ in this expression for the derivative.

$$\frac{dq}{dp} = -\frac{(q^2 + 100q)^2}{100,000\,(2q + 100)}$$

$$= -\frac{[200^2 + 100(200)]^2}{(100,000)[2(200) + 100]}$$

$$= -72$$

When the price is 200, the rate of change of demand with respect to price is -72 units per unit change in price.

6.4   Warmup Exercises

W1. $f(x) = x\ln(x^2 + 1)$

Use the product rule and write the factor

$\ln(x^2 + 1)$ as $h(g(x))$ with $h(x) = \ln x$ and

$g(x) = x^2 + 1$ and apply the chain rule.

$$f'(x) = (1)\ln(x^2 + 1) + (x)\frac{2x}{x^2 + 1}$$

$$= \ln(x^2 + 1) + \frac{2x^2}{x^2 + 1}$$

W2.   Use the quotient and chain rules, writing

e^{x^3} as $h(g(x))$ with $h(x) = e^x$ and

$g(x) = x^3$.

$$f(x) = \frac{e^{x^3}}{x^2}$$

$$f'(x) = \frac{\left(3x^2 e^{x^3}\right)\left(x^2\right) - \left(e^{x^3}\right)(2x)}{\left(x^2\right)^2}$$

$$= \frac{e^{x^3}\left(3x^4 - 2x\right)}{x^4}$$

$$= \frac{e^{x^3}\left(3x^3 - 2\right)}{x^3}$$

6.4   Exercises

1. $6x^2 + 5y^2 = 36$

$$\frac{d}{dx}(6x^2 + 5y^2) = \frac{d}{dx}(36)$$

$$\frac{d}{dx}(6x^2) + \frac{d}{dx}(5y^2) = \frac{d}{dx}(36)$$

$$12x + 5 \cdot 2y\frac{dy}{dx} = 0$$

$$10y\frac{dy}{dx} = -12x$$

$$\frac{dy}{dx} = -\frac{6x}{5y}$$

3. $8x^2 - 10xy + 3y^2 = 26$

$$\frac{d}{dx}(8x^2 - 10xy + 3y^2) = \frac{d}{dx}(26)$$

$$16x - \frac{d}{dx}(10xy) + \frac{d}{dx}(3y^2) = 0$$

$$16x - 10x\frac{dy}{dx} - y\frac{d}{dx}(10x) + 6y\frac{dy}{dx} = 0$$

$$16x - 10x\frac{dy}{dx} - 10y + 6y\frac{dy}{dx} = 0$$

$$(-10x + 6y)\frac{dy}{dx} = -16x + 10y$$

$$\frac{dy}{dx} = \frac{-16x + 10y}{-10x + 6y}$$

$$\frac{dy}{dx} = \frac{8x - 5y}{5x - 3y}$$

5. $5x^3 = 3y^2 + 4y$

$$\frac{d}{dx}(5x^3) = \frac{d}{dx}(3y^2 + 4y)$$

$$15x^2 = \frac{d}{dx}(3y^2) + \frac{d}{dx}(4y)$$

$$15x^2 = 6y\frac{dy}{dx} + 4\frac{dy}{dx}$$

$$\frac{15x^2}{6y + 4} = \frac{dy}{dx}$$

7. $3x^2 = \dfrac{2-y}{2+y}$

$$\frac{d}{dx}(3x^2) = \frac{d}{dx}\left(\frac{2-y}{2+y}\right)$$

$$6x = \frac{(2+y)\frac{d}{dx}(2-y) - (2-y)\frac{d}{dx}(2+y)}{(2+y)^2}$$

$$6x = \frac{(2+y)\left(-\frac{dy}{dx}\right) - (2-y)\frac{dy}{dx}}{(2+y)^2}$$

$$6x = \frac{-4\frac{dy}{dx}}{(2+y)^2}$$

$$6x(2+y)^2 = -4\frac{dy}{dx}$$

$$-\frac{3x(2+y)^2}{2} = \frac{dy}{dx}$$

9. $2\sqrt{x} + 4\sqrt{y} = 5y$

$$\frac{d}{dx}(2x^{1/2} + 4y^{1/2}) = \frac{d}{dx}(5y)$$

$$x^{-1/2} + 2y^{-1/2}\frac{dy}{dx} = 5\frac{dy}{dx}$$

$$(2y^{-1/2} - 5)\frac{dy}{dx} = -x^{-1/2}$$

$$\frac{dy}{dx} = \frac{x^{-1/2}}{5 - 2y^{-1/2}}\left(\frac{x^{1/2}y^{1/2}}{x^{1/2}y^{1/2}}\right)$$

$$= \frac{y^{1/2}}{x^{1/2}(5y^{1/2} - 2)}$$

$$= \frac{\sqrt{y}}{\sqrt{x}(5\sqrt{y} - 2)}$$

11. $x^4y^3 + 4x^{3/2} = 6y^{3/2} + 5$

$$\frac{d}{dx}(x^4y^3 + 4x^{3/2}) = \frac{d}{dx}(6y^{3/2} + 5)$$

$$\frac{d}{dx}(x^4y^3) + \frac{d}{dx}(4x^{3/2}) = \frac{d}{dx}(6y^{3/2}) + \frac{d}{dx}(5)$$

$$4x^3y^3 + x^4 \cdot 3y^2\frac{dy}{dx} + 6x^{1/2} = 9y^{1/2}\frac{dy}{dx} + 0$$

$$4x^3y^3 + 6x^{1/2} = 9y^{1/2}\frac{dy}{dx} - 3x^4y^2\frac{dy}{dx}$$

$$4x^3y^3 + 6x^{1/2} = (9y^{1/2} - 3x^4y^2)\frac{dy}{dx}$$

$$\frac{4x^3y^3 + 6x^{1/2}}{9y^{1/2} - 3x^4y^2} = \frac{dy}{dx}$$

13. $e^{x^2y} = 5x + 4y + 2$

$$\frac{d}{dx}(e^{x^2y}) = \frac{d}{dx}(5x + 4y + 2)$$

$$e^{x^2y}\frac{d}{dx}(x^2y) = \frac{d}{dx}(5x) + \frac{d}{dx}(4y) + \frac{d}{dx}(2)$$

$$e^{x^2y}\left(2xy + x^2\frac{dy}{dx}\right) = 5 + 4\frac{dy}{dx} + 0$$

$$2xye^{x^2y} + x^2e^{x^2y}\frac{dy}{dx} = 5 + 4\frac{dy}{dx}$$

$$x^2e^{x^2y}\frac{dy}{dx} - 4\frac{dy}{dx} = 5 - 2xye^{x^2y}$$

$$(x^2e^{x^2y} - 4)\frac{dy}{dx} = 5 - 2xye^{x^2y}$$

$$\frac{dy}{dx} = \frac{5 - 2xye^{x^2y}}{x^2e^{x^2y} - 4}$$

15. $x + \ln y = x^2y^3$

$$\frac{d}{dx}(x + \ln y) = \frac{d}{dx}(x^2y^3)$$

$$1 + \frac{1}{y}\frac{dy}{dx} = 2xy^3 + 3x^2y^2\frac{dy}{dx}$$

$$\frac{1}{y}\frac{dy}{dx} - 3x^2y^2\frac{dy}{dx} = 2xy^3 - 1$$

$$\left(\frac{1}{y} - 3x^2y^2\right)\frac{dy}{dx} = 2xy^3 - 1$$

$$\frac{dy}{dx} = \frac{2xy^3 - 1}{\frac{1}{y} - 3x^2y^2}$$

$$= \frac{y(2xy^3 - 1)}{1 - 3x^2y^3}$$

17. $x^2 + y^2 = 25$; tangent at $(-3, 4)$

$$\frac{d}{dx}(x^2 + y^2) = \frac{d}{dx}(25)$$

$$2x + 2y\frac{dy}{dx} = 0$$

$$2y\frac{dy}{dx} = -2x$$

$$\frac{dy}{dx} = -\frac{x}{y}$$

$$m = -\frac{x}{y} = -\frac{-3}{4} = \frac{3}{4}$$

$$y - y_1 = m(x - x_1)$$

$$y - 4 = \frac{3}{4}[x - (-3)]$$

$$4y - 16 = 3x + 9$$

$$4y = 3x + 25$$

$$y = \frac{3}{4}x + \frac{25}{4}$$

19. $x^2y^2 = 1$; tangent at $(-1, 1)$

$$\frac{d}{dx}(x^2y^2) = \frac{d}{dx}(1)$$

$$x^2\frac{d}{dx}(y^2) + y^2\frac{d}{dx}(x^2) = 0$$

$$x^2(2y)\frac{dy}{dx} + y^2(2x) = 0$$

$$2x^2y\frac{dy}{dx} = -2xy^2$$

$$\frac{dy}{dx} = \frac{-2xy^2}{2x^2y} = -\frac{y}{x}$$

$$m = -\frac{y}{x} = -\frac{1}{-1} = 1$$

$$y - 1 = 1[x - (-1)]$$

$$y = x + 1 + 1$$

$$y = x + 2$$

21. $2y^2 - \sqrt{x} = 4$; tangent at $(16, 2)$

$$\frac{d}{dx}(2y^2 - \sqrt{x}) = \frac{d}{dx}(4)$$

$$4y\frac{dy}{dx} - \frac{1}{2}x^{-1/2} = 0$$

$$4y\frac{dy}{dx} = \frac{1}{2x^{1/2}}$$

$$\frac{dy}{dx} = \frac{1}{8yx^{1/2}}$$

$$m = \frac{1}{8yx^{1/2}} = \frac{1}{8(2)(16)^{1/2}}$$

$$= \frac{1}{8(2)(4)} = \frac{1}{64}$$

$$y - 2 = -\frac{1}{4}(x - 4)$$

$$y = -\frac{1}{4}x + 3$$

$$y - 2 = \frac{1}{64}(x - 16)$$

$$64y - 128 = x - 16$$

$$64y = x + 112$$

$$y = \frac{x}{64} + \frac{7}{4}$$

23. $e^{x^2+y^2} = xe^{5y} - y^2e^{5x/2}$; tangent at $(2,1)$

$$\frac{d}{dx}(e^{x^2+y^2}) = \frac{d}{dx}(xe^{5y} - y^2e^{5x/2})$$

$$e^{x^2+y^2} \cdot \frac{d}{dx}(x^2 + y^2) = e^{5y} + x\frac{d}{dx}(e^{5y}) - \left[2y\frac{dy}{dx}e^{5x/2} + y^2e^{5x/2}\frac{d}{dx}\left(\frac{5x}{2}\right)\right]$$

$$e^{x^2+y^2}\left(2x + 2y\frac{dy}{dx}\right) = e^{5y} + x \cdot 5e^{5y}\frac{dy}{dx} - 2ye^{5x/2}\frac{dy}{dx} - \frac{5}{2}y^2e^{5x/2}$$

$$\left(2ye^{x^2+y^2} - 5xe^{5y} + 2ye^{5x/2}\right)\frac{dy}{dx} = -2xe^{x^2+y^2} + e^{5y} - \frac{5}{2}y^2e^{5x/2}$$

$$\frac{dy}{dx} = \frac{-2xe^{x^2+y^2} + e^{5y} - \frac{5}{2}y^2e^{5x/2}}{2ye^{x^2+y^2} - 5xe^{5y} + 2ye^{5x/2}}$$

$$m = \frac{-4e^5 + e^5 - \frac{5}{2}e^5}{2e^5 - 10e^5 + 2e^5} = \frac{-\frac{11}{2}e^5}{-6e^5} = \frac{11}{12}$$

$$y - 1 = \frac{11}{12}(x - 2)$$

$$y = \frac{11}{12}x - \frac{5}{6}$$

25. $\ln(x + y) = x^3y^2 + \ln(x^2 + 2) - 4$; tangent at $(1, 2)$

$$\frac{d}{dx}[\ln(x + y)] = \frac{d}{dx}[x^3y^2 + \ln(x^2 + 2) - 4]$$

$$\frac{1}{x + y} \cdot \frac{d}{dx}(x + y) = 3x^2y^2 + x^3 \cdot 2y\frac{dy}{dx} + \frac{1}{x^2 + 2} \cdot \frac{d}{dx}(x^2 + 2) - \frac{d}{dx}(4)$$

$$\left(\frac{1}{x + y} - 2x^3y\right)\frac{dy}{dx} = 3x^2y^2 + \frac{2x}{x^2 + 2} - \frac{1}{x + y}$$

$$\frac{dy}{dx} = \frac{3x^2y^2 + \frac{2x}{x^2+2} - \frac{1}{x+y}}{\frac{1}{x+y} - 2x^3y}$$

$$m = \frac{3 \cdot 1 \cdot 4 + \frac{2 \cdot 1}{3} - \frac{1}{3}}{\frac{1}{3} - 2 \cdot 1 \cdot 2} = \frac{\frac{37}{3}}{\frac{-11}{3}} = -\frac{37}{11}$$

$$y - 2 = -\frac{37}{11}(x - 1)$$

$$y = -\frac{37}{11}x + \frac{59}{11}$$

27. $y^3 + xy - y = 8x^4$; $x = 1$

First, find the y-value of the point.

$$y^3 + (1)y - y = 8(1)^4$$

$$y^3 = 8$$

$$y = 2$$

The point is $(1, 2)$.

Find $\frac{dy}{dx}$.

$$3y^2\frac{dy}{dx} + x\frac{dy}{dx} + y - \frac{dy}{dx} = 32x^3$$

$$(3y^2 + x - 1)\frac{dy}{dx} = 32x^3 - y$$

$$\frac{dy}{dx} = \frac{32x^3 - y}{3y^2 + x - 1}$$

At $(1, 2)$,

$$\frac{dy}{dx} = \frac{32(1)^3 - 2}{3(2)^2 + 1 - 1} = \frac{30}{12} = \frac{5}{2}.$$

$$y - 2 = \frac{5}{2}(x - 1)$$

$$y - 2 = \frac{5}{2}x - \frac{5}{2}$$

$$y = \frac{5}{2}x - \frac{1}{2}$$

29. $y^3 + xy^2 + 1 = x + 2y^2$; $x = 2$

Find the y-value of the point.

$$y^3 + 2y^2 + 1 = 2 + 2y^2$$

$$y^3 + 1 = 2$$

$$y^3 = 1$$

$$y = 1$$

The point is $(2, 1)$.

Find $\frac{dy}{dx}$.

$$3y^2\frac{dy}{dx} + x2y\frac{dy}{dx} + y^2 = 1 + 4y\frac{dy}{dx}$$

$$3y^2\frac{dy}{dx} + 2xy\frac{dy}{dx} - 4y\frac{dy}{dx} = 1 - y^2$$

$$(3y^2 + 2xy - 4y)\frac{dy}{dx} = 1 - y^2$$

$$\frac{dy}{dx} = \frac{1 - y^2}{3y^2 + 2xy - 4y}$$

At $(2, 1)$,

$$\frac{dy}{dx} = \frac{1 - 1^2}{3(1)^2 + 2(2)(1) - 4(1)} = 0.$$

$$y - 0 = 0(x - 2)$$

$$y = 1$$

31. $2y^3(x - 3) + x\sqrt{y} = 3$; $x = 3$

Find the y-value of the point.

$$2y^3(3 - 3) + 3\sqrt{y} = 3$$

$$3\sqrt{y} = 3$$

$$\sqrt{y} = 1$$

$$y = 1$$

The point is $(3, 1)$

Find $\frac{dy}{dx}$.

$$2y^3(1) + 6y^2(x - 3)\frac{dy}{dx}$$

$$+ x\left(\frac{1}{2}\right)y^{-1/2}\frac{dy}{dx} + \sqrt{y} = 0$$

$$6y^2(x - 3)\frac{dy}{dx} + \frac{x}{2\sqrt{y}}\frac{dy}{dx} = -2y^3 - \sqrt{y}$$

$$\left[6y^2(x - 3) + \frac{x}{2\sqrt{y}}\right]\frac{dy}{dx} = -2y^3 - \sqrt{y}$$

$$\frac{dy}{dx} = \frac{-2y^3 - \sqrt{y}}{6y^2(x - 3) + \frac{x}{2\sqrt{y}}}$$

$$= \frac{-4y^{7/2} - 2y}{12y^{5/2}(x - 3) + x}$$

At $(3, 1)$,

$$\frac{dy}{dx} = \frac{-4(1) - 2}{12(1)(3 - 3) + 3} = \frac{-6}{3} = -2.$$

$$y - 1 = -2(x - 3)$$

$$y - 1 = -2x + 6$$

$$y = -2x + 7$$

33. $x^{2/3} + y^{2/3} = 2$; $(1, 1)$

Find $\frac{dy}{dx}$.

$$\frac{2}{3}x^{-1/3} + \frac{2}{3}y^{-1/3}\frac{dy}{dx} = 0$$

$$\frac{2}{3}y^{-1/3}\frac{dy}{dx} = -\frac{2}{3}x^{-1/3}$$

$$\frac{dy}{dx} = \frac{-\frac{2}{3}x^{-1/3}}{\frac{2}{3}y^{-1/3}}$$

$$= -\frac{y^{1/3}}{x^{1/3}}$$

At $(1, 1)$

$$\frac{dy}{dx} = -\frac{1^{1/3}}{1^{1/3}} = -1$$

$$y - 1 = -1(x - 1)$$

$$y - 1 = -x + 1$$

$$y = -x + 2$$

35. $y^2(x^2 + y^2) = 20x^2$; (1, 2)

Find $\dfrac{dy}{dx}$.

$$2y(x^2 + y^2)\dfrac{dy}{dx} + y^2\left(2x + 2y\dfrac{dy}{dx}\right) = 40x$$

$$2x^2 y\dfrac{dy}{dx} + 2y^3\dfrac{dy}{dx} + 2xy^2 + 2y^3\dfrac{dy}{dx} = 40x$$

$$2x^2 y\dfrac{dy}{dx} + 4y^3\dfrac{dy}{dx} = -2xy^2 + 40x$$

$$(2x^2 y + 4y^3)\left(\dfrac{dy}{dx}\right) = -2xy^2 + 40x$$

$$\dfrac{dy}{dx} = \dfrac{-2xy^2 + 40x}{2x^2 y + 4y^3}$$

At (1, 2),

$$\dfrac{dy}{dx} = \dfrac{-2(1)(2)^2 + 40(1)}{2(1)^2(2) + 4(2)^3}$$

$$= \dfrac{32}{36} = \dfrac{8}{9}$$

$$y - 2 = \dfrac{8}{9}(x - 1)$$

$$y - 2 = \dfrac{8}{9}x - \dfrac{8}{9}$$

$$y = \dfrac{8}{9}x + \dfrac{10}{9}$$

37. $x^2 + y^2 = 100$

(a) Lines are tangent at points where $x = 6$. By substituting $x = 6$ in the equation, we find that the points are $(6, 8)$ and $(6, -8)$.

$$\dfrac{d}{dx}(x^2 + y^2) = \dfrac{d}{dx}(100)$$

$$2x + 2y\dfrac{dy}{dx} = 0$$

$$2y\dfrac{dy}{dx} = -2x$$

$$dy = -\dfrac{x}{y}$$

$$m_1 = -\dfrac{x}{y} = -\dfrac{6}{8} = -\dfrac{3}{4}$$

$$m_2 = -\dfrac{x}{y} = -\dfrac{6}{-8} = \dfrac{3}{4}$$

First tangent:

$$y - 8 = -\dfrac{3}{4}(x - 6)$$

$$y = -\dfrac{3}{4}x + \dfrac{25}{2}$$

Second tangent:

$$y - (-8) = \dfrac{3}{4}(x - 6)$$

$$y + 8 = \dfrac{3}{4}x - \dfrac{18}{4}$$

$$y = \dfrac{3}{4}x - \dfrac{25}{2}$$

(b)

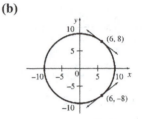

39. (a) $\sqrt{u} + \sqrt{2v + 1} = 5$

$$\dfrac{du}{dv}(\sqrt{u} + \sqrt{2v + 1}) = \dfrac{du}{dv}(5)$$

$$\dfrac{1}{2}u^{-1/2}\dfrac{du}{dv} + \dfrac{1}{2}(2v + 1)^{-1/2}(2) = 0$$

$$\dfrac{1}{2}u^{-1/2}\dfrac{du}{dv} = -\dfrac{1}{(2v + 1)^{1/2}}$$

$$\dfrac{du}{dv} = -\dfrac{2u^{1/2}}{(2v + 1)^{1/2}}$$

(b) $\sqrt{u} + \sqrt{2v + 1} = 5$

$$\dfrac{dv}{du}(\sqrt{u} + \sqrt{2v + 1}) = \dfrac{dv}{du}(5)$$

$$\dfrac{1}{2}u^{-1/2} + \dfrac{1}{2}(2v + 1)^{-1/2}(2)\dfrac{dv}{du} = 0$$

$$(2v + 1)^{-1/2}\dfrac{dv}{du} = -\dfrac{1}{2}u^{-1/2}$$

$$\dfrac{dv}{du} = -\dfrac{(2v + 1)^{1/2}}{2u^{1/2}}$$

The derivatives are reciprocals.

41. $x^2 + y^2 + 1 = 0$

$$\frac{d}{dx}(x^2 + y^2) = \frac{d}{dx}(-1)$$

$$2x + 2y\frac{dy}{dx} = 0$$

$$\frac{dy}{dx} = \frac{-2x}{2y} = -\frac{x}{y}$$

If x and y are real numbers, x^2 and y^2 are nonnegative; 1 plus a nonnegative number cannot equal zero, so there is no function $y = f(x)$ that satisfies $x^2 + y^2 + 1 = 0$.

43. $C^2 = x^2 + 100\sqrt{x} + 50$

(a) $2C\frac{dC}{dx} = 2x + \frac{1}{2}(100)x^{-1/2}$

$$\frac{dC}{dx} = \frac{2x + 50x^{-1/2}}{2C}$$

$$\frac{dC}{dx} = \frac{x + 25x^{-1/2}}{C} \cdot \frac{x^{1/2}}{x^{1/2}}$$

$$\frac{dC}{dx} = \frac{x^{3/2} + 25}{Cx^{1/2}}$$

When $x = 5$, the approximate increase in cost in dollars of an additional unit is

$$\frac{(5)^{3/2} + 25}{(5^2 + 100\sqrt{5} + 50)^{1/2}(5)^{1/2}} = \frac{36.18}{(17.28)\sqrt{5}}$$

$$\approx 0.94.$$

(b) $900(x - 5)^2 + 25R^2 = 22,500$

$$R^2 = 900 - 36(x - 5)^2$$

$$2R\frac{dR}{dx} = -72(x - 5)$$

$$\frac{dR}{dx} = \frac{-36(x - 5)}{R} = \frac{180 - 36x}{R}$$

When $x = 5$, the approximate change in revenue for a unit increase in sales is

$$\frac{180 - 36(5)}{R} = \frac{0}{R} = 0.$$

45. (a) $\ln q = D - 0.44\ln p$

$$\frac{1}{q}\frac{dq}{dp} = -\frac{0.44}{p}$$

$$\frac{dp}{dq} = -0.44\frac{q}{p}$$

$$E = -\frac{p}{q}\frac{dq}{dp}$$

$$= -\frac{p}{q}\left(-0.44\frac{q}{p}\right)$$

$$= 0.44$$

E is less than 1, so the demand is inelastic.

(b) Solving for q first:

$$\ln q = D - 0.44\ln p$$

$$e^{\ln q} = e^{D - 0.44\ln p}$$

$$q = e^D p^{-0.44}$$

$$\frac{dq}{dp} = e^D(-0.678)p^{-1.44}$$

$$E = -\frac{p}{q}\frac{dq}{dp}$$

$$= \left(-\frac{p}{e^D p^{-0.44}}\right)(e^D(-0.44)p^{-1.44})$$

$$= 0.44$$

This is the same answer as found in part (a).

47. First note that

if $\quad \log R(w) = 1.83 - 0.43\log(w)$

then $\quad R(w) = 10^{1.83 - 0.43\log(w)}$

$$= 10^{1.83}10^{-0.43\log(w)}$$

$$= 10^{1.83}[10^{\log(w)}]^{-0.43}$$

$$= 10^{1.83}w^{-0.43}$$

(a) $\dfrac{d}{dw}[\log R(w)] = \dfrac{d}{dw}[1.83 - 0.43\log(w)]$

$$\frac{1}{\ln 10}\frac{1}{R(w)}\frac{dR}{dw} = 0 - 0.43\frac{1}{\ln 10}\frac{1}{w}$$

$$\frac{dR}{dw} = -0.43\frac{R(w)}{w}$$

$$= -0.43\frac{10^{1.83}w^{-0.43}}{w}$$

$$\approx -29.0716w^{-1.43}$$

(b) $\quad R(w) = 10^{1.83}w^{-0.43}$

$$\frac{d}{dw}[R(w)] = \frac{d}{dw}[10^{1.83}w^{-0.43}]$$

$$\frac{dR}{dw} = 10^{1.83}(-0.43)w^{-1.43}$$

$$\approx -29.0716w^{-1.43}$$

49. $b - a = (b + a)^3$

$$\frac{d}{db}(b - a) = \frac{d}{db}[(b + a)^3]$$

$$1 - \frac{da}{db} = 3(b + a)^2 \frac{d}{db}(b + a)$$

$$1 - \frac{da}{db} = 3(b + a)^2 \left(1 + \frac{da}{db}\right)$$

$$1 - \frac{da}{db} = 3(b + a)^2 + 3(b + a)^2 \frac{da}{db}$$

$$-\frac{da}{db} - 3(b + a)^2 \frac{da}{db} = 3(b + a)^2 - 1$$

$$[-1 - 3(b + a)^2]\frac{da}{db} = 3(b + a)^2 - 1$$

$$\frac{da}{db} = \frac{3(b + a)^2 - 1}{-1 - 3(b + a)^2}$$

$$\frac{da}{db} = 0$$

$$3(b + a)^2 - 1 = 0$$

$$b + a = \frac{1}{\sqrt{3}}$$

Since $b - a = (b + a)^3 = \left(\frac{1}{\sqrt{3}}\right)^3 = \frac{1}{3\sqrt{3}}$.

$$b + a = \frac{1}{\sqrt{3}}$$

$$\underline{-(b - a) = -\frac{1}{3\sqrt{3}}}$$

$$2a = \frac{2}{3\sqrt{3}}$$

$$a = \frac{1}{3\sqrt{3}}$$

51. Differentiate $f(R)$ twice, treating t as a function of R.

$$f''(R)$$

$$= \frac{C}{(1 + hcR)^3}$$

$$\times \left| \frac{d^2t}{dR^2}\left(R + 2R^2hC + R^3h^2C^2\right) \right.$$
$$\left. + \frac{dt}{dR}2(1 + hcR) - 2thC \right|$$

$$= \frac{C}{(1 + hcR)^3}$$

$$\times \left| \frac{d^2t}{dR^2}R(1 + hcR)^2 \right.$$
$$\left. + \frac{dt}{dR}2(1 + hcR) - 2thC \right|$$

For the graph of $f(R)$ to be concave down, $f(R)$ must be negative, which will be true when the expression in brackets above is negative. This is what the exercise claims.

53. $s^3 - 4st + 2t^3 - 5t = 0$

$$3s^2\frac{ds}{dt} - \left(4t\frac{ds}{dt} + 4s\right) + 6t^2 - 5 = 0$$

$$3s^2\frac{ds}{dt} - 4t\frac{ds}{dt} - 4s + 6t^2 - 5 = 0$$

$$\frac{ds}{dt}(3s^2 - 4t) = 4s - 6t^2 + 5$$

$$\frac{ds}{dt} = \frac{4s - 6t^2 + 5}{3s^2 - 4t}$$

6.5   Related Rates

Your Turn 1

$x^3 + 2xy + y^2 = 1$, where both x and y are functions of t. Given $x = 1, y = -2$, and $dx/dy = 6$, find dy/dt.

$$\frac{d}{dt}\left(x^3 + 2xy + y^2\right) = \frac{d}{dt}(1)$$

$$3x^2\frac{dx}{dt} + 2x\frac{dy}{dt} + 2y\frac{dx}{dt} + 2y\frac{dy}{dt} = 0$$

$$52x\frac{dy}{dt} + 2y\frac{dy}{dt} = -\left(3x^2\frac{dx}{dt} + 2y\frac{dx}{dt}\right)$$

$$\frac{dy}{dt} = \frac{dx}{dt}\left(-\frac{3x^2 + 2y}{2x + 2y}\right)$$

Now substitute the given values.

$$\frac{dy}{dt} = \frac{dx}{dt}\left(-\frac{3x^2 + 2y}{2x + 2y}\right)$$

$$= 6\left(-\frac{3(1^2) + 2(-2)}{2(1) + 2(-2)}\right)$$

$$= 6\left(-\frac{-1}{-2}\right)$$

$$= -3$$

Your Turn 2

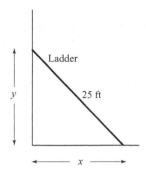

$x^2 + y^2 = 25^2$, where both x and y are functions of t. We are interested in what happens when $x = 7$ ft. At this time, $y = \sqrt{25^2 - 7^2} = 24$, and since the bottom of the ladder is slipping away from the building at 3 ft/min, $dx/dt = 3$.

$$\frac{d}{dt}\left(x^2 + y^2\right) = \frac{d}{dt}(25^2)$$

$$2x\frac{dx}{dt} + 2y\frac{dy}{dt} = 0$$

Now substitute the known values.

$$2(7)(3) + 2(24)\frac{dy}{dt} = 0$$

$$\frac{dy}{dt} = \frac{-2(7)(3)}{2(24)}$$

$$= -\frac{7}{8}$$

The latter is sliding *down* the side of the building at 7/8 ft/min.

Your Turn 3

In Example 4, differentiating the formula for the volume of a cone gives the following result:

$$\frac{dV}{dt} = \frac{1}{3}\pi\left[r^2\frac{dh}{dt} + (h)(2r)\frac{dr}{dt}\right]$$

For this problem,

$$\frac{dV}{dt} = -10, \quad \frac{dr}{dt} = -0.4, \quad r = 4, \quad \text{and } h = 20.$$

Substitute these values above and solve for dh/dt.

$$\frac{dV}{dt} = \frac{1}{3}\pi\left[r^2\frac{dh}{dt} + (h)(2r)\frac{dr}{dt}\right]$$

$$-10 = \frac{1}{3}\pi\left[(4^2)\frac{dh}{dt} + (20)(2)(4)(-0.4)\right]$$

$$-\frac{30}{\pi} + 64 = 16\frac{dh}{dt}$$

$$\frac{dh}{dt} = \frac{-\frac{30}{\pi} + 64}{16}$$

$$\frac{dh}{dt} \approx 3.4$$

The length increases at a rate of 3.4 cm per hour.

Your Turn 4

The revenue equation is

$$R = qp = q\left(2000 - \frac{q^2}{100}\right) = 2000q - \frac{q^3}{100}.$$

Differentiate with respect to t.

$$\frac{dR}{dt} = 2000\frac{dq}{dt} - \frac{3q^2}{100}\frac{dq}{dt}$$

$$= \left(2000 - \frac{3q^2}{100}\right)\frac{dq}{dt}$$

Substitute the know values for q and dq/dt, which are the same as in Example 5: $q = 200$, $dq/dt = 50$.

$$\frac{dR}{dt} = \left(2000 - \frac{3q^2}{100}\right)\frac{dq}{dt}$$

$$= \left(2000 - \frac{3(200^2)}{100}\right)(50)$$

$$= (800)(50) = 40,000$$

Revenue is increasing at the rate of $40,000 per day.

6.5 Warmup Exercises

W1. $x^3y + y^4x = 5$

Differentiate both sides with respect to x.

$$\left[x^3\frac{dy}{dx} + 3x^2y\right] + \left[x\left(4y^3\frac{dy}{dx}\right) + y^4\right] = 0$$

$$\frac{dy}{dx}\left(x^3 + 4xy^3\right) = -\left(3x^2y + y^4\right)$$

$$\frac{dy}{dx} = -\frac{3x^2y + y^4}{x^3 + 4xy^3}$$

W2. $x^2 + y^2 = 3xy^3$

Differentiate both sides with respect to x.

$$2x + 2y\frac{dy}{dx} = (3x)\left(3y^2\frac{dy}{dx}\right) + 3y^3$$

$$\frac{dy}{dx}\left(2y - 9xy^2\right) = 3y^3 - 2x$$

$$\frac{dy}{dx} = \frac{3y^3 - 2x}{2y - 9xy^2}$$

6.5 Exercises

1. $y^2 - 8x^3 = -55; \dfrac{dx}{dt} = -4, x = 2, y = 3$

$$2y\frac{dy}{dt} - 24x^2\frac{dx}{dt} = 0$$

$$y\frac{dy}{dt} = 12x^2\frac{dx}{dt}$$

$$3\frac{dy}{dt} = 48(-4)$$

$$\frac{dy}{dt} = -64$$

3. $2xy - 5x + 3y^3 = -51; \dfrac{dx}{dt} = -6,$

$$x = 3, y = -2$$

$$2x\frac{dy}{dt} + 2y\frac{dx}{dt} - 5\frac{dx}{dt} + 9y^2\frac{dx}{dt} = 0$$

$$(2x + 9y^2)\frac{dy}{dt} + (2y - 5)\frac{dx}{dt} = 0$$

$$(2x + 9y^2)\frac{dy}{dt} = (5 - 2y)\frac{dx}{dt}$$

$$\frac{dy}{dt} = \frac{5 - 2y}{2x + 9y^2} \cdot \frac{dx}{dt}$$

$$= \frac{5 - 2(-2)}{2(3) + 9(-2)^2} \cdot (-6)$$

$$= \frac{9}{42} \cdot (-6) = \frac{-54}{42}$$

$$= -\frac{9}{7}$$

5. $\dfrac{x^2 + y}{x - y} = 9; \dfrac{dx}{dt} = 2, x = 4, y = 2$

$$\frac{(x - y)\left(2x\frac{dx}{dt} + \frac{dy}{dt}\right) - (x^2 + y)\left(\frac{dx}{dt} - \frac{dy}{dt}\right)}{(x - y)^2} = 0$$

$$\frac{2x(x - y)\frac{dx}{dt} + (x - y)\frac{dy}{dt} - (x^2 + y)\frac{dx}{dt} + (x^2 + y)\frac{dy}{dt}}{(x - y)^2} = 0$$

$$[2x(x - y) - (x^2 + y)]\frac{dx}{dt} + [(x - y) + (x^2 + y)]\frac{dy}{dt} = 0$$

$$\frac{dy}{dt} = \frac{[(x^2 + y) - 2x(x - y)]\frac{dx}{dt}}{(x - y) + (x^2 + y)}$$

$$\frac{dy}{dt} = \frac{(-x^2 + y + 2xy)\frac{dx}{dt}}{x + x^2}$$

$$= \frac{[-(4)^2 + 2 + 2(4)(2)](2)}{4 + 4^2}$$

$$= \frac{4}{20} = \frac{1}{5}$$

7. $xe^y = 3 + \ln x; \dfrac{dx}{dt} = 6, x = 2, y = 0$

$$e^y\frac{dx}{dt} + xe^y\frac{dy}{dt} = 0 + \frac{1}{x}\frac{dx}{dt}$$

$$xe^y\frac{dy}{dt} = \left(\frac{1}{x} - e^y\right)\frac{dx}{dt}$$

$$\frac{dy}{dt} = \frac{\left(\frac{1}{x} - e^y\right)\frac{dx}{dt}}{xe^y}$$

$$= \frac{(1 - xe^y)\frac{dx}{dt}}{x^2e^y}$$

$$= \frac{[1 - (2)e^0](6)}{2^2e^0}$$

$$= \frac{-6}{4} = -\frac{3}{2}$$

9. $C = 0.2x^2 + 10,000; x = 80, \dfrac{dx}{dt} = 12$

$$\frac{dC}{dt} = 0.2(2x)\frac{dx}{dt} = 0.2(160)(12) = 384$$

The cost is changing at a rate of $384 per month.

11. $R = 50x - 0.4x^2; C = 5x + 15;$

$x = 40; \dfrac{dx}{dt} = 10$

(a) $\dfrac{dR}{dt} = 50\dfrac{dx}{dt} - 0.8x\dfrac{dx}{dt}$

$\qquad = 50(10) - 0.8(40)(10)$

$\qquad = 500 - 320$

$\qquad = 180$

Revenue is increasing at a rate of \$180 per day.

(b) $\dfrac{dC}{dt} = 5\dfrac{dx}{dt} = 5(10) = 50$

Cost is increasing at a rate of \$50 per day.

(c) Profit = Revenue − Cost

$\qquad P = R - C$

$\qquad \dfrac{dP}{dt} = \dfrac{dR}{dt} - \dfrac{dC}{dt} = 180 - 50 = 130$

Profit is increasing at a rate of \$130 per day.

13. $pq = 8000; p = 3.50, \dfrac{dp}{dt} = 0.15$

$\qquad\qquad pq = 8000$

$\qquad p\dfrac{dq}{dt} + q\dfrac{dp}{dt} = 0$

$\qquad\qquad \dfrac{dq}{dt} = \dfrac{-q\frac{dp}{dt}}{p}$

$\qquad\qquad\quad = \dfrac{-\left(\frac{8000}{3.50}\right)(0.15)}{3.50}$

$\qquad\qquad\quad \approx -98$

Demand is decreasing at a rate of approximately 98 units per unit time.

15. $V = k(R^2 - r^2); k = 555.6, R = 0.02$ mm,

$\dfrac{dR}{dt} = 0.003$ mm per minute; r is constant.

$\qquad V = k(R^2 - r^2)$

$\qquad V = 555.6(R^2 - r^2)$

$\qquad \dfrac{dV}{dt} = 555.6\left(2R\dfrac{dR}{dt} - 0\right)$

$\qquad\qquad = 555.6(2)(0.02)(0.003)$

$\qquad\qquad = 0.067$ mm/min

17. $b = 0.22m^{0.87}$

$\dfrac{db}{dt} = 0.22(0.87)m^{-0.13}\dfrac{dm}{dt}$

$\qquad = 0.1914m^{-0.13}\dfrac{dm}{dt}$

$\dfrac{dm}{dt} = \dfrac{m^{0.13}}{0.1914}\dfrac{db}{dt}$

$\qquad = \dfrac{25^{0.13}}{0.1914}(0.25)$

$\qquad \approx 1.9849$

The rate of change of the total weight is about 1.9849 g/day.

19. $r = 140.2m^{0.75}$

(a) $\dfrac{dr}{dt} = 140.2(0.75)m^{-0.25}\dfrac{dm}{dt}$

$\qquad = 105.15m^{-0.25}\dfrac{dm}{dt}$

(b) $\dfrac{dr}{dt} = 105.15(250)^{-0.25}(2)$

$\qquad \approx 52.89$

The rate of change of the average daily metabolic rate is about 52.89 kcal/day^2.

21. $C = \dfrac{1}{10}(T - 60)^2 + 100$

$\dfrac{dC}{dt} = \dfrac{1}{5}(T - 60)\dfrac{dT}{dt}$

If $T = 76°$ and $\dfrac{dT}{dt} = 8$,

$\dfrac{dC}{dt} = \dfrac{1}{5}(76 - 60)(8) = \dfrac{1}{5}(16)(8)$

$\qquad = 25.6.$

The crime rate is rising at the rate of 25.6 crimes/month.

23. Let $x =$ The distance of the base of the ladder from the base of the building

$\qquad y =$ The distance up the side of the building to the top of the ladder

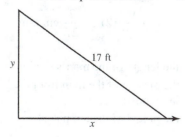

Find $\frac{dy}{dt}$ when $x = 8$ ft and $\frac{dx}{dt} = 9$ ft/min.

Since $y = \sqrt{17^2 - x^2}$, when $x = 8$,
$y = 15$.

By the Pythagorean theorem,

$$x^2 + y^2 = 17^2.$$

$$\frac{d}{dx}(x^2 + y^2) = \frac{d}{dt}(17^2)$$

$$2x\frac{dx}{dt} + 2y\frac{dy}{dt} = 0$$

$$2y\frac{dy}{dt} = -2x\frac{dx}{dt}$$

$$\frac{dy}{dt} = \frac{-2x}{2y} \cdot \frac{dx}{dt} = -\frac{x}{y} \cdot \frac{dx}{dt}$$

$$= -\frac{8}{15}(9)$$

$$= -\frac{24}{5}$$

The ladder is sliding down the building at the rate of $\frac{24}{5}$ ft/min.

25. Let $r =$ the raius of the circle formed by the ripple.

Find $\frac{dA}{dt}$ when $r = 4$ ft and $\frac{dr}{dt} = 2$ ft/min.

$$A = \pi r^2$$

$$\frac{dA}{dt} = 2\pi r\frac{dr}{dt}$$

$$= 2\pi(4)(2) = 16\pi$$

The area is changing at the rate of 16π ft^2/min.

27. $V = x^3$, $x = 3$ cm, and $\frac{dV}{dt} = 2$ cm^3/min

$$\frac{dV}{dt} = 3x^2\frac{dx}{dt}$$

$$\frac{dx}{dt} = \frac{1}{3x^2}\frac{dV}{dt}$$

$$= \frac{1}{3 \cdot 3^2}(2) = \frac{2}{27} \text{ cm/min}$$

29. Let $y =$ the length of the man's shadow;
$x =$ the distane of the man from the lamp post;
$h =$ the height of the lamp post.

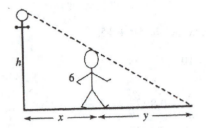

$$\frac{dx}{dt} = 50 \text{ ft/min}$$

Find $\frac{dy}{dt}$ when $x = 25$ ft.

Now $\frac{h}{x+y} = \frac{6}{y}$, by similar triangles.

When $x = 8$, $y = 10$,

$$\frac{h}{18} = \frac{6}{10}$$

$$h = 10.8.$$

$$\frac{10.8}{x+y} = \frac{6}{y},$$

$$10.8y = 6x + 6y$$

$$4.8y = 6x$$

$$y = 1.25x$$

$$\frac{dy}{dt} = 1.25\frac{dx}{dt}$$

$$= 1.25(50)$$

$$\frac{dy}{dt} = 62.5$$

The length of the shadow is increasing at the rate of 62.5 ft/min.

31. Let $x =$ the distance from the docks
$s =$ the length of the rope.

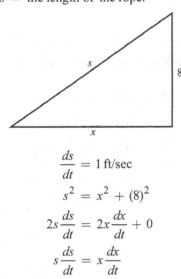

$$\frac{ds}{dt} = 1 \text{ ft/sec}$$

$$s^2 = x^2 + (8)^2$$

$$2s\frac{ds}{dt} = 2x\frac{dx}{dt} + 0$$

$$s\frac{ds}{dt} = x\frac{dx}{dt}$$

If $x = 8$,

$$s = \sqrt{(8)^2 + (8)^2} = \sqrt{128} = 8\sqrt{2}.$$

Then,

$$8\sqrt{2}(1) = 8\frac{dx}{dt}$$

$$\frac{dx}{dt} = \sqrt{2} \approx 1.41$$

The boat is approaching the dock at $\sqrt{2} \approx 1.41$ ft/sec.

6.6 Differentials: Linear Approximation

Your Turn 1

$y = 300x^{-2/3}$, $x = 8$, $dx = 0.05$

$$\frac{dy}{dx} = \left(-\frac{2}{3}\right)(300)\, x^{-5/3} = -200x^{-5/3}$$

$$dy = -200x^{-5/3}dx$$

$$= (-200)(8^{-5/3})(0.05)$$

$$= (-200)\left(\frac{1}{32}\right)\left(\frac{1}{20}\right)$$

$$= -\frac{5}{16}$$

Your Turn 2

Use the approximation formula for $f(x) = \sqrt{x}$ developed in Example 2:

$$f(x + \Delta x) \approx \sqrt{x} + \frac{1}{2\sqrt{x}}\,dx$$

For this problem, $x = 100$ and $dx = -1$.

$$f(99) = f(100 - 1)$$

$$\approx \sqrt{100} + \frac{1}{2\sqrt{100}}(-1)$$

$$= 10 - \frac{1}{20}$$

$$= 9.95$$

Your Turn 3

Use the approximation derived in Example 5:

$$dV = 4\pi r^2 dr$$

For this problem, $r = 1.25$ and $dr = \Delta r = \pm 0.025$.

$$dV = 4\pi(1.25)^2(\pm 0.025)$$

$$\approx 0.491$$

$$\approx 0.5$$

The maximum error in the volume is about 0.5 mm^3.

6.6 Warmup Exercises

W1. Use the chain rule, writing $f(x) = \sqrt{x^4 + 2}$ as $h(g(x))$ with $h(x) = x^{1/2}$ and $g(x) = x^4 + 2$.

$$f'(x) = f'(g(x))g'(x)$$

$$= \frac{1}{2}g(x)^{-1/2}\left(4x^3\right)$$

$$= \frac{2x^3}{\sqrt{x^4 + 2}}$$

W2. Use the chain rule, writing $f(x) = \ln(e^{2x} + 1)$ as $h(g(x))$ with $h(x) = \ln x$ and $g(x) = e^{2x} + 1$.

$$f'(x) = f'(g(x))g'(x)$$

$$= \frac{1}{g(x)}\left(2e^{2x}\right)$$

$$= \frac{2e^{2x}}{e^{2x} + 1}$$

6.6 Exercises

1. $y = 2x^3 - 5x$; $x = -2$, $\Delta x = 0.1$

   $$dy = (6x^2 - 5)dx$$

   $$\Delta y \approx (6x^2 - 5)\Delta x \approx [6(-2)^2 - 5](0.1) \approx 1.9$$

3. $y = x^3 - 2x^2 + 3$, $x = 1$, $\Delta x = -0.1$

   $$dy = (3x^2 - 4x)dx$$

   $$\Delta y \approx (3x^2 - 4x)\Delta x$$

   $$= [3(1^2) - 4(1)](-0.1)$$

   $$= 0.1$$

5. $y = \sqrt{3x + 2}$, $x = 4$, $\Delta x = 0.15$

   $$dy = 3\left[\frac{1}{2}(3x + 2)^{-1/2}\right]dx$$

   $$\Delta y \approx \frac{3}{2\sqrt{3x + 2}}\Delta x \approx \frac{3}{2(3.74)}(0.15) \approx 0.060$$

7. $y = \dfrac{2x - 5}{x + 1}$; $x = 2$, $\Delta x = -0.03$

$$dy = \dfrac{(x + 1)(2) - (2x - 5)(1)}{(x + 1)^2}\, dx$$

$$= \dfrac{7}{(x + 1)^2}\, dx$$

$$\Delta y \approx \dfrac{7}{(x + 1)^2}\, \Delta x$$

$$= \dfrac{7}{(2 + 1)^2}(-0.03)$$

$$= -0.023$$

9. $\sqrt{145}$

We know $\sqrt{144} = 12$, so $f(x) = \sqrt{x}$, $x = 144$, $dx = 1$.

$$\dfrac{dy}{dx} = \dfrac{1}{2}x^{-1/2}$$

$$dy = \dfrac{1}{2\sqrt{x}}\, dx$$

$$dy = \dfrac{1}{2\sqrt{144}}(1) = \dfrac{1}{24}$$

$$\sqrt{145} \approx f(x) + dy = 12 + \dfrac{1}{24}$$

$$\approx 12.0417$$

By calculator, $\sqrt{145} \approx 12.0416$.

The difference is $|12.0417 - 12.0416| = 0.0001$.

11. $\sqrt{0.99}$

We know $\sqrt{1} = 1$, so $f(x) = \sqrt{x}$, $x = 1$, $dx = -0.01$.

$$\dfrac{dy}{dx} = \dfrac{1}{2}x^{-1/2}$$

$$dy = \dfrac{1}{2\sqrt{x}}\, dx$$

$$dy = \dfrac{1}{2\sqrt{1}}(-0.01) = -0.005$$

$$\sqrt{0.99} \approx f(x) + dy = 1 - 0.005$$

$$= 0.995$$

By calculator, $\sqrt{0.99} \approx 0.9950$.

The difference is $|0.995 - 0.9950| = 0$.

13. $e^{0.01}$

We know $e^0 = 1$, so $f(x) = e^x$, $x = 0$, $dx = 0.01$.

$$\dfrac{dy}{dx} = e^x$$

$$dy = e^x dx$$

$$dy = e^0(0.01) = 0.01$$

$$e^{0.01} \approx f(x) + dy = 1 + 0.01 = 1.01$$

By calculator, $e^{0.01} \approx 1.0101$.

The difference is $|1.01 - 1.0101| = 0.0001$.

15. $\ln 1.05$

We know $\ln 1 = 0$, so $f(x) = \ln x$, $x = 1$, $dx = 0.05$.

$$\dfrac{dy}{dx} = \dfrac{1}{x}$$

$$dy = \dfrac{1}{x}\, dx$$

$$dy = \dfrac{1}{1}(0.05) = 0.05$$

$$\ln 1.05 \approx f(x) + dy = 0 + 0.05 = 0.05$$

By calculator, $\ln 1.05 \approx 0.0488$.

The difference is $|0.05 - 0.0488| = 0.0012$.

17. Let D = the demand in thousands of pounds; x = the price in dollars.

$$D(q) = -3q^3 - 2q^2 + 1500$$

(a) $q = 2$, $\Delta q = 0.10$

$$dD = (-9q^2 - 4q)dq$$

$$\Delta D \approx (-9q^2 - 4q)\, \Delta q$$

$$\approx [-9(4) - 4(2)](0.10)$$

$$\approx -4.4 \text{ thousand pounds}$$

(b) $q = 6$, $\Delta q = 0.15$

$$\Delta D \approx [-9(36) - 4(6)](0.15)$$

$$\approx -52.2 \text{ thousand pounds}$$

19. $R(x) = 12,000 \ln(0.01x + 1)$

$x = 100, \; \Delta x = 1$

$$dR = \frac{12,000}{0.01x + 1}(0.01)dx$$

$$\Delta R \approx \frac{120}{0.01x + 1}\Delta x$$

$$\approx \frac{120}{0.01(100) + 1}(1)$$

$$\approx \$60$$

21. If a cube is given a coating 0.1 in. thick, each edge increases in length by twice that amount, or 0.2 in. because there is a face at both ends of the edge.

$$V = x^3, \; x = 4, \; \Delta x = 0.2$$

$$dV = 3x^2 dx$$

$$\Delta V \approx 3x^2 \Delta x$$

$$= 3(4^2)(0.2)$$

$$= 9.6$$

For 1000 cubes $9.6(1000) = 9600 \text{ in.}^3$ of coating should be ordered.

23. (a) $A(x) = y = 0.0036311x^3 - 0.03746x^2$
$$+ 0.1012x + 0.009$$

Let $x = 1, \; dx = 0.2$.

$$\frac{dy}{dx} = 0.010893x^2 - 0.07492x + 0.1012$$

$$dy = (0.010893x^2 - 0.07492x + 0.1012)dx$$

$$\Delta y \approx (0.010893x^2 - 0.07492x + 0.1012)\Delta x$$

$$\approx (0.010893 \cdot 1^2 - 0.07492 \cdot 1 + 0.1012) \cdot 0.2$$

$$\approx 0.007435$$

The alcohol concentration increases by about 0.74 percent.

(b)

$$\Delta y \approx (0.010893 \cdot 3^2 - 0.07492 \cdot 3 + 0.1012) \cdot 0.2$$

$$\approx -0.005105$$

The alcohol concentration decreases by about 0.51 percent.

25. $P(t) = \dfrac{25t}{8 + t^2}$

$$dP = \frac{(8 + t^2)(25) - 25t(2t)}{(8 + t^2)^2}dt$$

$$\approx \frac{(8 + t^2)(25) - 25t(2t)}{(8 + t^2)^2}\Delta t$$

(a) $t = 2, \; \Delta t = 0.5$

$$dP = \frac{[(8 + 4)(25) - (25)(2)(4)](0.5)}{(8 + 4)^2}$$

$$\approx 0.347 \text{ million}$$

(b) $t = 3, \; \Delta t = 0.25$

$$dP = \frac{[(8 + 9)(25) - 25(3)(6)]0.25}{(8 + 9)^2}$$

$$\approx -0.022 \text{ million}$$

27. r changes from 14 mm to 16 mm, so $\Delta r = 2$.

$$V = \frac{4}{3}\pi r^3$$

$$dV = \frac{4}{3}(3)\pi r^2 \, dr$$

$$\Delta V \approx 4\pi r^2 \, \Delta r$$

$$= 4\pi(14)^2(2)$$

$$= 1568\pi \text{ mm}^3$$

29. r increases from 20 mm to 22 mm, so $\Delta r = 2$.

$$A = \pi r^2$$

$$dA = 2\pi r \, dr$$

$$\Delta A \approx 2\pi r \, \Delta r$$

$$= 2\pi(20)(2)$$

$$= 80\pi \text{ mm}^2$$

31. $W(t) = -3.5 + 197.5e^{-e^{-0.01394(t - 108.4)}}$

(a)

$$dW = 197.5e^{-e^{-0.01394(t - 108.4)}}(-1)e^{-0.01394(t - 108.4)}(-0.01394)dt$$

$$= 2.75315e^{-e^{-0.01394(t - 108.4)}}e^{-0.01394(t - 108.4)}dt$$

We are given $t = 80$ and $dt = 90 - 80$
$= 10$.

$$dW \approx 9.258$$

The pig will gain about 9.3 kg.

(b) The actual weight gain is calculated as

$$W(90) - W(80) \approx 50.736 - 41.202$$

$$= 9.534$$

or about 9.5 kg.

33. $r = 3$ cm, $\Delta r = -0.2$ cm

$$V = \frac{4}{3}\pi r^3$$

$$dV = 4\pi r^2 dr$$

$$\Delta V \approx 4\pi r^2 \Delta r$$

$$= 4\pi(9)(-0.2)$$

$$= -7.2\pi \text{ cm}^3$$

35. $V = \frac{1}{3}\pi r^2 h; h = 13, dh = 0.2$

$$V = \frac{1}{3}\pi\left(\frac{h}{15}\right)^2 h$$

$$= \frac{\pi}{775}h^3$$

$$dV = \frac{\pi}{775} \cdot 3h^2 dh$$

$$= \frac{\pi}{225}h^2 dh$$

$$\Delta V \approx \frac{\pi}{225}h^2 \Delta h$$

$$\approx \frac{\pi}{225}(13^2)(0.2)$$

$$\approx 0.472 \text{ cm}^3$$

37. $A = x^2; x = 4, dA = 0.01$

$$dA = 2x\, dx$$

$$\Delta A \approx 2x\, \Delta x$$

$$\Delta x \approx \frac{\Delta A}{2x} \approx \frac{0.01}{2(4)} \approx 0.00125 \text{ cm}$$

39. $V = \frac{4}{3}\pi r^3; r = 5.81, \Delta r = \pm 0.003$

$$dV = \frac{4}{3}\pi(3r^2)dr$$

$$\Delta V \approx \frac{4}{3}\pi(3r^2)\Delta r$$

$$= 4\pi(5.81)^2(\pm 0.003)$$

$$= \pm 0.405\pi \approx \pm 1.273 \text{ in.}^3$$

41. $h = 7.284$ in., $r = 1.09 \pm 0.007$ in.

$$V = \frac{1}{3}\pi r^2 h$$

$$dV = \frac{2}{3}\pi rh\, dr$$

$$\Delta V \approx \frac{2}{3}\pi rh\, \Delta r$$

$$= \frac{2}{3}\pi(1.09)(7.284)(0.007)$$

$$= \pm 0.116 \text{ in.}^3$$

Chapter 6   Review Exercises

1. False: The absolute maximum might occur at the endpoint of the interval of interest.

2. True

3. False: It could have either. For example $f(x) = 1/(1 - x^2)$ has an absolute minimum of 1 on $(-1, 1)$.

4. True

5. True

6. True

7. True

8. True

9. True

10. True

11. $f(x) = -x^3 + 6x^2 + 1; [-1, 6]$

$f'(x) = -3x^2 + 12x = 0$ when $x = 0, 4$.

$$f(-1) = 8$$
$$f(0) = 1$$
$$f(4) = 33$$
$$f(6) = 1$$

Absolute maximum of 33 at 4; absolute minimum of 1 at 0 and 6.

13. $f(x) = x^3 + 2x^2 - 15x + 3; [-4, 2]$

$f'(x) = 3x^2 + 4x - 15 = 0$ when

$(3x - 5)(x + 3) = 0$

$x = \dfrac{5}{3} \quad \text{or} \quad x = -3.$

$f(-4) = 31$

$f(-3) = 39$

$f\left(\dfrac{5}{3}\right) = -\dfrac{319}{27}$

$f(2) = -11$

Absolute maximum of 39 at -3; absolute minimum of $-\dfrac{319}{27}$ at $\dfrac{5}{3}$.

17. (a) $f(x) = \dfrac{2 \ln x}{x^2}; [1, 4]$

$f'(x) = \dfrac{x^2 \left(\dfrac{2}{x}\right) - (2 \ln x)(2x)}{x^4}$

$= \dfrac{2x - 4x \ln x}{x^4}$

$= \dfrac{2 - 4 \ln x}{x^3}$

$f'(x) = 0$ when

$2 - 4 \ln x = 0$

$2 = 4 \ln x$

$0.5 = \ln x$

$e^{0.5} = x$

$x \approx 1.6487.$

x	$f(x)$
1	0
$e^{0.5}$	0.36788
4	0.17329

Maximum is 0.37; minimum is 0.

(b) $[2, 5]$

Note that the critical number of f is not in the domain, so we only test the endpoints.

x	$f(x)$
2	0.34657
5	0.12876

Maximum is 0.35, minimum is 0.13.

21. $x^2 - 4y^2 = 3x^3 y^4$

$\dfrac{d}{dx}(x^2 - 4y^2) = \dfrac{d}{dx}(3x^3 y^4)$

$2x - 8y\dfrac{dy}{dx} = 9x^2 y^4 + 3x^3 \cdot 4y^3 \dfrac{dy}{dx}$

$(-8y - 3x^3 \cdot 4y^3)\dfrac{dy}{dx} = 9x^2 y^4 - 2x$

$\dfrac{dy}{dx} = \dfrac{2x - 9x^2 y^4}{8y + 12x^3 y^3}$

23. $2\sqrt{y - 1} = 9x^{2/3} + y$

$\dfrac{d}{dx}[2(y - 1)^{1/2}] = \dfrac{d}{dx}(9x^{2/3} + y)$

$2 \cdot \dfrac{1}{2} \cdot (y - 1)^{-1/2} \dfrac{dy}{dx} = 6x^{-1/3} + \dfrac{dy}{dx}$

$[(y - 1)^{-1/2} - 1]\dfrac{dy}{dx} = 6x^{-1/3}$

$\dfrac{1 - \sqrt{y - 1}}{\sqrt{y - 1}} \cdot \dfrac{dy}{dx} = \dfrac{6}{x^{1/3}}$

$\dfrac{dy}{dx} = \dfrac{6\sqrt{y - 1}}{x^{1/3}(1 - \sqrt{y - 1})}$

25. $\dfrac{6 + 5x}{2 - 3y} = \dfrac{1}{5x}$

$5x(6 + 5x) = 2 - 3y$

$30x + 25x^2 = 2 - 3y$

$\dfrac{d}{dx}(30x + 25x^2) = \dfrac{d}{dx}(2 - 3y)$

$30 + 50x = -3\dfrac{dy}{dx}$

$-\dfrac{30 + 50x}{3} = \dfrac{dy}{dx}$

27. $\ln(xy + 1) = 2xy^3 + 4$

$$\frac{d}{dx}[\ln(xy + 1)] = \frac{d}{dx}(2xy^3 + 4)$$

$$\frac{1}{xy + 1} \cdot \frac{d}{dx}(xy + 1) = 2y^3 + 2x \cdot 3y^2\frac{dy}{dx} + \frac{d}{dx}(4)$$

$$\frac{1}{xy + 1}\left(y + x\frac{dy}{dx} + \frac{d}{dx}(1)\right) = 2y^3 + 6xy^2\frac{dy}{dx}$$

$$\frac{y}{xy + 1} + \frac{x}{xy + 1} \cdot \frac{dy}{dx} = 2y^3 + 6xy^2\frac{dy}{dx}$$

$$\left(\frac{x}{xy + 1} - 6xy^2\right)\frac{dy}{dx} = 2y^3 - \frac{y}{xy + 1}$$

$$\frac{dy}{dx} = \frac{2y^3 - \dfrac{y}{xy + 1}}{\dfrac{x}{xy + 1} - 6xy^2}$$

$$= \frac{2y^3(xy + 1) - y}{x - 6xy^2(xy + 1)}$$

$$= \frac{2xy^4 + 2y^3 - y}{x - 6x^2y^3 - 6xy^2}$$

29. $\sqrt{2y} - 4xy = -22$, tangent line at $(3, 2)$.

$$\frac{d}{dx}\left(\sqrt{2y} - 4xy\right) = \frac{d}{dx}(-22)$$

$$\frac{1}{2}(2)(2y)^{-1/2}\frac{dy}{dx} - \left(4y + 4x\frac{dy}{dx}\right) = 0$$

$$((2y)^{-1/2} - 4x)\frac{dy}{dx} = 4y$$

$$\frac{dy}{dx} = \frac{4y}{\dfrac{1}{\sqrt{2y}} - 4x}$$

To find the slope m of the tangent line, substitute 3 for x and 2 for y.

$$m = \frac{4y}{\dfrac{1}{2\sqrt{2y}} - 4x}$$

$$= \frac{4(2)}{\dfrac{1}{\sqrt{2(2)}} - 4(3)}$$

$$= \frac{8}{\dfrac{1}{2} - 12}$$

$$= \frac{16}{1 - 24} = -\frac{16}{23}$$

The equation of the tangent line is

$$y - y_1 = m(x - x_1)$$

$$y - 2 = -\frac{16}{23}(x - 3)$$

$$y - 2 - \frac{48}{23} = -\frac{16}{23}x$$

$$y = -\frac{16}{23}x + \frac{94}{23}.$$

We can also write this equation as $16x + 23y = 94$.

33. $y = 8x^3 - 7x^2$, $\frac{dx}{dt} = 4$, $x = 2$

$$\frac{dy}{dt} = \frac{d}{dt}(8x^3 - 7x^2)$$

$$= 24x^2\frac{dx}{dt} - 14x\frac{dx}{dt}$$

$$= 24(2)^2(4) - 14(2)(4)$$

$$= 272$$

35. $y = \frac{1 + \sqrt{x}}{1 - \sqrt{x}}$, $\frac{dx}{dt} = -4$, $x = .4$

$$\frac{dy}{dt} = \frac{d}{dt}\left[\frac{1 + \sqrt{x}}{1 - \sqrt{x}}\right]$$

$$= \frac{\left[\left(1 - \sqrt{x}\right)\left(\frac{1}{2}x^{-1/2}\frac{dx}{dt}\right) - 1\left(1 + \sqrt{x}\right)\left(-\frac{1}{2}\right)\left(x^{-1/2}\frac{dx}{dt}\right)\right]}{\left(1 - \sqrt{x}\right)^2}$$

$$= \frac{\left[(1 - 2)\left(\frac{1}{2 \cdot 2}\right)(-4) - (1 + 2)\left(\frac{-1}{2 \cdot 2}\right)(-4)\right]}{(1 - 2)^2}$$

$$= \frac{1 - 3}{1} = -2$$

37. $y = xe^{3x}$; $\frac{dx}{dt} = -2$, $x = 1$

$$\frac{dy}{dt} = \frac{d}{dt}(xe^{3x})$$

$$= \frac{dx}{dt} \cdot e^{3x} + x \cdot \frac{d}{dt}(e^{3x})$$

$$= \frac{dx}{dt} \cdot e^{3x} + xe^{3x} \cdot 3\frac{dx}{dt}$$

$$= (1 + 3x)e^{3x}\frac{dx}{dt}$$

$$= (1 + 3 \cdot 1)e^{3(1)}(-2) = -8e^3$$

41. $y = \dfrac{3x - 7}{2x + 1}; x = 2, \Delta x = 0.003$

$dy = \dfrac{(3)(2x + 1) - (2)(3x - 7)}{(2x + 1)^2} dx$

$dy = \dfrac{17}{(2x + 1)^2} dx$

$\Delta y \approx \dfrac{17}{(2x + 1)^2} \Delta x$

$\quad = \dfrac{17}{(2[2] + 1)^2}(0.003)$

$\quad = 0.00204$

43. $-12x + x^3 + y + y^2 = 4$

$\dfrac{dy}{dx}(-12x + x^3 + y + y^2) = \dfrac{d}{dx}(4)$

$-12 + 3x^2 + \dfrac{dy}{dx} + 2y\dfrac{dy}{dx} = 0$

$(1 + 2y)\dfrac{dy}{dx} = 12 - 3x^2$

$\dfrac{dy}{dx} = \dfrac{12 - 3x^2}{1 + 2y}$

(a) If $\dfrac{dy}{dx} = 0$,

$12 - 3x^2 = 0$

$12 = 3x^2$

$\pm 2 = x.$

$x = 2$:

$-24 + 8 + y + y^2 = 4$

$y + y^2 = 20$

$y^2 + y - 20 = 0$

$(y + 5)(y - 4) = 0$

$y = -5 \text{ or } y = 4$

$(2, -5)$ and $(2, 4)$ are critical points.

$x = -2$:

$24 - 8 + y + y^2 = 4$

$y + y^2 = -12$

$y^2 + y + 12 = 0$

$y = \dfrac{-1 \pm \sqrt{1^2 - 48}}{2}$

This leads to imaginary roots.

$x = -2$ does not produce critical points.

(b)

x	y_1	y_2
1.9	-4.99	3.99
2	-5	4
2.1	-4.99	3.99

The point $(2, -5)$ is a relative minimum.

The point $(2, 4)$ is a relative maximum.

(c) There is no absolute maximum or minimum for x or y.

45. (a) $P(x) = -x^3 + 10x^2 - 12x$

$P'(x) = -3x^2 + 20x - 12 = 0$

$3x^2 - 20x + 12 = 0$

$(3x - 2)(x - 6) = 0$

$3x - 2 = 0 \quad \text{or} \quad x - 6 = 0$

$x = \dfrac{2}{3} \quad \text{or} \quad x = 6$

$P''(x) = -6x + 20$

$P''\left(\dfrac{2}{3}\right) = 16,$

which implies that $x = \dfrac{2}{3}$ is the location of the minimum.

$P''(6) = -16,$

which implies that $x = 6$ is the location of the maximum. Thus, 600 boxes will produce a maximum profit.

(b) Maximum profit $= P(6)$

$= -(6)^3 + 10(6)^2 - 12(6) = 72$

The maximum profit is \$720.

47. Volume of cylinder $= \pi r^2 h$

Surface area of cylinder open at one end

$= 2\pi rh + \pi r^2.$

$V = \pi r^2 h = 27\pi$

$h = \dfrac{27\pi}{\pi r^2} = \dfrac{27}{r^2}$

$A = 2\pi r\left(\dfrac{27}{r^2}\right) + \pi r^2$

$\quad = 54\pi r^{-1} + \pi r^2$

$A' = -54\pi r^{-2} + 2\pi r$

If $A' = 0$,

$$2\pi r = \frac{54\pi}{r^2}$$
$$r^3 = 27$$
$$r = 3.$$

If $r = 3$,

$$A'' = 108\pi r^{-3} + 2\pi > 0,$$

so the value at $r = 3$ is a minimum.

For the minimum cost, the radius of the bottom should be 3 inches.

49. Here $k = 0.15$, $M = 20,000$, and $f = 12$. We have

$$q = \sqrt{\frac{2fM}{k}} = \sqrt{\frac{2(12)20,000}{0.15}}$$
$$= \sqrt{3,200,000} \approx 1789$$

Ordering 1789 rolls each time minimizes annual cost.

51. Use equation (3) from Section 6.3 with $k = 1$, $M = 128,000$, and $f = 10$.

$$q = \sqrt{\frac{2fM}{k}} = \sqrt{\frac{2(10)(128,000)}{1}}$$
$$= \sqrt{2,560,000} = 1600$$

The number of lots that should be produced annually is

$$\frac{M}{q} = \frac{128,000}{1600} = 80.$$

53. $\ln q = D - 0.447\ln p$

$$\frac{1}{q}\frac{dq}{dp} = -\frac{0.47}{p}$$

$$\frac{dp}{dq} = -0.47\frac{q}{p}$$

$$E = -\frac{p}{q}\frac{dq}{dp}$$

$$= -\frac{p}{q}\left(-0.47\frac{q}{p}\right)$$

$$= 0.47$$

E is less than 1, so the demand is inelastic.

55. $A = \pi r^2$; $\frac{dr}{dt} = 4\,\text{ft/min}$, $r = 7\,\text{ft}$

$$\frac{dA}{dt} = 2\pi r\frac{dr}{dt}$$
$$\frac{dA}{dt} = 2\pi(7)(4)$$
$$\frac{dA}{dt} = 56\pi$$

The rate of change of the area is 56π ft^2/min.

57. (a)

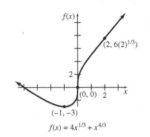

$f(x) = 4x^{1/3} + x^{4/3}$

(b) We use a graphing calculator to graph

$$M'(t) = -0.4321173 + 0.1129024t$$
$$- 0.0061518t^2 + 0.0001260t^3$$
$$- 0.0000008925t^4$$

on $[5,51]$ by $[0,7.5]$. We find the maximum value of $M'(t)$ on this graph at about 15.41, or on about the 15th day.

59. Let $x =$ the distance from the base of the ladder to the building;

$y =$ the height on the building at the top of the ladder.

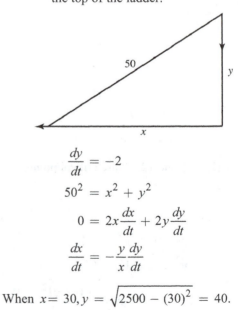

$$\frac{dy}{dt} = -2$$
$$50^2 = x^2 + y^2$$
$$0 = 2x\frac{dx}{dt} + 2y\frac{dy}{dt}$$
$$\frac{dx}{dt} = -\frac{y}{x}\frac{dy}{dt}$$

When $x = 30$, $y = \sqrt{2500 - (30)^2} = 40$.

So

$$\frac{dx}{dt} = \frac{-40}{30}(-2) = \frac{80}{30} = \frac{8}{3}$$

The base of the ladder is slipping away from the building at a rate of $\frac{8}{3}$ ft/min.

61. Let x = one-half of the width of the triangular cross section;
h = the height of the water;
V = the volume of the water.

$$\frac{dV}{dt} = 3.5 \text{ ft}^3/\text{min}.$$

Find $\frac{dV}{dt}$ when $h = \frac{1}{3}$.

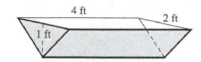

$$V = \begin{pmatrix} \text{Area of} \\ \text{triangular} \\ \text{side} \end{pmatrix}(\text{length})$$

Area of triangular cross section

$$= \frac{1}{2}(\text{base})(\text{altitude})$$

$$= \frac{1}{2}(2x)(h) = xh$$

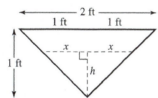

By similar triangles, $\frac{2x}{h} = \frac{2}{1}$, so $x = h$.

$$V = (xh)(4)$$

$$= h^2 \cdot 4$$

$$= 4h^2$$

$$\frac{dV}{dt} = 8h\frac{dh}{dt}$$

$$\frac{1}{8h} \cdot \frac{dV}{dt} = \frac{dh}{dt}$$

$$\frac{1}{8\left(\frac{1}{3}\right)}(3.5) = \frac{dh}{dt}$$

$$\frac{dh}{dt} = \frac{21}{16} = 1.3125$$

The depth of water is changing at the rate of 1.3125 ft/min.

63. $A = s^2; \; s = 9.2, \; \Delta s = \pm 0.04$

$$ds = 2s \, ds$$

$$\Delta A \approx 2s\Delta s$$

$$= 2(9.2)(\pm 0.04)$$

$$= \pm 0.736 \text{ in.}^2$$

65. We need to minimize y. Note that $x > 0$.

$$\frac{dy}{dx} = \frac{x}{8} - \frac{2}{x}$$

Set the derivative equal to 0.

$$\frac{x}{8} - \frac{2}{x} = 0$$

$$\frac{x}{8} = \frac{2}{x}$$

$$x^2 = 16$$

$$x = 4$$

Since $\lim\limits_{x \to 0} y = \infty$, $\lim\limits_{x \to \infty} y = \infty$, and $x = 4$ is the only critical value in $(0, \infty)$, $x = 4$ produces a minimum value.

$$y = \frac{4^2}{16} - 2\ln 4 + \frac{1}{4} + 2\ln 6$$

$$= 1.25 + 2(\ln 6 - \ln 4)$$

$$= 1.25 + 2\ln 1.5$$

The y coordinate of the Southern most point of the second boat's path is $1.25 + 2 \ln 1.5$.

67. Distance on shore: $40 - x$ feet

Speed on shore: 5 feet per second

Distance in water: $\sqrt{x^2 + 40^2}$ feet

Speed in water: 3 feet for second

The total travel time t is

$$t = t_1 + t_2 = \frac{d_1}{v_1} + \frac{d_2}{v_2}.$$

$$t(x) = \frac{40 - x}{5} + \frac{\sqrt{x^2 + 40^2}}{3}$$

$$= 8 - \frac{x}{5} + \frac{\sqrt{x^2 + 1600}}{3}$$

$$t'(x) = -\frac{1}{5} + \frac{1}{3} \cdot \frac{1}{2}(x^2 + 1600)^{-1/2}(2x)$$

$$= -\frac{1}{5} + \frac{x}{3\sqrt{x^2 + 1600}}$$

Minimize the travel time $t(x)$. If $t'(x) = 0$:

$$\frac{x}{3\sqrt{x^2 + 1600}} = \frac{1}{5}$$

$$5x = 3\sqrt{x^2 + 1600}$$

$$\frac{5x}{3} = \sqrt{x^2 + 1600}$$

$$\frac{25}{9}x^2 = x^2 + 1600$$

$$x^2 = \frac{1600 \cdot 9}{16}$$

$$x = \frac{40 \cdot 3}{4} = 30$$

(Discard the negative solution.)

To minimize the time, he should walk
$40 - x = 40 - 30 = 10$ ft along the shore
before paddling toward the desired destination.
The minimum travel time is

$$\frac{40 - 30}{5} + \frac{\sqrt{30^2 + 40^2}}{3} \approx 18.67 \text{ seconds.}$$

INTEGRATION

7.1   Antiderivatives

Your Turn 1

Find an antiderivative for the function $f(x) = 8x^7$.

Since the derivative of x^n is nx^{n-1}, the derivative of x^8 is $8x^7$. Thus x^8 is an antiderivative of $8x^7$. The general antiderivative is $x^8 + C$.

Your Turn 2

Find $\int \dfrac{1}{t^4}\, dt$.

Use the power rule with $n = -4$.

$$\int \frac{1}{t^4}\, dt = \int t^{-4}\, dt$$

$$= \frac{t^{-4+1}}{-4+1} + C$$

$$= \frac{t^{-3}}{-3} + C$$

$$= -\frac{1}{3t^3} + C$$

Your Turn 3

Find $\int (6x^2 + 8x - 9)\, dx$.

Use the sum or difference rule and the constant multiple rule.

$$\int (6x^2 + 8x - 9)\, dx = \int 6x^2\, dx + \int 8x\, dx - \int 9\, dx$$

$$= 6\int x^2\, dx + 8\int x\, dx - 9\int dx$$

Now use the power rule on each term.

$$6\int x^2\, dx + 8\int x\, dx - 9\int dx$$

$$= 6\left(\frac{x^3}{3}\right) + 8\left(\frac{x^2}{2}\right) - 9x + C$$

$$= 2x^3 + 4x^2 - 9x + C$$

Your Turn 4

$$\int \frac{x^3 - 2}{\sqrt{x}}\, dx = \int \left(\frac{x^3}{\sqrt{x}} - \frac{2}{\sqrt{x}}\right)$$

$$= \int \frac{x^3}{x^{1/2}}\, dx - \int \frac{2}{x^{1/2}}\, dx$$

$$= \int x^{5/2}\, dx - 2\int x^{-1/2}\, dx$$

$$= \frac{2}{7}x^{7/2} - 2\left(\frac{2}{1}x^{1/2}\right) + C$$

$$= \frac{2}{7}x^{7/2} - 4x^{1/2} + C$$

Your Turn 5

$$\int \left(\frac{3}{x} + e^{-3x}\right) dx = \int \frac{3}{x}\, dx + \int e^{-3x}\, dx$$

$$= 3\int \frac{1}{x}\, dx + \int e^{-3x}\, dx$$

$$= 3\ln|x| - \frac{1}{3}e^{-3x} + C$$

Your Turn 6

Suppose an object is thrown down from the top of the 2717-ft tall Burj Khalifa with an initial velocity of -20 ft/sec. Find when it hits the ground and how fast it is traveling when it hits the ground.

In Example 11 (b) we derived the formulas

$$v(t) = -32t - 20$$

$$s(t) = -16t^2 - 20t + 1100$$

for the velocity v and distance above the ground s for an object thrown down from the Willis Tower. The only change required for the new problem is to change 1100 to the height of the Burj Khalifa, 2717, so we have

$$s(t) = -16t^2 - 20t + 2717.$$

To find when the object hits the ground we solve

$$s(t) = -16t^2 - 20t + 2717 = 0.$$

Use the quadratic formula.

$$t = \frac{20 \pm \sqrt{20^2 - 4(-16)(2717)}}{2(-16)}$$

$$t \approx -13.62 \quad \text{or} \quad t \approx 12.42$$

Only the positive root is relevant. To find the speed on impact, substitute $t = 12.42$ into the formula for v.

$$v(12.42) = -32(12.42) - 20$$
$$\approx -417$$

The object hits the ground after 12.42 sec, traveling downward at 417 ft/sec.

Your Turn 7

$$f'(x) = 3x^{1/2} + 4$$

$$f(x) = \int f'(x)\, dx$$

$$= \int (3x^{1/2} + 4)\, dx$$

$$= 3\int x^{1/2}\, dx + 4\int dx$$

$$= 3\left(\frac{2}{3}x^{3/2}\right) + 4x + C$$

$$= 2x^{3/2} + 4x + C$$

Since the graph of f is to go through the point $(1, -2)$, $f(1) = -2$.

$$2(1)^{3/2} + 4(1) + C = -2$$
$$2 + 4 + C = -2$$
$$C = -8$$

Thus,

$$f(x) = 2x^{3/2} + 4x - 8.$$

7.1 Warmup Exercises

W1. $f(x) = 5x^4 + 6\sqrt{x} = 5x^4 + 6x^{1/2}$

$$f'(x) = (5)\left(4x^3\right) + (6)\left(\frac{1}{2}x^{-1/2}\right)$$

$$= 20x^3 + \frac{3}{\sqrt{x}}$$

W2. $f(x) = 7e^{3x} - 8x^{5/2}$

$$f'(x) = (7)\left(3e^{3x}\right) - (8)\left(\frac{5}{2}x^{(5/2)-1}\right)$$

$$= 21e^{3x} - 20x^{3/2}$$

7.1 Exercises

1. If $F(x)$ and $G(x)$ are both antiderivatives of $f(x)$, then there is a constant C such that

$$F(x) - G(x) = C.$$

The two functions can differ only by a constant.

5. $\displaystyle\int 6\, dk = 6\int 1\, dk$

$$= 6\int k^0\, dk$$

$$= 6\cdot\frac{1}{1}k^{0+1} + C$$

$$= 6k + C$$

7. $\displaystyle\int (2z + 3)\, dz$

$$= 2\int z\, dz + 3\int z^0\, dz$$

$$= 2\cdot\frac{1}{1+1}z^{1+1} + 3\cdot\frac{1}{0+1}z^{0+1} + C$$

$$= z^2 + 3z + C$$

9. $\displaystyle\int (6t^2 - 8t + 7)\, dt$

$$= 6\int t^2\, dt - 8\int t\, dt + 7t\int t^0\, dt$$

$$= \frac{6t^3}{3} - \frac{8t^2}{2} + 7t + C$$

$$= 2t^3 - 4t^2 + 7t + C$$

11. $\displaystyle\int (4z^3 + 3z^2 + 2z - 6)\, dz$

$$= 4\int z^3\, dz + 3\int z^2\, dz + 2\int z\, dz$$

$$-6\int z^0\, dz$$

$$= \frac{4z^4}{4} + \frac{3z^3}{3} + \frac{2z^2}{2} - 6z + C$$

$$= z^4 + z^3 + z^2 - 6z + C$$

13. $\displaystyle\int (5\sqrt{z} + \sqrt{2})\, dz = 5\int z^{1/2}\, dz + \sqrt{2}\int dz$

$$= \frac{5z^{3/2}}{\frac{3}{2}} + \sqrt{2}z + C$$

$$= 5\left(\frac{2}{3}\right)z^{3/2} + \sqrt{2}z + C$$

$$= \frac{10z^{3/2}}{3} + \sqrt{2}z + C$$

15. $\displaystyle\int 5x(x^2 - 8)\,dx = \int (5x^3 - 40x)\,dx$

$$= \frac{5x^4}{4} - \frac{40x^2}{2} + C$$

$$= \frac{5x^4}{4} - 20x^2 + C$$

17. $\displaystyle\int (4\sqrt{v} - 3v^{3/2})\,dv$

$$= 4\int v^{1/2}\,dv - 3\int v^{3/2}\,dv$$

$$= \frac{4v^{3/2}}{\frac{3}{2}} - \frac{3v^{5/2}}{\frac{5}{2}} + C$$

$$= \frac{8v^{3/2}}{3} - \frac{6v^{5/2}}{5} + C$$

19. $\displaystyle\int (10u^{3/2} - 14u^{5/2})\,du$

$$= 10\int u^{3/2}\,du - 14\int u^{5/2}\,du$$

$$= \frac{10u^{5/2}}{\frac{5}{2}} - \frac{14u^{7/2}}{\frac{7}{2}} + C$$

$$= 10\left(\frac{2}{5}\right)u^{5/2} - 14\left(\frac{2}{7}\right)u^{7/2} + C$$

$$= 4u^{5/2} - 4u^{7/2} + C$$

21. $\displaystyle\int \left(\frac{7}{z^2}\right)dz = \int 7z^{-2}\,dz$

$$= 7\int z^{-2}\,dz$$

$$= 7\left(\frac{z^{-2+1}}{-2+1}\right) + C$$

$$= \frac{7z^{-1}}{-1} + C$$

$$= -\frac{7}{z} + C$$

23. $\displaystyle\int \left(\frac{\pi^3}{y^3} - \frac{\sqrt{\pi}}{\sqrt{y}}\right)dy$

$$= \int \pi^3 y^{-3}\,dy - \int \sqrt{\pi}\,y^{-1/2}\,dy$$

$$= \pi^3 \int y^{-3}\,dy - \sqrt{\pi}\int y^{-1/2}\,dy$$

$$= \pi^3\left(\frac{y^{-2}}{-2}\right) - \sqrt{\pi}\left(\frac{y^{-1/2}}{\frac{1}{2}}\right) + C$$

$$= -\frac{\pi^3}{2y^2} - 2\sqrt{\pi y} + C$$

25. $\displaystyle\int (-9t^{-2.5} - 2t^{-1})\,dt$

$$= -9\int t^{-2.5}\,dt - 2\int t^{-1}\,dt$$

$$= \frac{-9t^{-1.5}}{-1.5} - 2\int \frac{dt}{t}$$

$$= 6t^{-1.5} - 2\ln|t| + C$$

27. $\displaystyle\int \frac{1}{3x^2}\,dx = \int \frac{1}{3}x^{-2}\,dx$

$$= \frac{1}{3}\int x^{-2}\,dx$$

$$= \frac{1}{3}\left(\frac{x^{-1}}{-1}\right) + C$$

$$= -\frac{1}{3}x^{-1} + C$$

$$= -\frac{1}{3x} + C$$

29. $\displaystyle\int 3e^{-0.2x}\,dx = 3\int e^{-0.2x}\,dx$

$$= 3\left(\frac{1}{-0.2}\right)e^{-0.2x} + C$$

$$= \frac{3(e^{-0.2x})}{-0.2} + C$$

$$= -15e^{-0.2x} + C$$

31. $\displaystyle\int \left(-\frac{3}{x} + 4e^{-0.4x} + e^{0.1}\right)dx$

$$= -3\int \frac{dx}{x} + 4\int e^{-0.4x}\,dx + e^{0.1}\int dx$$

$$= -3\ln|x| + \frac{4e^{-0.4x}}{-0.4} + e^{0.1}x + C$$

$$= -3\ln|x| - 10e^{-0.4x} + e^{0.1}x + C$$

33. $\displaystyle\int\left(\frac{1+2t^3}{4t}\right)dt = \int\left(\frac{1}{4t}+\frac{t^2}{2}\right)dt$

$\displaystyle\qquad = \frac{1}{4}\int\frac{1}{t}\,dt + \frac{1}{2}\int t^2\,dt$

$\displaystyle\qquad = \frac{1}{4}\ln|t| + \frac{1}{2}\left(\frac{t^3}{3}\right) + C$

$\displaystyle\qquad = \frac{1}{4}\ln|t| + \frac{t^3}{6} + C$

35. $\displaystyle\int(e^{2u}+4u)\,du = \frac{e^{2u}}{2}+\frac{4u^2}{2}+C$

$\displaystyle\qquad = \frac{e^{2u}}{2}+2u^2+C$

37. $\displaystyle\int(x+1)^2\,dx = \int(x^2+2x+1)\,dx$

$\displaystyle\qquad = \frac{x^3}{3}+\frac{2x^2}{2}+x+C$

$\displaystyle\qquad = \frac{x^3}{3}+x^2+x+C$

39. $\displaystyle\int\frac{\sqrt{x}+1}{\sqrt[3]{x}}\,dx = \int\left(\frac{\sqrt{x}}{\sqrt[3]{x}}+\frac{1}{\sqrt[3]{x}}\right)dx$

$\displaystyle\qquad = \int\left(x^{(1/2-1/3)}+x^{-1/3}\right)dx$

$\displaystyle\qquad = \int x^{1/6}\,dx + \int x^{-1/3}\,dx$

$\displaystyle\qquad = \frac{x^{7/6}}{\frac{7}{6}}+\frac{x^{2/3}}{\frac{2}{3}}+C$

$\displaystyle\qquad = \frac{6x^{7/6}}{7}+\frac{3x^{2/3}}{2}+C$

41. $\displaystyle\int 10^x\,dx = \frac{10^x}{\ln 10}+C$

43. Find $f(x)$ such that $f'(x) = x^{2/3}$, and $\left(1,\frac{3}{5}\right)$ is on the curve.

$\displaystyle\int x^{2/3}\,dx = \frac{x^{5/3}}{\frac{5}{3}}+C$

$\displaystyle f(x) = \frac{3x^{5/3}}{5}+C$

Since $\left(1,\frac{3}{5}\right)$ is on the curve,

$f(1) = \dfrac{3}{5}.$

$f(1) = \dfrac{3(1)^{5/3}}{5}+C = \dfrac{3}{5}$

$\dfrac{3}{5}+C = \dfrac{3}{5}$

$C = 0.$

Thus,

$f(x) = \dfrac{3x^{5/3}}{5}.$

45. $C'(x) = 4x - 5$; fixed cost is $8.

$\displaystyle C(x) = \int(4x-5)\,dx$

$\displaystyle\qquad = \frac{4x^2}{2}-5x+k$

$\displaystyle\qquad = 2x^2-5x+k$

$C(0) = 2(0)^2-5(0)+k = k$

Since $C(0) = 8,\ k = 8.$

Thus,

$C(x) = 2x^2-5x+8.$

47. $C'(x) = 0.03e^{0.01x}$; fixed cost $8.

$\displaystyle C(x) = \int 0.03e^{0.01x}\,dx$

$\displaystyle\qquad = 0.03\int e^{0.01x}\,dx$

$\displaystyle\qquad = 0.03\left(\frac{1}{0.01}e^{0.01x}\right)+k$

$\displaystyle\qquad = 3e^{0.01x}+k$

$C(0) = 3e^{0.01(0)}+k = 3(1)+k$

$\displaystyle\qquad = 3+k$

Since $C(0) = 8,\ 3+k = 8,$ and $k = 5.$

Thus,

$C(x) = 3e^{0.01x}+5.$

49. $C'(x) = x^{2/3} + 2$; 8 units cost $58.

$$C(x) = \int (x^{2/3} + 2)\,dx$$

$$= \frac{3x^{5/3}}{5} + 2x + k$$

$$C(8) = \frac{3(8)^{5/3}}{5} + 2(8) + k$$

$$= \frac{3(32)}{5} + 16 + k$$

Since $C(8) = 58$,

$$58 - 16 - \frac{96}{5} = k$$

$$\frac{114}{5} = k.$$

Thus,

$$C(x) = \frac{3x^{5/3}}{5} + 2x + \frac{114}{5}.$$

51. $C'(x) = 5x - \dfrac{1}{x}$; 10 units cost $94.20, so

$C(10) = 94.20.$

$$C(x) = \int \left(5x - \frac{1}{x}\right) dx = \frac{5x^2}{2} - \ln|x| + k$$

$$C(10) = \frac{5(10)^2}{2} - \ln(10) + k$$

$$= 250 - 2.30 + k.$$

Since $C(10) = 94.20$,

$$94.20 = 247.70 + k$$

$$-153.50 = k.$$

Thus, $C(x) = \dfrac{5x^2}{2} - \ln|x| - 153.50.$

53. $R'(x) = 175 - 0.02x - 0.03x^2$

$$R = \int (175 - 0.02x - 0.03x^2)\,dx$$

$$= 175x - 0.01x^2 - 0.01x^3 + C.$$

If $x = 0$, then $R = 0$ (no items sold means no revenue), and

$$0 = 175(0) - 0.01(0)^2 - 0.01(0)^3 + C$$

$$0 = C.$$

Thus, $R = 175x - 0.01x^2 - 0.01x^3$ gives the revenue function. Now, recall that $R = xp$, where p is the demand function. Then

$$175x - 0.01x^2 - 0.01x^3 = xp$$

$$175 - 0.01x - 0.01x^2 = p,$$

the demand function.

55. $R'(x) = 500 - 0.15\sqrt{x}$

$$R = \int (500 - 0.015\sqrt{x})\,dx$$

$$= 500x - 0.1x^{3/2} + C.$$

If $x = 0$, $R = 0$ (no items sold means no revenue), and

$$0 = 500(0) - 0.1(0)^{3/2} + C$$

$$0 = C.$$

Thus, $R = 500x - 0.1x^{3/2}$ gives the revenue function. Now, recall that $R = xp$, where p is the demand function. Then

$$500x - 0.1x^{3/2} = xp$$

$$500 - 0.1\sqrt{x} = p,$$ the demand function.

57. (a) $p'(t) = 43.14t - 143.5$

$$p(t) = \int p(t)\,dt$$

$$= (43.14)\left(\frac{1}{2}t^2\right) - 143.5t$$

$$= 21.57t^2 - 143.5t + C$$

In 2008, when $t = 8$, there were 828,328 patent applications, or 828.328 thousand.

$$p(8) = 828.328$$

$$21.57(8)^2 - 143.5(8) + C = 828.328$$

$$232.48 + C = 828.328$$

$$C = 828,328 - 232.48 = 595.848 \approx 595.85$$

Thus $p(t) = 21.57t^2 - 143.5t + 595.85$

(b) In 2013, $t = 13$.

$$p(13) = 21.57(13)^2 - 143.5(13) + 595.85$$

$$= 2375.68 \text{ thousand or about } 2,376,000$$

59. (a) $P'(x) = 50x^3 + 30x^2$; profit is -40 when no cheese is sold.

$$P(x) = \int (50x^3 + 30x^2)\,dx$$

$$= \frac{25x^4}{2} + 10x^3 + k$$

$$P(0) = \frac{25(0)^4}{2} + 10(0)^3 + k$$

Since
$$P(0) = -40,$$
$$-40 = k.$$
Thus,
$$P(x) = \frac{25x^4}{2} + 10x^3 - 40.$$

(b) $P(2) = \dfrac{25(2)^4}{2} + 10(2)^3 - 40 = 240$

The profit from selling 200 lbs of Brie cheese is $240.

61. $\displaystyle \int \frac{g(x)}{x}\,dx = \int \frac{a - bx}{x}\,dx$

$$= \int \left(\frac{a}{x} - b \right) dx$$

$$= a \int \frac{dx}{x} - b \int dx$$

$$= a \ln|x| - bx + C$$

Since x represents a positive quantity, the absolute value sign can be dropped.

$$\int \frac{g(x)}{x}\,dx = a \ln x - bx + C$$

63. $N'(t) = Ae^{kt}$

(a) $N(t) = \dfrac{A}{k} e^{kt} + C$

$A = 50$, $N(t) = 300$ when $t = 0$.

$$N(0) = \frac{50}{k} e^0 + C = 300$$
$$N'(5) = 250$$

Therefore,

$$N'(5) = 50e^{5k} = 250$$
$$e^{5k} = 5$$
$$5k = \ln 5$$
$$k = \frac{\ln 5}{5}.$$

$$N(0) = \frac{50}{\frac{\ln 5}{5}} + C = 300$$

$$\frac{250}{\ln 5} + C = 300$$

$$C = 300 - \frac{250}{\ln 5} \approx 144.67$$

$$N(t) = \frac{50}{\frac{\ln 5}{5}} e^{(\ln 5/5)t} + 144.67$$

$$= 155.3337 e^{0.321888t} + 144.67$$

$$\approx 155.3 e^{0.3219t} + 144.7$$

(b) $N(12) = 155.3337 e^{0.321888(12)} + 144.67$

$$\approx 7537$$

There are 7537 cells present after 12 days.

65. $B'(t) = 0.04764t^2 - 0.9148t + 13.20$

(a)

$B(t)$

$$= \int (0.04764t^2 - 0.9148t + 13.20)\,dt$$

$$= \frac{0.04764}{3} t^3 - \frac{0.9148}{2} t^2 + 13.20t + C$$

$$= 0.01588t^3 - 0.4574t^2 + 13.20t + C$$

In 1970, when $t = 0$, 792,316 or about 792.3 thousand degrees were conferred, so $B(0) = 792.3$ and thus $C = 792.3$ and the formula for B is

$$B(t) = 0.01588t^3 - 0.4574t^2$$
$$+ 13.20t + 792.3$$

(b) To project the number of bachelor's degrees conferred in 2020 we set t equal to 50 and evaluate $B(50)$.

$$B(50) = 0.01588(50)^3 - 0.4574(50)^2$$
$$+ 13.20(50) + 792.3$$
$$\approx 2294$$

The formula predicts that 2294 thousand or about 2,294,000 bachelor's degrees will be conferred in 2020.

67. $a(t) = 5t^2 + 4$

$$v(t) = \int (5t^2 + 4)\,dt$$

$$= \frac{5t^3}{3} + 4t + C$$

$$v(0) = \frac{5(0)^3}{3} + 4(0) + C$$

Since $v(0) = 6, C = 6.$

$$v(t) = \frac{5t^3}{3} + 4t + 6$$

69. $a(t) = -32$

$$v(t) = \int -32\, dt = -32t + C_1$$

$$v(0) = -32(0) + C_1$$

Since $v(0) = 0, C_1 = 0$.

$$v(t) = -32t$$

$$s(t) = \int -32t\, dt$$

$$= \frac{-32t^2}{2} + C_2$$

$$= -16t^2 + C_2$$

At $t = 0$, the plane is at 6400 ft.

That is, $s(0) = 6400$.

$$s(0) = -16(0)^2 + C_2$$
$$6400 = 0 + C_2$$
$$C_2 = 6400$$
$$s(t) = -16t^2 + 6400$$

When the object hits the ground, $s(t) = 0$.

$$-16t^2 + 6400 = 0$$
$$-16t^2 = -6400$$
$$t^2 = 400$$
$$t = \pm 20$$

Discard -20 since time must be positive.
The object hits the ground in 20 sec.

71. $a(t) = \dfrac{15}{2}\sqrt{t} = 3e^{-t}$

$$v(t) = \int \left(\frac{15}{2}\sqrt{t} + 3e^{-t} \right) dt$$

$$= \int \left(\frac{15}{2}t^{1/2} + 3e^{-t} \right) dt$$

$$= \frac{15}{2}\left(\frac{t^{3/2}}{\frac{3}{2}} \right) + 3\left(\frac{1}{-1}e^{-t} \right) + C_1$$

$$= 5t^{3/2} - 3e^{-t} + C_1$$

$$v(0) = 5(0)^{3/2} - 3e^{-0} + C_1 = -3 + C_1$$

Since $v(0) = -3, C_1 = 0$.

$$v(t) = 5t^{3/2} - 3e^{-t}$$

$$s(t) = \int (5t^{3/2} - 3e^{-t})\, dt$$

$$= 5\left(\frac{t^{5/2}}{\frac{5}{2}} \right) - 3\left(-\frac{1}{1}e^{-t} \right) + C_2$$

$$= 2t^{5/2} + 3e^{-t} + C_2$$

$$s(0) = 2(0)^{5/2} + 3e^{-0} + C_2 = 3 + C_2$$

Since $s(0) = 4, C_2 = 1$.

Thus,

$$s(t) = 2t^{5/2} + 3e^{-t} + 1.$$

73. First find $v(t)$ by integrating $a(t)$:

$$v(t) = \int (-32)dt = -32t + k.$$

When $t = 5, v(t) = 0$:

$$0 = -32(5) + k$$
$$160 = k$$

and

$$v(t) = -32t + 160.$$

Now integrate $v(t)$ to find $h(t)$.

$$h(t) = \int (-32t + 160)dt = -16t^2 + 160t + C$$

Since $h(t) = 412$ when $t = 5$, we can substitute these values into the equation for $h(t)$ to get $C = 12$ and

$$h(t) = -16t^2 + 160t + 12.$$

Therefore, from the equation given in Exercise 72, the initial velocity v_0 is 160 ft/sec and the initial height of the rocket h_0 is 12 ft.

7.2   Substitution

Your Turn 1

Find $\displaystyle\int 8x(4x^2 + 8)^6\, dx$.

Let $u = 4x^2 + 8$.

Then $du = 8x\, dx$.

Now substitute.

$$\int 8x(4x^2 + 8)^6 \, dx = \int (4x^2 + 8)^6 \, (8x \, dx)$$

$$= \int u^6 \, du$$

$$= \frac{1}{7}u^7 + C$$

Now replace u with $4x^2 + 8$.

$$\int 8x(4x^2 + 8)^6 \, dx = \frac{1}{7}u^7 + C$$

$$= \frac{1}{7}(4x^2 + 8)^7 + C$$

Your Turn 2

Find $\int x^3 \sqrt{3x^4 + 10} \, dx$.

Let $\quad u = 3x^4 + 10$.

Then $du = 12x^3 \, dx$.

Multiply the integral by $\frac{12}{12}$ to introduce the factor of 12 needed for du, and then substitute.

$$\int x^3 \sqrt{3x^4 + 10} \, dx = \frac{1}{12} \int 12x^3 \sqrt{3x^4 + 10} \, dx$$

$$= \frac{1}{12} \int \sqrt{3x^4 + 10} \, (12x^3 \, dx)$$

$$= \frac{1}{12} \int u^{1/2} \, du$$

$$= \frac{1}{12}\left(\frac{2}{3}u^{3/2} + C\right)$$

$$= \frac{1}{18}u^{3/2} + C$$

(Note that $(1/12) \, C$ is just a different constant, which we can also call C.) Now replace u with $3x^4 + 10$.

$$\int x^3 \sqrt{3x^4 + 10} \, dx = \frac{1}{18}u^{3/2} + C$$

$$= \frac{1}{18}(3x^4 + 10)^{3/2} + C$$

Your Turn 3

Find $\int \frac{x + 1}{(4x^2 + 8x)^3} \, dx$.

Let $\quad u = 4x^2 + 8x$.

Then $du = (8x + 8)\, dx = 8(x + 1)\, dx$

Multiply the integral by $\frac{8}{8}$ to introduce the factor of 8 needed for du, and then substitute.

$$\int \frac{x + 1}{(4x^2 + 8x)^3} \, dx = \frac{1}{8} \int \frac{8(x + 1)}{(4x^2 + 8x)^3} \, dx$$

$$= \frac{1}{8} \int \frac{1}{u^3} \, du$$

$$= \frac{1}{8} \int u^{-3} \, du$$

$$= \frac{1}{8}\left(-\frac{1}{2}u^{-2} + C\right)$$

$$= -\frac{1}{16u^2} + C$$

Now replace u with $4x^2 + 8x$.

$$\int \frac{x + 1}{(4x^2 + 8x)^3} \, dx = -\frac{1}{16u^2} + C$$

$$= -\frac{1}{16(4x^2 + 8x)^2} + C$$

Your Turn 4

Find $\int \frac{x + 3}{x^2 + 6x} \, dx$.

Let $\quad u = x^2 + 6x$.

Then $du = 2x + 6\, dx = 2(x + 3)\, dx$

Multiply the integral by $\frac{2}{2}$ to introduce the factor of 2 needed for du, and then substitute.

$$\int \frac{x + 3}{x^2 + 6x} \, dx = \frac{1}{2} \int \frac{2(x + 3)}{x^2 + 6x} \, dx$$

$$= \frac{1}{2} \int \frac{1}{u} \, du$$

$$= \frac{1}{2}(\ln|u| + C)$$

$$= \frac{1}{2}\ln|u| + C$$

Now replace u with $x^2 + 6x$.

$$\int \frac{x + 3}{x^2 + 6x} \, dx = \frac{1}{2}\ln|u| + C$$

$$= \frac{1}{2}|x^2 + 6x| + C$$

Your Turn 5

Find $\int x^3 e^{x^4} \, dx$.

Let $\quad u = x^4$.

Then $du = 4x^3 \, dx$.

Multiply the integral by $\frac{4}{4}$ to introduce the factor of 4 needed for du, and then substitute.

$$\int x^3 e^{x^4} \, dx = \frac{1}{4} \int 4x^3 e^{x^4} \, dx$$

$$= \frac{1}{4} \int e^{x^4} (4x^3 \, dx)$$

$$= \frac{1}{4} \int e^u \, du$$

$$= \frac{1}{4} e^u + C$$

Now replace u with x^4.

$$\int x^3 e^{x^4} \, dx = \frac{1}{4} e^u + C = \frac{1}{4} e^{x^4} + C$$

Your Turn 6

Find $\int x\sqrt{3 + x} \, dx$.

Let $u = 3 + x$.
Then $du = dx$ and $x = u - 3$.

Now substitute.

$$\int x\sqrt{3 + x} \, dx = \int (u - 3)\sqrt{u} \, du$$

$$= \int u\sqrt{u} \, du - 3 \int \sqrt{u} \, du$$

$$= \int u^{3/2} \, du - 3 \int u^{1/2} \, du$$

$$= \frac{2}{5} u^{5/2} - 3\left(\frac{2}{3} u^{3/2}\right) + C$$

$$= \frac{2}{5} u^{5/2} - 2u^{3/2} + C$$

Now replace u with $3 + x$.

$$\int x\sqrt{3 + x} \, dx = \frac{2}{5} u^{5/2} - 2u^{3/2} + C$$

$$= \frac{2}{5} (3 + x)^{5/2} - 2(3 + x)^{3/2} + C$$

7.2 Warmup Exercises

W1. $\displaystyle \int 10x^4 + 6\sqrt{x} \, dx = \int 10x^4 + 6x^{1/2} \, dx$

$$= 10\left(\frac{1}{5} x^5\right) + 6\left(\frac{1}{3/2} x^{3/2}\right) + C$$

$$= 2x^5 + 4x^{3/2} + C$$

W2. $\displaystyle \int \frac{5}{x} + \frac{4}{x^3} \, dx = \int \frac{5}{x} + 4x^{-3} \, dx$

$$= 5\ln|x| - 4\left(-\frac{1}{2} x^{-2}\right) + C$$

$$= 5\ln|x| - \frac{2}{x^2} + C$$

W3. $\displaystyle \int e^{6x} \, dx = \frac{1}{6} e^{6x} + C = \frac{e^{6x}}{6} + C$

7.2 Exercises

3. $\displaystyle \int 4(2x + 3)^4 \, dx = 2 \int 2(2x + 3)^4 \, dx$

Let $u = 2x + 3$, so that $du = 2 \, dx$.

$$= 2 \int u^4 \, du$$

$$= \frac{2 \cdot u^5}{5} + C$$

$$= \frac{2(2x + 3)^5}{5} + C$$

5. $\displaystyle \int \frac{2 \, dm}{(2m + 1)^3} = \int 2(2m + 1)^{-3} \, dm$

Let $u = 2m + 1$, so that $du = 2 \, dm$.

$$= \int u^{-3} \, du$$

$$= \frac{u^{-2}}{-2} + C = \frac{-(2m + 1)^{-2}}{2} + C$$

7. $\displaystyle \int \frac{2x + 2}{(x^2 + 2x - 4)^4} \, dx$

$$= \int (2x + 2)(x^2 + 2x - 4)^{-4} \, dx$$

Let $w = x^2 + 2x - 4$, so that
$dw = (2x + 2) \, dx$.

$$= \int w^{-4} \, dw$$

$$= \frac{w^{-3}}{-3} + C$$

$$= -\frac{(x^2 + 2x - 4)^{-3}}{3} + C$$

$$= -\frac{1}{3(x^2 + 2x - 4)^3} + C$$

9. $\displaystyle\int z\sqrt{4z^2 - 5}\,dz = \int z(4z^2 - 5)^{1/2}dz$

$$= \frac{1}{8}\int 8z(4z^2 - 5)^{1/2}\,dz$$

Let $u = 4z^2 - 5$, so that $du = 8z\,dz$.

$$= \frac{1}{8}\int u^{1/2}du$$

$$= \frac{1}{8}\cdot\frac{u^{3/2}}{\frac{3}{2}} + C$$

$$= \frac{1}{8}\cdot\left(\frac{2}{3}\right)u^{3/2} + C$$

$$= \frac{(4z^2 - 5)^{3/2}}{12} + C$$

11. $\displaystyle\int 3x^2 e^{2x^3}\,dx = \frac{1}{2}\int 2\cdot 3x^2 e^{2x^3}\,dx$

Let $u = 2x^3$, so that $du = 6x^2 dx$.

$$= \frac{1}{2}\int e^u\,du$$

$$= \frac{1}{2}e^u + C$$

$$= \frac{e^{2x^3}}{2} + C$$

13. $\displaystyle\int (1 - t)e^{2t - t^2}\,dt$

$$= \frac{1}{2}\int 2(1 - t)e^{2t - t^2}\,dt$$

Let $u = 2t - t^2$, so that $du = (2 - 2t)dt$.

$$= \frac{1}{2}\int e^u\,du$$

$$= \frac{e^u}{2} + C = \frac{e^{2t - t^2}}{2} + C$$

15. $\displaystyle\int\frac{e^{1/z}}{z^2}\,dz = -\int e^{1/z}\cdot\frac{-1}{z^2}\,dz$

Let $u = \frac{1}{z}$, so that $du = \frac{-1}{z^2}\,dx$.

$$\int\frac{e^{1/z}}{z^2}\,dz = -\int e^u\,du$$

$$= -e^u + C$$

$$= -e^{1/z} + C$$

17. $\displaystyle\int\frac{t}{t^2 + 2}\,dt$

Let $t^2 + 2 = u$, so that $2t\,dt = du$.

$$= \frac{1}{2}\int\frac{du}{u}$$

$$= \frac{1}{2}\ln|u| + C$$

$$= \frac{\ln(t^2 + 2)}{2} + C$$

19. $\displaystyle\int\frac{x^3 + 2x}{x^4 + 4x^2 + 7}\,dx$

Let $u = x^4 + 4x^2 + 7$.

Then $du = (4x^3 + 8x)dx = 4(x^3 + 2x)dx$.

$$\int\frac{x^3 + 2x}{x^4 + 4x^2 + 7}\,dx = \frac{1}{4}\int\frac{(4x^3 + 2x)}{x^4 + 4x^2 + 7}\,dx$$

$$= \frac{1}{4}\int\frac{1}{u}\,du = \frac{1}{4}\ln|u| + C$$

$$= \frac{1}{4}\ln(x^4 + 4x^2 + 7) + C$$

Since $x^4 + 4x^2 + 7 > 0$ for all x, we can write this answer as $\frac{1}{4}\ln(x^4 + 4x^2 + 7) + C$.

21. $\displaystyle\int\frac{2x + 1}{(x^2 + x)^3}\,dx$

$$= \int(2x + 1)(x^2 + x)^{-3}\,dx$$

Let $u = x^2 + x$, so that $du = (2x + 1)\,dx$.

$$= \int u^{-3}\,du = \frac{u^{-2}}{-2} + C$$

$$= \frac{-1}{2u^2} + C = \frac{-1}{2(x^2 + x)^2} + C$$

23. $\displaystyle\int p(p + 1)^5\,dp$

Let $u = p + 1$, so that $du = dp$; also, $p = u - 1$.

$$= \int (u - 1)u^5 du$$

$$= \int (u^6 - u^5) du$$

$$= \frac{u^7}{7} - \frac{u^6}{6} + C$$

$$= \frac{(p + 1)^7}{7} - \frac{(p + 1)^6}{6} + C$$

25. $\int \dfrac{u}{\sqrt{u - 1}} du$

$$= \int u(u - 1)^{-1/2} du$$

Let $w = u - 1$, so that $dw = du$ and $u = w + 1$.

$$= \int (w + 1)w^{-1/2} dw$$

$$= \int (w^{1/2} + w^{-1/2}) dw$$

$$= \frac{w^{3/2}}{\frac{3}{2}} + \frac{w^{1/2}}{\frac{1}{2}} + C$$

$$= \frac{2(u - 1)^{3/2}}{3} + 2(u - 1)^{1/2} + C$$

27. $\int \left(\sqrt{x^2 + 12x} \right)(x + 6)\, dx$

$$= \int (x^2 + 12x)^{1/2}(x + 6)\, dx$$

Let $x^2 + 12x = u$, so that

$$(2x + 12)\, dx = du$$
$$2(x + 6)\, dx = du.$$

$$= \frac{1}{2} \int u^{1/2} du = \frac{1}{2}\left(\frac{2}{3}\right)u^{3/2} + C$$

$$= \frac{(x^2 + 12x)^{3/2}}{3} + C$$

29. $\int \dfrac{3(1 + 3 \ln x)^2}{x} dx$

Let $u = 1 + 3 \ln x$, so that $du = \frac{3}{x} dx$.

$$= \frac{1}{3} \int \frac{3(1 + 3 \ln x)^2}{x} dx$$

$$= \frac{1}{3} \int u^2 du$$

$$= \frac{1}{3} \cdot \frac{u^3}{3} + C$$

$$= \frac{(1 + 3 \ln x)^3}{9} + C$$

31. $\int \dfrac{e^{2x}}{e^{2x} + 5} dx$

Let $u = e^{2x} + 5$, so that $du = 2e^{2x}\, dx$.

$$= \frac{1}{2} \int \frac{du}{u}$$

$$= \frac{1}{2} \ln |u| + C$$

$$= \frac{1}{2} \ln |e^{2x} + 5| + C$$

$$= \frac{1}{2} \ln (e^{2x} + 5) + C$$

33. $\int \dfrac{\log x}{x} dx$

Let $u = \log x$, so that $du = \frac{1}{(\ln 10)x} dx$.

$$\int \frac{\log x}{x} dx = (\ln 10) \int \frac{\log x}{(\ln 10)x} dx = (\ln 10) \int u\, du$$

$$= (\ln 10)\left(\frac{u^2}{2}\right) + C$$

$$= \frac{(\ln 10)(\log x)^2}{2} + C$$

35. $\int x 8^{3x^2 + 1} dx$

Let $u = 3x^2 + 1$, so that $du = 6x\, dx$.

$$= \frac{1}{6} \int 6x \cdot 8^{3x^2 + 1} dx$$

$$= \frac{1}{6} \int 8^u du$$

$$= \frac{1}{6}\left(\frac{8^u}{\ln 8}\right) + C$$

$$= \frac{8^{3x^2 + 1}}{6 \ln 8} + C$$

39. **(a)** $R'(x) = 4x(x^2 + 27,000)^{-2/3}$

$$R(x) = \int 4x(x^2 + 27,000)^{-2/3}\,dx$$

$$= 2\int 2x(x^2 + 27,000)^{-2/3}\,dx$$

Let $u = x^2 + 27,000$, so that $du = 2x\,dx$.

$$R = 2\int u^{-2/3}\,du$$

$$= 2 \cdot 3u^{1/3} + C$$

$$= 6(x^2 + 27,000)^{1/3} + C$$

$$R(125) = 6(125^2 + 27,000)^{1/3} + C$$

Since $R(125) = 29.591$,

$$6(125^2 + 27,000)^{1/3} + C = 29.591$$

$$C = -180$$

Thus,

$$R(x) = 6(x^2 + 27,000)^{1/3} - 180.$$

(b) $R(x) = 6(x^2 + 27,000)^{1/3} - 180 \geq 40$

$$6(x^2 + 27,000)^{1/3} \geq 220$$

$$(x^2 + 27,000)^{1/3} \geq 36.6667$$

$$x^2 + 27,000 \geq 49,296.43$$

$$x^2 \geq 22,296.43$$

$$x \geq 149.4$$

For a revenue of at least $40,000, 150 players must be sold

41. $C'(x) = \dfrac{60x}{5x^2 + e}$

(a) Let $u = 5x^2 + e$, so that $du = 10x\,dx$.

$$C(x) = \int C'(x)\,dx$$

$$= \int \frac{60x}{5x^2 + e}\,dx$$

$$= 6\int \frac{du}{u} = 6\ln|u| + C$$

$$= 6\ln(5x^2 + e) + C$$

Since $C(0) = 10$, $C = 4$. Therefore,

$$C(x) = 6\ln|5x^2 + e| + 4$$

$$= 6\ln(5x^2 + e) + 4.$$

(b) $C(5) = 6\ln(5 \cdot 5^2 + e) + 4 \approx 33.099$

Since this represents $33,099 dollars which is greater than $20,000, a new source of investment income should be sought.

43. $f'(t) = 4.0674 \cdot 10^{-4} t(t - 1970)^{0.4}$

(a) Let $u = t - 1970$. To get the t outside the parentheses in terms of u, solve $u = t - 1970$ for t to get $t = u + 1970$. Then $dt = du$ and we can substitute as follows.

$$f(t) = \int f'(t)dt = \int 4.0674 \cdot 10^{-4} t(t - 1970)^{0.3} \, dt$$

$$= \int 4.0674 \cdot 10^{-4}(u + 1970)(u)^{0.3} \, du$$

$$= 4.0674 \cdot 10^{-4} \int (u + 1970)(u)^{0.3} \, du$$

$$= 4.0674 \cdot 10^{-4} \int (u^{1.3} + 1970u^{0.3})du$$

$$= 4.0674 \cdot 10^{-4}\left(\frac{u^{2.3}}{2.3} + \frac{1970u^{1.3}}{1.3}\right) + C$$

$$= 4.0674 \cdot 10^{-4}\left[\frac{(t - 1970)^{2.3}}{2.3} + \frac{1970(t - 1970)^{1.3}}{1.3}\right] + C$$

Since $f(1970) = 61.298, C = 61.298$.

Therefore, $f(t) = 4.0674 \cdot 10^{-4}\left[\dfrac{(t - 1970)^{2.3}}{2.3} + \dfrac{1970(t - 1970)^{1.3}}{1.3}\right] + 61.298$.

(b) $f(2020) = 4.0674 \cdot 10^{-4}\left[\dfrac{(2020 - 1970)^{2.3}}{2.3} + \dfrac{1970(2020 - 1970)^{1.3}}{1.3}\right] + 61.298 \approx 162.0$

In the year 2020, there will be about 162,000 local transit vehicles.

45. (a) $P'(t) = 500te^{-t^2/5}$

$$P(t) = \int P'(t) \, dt = \int 500te^{-t^2/5} \, dt$$

Let $u = -t^2/5$; then $du = (-2/5)t \, dt$.

$$\int 500te^{-t^2/5} \, dt$$

$$= (-1250)\int e^{-t^2/5}\left((-2/5)t \, dt\right)$$

$$= -1250 \int e^u \, du$$

$$= -1250e^u + C$$

$$= -1250e^{-t^2/5} + C$$

Since $P(2000)$ is 0,

$$2000 = -1250e^{-(0)^2/5} + C$$

$$C = 3250$$

$$P(t) = 3250 - 1250e^{-t^2/5}$$

(b) $P(3) = 3250 - 1250e^{-(3)^2/5}$

$$= 3043 \text{ to the nearest integer}$$

7.3   Area and the Definite Integral

Your Turn 1

Approximate $\displaystyle\int_1^5 4x\,dx$ using four rectangles.

Find the area of the shaded region:

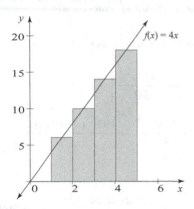

Build a table giving the heights of the rectangles, which are the values of $f(x) = 4x$ at the midpoint of each interval.

i	x_i	$f(x_i)$
1	1.5	6.0
2	2.5	10.0
3	3.5	14.0
4	4.5	18.0

For each interval, $\Delta x = 1$. The sum of the areas of the rectangles is

$$\sum_{i=1}^{4} f(x_i)\,\Delta x_i = f(1.5)\Delta x + f(2.5)\Delta x + f(3.5)\Delta x + f(4.5)\Delta x$$

$$= 1(6) + 1(10) + 1(14) + 1(18)$$

$$= 48.$$

Thus our approximation to the integral is 48. In this case the approximation is exact.

Your Turn 2

A driver has the following velocities at various times:

Time (hr)	0	0.5	1	1.5	2
Velocity (mph)	0	50	56	40	48

Approximate the total distance traveled during the 2-hour period.

Using left endpoints:

distance $= 0(0.5) + 50(0.5) + 56(0.5) + 40(0.5)$

$= 73$ miles

Using right endpoints:

distance $= 50(0.5) + 56(0.5) + 40(0.5) + 48(0.5)$

$= 97$ miles

Averaging these two estimates:

$$\text{distance} = \frac{73 + 97}{2} = 85 \text{ miles}$$

7.3 Exercises

3. $f(x) = 2x + 5$, $x_1 = 0$, $x_2 = 2$, $x_3 = 4$, $x_4 = 6$, and $\Delta x = 2$

(a) $\displaystyle\sum_{i=1}^{4} f(x_i)\Delta x$

$= f(x_1)\Delta x + f(x_2)\Delta x + f(x_3)\Delta x + f(x_4)\Delta x$

$= f(0)(2) + f(2)(2) + f(4)(2) + f(6)(2)$

$= [2(0) + 5](2) + [2(2) + 5](2)$

$\quad + [2(4) + 5](2) + [2(6) + 5](2)$

$= 10 + 9(2) + 13(2) + 17(2)$

$= 88$

(b)

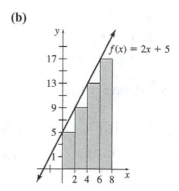

The sum of these rectangles approximates

$$\int_0^8 (2x + 5)\, dx.$$

5. $f(x) = 2x + 5$ from $x = 2$ to $x = 4$

For $n = 4$ rectangles:

$$\Delta x = \frac{4 - 2}{4} = 0.5$$

(a) Using the left endpoints:

i	x_i	$f(x_i)$
1	2	9
2	2.5	10
3	3	11
4	3.5	12

$A = \displaystyle\sum_{1}^{4} f(x_i)\Delta x$

$= 9(0.5) + 10(0.5) + 11(0.5) + 12(0.5)$

$= 21$

(b) Using the right endpoints:

i	x_i	$f(x_i)$
1	2.5	10
2	3	11
3	3.5	12
4	4	13

$A = 10(0.5) + 11(0.5) + 12(0.5) + 13(0.5) = 23$

(c) Average $= \dfrac{21 + 23}{2} = \dfrac{44}{2} = 22$

(d) Using the midpoints:

i	x_i	$f(x_i)$
1	2.25	9.5
2	2.75	10.5
3	3.25	11.5
4	3.75	12.5

$A = \displaystyle\sum_{1}^{4} f(x_i)\Delta x = 9.5(0.5) + 10.5(0.5)$

$\quad + 11.5(0.5) + 12.5(0.5) = 22$

7. $f(x) = -x^2 + 4$ from $x = -2$ to $x = 2$

For $n = 4$ rectangles:

$$\Delta x = \frac{2 - (-2)}{4} = 1$$

(a) Using the left endpoints:

i	x_i	$f(x_i)$
1	-2	$-(-2)^2 + 4 = 0$
2	-1	$-(-1)^2 + 4 = 3$
3	0	$-(0)^2 + 4 = 4$
4	1	$-(1)^2 + 4 = 3$

$$A = \sum_{i=1}^{4} f(x_i)\Delta x$$
$$= (0)(1) + (3)(1) + (4)(1) + (3)(1)$$
$$= 10$$

(b) Using the right endpoints:

i	x_i	$f(x_i)$
1	-1	3
2	0	4
3	1	3
4	2	0

Area $= 1(3) + 1(4) + 1(3) + 1(0) = 10$

(c) Average $= \dfrac{10 + 10}{2} = 10$

(d) Using the midpoints:

i	x_i	$f(x_i)$
1	$-\dfrac{3}{2}$	$\dfrac{7}{4}$
2	$-\dfrac{1}{2}$	$\dfrac{15}{4}$
3	$\dfrac{1}{2}$	$\dfrac{15}{4}$
4	$\dfrac{3}{2}$	$\dfrac{7}{4}$

$$A = \sum_{i=1}^{4} f(x_i)\Delta x$$
$$= \frac{7}{4}(1) + \frac{15}{4}(1) + \frac{15}{4}(1) + \frac{7}{4}(1)$$
$$= 11$$

9. $f(x) = e^x + 1$ from $x = -2$ to $x = 2$

For $n = 4$ rectangles:

$$\Delta x = \frac{2 - (-2)}{4} = 1$$

(a) Using the left endpoints:

i	x_i	$f(x_i)$
1	-2	$e^{-2} + 1$
2	-1	$e^{-1} + 1$
3	0	$e^0 + 1 = 2$
4	1	$e^1 + 1$

$$A = \sum_{i=1}^{4} f(x_i)\Delta x = \sum_{i=1}^{4} f(x_i)(1) = \sum_{i=1}^{4} f(x_i)$$
$$= (e^{-2} + 1) + (e^{-1} + 1) + 2 + e^1 + 1$$
$$\approx 8.2215 \approx 8.22$$

(b) Using the right endpoints:

i	x_i	$f(x_i)$
1	-1	$e^{-1} + 1$
2	0	2
3	1	$e + 1$
4	2	$e^2 + 1$

Area $= 1(e^{-1} + 1) + 1(2) + 1(e + 1) + 1(e^2 + 1)$
$$\approx 15.4752 \approx 15.48$$

(c) Average $= \dfrac{8.2215 + 15.4752}{2}$
$$= 11.84835$$
$$\approx 11.85$$

(d) Using the midpoints:

i	x_i	$f(x_i)$
1	$-\dfrac{3}{2}$	$e^{-3/2} + 1$
2	$-\dfrac{1}{2}$	$e^{-1/2} + 1$
3	$\dfrac{1}{2}$	$e^{1/2} + 1$
4	$\dfrac{3}{2}$	$e^{3/2} + 1$

$$A = \sum_{i=1}^{4} f(x_i)\Delta x$$
$$= (e^{-3/2} + 1)(1) + (e^{-1/2} + 1)(1)$$
$$\quad + (e^{1/2} + 1)(1) + (e^{3/2} + 1)(1)$$
$$\approx 10.9601 \approx 10.96$$

11. $f(x) = \dfrac{2}{x}$ from $x = 1$ to $x = 9$

For $n = 4$ rectangles:

$$\Delta x = \frac{9-1}{4} = 2$$

(a) Using the left endpoints:

i	x_i	$f(x_i)$
1	1	$\frac{2}{1} = 2$
2	3	$\frac{2}{3}$
3	5	$\frac{2}{5} = 0.4$
4	7	$\frac{2}{7}$

$$A = \sum_{i=1}^{4} f(x_i)\Delta x$$

$$= (2)(2) + \frac{2}{3}(2) + (0.4)(2) + \left(\frac{2}{7}\right)(2)$$

$$\approx 6.7048 \approx 6.70$$

(b) Using the right endpoints:

i	x_i	$f(x_i)$
1	3	$\frac{2}{3}$
2	5	$\frac{2}{5}$
3	7	$\frac{2}{7}$
4	9	$\frac{2}{9}$

$$\text{Area} = 2\left(\frac{2}{3}\right) + 2\left(\frac{2}{5}\right) + 2\left(\frac{2}{7}\right) + 2\left(\frac{2}{9}\right)$$

$$= \frac{4}{3} + \frac{4}{5} + \frac{4}{7} + \frac{4}{9} \approx 3.1492 \approx 3.15$$

(c) Average $= \dfrac{6.7 + 3.15}{2} = 4.93$

(d) Using the midpoints:

i	x_i	$f(x_i)$
1	2	1
2	4	$\frac{1}{2}$
3	6	$\frac{1}{3}$
4	8	$\frac{1}{4}$

$$A = \sum_{i=1}^{4} f(x_i)\Delta x$$

$$= 1(2) + \frac{1}{2}(2) + \frac{1}{3}(2) + \frac{1}{4}(2)$$

$$\approx 4.1667 \approx 4.17$$

13. **(a)** Width $= \dfrac{4-0}{4} = 1$; $f(x) = \dfrac{x}{2}$

$$\text{Area} = 1 \cdot f\left(\frac{1}{2}\right) + 1 \cdot f\left(\frac{3}{2}\right)$$

$$+ 1 \cdot f\left(\frac{5}{2}\right) + 1 \cdot f\left(\frac{7}{2}\right)$$

$$= \frac{1}{4} + \frac{3}{4} + \frac{5}{4} + \frac{7}{4} = \frac{16}{4} = 4$$

(b)

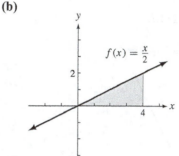

$$\int_0^4 f(x)dx = \int_0^4 \frac{x}{2}dx = \frac{1}{2}(\text{base})(\text{height})$$

$$= \frac{1}{2}(4)(2) = 4$$

15. **(a)** Area of triangle is $\frac{1}{2} \cdot$ base $\cdot$ height.

The base is 4; the height is 2.

$$\int_0^4 f(x)\,dx = \frac{1}{2} \cdot 4 \cdot 2 = 4$$

(b) The larger triangle has an area of
$\frac{1}{2} \cdot 3 \cdot 3 = \frac{9}{2}$. The smaller triangle has an
area of $\frac{1}{2} \cdot 1 \cdot 1 = \frac{1}{2}$. The sum is
$\frac{9}{2} + \frac{1}{2} = \frac{10}{2} = 5$.

17. $\displaystyle\int_{-4}^{0} \sqrt{16 - x^2}\,dx$

Graph $y = \sqrt{16 - x^2}$.

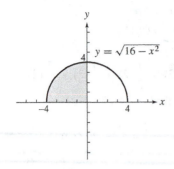

$y = \sqrt{16 - x^2}$

$\displaystyle\int_{-4}^{0} \sqrt{16 - x^2}\, dx$ is the area of the portion of the circle in the second quadrant, which is one-fourth of a circle. The circle has radius 4.

$$\text{Area} = \frac{1}{4}\pi r^2 = \frac{1}{4}\pi(4)^2 = 4\pi$$

19. $\displaystyle\int_{2}^{5} (1 + 2x)\, dx$

Graph $y = 1 + 2x$.

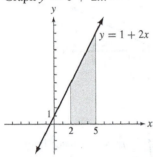

$y = 1 + 2x$

$\displaystyle\int_{2}^{5} (1 + 2x)\, dx$ is the area of the trapezoid with $B = 11, b = 5,$ and $h = 3.$ The formula for the area is

$$A = \frac{1}{2}(B + b)h,$$

so we have

$$A = \frac{1}{2}(11 + 5)(3) = 24.$$

21. (a) With $n = 10, \Delta x = \frac{1-0}{10} = 0.1,$ and

$x_1 = 0 + 0.1 = 0.1,$ use the command

seq $(X^2,$ X, 0.1, 1, 0.1) $\to$L1. The resulting screen is:

```
seq(X²,X,.1,1,.1
)→L1
{.01 .04 .09 .1…
```

(b) Since $\sum_{i=1}^{n} f(x_i)\Delta x = \Delta x\left(\sum_{i=1}^{n} f(x_i)\right),$ use the command 0.1*sum (L1) to approximate $\displaystyle\int_{0}^{1} x^2 dx.$ The resulting screen is:

```
.1*sum(L1)
                .385
```

$$\int_{0}^{1} x^2 dx \approx 0.385$$

(c) With $n = 100, \Delta x = \frac{1-0}{100} = 0.01$ and

$x_1 = 0 + 0.01 = 0.01,$ use the command

seq $(X^2,$ X, 0.01, 1, 0.01) $\to$ L1. The resulting screen is:

```
seq(X²,X,.01,1,.
01)→L1
{1E-4 4E-4 9E-4…
```

Use the command 0.01*sum(L1) to approximate $\displaystyle\int_{0}^{1} x^2 dx.$ The resulting screen is:

```
.01*sum(L1)
              .33835
```

$$\int_{0}^{1} x^2 dx \approx 0.33835$$

(d) With $n = 500, \Delta x = \frac{1-0}{500} = 0.002,$ and

$x_1 = 0 + 0.002 = 0.002,$ use the command

seq $(X^2,$ X, 0.002, 1, 0.002) $\to$ L1. The resulting screen is:

```
seq(X²,X,.002,1,
.002)→L1
{4E-6 1.6E-5 3.…
```

Use the command 0.002*sum(L1) to approximate $\int_0^1 x^2 dx$. The resulting screen is:

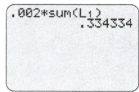

.002*sum(L1)
          .334334

$$\int_0^1 x^2 dx \approx 0.334334$$

(e) As n gets larger the approximation for $\int_0^1 x^2 dx$ seems to be approaching 0.333333 or $\frac{1}{3}$. We estimate $\int_0^1 x^2 dx = \frac{1}{3}$.

For Exercises 24–34, reading on the graphs and answers may vary.

25. Left endpoints:

Read values of the function on the graph every three years from 2001 to 2010, that is, at $x_0 = 2001$, $x_1 = 2004$, $x_2 = 2007$, and $x_3 = 2010$. These values give us the heights of four rectangles. The width of each rectangle is $\Delta x = 3$. We estimate the area under the curve as follows:

$$A = \sum_{i=0}^{3} f(x_i)\Delta x$$
$$= 70(3) + 142(3) + 341(3) + 923(3)$$
$$= 4428$$

Right endpoints:

Read values of the function on the graph every three years from 2004 to 2013, that is, discard the point $x_0 = 2001$ and include the point $x_4 = 2010$. We estimate the area under the curve as follows:

$$A = \sum_{i=1}^{4} f(x_i)\Delta x$$
$$= 142(3) + 341(3) + 923(3) + 1595(3)$$
$$= 9003$$

Average:
$$\frac{4428 + 9003}{2} = 6715.5 \text{ trillion BTUs}$$

We estimate the total wind energy consumption over the 12-year period from 2001 to 2013 as 6715.5 trillion BTUs.

27. The demand function is

$$D(q) = \frac{2}{125}q^2 - \frac{4}{5}q + 15$$

Left endpoints:

Compute values of the demand function every 5 liters from 0 to 20, that is, at $q_i = 5 \cdot i$ for $i = 0, 1, 2, 3, 4$. These values give us the heights of five rectangles. The width of each rectangle is $\Delta x = 5$. We estimate the area under the demand curve as follows:

$$A = \sum_{i=0}^{4} D(q_i)\Delta x = \Delta x \sum_{i=0}^{4} D(q_i)$$
$$= (5)(15 + 11.4 + 8.6 + 6.6 + 5.4)$$
$$= 235 \text{ or a total revenue of } \$235$$

Right endpoints:

Compute values of the demand function every 10 liters from 5 to 25, that is, at $q_i = 5 \cdot i$ for $i = 1, 2, 3, 4, 5$. We estimate the area under the demand curve as follows:

$$A = \Delta x \sum_{i=1}^{5} D(q_i)$$
$$= (5)(11.4 + 8.6 + 6.6 + 5.4 + 5)$$
$$= 185 \text{ or a total revenue of } \$185$$

Average: $\frac{235 + 185}{2} = 210$ or $210

We estimate the total as $210.

29. First read approximate data values from the graph. These readings are just estimates, and you may get different answers if your estimated readings differ from these. Month 1 represents mid-February.

Cows		Pigs	
Month	Cases	Month	Cases
1	3000	1	2000
2	165,000	2	62,000
3	267,000	3	68,000
4	54,000	4	3000
5	44,000	5	1000
6	21,000	6	9000
7	16,500	7	1000
8	11,500	8	0
9	1000	9	0

(a) Left endpoints:

Add up the values corresponding to months 1 through 8 in the Cows table. The total is 582,000 cases.

Right endpoints:

Add up the values corresponding to months
2 through 9. The total is 580,000 cases.

The average of these two values is 581,000
cases.

(b) Left endpoints:

Add up the values corresponding to months
1 through 8 in the Pigs table. The total is
146,000 cases.

Right endpoints:

Add up the values corresponding to months
2 through 9. The total is 144,000 cases.
The average of these two values is 145,000
cases.

31. Read the value of the function at $x = 1.5$ sec, 4.5 sec, 7.5 sec and 10 sec. These are the midpoints of rectangles with widths Δx equal to 3, 3, 3 and 2. We estimate the function values as 22, 76, 106 and 122, all in mph, which we will convert to feet per second after taking the sum, which will give us an answer in feet.

$$\sum_{i=1}^{4} f(x_i)\Delta x \approx 22(3) + 76(3) + 106(3) + 122(2) = 856$$

$$\frac{856}{3600}(5280) \approx 1300$$

The Lamborghini traveled about 1300 ft.

33. This problem is best solved using a calculator that can handle lists and summations. Extend the given table with a speed of 0 mph at 0 seconds and then make a list of the time differences:

$d_1 = 1.7 - 0 = 1.7$

$d_2 = 2.3 - 1.7 = 0.6$

$d_3 = 3.1 - 2.3 = 0.8$

$\vdots$

$d_{14} = 27.0 - 21.7 = 5.3$

These are the widths of the 14 time intervals. There are 15 speeds, including the first speed, 0 mph and the final speed, 160 mph. List these speeds:

$s_1 = 0$

$s_2 = 30$

$s_3 = 40$

$\vdots$

$s_{15} = 160$

For left-endpoint rectangles, we pair d_1 through d_{14} with s_1 through s_{14}. Noting that 1 second is 1/3600 hr and 1 mile is 5280 feet, the left endpoint estimate in feet is

$$\frac{5280}{3600}\sum_{k=1}^{14} d_k \cdot s_k = 4160.933 \approx 4161 \text{ ft}$$

For the right endpoint estimate we pair we pair d_1 through d_{14} with s_2 through s_{15}. The right endpoint estimate is

$$\frac{5280}{3600}\sum_{k=1}^{14} d_k \cdot s_{k+1} = 4606.800 \approx 4607 \text{ ft}$$

The average of these two estimates is

$$\frac{4160.933 + 4606.800}{2} = 4383.867$$

$$\approx 4384 \text{ ft,}$$

so our best estimate of the distance traveled by the Mercedes Benz C63 AMG Edition 507 is 4384 ft.

35. **(a)** Read values of the function on the plain glass graph every 2 hr from 6 to 6. These are at midpoints of the widths $\Delta x = 2$ and represent the heights of the rectangles.

$$\sum f(x_i)\Delta x = 80(2) + 110(2) + 80(2) + 26(2) + 21(2) + 18(2) + 8(2) = 684$$

The total heat gain was about 680 BTUs per square foot.

(b) Read values on the triple-glazed graph every 2 hr from 6 to 6.

$$\sum f(x_i)\Delta x = 35(2) + 54(2) + 38(2) + 12(2) + 10(2) + 8(2) + 3(2) = 320$$

37. **(a)** Then area of a trapezoid is

$$A = \frac{1}{2}h(b_1 + b_2) = \frac{1}{2}(6)(1 + 2) = 9.$$

Car A has traveled 9 ft.

(b) Car A is furthest ahead of car B at 2 sec. Notice that from $t = 0$ to $t = 2$, $v(t)$ is larger for car A than for car B. For $t > 2$, $v(t)$ is larger for car B than for car A.

(c) As seen in part (a), car A drove 9 ft in 2 sec. The distance of car B can be calculated as follows:

$$\frac{2 - 0}{4} = \frac{1}{2} = \text{width}$$

$$\text{Distance} = \frac{1}{2} \cdot v(0.25) + \frac{1}{2}v(0.75) + \frac{1}{2}v(1.25) + \frac{1}{2}v(1.75)$$

$$= \frac{1}{2}(0.2) + \frac{1}{2}(1) + \frac{1}{2}(2.6) + \frac{1}{2}(5)$$

$$= 4.4$$

$$9 - 4.4 = 4.6$$

The furthest car A can get ahead of car B is about 4.6 ft.

(d) At $t = 3$, car A travels $\frac{1}{2}(6)(2 + 3) = 15$ ft and car B travels approximately 13 ft.

At $t = 3.5$, car A travels $\frac{1}{2}(6)(2.5 + 3.5) = 18$ ft and car B travels approximately 18.25 ft. Therefore, car B catches up with car A between 3 and 3.5 sec.

39.   Using the left endpoints:

$$\text{Distance} = v_0(1) + v_1(1) + v_2(1)$$

$$= 10 + 6.5 + 6 = 22.5 \text{ ft}$$

Using the right endpoints:

$$\text{Distance} = v_1(1) + v_2(1) + v_3(1)$$

$$= 6.5 + 6 + 5.5 = 18 \text{ ft}$$

7.4 The Fundamental Theorem of Calculus

Your Turn 1

Find $\displaystyle\int_1^3 3x^2 \, dx$.

The indefinite integral is $\displaystyle\int 3x^2 \, dx = x^3 + C$.

By the Fundamental Theorem,

$$\int_1^3 3x^2 \, dx = x^3 \Big|_1^3$$

$$= 3^3 - 1^3$$

$$= 27 - 1$$

$$= 26$$

Your Turn 2

Find $\displaystyle\int_3^5 (2x^3 - 3x + 4) \, dx$.

$$\int_3^5 (2x^3 - 3x + 4) \, dx$$

$$= 2\int_3^5 x^3 \, dx - 3\int_3^5 x \, dx + 4\int_3^5 dx$$

$$= \frac{2}{4}x^4 \Big|_3^5 - \frac{3}{2}x^2 \Big|_3^5 + 4x \Big|_3^5$$

$$= \frac{1}{2}(5^4 - 3^4) - \frac{3}{2}(5^2 - 3^2) + 4(5 - 3)$$

$$= 256$$

Your Turn 3

Find $\displaystyle\int_1^3 \frac{2}{y} \, dy$.

$$\int_1^3 \frac{2}{y} \, dy = 2\int_1^3 \frac{1}{y} \, dy$$

$$= 2 \ln |y| \Big|_1^3$$

$$= 2(\ln 3 - \ln 1)$$

$$= 2 \ln 3 \text{ or } \ln 3^2 = \ln 9$$

Your Turn 4

Evaluate $\displaystyle\int_0^4 2x\sqrt{16 - x^2} \, dx$.

Using Method 1:

Let $u = 16 - x^2$.

Then $du = -2x$.

If $x = 4$, then $u = 16 - 4^2 = 0$.

If $x = 0$, then $u = 16 - 0^2 = 16$.

Now substitute.

$$\int_0^4 2x\sqrt{16 - x^2} \, dx$$

$$= \int_0^{16} \sqrt{u} \, du$$

$$= \frac{2}{3}u^{3/2} \Big|_0^{16} = \frac{2}{3}(64 - 0)$$

$$= \frac{128}{3}$$

Your Turn 5

Find the area between the graph of the function $f(x) = x^2 - 9$ and the x-axis from $x = 0$ to $x = 6$. Here is a graph of the function and the area to be found.

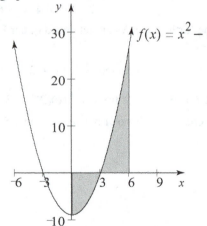

Since the curve is below the x-axis on the interval $(0, 3)$, the definite integral will count this areas as negative. The total positive area is thus

$$\left| \int_0^3 (x^2 - 9)\, dx \right| + \int_3^6 (x^2 - 9)\, dx$$

$$= \left| \left(\frac{1}{3}x^3 - 9x \right) \Big|_0^3 \right| + \left(\frac{1}{3}x^3 - 9x \right) \Big|_3^6$$

$$|9 - 27| + (27 - 54 - (9 - 27))$$

$$= 18 + 18 + 18 = 54.$$

7.4 Warmup Exercises

W1. $\displaystyle \int 6x\sqrt{x^2 + 4}\, dx$

Let $u = x^2 + 4$; then $du = 2x\, dx$.

$$\int 6x\sqrt{x^2 + 4}\, dx$$

$$= 3 \int u^{1/2}\, du$$

$$= 3\left(\frac{2}{3} \right) u^{3/2} + C$$

$$= 2\left(x^2 + 4 \right)^{3/2} + C$$

W2. $\displaystyle \int \frac{4x}{x^2 + 2}\, dx$

Let $u = x^2 + 2$; then $du = 2x\, dx$.

$$\int \frac{4x}{x^2 + 2}\, dx$$

$$= 2 \int \frac{1}{u}\, du$$

$$= 2\ln|u| + C$$

$$= 2\ln\left|x^2 + 2\right| + C$$

$$= 2\ln\left(x^2 + 2\right) + C \text{ since } x^2 + 2 \text{ is}$$
always positive

W3. $\displaystyle \int 15x^2 e^{x^3}\, dx$

Let $u = x^3$; then $du = 3x^2\, dx$.

$$\int 15x^2 e^{x^3}\, dx$$

$$= 5 \int e^u\, du$$

$$= 5e^u + C$$

$$= 5e^{x^3} + C$$

7.4 Exercises

1. $\displaystyle \int_{-2}^{4} (-3)\, dp = -3 \int_{-2}^{4} dp = -3 \cdot p \Big|_{-2}^{4}$

$$= -3[4 - (-2)]$$

$$= -18$$

3. $\displaystyle \int_{-1}^{2} (5t - 3)\, dt = 5 \int_{-1}^{2} t\, dt - 3 \int_{-1}^{2} dt$

$$= \frac{5}{2}t^2 \Big|_{-1}^{2} - 3t \Big|_{-1}^{2}$$

$$= \frac{5}{2}[2^2 - (-1)^2] - 3[2 - (-1)]$$

$$= \frac{5}{2}(4 - 1) - 3(2 + 1)$$

$$= \frac{15}{2} - 9$$

$$= \frac{15}{2} - \frac{18}{2} = -\frac{3}{2}$$

5. $\displaystyle \int_0^2 (5x^2 - 4x + 2)\, dx$

$$= 5 \int_0^2 x^2\, dx - 4 \int_0^2 x\, dx + 2 \int_0^2 dx$$

$$= \frac{5x^3}{3} \Big|_0^2 - 2x^2 \Big|_0^2 + 2x \Big|_0^2$$

$$= \frac{5}{3}(2^3 - 0^3) - 2(2^2 - 0^2) + 2(2 - 0)$$

$$= \frac{5}{3}(8) - 2(4) + 2(2) = \frac{40 - 24 + 12}{3}$$

$$= \frac{28}{3}$$

7. $\displaystyle \int_0^2 3\sqrt{4u + 1}\, du$

Let $4u + 1 = x$, so that $4\, du = dx$.

When $u = 0$, $x = 4(0) + 1 = 1$.

When $u = 2$, $x = 4(2) + 1 = 9$.

$$\int_0^2 3\sqrt{4u+1}\,du$$

$$= \frac{3}{4}\int_0^2 \sqrt{4u+1}\,(4\,du)$$

$$= \frac{3}{4}\int_1^9 x^{1/2}\,dx$$

$$= \frac{3}{4}\cdot\frac{x^{3/2}}{3/2}\Big|_1^9$$

$$= \frac{3}{4}\cdot\frac{2}{3}(9^{3/2}-1^{3/2})$$

$$= \frac{1}{2}(27-1) = \frac{26}{2} = 13$$

9. $\displaystyle\int_0^4 2(t^{1/2}-t)\,dt = 2\int_0^4 t^{1/2}\,dt - 2\int_0^4 t\,dt$

$$= 2\cdot\frac{t^{3/2}}{\frac{3}{2}}\Big|_0^4 - 2\cdot\frac{t^2}{2}\Big|_0^4$$

$$= \frac{4}{3}(4^{3/2}-0^{3/2}) - (4^2-0^2)$$

$$= \frac{32}{3} - 16 = -\frac{16}{3}$$

11. $\displaystyle\int_1^4 (5y\sqrt{y}+3\sqrt{y})\,dy$

$$= 5\int_1^4 y^{3/2}\,dy + 3\int_1^4 y^{1/2}\,dy$$

$$= 5\left(\frac{y^{5/2}}{\frac{5}{2}}\right)\Big|_1^4 + 3\left(\frac{y^{3/2}}{\frac{3}{2}}\right)\Big|_1^4$$

$$= 2y^{5/2}\Big|_1^4 + 2y^{3/2}\Big|_1^4$$

$$= 2(4^{5/2}-1) + 2(4^{3/2}-1)$$

$$= 2(32-1) + 2(8-1)$$

$$= 62 + 14$$

$$= 76$$

13. $\displaystyle\int_4^6 \frac{2}{(2x-7)^2}\,dx$

Let $u = 2x-7$, so that $du = 2\,dx$.

When $x=6$, $u = 2\cdot 6 - 7 = 5$.

When $x=4$, $u = 2\cdot 4 - 7 = 1$.

$$\int_4^6 \frac{2}{(2x-7)^2}\,dx = \int_1^5 u^{-2}\,du$$

$$= \frac{u^{-1}}{-1}\Big|_1^5$$

$$= -u^{-1}\Big|_1^5$$

$$= -\left(\frac{1}{5}-1\right)$$

$$= -\left(-\frac{4}{5}\right)$$

$$= \frac{4}{5}$$

15. $\displaystyle\int_1^5 (6n^{-2}-n^{-3})\,dn$

$$= 6\int_1^5 n^{-2}\,dn - \int_1^5 n^{-3}\,dn$$

$$= 6\cdot\frac{n^{-1}}{-1}\Big|_1^5 - \frac{n^{-2}}{-2}\Big|_1^5$$

$$= \frac{-6}{n}\Big|_1^5 + \frac{1}{2n^2}\Big|_1^5$$

$$= \frac{-6}{5} - \left(\frac{-6}{1}\right) + \left[\frac{1}{2(25)} - \frac{1}{2(1)}\right]$$

$$= \frac{-6}{5} - \frac{6}{1} + \frac{1}{50} - \frac{1}{2}$$

$$= \frac{108}{25}$$

17. $\displaystyle\int_{-3}^{-2}\left(2e^{-0.1y}+\frac{3}{y}\right)\,dy$

$$= 2\int_{-3}^{-2} e^{-0.1y}\,dy + \int_{-3}^{-2}\frac{3}{y}\,dy$$

$$= 2\cdot\frac{e^{-0.1y}}{-0.1}\Big|_{-3}^{-2} + 3\ln|y|\Big|_{-3}^{-2}$$

$$= -20e^{-0.1y}\Big|_{-3}^{-2} + 3\ln|y|\Big|_{-3}^{-2}$$

$$= 20e^{0.3} - 20e^{0.2} + 3\ln 2 - 3\ln 3$$

$$\approx 1.353$$

19. $\displaystyle\int_1^2 \left(e^{4u} - \frac{1}{(u+1)^2} \right) du$

$\displaystyle = \int_1^2 e^{4u}\, du - \int_1^2 \frac{1}{(u+1)^2}\, du$

$\displaystyle = \frac{e^{4u}}{4}\bigg|_1^2 - \frac{-1}{u+1}\bigg|_1^2$

$\displaystyle = \frac{e^8}{4} - \frac{e^4}{4} + \frac{1}{2+1} - \frac{1}{1+1}$

$\displaystyle = \frac{e^8}{4} - \frac{e^4}{4} - \frac{1}{6}$

≈ 731.4

21. $\displaystyle\int_{-1}^0 y(2y^2 - 3)^5\, dy$

Let $u = 2y^3 - 3$, so that

$\displaystyle du = 4y\, dy \text{ and } \frac{1}{4}\, du = y\, dy.$

When $y = -1$, $u = 2(-1)^2 - 3 = -1$.

When $y = 0$, $u = 2(0)^2 - 3 = -3$.

$\displaystyle \frac{1}{4}\int_{-1}^{-3} u^5 du = \frac{1}{4} \cdot \frac{u^6}{6}\bigg|_{-1}^{-3}$

$\displaystyle = \frac{1}{24} u^6 \bigg|_{-1}^{-3}$

$\displaystyle = \frac{1}{24}(-3)^6 - \frac{1}{24}(-1)^6$

$\displaystyle = \frac{729}{24} - \frac{1}{24}$

$\displaystyle = \frac{728}{24} = \frac{91}{3}$

23. $\displaystyle\int_1^{64} \frac{\sqrt{z} - 2}{\sqrt[3]{z}}\, dz$

$\displaystyle = \int_1^{64} \left(\frac{z^{1/2}}{z^{1/2}} - 2z^{-1/3} \right) dz$

$\displaystyle = \int_1^{64} z^{1/6} dz - 2\int_1^{64} z^{-1/3} dz$

$\displaystyle = \frac{z^{7/6}}{\frac{7}{6}}\bigg|_1^{64} - 2 \frac{z^{2/3}}{\frac{2}{3}}\bigg|_1^{64}$

$\displaystyle = \frac{6z^{7/6}}{7}\bigg|_1^{64} - 3z^{2/3}\bigg|_1^{64}$

$\displaystyle = \frac{6(64)^{7/6}}{7} - \frac{6(1)^{7/6}}{7}$

$\displaystyle \qquad - 3(64^{2/3} - 1^{2/3})$

$\displaystyle = \frac{6(128)}{7} - \frac{6}{7} - 3(16 - 1)$

$\displaystyle = \frac{768 - 6 - 315}{7} = \frac{447}{7} \approx 63.86$

25. $\displaystyle\int_1^2 \frac{\ln x}{x}\, dx$

Let $u = \ln x$, so that

$\displaystyle du = \frac{1}{x}\, dx.$

When $x = 1$, $u = \ln 1 = 0$.

When $x = 2$, $u = \ln 2$.

$\displaystyle \int_0^{\ln 2} u\, du = \frac{u^2}{2}\bigg|_0^{\ln 2}$

$\displaystyle = \frac{(\ln 2)^2}{2} - 0$

$\displaystyle = \frac{(\ln 2)^2}{2}$

≈ 0.2402

27. $\displaystyle\int_0^8 x^{1/3}\sqrt{x^{4/3} + 9}\, dx$

Let $u = x^{4/3} + 9$, so that

$\displaystyle du = \frac{4}{3}x^{1/3} dx \text{ and } \frac{3}{4}\, du = x^{1/3} dx.$

When $x = 0$, $u = 0^{4/3} + 9 = 9$.

When $x = 8$, $u = 8^{4/3} + 9 = 25$.

$\displaystyle \frac{3}{4}\int_9^{25} \sqrt{u}\, du = \frac{3}{4}\int_9^{25} u^{1/2} du$

$\displaystyle = \frac{3}{4} \cdot \frac{u^{3/2}}{\frac{3}{2}}\bigg|_9^{25}$

$\displaystyle = \frac{1}{2} u^{3/2}\bigg|_9^{25}$

$\displaystyle = \frac{1}{2}(25)^{3/2} - \frac{1}{2}(9)^{3/2}$

$\displaystyle = \frac{125}{2} - \frac{27}{2} = 49$

29. $\displaystyle\int_0^1 \frac{e^{2t}}{(3+e^{2t})^2}\,dt$

Let $u = 3 + e^{2t}$, so that $du = 2e^{2t}\,dt$.

When $x = 1$, $u = 3 + e^{2\cdot1} = 3 + e^2$.

When $x = 0$, $u = 3 + e^{2\cdot0} = 4$.

$$\int_0^1 \frac{e^{2t}}{(3+e^{2t})^2}\,dt = \frac{1}{2}\int_4^{3+e^2} u^{-2}\,du$$

$$= \frac{1}{2}\cdot\frac{u^{-1}}{-1}\Big|_4^{3+e^2} = \frac{-1}{2u}\Big|_4^{3+e^2}$$

$$= \frac{1}{8} - \frac{1}{2(3+e^2)}$$

$$\approx 0.07687$$

31. $f(x) = 2x - 14;\ [6, 10]$

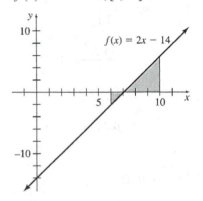

The graph crosses the x-axis at

$$0 = 2x - 14$$
$$2x = 14$$
$$x = 7.$$

This location is in the interval. The area of the region is

$$\left|\int_6^7 (2x - 14)\,dx\right| + \int_7^{10} (2x - 14)\,dx$$

$$= \left|(x^2 - 14x)\right|_6^7 + (x^2 - 14x)\Big|_7^{10}$$

$$= |(7^2 - 98) - (6^2 - 84)|$$
$$\quad + (10^2 - 140) - (7^2 - 98)$$

$$= |-1| + (-40) - (-49)$$

$$= 10.$$

33. $f(x) = 2 - 2x^2;\ [0, 5]$

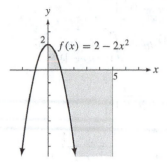

Find the points where the graph crosses the x-axis by solving $2 - 2x^2 = 0$.

$$2 - 2x^2 = 0$$
$$2x^2 = 2$$
$$x^2 = 1$$
$$x = \pm 1.$$

The only solution in the interval $[0, 5]$ is 1. The total area is

$$\int_0^1 (2 - 2x^2)\,dx + \left|\int_2^5 (2 - 2x^2)\,dx\right|$$

$$= \left(2x - \frac{2x^3}{3}\right)\Big|_0^1 + \left|\left(2x - \frac{2x^3}{3}\right)\Big|_1^5\right|$$

$$= 2 - \frac{2}{3} + \left|10 - \frac{2(5^3)}{3} - 2 + \frac{2}{3}\right|$$

$$= \frac{4}{3} + \left|\frac{-224}{3}\right|$$

$$= \frac{228}{3}$$

$$= 76.$$

35. $f(x) = x^3;\ [-1, 3]$

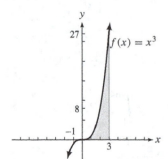

The solution

$$x^3 = 0$$
$$x = 0$$

indicates that the graph crosses the x-axis at 0 in the given interval $[-1, 3]$.

The total area is

$$\left|\int_{-1}^{0} x^3 \, dx\right| + \int_{0}^{3} x^3 \, dx$$

$$= \left|\frac{x^4}{4}\right|_{-1}^{0} + \left|\frac{x^4}{4}\right|_{0}^{3}$$

$$= \left|\left(0 - \frac{1}{4}\right)\right| + \left(\frac{3^4}{4} - 0\right)$$

$$= \frac{1}{4} + \frac{81}{4} = \frac{82}{4}$$

$$= \frac{41}{2}.$$

37. $f(x) = e^x - 1; \; [-1, 2]$

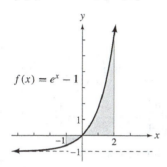

Solve

$$e^x - 1 = 0.$$

$$e^x = 1$$

$$x \ln e = \ln 1$$

$$x = 0$$

The graph crosses the x-axis at 0 in the given interval $[-1, 2]$.

The total area is

$$\left|\int_{-1}^{0} (e^x - 1) \, dx\right| + \int_{0}^{2} (e^x - 1) \, dx$$

$$= \left|(e^x - x)\right|_{-1}^{0} + (e^x - x)\Big|_{0}^{2}$$

$$= |(1 - 0) - (e^{-1} + 1)| + (e^2 - 2) - (1 - 0)$$

$$= |1 - e^{-1} - 1| + e^2 - 2 - 1$$

$$= \frac{1}{e} + e^2 - 3$$

$$\approx 4.757.$$

39. $f(x) = \frac{1}{x} - \frac{1}{e}; \; [1, e^2]$

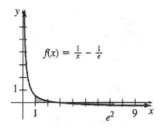

The graph crosses the x-axis at

$$0 = \frac{1}{x} - \frac{1}{e}$$

$$\frac{1}{x} = \frac{1}{e}$$

$$x = e.$$

This location is in the interval. The area of the region is

$$\int_{1}^{e}\left(\frac{1}{x} - \frac{1}{e}\right) dx + \left|\int_{e}^{e^2}\left(\frac{1}{x} - \frac{1}{e}\right) dx\right|$$

$$= \left|\ln|x| - \frac{x}{e}\right|_{1}^{e} + \left|\left(\ln|x| - \frac{x}{e}\right)\right|_{e}^{e^2}\right|$$

$$= 0 - \left(-\frac{1}{e}\right) + |(2 - e) - 0|$$

$$= \frac{1}{e} + |2 - e|$$

$$= e - 2 + \frac{1}{e}.$$

41. $y = 4 - x^2; \; [0, 3]$

From the graph, we see that the total area is

$$\int_0^2 (4 - x^2)\, dx + \left| \int_2^3 (4 - x^2)\, dx \right|$$

$$= \left(4x - \frac{x^3}{3} \right) \Bigg|_0^2 + \left| \left(4x - \frac{x^3}{3} \right) \Bigg|_2^3 \right|$$

$$= \left[\left(8 - \frac{8}{3} \right) - 0 \right]$$

$$\quad + \left| \left[(12 - 9) - \left(8 - \frac{8}{3} \right) \right] \right|$$

$$= \frac{16}{3} + \left| 3 - \frac{16}{3} \right|$$

$$= \frac{16}{3} + \frac{7}{3}$$

$$= \frac{23}{3}$$

43. $y = e^x - e;\ [0, 2]$

From the graph, we see that total area is

$$\left| \int_0^1 (e^x - e)\, dx \right| + \int_1^2 (e^x - e)\, dx$$

$$= \left| (e^x - xe) \Big|_0^1 \right| + (e^x - xe) \Big|_1^2$$

$$= |(e^1 - e) - (e^0 + 0)| + (e^2 - 2e) - (e^1 - e)$$

$$= |-1| + e^2 - 2e$$

$$= 1 + e^2 - 2e \approx 2.952.$$

45. $\displaystyle \int_a^c f(x)\, dx = \int_a^b f(x)\, dx + \int_b^c f(x)\, dx$

47. $\displaystyle \int_0^{16} f(x)\, dx = \int_0^2 f(x)\, dx + \int_2^5 f(x)\, dx$

$$\quad + \int_5^8 f(x)\, dx + \int_8^{16} f(x)\, dx$$

$$= \frac{1}{2} \cdot 2(1 + 3) + \frac{\pi(3^2)}{4}$$

$$\quad - \frac{\pi(3^2)}{4} - \frac{1}{2}(3)(8)$$

$$= 4 + \frac{9}{4}\pi - \frac{9}{4}\pi - 12 = -8$$

49. Prove: $\displaystyle \int_a^b f(x)\, dx$

$$= \int_a^c f(x)\, dx + \int_c^b f(x)\, dx.$$

Let $F(x)$ be an antiderivative of $f(x)$.

$$\int_a^c f(x)\, dx + \int_c^b f(x)\, dx$$

$$= F(x) \Big|_a^c + F(x) \Big|_c^b$$

$$= [F(c) - F(a)] + [F(b) - F(c)]$$

$$= F(c) - F(a) + F(b) - F(c)$$

$$= F(b) - F(a)$$

$$= \int_a^b f(x)\, dx$$

51. $\displaystyle \int_{-1}^4 f(x)\, dx$

$$= \int_{-1}^0 (2x + 3) + dx \int_0^4 \left(-\frac{x}{4} - 3 \right) dx$$

$$= (x^2 + 3x) \Big|_{-1}^0 + \left(-\frac{x^2}{8} - 3x \right) \Big|_0^4$$

$$= -(1 - 3) + (-2 - 12)$$

$$= 2 - 14$$

$$= -12$$

53. (a) $g(t) = t^4$ and $c = 1$, use substitution.

$$f(x) = \int_c^x g(t)\, dt$$

$$= \int_1^x t^4\, dt$$

$$= \frac{t^5}{5} \Big|_1^x$$

$$= \frac{x^5}{5} - \frac{(1)^5}{5}$$

$$= \frac{x^5}{5} - \frac{1}{5}$$

(b) $f'(x) = \dfrac{d}{dx}(f(x))$

$\quad = \dfrac{d}{dx}\left(\dfrac{x^5}{5} - \dfrac{1}{5}\right)$

$\quad = \dfrac{1}{5} \cdot \dfrac{d}{dx}(x^5) - \dfrac{d}{dx}\left(\dfrac{1}{5}\right)$

$\quad = \dfrac{1}{5} \cdot 5x^4 - 0 = x^4$

Since $g(t) = t^4$, then $g(x) = x^4$ and we see $f'(x) = g(x)$.

(c) Let $g(t) = e^{t^2}$ and $c = 0$, then

$$f(x) = \int_0^x e^{t^2}\,dt.$$

$f(1) = \displaystyle\int_0^1 e^{t^2}\,dt$ and $f(1.01) = \displaystyle\int_0^{1.01} e^{t^2}\,dt.$

Use the fnInt command in the Math menu of your calculator to find $\displaystyle\int_0^1 e^{x^2}\,dx$ and

$\displaystyle\int_0^{1.01} e^{x^2}\,dx.$ The resulting screens are:

$f(1) \approx 1.46265$

$f(1.01) \approx 1.49011$

Use $\dfrac{f(1+h) - f(1)}{h}$ to approximate

$f'(1)$ with $h = 0.01$

$\dfrac{f(1+h) - f(1)}{h} = \dfrac{f(1.01) - f(1)}{0.01}$

$\quad \approx \dfrac{1.49011 - 1.46265}{0.01}$

$\quad = 2.746$

So $f'(1) \approx 2.746$, and $g(1) = e^{1^2} = e \approx 2.718.$

55. $P'(t) = (3t + 3)(t^2 + 2t + 2)^{1/3}$

(a) $\displaystyle\int_0^3 3(t + 1)(t^2 + 2t + 2)^{1/3}\,dt$

Let $u = t^2 + 2t + 2$, so that

$du = (2t + 2)\,dt$ and $\tfrac{1}{2}du = (t + 1)\,dt.$

When $t = 0$, $u = 0^2 + 2 \cdot 0 + 2 = 2.$

When $t = 3$, $u = 3^2 + 2 \cdot 3 + 2 = 17.$

$\dfrac{3}{2}\displaystyle\int_2^{17} u^{1/3}\,du = \dfrac{3}{2} \cdot \left.\dfrac{u^{4/3}}{\frac{4}{3}}\right|_2^{17}$

$\quad = \dfrac{9}{8}u^{4/3}\Big|_2^{17}$

$\quad = \dfrac{9}{8}(17)^{4/3} - \dfrac{9}{8}(2)^{4/3}$

$\quad \approx 46.341$

Total profits for the first 3 yr were

$\dfrac{9000}{8}(17^{4/3} - 2^{4/3}) \approx \$46,341.$

(b) $\displaystyle\int_3^4 3(t + 1)(t^2 + 2t + 2)^{1/3}\,dt$

Let $u = t^2 + 2t + 2$, so that

$du = (2t + 2)\,dt = 2(t + 1)\,dt$ and

$\dfrac{3}{2}du = 3(t + 1)\,dt.$

When $t = 3$, $u = 3^2 + 2 \cdot 3 + 2 = 17.$

When $t = 4$, $u = 4^2 + 2 \cdot 4 + 2 = 26.$

$\dfrac{3}{2}\displaystyle\int_{17}^{26} u^{1/3}\,du = \dfrac{9}{8}u^{4/3}\Big|_{17}^{26}$

$\quad = \dfrac{9}{8}(26)^{4/3} - \dfrac{9}{8}(17)^{4/3} \approx 37.477$

Profit in the fourth year was

$\dfrac{9000}{8}(26^{4/3} - 17^{4/3}) \approx \$37,477.$

(c) $\displaystyle\lim_{x \to \infty} P'(t)$

$\quad = \displaystyle\lim_{x \to \infty}(3t + 3)(t^2 + 2t + 2)^{1/3}$

$\quad = \infty$

The annual profit is slowly increasing without bound.

57. $P'(t) = 140t^{5/2}$

$$\int_0^4 140t^{5/2}\,dt = 140 \cdot \frac{t^{7/2}}{\frac{7}{2}}\Big|_0^4$$

$$= 40t^{7/2}\Big|_0^4$$

$$= 5120$$

Since 5120 is above the total level of acceptable pollution (4850), the factory cannot operate for 4 years without killing all the fish in the lake.

59. Growth rate is $0.6 + \frac{4}{(t+1)^3}$ ft/yr.

(a) Total growth in the second year is

$$\int_1^2 \left[0.6 + \frac{4}{(t+1)^3}\right]dt$$

$$= \left[0.6t + \frac{4}{-2(t+1)^2}\right]\Big|_1^2$$

$$= \left[0.6(2) - \frac{2}{(2+1)^2}\right]$$

$$\quad - \left[0.6(1) - \frac{2}{(1+1)^2}\right]$$

$$= \frac{44}{45} - \frac{1}{10}$$

$$\approx 0.8778\,\text{ft}.$$

(b) Total growth in the third year is

$$\int_2^3 \left[0.6 + \frac{4}{(t+1)^3}\right]dt$$

$$= \left[0.6t + \frac{4}{-2(t+1)^2}\right]\Big|_2^3$$

$$= \left[0.6(3) - \frac{2}{(3+1)^2}\right]$$

$$\quad - \left[0.6(2) - \frac{2}{(2+1)^2}\right]$$

$$= \frac{67}{40} - \frac{44}{45}$$

$$\approx 0.6972\,\text{ft}.$$

61. $R'(t) = \frac{5}{t+1} + \frac{2}{\sqrt{t+1}}$

(a) Total reaction from $t = 1$ to $t = 12$ is

$$\int_1^{12}\left(\frac{5}{t+1} + \frac{2}{\sqrt{t+1}}\right)dt$$

$$= \left[5\ln(t+1) + 4\sqrt{t+1}\right]\Big|_1^{12}$$

$$= (5\ln 13 + 4\sqrt{13}) - (5\ln 2 + 4\sqrt{2})$$

$$\approx 18.12.$$

(b) Total reaction from $t = 12$ to $t = 24$ is

$$\int_{12}^{24}\left(\frac{5}{t+1} + \frac{2}{\sqrt{t+1}}\right)dt$$

$$= \left[5\ln(t+1) + 4\sqrt{t+1}\right]\Big|_{12}^{24}$$

$$= (5\ln 25 + 4\sqrt{25}) - (5\ln 13 + 4\sqrt{13})$$

$$\approx 8.847.$$

63. (b) $\int_0^{60} n(x)\,dx$

(c) $\int_5^{10}\sqrt{5x+1}\,dx$

Let $u = 5x + 1$. Then $du = 5\,dx$.

When $x = 5$, $u = 26$; when $x = 10$, $u = 51$.

$$\frac{1}{5}\int_{26}^{51} u^{1/2}\,du$$

$$= \frac{1}{5} \cdot \frac{u^{3/2}}{\frac{3}{2}}\Big|_{26}^{51}$$

$$= \frac{2}{15}u^{3/2}\Big|_{26}^{51}$$

$$= \frac{2}{15}(51^{3/2} - 26^{3/2})$$

$$\approx 30.89 \text{ million}$$

65. $v = k(R^2 - r^2)$

(a) $Q(R) = \int_0^R 2\pi v r\,dr$

$$= \int_0^R 2\pi k (R^2 - r^2) r \, dr$$

$$= 2\pi k \int_0^R (R^2 r - r^2) \, dr$$

$$= 2\pi k \left(\frac{R^2 r^2}{2} - \frac{r^4}{4} \right) \Big|_0^R$$

$$= 2\pi k \left(\frac{R^4}{2} - \frac{R^4}{4} \right)$$

$$= 2\pi k \left(\frac{R^4}{4} \right)$$

$$= \frac{\pi k R^4}{2}$$

(b) $\quad Q(0.4) = \dfrac{\pi k (0.4)^4}{2}$

$$= 0.04k \text{ mm/min}$$

67. $\quad E(t) = 753 t^{-0.1321}$

(a) Since t is the age of the beagle in years, to convert the formula to days, let $T = 365t$, or $t = \dfrac{T}{365}$.

$$E(T) = 753 \left(\frac{T}{365} \right)^{-0.1321}$$

$$\approx 1642 T^{-0.1321}$$

Now, replace T with t.
$$E(t) = 1642 t^{-0.1321}$$

(b) The beagle's age in days after one year is 365 days and after 3 years she is 1095 days old.

$$\int_{365}^{1095} 1642 t^{-0.1321} \, dt$$

$$= 1642 \frac{1}{0.8679} t^{0.8679} \Big|_{365}^{1095}$$

$$\approx 1892 (1,095^{0.8679} - 365^{0.8679})$$

$$\approx 505,155$$

The beagle's total energy requirements are about $505,000 \text{ kJ/W}^{0.67}$, where W represents weight.

69. **(a)** $\quad f(x) = 40.2 + 3.50x - 0.897x^2$

$$\int_0^9 (40.2 + 3.50x - 0.897x^2) \, dx$$

$$= (40.2x + 1.74x^2 - 0.299x^3) \Big|_0^9$$

$$\approx 286$$

The integral represents the population aged 0 to 90, which is about 286 million.

(b) $\quad \displaystyle\int_{4.5}^{6.5} (40.2 + 3.50x - 0.897x^2) \, dx$

$$= (40.2x + 1.74x^2 - 0.299x^3) \Big|_{4.5}^{6.5}$$

$$\approx 64$$

The number of baby boomers is about 64 million.

71. $\quad c'(t) = ke^{rt}$

(a) $\quad c'(t) = 1.2 \, e^{0.04t}$

(b) The amount of oil that the company will sell in the next ten years is given by the integral

$$\int_0^{10} 1.2 e^{0.04t} \, dt.$$

(c) $\quad \displaystyle\int_0^{10} 1.2 e^{0.04t} \, dx = \frac{1.2 e^{0.04t}}{0.04} \Big|_0^{10}$

$$= 30 e^{0.04t} \Big|_0^{10}$$

$$= 30 e^{0.4} - 30$$

$$\approx 14.75$$

This represents about 14.75 billion barrels of oil.

(d) $\quad \displaystyle\int_0^T 1.2 e^{0.04t} \, dt = 30 e^{0.04t} \Big|_0^T$

$$= 30 e^{0.04T} - 30$$

Solve

$$20 = 30 e^{0.04T} - 30.$$

$$50 = 30 e^{0.04T}$$

$$\frac{5}{3} = e^{0.04T}$$

$$\ln \frac{5}{3} = 0.04T \ln e$$

$$T = \frac{\ln \frac{5}{3}}{0.04}$$

$$\approx 12.8$$

The oil will last about 12.8 years.

(e) $\displaystyle\int_0^T 1.2e^{0.02t}\, dt = 60e^{0.02t}\Big|_0^T$

$$= 60e^{0.02T} - 60$$

Solve

$$20 = 60e^{0.02T} - 60.$$

$$80 = 60e^{0.02T}$$

$$\frac{4}{3} = e^{0.02T}$$

$$\ln\frac{4}{3} = 0.02T \ln e$$

$$T = \frac{\ln\frac{4}{3}}{0.02} \approx 14.4$$

The oil will last about 14.4 years.

7.5 The Area Between Two Curves

Your Turn 1

Find the area bounded by $f(x) = 4 - x^2$, $g(x) = x + 2$, $x = -2$, and $x = 1$. A sketch such as the one below shows that the two graphs intersect at the points $(-2, 0)$ and $(1, 3)$.

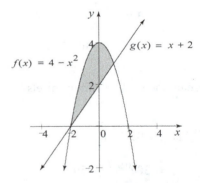

Over the interval $[-2, 1]$, $f(x) \geq g(x)$, so the area will be given by $\displaystyle\int_{-2}^1 [f(x) - g(x)]\, dx$.

$$\int_{-2}^1 [f(x) - g(x)]\, dx = \int_{-2}^1 [4 - x^2 - (x + 2)]\, dx$$

$$= \int_{-2}^1 (2 - x - x^2)\, dx$$

$$= \left(2x - \frac{1}{2}x^2 - \frac{1}{3}x^3\right)\Big|_{-2}^1$$

$$= \left(2 - \frac{1}{2} - \frac{1}{3}\right) - \left(-4 - 2 + \frac{8}{3}\right)$$

$$= \frac{9}{2}$$

Your Turn 2

Find the area between the curves $y = x^{1/4}$ and $y = x^2$. First find where these two curves intersect by setting the two righthand sides equal.

$$x^{1/4} = x^2$$

$$x = x^8$$

$$x^8 - x = 0$$

$$x(x^7 - 1) = 0$$

$$x = 0 \text{ or } x^7 - 1 = 0$$

$$x = 0 \text{ or } x = 1.$$

The corresponding y values are 0 and 1, respectively, so the curves intersect at $(0, 0)$ and $(1, 1)$, as shown in the graph below. Here $f(x) = x^{1/4}$ and $g(x) = x^2$.

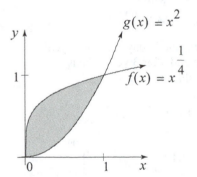

Over the interval $[0, 1]$, $f(x) \geq g(x)$, so the area is given by the integral $\displaystyle\int_0^1 [f(x) - g(x)]\, dx$.

$$\int_0^1 [f(x) - g(x)]\,dx = \int_0^1 (x^{1/4} - x^2)\,dx$$

$$= \left(\frac{4}{5}x^{5/2} - \frac{1}{3}x^3\right)\Big|_0^1$$

$$= \left(\frac{4}{5} - \frac{1}{3}\right) - (0 - 0) = \frac{7}{15}$$

Your Turn 3

Find the area enclosed by $y = x^2 - 3x$ and $y = 2x$ on $[0, 6]$. First find where the two graphs intersect.

$$x^2 - 3x = 2x$$
$$x^2 - 5x = 0$$
$$x(x - 5) = 0$$
$$x = 0 \quad \text{or} \quad x = 5.$$

The intersection points are $(0, 0)$ and $(5, 10)$, so we will need to use two integrals. On $(0, 5)$, $2x$ is the larger function and on $(5, 6)$, $x^2 - 3x$ is the larger function, as illustrated in the following graph.

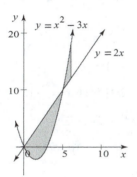

$$\text{Area} = \int_0^5 [2x - (x^2 - 3x)]\,dx + \int_5^6 [(x^2 - 3x) - 2x]\,dx$$

$$= \int_0^5 (5x - x^2)\,dx + \int_5^6 (x^2 - 5x)\,dx$$

$$= \left(\frac{5}{2}x^2 - \frac{1}{3}x^3\right)\Big|_0^5 + \left(\frac{1}{3}x^3 - \frac{5}{2}x^2\right)\Big|_5^6$$

$$= \left(\frac{125}{2} - \frac{125}{3} - 0\right) + \left(\frac{216}{3} - \frac{180}{2} - \frac{125}{3} + \frac{125}{2}\right)$$

$$= \frac{71}{3}$$

Your Turn 4

Find the consumers' surplus and the producers' surplus for oat bran when the price in dollars per ton is $D(q) = 600 - e^{q/3}$ when the demand is q tons, and the price

in dollars per ton is $S(q) = e^{q/3} - 100$ when the demand is q tons.

First find the equilibrium quantity.

$$e^{q/3} - 100 = 600 - e^{q/3}$$
$$2e^{q/3} = 700$$
$$e^{q/3} = 350$$
$$\frac{q}{3} = \ln 350$$
$$q = 3\ln 350$$
$$q \approx 17.57380$$

The equilibrium price is

$$S(17.57380) = e^{17.57380/3} - 100$$
$$\approx 250.00$$

The consumers' surplus is given by the following integral:

$$\int_0^{17.57380} (600 - e^{q/3} - 250)\,dq$$

$$= \left(350q - 3e^{q/3}\right)\Big|_0^{17.57380}$$

$$= (350(17.57380) - 3e^{17.57380/3}) - (0 - 3)$$

$$\approx 5103.83$$

The consumers' surplus is \$5103.83. As in Example 5, the producers' surplus has the same value, \$5103.83.

7.5 Warmup Exercises

W1.

$$\int_1^8 x^{1/3} + x^{4/3}\,dx$$

$$= \left(\frac{3}{4}x^{4/3} + \frac{3}{7}x^{7/3}\right)\Big|_1^8$$

$$= \left(\frac{3}{4}8^{4/3} + \frac{3}{7}8^{7/3}\right) - \left(\frac{3}{4}1^{4/3} + \frac{3}{7}1^{7/3}\right)$$

$$= \frac{3}{4}16 + \frac{3}{7}128 - \frac{3}{4} - \frac{3}{7}$$

$$= \frac{1839}{28} \approx 65.68$$

W2. $\displaystyle \int_0^3 e^{2x}\,dx = \frac{1}{2}e^{2x}\Big|_0^3$

$$= \frac{1}{2}\left(e^6 - e^0\right)$$

$$= \frac{e^6 - 1}{2} \approx 201.2$$

7.5 Exercises

1. $x = -2, x = 1, y = 2x^2 + 5, y = 0$

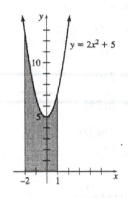

$$\int_{-2}^{1} [(2x^2 + 5) - 0] = \left(\frac{2x^3}{3} + 5x \right) \Bigg|_{-2}^{1}$$

$$= \left(\frac{2}{3} + 5 \right) - \left(-\frac{16}{3} - 10 \right)$$

$$= 21$$

3. $x = -3, x = 1, y = x^3 + 1, y = 0$

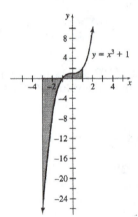

To find the points of intersection of the graphs, substitute for y.

$$x^3 + 1 = 0$$
$$x^3 = -1$$
$$x = -1$$

The region is composed of two separate regions because $y = x^3 + 1$ intersects $y = 0$ at $x = -1$.

Let $f(x) = x^3 + 1, g(x) = 0$.

In the interval $[-3, -1]$, $g(x) \geq f(x)$.

In the interval $[-1, 1]$, $f(x) \geq g(x)$.

$$\int_{-3}^{-1} [0 - (x^3 + 1)\, dx] + \int_{-1}^{1} [(x^3 + 1) - 0]\, dx$$

$$= \left(\frac{-x^4}{4} - x \right) \Bigg|_{-3}^{-1} + \left(\frac{x^4}{4} + x \right) \Bigg|_{-1}^{1}$$

$$= \left(-\frac{1}{4} + 1 \right) - \left(-\frac{81}{4} + 3 \right) + \left(\frac{1}{4} + 1 \right) - \left(\frac{1}{4} - 1 \right)$$

$$= 20$$

5. $x = -2, x = 1, y = 2x, y = x^2 - 3$

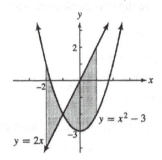

Find the points of intersection of the graphs of $y = 2x$ and $y = x^2 - 3$ by substituting for y.

$$2x = x^2 - 3$$
$$0 = x^2 - 2x - 3$$
$$0 = (x - 3)(x + 1)$$

The only intersection in $[-2, 1]$ is at $x = -1$.

In the interval $[-2, -1]$, $(x^2 - 3) \geq 2x$.

In the interval $[-1, 1]$, $2x \geq (x^2 - 3)$.

$$\int_{-2}^{-1} [(x^2 - 3) - (2x)]\, dx + \int_{-1}^{1} [(2x) - (x^2 - 3)]\, dx$$

$$= \int_{-2}^{-1} (x^2 - 3 - 2x)\, dx + \int_{-1}^{1} (2x - x^2 + 3)\, dx$$

$$= \left(\frac{x^3}{3} - 3x - x^2 \right) \Bigg|_{-2}^{-1} + \left(x^2 - \frac{x^3}{3} + 3x \right) \Bigg|_{-1}^{1}$$

$$= -\frac{1}{3} + 3 - 1 - \left(-\frac{8}{3} + 6 - 4 \right) + 1 - \frac{1}{3} + 3$$

$$- \left(1 + \frac{1}{3} - 3 \right)$$

$$= \frac{5}{3} + 6 = \frac{23}{3}$$

7. $y = x^2 - 30$
   $y = 10 - 3x$

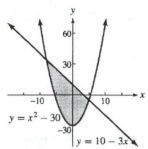

Find the points of intersection.

$$x^2 - 30 = 10 - 3x$$
$$x^2 + 3x - 40 = 0$$
$$(x + 8)(x - 5) = 0$$
$$x = -8 \quad \text{or} \quad x = 5$$

Let $f(x) = 10 - 3x$ and $g(x) = x^2 - 30$.

The area between the curves is given by

$$\int_{-8}^{5} [f(x) - g(x)] \, dx$$

$$= \int_{-8}^{5} [(10 - 3x) - (x^2 - 30)] \, dx$$

$$= \int_{-8}^{5} (-x^2 - 3x + 40) \, dx$$

$$= \left(\frac{-x^3}{3} - \frac{3x^3}{2} + 40x \right) \Bigg|_{-8}^{5}$$

$$= \frac{-5^3}{3} - \frac{3(5)^2}{2} + 40(5)$$

$$\quad - \left[\frac{-(-8)^3}{3} - \frac{3(-8)^2}{2} + 40(-8) \right]$$

$$= \frac{-125}{3} - \frac{75}{2} + 200 - \frac{512}{3} + \frac{192}{2} + 320$$

$$\approx 366.1667.$$

9. $y = x^2, y = 2x$

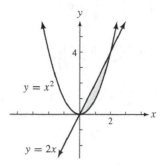

Find the points of intersection.

$$x^2 = 2x$$
$$x^2 - 2x = 0$$
$$x(x - 2) = 0$$
$$x = 0 \quad \text{or} \quad x = 2$$

Let $f(x) = 2x$ and $g(x) = x^2$.

The area between the curves is given by

$$\int_0^2 [f(x) - g(x)] \, dx = \int_0^2 (2x - x^2) \, dx$$

$$= \left(\frac{2x^2}{2} - \frac{x^3}{3} \right) \Bigg|_0^2$$

$$= 4 - \frac{8}{2} = \frac{4}{3}.$$

11. $x = 1, x = 6, y = \frac{1}{x}, y = \frac{1}{2}$

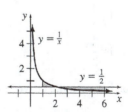

To find the points of intersection of the graphs, substitute for y.

$$\frac{1}{x} = \frac{1}{2}$$
$$x = 2$$

The region is composed of two separate regions because $y = \frac{1}{x}$ intersects $y = \frac{1}{2}$ at $x = 2$.

Let $f(x) = \frac{1}{x}, g(x) = \frac{1}{2}$.

In the interval $[1, 2]$, $f(x) \geq g(x)$.

In the interval $[2, 6]$, $g(x) \geq f(x)$.

$$\int_1^2 \left(\frac{1}{x} - \frac{1}{2} \right) dx + \int_2^6 \left(\frac{1}{2} - \frac{1}{x} \right) dx$$

$$= \left(\ln|x| - \frac{x}{2} \right)\Big|_1^2 + \left(\frac{x}{2} - \ln|x| \right)\Big|_2^6$$

$$= (\ln 2 - 1) - \left(0 - \frac{1}{2} \right) + (3 - \ln 6) - (1 - \ln 2)$$

$$= 2 \ln 2 - \ln 6 + \frac{3}{2} \approx 1.095$$

13. $x = -1, x = 1, y = e^x, y = 3 - e^x$

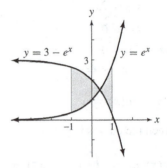

To find the point of intersection, set
$e^x = 3 - e^x$ and solve for x.

$$e^x = 3 - e^x$$

$$2e^x = 3$$

$$e^x = \frac{3}{2}$$

$$\ln e^x = \ln \frac{3}{2}$$

$$x \ln e = \ln \frac{3}{2}$$

$$x = \ln \frac{3}{2}$$

The area of the region between the curves from $x = -1$ to $x = 1$ is

$$\int_{-1}^{\ln 3/2} [(3 - e^x) - e^x] \, dx$$

$$+ \int_{\ln 3/2}^{1} [e^x - (3 - e^x)] \, dx$$

$$= \int_{-1}^{\ln 3/2} (3 - 2e^x) \, dx + \int_{\ln 3/2}^{1} (2e^x - 3) \, dx$$

$$= (3x - 2e^x)\Big|_{-1}^{\ln 3/2} + (2e^x - 3x)\Big|_{\ln 3/2}^{1}$$

$$= \left[\left(3 \ln \frac{3}{2} - 2e^{\ln 3/2} \right) - [3(-1) - 2e^{-1}] \right]$$

$$+ \left[2e^1 - 3(1) - \left(2e^{\ln 3/2} - 3 \ln \frac{3}{2} \right) \right]$$

$$= \left[\left(3 \ln \frac{3}{2} - 3 \right) - \left(-3 - \frac{2}{e} \right) \right]$$

$$+ \left[2e - 3 - \left(3 - 3 \ln \frac{3}{2} \right) \right]$$

$$= 6 \ln \frac{3}{2} + \frac{2}{e} + 2e - 6 \approx 2.605.$$

15. $x = -1, x = 2, y = 2e^{2x}, y = e^{2x} + 1$

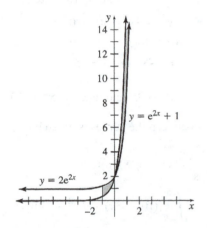

To find the points of intersection of the graphs, substitute for y.

$$2e^{2x} = e^{2x} + 1$$

$$e^{2x} = 1$$

$$2x = 0$$

$$x = 0$$

The region is composed of two separate regions because $y = 2e^{2x}$ intersects $y = e^{2x} + 1$ at $x = 0$.

Let $f(x) = 2e^{2x}, g(x) = e^{2x} + 1$.

In the interval $[-1, 0]$, $g(x) \geq f(x)$.

In the interval $[0, 2]$, $f(x) \geq g(x)$.

$$\int_{-1}^{0} (e^{2x} + 1 - 2e^{2x}) \, dx$$

$$+ \int_{0}^{2} [2e^{2x} - (e^{2x} + 1)] \, dx$$

$$= \left(-\frac{e^{2x}}{2} + x \right)\Big|_{-1}^{0} + \left(\frac{e^{2x}}{2} - x \right)\Big|_{0}^{2}$$

$$= \left(-\frac{1}{2} + 0 \right) - \left(-\frac{e^{-2}}{2} - 1 \right)$$

$$+ \left(\frac{e^4}{2} - 2 \right) - \left(\frac{1}{2} - 0 \right)$$

$$= \frac{e^{-2} + e^4}{2} - 2 \approx 25.37$$

17. $y = x^3 - x^2 + x + 1, \; y = 2x^2 - x + 1$

Find the points of intersection.

$$x^3 - x^2 + x + 1 = 2x^2 - x + 1$$
$$x^3 - 3x^2 + 2x = 0$$
$$x(x^2 - 3x + 2) = 0$$
$$x(x - 2)(x - 1) = 0$$

The points of intersection are at $x = 0$, $x = 1$, and $x = 2$.

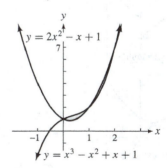

Area between the curves is

$$\int_{0}^{1} [(x^3 - x^2 + x + 1) - (2x^2 - x + 1)] \, dx$$

$$+ \int_{0}^{2} [(2x^2 - x + 1) - (x^3 - x^2 + x + 1)] \, dx$$

$$= \int_{0}^{1} (x^3 - 3x^2 + 2x) \, dx + \int_{1}^{2} (-x^3 + 3x^2 - 2x)] \, dx$$

$$= \left(\frac{x^4}{4} - x^3 + x^2 \right)\Big|_{0}^{1} + \left(\frac{-x^4}{4} + x^3 - x^2 \right)\Big|_{1}^{2}$$

$$= \left[\left(\frac{1}{4} - 1 + 1 \right) - (0) \right]$$

$$+ \left[(-4 + 8 - 4) - \left(-\frac{1}{4} + 1 - 1 \right) \right]$$

$$= \frac{1}{4} + \frac{1}{4}$$

$$= \frac{1}{2}.$$

19. $y = x^4 + \ln (x + 10),$
        $y = x^3 + \ln (x + 10)$

Find the points of intersection.

$$x^4 + \ln (x + 10) = x^3 + \ln (x + 10)$$
$$x^4 - x^3 = 0$$
$$x^3(x - 1) = 0$$
$$x = 0 \text{ or } x = 1$$

The points of intersection are at $x = 0$ and $x = 1$.

The area between the curves is

$$\int_{0}^{1} [(x^3 + \ln (x + 10)) - (x^4 + \ln (x + 10))] \, dx$$

$$= \int_{0}^{1} (x^3 - x^4) \, dx$$

$$= \left(\frac{x^4}{4} - \frac{x^5}{5} \right)\Big|_{0}^{1}$$

$$= \left(\frac{1}{4} - \frac{1}{5} \right) - (0) = \frac{1}{20}.$$

21. $y = x^{4/3}, \; y = 2x^{1/3}$

Find the points of intersection.

$$x^{4/3} = 2x^{1/3}$$
$$x^{4/3} - 2x^{1/3} = 0$$
$$x^{1/3}(x - 2) = 0$$
$$x = 0 \text{ or } x = 2$$

The points of intersection are at $x = 0$ and $x = 2$.

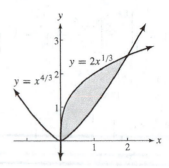

The area between the curves is

$$\int_0^2 (2x^{1/3} - x^{4/3})\, dx = 2\frac{x^{4/3}}{\frac{4}{3}} - \frac{x^{7/3}}{\frac{7}{3}}\Bigg|_0^2$$

$$= \frac{3}{2}x^{4/3} - \frac{3}{7}x^{7/3}\Bigg|_0^2$$

$$= \left[\frac{3}{2}(2)^{4/3} - \frac{3}{7}(2)^{7/3}\right] - 0$$

$$= \frac{3(2^{4/3})}{2} - \frac{3(2^{7/3})}{7}$$

$$\approx 1.62.$$

23. $x = 0,\ x = 3,\ y = 2e^{3x},\ y = e^{3x} + e^6$

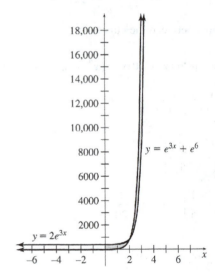

To find the points of intersection of the graphs, substitute for y.

$$2e^{3x} = e^{3x} + e^6$$

$$e^{3x} = e^6$$

$$3x = 6$$

$$x = 2$$

The region is composed of two separate regions because $y = 2e^{3x}$ intersects $y = e^{3x} + e^6$ at $x = 2$.

Let $f(x) = 2e^{3x}$, $g(x) = e^{3x} + e^6$.

In the interval $[0, 2]$, $g(x) \geq f(x)$.

In the interval $[2, 3]$, $f(x) \geq g(x)$.

$$\int_0^2 (e^{3x} + e^6 - 2e^{3x})\, dx + \int_2^3 [2e^{3x} - (e^{3x} + e^6)]\, dx$$

$$= \left(-\frac{e^{3x}}{3} + e^6 x\right)\Bigg|_0^2 + \left(\frac{e^{3x}}{3} - e^6 x\right)\Bigg|_2^3$$

$$= \left(-\frac{e^6}{3} + 2e^6\right) - \left(-\frac{1}{3} + 0\right)$$

$$+ \left(\frac{e^9}{3} - 3e^6\right) - \left(\frac{e^6}{3} - 2e^6\right)$$

$$= \frac{e^9 + e^6 + 1}{3}$$

$$\approx 2836$$

25. Graph $y_1 = e^x$ and $y_2 = -x^2 - 2x$ on your graphing calculator. Use the intersect command to find the two intersection points. The resulting screens are:

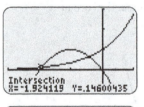

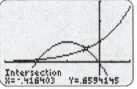

These screens show that $e^x = -x^2 - 2x$ when $x \approx -1.9241$ and $x \approx -0.4164$.

In the interval $[-1.9241, -0.4164]$,

$$e^x < -x^2 - 2x.$$

The area between the curves is given by

$$\int_{-1.9241}^{-0.4164} [(-x^2 - 2x) - e^x]\, dx.$$

Use the fnInt command to approximate this definite integral.

The resulting screen is:

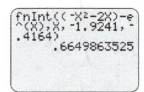

The last screen shows that the area is approximately 0.6650.

27. **(a)** It is profitable to use the machine until $S'(t) = C'(t)$.

$$150 - t^2 = t^2 + \frac{11}{4}t$$

$$2t^2 + \frac{11}{4}t - 150 = 0$$

$$8t^2 + 11t - 600 = 0$$

$$t = \frac{-11 \pm \sqrt{121 - 4(8)(-600)}}{16}$$

$$= \frac{-11 \pm 139}{16}$$

$$t = 8 \quad \text{or} \quad t = -9.375$$

It will be profitable to use this machine for 8 years. Reject the negative solution.

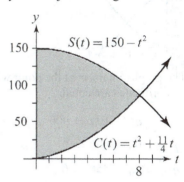

(b) Since $150 - t^2 > t^2 + \frac{11}{4}t$, in the interval $[0, 8]$, the net total saving in the first year are

$$\int_0^1 \left[(150 - t^2) - \left(t^2 + \frac{11}{4}t\right)\right] dt$$

$$= \int_0^1 \left(-2t^2 - \frac{11}{4}t + 150t\right) dt$$

$$= \left(\frac{-2t^3}{3} - \frac{11t^2}{8} + 150t\right)\Bigg|_0^1$$

$$= -\frac{2}{3} - \frac{11}{8} + 150 \approx \$148.$$

(c) The net total savings over the entire period of use are

$$\int_0^8 \left[(150 - t^2) - \left(t^2 + \frac{11}{4}t\right)\right] dt$$

$$= \left(\frac{-2t^3}{3} - \frac{11t^2}{8} + 150t\right)\Bigg|_0^8$$

$$= \frac{-2(8^3)}{3} - \frac{11(8^2)}{8} + 150(8)$$

$$= \frac{-1024}{3} - \frac{704}{8} + 1200 \approx \$771.$$

29. **(a)** $E'(x) = e^{0.1x}$ and $I'(x) = 98.8 - e^{0.1x}$

To find the point of intersection, where profit will be maximized, set the functions equal to each other and solve for x.

$$e^{0.1x} = 98.8 - e^{0.1x}$$

$$2e^{0.1x} = 98.8$$

$$e^{0.1x} = 49.4$$

$$0.1x = \ln 49.4$$

$$x = \frac{\ln 49.4}{0.1} \approx 39$$

The optimum number of days for the job to last is 39.

(b) The total income for 39 days is

$$\int_0^{39} (98.8 - e^{0.1x}) \, dx$$

$$= \left(98.8x - \frac{e^{0.1x}}{0.1}\right)\Bigg|_0^{39}$$

$$= \left(98.8x - 10e^{0.1x}\right)\Bigg|_0^{39}$$

$$= [98.8(39) - 10e^{3.9}] - (0 - 10)$$

$$= \$3369.18.$$

(c) The total expenditure for 39 days is

$$\int_0^{39} e^{0.1x} dx = \frac{e^{0.1x}}{0.1}\Bigg|_0^{39}$$

$$= 10e^{0.1x}\Big|_0^{39}$$

$$= 10e^{3.9} - 10$$

$$= \$484.02.$$

(d) Profit $=$ Income $-$ Expense
$$= 3369.18 - 484.02 = \$2885.16$$

31. $S(q) = q^{5/2} + 2q^{3/2} + 50$; $q = 16$ is the equilibrium quantity.

Producers surplus $= \displaystyle\int_0^{q_0} [p_0 - S(q)]\,dq$,

where p_0 is the equilibrium price and q_0 is equilibrium supply.

$$p_0 = S(16) = (16)^{5/2} + 2(16)^{3/2} + 50$$
$$= 1202$$

Therefore, the producers' surplus is

$$\int_0^{16} [1202 - (q^{5/2} + 2q^{3/2} + 50)]\,dq$$

$$= \int_0^{16} (1152 - q^{5/2} - 2q^{3/2})\,dq$$

$$= \left(1152q - \frac{2}{7}q^{7/2} - \frac{4}{5}q^{5/2}\right)\Big|_0^{16}$$

$$= 1152(16) - \frac{2}{7}(16)^{7/2} - \frac{4}{5}(16)^{5/2}$$

$$= 18,432 - \frac{32,768}{7} - \frac{4096}{5}$$

$$= 12,931.66.$$

The producers' surplus is $12,931.66.

33. $D(q) = \dfrac{200}{(3q + 1)^2}$; $q = 3$ is the equilibrium quantity.

Consumers' surplus $= \displaystyle\int_0^{q_0} |D(q) - p_0|\,dq$

$$p_0 = D(3) = 2$$

Therefore, the consumers' surplus is

$$\int_0^3 \left|\frac{200}{(3q + 1)^2} - 2\right| dq$$

$$= \int_0^3 \frac{200}{(3q + 1)^2}\,dq - \int_0^3 2\,dq.$$

Let $u = 3q + 1$, so that

$$du = 3\,dq \text{ and } \frac{1}{3}\,du = dq.$$

$$\int_0^3 \frac{200}{(3q + 1)^2}\,dq - \int_0^3 2\,dq$$

$$= \frac{1}{3}\int_1^{10} \frac{200}{u^2}\,du - \int_0^3 2\,dq$$

$$= \frac{200}{3}\int_1^{10} u^{-2}\,du - \int_0^3 2\,dq$$

$$= \frac{200}{3} \cdot \frac{u^{-1}}{-1}\Big|_1^{10} - 2q\Big|_0^3$$

$$= -\frac{200}{3u}\Big|_1^{10} - 6$$

$$= -\frac{200}{30} + \frac{200}{3} - 6$$

$$= \$54$$

35. $S(q) = q^2 + 10q$

$D(q) = 900 - 20q - q^2$

(a) The graphs of the supply and demand functions are parabolas with vertices at $(-5, -25)$ and $(-10, 1900)$, respectively.

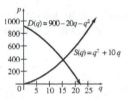

(b) The graphs intersect at the point where the y-coordinates are equal.

$$q^2 + 10q = 900 - 20q - q^2$$
$$2q^2 + 30q - 900 = 0$$
$$q^2 + 15q - 450 = 0$$
$$(q + 30)(q - 15) = 0$$
$$q = -30 \text{ or } q = 15$$

Disregard the negative solution.

The supply and demand functions are in equilibrium when $q = 15$.

$$S(15) = 15^2 + 10(15) = 375$$

The point is $(15, 375)$.

(c) Find the consumers' surplus.

$$\int_0^{q_0} [D(q) - p_0]\,dq$$

$$p_0 = D(15) = 375$$

$$\int_0^{15} [(900 - 20q - q^2) - 375]\, dq$$

$$= \int_0^{15} (525 - 20q - q^2)\, dq$$

$$= \left(525q - 10q^2 - \frac{1}{3}q^3 \right)\Big|_0^{15}$$

$$= \left[525(15) - 10(15)^2 - \frac{1}{3}(15)^3 \right] - 0 = 4500$$

The consumer's surplus is $4500.

(d) Find the producers' surplus.

$$\int_0^{q_0} [p_0 - S(q)]\, dq$$

$$p_0 = S(15) = 375$$

$$\int_0^{15} [375 - (q^2 + 10q)]\, dq$$

$$= \int_0^{15} (375 - q^2 - 10q)\, dq$$

$$= \left(375q - \frac{1}{3}q^3 - 5q^2 \right)\Big|_0^{15}$$

$$= \left[375(15) - \frac{1}{3}(15)^3 - 5(15)^2 \right] - 0$$

$$= 3375$$

The producer's surplus is $3375.

37. **(a)** $S(q) = q^2 + 10q$; $S(q) = 264$ is the price the government set.

$$264 = q^2 + 10q$$

$$0 = q^2 + 10q - 264$$

$$0 = (q - 12)(q + 22)$$

$$q = 12 \quad \text{or} \quad q = -22$$

Only 12 is a meaningful solution here. Thus, 12 units of oil will be produced.

(b) The consumers' surplus is given by

$$\int_0^{12} (900 - 20q - q^2 - 264)\, dq$$

$$= \int_0^{12} (636 - 20q - q^2)\, dq$$

$$= \left(636q - 10q^2 - \frac{1}{3}q^3 \right)\Big|_0^{12}$$

$$= 636(12) - 10(12)^2 - \frac{1}{3}(12)^3 - 0$$

$$= 5616$$

Here the consumer' surplus is $5616. In this case, the consumers' surplus is $5616 - 4500 = \$1116$ larger.

(c) The producers' surplus is given by

$$\int_0^{12} [264 - (q^2 + 10q)]\, dq$$

$$= \int_0^{12} (264 - q^2 - 10q)\, dq$$

$$= \left(264q - \frac{1}{3}q^3 - 5q^2 \right)\Big|_0^{12}$$

$$= 264(12) - \frac{1}{3}(12)^3 - 5(12)^2 - 0$$

$$= 1872$$

Here the producers' surplus is $1872. In this case, the producers' surplus is $3375 - 1872 = \$1503$ smaller.

(d) For the equilibrium price, the total consumers' and producers' surplus is

$$4500 + 3375 = \$7875$$

For the government price, the total consumers' and producers' surplus is

$$5616 + 1872 = \$7488.$$

The difference is

$$7875 - 7488 = \$387.$$

39. **(a)** The pollution level in the lake is changing at the rate $f(t) - g(t)$ at any time t. We find the amount of pollution by integrating.

$$\int_0^{12} [f(t) - g(t)]\, dt$$

$$= \int_0^{12} [10(1 - e^{-0.5t}) - 0.4t]\, dt$$

$$= \left(10t - 10 \cdot \frac{1}{-0.5} e^{-0.5t} - 0.4 \cdot \frac{1}{2}t^2 \right)\Big|_0^{12}$$

$$= (20e^{-0.5t} + 10t - 0.2t^2)\big|_0^{12}$$

$$= [20e^{-0.5(12)} + 10(12) - 0.2(12)^2]$$

$$\quad - [20e^{-0.5(0)} + 10(0) - 0.2(0)^2]$$

$$= (20e^{-6} + 91.2) - (20)$$

$$= 20e^{-6} + 71.2 \approx 71.25$$

After 12 hours, there are about 71.25 gallons.

(b) The graphs of the functions intersect at about 25.00. So the rate that pollution enters the lake equals the rate the pollution is removed at about 25 hours.

(c) $\displaystyle\int_0^{25} [f(t) - g(t)]\,dt$

$$= (20e^{-0.5t} + 10t - 0.2t^2)\Big|_0^{25}$$

$$= [20e^{-0.5(25)} + 10(25) - 0.2(25)^2)] - 20$$

$$= 20e^{-12.5} + 105 \approx 105$$

After 25 hours, there are about 105 gallons.

(d) For $t > 25$, $g(t) > f(t)$, and pollution is being removed at the rate $g(t) - f(t)$. So, we want to solve for c, where

$$\int_0^c [f(t) - g(t)]\,dt = 0.$$

Alternatively, we could solve for c in

$$\int_{25}^c [g(t) - f(t)]\,dt = 105.$$

One way to do this with a graphing calculator is to graph the function

$$y = \int_0^x [f(t) - g(t)]\,dt$$

and determine the values of x for which $y = 0$. The first window shows how the function can be defined.

A suitable window for the graph is $[0, 50]$ by $[0, 110]$.

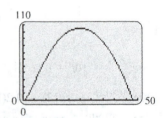

Use the calculator's features to approximate where the graph intersects the x-axis. These are at 0 and about 47.91. Therefore, the pollution will be removed from the lake after about 47.91 hours.

41. $I(x) = 0.9x^2 + 0.1x$

(a) $I(0.1) = 0.9(0.1)^2 + 0.1(0.1)$

$$= 0.019$$

The lower 10% of income producers earn 1.9% of total income of the population.

(b) $I(0.4) = 0.9(0.4)^2 + 0.1(0.4) = 0.184$

The lower 40% of income producers earn 18.4% of total income of the population.

(c)   The graph of $I(x) = x$ is a straight line through the points $(0, 0)$ and $(1, 1)$. The graph of $I(x) = 0.9x^2 + 0.1x$ is a parabola with vertex $\left(-\frac{1}{18}, -\frac{1}{360}\right)$. Restrict the domain to $0 \le x \le 1$.

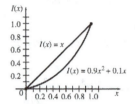

(d)   To find the points of intersection, solve

$$x = 0.9x^2 + 0.1x.$$

$$0.9x^2 - 0.9x = 0$$

$$0.9x(x - 1) = 0$$

$$x = 0 \quad \text{or} \quad x = 1$$

The area between the curves is given by

$$\int_0^1 [x - (0.9x^2 + 0.1x)]\,dx$$

$$= \int_0^1 (0.9x - 0.9x^2)\,dx$$

$$= \left(\frac{0.9x^2}{2} - \frac{0.9x^3}{3}\right)\Big|_0^1$$

$$= \frac{0.9}{2} - \frac{0.9}{3} = 0.15.$$

(e)   Income is distributed less equally in 2012 than in 1968.

7.6 Numerical Integration

Your Turn 1

Use the trapezoidal rule with $n = 4$ to approximate

$$\int_1^3 \sqrt{x^2 + 3}\, dx.$$

Here $f(x) = \sqrt{x^2 + 3}$, $a = 1$, $b = 3$, and $n = 4$. The subintervals have length $(3 - 1)/4 = 1/2$. The following table summarizes the information required.

i	x_i	$f(x_i)$
0	1	$f(1) = 2$
1	3/2	$f(3/2) \approx 2.29129$
2	2	$f(2) \approx 2.64575$
3	5/2	$f(5/2) \approx 3.04138$
4	3	$f(3) \approx 3.46410$

The trapezoidal rule gives

$$\int_1^3 \sqrt{x^2 + 3}\, dx$$

$$\approx \frac{3-1}{2}\left[\frac{1}{2}(2) + 2.29129 + 2.64757 \right.$$

$$\left. + 3.04138 + \frac{1}{2}(3.46410)\right]$$

$$\approx 5.3552.$$

Your Turn 2

Use Simpson's rule with $n = 4$ to approximate

$$\int_1^3 \sqrt{x^2 + 3}\, dx.$$

Here $f(x) = \sqrt{x^2 + 3}$, $a = 1$, $b = 3$, and $n = 4$. The subintervals have length $(3 - 1)/4 = 1/2$. The following table summarizes the information required; it is the same as the table used in Your Turn 1.

i	x_i	$f(x_i)$
0	1	$f(1) = 2$
1	3/2	$f(3/2) \approx 2.29129$
2	2	$f(2) \approx 2.64575$
3	5/2	$f(5/2) \approx 3.04138$
4	3	$f(3) \approx 3.46410$

For Simpson's rule, the factor in front is $(b - a)/3n$ $= (3 - 1)/12 = 1/6$. Simpson's rule thus gives

$$\int_1^3 \sqrt{x^2 + 3}\, dx$$

$$\approx \frac{1}{6}\left[2 + 4(2.29129) + 2(2.64757) + 4(3.04138) + 3.46410\right]$$

$$\approx 5.3477.$$

7.6 Exercises

1. $\displaystyle\int_0^2 (3x^2 + 2)\, dx$

    $n = 4$, $b = 2$, $a = 0$, $f(x) = 3x^2 + 2$

i	x_i	$f(x_i)$
0	0	2
1	$\dfrac{1}{2}$	2.75
2	1	5
3	$\dfrac{3}{2}$	8.75
4	2	14

(a) Trapezoidal rule:

$$\int_0^2 (3x^2 + 2)\, dx$$

$$\approx \frac{2-0}{4}\left[\frac{1}{2}(2) + 2.75 + 5 + 8.75 + \frac{1}{2}(14)\right]$$

$$= 0.5(24.5)$$

$$= 12.25$$

(b) Simpson's rule:

$$\int_0^2 (3x^2 + 2)\, dx$$

$$\approx \frac{2-0}{3(4)}[2 + 4(2.75) + 2(5) + 4(8.75) + 14]$$

$$= \frac{2}{12}(72)$$

$$= 12$$

(c) Exact value:

$$\int_0^2 (3x^2 + 2)\, dx = (x^3 + 2x)\Big|_0^2$$

$$= (8 + 4) - 0$$

$$= 12$$

3. $\displaystyle\int_{-1}^{3} \frac{3}{5-x}\,dx$

$n = 4, b = 3, a = -1, f(x) = \dfrac{3}{5-x}$

i	x_i	$f(x_i)$
0	-1	0.5
1	0	0.6
2	1	0.75
3	2	1
4	3	1.5

(a) Trapezoidal rule:

$\displaystyle\int_{-1}^{3}\frac{3}{5-x}\,dx$

$\approx \dfrac{3-(-1)}{4}\left[\dfrac{1}{2}(0.5)+0.6+0.75+1+\dfrac{1}{2}(1.5)\right]$

$= 1(3.35)$

$= 3.35$

(b) Simpson's rule:

$\displaystyle\int_{-1}^{3}\frac{3}{5-x}\,dx$

$\approx \dfrac{3-(-1)}{3(4)}[0.5+4(0.6)+2(0.75)+4(1)+1.5]$

$= \dfrac{1}{3}\left(\dfrac{99}{10}\right)$

$= \dfrac{33}{10} \approx 3.3$

(c) Exact value:

$\displaystyle\int_{-1}^{3}\frac{3}{5-x}\,dx = -3\ln|5-x|\,\Big|_{-1}^{3}$

$= -3(\ln|2|-\ln|6|)$

$= 3\ln 3 \approx 3.296$

5. $\displaystyle\int_{-1}^{2}(2x^3+1)\,dx$

$n = 4, b = 2, a = -1, f(x) = 2x^3+1$

i	x_i	$f(x)$
0	-1	-1
1	$-\dfrac{1}{4}$	$\dfrac{31}{32}$
2	$\dfrac{1}{2}$	$\dfrac{5}{4}$
3	$\dfrac{5}{4}$	$\dfrac{157}{32}$
4	2	17

(a) Trapezoidal rule:

$\displaystyle\int_{-1}^{2}(2x^3+1)\,dx$

$\approx \dfrac{2-(-1)}{4}\left[\dfrac{1}{2}(-1)+\dfrac{31}{32}+\dfrac{5}{4}\right.$

$\left.\qquad\qquad +\dfrac{157}{32}+\dfrac{1}{2}(17)\right]$

$= 0.75(15.125)$

≈ 11.34

(b) Simpson's rule:

$\displaystyle\int_{-1}^{2}(2x^3+1)\,dx$

$\approx \dfrac{2-(-1)}{3(4)}\left[-1+4\left(\dfrac{31}{32}\right)+2\left(\dfrac{5}{4}\right)+4\left(\dfrac{157}{32}\right)+17\right]$

$= \dfrac{1}{4}(42) = 10.5$

(c) Exact value:

$\displaystyle\int_{-1}^{2}(2x^3+1)\,dx$

$= \left(\dfrac{x^4}{2}+x\right)\Big|_{-1}^{2}$

$= (8+2)-\left(\dfrac{1}{2}-1\right)$

$= \dfrac{21}{2}$

$= 10.5$

7. $\displaystyle\int_1^5 \frac{1}{x^2}\,dx$

$n = 4, b = 5, a = 1, f(x) = \dfrac{1}{x^2}$

i	x_i	$f(x_i)$
0	1	1
1	2	0.25
2	3	0.1111
3	4	0.0625
4	5	0.04

(a) Trapezoidal rule:

$$\int_1^5 \frac{1}{x^2}\,dx$$

$$\approx \frac{5-1}{4}\left[\frac{1}{2}(1) + 0.25 + 0.1111\right.$$

$$\left. + 0.0625 + \frac{1}{2}(0.04)\right]$$

$$\approx 0.9436$$

(b) Simpson's rule:

$$\int_1^5 \frac{1}{x^2}\,dx$$

$$\approx \frac{5-1}{12}[1 + 4(0.25) + 2(0.1111)$$

$$+ 4(0.0625) + 0.04]$$

$$\approx 0.8374$$

(c) Exact value:

$$\int_1^5 x^{-2}\,dx = -x^{-1}\Big|_1^5$$

$$= -\frac{1}{5} + 1$$

$$= \frac{4}{5} = 0.8$$

9. $\displaystyle\int_0^1 4xe^{-x^2}\,dx$

$n = 4, b = 1, a = 0, f(x) = 4xe^{-x^2}$

i	x_i	$f(x_i)$
0	0	0
1	$\dfrac{1}{4}$	$e^{-1/16}$
2	$\dfrac{1}{2}$	$2e^{-1/4}$
3	$\dfrac{3}{4}$	$3e^{-9/16}$
4	1	$4e^{-1}$

(a) Trapezoidal rule:

$$\int_0^1 4xe^{-x^2}\,dx$$

$$\approx \frac{1-0}{4}\left[\frac{1}{2}(0) + e^{-1/16} + 2e^{-1/4}\right.$$

$$\left. + 3e^{-9/16} + \frac{1}{2}(4e^{-1})\right]$$

$$= \frac{1}{4}(e^{-1/16} + 2e^{-1/4} + 3e^{-9/16} + 2e^{-1})$$

$$\approx 1.236$$

(b) Simpson's rule:

$$\int_0^1 4xe^{-x^2}\,dx$$

$$\approx \frac{1-0}{3(4)}[0 + 4(e^{-1/16}) + 2(2e^{-1/4})$$

$$+ 4(3e^{-9/16}) + 4e^{-1}]$$

$$= \frac{1}{12}(4e^{-1/16} + 4e^{-1/4} + 12e^{-9/16} + 4e^{-1})$$

$$\approx 1.265$$

(c) Exact value:

$$\int_0^1 4xe^{-x^2}\,dx = -2e^{-x^2}\Big|_0^1$$

$$= (-2e^{-1}) - (-2)$$

$$= 2 - 2e^{-1} \approx 1.264$$

11. $y = \sqrt{4 - x^2}$

$n = 8, b = 2, a = -2, f(x) = \sqrt{4 - x^2}$

i	x_i	y
0	-2.0	0
1	-1.5	1.32289
2	-1.0	1.73205
3	-0.5	1.93649
4	0	2
5	0.5	1.93649
6	1.0	1.73205
7	1.5	1.32289
8	2.0	0

(a) Trapezoidal rule:

$$\int_{-2}^{2} \sqrt{4 - x^2}\, dx$$

$$\approx \frac{2 - (-2)}{8}$$

$$\cdot \left[\frac{1}{2}(0) + 1.32289 + 1.73205 + \cdots + \frac{1}{2}(0)\right]$$

$$\approx 5.991$$

(b) Simpson's rule:

$$\int_{-2}^{2} \sqrt{4 - x^2}\, dx$$

$$\approx \frac{2 - (-2)}{3(8)}$$

$$\cdot [0 + 4(1.32289) + 2(1.73205) + 4(1.93649) + 2(2) + 4(1.93649) + 2(1.73205) + 4(1.32289) + 0]$$

$$\approx 6.167$$

(c) Area of semicircle $= \frac{1}{2}\pi r^2 = \frac{1}{2}\pi(2)^2$

$$\approx 6.283$$

Simpson's rule is more accurate.

13. Since $f(x) > 0$ and $f''(x) > 0$ for all x between a and b, we know the graph of $f(x)$ on the interval from a to b is concave upward. Thus, the trapezoid that approximates the area will have an area greater than the actual area Thus,

$$T > \int_{a}^{b} f(x)\, dx.$$

The correct choice is (b).

15. **(a)** $\int_{0}^{1} x^4\, dx = \left(\frac{1}{5}\right)x^5 \Big|_{0}^{1}$

$$= \frac{1}{5}$$

$$= 0.2$$

(b) $n = 4, b = 1, a = 0, f(x) = x^4$

$$\int_{0}^{1} x^4\, dx \approx \frac{1 - 0}{4}\left[\frac{1}{2}(0) + \frac{1}{256} + \frac{1}{16} + \frac{81}{256} + \frac{1}{2}(1)\right]$$

$$= \frac{1}{4}\left(\frac{226}{256}\right)$$

$$\approx 0.220703$$

$n = 8, b = 1, a = 0, f(x) = x^4$

$$\int_{0}^{1} x^4\, dx \approx \frac{1 - 0}{8}\left[\frac{1}{2}(0) + \frac{1}{4096} + \frac{1}{256} + \frac{81}{4096}\right.$$

$$\left. + \frac{1}{16} + \frac{625}{4096} + \frac{81}{256} + \frac{2401}{4096} + \frac{1}{2}(1)\right]$$

$$= \frac{1}{8}\left(\frac{6724}{4096}\right)$$

$$\approx 0.205200$$

$n = 16, b = 1, a = 0, f(x) = x^4$

$$\int_{0}^{1} x^4\, dx \approx \frac{1 - 0}{16}\left[\frac{1}{2}(0) + \frac{1}{65,536} + \frac{1}{4096}\right.$$

$$+ \frac{81}{65,536} + \frac{1}{256} + \frac{625}{65,536}$$

$$+ \frac{81}{4096} + \frac{2401}{65,536} + \frac{1}{16}$$

$$+ \frac{6561}{65,536} + \frac{625}{4096} + \frac{14,641}{65,536}$$

$$+ \frac{81}{256} + \frac{28,561}{65,536} + \frac{2401}{4096}$$

$$\left. + \frac{50,625}{65,536} + \frac{1}{2}(1)\right]$$

$$\approx \frac{1}{16}\left(\frac{211,080}{65,536}\right)$$

$$\approx 0.201302$$

$n = 32, b = 1, a = 0, f(x) = x^4$

$$\int_0^1 x^4\,dx$$

$$\approx \frac{1-0}{32}\left[\frac{1}{2}(0) + \frac{1}{1,048,576} + \frac{1}{65,536}\right.$$

$$+ \frac{81}{1,048,576} + \frac{1}{4096} + \frac{625}{1,048,576}$$

$$+ \frac{81}{65,536} + \frac{2401}{1,048,576} + \frac{1}{256} + \frac{6561}{1,048,576}$$

$$+ \frac{625}{65,536} + \frac{14,641}{1,048,576} + \frac{81}{4096} + \frac{28,561}{1,048,576}$$

$$+ \frac{2401}{65,536} + \frac{50,625}{1,048,576} + \frac{1}{16} + \frac{83,521}{1,048,576}$$

$$+ \frac{6561}{65,536} + \frac{130,321}{1,048,576} + \frac{625}{4096} + \frac{194,481}{1,048,576}$$

$$+ \frac{14,641}{65,536} + \frac{279,841}{1,048,576} + \frac{81}{256} + \frac{390,625}{1,048,576}$$

$$+ \frac{28,561}{65,536} + \frac{531,441}{1,048,576} + \frac{2401}{4096} + \frac{707,281}{1,048,576}$$

$$+ \left.\frac{50,625}{65,536} + \frac{923,521}{1,048,576} + \frac{1}{2}(1)\right]$$

$$\approx \frac{1}{32}\left(\frac{6,721,808}{1,048,576}\right) \approx 0.200325$$

To find error for each value of n, subtract as indicated.

$n = 4$: $(0.220703 - 0.2) = 0.020703$

$n = 8$: $(0.205200 - 0.2) = 0.005200$

$n = 16$: $(0.201302 - 0.2) = 0.001302$

$n = 32$: $(0.200325 - 0.2) = 0.000325$

(c) $p = 1$

$$4^1(0.020703) = 4(0.020703)$$
$$= 0.082812$$

$$8^1(0.005200) = 8(0.005200)$$
$$= 0.0416$$

Since these are not the same, try $p = 2$.

$p = 2$:

$$4^2(0.020703) = 16(0.020703)$$
$$= 0.331248$$

$$8^2(0.005200) = 64(0.005200) = 0.3328$$

$$16^2(0.001302) = 256(0.001302)$$
$$= 0.333312$$

$$32^2(0.000325) = 1024(0.000325)$$
$$= 0.3328$$

Since these values are all approximately the same, the correct choice is $p = 2$.

17. **(a)** $\displaystyle\int_0^1 x^4\,dx = \frac{1}{5}x^5\Big|_0^1$

$$= \frac{1}{5}$$

$$= 0.2$$

(b) $n = 4, b = 1, a = 0, f(x) = x^4$

$$\int_0^1 x^4\,dx \approx \frac{1-0}{3(4)}\left[0 + 4\left(\frac{1}{256}\right) + 2\left(\frac{1}{16}\right)\right.$$

$$\left. + 4\left(\frac{81}{256}\right) + 1\right]$$

$$= \frac{1}{12}\left(\frac{77}{32}\right)$$

$$\approx 0.2005208$$

$n = 8, b = 1, a = 0, f(x) = x^4$

$$\int_0^1 x^4\,dx \approx \frac{1-0}{3(8)}\left[0 + 4\left(\frac{1}{4096}\right) + 2\left(\frac{1}{256}\right)\right.$$

$$+ 4\left(\frac{81}{4096}\right) + 2\left(\frac{1}{16}\right) + 4\left(\frac{625}{4096}\right)$$

$$\left. + 2\left(\frac{18}{256}\right) + 4\left(\frac{2401}{4096}\right) + 1\right]$$

$$= \frac{1}{24}\left(\frac{4916}{1024}\right)$$

$$\approx 0.2000326$$

$n = 16, b = 1, a = 0, f(x) = x^4$

$$\int_0^1 x^4 dx$$

$$\approx \frac{1-0}{3(16)}\left[0 + 4\left(\frac{1}{65,536}\right) + 2\left(\frac{1}{4096}\right)\right.$$

$$+ 4\left(\frac{81}{65,536}\right) + 2\left(\frac{1}{256}\right) + 4\left(\frac{625}{65,536}\right)$$

$$+ 2\left(\frac{81}{4096}\right) + 4\left(\frac{2401}{65,536}\right) + 2\left(\frac{1}{16}\right)$$

$$+ 4\left(\frac{6561}{65,536}\right) + 2\left(\frac{625}{4096}\right) + 4\left(\frac{14,641}{65,536}\right)$$

$$+ 2\left(\frac{81}{256}\right) + 4\left(\frac{28,561}{65,536}\right) + 2\left(\frac{2401}{4096}\right)$$

$$\left. + 4\left(\frac{50,625}{65,536} + 1\right)\right]$$

$$= \frac{1}{48}\left(\frac{157,288}{16,384}\right) \approx 0.2000020$$

$$n = 32, b = 1, a = 0, f(x) = x^4$$

$$\int_0^1 x^4 dx$$

$$\approx \frac{1-0}{3(32)}\left[0 + 4\left(\frac{1}{1,048,576}\right) + 2\left(\frac{1}{65,536}\right)\right.$$

$$+ 4\left(\frac{81}{1,048,576}\right) + 2\left(\frac{1}{4096}\right) + 4\left(\frac{625}{1,048,576}\right)$$

$$+ 2\left(\frac{625}{65,536}\right) + 4\left(\frac{14,641}{1,048,576}\right) + 2\left(\frac{81}{4096}\right)$$

$$+ 4\left(\frac{28,561}{1,048,576}\right) + 2\left(\frac{2401}{65,536}\right) + 4\left(\frac{50,625}{1,048,576}\right)$$

$$+ 2\left(\frac{1}{16}\right) + 4\left(\frac{83,521}{1,048,576}\right) + 2\left(\frac{6561}{65,536}\right)$$

$$+ 4\left(\frac{130,321}{1,048,576}\right) + 2\left(\frac{625}{4096}\right) + 4\left(\frac{194,481}{1,048,576}\right)$$

$$+ 2\left(\frac{14,641}{65,536}\right) + 4\left(\frac{279,841}{1,048,576}\right) + 2\left(\frac{81}{256}\right)$$

$$+ 4\left(\frac{390,625}{1,048,576}\right) + 2\left(\frac{28,561}{65,536}\right) + 4\left(\frac{531,441}{1,048,576}\right)$$

$$+ 2\left(\frac{2401}{4096}\right) + 4\left(\frac{707,281}{1,048,576}\right) + 2\left(\frac{50,625}{65,536}\right)$$

$$\left. + 4\left(\frac{923,521}{1,048,576}\right) + 1\right]$$

$$= \frac{1}{96}\left(\frac{50,033,168}{262,144}\right) \approx 0.2000001$$

To find error for each value of n, subtract as indicated.

$n = 4$: $(0.2005208 - 0.2) = 0.0005208$

$n = 8$: $(0.2000326 - 0.2) = 0.0000326$

$n = 16$: $(0.2000020 - 0.2) = 0.0000020$

$n = 32$: $(0.2000001 - 0.2) = 0.0000001$

(c) $p = 1$:

$4^1(0.0005208) = 4(0.0005208) = 0.0020832$

$8^1(0.0000326) = 8(0.0000326) = 0.0002608$

Try $p = 2$:

$4^2(0.0005208) = 16(0.0005208) = 0.0083328$

$8^2(0.0000326) = 64(0.0000326) = 0.0020864$

Try $p = 3$:

$4^3(0.0005208) = 64(0.0005208) = 0.0333312$

$8^3(0.0000326) = 512(0.0000326) = 0.0166912$

Try $p = 4$:

$4^4(0.0005208) = 256(0.0005208) = 0.1333248$

$8^4(0.0000326) = 4096(0.0000326) = 0.1335296$

$16^4(0.0000020) = 65536(0.0000020) = 0.131072$

$32^4(0.0000001) = 1048576(0.0000001) = 0.1048576$

These are the closest values we can get; thus, $p = 4$.

19. Midpoint rule:

$$n = 4, b = 5, a = 1, f(x) = \frac{1}{x^2}, \Delta x = 1$$

i	x_i	$f(x_i)$
0	$\frac{3}{2}$	$\frac{4}{9}$
2	$\frac{5}{2}$	$\frac{4}{25}$
3	$\frac{7}{2}$	$\frac{4}{49}$
4	$\frac{9}{2}$	$\frac{4}{81}$

$$\int_1^5 \frac{1}{x^2} dx \approx \sum_{i=1}^{4} f(x_i)\Delta x$$

$$= \frac{4}{9}(1) + \frac{4}{25}(1) + \frac{4}{49}(1) + \frac{4}{81}(1)$$

$$\approx 0.7355$$

Simpson's rule:

$$m = 8, b = 5, a = 1, f(x) = \frac{1}{x^2}$$

i	x_i	$f(x_i)$
0	1	1
1	$\dfrac{3}{2}$	$\dfrac{4}{9}$
2	2	$\dfrac{1}{4}$
3	$\dfrac{5}{2}$	$\dfrac{4}{25}$
4	3	$\dfrac{1}{9}$
5	$\dfrac{7}{2}$	$\dfrac{4}{49}$
6	4	$\dfrac{1}{16}$
7	$\dfrac{9}{2}$	$\dfrac{4}{81}$
8	5	$\dfrac{1}{25}$

$$\int_1^5 \frac{1}{x^2}\, dx$$

$$\approx \frac{5-1}{3(8)}\left[1 + 4\left(\frac{4}{9}\right) + 2\left(\frac{1}{4}\right) + 4\left(\frac{4}{25}\right)\right.$$

$$+ 2\left(\frac{1}{9}\right) + 4\left(\frac{4}{49}\right) + 2\left(\frac{1}{16}\right)$$

$$\left. + 4\left(\frac{4}{81}\right) + \frac{1}{25}\right]$$

$$\approx \frac{1}{6}(4.82906)$$

$$\approx 0.8048$$

From #7 part a, $T \approx 0.9436$, when $n = 4$.

To verify the formula evaluate $\frac{2M+T}{3}$.

$$\frac{2M+T}{3} \approx \frac{2(0.7355) + 0.9436}{3}$$

$$\approx 0.8048$$

21. **(a)** Using the trapezoidal rule, the total wind energy consumption is

$$\frac{2013 - 2001}{4}\left(\begin{array}{l} \frac{1}{2} \cdot 70 + 142 + 341 \\[4pt] \quad + 923 + \frac{1}{2} \cdot 1595 \end{array}\right)$$

$$= 6715.5 \text{ or } 6715.5 \text{ trillion BTUs.}$$

(b) Using Simpson's rule, the total wind energy consumption is

$$\frac{2013 - 2001}{3 \cdot 4}\left(\begin{array}{l} 70 + 4 \cdot 142 + 2 \cdot 341 \\[4pt] \quad + 4 \cdot 923 + 1595 \end{array}\right)$$

$$= 6607 \text{ or } 6607 \text{ trillion BTUs.}$$

23. $y = e^{-t^2} + \dfrac{1}{t+1}$

The total reaction is

$$\int_1^9 \left(e^{-t^2} + \frac{1}{t+1}\right) dt.$$

$n = 8, b = 9, a = 1, f(t) = e^{-t^2} + \frac{1}{t+1}$

i	x_i	$f(x_i)$
0	1	0.8679
1	2	0.3516
2	3	0.2501
3	4	0.2000
4	5	0.1667
5	6	0.1429
6	7	0.1250
7	8	0.1111
8	9	0.1000

(a) Trapezoidal rule:

$$\int_1^9 \left(e^{-t^2} + \frac{1}{t+1}\right) dt$$

$$\approx \frac{9-1}{8}\left[\frac{1}{2}(0.8679) + 0.3516 + 0.2501\right.$$

$$\left. + \cdots + \frac{1}{2}(0.1000)\right]$$

$$\approx 1.831$$

(b) Simpson's rule:

$$\int_1^9 \left(e^{-t^2} + \frac{1}{t+1}\right) dt$$

$$\approx \frac{9-1}{3(8)}[0.8679 + 4(0.3516) + 2(0.2501)$$

$$+ 4(0.2000) + 2(0.1667) + 4(0.1429)$$

$$+ 2(0.1250) + 4(0.1111) + 0.1000]$$

$$= \frac{1}{3}(5.2739)$$

$$\approx 1.758$$

25. Note that heights may differ depending on the readings of the graph. Thus, answers may vary.

$n = 10, \ b = 20, \ a = 0$

i	x_i	$f(x_i)$
0	1	0
1	2	5
2	4	3
3	6	2
4	8	1.5
5	10	1.2
6	12	1
7	14	0.5
8	16	0.3
9	18	0.2
10	20	0.2

Area under curve for Formulation A

$$= \frac{20 - 0}{10}\left[\frac{1}{2}(0) + 5 + 3 + 2 + 1.5 + 1.2 \right.$$

$$\left. + 1 + 0.5 + 0.3 + 0.2 + \frac{1}{2}(0.2)\right]$$

$$= 2(14.8)$$

$$\approx 30 \text{ mcg(h)/ml}$$

This represents the total amount of drug available to the patient for each ml of blood.

27. As in Exercise 25, readings on the graph may vary, so answers may vary. The area both under the curve for Formulation A and above the minimum effective concentration line is on the interval $\left[\frac{1}{2}, 6\right]$.

Area under curve for Formulation A on $\left[\frac{1}{2}, 1\right]$, with $n = 1$

$$= \frac{1 - \frac{1}{2}}{1}\left[\frac{1}{2}(2 + 6)\right]$$

$$= \frac{1}{2}(4) = 2$$

Area under curve for Formulation A on $[1, 6]$, with $n = 5$

$$= \frac{6 - 1}{5}\left[\frac{1}{2}(6) + 5 + 4 + 3 + 2.4 + \frac{1}{2}(2)\right]$$

$$= 18.4$$

Area under minimum effective concentration line $\left[\frac{1}{2}, 6\right]$

$$= 5.5(2) = 11.0$$

Area under the curve for Formulation A and above minimum effective concentration line

$$= 2 + 18.4 - 110$$

$$\approx 9 \text{ mcg(h)/ml}$$

This represents the total erective amount of drug available to the patient for each ml of blood.

29. $y = b_0 w^{b_1} e^{-b_2 w}$

(a) if $t = 7w$ then $w = \dfrac{t}{7}$.

$$y = b_0 \left(\frac{t}{7}\right)^{b_1} e^{-b_2 t/7}$$

(b) Replacing the constants with the given values, we have

$$y = 5.955\left(\frac{t}{7}\right)^{0.233} e^{-0.027t/7} dt$$

In 25 weeks, there are 175 days.

$$\int_0^{175} 5.955\left(\frac{t}{7}\right)^{0.233} e^{-0.027t/7} dt$$

$n = 10, b = 175, a = 0,$

$$f(t) = 5.955\left(\frac{t}{7}\right)^{0.233} e^{-0.027t/7}$$

i	t_i	$f(t_i)$
0	0	0
1	1.75	6.89
2	35	7.57
3	52.5	7.78
4	70	7.77
5	87.5	7.65
6	105	7.46
7	122.5	7.23
8	140	6.97
9	157.5	6.70
10	175	6.42

Trapezoidal rule:

$$\int_0^{175} 5.955\left(\frac{t}{7}\right)^{0.233} e^{-0.027t/7} dt$$

$$\approx \frac{175-0}{10}\left[\frac{1}{2}(0) + 6.89 + 7.57 + 7.78 + 7.77\right.$$

$$\left. + 7.65 + 7.46 + 7.23 + 6.97 + 6.70 + \frac{1}{2}(6.42)\right]$$

$$= 17.5(69.23)$$

$$= 1211.525$$

The total milk consumed is about 1212 kg.

Simpson's rule:

$$\int_0^{175} 5.955\left(\frac{t}{7}\right)^{0.233} e^{-0.027t/7} dt$$

$$\approx \frac{175-0}{3(10)}[0 + 4(6.89) + 2(7.57) + 4(7.78)$$

$$+ 2(7.77) + 4(7.65) + 2(7.46) + 4(7.23)$$

$$+ 2(6.97) + 4(6.70) + 6.42]$$

The total milk consumed is about 1231 kg.

(c) Replacing the constants with the given values, we have

$$y = 8.409\left(\frac{t}{7}\right)^{0.143} e^{-0.037t/7}.$$

In 25 weeks, there are 175 days.

$$\int_0^{175} 8.409\left(\frac{t}{7}\right)^{0.143} e^{-0.037t/7} dt$$

$$n = 10, b = 175, a = 0,$$

$$f(t) = 8.409\left(\frac{t}{7}\right)^{0.143} e^{-0.037t/7}$$

i	t_i	$f(t_i)$
0	0	0
1	17.5	8.74
2	35	8.80
3	52.5	8.50
4	70	8.07
5	87.5	7.60
6	105	7.11
7	122.5	6.63
8	140	6.16
9	157.5	5.71
10	175	5.28

Trapezoidal rule:

$$\int_0^{175} 8.409\left(\frac{t}{7}\right)^{0.143} e^{-0.037t/7} dt$$

$$\approx \frac{175-0}{10}\left[\frac{1}{2}(0) + 8.74 + 8.80 + 8.50\right.$$

$$+ 8.07 + 7.60 + 7.11 + 6.63$$

$$\left. + 6.16 + 5.71 + \frac{1}{2}(5.28)\right]$$

$$= 17.5(69.96)$$

$$= 1224.30$$

The total milk consumed is about 1224 kg.

Simpson's rule:

$$\int_0^{175} 8.409\left(\frac{t}{7}\right)^{0.143} e^{-0.037t/7} dt$$

$$\approx \frac{175-0}{3(10)}[0 + 4(8.74) + 2(8.80) + 4(8.50)]$$

$$+ 2(8.07) + 4(7.60) + 2(7.11) + 4(6.63)$$

$$+ 2(6.16) + 4(5.71) + 5.28]$$

$$= \frac{35}{6}(214.28)$$

$$= 1249.97$$

The total milk consumed is about 1250 kg.

31. (a) $\frac{7-1}{6}\frac{1}{2}$

$$\left[(4) + 7 + 11 + 9 + 15 + 16 + \frac{1}{2}(23)\right]$$

$$= 71.5$$

(b) $\frac{7-1}{3(6)}[4 + 4(7) + 2(11) + 4(9)$

$$+ 2(15) + 4(16) + 23] = 69.0$$

33. We need to evaluate

$$\int_{12}^{36} (105e^{0.01x} + 32)\, dx.$$

Using a calculator program for Simpson's rule with $n = 20$, we obtain 3979.242 as the value of this integral. This indicates that the total revenue between the twelfth and thirty-sixth months is about 3979.

35. Use a calculator program for Simpson's rule with $n = 20$ to evaluate each of the integrals in this exercise.

(a) $\displaystyle\int_{-1}^{1}\left(\frac{1}{\sqrt{2\pi}}e^{-x^2/2}\right)dx \approx 0.6827$

The probability that a normal random variable is within 1 standard deviation of the mean is about 0.6827.

(b) $\displaystyle\int_{-2}^{2}\left(\frac{1}{\sqrt{2\pi}}e^{-x^2/2}\right)dx \approx 0.9545$

The probability that a normal random variable is within 2 standard deviation of the mean is about 0.9545.

(c) $\displaystyle\int_{-3}^{3}\left(\frac{1}{\sqrt{2\pi}}e^{-x^2/2}\right)dx \approx 0.9973$

The probability that a normal random variable is within 3 standard deviations of the mean is about 0.9973.

Chapter 7   Review Exercises

1. True

2. False: The statement is false for $n = -1$.

3. False: For example, if $f(x) = 1$ the first expression is equal to $x^2/2 + C$ and the second is equal to $x^2 + C$.

4. True

5. True

6. False: The derivative gives the instantaneous rate of change.

7. False: If the function is positive over the interval of integration the definite integral gives the exact area.

8. True

9. True

10. False: The definite integral may be positive, negative, or zero.

11. True

12. False: Sometimes true, but not in general.

13. False: The trapezoidal rule allows any number of intervals.

14. True

19. $\displaystyle\int (2x + 3)\,dx = \frac{2x^2}{2} + 3x + C$

$= x^2 + 3x + C$

21. $\displaystyle\int (x^2 - 3x + 2)\,dx$

$= \dfrac{x^3}{3} - \dfrac{3x^2}{2} + 2x + C$

23. $\displaystyle\int 3\sqrt{t}\,dt = 3\int t^{1/2}\,dt$

$= \dfrac{3t^{3/2}}{\frac{3}{2}} + C$

$= 2t^{3/2} + C$

25. $\displaystyle\int (x^{1/2} + 3x^{-2/3})\,dx$

$= \dfrac{x^{3/2}}{\frac{3}{2}} + \dfrac{3x^{1/3}}{\frac{1}{3}} + C$

$= \dfrac{2x^{3/2}}{3} + 9x^{1/3} + C$

27. $\displaystyle\int \frac{-4}{y^3}\,dy = \int -4y^{-3}\,dy$

$= \dfrac{-4y^{-2}}{-2} + C$

$= 2y^{-2} + C$

29. $\displaystyle\int -3e^{2x}\,dx = \dfrac{-3e^{2x}}{2} + C$

31. $\displaystyle\int xe^{3x^2}\,dx = \frac{1}{6}\int 6xe^{3x^2}\,dx$

Let $u = 3x^2$, so that $du = 6x\,dx$.

$= \dfrac{1}{6}\int e^u\,du$

$= \dfrac{1}{6}e^u + C$

$= \dfrac{e^{3x^2}}{6} + C$

33. $\displaystyle\int \frac{3u}{u^2 - 1}\,du = 3\left(\frac{1}{2}\right)\int \frac{2u\,du}{u^2 - 1}$

Let $w = u^2 - 1$, so that $dw = 2u\,du$.

$$= \frac{3}{2} \int \frac{dw}{w}$$

$$= \frac{3}{2} \ln |w| + C$$

$$= \frac{3 \ln |u^2 - 1|}{2} + C$$

35. $\displaystyle\int \frac{x^2\,dx}{(x^3+5)^4} = \frac{1}{3} \int \frac{3x^2\,dx}{(x^3+5)^4}$

Let $u = x^3 + 5$, so that

$$du = 3x^2\,dx.$$

$$= \frac{1}{3} \int \frac{du}{u^4}$$

$$= \frac{1}{3} \int u^{-4}\,du$$

$$= \frac{1}{3} \left(\frac{u^{-3}}{-3} \right) + C$$

$$= \frac{-(x^3+5)^{-3}}{9} + C$$

37. $\displaystyle\int \frac{x^3}{e^{3x^4}}\,dx = \int x^3 e^{-3x^4}$

$$= -\frac{1}{12} \int -12x^3 e^{-3x^4}\,dx$$

Let $u = -3x^4$, so that $du = -12x^3\,dx$.

$$= -\frac{1}{12} \int e^u\,du$$

$$= -\frac{1}{12} \int e^u + C$$

$$= \frac{-e^{-3x^4}}{12} + C$$

39. $\displaystyle\int \frac{(3 \ln z + 2)^4}{z}\,dz$

Let $u = 3 \ln z + 2$ so that

$$du = \frac{3}{x}\,dz.$$

$$\int \frac{(3 \ln z + 2)^4}{z}\,dz = \frac{1}{3} \int \frac{3(3 \ln z + 2)^4}{z}\,dz$$

$$= \frac{1}{3} \int u^4\,du$$

$$= \frac{1}{3} \cdot \frac{u^5}{5} + C$$

$$= \frac{(3 \ln z + 2)^5}{15} + C$$

41. $f(x) = 3x + 1$, $x_1 = -1$, $x_2 = 0$, $x_3 = 1$,
$x_4 = 2$, $x_5 = 3$

$f(x_1) = -2$, $f(x_2) = 1$, $f(x_3) = 4$,
$f(x_4) = 7$, $f(x_5) = 10$

$$\sum_{i=1}^{5} f(x_i)$$

$$= f(1) + f(2) + f(3) + f(4) + f(5)$$

$$= -2 + 1 + 4 + 7 + 10$$

$$= 20$$

43. $f(x) = 2x + 3$, from $x = 0$ to $x = 4$

$$\Delta x = \frac{4 - 0}{4} = 1$$

i	x_i	$f(x_i)$
1	0	3
2	1	5
3	2	7
4	3	9

$$A = \sum_{i=1}^{4} f(x_i)\Delta x$$

$$= 3(1) + 5(1) + 7(1) + 9(1)$$

$$= 24$$

45. **(a)** Since $s(t)$ represents the odometer reading, the distance traveled between $t = 0$ and $t = T$ will be $s(T) - s(0)$.

(b) $\displaystyle\int_0^T v(t)\,dt = s(T) - s(0)$ is equivalent to the Fundamental Theorem of Calculus with $a = 0$, and $b = T$ because $s(t)$ is an antiderivative of $v(t)$.

47. $\displaystyle\int_1^2 (3x^2 + 5)\,dx = \left(\frac{3x^3}{3} + 5x \right)\Big|_1^2$

$$= (2^3 + 10) - (1 + 5)$$

$$= 18 - 6$$

$$= 12$$

49. $\int_1^5 (3x^{-1} + x^{-3})\, dx = \left(3\ln |x| + \dfrac{x^{-2}}{-2}\right)\Bigg|_1^5$

$$= \left(3\ln 5 - \dfrac{1}{50}\right) - \left(3\ln 1 - \dfrac{1}{2}\right)$$

$$= 3\ln 5 + \dfrac{12}{25} \approx 5.308$$

51. $\int_0^1 x\sqrt{5x^2 + 4}\, dx$

Let $u = 5x^2 + 4$, so that

$$du = 10x\, dx \text{ and } \dfrac{1}{10}\, du = x\, dx.$$

When $x = 0, u = 5(0^2) + 4 = 4$.

When $x = 1, u = 5(1^2) + 4 = 9$.

$$= \dfrac{1}{10}\int_4^9 \sqrt{u}\, du = \dfrac{1}{10}\int_4^9 u^{1/2}\, du$$

$$= \dfrac{1}{10} \cdot \dfrac{u^{3/2}}{3/2}\Bigg|_4^9 = \dfrac{1}{15} u^{3/2}\Bigg|_4^9$$

$$= \dfrac{1}{15}(9)^{3/2} - \dfrac{1}{15}(4)^{3/2}$$

$$= \dfrac{27}{15} - \dfrac{8}{15}$$

$$= \dfrac{19}{15}$$

53. $\int_0^2 3e^{-2w}\, dw = \dfrac{-3e^{-2w}}{2}\Bigg|_0^2$

$$= \dfrac{-3e^{-4}}{2} + \dfrac{3}{2}$$

$$= \dfrac{3(1 - e^{-4})}{2} \approx 1.473$$

55. $\int_0^{1/2} x\sqrt{1 - 16x^4}\, dx$

Let $u = 4x^2$. Then $du = 8x\, dx$.

When $x = 0, u = 0$, and when $x = \frac{1}{2}, u = 1$.

Thus,

$$\int_0^{1/2} x\sqrt{1 - 16x^4}\, dx = \dfrac{1}{8}\int_0^1 \sqrt{1 - u^2}\, du.$$

Note that this integral represents the area of right upper quarter of a circle centered at the origin with a radius of 1.

Area of circle $= \pi r^2 = \pi(1^2) = \pi$

$$\int_0^1 \sqrt{1 - u^2}\, du = \dfrac{\pi}{4}$$

$$\dfrac{1}{8}\int_0^1 \sqrt{1 - u^2}\, du = \dfrac{1}{8} \cdot \dfrac{\pi}{4} = \dfrac{\pi}{32}$$

57. $\int_1^{e^5} \dfrac{\sqrt{25 - (\ln x)^2}}{x}\, dx$

Let $u = \ln x$. Then $du = \frac{1}{x}\, dx$.

When $x = e^5, u = \ln (e^5) = 5$.

When $x = 1, u = \ln(1) = 0$.

Thus,

$$\int_1^{e^5} \dfrac{\sqrt{25 - (\ln x)^2}}{x}\, dx = \int_0^5 \sqrt{25 - u^2}\, du.$$

Note that this integral represents the area of a right upper quarter of a circle centered at the origin with a radius of 5.

Area of circle $= \pi r^2 = \pi(5)^2 = 25\pi$

$$\int_0^5 \sqrt{25 - u^2}\, du = \dfrac{25\pi}{4}$$

59. $f(x) = \sqrt{4x - 3};\ [1, 3]$

$$\text{Area} = \int_1^3 \sqrt{4x - 3}\, dx$$

$$= \int_1^3 (4x - 3)^{1/2}\, dx$$

$$= \dfrac{2}{3} \cdot \dfrac{1}{4} \cdot (4x - 3)^{3/2}\Bigg|_1^3$$

$$= \dfrac{1}{6}(9)^{3/2} - \dfrac{1}{6}(1)^{3/2}$$

$$= \dfrac{1}{6}(26)$$

$$= \dfrac{13}{3}$$

61. $f(x) = xe^{x^2};\ [0, 2]$

$$\text{Area} = \int_0^2 xe^{x^2}\, dx$$

$$= \dfrac{e^{x^2}}{2}\Bigg|_0^2$$

$$= \frac{e^4}{2} - \frac{1}{2}$$

$$= \frac{e^4 - 1}{2}$$

$$\approx 26.80$$

63. $f(x) = 5 - x^2$, $g(x) = x^2 - 3$

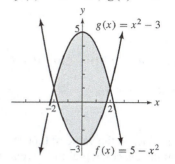

Points of intersection:

$$5 - x^2 = x^2 - 3$$

$$2x^2 - 8 = 0$$

$$2(x^2 - 4) = 0$$

$$x = \pm 2$$

Since $f(x) \geq g(x)$ in $[-2, 2]$, the area between the graphs is

$$\int_{-2}^{2} [f(x) - g(x)]\,dx = \int_{-2}^{2} [(5 - x^2) - (x^2 - 3)]\,dx$$

$$= \int_{-2}^{2} (-2x^2 + 8)\,dx$$

$$= \left(\frac{-2x^3}{3} + 8x \right) \Big|_{-2}^{2}$$

$$= -\frac{2}{3}(8) + 16 + \frac{2}{3}(-8) - 8(-2)$$

$$= \frac{-32}{3} + 32 = \frac{64}{3}.$$

65. $f(x) = x^2 - 4x$, $g(x) = x + 6$,
$x = -2$, $x = 4$

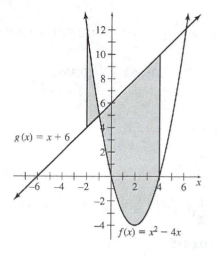

Points of intersection:

$$x^2 - 4x = x + 6$$

$$x^2 - 5x - 6 = 0$$

$$(x + 1)(x - 6) = 0$$

$$x = -1 \text{ or } x = 6$$

Thus, the area is

$$\int_{-2}^{-1} [x^2 - 4x - (x + 6)]\,dx$$

$$+ \int_{-1}^{4} [x + 6 - (x^2 - 4x)]\,dx$$

$$= \left(\frac{x^3}{3} - \frac{5x^2}{2} - 6x \right) \Big|_{-2}^{-1} + \left(-\frac{x^3}{3} + \frac{5x^2}{2} + 6x \right) \Big|_{-1}^{4}$$

$$= \left(\frac{19}{6} + \frac{2}{3} \right) + \left(\frac{128}{3} + \frac{19}{6} \right) = \frac{149}{3}$$

67. $\int_{1}^{3} \frac{\ln x}{x}\,dx$

Trapezoidal Rule:

$n = 4$, $b = 3$, $a = 1$, $f(x) = \frac{\ln x}{x}$

i	x_1	$f(x_i)$
0	1	0
1	1.5	0.27031
2	2	0.34657
3	2.5	0.36652
4	3	0.3662

$$\int_1^3 \frac{\ln x}{x}\,dx \approx \frac{3-1}{4}\left[\frac{1}{2}(0) + 0.27031 + 0.34657\right.$$

$$\left. + 0.36652 + \frac{1}{2}(0.3662)\right]$$

$$= 0.5833$$

Exact Value:

$$\int_1^3 \frac{\ln x}{x}\,dx$$

$$= \frac{1}{2}(\ln x)^2\Big|_1^3 = \frac{1}{2}(\ln 3)^2 - \frac{1}{2}(\ln 1)^2$$

$$\approx 0.6035$$

69. $\displaystyle\int_0^1 e^x\sqrt{e^x + 4}\,dx$

Trapezoidal Rule:

$$n = 4, b = 1, a = 0, f(x) = e^x\sqrt{e^x + 4}$$

i	x_i	$f(x_i)$
0	0	2.236
1	0.25	2.952
2	0.5	3.919
3	0.75	5.236
4	1	7.046

$$\int_0^1 e^x\sqrt{e^x + 4}\,dx$$

$$= \frac{1-0}{4}\left[\frac{1}{2}(2.236) + 2.952\right.$$

$$\left. + 3.919 + 5.236 + \frac{1}{2}(7.046)\right]$$

$$\approx 4.187$$

Exact value:

$$\int_0^1 e^x\sqrt{e^x + 4}\,dx = \int_0^1 e^x(e^x + 4)^{1/2}\,dx$$

$$= \frac{2}{3}(e^x + 4)^{3/2}\Big|_0^1$$

$$= \frac{2}{3}(e + 4)^{3/2} - \frac{2}{3}(5)^{3/2}$$

$$\approx 4.155$$

71. $\displaystyle\int_1^3 \frac{\ln x}{x}\,dx$

Simpson's rule:

$$n = 4, b = 3, a = 1, f(x) = \frac{\ln x}{x}$$

i	x_i	$f(x_i)$
0	1	0
1	1.5	0.27031
2	2	0.34657
3	2.5	0.36652
4	3	0.3662

$$\int_1^3 \frac{\ln x}{x}\,dx$$

$$\approx \frac{3-1}{3(4)}\left[0 + 4(0.27031) + 2(0.34657)\right.$$

$$\left. + 4(0.36652) + 0.3662\right]$$

$$\approx 0.6011$$

This answer is close to the value of 0.6035 obtained from the exact integral in Exercise 67.

73. $\displaystyle\int_0^1 e^x\sqrt{e^x + 4}\,dx$

Simpson's rule:

$$n = 4, b = 1, a = 0, f(x) = e^x\sqrt{e^x + 4}$$

i	x_i	$f(x_i)$
0	0	2.236
1	0.25	2.952
2	0.5	3.919
3	0.75	5.236
4	1	7.046

$$\int_0^1 e^x\sqrt{e^x + 4}\,dx$$

$$= \frac{1-0}{3(4)}[2.236 + 4(2.952) + 2(3.919)$$

$$+ 4(5.236) + 7.046]$$

$$\approx 4.156$$

This answer is close to the answer of 4.155 obtained from the exact integral in Exercise 69.

75. (a) $\displaystyle\int_1^5 \left[\sqrt{x-1} - \left(\frac{x-1}{2}\right)\right]dx$

$$= \int_1^5 \left(\sqrt{x-1} - \frac{x}{2} + \frac{1}{2}\right)dx$$

$$= \left(\frac{2}{3}(x-1)^{3/2} - \frac{x^2}{4} + \frac{x}{2}\right)\Big|_1^5$$

$$= \left(\frac{16}{3} - \frac{25}{4} + \frac{5}{2}\right) - \left(0 - \frac{1}{4} + \frac{1}{2}\right)$$

$$= \frac{16}{2} - 6 + 2 = \frac{4}{3}$$

(b) $n = 4, b = 5, a = 1,$

$$f(x) = \sqrt{x - 1} - \frac{x}{2} + \frac{1}{2}$$

i	x_i	$f(x_i)$
0	1	0
1	2	0.5
2	3	0.41421
3	4	0.23205
4	5	0

$$\int_1^5 \left(\sqrt{x - 1} - \frac{x}{2} + \frac{1}{2} \right) dx$$

$$= \left(\frac{5 - 1}{4} \right) \left[\frac{1}{2}(0) + 0.5 + 0.41421 \right.$$

$$\left. + 0.23205 + \frac{1}{2}(0) \right]$$

$$= 1.146$$

(c) $\displaystyle\int_1^5 \left(\sqrt{x - 1} - \frac{x}{2} + \frac{1}{2} \right) dx$

$$= \left(\frac{5 - 1}{3(4)} \right) [0 + 4(0.5) + 2(0.41421)$$

$$+ 4(0.23205) + 0]$$

$$= \left(\frac{1}{3} \right)(3.75662)$$

$$= 1.252$$

77. $\displaystyle\int_{-2}^2 [x(x - 1)(x + 1)(x - 2)(x + 2)]^2 dx$

(a) Trapezoidal Rule:

$n = 4, b = -2, a = 2,$

$$f(x) = [x(x - 1)(x + 1)(x - 2)(x + 2)]^2$$

i	x_i	$f(x_i)$
0	-2	0
1	-1	0
2	0	0
3	1	0
4	2	0

$$\int_{-2}^2 [x(x - 1)(x + 1)(x - 2)(x + 2)]^2 dx$$

$$\approx \frac{2 - (-2)}{4} \left[\frac{1}{2}(0) + 0 + 0 + 0 + \frac{1}{2}(0) \right]$$

$$= 0$$

(b) Simpson's Rule:

$n = 4, b = 2, a = 2,$

$$f(x) = [x(x - 1)(x + 1)(x - 2)(x + 2)]^2$$

i	x_i	$f(x_i)$
0	-2	0
1	-1	0
2	0	0
3	1	0
4	2	0

$$\int_{-2}^2 [x(x - 1)(x + 1)(x - 2)(x + 2)]^2 dx$$

$$\approx \frac{2 - (-2)}{3(4)} [0 + 4(0) + 2(0) + 4(0) + 0]$$

$$= 0$$

79. $C'(x) = 3\sqrt{2x - 1};$ 13 units cost $270.

$$C(x) = \int 3(2x - 1)^{1/2} dx$$

$$= \frac{3}{2} \int 2(2x - 1)^{1/2} dx$$

Let $u = 2x - 1,$ so that

$$du = 2dx.$$

$$= \frac{3}{2} \int u^{1/2} du$$

$$= \frac{3}{2} \left(\frac{u^{3/2}}{3/2} \right) + C$$

$$= (2x - 1)^{3/2} + C$$

$$C(13) = [2(13) - 1]^{3/2} + C$$

Since $C(13) = 270,$

$$270 = 25^{3/2} + C$$

$$270 = 125 + C$$

$$C = 145.$$

Thus,

$$C(x) = (2x - 1)^{3/2} + 145.$$

81. The deficit in the years 2004, 2006, 2008, 2010 and 2012 is approximately 400, 250, 450, 1300, and 1100 in trillions of dollars. Thus the total amount of debt accumulated from 2003 to 2013 is 2(400 + 250 + 450 + 1300 + 1100) = $7000 trillion.

83. $S'(x) = 3\sqrt{2x + 1} + 3$

$$S(x) = \int_0^4 (3\sqrt{2x + 1} + 3)\, dx$$

$$= [(2x + 1)^{3/2} + 3x]\Big|_0^4$$

$$= (27 + 12) - (1 + 0) = 38$$

Total sales = \$38,000.

85. $S(q) = q^2 + 5q + 100$

$D(q) = 350 - q^2$

$S(q) = D(q)$ at the equilibrium point.

$$q^2 + 5q + 100 = 350 - q^2$$

$$2q^2 + 5q - 250 = 0$$

$$(-2q + 25)(q - 10) = 0$$

$$q = -\frac{25}{2} \quad \text{or} \quad q = 10$$

Since the number of units produced would not be negative, the equilibrium point occurs when $q = 10$.

Equilibrium supply

$$= (10)^2 + 5(10) + 100 = 250$$

Equilibrium demand

$$= 350 - (10)^2 = 250$$

(a) Producers' surplus

$$= \int_0^{10} \left[250 - (q^2 + 5q + 100)\right] dq$$

$$= \int_0^{10} (-q^2 - 5q + 150)\, dq$$

$$= \left(\frac{-q^3}{3} - \frac{5q^2}{2} + 150q\right)\Big|_0^{10}$$

$$= \frac{-1000}{3} - \frac{500}{2} + 1500$$

$$= \frac{\$2750}{3} \approx \$916.67$$

(b) Consumers' surplus

$$= \int_0^{10} [(350 - q^2) - 250]\, dq$$

$$= \int_0^{10} (100 - q^2)\, dq$$

$$= \left(100q - \frac{q^3}{3}\right)\Big|_0^{10}$$

$$= 1000 - \frac{1000}{3} = \frac{\$2000}{3} \approx \$666.67$$

87. **(a)** Total amount $\approx \frac{1}{2}(1.986) + 1.891$

$$+ 1.857 + 1.853$$

$$+ 1.825 + 1.954$$

$$+ 1.997 + 2.063$$

$$+ 2.368 + \frac{1}{2}(2.716)$$

$$= 18.159$$

The estimate is 18.159 billion barrels.

(b) The left endpoint sum is 17.794 and the right endpoint sum is 18.524. Their average is

$$\frac{17.794 + 18.524}{2} = 18.159.$$

The estimate is 181.59 billion barrels.

(d) The line of best fit has the equation $y = 0.06970t + 1.459$.

$$\int_4^{13} (0.06970t + 1.459)\, dt \approx 18.46$$

The integral yields an estimate of 18.46 billion barrels.

89. $f(t) = 100 - t\sqrt{0.4t^2 + 1}$

The total number of additional spiders in the first ten months is

$$\int_0^{10} (100 - t\sqrt{0.4t^2 + 1})\, dt,$$

where t is the time in months.

$$= \int_0^{10} 100\, dt - \int_0^{10} t\sqrt{0.4t^2 + 1}\, dt.$$

Let $u = 0.4t^2 + 1$, so that

$$du = 0.8t\, dt \text{ and } \tfrac{1}{0.8}\, du = t\, dt.$$

When $t = 10, u = 41$.

When $t = 0, u = 1$.

$$= \int_0^{10} 100\, dt - \frac{1}{0.8} \int_1^{41} u^{1/2}\, du$$

$$= 100t\Big|_0^{10} - \frac{5}{4} \cdot \frac{u^{3/2}}{\frac{3}{2}}\Big|_1^{41} = 1000 - \frac{5}{6}u^{3/2}\Big|_1^{41}$$

$$\approx 782$$

The total number of additional spiders in the first 10 months is about 782.

91. **(a)** The total area is the area of the triangle on [0, 12] with height 0.024 plus the area of the rectangle on [12, 17.6] with height 0.024.

$$A = \frac{1}{2}(12 - 0)(0.024) + (17.6 - 12)(0.024)$$

$$= 0.144 + 0.1344 = 0.2784$$

(b) On [0, 12] we define the function $f(x)$ with slope $\frac{0.024 - 0}{12 - 0} = 0.002$ and y-intercept 0.

$$f(x) = 0.002x$$

On [12, 17.6], define $g(x)$ as the constant value.

$$g(x) = 0.024.$$

The area is the sum of the integrals of these two functions.

$$A = \int_0^{12} 0.002x\,dx + \int_{12}^{17.6} 0.024\,dx$$

$$= 0.001x^2\Big|_0^{12} + 0.024x\Big|_{12}^{17.6}$$

$$= 0.001(12^2 - 0^2) + 0.024(17.6 - 12)$$

$$= 0.144 + 0.1344 = 0.2784$$

93. **(a)** $\int_0^{321} 1.87t^{1.49}e^{-0.189(\ln t)^2}\,dt$

Trapezoidal rule:

$n = 8, b = 321, a = 1,$

$$f(t) = 1.87t^{1.49}e^{-0.189(\ln t)^2}$$

i	x_i	$f(t_i)$
0	1	1.87
1	41	34.9086
2	81	33.9149
3	121	30.7147
4	161	27.5809
5	161	24.8344
6	201	22.4794
7	281	20.4622
8	321	18.7255

$$\text{Total amount} \approx \frac{321-1}{8}\left[\begin{array}{l}\frac{1}{2}(1.87) + 34.9086 + 33.9149 + 30.7147 + 27.5809 \\[2mm] + 24.8344 + 22.4794 + 20.4622 + \frac{1}{2}(18.7255)\end{array}\right]$$

$$\approx 8208$$

The total milk production from $t = 1$ to $t = 321$ is approximately 8208 kg.

(b) Simpson's rule:

$$n = 8, b = 321, a = 1, \ f(t) = 1.87t^{1.49}e^{-0.189(\ln t)^2}$$

i	t_i	$f(t_i)$
0	1	1.87
1	41	34.9086
2	81	33.9149
3	121	30.7147
4	161	27.5809
5	201	24.8344
6	241	22.4794
7	281	20.4622
8	321	18.7255

$$\text{Total amount} \approx \frac{321-1}{8(3)}\left[\begin{array}{l}1.87 + 4(34.9086) + 2(33.9149) + 4(30.7147) + 2(27.5809) \\[2mm] + 4(24.8344) + 2(22.4794) + 4(20.4622) + 18.7255\end{array}\right]$$

$$\approx 8430$$

The total milk production from $t = 1$ to $t = 321$ is approximately 8430 kg.

(c) Numerical evaluation gives $\displaystyle\int_0^{321} 1.87t^{1.49}e^{-0.189(\ln t)^2} \, dt \approx 8558$, or 8558 kg

95. $v(t) = t^2 - 2t$

$$s(t) = \int_0^t (t^2 - 2t) \, dt$$

$$s(t) = \frac{t^3}{3} - t^2 + s_0$$

If $t = 3, s = 8$.

$$8 = 9 - 9 + s_0$$
$$8 = s_0$$

Thus,

$$s(t) = \frac{t^3}{3} - t^2 + 8.$$

FURTHER TECHNIQUES AND APPLICATIONS OF INTEGRATION

8.1 Integration by Parts

Your Turn 1

$$\int xe^{-2x}dx$$

Let $u = x$ and $dv = e^{-2x}dx$.

Then $du = dx$ and $v = \int e^{-2x}dx$

$$= -\frac{1}{2}e^{-2x}.$$

$$\int u\,dv = uv - \int v\,du$$

$$\int xe^{-2x}dx = -\frac{1}{2}xe^{-2x} - \int\left(-\frac{1}{2}e^{-2x}\right)dx$$

$$= -\frac{1}{2}xe^{-2x} + \frac{1}{2}\int e^{-2x}dx$$

$$= -\frac{1}{2}xe^{-2x} + \frac{1}{2}\left(-\frac{1}{2}\right)e^{-2x} + C$$

$$= -\frac{1}{2}xe^{-2x} - \frac{1}{4}e^{-2x} + C$$

$$= -\frac{1}{4}e^{-2x}(2x + 1) + C$$

Your Turn 2

$$\int \ln 2x\,dx$$

Let $u = \ln 2x$ and $dv = dx$.

Then $du = \frac{1}{2x}(2) = \frac{1}{x}$ and $v = \int dx = x$.

$$\int u\,dv = uv - \int v\,du$$

$$\int \ln 2x\,dx = (\ln 2x)(x) - \int x \cdot \frac{1}{x}dx$$

$$= x\ln 2x - \int dx$$

$$= x\ln 2x - x + C$$

Your Turn 3

$$\int (3x^2 + 4)e^{2x}dx$$

Choose $3x^2 + 4$ as the part to be differentiated and e^{2x} as the part to be integrated.

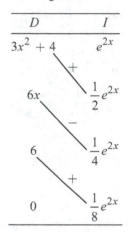

D	I
$3x^2 + 4$	e^{2x}
$6x$	$\frac{1}{2}e^{2x}$
6	$\frac{1}{4}e^{2x}$
0	$\frac{1}{8}e^{2x}$

$$\int (3x^2 + 4)e^{2x}dx$$

$$= (3x^2 + 4)\left(\frac{1}{2}e^{2x}\right) - (6x)\left(\frac{1}{4}e^{2x}\right) + \left(\frac{1}{8}e^{2x}\right) + C$$

$$= \frac{1}{8}e^{2x}[4(3x^2 + 4) - 2(6x) + 6] + C$$

$$= \frac{1}{8}e^{2x}(12x^2 + 16 - 12x + 6) + C$$

$$= \frac{1}{4}e^{2x}(6x^2 - 6x + 11) + C$$

Your Turn 4

$$\int_1^e x^2 \ln x\,dx$$

Let $u = \ln x$ and $dv = x^2dx$.

Then $du = \frac{1}{x}dx$ and $v = \int x^2dx = \frac{x^3}{3}$.

$$\int u\,dv = uv - \int v\,du$$

$$\int x^2 \ln x \, dx = \frac{x^3}{3} \ln x - \int \left(\frac{x^3}{3} \cdot \frac{1}{x} \right) dx$$

$$= \frac{x^3}{3} \ln x - \frac{1}{3} \int x^2 dx$$

$$= \frac{x^3}{3} \ln x - \frac{1}{3} \left(\frac{x^3}{3} \right) + C$$

$$= \frac{x^3}{3} \left(\ln x - \frac{1}{3} \right) + C$$

$$\int_1^e x^2 \ln x \, dx = \frac{x^3}{3} \left(\ln x - \frac{1}{3} \right) \Big|_1^e$$

$$= \frac{e^3}{3} \left(1 - \frac{1}{3} \right) - \frac{1}{3} \left(0 - \frac{1}{3} \right)$$

$$= \frac{2}{9} e^3 + \frac{1}{9}$$

$$= \frac{2e^3 + 1}{9}$$

Your Turn 5

$$\int \frac{1}{x\sqrt{4 + x^2}} dx$$

Use formula 10 from the table of integrals with $a = 2$.

$$\int \frac{1}{x\sqrt{4 + x^2}} dx = \int \frac{1}{x\sqrt{2^2 + x^2}} dx$$

$$= -\frac{1}{2} \ln \left| \frac{2 + \sqrt{4 + x^2}}{x} \right| + C$$

8.1 Warmup

W1. $\ln(3x) + \ln\left(x^2\right) = \ln 3 + \ln x + 2\ln x$

$$= \ln 3 + 3\ln x$$

$$\ln(3x) + \ln\left(x^2\right) = \ln 3 + \ln x + 2\ln x$$

$$= \ln 3 + 3\ln x$$

W2. $\int \left(3x^5 - \frac{2}{x^3} \right) dx$

$$= \int \left(3x^5 - 2x^{-3} \right) dx$$

$$= 3\left(\frac{1}{6}x^6 \right) - 2\left(-\frac{1}{2}x^{-2} \right) + C$$

$$= \frac{x^6}{2} + \frac{1}{x^2} + C$$

W3. $\int \left(3x^5 - \frac{2}{x^3} \right) dx$

$$= \int \left(3x^5 - 2x^{-3} \right) dx$$

$$= 3\left(\frac{1}{6}x^6 \right) - 2\left(-\frac{1}{2}x^{-2} \right) + C$$

$$= \frac{x^6}{2} + \frac{1}{x^2} + C$$

W4. $\int \left(e^{2x} + \frac{5}{e^x} \right) dx$

$$= \int \left(e^{2x} + 5e^{-x} \right) dx$$

$$= \frac{1}{2}e^{2x} + 5\left(-e^{-x} \right) + C$$

$$= \frac{e^{2x}}{2} - \frac{5}{e^x} + C$$

W5. Let $u = x^5$; then $du = 5x^4$.

$$\int x^4 e^{x^5} dx = \frac{1}{5} \int 5x^4 e^{x^5} dx$$

$$= \frac{1}{5} \int e^u du = \frac{1}{5}e^u + C$$

$$= \frac{e^{x^5}}{5} + C$$

W6. $\int_0^3 \left(6x^2 - 5 + e^x \right) dx$

$$= \left(2x^3 - 5x + e^x \right) \Big|_0^3$$

$$= \left[\left(54 - 15 + e^3 \right) - (0 - 0 + 1) \right]$$

$$= 38 + e^3 \approx 58.09$$

8.1 Exercises

1. $\int xe^x dx$

   Let $dv = e^x dx$ and $u = x$.

   Then $v = \int e^x dx$ and $du = dx$.

   $$v = e^x$$

   Use the formula

   $$\int u \, dv = uv - \int v \, du.$$

$$\int xe^x\,dx = xe^x - \int e^x\,dx = xe^x - e^x + C$$

3. $\displaystyle\int (4x - 12)e^{-8x}\,dx$

Let $dv = e^{-8x}\,dx$ and $u = 4x - 12$

Then $v = \displaystyle\int e^{-8x}\,dx$ and $du = 4\,dx$.

$$v = \frac{e^{-8x}}{-8}$$

$$\int (4x - 12)e^{-8x}\,dx$$

$$= (4x - 12)\left(\frac{e^{-8x}}{-8}\right) - \int \left(\frac{e^{-8x}}{-8}\right)\cdot 4\,dx$$

$$= -\frac{4x}{8}e^{-8x} + \frac{12}{8}e^{-8x} - \left(-\frac{4}{8}\cdot\frac{e^{-8x}}{-8}\right) + C$$

$$= -\frac{x}{2}e^{-8x} + \frac{3}{2}e^{-8x} - \frac{1}{16}e^{-8x} + C$$

$$= \left(-\frac{x}{2} + \frac{23}{16}\right)e^{-8x} + C$$

5. $\displaystyle\int x\ln\,dx$

Let $dv = x\,dx$ and $u = \ln x$.

Then $v = \frac{x^2}{2}$ and $du = \frac{1}{x}\,dx$.

$$\int x\ln\,dx = \frac{x^2}{2}\ln x - \int \frac{x}{2}\,dx$$

$$= \frac{x^2\ln x}{2} - \frac{x^2}{4} + C$$

7. $\displaystyle\int_0^1 \frac{2x+1}{e^x}\,dx$

$$= \int_0^1 (2x+1)e^{-x}\,dx$$

Let $dv = e^{-x}\,dx$ and $u = 2x + 1$.

Then $v = \displaystyle\int e^{-x}\,dx$ and $du = 2\,dx$.

$$v = -e^{-x}$$

$$\int \frac{2x+1}{e^x}\,dx$$

$$= -(2x+1)e^{-x} + \int 2e^{-x}\,dx$$

$$= -(2x+1)e^{-x} - 2e^{-x}$$

$$\int_0^1 \frac{2x+1}{e^x}\,dx$$

$$= \left[-(2x+1)e^{-x} - 2e^{-x}\right]\Big|_0^1$$

$$= \left[-(3)e^{-1} - 2e^{-1}\right] - (-1 - 2)$$

$$= -5e^{-1} + 3$$

$$\approx 1.161$$

9. $\displaystyle\int \ln 3x\,dx$

Let $dv = dx$ and $u = \ln 3x$.

Then $v = x$ and $du = \frac{1}{x}\,dx$.

$$\int \ln 3x\,dx = x\ln 3x - \int dx$$

$$= x\ln 3x - x$$

$$\int \ln 3x\,dx = (x\ln 3x - x)\Big|_1^9$$

$$= (9\ln 27 - 9) - (\ln 3 - 1)$$

$$= 9\ln 3^3 - 9 - \ln 3 + 1$$

$$= 27\ln 3 - \ln 3 - 8$$

$$= 26\ln 3 - 8 \approx 20.56$$

11. The area is $\displaystyle\int_2^4 (x - 2)e^x\,dx$.

Let $dv = e^x\,dx$ and $u = x - 2$.

Then $v = e^x$ and $du = dx$.

$$\int (x - 2)e^x\,dx = (x - 2)e^x - \int e^x\,dx$$

$$\int_1^4 (x - 2)e^x\,dx = \left[(x - 2)e^x - e^x\right]\Big|_2^4$$

$$= (2e^4 - e^4) - (0 - e^2)$$

$$= e^4 + e^2 \approx 61.99$$

13. $\displaystyle\int x^2 e^{2x}\,dx$

Let $u = x^2$ and $dv = e^{2x}\,dx$.

Use column integration.

D		I
x^2	$+$	e^{2x}
$2x$	$-$	$\dfrac{e^{2x}}{2}$
2	$+$	$\dfrac{e^{2x}}{4}$
0		$\dfrac{e^{2x}}{8}$

$$\int x^2 e^{2x}\, dx = x^2\left(\frac{e^{2x}}{2}\right) - 2x\left(\frac{e^{2x}}{4}\right) + \frac{2e^{2x}}{8} + C$$

$$= \frac{x^2 e^{2x}}{2} - \frac{xe^{2x}}{2} + \frac{e^{2x}}{4} + C$$

15. $\displaystyle\int x^2\sqrt{x+4}\, dx$

Let $u = x^2$ and $dv = (x+4)^{1/2}$. Use column integration.

D		I
x^2	$+$	$(x+4)^{1/2}$
$2x$	$-$	$\frac{2}{3}(x+4)^{3/2}$
2	$+$	$\left(\frac{2}{3}\right)\left(\frac{2}{5}\right)(x+4)^{5/2}$
0		$\left(\frac{2}{3}\right)\left(\frac{2}{5}\right)\left(\frac{2}{7}\right)(x+4)^{7/2}$

$$\int x^2\sqrt{x+4}\, dx$$

$$= x^2(x+4)^{3/2}\left(\frac{2}{3}\right) - 2x(x+4)^{5/2}\left(\frac{2}{3}\right)\left(\frac{2}{5}\right)$$

$$\quad + 2(x+4)^{7/2}\left(\frac{2}{3}\right)\left(\frac{2}{5}\right)\left(\frac{2}{7}\right) + C$$

$$= \frac{2}{3}x^2(x+4)^{3/2} - \frac{8}{15}x(x+4)^{5/2}$$

$$\quad + \frac{16}{105}(x+4)^{7/2} + C$$

17. $\displaystyle\int (8x+10)\ln(5x)\, dx$

Let $dv = (8x+10)\, dx$ and $u = \ln(5x)$.

Then $v = 4x^2 + 10x$ and $du = \dfrac{1}{x}\, dx$.

$$\int (8x+10)\ln(5x)\, dx$$

$$= (4x^2 + 10x)\ln(5x) - \int (4x^2 + 10x)\left(\frac{1}{x}\right) dx$$

$$= (4x^2 + 10x)\ln(5x) - \int (4x + 10)\, dx$$

$$= (4x^2 + 10x)\ln(5x) - 2x^2 - 10x + C$$

19. $\displaystyle\int_1^2 (1 - x^2)e^{2x}\, dx$

Let $u = 1 - x^2$ and $dv = e^{2x}\, dx$.

Use column integration.

D		I
$1 - x^2$	$+$	e^{2x}
$-2x$	$-$	$\dfrac{e^{2x}}{2}$
-2	$+$	$\dfrac{e^{2x}}{4}$
0		$\dfrac{e^{2x}}{8}$

$$\int (1 - x^2)e^{2x}\, dx$$

$$= \frac{(1 - x^2)e^{2x}}{2} - \frac{(-2x)e^{2x}}{4} + \frac{(-2)e^{2x}}{8}$$

$$= \frac{(1 - x^2)e^{2x}}{2} + \frac{xe^{2x}}{2} - \frac{e^{2x}}{4}$$

$$= \frac{e^{2x}}{2}\left(1 - x^2 + x - \frac{1}{2}\right) = \frac{e^{2x}}{2}\left(\frac{1}{2} - x^2 + x\right)$$

$$\int_1^2 (1 - x^2)e^{2x}\, dx = \frac{e^{2x}}{2}\left(\frac{1}{2} - x^2 + x\right)\Bigg|_1^2$$

$$= \frac{e^4}{2}\left(-\frac{3}{2}\right) - \frac{e^2}{2}\left(\frac{1}{2}\right)$$

$$= -\frac{e^2}{4}(3e^2 + 1) \approx -42.80$$

21. $\displaystyle\int_0^1 \frac{x^3\, dx}{\sqrt{3+x^2}} = \int_0^1 x^3(3x + x^2)^{-1/2}\, dx$

Let $dv = x(3 + x^2)^{-1/2}\, dx$ and $u = x^2$.

Then $v = \dfrac{2(3+x^2)^{1/2}}{2}$

$\quad v = (3 + x^2)^{1/2}$ and $du = 2x\, dx$.

$$\int \frac{x^3 dx}{\sqrt{3 + x^2}}$$

$$= x^2(3 + x^2)^{1/2} - \int 2x(3 + x^2)^{1/2} \, dx$$

$$= x^2(3 + x^2)^{1/2} - \frac{2}{3}(3 + x^2)^{3/2}$$

$$\int_0^1 \frac{x^3 dx}{\sqrt{3 + x^2}}$$

$$= \left[x^2(3 + x^2)^{1/2} - \frac{2}{3}(3 + x^2)^{3/2} \right]\Big|_0^1$$

$$= 4^{1/2} - \frac{2}{3}(4^{3/2}) - 0 + \frac{2}{3}(3^{3/2})$$

$$= 2 - \frac{2}{3}(8) + \frac{2}{3}(3^{3/2})$$

$$= -\frac{10}{3} + 2\sqrt{3}$$

$$\approx 0.1308$$

23. $\displaystyle\int \frac{16}{\sqrt{x^2 + 16}} \, dx$

Use formula 5 from the table of integrals with $a = 4$.

$$\int \frac{16}{\sqrt{x^2 + 16}} \, dx = 16\int \frac{1}{\sqrt{x^2 + 4^2}} \, dx$$

$$= 16\ln\left| x + \sqrt{x^2 + 16} \right| + C$$

25. $\displaystyle\int \frac{3}{x\sqrt{121 - x^2}} \, dx = 3\int \frac{dx}{x\sqrt{11^2 - x^2}}$

If $a = 11$, this integral matches formula 9 in the table.

$$= 3\left(-\frac{1}{11} \ln\left| \frac{11 + \sqrt{121 - x^2}}{x} \right| \right) + C$$

$$= -\frac{3}{11} \ln\left| \frac{11 + \sqrt{121 - x^2}}{x} \right| + C$$

27. $\displaystyle\int \frac{-6}{x(4x + 6)^2} \, dx$

Use formula 14 from the table of integrals with $a = 4$ and $b = 6$.

$$\int \frac{-6}{x(4x + 6)^2} \, dx$$

$$= -6\int \frac{1}{x(4x + 6)^2} \, dx$$

$$= -6\left[\frac{1}{6(4x + 6)} + \frac{1}{6^2} \ln\left| \frac{x}{4x + 6} \right| \right] + C$$

$$= \frac{-1}{(4x + 6)} - \frac{1}{6} \ln\left| \frac{x}{4x + 6} \right| + C$$

31. First find the indefinite integral using integration by parts.

$$\int u \, dv = uv - \int v \, du$$

Now substitute the given values.

$$\int_0^1 u \, dv = uv\Big|_0^1 - \int_0^1 v \, du$$

$$= [u(1)v(1) - u(0)v(0)] - 4$$

$$= (3)(-4) - (2)(1) - 4 = -18$$

33. $\displaystyle\int r \, ds = rs - \int s \, dr$

$$\int_0^2 r \, ds = rs\Big|_0^2 - \int_0^2 s \, dr$$

$$10 = r(2)s(2) - r(0)s(0) - 5$$

$$15 = r(s)s(2)$$

35. $\displaystyle\int x^n \cdot \ln|x| \, dx, \; n \neq -1$

Let $u = \ln|x|$ and $dv = x^n \, dx$.
Use column integration.

D		I
$\ln|x|$	$+$	x^n
$\frac{1}{x}$	$-$	$\frac{1}{n+1}x^{n+1}$

$$\int x^n \cdot \ln|x| \, dx$$

$$= \frac{1}{n+1}x^{n+1} \ln|x| - \int\left[\frac{1}{x} \cdot \frac{1}{n+1}x^{n+1} \right] dx$$

$$= \frac{1}{n+1}x^{n+1} \ln|x| - \int \frac{1}{n+1}x^n \, dx$$

$$= \frac{1}{n+1}x^{n+1} \ln|x| - \frac{1}{(n+1)^2}x^{n+1} + C$$

$$= x^{n+1}\left[\frac{\ln|x|}{n+1} - \frac{1}{(n+1)^2} \right] + C$$

37. $\int x\sqrt{x+1}\,dx$

(a) Let $u = x$ and $dv = \sqrt{x+1}\,dx$.
Use column integration.

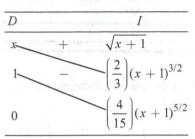

D	I
x $+$	$\sqrt{x+1}$
1 $-$	$\left(\dfrac{2}{3}\right)(x+1)^{3/2}$
0	$\left(\dfrac{4}{15}\right)(x+1)^{5/2}$

$$\int x\sqrt{x+1}\,dx$$
$$= \left(\frac{2}{3}\right)x(x+1)^{3/2} - \left(\frac{4}{15}\right)(x+1)^{5/2} + C$$

(b) Let $u = x + 1$; then $u - 1 = x$
and $du = dx$.

$$\int x\sqrt{x+1}\,dx$$
$$= \int (u-1)u^{1/2}\,du = \int (u^{3/2} - u^{1/2})\,du$$
$$= \frac{2}{5}u^{5/2} - \frac{2}{3}u^{3/2} + C$$
$$= \frac{2}{5}(x+1)^{5/2} - \frac{2}{3}(x+1)^{3/2} + C$$

(c) Both results factor as
$\frac{2}{15}(x+1)^{3/2}(3x-2) + C$, so they are
equivalent.

39. $R = \displaystyle\int_0^{12} (x+1)\ln(x+1)\,dx$

Let $u = \ln(x+1)$ and $dv = (x+1)\,dx$.

Then $du = \frac{1}{x+1}\,dx$ and $v = \frac{1}{2}(x+1)^2$.

$$\int (x+1)\ln(x+1)\,dx$$
$$= \frac{1}{2}(x+1)^2 \ln(x+1)$$
$$\quad - \int \left[\frac{1}{2}(x+1)^2 \cdot \frac{1}{x+1}\right]dx$$
$$= \frac{1}{2}(x+1)^2 \ln(x+1) - \int \frac{1}{2}(x+1)\,dx$$
$$= \frac{1}{2}(x+1)^2 \ln(x+1) - \frac{1}{4}(x+1)^2 + C$$

$$\int_0^{12} (x+1)\ln(x+1)\,dx$$
$$= \left[\frac{1}{2}(x+1)^2 \ln(x+1) - \frac{1}{4}(x+1)^2\right]\Big|_0^{12}$$
$$= \frac{169}{2}\ln 13 - 42 \approx \$174.74$$

41. The total accumulated growth of the microbe
population during the first 2 days is given by

$$\int_0^2 27t\,e^{3t}\,dt.$$

Let $dv = e^{3t}\,dt$ and $u = 27t$.

Then $v = \dfrac{e^{3t}}{3}$ and $du = 27\,dt$.

$$\int 27t e^{3t}\,dt = 27t \cdot \frac{e^{3t}}{3} - \int \frac{e^{3t}}{3} \cdot 27\,dt$$
$$= 9t e^{3t} - 3 e^{3t}$$

$$\int_0^2 27t e^{3t}\,dt = \left(9te^{3t} - 3e^{3t}\right)\Big|_0^2$$
$$= (18e^6 - 3e^6) - (0 - 3)$$
$$= 15e^6 + 3 \approx 6054$$

43. $\displaystyle\int_0^6 (-10.28 + 175.9te^{-t/1.3})\,dt$

$$= -10.28t + 175.9\int te^{-t/1.3}\,dt$$

Evaluate this integral using integration by parts.

Let $u = t$ and $dv = e^{-t/1.3}\,dt$.

Then $du = dt$ and $v = -1.3e^{-t/1.3}$.

$$\int te^{-t/1.3}\,dt$$
$$= (t)(-1.3e^{-t/1.3}) - \int (-1.3e^{-t/1.3})\,dt$$
$$= -1.3te^{-t/1.3} - 1.69e^{-t/1.3} + C$$

Substitute this expression in the earlier expression.

$$-10.28t + 175.9(-1.3te^{-t/1.3} - 1.69e^{-t/1.3})\Big|_0^6$$

$$= -10.28t - 228.67te^{-t/1.3} - 297.271e^{-t/1.3}\Big|_0^6$$

$$= (-61.68 - 1669.291e^{-6/1.3}) - (-297.271)$$

$$\approx 219.07$$

The total thermic energy is about 219 kJ.

8.2 Volume and Average Value

Your Turn 1

Find the volume of the solid of revolution formed by rotating about the x-axis the region bounded by $y = x^2 + 1$, $y = 0$, $x = -1$, and $x = 1$.

The region and the solid are shown below.

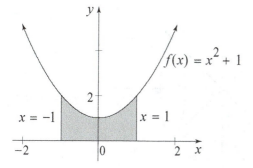

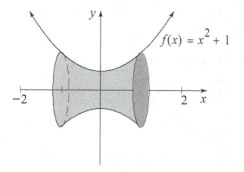

$$V = \int_{-1}^{1} \pi(x^2 + 1)^2\, dx$$

$$= \pi \int_{-1}^{1} (x^4 + 2x^2 + 1)\, dx$$

$$= \pi \left(\frac{x^5}{5} + \frac{2x^3}{3} + x \right)\Big|_{-1}^{1}$$

$$= \pi \left[\left(\frac{1}{5} + \frac{2}{3} + 1 \right) - \left(-\frac{1}{5} - \frac{2}{3} - 1 \right) \right]$$

$$= \pi \left(\frac{56}{15} \right) = \frac{56\pi}{15}$$

Your Turn 2

Find the average value of the function $f(x) = x + \sqrt{x}$ on the interval $[1, 4]$.

Use the formula for average value with $a = 1$ and $b = 4$.

$$\frac{1}{4-1}\int_1^4 (x + \sqrt{x})\, dx = \frac{1}{3}\left(\frac{x^2}{2} + \frac{2}{3}x^{3/2} \right)\Big|_1^4$$

$$= \frac{1}{3}\left(8 + \frac{16}{3} - \frac{1}{2} - \frac{2}{3} \right)$$

$$= \frac{1}{3}\left(\frac{73}{6} \right) = \frac{73}{18}$$

8.2 Warmup Exercises

W1. $\displaystyle\int_0^2 (x + 5)^2\, dx$

$$= \frac{1}{3}(x + 5)^3\Big|_0^2$$

$$= \frac{7^3}{3} - \frac{5^3}{3} = \frac{343 - 125}{3}$$

$$= \frac{218}{3}$$

W2. $\displaystyle\int_0^3 \left(6 - x^2\right)^2 dx$

$$= \int_0^3 \left(36 - 12x^2 + x^4\right) dx$$

$$= \left(36x - 4x^3 + \frac{1}{5}x^5\right)\Big|_0^3$$

$$= \left(108 - 108 + \frac{243}{5}\right) - (0)$$

$$= \frac{243}{5}$$

W3. $\displaystyle\int_{-2}^1 \sqrt{x + 3}\; dx$

$$= \frac{2}{3}(x + 3)^{3/2}\Big|_{-2}^1$$

$$= \frac{2 \cdot 4^{3/2}}{3} - \frac{2 \cdot 1^{3/2}}{3}$$

$$= \frac{16}{3} - \frac{2}{3} = \frac{14}{3}$$

W4. $\displaystyle\int_2^7 \frac{7}{\sqrt{x + 2}}\, dx = 7\int_2^7 (x + 2)^{-1/2}\, dx$

$$= 7\left(2(x + 2)^{1/2}\right)\Big|_2^7$$

$$= 14\left(9^{1/2} - 4^{1/2}\right)$$

$$= 14$$

W5. $\displaystyle\int_0^1 \left(e^x\right)^2 dx = \int_0^1 e^{2x} dx$

$$= \frac{e^{2x}}{2}\Big|_0^1 = \frac{1}{2}\left(e^2 - e^0\right)$$

$$= \frac{e^2 - 1}{2} \approx 3.1945$$

W6. $\displaystyle\int_1^3 x \ln x\, dx$

Let $v = \ln x$ and $du = x\, dx$.

Then $dv = \dfrac{1}{x} dx$ and $u = \dfrac{x^2}{2}$.

$$\int x \ln x\, dx = \int v\, du$$

$$= uv - \int u\, dv$$

$$= \frac{x^2 \ln x}{2} - \int \frac{x}{2}\, dx$$

$$= \frac{x^2 \ln x}{2} - \frac{x^2}{4} + C$$

$$\int_1^3 x \ln x\, dx = \frac{x^2}{2}\left(\ln x - \frac{1}{2}\right)\Big|_1^3$$

$$= \left(\frac{9}{2}\right)\left(\ln 3 - \frac{1}{2}\right) - \left(\frac{1}{2}\right)\left(\ln 1 - \frac{1}{2}\right)$$

$$= \frac{9 \ln 3}{2} - 2 \approx 2.9438$$

8.2 Exercises

1. $f(x) = x, y = 0, x = 0, x = 3$

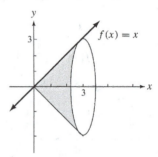

$$V = \pi\int_0^3 x^2\, dx = \frac{\pi x^3}{2}\Big|_0^3 = \frac{\pi(27)}{3} - 0 = 9\pi$$

3. $f(x) = 2x + 1, y = 0, x = 0, x = 4$

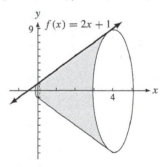

$$V = \pi\int_0^4 (2x + 1)^2\, dx$$

Let $u = 2x + 1$. Then $du = 2\, dx$.

If $x = 4$, $u = 9$. If $x = 0$, $u = 1$.

$$V = \frac{1}{2}\pi \int_0^4 2(2x + 1)^2 \, dx$$

$$= \frac{1}{2}\pi \int_1^9 u^2 \, du$$

$$= \frac{\pi}{2} \left(\frac{u^3}{3} \right) \Big|_1^9$$

$$= \frac{\pi}{2} \left(\frac{729}{3} - \frac{1}{3} \right)$$

$$= \frac{728\pi}{6}$$

$$= \frac{364\pi}{3}$$

5. $f(x) = \frac{1}{3}x + 2,\ y = 0,\ x = 1,\ x = 3$

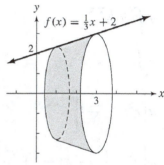

$$V = \pi \int_1^3 \left(\frac{1}{3}x + 2 \right)^2 dx$$

$$= 3\pi \int_1^3 \frac{1}{3} \left(\frac{1}{3}x + 2 \right)^2 dx$$

$$= 3\pi \frac{\left(\frac{1}{3}x + 2 \right)^3}{3} \Big|_1^3$$

$$= \pi \left(\frac{1}{3}x + 2 \right)^3 \Big|_1^3$$

$$= 27\pi - \frac{343\pi}{27} = \frac{386\pi}{27}$$

7. $f(x) = \sqrt{x},\ y = 0,\ x = 1,\ x = 4$

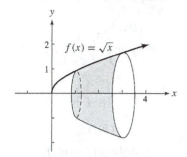

$$V = \pi \int_1^4 (\sqrt{x})^2 \, dx = \pi \int_1^4 x \, dx$$

$$= \frac{\pi x^2}{2} \Big|_1^4$$

$$= 8\pi - \frac{\pi}{2} = \frac{15\pi}{2}$$

9. $f(x) = \sqrt{2x + 1},\ y = 0,\ x = 1,\ x = 4$

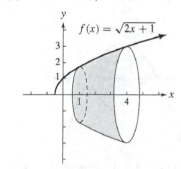

$$V = \pi \int_1^4 (\sqrt{2x + 1})^2 \, dx$$

$$= \pi \int_1^4 (2x + 1) \, dx$$

$$= \pi \left(\frac{2x^2}{2} + x \right) \Big|_1^4$$

$$= \pi[(16 + 4) - 2]$$

$$= 18\pi$$

11. $f(x) = e^x;\ y = 0,\ x = 0,\ x = 2$

$$V = \pi \int_0^2 e^{2x} \, dx = \frac{\pi e^{2x}}{2} \Big|_0^2$$

$$= \frac{\pi e^4}{2} - \frac{\pi}{2}$$

$$= \frac{\pi}{2}(e^4 - 1)$$

$$\approx 84.19$$

13. $f(x) = \dfrac{2}{\sqrt{x}}, y = 0, x = 1, x = 3$

$$V = \pi \int_1^3 \left(\frac{2}{\sqrt{x}}\right)^2 dx$$

$$= \pi \int_1^3 \frac{4}{x} dx$$

$$= 4\pi \ln|x| \Big|_1^3$$

$$= 4\pi (\ln 3 - \ln 1)$$

$$= 4\pi \ln 3 \approx 13.81$$

15. $f(x) = x^2, y = 0, x = 1, x = 5$

$$V = \pi \int_1^5 x^4 dx = \frac{\pi x^5}{5}\Big|_1^5 = 625\pi - \frac{\pi}{5} = \frac{3124\pi}{5}$$

17. $f(x) = 1 - x^2, y = 0$

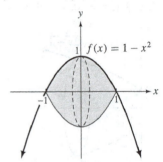

Since $f(x) = 1 - x^2$ intersects $y = 0$ where

$$1 - x^2 = 0$$
$$x = \pm 1,$$
$$a = -1 \text{ and } b = 1.$$

$$V = \pi \int_{-1}^1 (1 - x^2)^2 dx$$

$$= \pi \int_{-1}^1 (1 - 2x^2 + x^4) dx$$

$$= \pi \left(x - \frac{2x^3}{3} + \frac{x^5}{5}\right)\Big|_{-1}^1$$

$$= \pi \left(1 - \frac{2}{3} + \frac{1}{5}\right) - \pi\left(-1 + \frac{2}{3} - \frac{1}{5}\right)$$

$$= 2\pi - \frac{4\pi}{3} + \frac{2\pi}{5}$$

$$= \frac{16\pi}{15}$$

19. $f(x) = \sqrt{1 - x^2}$

$\quad r = \sqrt{1} = 1$

$$V = \pi \int_{-1}^1 (\sqrt{1 - x^2})^2 dx$$

$$= \pi \int_{-1}^1 (1 - x^2) dx$$

$$= \pi \left(x - \frac{x^3}{3}\right)\Big|_{-1}^1$$

$$= \pi\left(1 - \frac{1}{3}\right) - \pi\left(-1 + \frac{1}{3}\right)$$

$$= 2\pi - \frac{2}{3}\pi = \frac{4\pi}{3}$$

21. $f(x) = \sqrt{r^2 - x^2}$

$$V = \pi \int_{-r}^r (\sqrt{r^2 - x^2})^2 dx$$

$$= \pi \int_{-r}^r (r^2 - x^2) dx$$

$$= \pi \left(r^2 x - \frac{x^3}{3}\right)\Big|_{-r}^r$$

$$= \pi\left(r^3 - \frac{r^3}{3}\right) - \pi\left(-r^3 + \frac{r^3}{3}\right)$$

$$= 2r^3\pi - \left(\frac{2r^3\pi}{3}\right)$$

$$= \frac{4\pi r^3}{3}$$

23. $f(x) = r, x = 0, x = h$

Graph $f(x) = r$; then show the solid of revolution formed by rotating about the x-axis the region bounded by $f(x), x = 0, x = h$.

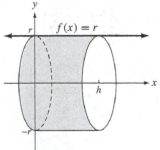

$$\int_0^h \pi r^2 dx = \pi r^2 x \Big|_0^h$$

$$= \pi r^2 h - 0$$

$$= \pi r^2 h$$

25. $f(x) = x^2 - 4; [0, 5]$

Average value

$$= \frac{1}{5 - 0} \int_0^5 (x^2 - 4)dx$$

$$= \frac{1}{5} \left(\frac{x^3}{3} - 4x \right) \Big|_0^5$$

$$= \frac{1}{5} \left[\left(\frac{125}{3} - 20 \right) - 0 \right]$$

$$= \frac{13}{3} \approx 4.333$$

27. $f(x) = \sqrt{x + 1}; [3, 8]$

Average value

$$= \frac{1}{8 - 3} \int_3^8 \sqrt{x + 1}\, dx$$

$$= \frac{1}{5} \int_3^8 (x + 1)^{1/2} dx$$

$$= \frac{1}{5} \cdot \frac{2}{3} (x + 1)^{3/2} \Big|_3^8$$

$$= \frac{2}{15} (9^{3/2} - 4^{3/2})$$

$$= \frac{2}{15} (27 - 8) = \frac{38}{15} \approx 2.533$$

29. $f(x) = e^{x/7}; [0, 7]$

Average value

$$= \frac{1}{7 - 0} \int_0^7 e^{x/7} dx$$

$$= \frac{1}{7} \cdot 7e^{x/7} \Big|_0^7$$

$$= e^{x/7} \Big|_0^7 = e^1 - e^0$$

$$e - 1 \approx 1.718$$

31. $f(x) = x^2 e^{2x}; [0, 2]$

Average value $= \dfrac{1}{2 - 0} \displaystyle\int_0^2 x^2 e^{2x}\, dx$

Let $u = x^2$ and $dv = e^{2x}\, dx$.

Use column integration.

D	I
x^2 $+$	e^{2x}
$2x$ $-$	$\frac{1}{2}e^{2x}$
2 $+$	$\frac{1}{4}e^{2x}$
0	$\frac{1}{8}e^{2x}$

$$\frac{1}{2 - 0} \int_0^2 x^2 e^{2x}\, dx$$

$$= \frac{1}{2} \left[(x^2)\left(\frac{1}{2}\right)e^{2x} - (2x)\left(\frac{1}{4}\right)e^{2x} + 2\left(\frac{1}{8}\right)e^{2x} \right]\Big|_0^2$$

$$= \frac{1}{2} \left(2e^4 - e^4 + \frac{1}{4}e^4 - \frac{1}{4} \right)$$

$$= \frac{5e^4 - 1}{8} \approx 34.00$$

33. $f(x) = e^{-x^2}, y = 0, x = -1, x = 1$

$$V = \pi \int_{-1}^1 (e^{-x^2})^2 dx = \pi \int_{-2}^2 e^{-2x^2} dx$$

Using an integration feature on a graphing calculator to evaluate the integral, we obtain $3.758249634 \approx 3.758$.

35. Use the formula for average value with $a = 0$ and $b = 6$.

$$\frac{1}{6 - 0} \int_0^6 (37 + 6e^{-0.03t})dt$$

$$= \frac{1}{6} \left(37t + \frac{6}{-0.03}e^{-0.03t} \right)\Big|_0^6$$

$$= \frac{1}{6}(37t - 200e^{-0.03t})\Big|_0^6$$

$$= \frac{1}{6}[(222 - 200e^{-0.18}) - (0 - 200)]$$

$$= \frac{1}{6}(422 - 200e^{-0.18})$$

$$\approx 42.49$$

The average price is $42.49.

37. Use the formula for average value with $a = 0$ and $b = 6$. The average price is

$$\frac{1}{30 - 0} \int_0^{30} (600 - 20\sqrt{30t})\, dt$$

$$= \frac{1}{30}\left(600t - 20\sqrt{30} \cdot \frac{2}{3} t^{3/2} \right)\Big|_0^{30}$$

$$= \frac{1}{30}\left(600t - \frac{40\sqrt{30}}{3} t^{3/2} \right)\Big|_0^{30}$$

$$= \frac{1}{30}(18{,}000 - 12{,}000)$$

$$= 200 \text{ cases}$$

39. $\displaystyle \int p(t)\, dt = \int 1.757(1.0248)^t\, dt$

$$= \frac{1.757}{\ln(1.0248)} \cdot (1.0248)^t + C$$

(a) The average corn production from 1930 to 1950 is

$$\frac{\left(\dfrac{1.757}{\ln(1.0248)} \cdot (1.0248)^t \right)\Big|_0^{20}}{20 - 0}$$

$$\approx 2.2672$$

$$\approx 2.27 \text{ billion bushels.}$$

(b) The average corn production from 2000 to 2010 is

$$\frac{\left(\dfrac{1.757}{\ln(1.0248)} \cdot (1.0248)^t \right)\Big|_{70}^{80}}{80 - 70}$$

$$\approx 11.0609$$

$$\approx 11.06 \text{ billion bushels.}$$

41. $R(t) = te^{-0.1t}$

"During the nth hour" corresponds to the interval $(n - 1, n)$.

The average intensity during nth hour is

$$\frac{1}{n - (n - 1)} \int_{n-1}^{n} te^{-0.1t}\, dt = \int_{n-1}^{n} te^{-0.1t}\, dt$$

Let $u = t$ and $dv = e^{-0.1t}\, dt$.

D		I
t	$+$	$e^{-0.1t}$
1	$-$	$-10e^{-0.1t}$
0		$100e^{-0.1t}$

$$\int_{n-1}^{n} te^{-0.1t}\, dt$$

$$= (-10te^{-0.1t} - 100e^{-0.1t})\Big|_{n-1}^{n}$$

(a) Second hour, $n = 2$

Average intensity

$$= -10e^{-0.2}(12) + 10e^{-0.1}(11)$$

$$= 110e^{-0.1} - 120e^{-0.2}$$

$$\approx 1.284$$

(b) Twelfth hour, $n = 12$

Average intensity

$$= -10e^{-1.2}(12 + 10) + 10e^{-1.1}(11 + 10)$$

$$= 210e^{-1.1} - 220e^{-1.2} \approx 3.640$$

(c) Twenty-fourth hour, $n = 24$

Average intensity

$$= -10e^{-2.4}(24 + 10) + 10e^{-2.3}(23 + 10)$$

$$= 330e^{-2.3} - 340e^{-2.4} \approx 2.241$$

43. For each part below, use

Average value

$$= \frac{1}{b - a} \int_a^b 45\ln(t + 1)\, dt$$

$$= \frac{45}{b - a} \int_a^b \ln(t + 1)\, dt.$$

Evaluate the integral using integration by parts.

Let $\quad u = \ln(t + 1) \quad$ and $\quad dv = dt$.

Then $\quad du = \dfrac{1}{t + 1}\, dt \quad$ and $\quad v = t$.

$$\int \ln(t + 1)\, dt$$

$$= t\ln(t + 1) - \int \frac{t}{t + 1}\, dt$$

$$= t\ln(t + 1) - \int \left(1 - \frac{1}{t + 1} \right) dt$$

$$= t\ln(t + 1) - t + \ln(t + 1) + C$$

$$= (t + 1)\ln(t + 1) - t + C$$

Therefore

Average value

$$= \frac{1}{b - a} \int_a^b 45\ln(t + 1)\, dt$$

$$= \frac{45}{b - a}[(t + 1)\ln(t + 1) - t]\Big|_a^b.$$

(a) The average number of items produced daily after 5 days is

$$\frac{45}{5-0}[(t+1)\ln(t+1)-t]\Big|_0^5$$

$$= 9[(6\ln 6 - 5) - (\ln 1 - 0)]$$

$$= 9(6\ln 6 - 5) \approx 51.76.$$

(b) The average number of items produced daily after 9 days is

$$\frac{45}{9-0}[(t+1)\ln(t+1)-t]\Big|_0^9$$

$$= 5(10\ln 10 - 9) \approx 70.13.$$

(c) The average number of items produced daily after 30 days is

$$\frac{45}{30-0}[(t+1)\ln(t+1)-t]\Big|_0^{30}$$

$$= \frac{3}{2}(31\ln 31 - 30) \approx 114.7.$$

45. From Exercise 22, the volume of an ellipsoid with horizontal axis of length $2a$ and vertical axis of length $2b$ is

$$V = \frac{4ab^2\pi}{3}.$$

For the Earth, $a = 6,356,752.3142$ and $b = 6,378,137$.

$$V = \frac{4(6,356,752.3142)(6,378,137)^2\pi}{3}$$

$$\approx 1.083 \times 10^{21}$$

The volume of the Earth is about 1.083×10^{21} cubic meters (m^3).

8.3 Continuous Money Flow

Your Turn 1

$$f(t) = 810e^{kt}$$

Use $f(1) = 797.94$ to find k.

$$f(1) = 810e^{k(1)}$$

$$797.94 = 810e^k$$

$$e^k = \frac{797.94}{810} \approx 0.9851$$

$$k \approx \ln 0.9851 \approx -0.015$$

$$f(t) = 810e^{-0.015t}$$

$$\text{Total income} = \int_0^2 810e^{-0.015t}\, dt$$

$$= -\frac{810}{0.015}e^{-0.015t}\Big|_0^2$$

$$= -54,000e^{-0.015t}\Big|_0^2$$

$$= -54,000(e^{-0.03} - 1)$$

$$\approx 1595.94$$

The total income is $1595.94

Your Turn 2

Find the present value of an income given by $f(t) = 50,000t$ over the next 5 years if the interest rate is 3.5%.

$$f(t) = 50,000t, \quad 0 \le t \le 5$$

$$P = \int_0^5 50,000te^{-0.035t}\, dt = 50,000\int_0^5 te^{-0.035t}\, dt$$

Use integration by parts.

Let $\quad u = t \quad$ and $\quad dv = e^{-0.035t}\, dt$.

Then $du = dt$ and $v = -\frac{1}{0.035}e^{-0.035t}$

$$= -28.57143e^{-0.035t}.$$

$$\int_0^5 te^{-0.035t}\, dt$$

$$= -28.57143te^{-0.035t} - \frac{28.57143}{0.035}e^{-0.035t} + C$$

$$= (-28.57143t - 816.32657)e^{-0.035t} + C$$

$$P = 50,000\int_0^5 te^{-0.035t}\, dt$$

$$= 50,000(-28.57143t - 816.32657)e^{-0.035t}\Big|_0^5$$

$$= 50,000[(-28.57143(5) - 816.32657)e^{-0.035(5)}$$
$$\quad - (0 - 816.32657)]$$

$$= 50,000(-805.19351 + 816.32657)$$

$$\approx 556,653$$

The present value of the income is $556,653.

Your Turn 3

Find the accumulated amount of money flow for an income given by $f(t) = 50,000t$ over the next 5 years if the interest rate is 3.5%.

$$A = e^{0.035(5)} \int_0^5 50{,}000te^{-0.035t}\, dt$$

The integral was computed in Your Turn 2. Using this value, the accumulated value is

$$e^{0.175}(556{,}653) \approx 663{,}111.$$

The accumulated amount of money flow is $663,111.

Your Turn 4

Find the present value at the end of 8 years of the continuous flow of money given by

$$f(t) = 200t^2 + 100t + 50$$

at 5% compounded continuously.

$$P = \int_0^8 (200t^2 + 100t + 50)e^{-0.05t}\, dt$$

Using integration by parts as in Example 5, you can verify that

$$\int (200t^2 + 100t + 50)e^{-0.05t}\, dt$$

$$= -1000e^{-0.05t}(4t^2 + 162t + 3241) + C.$$

Thus $P = -1000e^{-0.05t}(4t^2 + 162t + 3241)\Big|_0^8$

$$= -1000(e^{-0.4}(4793) - 3241)$$

$$\approx 28{,}156.02$$

The present value at the end of 8 years of this flow of money is $28,156.02.

8.3 Warmup Exercises

W1. $\displaystyle \int 300e^{0.01t}\, dt = 300\left(\frac{1}{0.01}e^{0.01t}\right) + C$

$$= 30{,}000e^{0.01t} + C$$

W2. $\displaystyle \int_0^5 100e^{0.04t}\, dt$

$$= \frac{100}{0.04}e^{0.04t}\Big|_0^5$$

$$= 2500\left(e^{0.2} - 1\right)$$

$$\approx 553.51$$

W3. $\displaystyle \int 400te^{-0.04t}\, dt = 400\int te^{-0.04t}\, dt$

Let $v = t$ and $du = e^{-0.04t}$.

Then $dv = dt$ and

$$u = -\frac{1}{0.04}e^{-0.04t} = -25e^{-0.04t}.$$

$$400\int te^{-0.04t}\, dt$$

$$= 400\int v\, du$$

$$= 400\left[uv - \int u\, dv\right]$$

$$= 400\left[-25te^{-0.04t} - \left(-25\int e^{-0.04t}\, dt\right)\right]$$

$$= -10{,}000\left(te^{-0.04t} + \frac{1}{0.04}e^{-0.04t}\right) + C$$

$$= \left(-10{,}000e^{-0.04t}\right)(t + 25) + C$$

W4. Using the result in W3,

$$1000\int te^{-0.04t}\, dt$$

$$= \left(\frac{5}{2}\right)\left(400\int te^{-0.04t}\, dt\right)$$

$$= \left(\frac{5}{2}\right)\left[\left(-10{,}000e^{-0.04t}\right)(t + 25) + C\right]$$

$$= \left(-25{,}000e^{-0.04t}\right)(t + 25) + C'$$

$$\int_0^5 1000te^{-0.04t}\, dt$$

$$= \left[\left(-25{,}000e^{-0.04t}\right)(t + 25)\right]\Big|_0^5$$

$$= -25{,}000\left[(5 + 25)e^{-0.2} - (0 + 25)e^{-0}\right]$$

$$\approx 10{,}951.94$$

8.3 Exercises

1. $f(t) = 1000$

  (a) $\displaystyle P = \int_0^{10} 1000e^{-0.08t}\, dt$

$$= \frac{1000}{-0.08}e^{-0.08t}\Big|_0^{10}$$

$$= -12{,}500(e^{-0.8} - e^0)$$

$$= -12{,}500(e^{-0.8} - 1)$$

$$\approx 6883.387949$$

(We will use this value for P in part (b).
Store it in your calculator without
rounding.)
The present value is $6883.39.

(b) $A = e^{0.08(10)} \int_0^{10} 1000e^{-0.08t}\,dt$

$= e^{0.8}P$

$\approx 15{,}319.26161$

The accumulated value is $15{,}319.26.

3. $f(t) = 500$

(a) $P = \int_0^{10} 500e^{-0.08t}\,dt$

$= \dfrac{500}{-0.08}e^{-0.08t}\Big|_0^{10}$

$= -6250(e^{-0.8} - e^0)$

≈ 3441.693974

The present value is $3441.69.

(b) $A = e^{0.08(10)} \int_0^{10} 500e^{-0.08t}\,dt$

$= e^{0.8}P \approx 7659.630803$

The accumulated value is $7659.63.

5. $f(t) = 400e^{0.03t}$

(a) $P = \int_0^{10} 400e^{0.03t}e^{-0.08t}\,dt$

$= 400\int_0^{10} e^{-0.05t}\,dt = \dfrac{400}{-0.05}e^{-0.05t}\Big|_0^{10}$

$= -8000(e^{-0.5} - e^0)$

≈ 3147.754722

The present value is $3147.75.

(b) $A = e^{0.08(10)} \int_0^{10} 400e^{0.03t}e^{-0.08t}\,dt$

$= e^{0.8}P \approx 7005.456967$

The accumulated value is $7005.46.

7. $f(t) = 5000e^{-0.01t}$

(a) $P = \int_0^{10} 5000e^{-0.01t}e^{-0.08t}\,dt$

$= 5000\int_0^{10} e^{-0.09t}\,dt$

$= \dfrac{5000}{-0.09}e^{-0.09t}\Big|_0^{10}$

$= -\dfrac{5000}{0.09}(e^{-0.9} - e^0)$

$\approx 32{,}968.35224$

The present value is $32{,}968.35.

(b) $A = e^{0.08(10)} \int_0^{10} 5000e^{-0.01t}e^{-0.08t}\,dt$

$= e^{0.8}P$

$\approx 73{,}372.41725$

The accumulated value is $73{,}372.42.

9. $f(t) = 25t$

(a) $P = \int_0^{10} 25te^{-0.08t}\,dt$

$= 25\int_0^{10} te^{-0.08t}\,dt$

Find the antiderivative using integration by
parts.

Let $u = t$ and $dv = e^{-0.08t}\,dt$.

Then $du = dt$ and $v = \dfrac{1}{-0.08}e^{-0.08t}$

$= -12.5e^{-0.08t}$.

$\displaystyle\int te^{-0.08t}\,dt$

$= t(-12.5e^{-0.08t}) - \int(-12.5e^{-0.08t})\,dt$

$= -12.5te^{-0.08t} + 12.5\int e^{-0.08t}\,dt$

$= -12.5te^{-0.08t} + \dfrac{12.5}{-0.08}e^{-0.08t} + C$

$= -(12.5t + 156.25)e^{-0.08t} + C$

Therefore

$P = 25[-(12.5t + 156.25)e^{-0.08t}]\Big|_0^{10}$

$= [-25(12.5t + 156.25)e^{-0.08t}]\Big|_0^{10}$

$= (-7031.25e^{-0.8}) - (-3906.25e^0)$

$\approx 746.9057211.$

The present value is $746.91.

(b) $A = e^{0.08(10)} \int_0^{10} 25te^{-0.08t}\,dt$

$= e^{0.8}P$

≈ 1662.269252

The accumulated value is $1662.27.

11. $f(t) = 0.01t + 100$

(a) $P = \int_0^{10} (0.01t + 100)e^{-0.08t}\,dt$

$= \int_0^{10} 0.01te^{-0.08t}\,dt$

$+ \int_0^{10} 100e^{-0.08t}\,dt$

$= 0.01\int_0^{10} te^{-0.08t}\,dt$

$+ 100\int_0^{10} e^{-0.08t}\,dt$

From Exercise 9, we know that

$\int te^{-0.08t}\,dt = -(12.5t + 156.25)e^{-0.08t} + C$

From Exercise 1, we know that

$\int e^{-0.08t}\,dt = -12.5e^{-0.08t} + C$

Substitute the given expressions and simplify.

$P = \{0.01[-(12.5t + 156.25)e^{-0.08t}]$

$+ 100(-12.5e^{-0.08t})\}\Big|_0^{10}$

$= [-(0.125t + 1251.5625)e^{-0.08t}]\Big|_0^{10}$

$= (-1252.8125e^{-0.8}) - (-1251.5625e^0)$

≈ 688.6375571

The present value is $688.64.

(b) $A = e^{0.08(10)} \int_0^{10} (0.01t + 100)e^{-0.08t}\,dt$

$= e^{0.8}P$

≈ 1532.591068

The accumulated value is $1532.59.

13. $f(t) = 1000t - 100t^2$

(a) $P = \int_0^{10} (1000t - 100t^2)e^{-0.08t}\,dt$

$= 1000\int_0^{10} te^{-0.08t}\,dt$

$- 100\int_0^{10} t^2 e^{-0.08t}\,dt$

From Exercise 9, we know that

$\int te^{-0.08t}\,dt = -(12.5t + 156.25)e^{-0.08t} + C$

Evaluate the antiderivative $\int t^2 e^{-0.08t}\,dt$ using column integration. (Note that $\frac{1}{-0.08} = -12.5$.)

D		I
t^2	$+$	$e^{-0.08t}$
$2t$	$-$	$12.5\,e^{-0.08t}$
2	$+$	$156.25\,e^{-0.08t}$
0		$-1953.125\,e^{-0.08t}$

Thus,

$\int_0^{10} t^2 e^{-0.08t}\,dt$

$= (t^2)(-12.5e^{-0.08t})$

$- (2t)(156.25e^{-0.08t})$

$+ (2)(-1953.125e^{-0.08t}) + C$

$= -(12.5t^2 + 312.5t$

$+ 3906.25)e^{-0.08t} + C.$

Therefore:

$P = \{1000[-(12.5t + 156.25)e^{-0.08t}]$

$- 100[-(12.5t^2 + 312.5t$

$+ 3906.25)e^{-0.08t}]\}\Big|_0^{10}$

Collect like terms and simplify.

$P = [(1250t^2 + 18,750t + 234,375)e^{-0.08t}]\Big|_0^{10}$

$= (546,875e^{-0.8}) - (234,375e^0)$

$\approx 11,351.77725$

The principal value is $11,351.78.

(b) $A = e^{0.08(10)} \displaystyle\int_0^{10} (1000t - 100t^2)e^{-0.08t}\,dt$

$ = e^{0.8}P$

$ \approx 25{,}263.84488$

The accumulated value is $25,263.84.

15. $A = e^{0.04(3)} \displaystyle\int_0^3 20{,}000e^{-0.04t}\,dt$

$ = e^{0.12}\left(\dfrac{20{,}000}{-0.04} e^{-0.04t} \right)\Big|_0^3$

$ = e^{0.12}\left(\dfrac{20{,}000}{-0.04} e^{-0.12} + \dfrac{20{,}000}{0.04} \right)$

$ \approx \$63{,}748.43$

17. **(a)** Present value

$ = \displaystyle\int_0^8 5000e^{-0.01t}e^{-0.08t}\,dt$

$ = \displaystyle\int_0^8 5000e^{-0.09t}\,dt$

$ = \left(\dfrac{5000}{-0.09} e^{-0.09t} \right)\Big|_0^8$

$ = \dfrac{5000e^{-0.72}}{-0.09} + \dfrac{5000}{0.09}$

$ = \$28{,}513.76$

(b) Final amount

$ = e^{0.08(8)} \displaystyle\int_0^8 5000e^{-0.01t}e^{-0.08t}\,dt$

$ \approx e^{0.64}(28{,}513.76)$

$ \approx \$54{,}075.81$

19. $P = \displaystyle\int_0^5 (1500 - 60t^2)e^{-0.05t}\,dt$

$ = \displaystyle\int_0^5 1500e^{-0.05t}\,dt - \int_0^5 60t^2 e^{-0.05t}\,dt$

$ = 1500 \displaystyle\int_0^5 e^{-0.05t}\,dt - 60 \int_0^5 t^2 e^{-0.05t}\,dt$

Find the second integral by column integration.

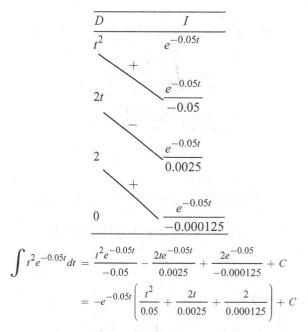

D	I
t^2	$e^{-0.05t}$
$2t$	$\dfrac{e^{-0.05t}}{-0.05}$
2	$\dfrac{e^{-0.05t}}{0.0025}$
0	$\dfrac{e^{-0.05t}}{-0.000125}$

$\displaystyle\int t^2 e^{-0.05t}\,dt = \dfrac{t^2 e^{-0.05t}}{-0.05} - \dfrac{2te^{-0.05t}}{0.0025} + \dfrac{2e^{-0.05}}{-0.000125} + C$

$\phantom{\int t^2 e^{-0.05t}\,dt} = -e^{-0.05t}\left(\dfrac{t^2}{0.05} + \dfrac{2t}{0.0025} + \dfrac{2}{0.000125} \right) + C$

Now add the first integral to this result.

$1500 \displaystyle\int_0^5 e^{-0.05t}\,dt - 60 \int_0^5 t^2 e^{-0.05t}\,dt$

$= \dfrac{1500}{-0.05} e^{-0.05t}\Big|_0^5$

$\; + 60e^{-0.05t}\left(\dfrac{t^2}{0.05} + \dfrac{2t}{0.0025} + \dfrac{2}{0.000125} \right)\Big|_0^5$

$= \dfrac{1500}{-0.05}(e^{-0.25} - 1)$

$\; + 60\left[e^{-0.25}\left(\dfrac{25}{0.05} + \dfrac{10}{0.0025} + \dfrac{2}{0.000125} \right) - \dfrac{2}{0.000125} \right]$

$\approx 6636.977 - 2075.037$

$= \$4560.94$

8.4 Improper Integrals

Your Turn 1

(a) $\displaystyle\int_8^{\infty} \dfrac{1}{x^{1/3}}\,dx = \lim_{b \to \infty} \int_8^b \dfrac{1}{x^{1/3}}\,dx$

$\phantom{(a)\;\int_8^{\infty} \dfrac{1}{x^{1/3}}\,dx} = \lim_{b \to \infty} \left(\dfrac{3}{2} x^{2/3} \right)\Big|_8^b$

$\phantom{(a)\;\int_8^{\infty} \dfrac{1}{x^{1/3}}\,dx} = \lim_{b \to \infty} \left(\dfrac{3}{2} b^{2/3} - \dfrac{3}{2}(4) \right)$

As $b \to \infty$, $b^{2/3} \to \infty$, so the limit above does not exist and the integral diverges.

(b) $\displaystyle\int_8^\infty \frac{1}{x^{4/3}}\,dx$

$$= \lim_{b\to\infty}\int_8^b \frac{1}{x^{4/3}}\,dx$$

$$= \lim_{b\to\infty}\left(-3x^{-1/3}\right)\Big|_8^b$$

$$= \lim_{b\to\infty}\left[-3b^{-1/3}-\left(-\frac{3}{8^{1/3}}\right)\right]$$

As $b\to\infty$, $b^{-1/3}=\dfrac{1}{b^{1/3}}\to 0$, so the integral is convergent and its value is $\dfrac{3}{8^{1/3}}=\dfrac{3}{2}$.

Your Turn 2

$$\int_0^\infty 5e^{-2x}\,dx = \lim_{b\to\infty}\int_0^b 5e^{-2x}\,dx$$

$$= \lim_{b\to\infty}\left(-\frac{5}{2}e^{-2x}\right)\Big|_0^b$$

$$= \lim_{b\to\infty}\left[-\frac{5}{2}e^{-2b}-\left(-\frac{5}{2}e^{-2(0)}\right)\right]$$

As $b\to\infty$, $e^{-2b}=\dfrac{1}{e^b}\to 0$, so the integral is convergent and its value is $\dfrac{5}{2}e^0=\dfrac{5}{2}$.

8.4 Warmup Exercises

W1. $\displaystyle\lim_{x\to\infty}\frac{3-x^2}{x^4}=\lim_{x\to\infty}\frac{3}{x^2}-\lim_{x\to\infty}\frac{x^2}{x^4}$
  where both limits exist
  $= 0+0=0$

W2. $\displaystyle\lim_{x\to\infty}\frac{4x^2+3x-7}{2x^5+3x-5}$

$$= \lim_{x\to\infty}\frac{4\cdot\dfrac{x^2}{x^5}+3\cdot\dfrac{x}{x^5}-7\cdot\dfrac{1}{x^5}}{2\cdot\dfrac{x^5}{x^5}+3\cdot\dfrac{x}{x^5}-5\cdot\dfrac{1}{x^5}}$$

$$= \frac{\lim\limits_{x\to\infty}\left(4\cdot\dfrac{1}{x^3}+3\cdot\dfrac{1}{x^4}-7\cdot\dfrac{1}{x^5}\right)}{\lim\limits_{x\to\infty}\left(2+3\cdot\dfrac{1}{x^4}-5\cdot\dfrac{1}{x^5}\right)}$$

$$= \frac{0}{2}=0$$

W3. $\displaystyle\lim_{x\to\infty}e^{-4x}=0$

For any given δ less than 1 we can make $0<e^{-4x}<\delta$ by taking $x>\dfrac{-\ln\delta}{4}$.

W4. $\displaystyle\lim_{x\to-\infty}e^{-0.2x}=\infty$

For any given M greater than 1 we can make $e^{-0.2x}>M$ by taking $x<-5\ln M$.

8.4 Exercises

1. $\displaystyle\int_3^\infty \frac{1}{x^2}\,dx = \lim_{b\to\infty}\int_3^b x^{-2}\,dx$

$$= \lim_{b\to\infty}-x^{-1}\Big|_3^b$$

$$= \lim_{b\to\infty}\left(-\frac{1}{b}+\frac{1}{3}\right)$$

$$= \lim_{b\to\infty}\left(-\frac{1}{b}\right)+\lim_{b\to\infty}\frac{1}{3}$$

As $b\to\infty$, $-\dfrac{1}{b}\to 0$. The integral is convergent.

$$\int_3^\infty \frac{1}{x^2}\,dx = 0+\frac{1}{3}=\frac{1}{3}$$

3. $\displaystyle\int_4^\infty \frac{2}{\sqrt{x}}\,dx = \lim_{b\to\infty}\int_4^b 2x^{-1/2}\,dx$

$$= \lim_{b\to\infty}4x^{1/2}\Big|_4^b$$

$$= \lim_{b\to\infty}\left(4\sqrt{b}-4\sqrt{4}\right)$$

$$= \lim_{b\to\infty}4\sqrt{b}-8$$

As $b\to\infty$, $4\sqrt{b}\to\infty$. The integral diverges.

5. $\displaystyle\int_{-\infty}^{-1}\frac{2}{x^3}\,dx = \int_{-\infty}^{-1}2x^{-3}\,dx$

$$= \lim_{a\to-\infty}\int_a^{-1}2x^{-3}\,dx$$

$$= \lim_{a\to-\infty}\left(\frac{2x^{-2}}{-2}\right)\Big|_a^{-1}$$

$$= \lim_{a\to-\infty}\left(-1+\frac{1}{a^2}\right)$$

As $a \to -\infty, \frac{1}{a^2} \to 0.$ The integral is convergent.

$$\int_{-\infty}^{-1} \frac{2}{x^3} dx = -1 + 0 = -1$$

7. $\int_{1}^{\infty} \frac{1}{x^{1.0001}} dx$

$$= \int_{1}^{\infty} x^{-1.0001} dx$$

$$= \lim_{b \to \infty} \int_{1}^{b} x^{-1.0001} dx$$

$$= \lim_{b \to \infty} \left(\frac{x^{-0.0001}}{-0.0001} \right) \Big|_{1}^{b}$$

$$= \lim_{b \to \infty} \left(-\frac{1}{(0.0001)b^{0.0001}} + \frac{1}{0.0001} \right)$$

As $b \to \infty, -\frac{1}{0.0001\, b^{0.0001}} \to 0.$

The integral is convergent.

$$\int_{1}^{\infty} \frac{1}{x^{1.0001}} dx = 0 + \frac{1}{0.0001} = 10,000$$

9. $\int_{-\infty}^{-10} x^{-2} dx = \lim_{a \to -\infty} \int_{a}^{-10} x^{-2} dx$

$$= \lim_{a \to -\infty} (-x^{-1}) \Big|_{a}^{-10}$$

$$= \lim_{a \to -\infty} \left(\frac{1}{10} + \frac{1}{a} \right)$$

$$= \frac{1}{10} + 0 = \frac{1}{10}$$

The integral is convergent and its value is $\frac{1}{10}$.

11. $\int_{-\infty}^{-1} x^{-8/3} dx = \lim_{a \to -\infty} \int_{a}^{-1} x^{-8/3} dx$

$$= \lim_{a \to -\infty} \left(-\frac{3}{5} x^{-5/3} \right) \Big|_{a}^{-1}$$

$$= \lim_{a \to -\infty} \left(\frac{3}{5} + \frac{3}{5a^{5/3}} \right)$$

$$= \frac{3}{5} + 0 = \frac{3}{5}$$

The integral is convergent, and its value is $\frac{3}{5}$.

13. $\int_{0}^{\infty} 8e^{-8x} dx = \lim_{b \to \infty} \int_{0}^{b} 8e^{-8x} dx$

$$= \lim_{b \to \infty} \left(\frac{8e^{-8x}}{-8} \right) \Big|_{0}^{b}$$

$$= \lim_{b \to \infty} (-e^{-8b} + 1)$$

$$= \lim_{b \to \infty} \left(-\frac{1}{e^{8b}} + 1 \right)$$

$$= 0 + 1 = 1$$

The integral is convergent, and its value is 1.

15. $\int_{-\infty}^{0} 1000e^{x} dx = \lim_{a \to -\infty} \int_{a}^{0} 1000e^{x} dx$

$$= \lim_{a \to -\infty} (1000e^{x}) \Big|_{a}^{0}$$

$$= \lim_{a \to -\infty} (1000 - 1000e^{a})$$

As $a \to \infty, -1000e^{a} \to 0.$ The integral is convergent.

$$\int_{-\infty}^{0} 1000e^{x} dx = 1000 - 0 = 1000$$

17. $\int_{-\infty}^{-1} \ln |x| dx = \lim_{a \to -\infty} \int_{a}^{-1} \ln |x| dx$

Let $u = \ln |x|$ and $dv = dx.$

Then $du = \frac{1}{x} dx$ and $v = x.$

$$\int \ln |x| dx = x \ln |x| - \int \frac{x}{x} dx$$

$$= x \ln |x| - x + C$$

$$\int_{-\infty}^{-1} \ln |x| dx = \lim_{a \to -\infty} (x \ln |x| - x) \Big|_{a}^{-1}$$

$$= \lim_{a \to -\infty} (-\ln 1 + 1 - a \ln |a| + a)$$

$$= \lim_{a \to -\infty} (1 + a - a \ln |a|)$$

The integral is divergent, since as $a \to -\infty.$

$(a - a \ln |a|) = -a(-1 + \ln |a|) \to \infty.$

19. $\displaystyle\int_0^\infty \frac{dx}{(x+1)^2}$

$\displaystyle = \lim_{b\to\infty} \int_0^b \frac{dx}{(x+1)^2}$ *Use substitution*

$\displaystyle = \lim_{b\to\infty} -(x+1)^{-1}\Big|_0^b$

$\displaystyle = \lim_{b\to\infty}\left(\frac{-1}{b+1}+1\right)$

As $b\to\infty$, $-\frac{1}{b+1}\to 0$. The integral is convergent.

$\displaystyle\int_0^\infty \frac{dx}{(x+1)^2} = 0+1 = 1$

21. $\displaystyle\int_{-\infty}^{-1} \frac{2x-1}{x^2-x}\,dx$

$\displaystyle = \lim_{a\to-\infty} \int_0^{-1} \frac{2x-1}{x^2-x}\,dx$ *Use substitution*

$\displaystyle = \lim_{a\to-\infty} \ln|x^2-x|\Big|_a^{-1}$

$\displaystyle = \lim_{a\to-\infty} (\ln 2 - \ln|a^2 - a|)$

As $a\to-\infty$, $\ln|a^2-a|\to\infty$. The integral is divergent.

23. $\displaystyle\int_2^\infty \frac{1}{x\ln x}\,dx$

$\displaystyle = \lim_{b\to\infty} \int_2^b \frac{1}{x\ln x}\,dx$ *Use substitution*

$\displaystyle = \lim_{b\to\infty}\left[\ln(\ln x)\Big|_2^b\right]$

$\displaystyle = \lim_{b\to\infty}\left[\ln(\ln b) - \ln(\ln 2)\right]$

As $b\to\infty$, $\ln(\ln b)\to\infty$. The integral is divergent.

25. $\displaystyle\int_0^\infty xe^{4x}\,dx = \lim_{b\to\infty}\int_0^b xe^{4x}\,dx$

Let $\quad dv = e^{4x}\,dx \quad$ and $\quad u = x$.

Then $\quad v = \frac{1}{4}e^{4x}\,dx \quad$ and $\quad du = dx$.

$\displaystyle\int xe^{4x}\,dx = \frac{x}{4}e^{4x} - \int \frac{1}{4}e^{4x}\,dx$

$\displaystyle = \frac{x}{4}e^{4x} - \frac{1}{16}e^{4x} + C$

$\displaystyle = \frac{1}{16}(4x-1)e^{4x} + C$

$\displaystyle\int_0^\infty xe^{4x}\,dx$

$\displaystyle = \lim_{b\to\infty}\left[\frac{1}{16}(4x-1)e^{4x}\right]\Big|_0^b$

$\displaystyle = \lim_{b\to\infty}\left[\frac{1}{16}(4b-1)e^{4b} - \frac{1}{16}(-1)(1)\right]$

$\displaystyle = \lim_{b\to\infty}\left[\frac{1}{16}(4b-1)e^{4b} + \frac{1}{16}\right]$

As $b\to\infty$, $\frac{1}{16}(4b-1)e^{4b}\to\infty$. The integral is divergent.

27. $\displaystyle\int_{-\infty}^\infty x^3 e^{-x^4}\,dx$

$\displaystyle = \int_{-\infty}^0 x^3 e^{-x^4}\,dx + \int_0^\infty x^3 e^{-x^4}\,dx$

We evaluate each of two improper integrals on the right.

$\displaystyle\int_{-\infty}^0 x^3 e^{-x^4}\,dx = \lim_{b\to-\infty}\int_b^0 x^3 e^{-x^4}\,dx$ *Use substitution*

$\displaystyle = \lim_{b\to-\infty}\left[-\frac{1}{4}e^{-x^4}\,\Big|_b^0\right]$

$\displaystyle = \lim_{b\to-\infty}\left[-\frac{1}{4} + \frac{1}{4e^{b^4}}\right]$

As $b\to-\infty$, $\frac{1}{4e^{b^4}}\to 0$. The integral is convergent.

$\displaystyle\int_{-\infty}^0 x^3 e^{-x^4}\,dx = -\frac{1}{4} + 0 = -\frac{1}{4}$

$\displaystyle\int_0^\infty x^3 e^{-x^4}\,dx = \lim_{b\to\infty}\int_0^b x^3 e^{-x^4}\,dx$ *Use substitution*

$\displaystyle = \lim_{b\to\infty}\left[-\frac{1}{4}e^{-x^4}\,\Big|_0^b\right]$

$\displaystyle = \lim_{b\to\infty}\left[\frac{1}{4e^{b^4}} + \frac{1}{4}\right]$

As $b \to -\infty, -\dfrac{1}{4e^{b^4}} \to 0$. The integral is

convergent.

$$\int_0^\infty x^3 e^{-x^4}\, dx = 0 + \frac{1}{4} = \frac{1}{4}$$

Since each of the improper integrals converges, the original improper integral converges.

$$\int_{-\infty}^\infty x^3 e^{-x^4}\, dx = -\frac{1}{4} + \frac{1}{4} = 0$$

29. $\displaystyle\int_{-\infty}^\infty \frac{x}{x^2 + 1}\, dx$

$$= \int_{-\infty}^0 \frac{x}{x^2 + 1}\, dx + \int_0^\infty \frac{x}{x^2 + 1}\, dx$$

We evaluate the first improper integrals on the right.

$$\int_{-\infty}^0 \frac{x}{x^2 + 1}\, dx$$

$$= \lim_{b \to -\infty} \int_b^0 \frac{x}{x^2 + 1}\, dx \quad \text{Use substitution}$$

$$= \lim_{b \to -\infty} \left[\frac{1}{2} \ln(x^2 + 1) \,\Big|_b^0 \right]$$

$$= \lim_{b \to -\infty} \left[0 - \frac{1}{2} \ln(b^2 + 1) \right]$$

As $b \to -\infty, \ln(b^2 + 1) \to \infty$. The integral is divergent. Since one of the two improper integrals on the right diverges, the original improper integral diverges.

31. $f(x) = \dfrac{1}{x - 1}$ for $(-\infty, 0]$

$$\int_{-\infty}^0 \frac{1}{x - 1}\, dx = \lim_{a \to -\infty} \int_a^0 \frac{dx}{x - 1} \quad \ln|x - 1|$$

$$= \lim_{a \to -\infty} \left(\ln|x - 1| \,\Big|_a^0 \right)$$

$$= \lim_{a \to -\infty} (\ln|-1| - \ln|a - 1|)$$

But $\displaystyle\lim_{a \to -\infty} (\ln|a - 1|) = \infty.$

The integral is divergent, so the area cannot be found.

33. $f(x) = \dfrac{1}{(x - 1)^2}$ for $(-\infty, 0]$

$$\int_{-8}^0 \frac{1}{(x - 1)^2}$$

$$= \lim_{a \to -\infty} \int_a^0 \frac{1}{(x - 1)^2} \quad \text{Use substitution}$$

$$= \lim_{a \to -\infty} -(x - 1)^{-1} \,\Big|_a^0$$

$$= \lim_{a \to -\infty} \left(-\frac{1}{-1} + \frac{1}{a - 1} \right)$$

As $a \to -\infty, \dfrac{1}{a-1} \to 0$. The integral is convergent.

$$= 1 + 0 = 1$$

Therefore, the area is 1.

35. $\displaystyle\int_{-\infty}^\infty xe^{-x^2}\, dx$

Let $u = -x^2$, so that $du = -2x\, dx.$

$$= \lim_{a \to -\infty} \left(-\frac{1}{2} \int_a^0 -2xe^{-x^2}\, dx \right)$$

$$+ \lim_{b \to \infty} \left(-\frac{1}{2} \int_0^b -2xe^{-x^2}\, dx \right)$$

$$= \lim_{a \to -\infty} \left(-\frac{1}{2} e^{-x^2} \right) \Big|_a^0$$

$$+ \lim_{b \to \infty} \left(-\frac{1}{2} e^{-x^2} \right) \Big|_0^b$$

$$= \lim_{a \to -\infty} \left(-\frac{1}{2} + \frac{1}{2e^{-a^2}} \right)$$

$$+ \lim_{b \to \infty} \left(-\frac{1}{2e^{b^2}} + \frac{1}{2} \right)$$

$$= -\frac{1}{2} + \frac{1}{2} = 0$$

37. $\displaystyle\int_1^\infty \frac{1}{x^p}\,dx$

Case 1a $p < 1$:

$$\int_1^\infty \frac{1}{x^p}\,dx$$

$$= \int_1^\infty x^{-p}\,dx$$

$$= \lim_{a\to\infty} \int_1^a x^{-p}\,dx$$

$$= \lim_{a\to\infty} \left[\frac{x^{-p+1}}{(-p+1)}\Big|_1^a\right]$$

$$= \lim_{a\to\infty} \left[\frac{1}{(-p+1)}(a^{-p+1}-1)\right]$$

$$= \lim_{a\to\infty} \left[\frac{1}{(-p+1)}a^{1-p} - \frac{1}{(-p+1)}\right]$$

Since $p < 1$, $1 - p$ is positive and, as $a \to \infty$, $a^{1-p} \to \infty$. The integral diverges.

Case 1b $p = 1$:

$$\int_1^\infty \frac{1}{x^p}\,dx = \int_1^\infty \frac{1}{x}\,dx$$

$$= \lim_{a\to\infty} \int_1^a \frac{1}{x}\,dx$$

$$= \lim_{a\to\infty} \left(\ln|x|\Big|_1^a\right)$$

$$= \lim_{a\to\infty} \left(\ln|a| - \ln 1\right)$$

$$= \lim_{a\to\infty} \ln|a|$$

As $a \to \infty$, $\ln|a| \to \infty$. The integral diverges.

Therefore, $\int_1^\infty \frac{1}{x^p}$ diverges when $p \le 1$.

Case 2 $p > 1$:

$$\int_1^\infty \frac{1}{x^p}\,dx = \lim_{x\to\infty} \int_1^a x^{-p}\,dx$$

$$= \lim_{a\to\infty} \left(\frac{x^{-p+1}}{-p+1}\Big|_1^a\right)$$

$$= \lim_{a\to\infty} \left[\frac{a^{-p+1}}{(-p+1)} - \frac{1}{(-p+1)}\right]$$

Since $p > 1$, $-p + 1 < 0$; thus as $a \to \infty$,

$$\frac{a^{-p+1}}{(-p+1)} \to 0.$$

Hence,

$$\lim_{a\to\infty} \left[\frac{a^{-p+1}}{(-p+1)} - \frac{1}{(-p+1)}\right] = 0 - \frac{1}{(-p+1)}$$

$$= \frac{-1}{-p+1}$$

$$= \frac{1}{p-1}.$$

The integral converges.

39. **(a)** Use the *fnInt* feature on a graphing utility to obtain

$$\int_1^{20} \frac{1}{\sqrt{1+x^2}}\,dx \approx 2.808;$$

$$\int_1^{50} \frac{1}{\sqrt{1+x^2}}\,dx \approx 3.724;$$

$$\int_1^{100} \frac{1}{\sqrt{1+x^2}}\,dx \approx 4.417;$$

$$\int_1^{1000} \frac{1}{\sqrt{1+x^2}}\,dx \approx 6.720;$$

$$\int_1^{10,000} \frac{1}{\sqrt{1+x^2}}\,dx \approx 9.022.$$

(b) Since the values of the integrals in part a do not appear to be approaching some fixed finite number but get bigger, the integral $\int_1^\infty \frac{1}{\sqrt{1+x^2}}\,dx$ appears to be divergent.

(c) Use the *fnInt* feature on a graphing utility to obtain

$$\int_1^{20} \frac{1}{\sqrt{1+x^4}}\,dx \approx 0.8770;$$

$$\int_1^{50} \frac{1}{\sqrt{1+x^4}}\,dx \approx 0.9070;$$

$$\int_1^{100} \frac{1}{\sqrt{1+x^4}}\,dx \approx 0.9170;$$

$$\int_1^{1000} \frac{1}{\sqrt{1+x^4}}\,dx \approx 0.9260;$$

$$\int_1^{10,000} \frac{1}{\sqrt{1+x^4}}\,dx \approx 0.9269.$$

(d) Since the values of the integrals in part c appear to be approaching some fixed finite number, the integral

$$\int_1^\infty \frac{1}{\sqrt{1+x^4}}\,dx$$

appears to be convergent.

(e) For large x, we may consider $1 + x^2 \approx x^2$ and $1 + x^4 \approx x^4$.

Thus,

$$\frac{1}{\sqrt{1+x^2}} \approx \frac{1}{\sqrt{x^2}} = \frac{1}{x} \text{ and}$$

$$\frac{1}{\sqrt{1+x^4}} \approx \frac{1}{\sqrt{x^4}} = \frac{1}{x^2}.$$

In Example 1(a) we showed that $\int_1^\infty \frac{1}{x}\,dx$ diverges. Thus, we might guess that $\int_1^\infty \frac{1}{\sqrt{1+x^2}}\,dx$ diverges as well. In Exercise 1, we saw that $\int_2^\infty \frac{1}{x^2}\,dx$ converges. Thus, we might guess that $\int_1^\infty \frac{1}{\sqrt{1+x^4}}\,dx$ converges as well.

41. **(a)** Use the *fnInt* feature on a graphing utility to obtain

$$\int_0^{10} e^{-.00001x}\,dx \approx 9.9995;$$

$$\int_0^{50} e^{-.00001x}\,dx \approx 49.9875;$$

$$\int_0^{100} e^{-.00001x}\,dx \approx 99.9500;$$

$$\int_0^{1000} e^{-.00001x}\,dx \approx 995.0166.$$

(b) Since the values of the integrals in part a do not appear to be approaching some fixed finite number, the integral $\int_0^\infty e^{-0.00001x}\,dx$ appears to be divergent.

(c) $$\int_0^\infty e^{-0.00001x}\,dx$$

$$= \lim_{b\to\infty} \int_0^b e^{-0.00001x}\,dx$$

$$= \lim_{b\to\infty} \left[\frac{e^{-0.00001x}}{-0.00001} \Big|_0^b \right]$$

$$= \lim_{b\to\infty} \left[-\frac{1}{0.00001 e^{0.00001b}} + \frac{1}{0.00001} \right]$$

$$= 0 + 100,000 = 100,000$$

43. $$\int_0^\infty 1,000,000 e^{-0.05t}\,dt$$

$$= \lim_{b\to\infty} \int_0^b 1,000,000 e^{-0.05t}\,dt$$

$$= \lim_{b\to\infty} \left(\frac{1,000,000}{-0.05} e^{-0.05t} \right)\Big|_0^b$$

$$= -20,000,000 \left[\lim_{b\to\infty} (e^{-0.05b}) - e^0 \right]$$

As $b \to \infty$, $e^{-0.05b} = \frac{1}{e^{0.05b}} \to 0$. The integral converges.

$$\int_0^\infty 1,000,000 e^{-0.05t}\,dt$$

$$= -20,000,000(0 - 1)$$

$$= 20,000,000$$

The capital value is $20,000,000.

45. $$\int_0^\infty 1200 e^{0.03t} e^{-0.07t}\,dt$$

$$= \lim_{b\to\infty} \int_0^b 1200 e^{-0.04t}\,dt$$

$$= \lim_{b\to\infty} \left(\frac{1200}{-0.04} e^{-0.04t} \right)\Big|_0^b$$

$$= -30,000 \left[\lim_{b\to\infty} (e^{-0.04b}) - e^0 \right]$$

As $b \to \infty$, $e^{-0.04b} = \frac{1}{e^{0.04b}} \to 0$. The integral converges.

$$\int_0^\infty 1200 e^{0.03t} e^{-0.07t}\,dt = -30,000(0 - 1)$$

$$= 30,000$$

The capital value is $30,000.

47. $$\int_0^\infty 4200 e^{-0.075t}\,dt$$

$$= \lim_{b\to\infty} \int_0^b 4200 e^{-0.075t}\,dt$$

$$= \lim_{b\to\infty} \frac{4200 e^{-0.075b}}{-0.075} \Big|_0^b$$

$$= \lim_{b\to\infty} \left(\frac{4200 e^{-0.075b}}{-0.075} + \frac{4200}{0.075} \right)$$

$$= 0 + 560,000$$

$$= \$560,000$$

49. $r'(x) = 2x^2 e^{-x}$

$$r(x) = \int_0^\infty 2x^2 e^{-x}\, dx = \lim_{b\to\infty} \int_0^b 2x^2 e^{-x}\, dx$$

Let $u = 2x^2$ and $dv = e^{-x} dx$.

Use column integration to obtain

$$\lim_{b\to\infty}\int_0^b 2x^2 e^{-x} dx$$

$$= \lim_{b\to\infty}\left[(-2x^2 e^{-x} - 4xe^{-x} - 4e^{-x})\Big|_0^b \right]$$

$$= \lim_{b\to\infty}[(-2b^2 e^{-b} - 4be^{-b} - 4e^{-b}) - (0 - 4\cdot 1)]$$

$$= -(-4) \quad \textit{Use hint to evaluate limits}$$

$$= 4.$$

51. $P = \int_0^\infty e^{-rt}(at + b)K\, dt$

$$= K \lim_{c\to\infty}\int_0^c (at+b)e^{-rt} dt$$

Evaluate $\int_0^c (at+b)e^{-rt} dt$ using integration by parts.

Let $u = at + b$ and $dv = e^{-rt} dt$.

Then $du = a\, dt$ and $v = -\frac{1}{r}e^{-rt}$.

$$\int_0^c (at+b)e^{-rt} dt$$

$$= \left[(at+b)\left(-\frac{1}{r}e^{-rt}\right) - \int\left(-\frac{1}{r}e^{-rt}\right)a\, dt \right]\Big|_0^c$$

$$= \left[-\frac{at+b}{r}e^{-rt} + \frac{a}{r}\int e^{-rt} dt \right]\Big|_0^c$$

$$= \left[-\frac{at+b}{r}e^{-rt} - \frac{a}{r^2}e^{-rt} \right]\Big|_0^c$$

$$= \left(-\frac{ac+b}{r}e^{-rc} - \frac{a}{r^2}e^{-rc} \right) - \left(-\frac{b}{r}e^0 - \frac{a}{r^2}e^0 \right)$$

Therefore,

$$K\lim_{c\to\infty}\int_0^c (at+b)e^{-rt} dt$$

$$= K\lim_{c\to\infty}\left(-\frac{ac+b}{r}e^{-rc} - \frac{a}{r^2}e^{-rc} + \frac{b}{r} + \frac{a}{r^2} \right)$$

$$= K\left(0 - 0 + \frac{b}{r} + \frac{a}{r^2} \right)$$

$$= \frac{K(a+br)}{r^2}$$

53. $\int_0^\infty 50e^{-0.04t} dt$

$$= 50\lim_{b\to\infty}\int_0^b e^{-0.04t} dt$$

$$= 50\lim_{b\to\infty}\frac{e^{-0.04t}}{-0.04}\Big|_0^b$$

$$= \frac{50}{-0.04}\lim_{b\to\infty}\left(\frac{1}{e^{0.04b}} - 1 \right)$$

As $b \to \infty$, $\frac{1}{e^{0.04b}} \to 0$.

$$\int_0^\infty 50e^{-0.04t} dt = -\frac{50}{-0.04}$$

$$= 1250$$

Chapter 8   Review Exercises

1. False: This integral is best evaluated by substitution.

2. True

3. False: Using the substitution $u = x^2$, $dv = xe^{-x^2}$ this integral requires only one integration by parts.

4. True

5. False: The integrand should be just $2x^2 + 3$.

6. False: The integrand should be $\pi(x^2 + 1)$.

7. True

8. True

9. True

10. False: We must write the integral as $\lim_{a\to-\infty}\int_a^c xe^{-2x} dx + \lim_{b\to\infty}\int_c^b xe^{-2x} dx$. The first of these integrals diverges so $\int_{-\infty}^\infty xe^{-2x} dx$ diverges.

15. $\int \frac{3x}{\sqrt{x-2}} dx = \int 3x(x-2)^{-1/2} dx$

Let $u = 3x$ and $dv = (x-2)^{-1/2} dx$.

Then $du = 3\, dx$ and $v = 2(x-2)^{1/2}$.

$$\int \frac{3x}{\sqrt{x-2}}\,dx$$

$$= 6x(x-2)^{1/2} - 6\int (x-2)^{1/2}\,dx$$

$$= 6x(x-2)^{1/2} - \frac{6(x-2)^{3/2}}{\frac{3}{2}} + C$$

$$= 6x(x-2)^{1/2} - 4(x-2)^{3/2} + C$$

17. $\displaystyle\int (3x+6)e^{-3x}\,dx$

Let $\quad u = 3x+6 \quad$ and $\quad dv = e^{-3x}\,dx.$

Then $\quad du = 3dx \quad$ and $\quad v = \frac{1}{-3}e^{-3x}.$

$$\int (3x+6)e^{-3x}\,dx$$

$$= (3x+6)\left(-\frac{1}{3}e^{-3x}\right) - \int\left(-\frac{1}{3}e^{-3x}\right)3\,dx$$

$$= -(x+2)e^{-3x} + \int e^{-3x}\,dx$$

$$= -(x+2)e^{-3x} - \frac{1}{3}e^{-3x} + C$$

19. $\displaystyle\int (x-1)\ln|x|\,dx$

Let $\quad u = \ln|x| \quad$ and $\quad dv = (x-1)\,dx.$

Then $\quad du = \frac{1}{x}dx \quad$ and $\quad v = \frac{x^2}{2} - x.$

$$\int (x-1)\ln|x|\,dx$$

$$= \left(\frac{x^2}{2} - x\right)\ln|x| - \int\left(\frac{x}{2} - 1\right)dx$$

$$= \left(\frac{x^2}{2} - x\right)\ln|x| - \frac{x^2}{4} + x + C$$

21. $\displaystyle\int \frac{x}{\sqrt{16+8x^2}}\,dx$

Use substitution.

Let $u = 16 + 8x^2.$ Then $du = 16x\,dx.$

$$\int \frac{x}{\sqrt{16+8x^2}}\,dx = \frac{1}{16}\int \frac{16x}{\sqrt{16+8x^2}}\,dx$$

$$= \frac{1}{16}\int \frac{1}{\sqrt{u}}\,du$$

$$= \frac{1}{16}\int u^{-1/2}\,du$$

$$= \frac{1}{16}(2)u^{1/2} + C$$

$$= \frac{1}{8}(16+8x^2)^{1/2} + C$$

$$= \frac{1}{8}\sqrt{16+8x^2} + C$$

23. $\displaystyle\int_0^1 x^2 e^{x/2}\,dx$

Let $u = x^2$ and $dv = e^{x/2}\,dx.$

Use column integration.

D		I
x^2	$+$	$e^{x/2}$
$2x$	$-$	$2e^{x/2}$
2	$+$	$4e^{x/2}$
0		$8e^{x/2}$

$$\int_0^1 x^2 e^{x/2}\,dx = \left(2x^2 e^{x/2} - 8xe^{x/2} + 16e^{x/2}\right)\Big|_0^1$$

$$= 2e^{1/2} - 8e^{1/2} + 16e^{1/2} - 16$$

$$= 10e^{1/2} - 16 \approx 0.4872$$

25. $\displaystyle A = \int_1^3 x^3(x^2-1)^{1/3}\,dx$

Let $\quad u = x^2$ and $dv = x(x^2-1)^{1/3}\,dx.$

Then $du = 2x\,dx$ and $v = \frac{3}{8}(x^2-1)^{4/3}.$

$$\int x^3(x^2 - 1)^{1/3}\,dx$$

$$= \frac{3x^2}{8}(x^2 - 1)^{4/3} - \frac{3}{4}\int x(x^2 - 1)^{4/3}\,dx$$

$$= \frac{3x^2}{8}(x^2 - 1)^{4/3} - \frac{3}{4}\left[\frac{1}{2}\cdot\frac{3}{7}(x^2 - 1)^{7/3}\right]$$

$$= \frac{3x^2}{8}(x^2 - 1)^{4/3} - \frac{9}{56}(x^2 - 1)^{7/3} + C$$

$$A = \left[\frac{3x^2}{8}(x^2 - 1)^{4/3} - \frac{9}{56}(x^2 - 1)^{7/3}\right]\Big|_0^3$$

$$= \frac{3}{8}(144) - \frac{9}{56}(128)$$

$$= 54 - \frac{144}{7} = \frac{234}{7} \approx 33.43$$

27.
$$-11 = \int_0^2 v\,du = (uv)\Big|_0^2 - \int_0^2 u\,dv$$

$$\int_0^2 u\,dv = \big(u(2)v(2) - u(0)v(0)\big) + 11$$

$$= \big((3)(-2) - (-6)(-4)\big) + 11$$

$$= -19$$

29. $f(x) = \sqrt{x - 4};\ y = 0;\ x = 13$

Since $f(x) = \sqrt{x - 4}$ intersects $y = 0$
at $x = 4$, the integral has lower bound $a = 4$.

$$V = \pi\int_4^{13}(\sqrt{x - 4})^2\,dx$$

$$= \pi\int_4^{13}(x - 4)\,dx$$

$$= \pi\left(\frac{x^2}{2} - 4x\right)\Big|_4^{13}$$

$$= \pi\left[\left(\frac{169}{2} - 52\right) - (8 - 16)\right]$$

$$= \pi\left(\frac{65}{2} + 8\right) = \frac{81}{2}\pi \approx 127.2$$

31. $f(x) = \dfrac{1}{\sqrt{x - 1}},\ y = 0,\ x = 2,\ x = 4$

$$V = \pi\int_2^4\left(\frac{1}{\sqrt{x - 1}}\right)^2\,dx$$

$$= \pi\int_2^4\frac{dx}{x - 1}$$

$$= \pi\left(\ln|x - 1|\right)\Big|_2^4$$

$$= \pi\ln 3 \approx 3.451$$

33. $f(x) = \dfrac{x^2}{4},\ y = 0,\ x = 4$

Since $f(x) = \frac{x^2}{4}$ intersects $y = 0$ at $x = 0$,
the integral has a lower bound, $a = 0$.

$$V = \pi\int_0^4\left(\frac{x^2}{4}\right)^2\,dx = \pi\int_0^4\frac{x^4}{16}$$

$$= \frac{\pi}{16}\left(\frac{x^5}{5}\right)\Big|_0^4 = \frac{\pi}{16}\left(\frac{1024}{5}\right)$$

$$= \frac{64\pi}{5} \approx 40.21$$

37.   Average value $= \dfrac{1}{2 - 0}\displaystyle\int_0^2 7x^2(x^3 + 1)^6\,dx$

$$= \frac{7}{2}\int_0^2 x^2(x^3 + 1)^6\,dx$$

Let $u = x^3 + 1$. Then $du = 3x^2\,dx$.

$$\int x^2(x^3 + 1)^6\,dx = \frac{1}{3}\int 3x^2(x^3 + 1)^6\,dx$$

$$= \frac{1}{3}\int u^6\,du$$

$$= \frac{1}{3}\cdot\frac{1}{7}u^7 + C$$

$$= \frac{1}{21}(x^3 + 1)^7 + C$$

$$\frac{7}{2}\int_0^2 x^2(x^3 + 1)^6\,dx = \frac{7}{2}\cdot\frac{1}{21}(x^3 + 1)^7\Big|_0^2$$

$$= \frac{1}{6}(9^7 - 1^7) = \frac{1}{6}(4{,}782{,}969 - 1)$$

$$= \frac{2{,}391{,}484}{3}$$

39. $\displaystyle\int_{-\infty}^{-5} x^{-2}dx = \lim_{a\to-\infty}\int_{a}^{-5}x^{-2}dx$

$$= \lim_{a\to-\infty}\left(\frac{x^{-1}}{-1}\right)\bigg|_{a}^{-5}$$

$$= \lim_{a\to-\infty}\left(-\frac{1}{x}\right)\bigg|_{a}^{-5}$$

$$= \frac{1}{5} + \lim_{a\to-\infty}\left(\frac{1}{a}\right)$$

As $a \to -\infty$, $\frac{1}{a} \to 0$. The integral converges.

$$\int_{-\infty}^{-5}x^{-2}dx = \frac{1}{5} + 0 = \frac{1}{5}$$

41. $\displaystyle\int_{1}^{\infty}6e^{-x}dx = \lim_{b\to\infty}\int_{1}^{b}6e^{-x}dx$

$$= \lim_{b\to\infty}-6e^{-x}\bigg|_{1}^{b}$$

$$= \lim_{b\to\infty}(-6e^{-b} + 6e^{-1})$$

$$= \lim_{b\to\infty}\left(\frac{-6}{e^{b}} + \frac{6}{e}\right)$$

As $b \to \infty$, $e^{b} \to \infty$, so $\frac{-6}{e^{b}} \to 0$. The integral

converges.

$$\int_{1}^{\infty}6e^{-x}dx = 0 + \frac{6}{e} = \frac{6}{e} \approx 2.207$$

43. $\displaystyle\int_{4}^{\infty}\ln(5x)dx = \lim_{b\to\infty}\int_{4}^{b}\ln(5x)dx$

Let $\quad u = \ln(5x) \quad$ and $\quad dv = dx$.

Then $\quad du = \frac{1}{x}dx \quad$ and $\quad v = x$.

$$\int \ln(5x)dx = x\ln(5x) - \int x\cdot\frac{1}{x}dx$$

$$= x\ln(5x) - \int dx$$

$$= x\ln(5x) - x + C$$

$$\lim_{b\to\infty}\int_{4}^{b}\ln(5x)dx$$

$$= \lim_{b\to\infty}\left[x\ln(5x) - x\right]\bigg|_{4}^{b}$$

$$= \lim_{b\to\infty}\left[b\ln(5b) - b\right] - (4\ln 20 - 4)$$

As $b \to \infty$, $b\ln(5b) - b \to \infty$. The integral

diverges.

45. $f(x) = 3e^{-x}$ for $[0, \infty)$

$$A = \int_{0}^{\infty}3e^{-x}dx$$

$$= \lim_{b\to\infty}\int_{0}^{b}3e^{-x}dx$$

$$= \lim_{b\to\infty}(-3e^{-x})\bigg|_{0}^{b}$$

$$= \lim_{b\to\infty}\left(\frac{-3}{e^{b}} + 3\right)$$

As $b \to \infty$, $\frac{-3}{e^{b}} \to 0$.

$$A = 0 + 3 = 3$$

47. $R' = x(x - 50)^{1/2}$

$$R = \int_{50}^{75}x(x - 50)^{1/2}dx$$

Let $\quad u = x$ and $dv = (x - 50)^{1/2}$.

Then $du = dx$ and $v = \frac{2}{3}(x - 50)^{3/2}$.

$$\int x(x - 50)^{1/2}dx$$

$$= \frac{2}{3}x(x - 50)^{3/2} - \frac{2}{3}\int(x - 50)^{3/2}dx$$

$$= \frac{2}{3}x(x - 50)^{3/2} - \frac{2}{3}\cdot\frac{2}{5}(x - 50)^{5/2}$$

$$R = \left[\frac{2}{3}x(x - 50)^{3/2} - \frac{4}{15}(x - 50)^{5/2}\right]\bigg|_{50}^{75}$$

$$= \frac{2}{3}(75)(25^{3/2}) - \frac{4}{15}(25^{5/2})$$

$$= 6250 - \frac{2500}{3}$$

$$= \frac{16,250}{3} \approx \$5416.67$$

49. $f(t) = 25{,}000; 12 \text{ yr}; 10\%$

$$P = \int_{0}^{12}25{,}000e^{-0.10t}dt$$

$$= 25{,}000\left(\frac{e^{-0.10t}}{-0.10}\right)\bigg|_{0}^{12}$$

$$\approx 250{,}000(-0.3012 + 1)$$

$$\approx \$174{,}701.45$$

51. $f(t) = 15t; 18$ mo; 8%

$$P = \int_0^{1.5} 15te^{-0.08t}\, dt$$

$$= 15 \int_0^{1.5} te^{-0.08t}\, dt$$

Find the antiderivative using integration by parts.

Let $u = t$ and $dv = e^{-0.08t}\, dt$.

Then $du = dt$ and $v = \dfrac{1}{-0.08} e^{-0.08t}$

$$= -12.5e^{-0.08t}$$

$$\int te^{-0.08t}\, dt = -12.5te^{-0.08t} - \int (-12.5e^{-0.08t})\, dt$$

$$= -12.5te^{-0.08t} - 156.25e^{-0.08t} + C$$

$$P = 15 \int_0^{1.5} te^{-0.08t}\, dt$$

$$= 15(-12.5te^{-0.08t} - 156.25e^{-0.08t})\Big|_0^{1.5}$$

$$= 15[(-18.75e^{-0.12} - 156.25e^{-0.12})$$
$$- (0 - 156.25)]$$

$$= 15(-175e^{-0.12} + 156.25)$$

$$\approx 15.58385362$$

The present value is $15.58.

53. $f(t) = 500e^{-0.04t}; 8$ yr; 10% per yr

$$A = e^{0.1(8)} \int_0^8 500e^{-0.04t} \cdot e^{-0.1t}\, dt$$

$$= e^{0.8} \int_0^8 500e^{-0.14t}\, dt$$

$$= e^{0.8}\left(\frac{500}{-0.14} e^{-0.14t} \right)\Big|_0^8$$

$$= e^{0.8}\left[\frac{500}{-0.14}(e^{-1.12} - 1) \right]$$

$$\approx 5354.971041$$

The accumulated value is $5354.97.

55. $f(t) = 1000 + 200t; 10$ yr; 9% per yr

$$e^{(0.09)(10)} \int_0^{10} (1000 + 200t)e^{-0.09t}\, dt$$

$$= e^{0.9}\left[\frac{1000}{-0.09} e^{0.09t} + \frac{200}{(0.09)^2}(-0.09t - 1)e^{-0.09t} \right]\Big|_0^{10}$$

$$= e^{0.9}\left[\frac{1000}{-0.09}(e^{-0.9} - 1) + \frac{200}{(0.09)^2}(-1.9e^{-0.9} + 1) \right]$$

$$\approx \$30{,}035.17$$

57. $e^{0.105(10)} \displaystyle\int_0^{10} 10{,}000e^{-0.105t}\, dt$

$$= e^{1.05}\left(\frac{10{,}000e^{-0.105t}}{-0.105} \right)\Big|_0^{10}$$

$$= \frac{10{,}000e^{1.05}}{-0.105}(e^{-1.05} - 1)$$

$$\approx -272{,}157.25(-0.65006)$$

$$\approx \$176{,}919.15$$

59. $\displaystyle\int_0^5 0.5te^{-t}\, dt = 0.5 \int_0^5 te^{-t}\, dt$

Let $u = t$ and $dv = e^{-t}\, dt$.

Then $du = dt$ and $v = \dfrac{e^{-t}}{-1}$.

$$\int te^{-t}\, dt = \frac{te^{-t}}{-1} + \int e^{-t}\, dt$$

$$= -te^{-t} + \frac{e^{-t}}{-1}$$

$$0.5 \int_0^5 te^{-t}\, dt = 0.5(-te^{-t} - e^{-t})\Big|_0^5$$

$$= 0.5(-5e^{-5} - e^{-5} + e^0)$$

$$\approx 0.4798$$

The total reaction over the first 5 hr is 0.4798.

61. **(a)** $\bar{T} = \dfrac{1}{10 - 0} \displaystyle\int_0^{10} (160 - 0.05x^2)\, dx$

$\qquad = \dfrac{1}{10}\left(160x - \dfrac{0.05x^3}{3}\right)\Big|_0^{10}$

$\qquad = \dfrac{1}{10}\left[160(10) - \dfrac{0.05}{3}(10)^3\right]$

$\qquad \approx \dfrac{1}{10}(1583.3) \approx 158.3°$

(b) $\bar{T} = \dfrac{1}{40 - 10} \displaystyle\int_{10}^{40} (160 - 0.05x^2)dx$

$\qquad = \dfrac{1}{30}\left(160x - \dfrac{0.05x^3}{3}\right)\Big|_{10}^{40}$

$\qquad = \dfrac{1}{30}\left[\left(160(40) - \dfrac{0.05(40)^3}{3}\right)\right.$

$\qquad\qquad \left. - \left(160(10) - \dfrac{0.05(10)^3}{3}\right)\right]$

$\qquad \approx \dfrac{1}{30}(5333.33 - 1583.33)$

$\qquad = 125°$

(c) $\bar{T} = \dfrac{1}{40 - 0} \displaystyle\int_0^{40} (160 - 0.05x^2)dx$

$\qquad = \dfrac{1}{40}\left(160x - \dfrac{0.05x^3}{3}\right)\Big|_0^{40}$

$\qquad = \dfrac{1}{40}\left[\left((160)(40) - \dfrac{(0.05)(40)^3}{3}\right)\right]$

$\qquad \approx \dfrac{1}{40}(5{,}333.33)$

$\qquad \approx 133.3°$

MULTIVARIABLE CALCULUS

9.1 Functions of Several Variables

Your Turn 1

$$f(x, y) = 4x^2 + 2xy + \frac{3}{y}$$

$$f(2, 3) = 4(2)^2 + 2(2)(3) + \frac{3}{3}$$

$$= 16 + 12 + 1$$

$$= 29$$

Your Turn 2

$$f(x, y, z) = 4xz - 3x^2y + 2z^2$$

$$f(1, 2, 3) = 4(1)(3) - 3(1)^2(2) + 2(3)^2$$

$$= 12 - 6 + 18$$

$$= 24$$

Your Turn 3

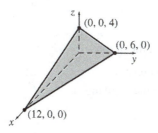

Your Turn 4

Use the Cobb-Douglas production function
$z = x^{1/4}y^{3/4}$.

$$27 = x^{1/4}y^{3/4}$$

$$\frac{27}{x^{1/4}} = y^{3/4}$$

$$\left(\frac{27}{x^{1/4}}\right)^4 = \left(y^{3/4}\right)^4$$

$$y^3 = \frac{(27)^4}{x}$$

$$\left(y^3\right)^{1/3} = \left(\frac{(27)^4}{x}\right)^{1/3}$$

$$y = \frac{(27)^{4/3}}{x^{1/3}} = \frac{81}{x^{1/3}}$$

9.1 Warmup Exercises

W1.

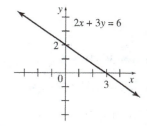

W2.

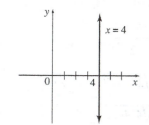

W3.

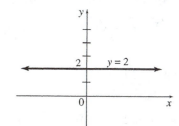

W4. $f(-2) = 2(-2)^2 + 3(-2) - 6$

$$= 8 - 6 - 6$$

$$= -4$$

W5. $f(3a) = 2(3a)^2 + 3(3a) - 6$

$$= 2\left(9a^2\right) + 9a - 6$$

$$= 18a^2 + 9a - 6$$

W6.

$$\frac{f(x + h) - f(x)}{h}$$

$$= \frac{\left[2(x + h)^2 + 3(x + h) - 6\right] - \left(2x^2 + 3x - 6\right)}{h}$$

$$\left(2x^2 + 4xh + 2h^2 + 3x + 3h - 6\right)$$

$$= \frac{- \left(2x^2 + 3x - 6\right)}{h}$$

$$= \frac{4xh + 2h^2 + 3h}{h} = 4x + 2h + 3$$

9.1 Exercises

1. $f(x, y) = 2x - 3y + 5$

    (a) $f(2, -1) = 2(2) - 3(-1) + 5 = 12$

    (b) $f(-4, 1) = 2(-4) - 3(1) + 5 = -6$

    (c) $f(-2, -3) = 2(-2) - 3(-3) + 5 = 10$

    (d) $f(0, 8) = 2(0) - 3(8) + 5 = -19$

3. $h(x, y) = \sqrt{x^2 + 2y^2}$

    (a) $h(5, 3) = \sqrt{25 + 2(9)} = \sqrt{43}$

    (b) $h(2, 4) = \sqrt{4 + 32} = 6$

    (c) $h(-1, -3) = \sqrt{1 + 18} = \sqrt{19}$

    (d) $h(-3, -1) = \sqrt{9 + 2} = \sqrt{11}$

5. $f(x, y) = e^x + \ln(x + y)$

    (a) $f(1, 0) = e^1 + \ln(1 + 0) = e$

    (b) $f(2, -1) = e^2 + \ln(2 + (-1)) = e^2$

    (c) $f(0, e) = e^0 + \ln(0 + e) = 1 + 1 = 2$

    (d)

    $f(0, e^2) = e^0 + \ln(0 + e^2) = 1 + 2 = 3$

7. $x + y + z = 9$

    If $x = 0$ and $y = 0$, $z = 9$.

    If $x = 0$ and $z = 0$, $y = 9$.

    If $y = 0$ and $z = 0$, $x = 9$.

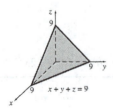

9. $2x + 3y + 4z = 12$

    If $x = 0$ and $y = 0$, $z = 3$.

    If $x = 0$ and $z = 0$, $y = 4$.

    If $y = 0$ and $z = 0$, $x = 6$.

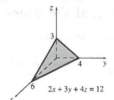

11. $x + y = 4$

    If $x = 0$, $y = 4$.

    If $y = 0$, $x = 4$.

    There is no z-intercept.

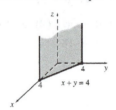

13. $x = 5$

    The point $(5, 0, 0)$ is on the graph.

    There are no y- or z-intercepts.

    The plane is parallel to the yz-plane.

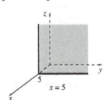

15. $3x + 2y + z = 24$

    For $z = 0$, $3x + 2y = 24$. Graph the line
    $3x + 2y = 24$ in the xy-plane.

    For $z = 2$, $3x + 2y = 22$. Graph the line
    $3x + 2y = 22$ in the plane $z = 2$.

    For $z = 4$, $3x + 2y = 20$. Graph the line
    $3x + 2y = 20$ in the plane $z = 4$.

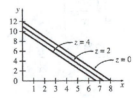

17. $y^2 - x = -z$

For $z = 0$, $x = y^2$. Graph $x = y^2$ in the xy-plane.

For $z = 2$, $x = y^2 + 2$. Graph $x = y^2 + 2$ in the plane $z = 2$.

For $z = 4$, $x = y^2 + 4$. Graph $x = y^2 + 4$ in the plane $z = 4$.

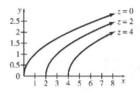

23. $z = x^2 + y^2$

The xz-trace is

$$z = x^2 + 0 = x^2.$$

The yz-trace is

$$z = 0 + y^2 = y^2.$$

Both are parabolas with vertices at the origin that open upward.

The xy-trace is

$$0 = x^2 + y^2.$$

This is a point, the origin.

The equation is represented by a paraboloid, as shown in (c).

25. $x^2 - y^2 = z$

The xz-trace is

$$x^2 = z,$$

which is a parabola with vertex at the origin that opens upward.

The yz-trace is

$$-y^2 = z,$$

which is a parabola with vertex at the origin that opens downward.

The xy-trace is

$$x^2 - y^2 = 0$$
$$x^2 = y^2$$
$$x = y \quad \text{or} \quad x = -y,$$

which are two lines that intersect at the origin.

The equation is represented by a hyperbolic paraboloid, as shown in (e).

27. $\dfrac{x^2}{16} + \dfrac{y^2}{25} + \dfrac{z^2}{4} = 1$

xz-trace:

$$\dfrac{x^2}{16} + \dfrac{z^2}{4} = 1, \text{ an ellipse}$$

yz-trace:

$$\dfrac{y^2}{25} + \dfrac{z^2}{4} = 1, \text{ an ellipse}$$

xy-trace:

$$\dfrac{x^2}{16} + \dfrac{y^2}{25} = 1, \text{ an ellipse}$$

The graph is an ellipsoid, as shown in (b).

29. $f(x, y) = 4x^2 - 2y^2$

(a) $\dfrac{f(x + h, y) - f(x, y)}{h}$

$$= \dfrac{[4(x + h)^2 - 2y^2] - [4x^2 - 2y^2]}{h}$$

$$= \dfrac{4x^2 + 8xh + 4h^2 - 2y^2 - 4x^2 + 2y^2}{h}$$

$$= \dfrac{h(8x + 4h)}{h} = 8x + 4h$$

(b) $\dfrac{f(x, y + h) - f(x, y)}{h}$

$$= \dfrac{[4x^2 - 2(y + h)^2] - [4x^2 - 2y^2]}{h}$$

$$= \dfrac{4x^2 - 2y^2 - 4yh - 2h^2 - 4x^2 + 2y^2}{h}$$

$$= \dfrac{h(-4y - 2h)}{h}$$
$$= -4y - 2h$$

(c) $\displaystyle\lim_{h \to 0} \dfrac{f(x + h, y) - f(x, y)}{h}$

$$= \lim_{h \to 0} (8x + 4h)$$

$$= 8x + 4(0) = 8x$$

(d) $\displaystyle\lim_{h \to 0} \dfrac{f(x, y + h) - f(x, y)}{h}$

$$= \lim_{h \to 0} (-4y - 2h)$$

$$= -4y - 2(0) = -4y$$

31. $f(x, y) = xye^{x^2 + y^2}$

(a) $\displaystyle\lim_{h \to 0} \frac{f(1 + h, 1) - f(1, 1)}{h}$

$\displaystyle = \lim_{h \to 0} \frac{(1 + h)(1)e^{1 + 2h + h^2 + 1} - (1)(1)e^{1 + 1}}{h}$

$\displaystyle = \lim_{h \to 0} \frac{(1 + h)e^{2 + 2h + h^2} - e^2}{h}$

$\displaystyle = e^2 \lim_{h \to 0} \frac{(1 + h)e^{2h + h^2} - 1}{h}$

X	Y1
.001	3.005
1E-4	3.0005
1E-5	3.0001
1E-6	3
-1E-5	3
-1E-4	2.9995
-.001	2.995

Y1◼((1+X)e^(2X+...

The graphing calculator indicates that

$\displaystyle\lim_{h \to 0} \frac{(1+h)e^{2h + h^2} - 1}{h} = 3$, thus

$\displaystyle\lim_{h \to 0} \frac{f(1+h, 1) - f(1,1)}{h} = 3e^2$.

The slope of the tangent line in the direction of x at $(1, 1)$ is $3e^2$.

(b) $\displaystyle\lim_{h \to 0} \frac{f(1, 1 + h) - f(1, 1)}{h}$

$\displaystyle = \lim_{h \to 0} \frac{(1)(1 + h)e^{1 + 1 + 2h + h^2} - (1)(1)e^{1 + 1}}{h}$

$\displaystyle = \lim_{h \to 0} \frac{(1 + h)e^{2 + 2h + h^2} - e^2}{h}$

$\displaystyle = e^2 \lim_{h \to 0} \frac{(1 + h)e^{2h + h^2} - 1}{h}$

So, this limit reduces to the exact same limit as in part a. Therefore, since

$$\lim_{h \to 0} \frac{(1 + h)e^{2h + h^2} - 1}{h} = 3,$$

then

$$\lim_{h \to 0} \frac{f(1, 1 + h) - f(1, 1)}{h} = 3e^2.$$

The slope of the tangent line in the direction of y at $(1, 1)$ is $3e^2$.

33. $P(x, y) = 100\left[\dfrac{3}{5} x^{-2/5} + \dfrac{2}{5} y^{-2/5}\right]^{-5}$

(a) $P(32, 1)$

$\displaystyle = 100\left[\frac{3}{5}(32)^{-2/5} + \frac{2}{5}(1)^{-2/5}\right]^{-5}$

$\displaystyle = 100\left[\frac{3}{5}\left(\frac{1}{4}\right) + \frac{2}{5}(1)\right]^{-5}$

$\displaystyle = 100\left(\frac{11}{20}\right)^{-5}$

$\displaystyle = 100\left(\frac{20}{11}\right)^{5}$

≈ 1986.95

The production is approximately 1987 cameras.

(b) $P(1, 32)$

$\displaystyle = 100\left[\frac{3}{5}(1)^{-2/5} + \frac{2}{5}(32)^{-2/5}\right]^{-5}$

$\displaystyle = 100\left[\frac{3}{5}(1) + \frac{2}{5}\left(\frac{1}{4}\right)\right]^{-5}$

$\displaystyle = 100\left(\frac{7}{10}\right)^{-5}$

$\displaystyle = 100\left(\frac{10}{7}\right)^{5} \approx 595$

The production is approximately 595 cameras.

(c)   32 work hours means that $x = 32$. 243 units of capital means that $y = 243$.

$P(32, 243)$

$\displaystyle = 100\left[\frac{3}{5}(32)^{-2/5} + \frac{2}{5}(243)^{-2/5}\right]^{-5}$

$\displaystyle = 100\left[\frac{3}{5}\left(\frac{1}{4}\right) + \frac{2}{5}\left(\frac{1}{9}\right)\right]^{-5}$

$\displaystyle = 100\left(\frac{7}{36}\right)^{-5}$

$\displaystyle = 100\left(\frac{36}{7}\right)^{5}$

$\approx 359{,}767.81$

The production is approximately 359,768 cameras.

35. $z = x^{0.6}y^{0.4}$ where $z = 500$

$$500 = x^{3/5}y^{2/5}$$

$$\frac{500}{x^{3/5}} = y^{2/5}$$

$$\left(\frac{500}{x^{3/5}}\right)^{5/2} = (y^{2/5})^{5/2}$$

$$y = \frac{(500)^{5/2}}{x^{3/2}}$$

$$y \approx \frac{5,590,170}{x^{3/2}}$$

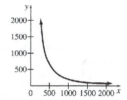

37. The cost function, C, is the sum of the products of the unit costs times the quantities x, y, and z.
Therefore,

$$C(x, y, z) = 250x + 150y + 75z.$$

39. $M = f(40, 0.06, 0.28)$

$$= \frac{(1 + 0.06)^{40}(1 - 0.28) + 0.28}{[1 + (1 - 0.28)(0.06)]^{40}}$$

$$= \frac{(1.06)^{40}(0.72) + 0.28}{[1 + (0.72)(0.06)]^{40}}$$

$$\approx 1.416$$

The multiplier is 1.416. Since $M > 1$, the IRA account grows faster.

41. $A = 0.024265h^{0.3964}m^{0.5378}$

(a) $A = 0.024265(178)^{0.3964}(72)^{0.5378}$

$\approx 1.89 \text{ m}^2$

(b) $A = 0.024265(140)^{0.3964}(65)^{0.5378}$

$\approx 1.62 \text{ m}^2$

(c) $A = 0.024265(160)^{0.3964}(70)^{0.5378}$

$\approx 1.78 \text{ m}^2$

(d) Answers will vary.

43. $P(W, R, A) = 48 - 2.43W - 1.81R - 1.22A$

(a) $P(5, 15, 0)$
$= 48 - 2.43(5) - 1.81(15) - 1.22(0)$
$= 8.7$

8.7% of fish will be intolerant to pollution.

(b) The maximum percentage will occur when the variable factors are a minimum, or when $W = 0$, $R = 0$, and $A = 0$.

$P(0, 0, 0) = 48 - 2.43(0) - 1.81(0) - 1.22(0)$
$= 48$

48% of fish will be intolerant to pollution.

(c) Any combination of values of W, R, and A that result in $P = 0$ is a scenario that will drive the percentage of fish intolerant to pollution to zero.

If $R = 0$ and $A = 0$:

$P(W, 0, 0) = 48 - 2.43W - 1.81(0) - 1.22(0)$
$= 48 - 2.43W.$

$$48 - 2.43W = 0$$

$$W = \frac{48}{2.43}$$

$$\approx 19.75$$

So $W = 19.75$, $R = 0$, $A = 0$ is one scenario.

If $W = 10$ and $R = 10$:

$P(10, 10, A) = 48 - 2.43(10) - 1.81(10) - 1.22A$
$= 5.6 - 1.22A$

$$5.6 - 1.22A = 0$$

$$A = \frac{5.6}{1.22}$$

$$\approx 4.59$$

So $W = 10$, $R = 10$, $A = 4.59$ is another scenario.

(d) Since the coefficient of W is greater than the coefficients of R and A, a change in W will affect the value of P more than an equal change in R or A. Thus, the percentage of wetland (W) has the greatest influence on P.

45. $A(L,T,U,C) = 53.02 + 0.383L + 0.0015T$
$$+ 0.0028U - 0.0003C$$

(a) $A(266,\ 107{,}484,\ 31{,}697,\ 24{,}870)$

$= 53.02 + 0.383(266) + 0.0015(107{,}484)$

$\qquad + 0.0028(31{,}697) - 0.0003(24{,}870)$

≈ 397

The estimated number of accidents is 397.

47. $\ln(T) = 5.49 - 3.00\ln(F) + 0.18\ln(C)$

(a) $e^{\ln(T)} = e^{5.49 - 3.00\ln(F) + 0.18\ln(C)}$

$T = e^{5.49} e^{-3.00\ln(F)} e^{0.18\ln(C)}$

$\quad = \dfrac{e^{5.49} e^{\ln(C^{0.18})}}{e^{\ln(F^3)}}$

$T \approx \dfrac{242.257 C^{0.18}}{F^3}$

(b) Replace F with 2 and C with 40 in the preceding formula.

$$T \approx \frac{242.257(40)^{0.18}}{(2)^3} \approx 58.82$$

T is about 58.8%. In other words, a tethered sow spends nearly 59% of the time doing repetitive behavior when she is fed 2 kg of food per day and neighboring sows spend 40% of the time doing repetitive behavior.

49. The girth is $2H + 2W$. Thus,

$$f(L,W,H) = L + 2H + 2W.$$

51. $f(H,D) = \sqrt{H^2 + D^2}$ with $D = 3.75$ in

(a) $\dfrac{H}{D} = \dfrac{3}{4}$

$H = \dfrac{3}{4}D$

$H = \dfrac{3}{4}(3.75)$

$H = 2.8125$

$f(2.8125,\ 3.75)$

$= \sqrt{(2.8125)^2 + (3.75)^2}$

≈ 4.69

The length of the ellipse is approximately 4.69 inches, and its width is 3.75 inches.

(b) $\dfrac{H}{D} = \dfrac{2}{5}$

$H = \dfrac{2}{5}D$

$H = \dfrac{2}{5}(3.75)$

$H = 1.5$

$f(1.5,\ 3.75)$

$= \sqrt{(1.5)^2 + (3.75)^2}$

≈ 4.04

The length of the ellipse is approximately 4.04 inches, and its width is 3.75 inches.

9.2 Partial Derivatives

Your Turn 1

$$f(x, y) = 2x^2 y^3 + 6x^5 y^4$$
$$f_x(x, y) = 4xy^3 + 30x^4 y^4$$
$$f_y(x, y) = 6x^2 y^2 + 24x^5 y^3$$

Your Turn 2

$$f(x, y) = e^{3x^2 y}$$
$$f_x(x, y) = e^{3x^2 y} \cdot \frac{\partial}{\partial x}(3x^2 y)$$
$$= 6xy e^{3x^2 y}$$
$$f_y(x, y) = e^{3x^2 y} \cdot \frac{\partial}{\partial y}(3x^2 y)$$
$$= 3x^2 e^{3x^2 y}$$

Your Turn 3

$$f(x, y) = xye^{x^2+y^3}$$

Use the product rule.

$$f_x(x, y) = xy \cdot \frac{\partial}{\partial x}\left(e^{x^2+y^3}\right) + \left(e^{x^2+y^3}\right)\frac{\partial}{\partial x}(xy)$$

$$= xy\left[2xe^{x^2+y^3}\right] + \left(e^{x^2+y^3}\right)y$$

$$= (2x^2y + y)e^{x^2+y^3}$$

$$f_x(2, 1) = [2(2)^2(1) + 1]e^{2^2+1^3}$$

$$= 9e^5$$

$$f_y(x, y) = xy \cdot \frac{\partial}{\partial y}\left(e^{x^2+y^3}\right) + \left(e^{x^2+y^3}\right)\frac{\partial}{\partial y}(xy)$$

$$= xy\left[3y^2e^{x^2+y^3}\right] + \left(e^{x^2+y^3}\right)x$$

$$= (3xy^3 + x)e^{x^2+y^3}$$

$$f_y(2, 1) = [3(2)(1)^3 + 2]e^{2^2+1^3}$$

$$= 8e^5$$

Your Turn 4

$$f(x, y) = x^2e^{7y} + x^4y^5$$

$$f_x(x, y) = 2xe^{7y} + 4x^3y^5$$

$$f_y(x, y) = 7x^2e^{7y} + 5x^4y^4$$

$$f_{xx}(x, y) = \frac{\partial}{\partial x}f_x(x, y)$$

$$= \frac{\partial}{\partial x}(2xe^{7y} + 4x^3y^5)$$

$$= 2e^{7y} + 12x^2y^5$$

$$f_{yy}(x, y) = \frac{\partial}{\partial y}f_y(x, y)$$

$$= \frac{\partial}{\partial y}(7x^2e^{7y} + 5x^4y^4)$$

$$= 49x^2e^{7y} + 20x^4y^3$$

$$f_{xy}(x, y) = \frac{\partial}{\partial y}f_x(x, y)$$

$$= \frac{\partial}{\partial y}(2xe^{7y} + 4x^3y^5)$$

$$= 14xe^{7y} + 20x^3y^4$$

$$f_{yx}(x, y) = \frac{\partial}{\partial x}f_y(x, y)$$

$$= \frac{\partial}{\partial x}(7x^2e^{7y} + 5x^4y^4)$$

$$= 14xe^{7y} + 20x^3y^4$$

Note that, as in Example 6, $f_{xy}(x, y) = f_{yx}(x, y)$.

9.2 Warmup Exercises

W1. $f(x) = 2x^3 + 7x - 12$

$$f'(x) = 2\left(3x^2\right) + 7(1) - 0$$

$$= 6x^2 + 7$$

W2.

$$f(x) = \sqrt{x} + \frac{5}{x^2} + 6e^{2x} = x^{1/2} + 5x^{-2} + 6e^{2x}$$

$$f'(x) = (1/2)\left(x^{-1/2}\right) + 5\left(-2x^{-3}\right) + 6\left(2e^{2x}\right)$$

$$= \frac{1}{2\sqrt{x}} - \frac{10}{x^3} + 12e^{2x}$$

W3. $f(x) = xe^x$ Use the product rule.

$$f'(x) = \left(\frac{d}{dx}x\right)\left(e^x\right) + (x)\left(\frac{d}{dx}e^x\right)$$

$$= e^x + xe^x$$

W4. $f(x) = \frac{3x^4}{x^2 - 5}$ Use the quotient rule.

$$f'(x) = \frac{\left(12x^3\right)\left(x^2 - 5\right) - \left(3x^4\right)(2x)}{\left(x^2 - 5\right)^2}$$

$$= \frac{6x^5 - 60x^3}{\left(x^2 - 5\right)^2}$$

W5. $f(x) = \sqrt{2x + 8}$ Use the chain rule.

$$f'(x) = \left(\frac{1}{2}\right)\left(\frac{1}{\sqrt{2x + 8}}\right)\left(\frac{d}{dx}(2x + 8)\right)$$

$$= \frac{1}{\sqrt{2x + 8}}$$

W6.

$$f(x) = 9\left(e^{x^2} + 5x\right)^3 \quad \text{Use the chain rule twice.}$$

$$f'(x) = 9\left[(3)\left(e^{x^2} + 5x\right)^2\left(\frac{d}{dx}\left(e^{x^2} + 5x\right)\right)\right]$$

$$= 27\left(e^{x^2} + 5x\right)^2\left[e^{x^2}\left(\frac{d}{dx}x^2\right) + 5\right]$$

$$= 27\left(e^{x^2} + 5x\right)^2\left(2xe^{x^2} + 5\right)$$

W7. $f(x) = \ln\left|2x^3 + 5\right| \quad \text{Use the chain rule.}$

$$f'(x) = \frac{1}{2x^3 + 5}\left(\frac{d}{dx}\left|2x^3 + 5\right|\right)$$

$$= \frac{6x^2}{2x^3 + 5}$$

W8.

$$f(x) = (x + 1)\ln x^2 \quad \text{Use the product rule} \\ \text{and the chain rule.}$$

$$f'(x) = \left(\frac{d}{dx}(x + 1)\right)\left(\ln x^2\right) + (x + 1)\left(\frac{d}{dx}\left(\ln x^2\right)\right)$$

$$= \ln x^2 + (x + 1)\left[\frac{1}{x^2}\frac{d}{dx}\left(x^2\right)\right]$$

$$= \ln x^2 + \frac{(x + 1)(2x)}{x^2} = \ln x^2 + \frac{2(x + 1)}{x}$$

9.2 Exercises

1. $z = f(x, y) = 6x^2 - 4xy + 9y^2$

   (a) $\dfrac{\partial z}{\partial x} = 12x - 4y$

   (b) $\dfrac{\partial z}{\partial y} = -4x + 18y$

   (c) $\dfrac{\partial f}{\partial x}(2, 3) = 12(2) - 4(3) = 12$

   (d) $f_y(1, -2) = -4(1) + 18(-2)$

$$= -40$$

3. $f(x, y) = -4xy + 6y^3 + 5$

   $f_x(x, y) = -4y$

   $f_y(x, y) = -4x + 18y^2$

   $f_x(2, -1) = -4(-1) = 4$

   $f_y(-4, 3) = -4(-4) + 18(3)^2$

$$= 16 + 18(9)$$

$$= 178$$

5. $f(x, y) = 5x^2y^3$

   $f_x(x, y) = 10xy^3$

   $f_y(x, y) = 15x^2y^2$

   $f_x(2, -1) = 10(2)(-1)^3 = -20$

   $f_y(-4, 3) = 15(-4)^2(3)^2 = 2160$

7. $f(x, y) = e^{x+y}$

   $f_x(x, y) = e^{x+y}$

   $f_y(x, y) = e^{x+y}$

   $f_x(2, -1) = e^{2-1}$

$$= e^1 = e$$

   $f_y(-4, 3) = e^{-4+3}$

$$= e^{-1}$$

$$= \frac{1}{e}$$

9. $f(x, y) = -6e^{4x-3y}$

   $f_x(x, y) = -24e^{4x-3y}$

   $f_y(x, y) = 18e^{4x-3y}$

   $f_x(2, -1) = -24e^{4(2)-3(-1)} = -24e^{11}$

   $f_y(-4, 3) = 18e^{4(-4)-3(3)} = 18e^{-25}$

11. $f(x, y) = \dfrac{x^2 + y^3}{x^3 - y^2}$

$$f_x(x, y) = \frac{2x(x^3 - y^2) - 3x^2(x^2 + y^3)}{(x^3 - y^2)^2}$$

$$= \frac{2x^4 - 2xy^2 - 3x^4 - 3x^2y^3}{(x^3 - y^2)^2}$$

$$= \frac{-x^4 - 2xy^2 - 3x^2y^3}{(x^3 - y^2)^2}$$

$$f_y(x, y) = \frac{3y^2(x^3 - y^2) - (-2y)(x^2 + y^3)}{(x^3 - y^2)^2}$$

$$= \frac{3x^3y^2 - 3y^4 + 2x^2y + 2y^4}{(x^3 - y^2)^2}$$

$$= \frac{3x^3y^2 - y^4 + 2x^2y}{(x^3 - y^2)^2}$$

$$f_x(2, -1) = \frac{-2^4 - 2(2)(-1)^2 - 3(2^2)(-1)^3}{[2^3 - (-1)^2]^2}$$

$$= -\frac{8}{49}$$

$$f_y(-4, 3) = \frac{3(-4)^3(3)^2 - 3^4 + 2(-4)^2(3)}{[(-4)^3 - 3^2]^2}$$

$$= -\frac{1713}{5329}$$

13. $f(x, y) = \ln|1 + 5x^3y^2|$

$$f_x(x, y) = \frac{1}{1 + 5x^3y^2} \cdot 15x^2y^2 = \frac{15x^2y^2}{1 + 5x^3y^2}$$

$$f_y(x, y) = \frac{1}{1 + 5x^3y^2} \cdot 10x^3y = \frac{10x^3y}{1 + 5x^3y^2}$$

$$f_x(2, -1) = \frac{15(2)^2(-1)^2}{1 + 5(2)^3(-1)^2} = \frac{60}{41}$$

$$f_y(-4, 3) = \frac{10(-4)^3(3)}{1 + 5(-4)^3(3)^2} = \frac{1920}{2879}$$

15. $f(x, y) = xe^{x^2y}$

$$f_x(x, y) = e^{x^2y} \cdot 1 + x(2xy)(e^{x^2y})$$

$$= e^{x^2y}(1 + 2x^2y)$$

$$f_y(x, y) = x^3 e^{x^2y}$$

$$f_x(2, -1) = e^{-4}(1 - 8) = -7e^{-4}$$

$$f_y(-4, 3) = -64e^{48}$$

17. $f(x, y) = \sqrt{x^4 + 3xy + y^4 + 10}$

$$f_x(x, y) = \frac{4x^3 + 3y}{2\sqrt{x^4 + 3xy + y^4 + 10}}$$

$$f_y(x, y) = \frac{3x + 4y^3}{2\sqrt{x^4 + 3xy + y^4 + 10}}$$

$$f_x(2, -1) = \frac{4(2)^3 + 3(-1)}{2\sqrt{2^4 + 3(2)(-1) + (-1)^4 + 10}}$$

$$= \frac{29}{2\sqrt{21}}$$

$$f_y(-4, 3) = \frac{3(-4) + 4(3)^3}{2\sqrt{(-4)^4 + 3(-4)(3) + 3^4 + 10}}$$

$$= \frac{48}{\sqrt{311}}$$

19.

$$f(x, y) = \frac{3x^2y}{e^{xy} + 2}$$

$$f_x(x, y) = \frac{6xy(e^{xy} + 2) - ye^{xy}(3x^2y)}{(e^{xy} + 2)^2}$$

$$= \frac{6xy(e^{xy} + 2) - 3x^2y^2e^{xy}}{(e^{xy} + 2)^2}$$

$$f_y(x, y) = \frac{3x^2(e^{xy} + 2) - xe^{xy}(3x^2y)}{(e^{xy} + 2)^2}$$

$$= \frac{3x^2(e^{xy} + 2) - 3x^3ye^{xy}}{(e^{xy} + 2)^2}$$

$$f_x(2, -1) = \frac{6(2)(-1)(e^{2(-1)} + 2) - 3(2)^2(-1)^2e^{2(-1)}}{(e^{2(-1)} + 2)^2}$$

$$= \frac{-12e^{-2} - 24 - 12e^{-2}}{(e^{-2} + 2)^2}$$

$$= \frac{-24(e^{-2} + 1)}{(e^{-2} + 2)^2}$$

$$f_y(-4, 3) = \frac{3(-4)^2\left(e^{(-4)(3)} + 2\right) - 3(-4)^3(3)e^{(-4)(3)}}{(e^{(-4)(3)} + 2)^2}$$

$$= \frac{48e^{-12} + 96 + 576e^{-12}}{(e^{-12} + 2)^2}$$

$$= \frac{624e^{-12} + 96}{(e^{-12} + 2)^2}$$

21. $f(x, y) = 4x^2y^2 - 16x^2 + 4y$

$f_x(x, y) = 8xy^2 - 32x$

$f_y(x, y) = 8x^2y + 4$

$f_{xx}(x, y) = 8y^2 - 32$

$f_{yy}(x, y) = 8x^2$

$f_{xy}(x, y) = f_{yx}(x, y) = 16xy$

23. $R(x, y) = 4x^2 - 5xy^3 + 12y^2x^2$

$R_x(x, y) = 8x - 5y^3 + 24y^2x$

$R_y(x, y) = -15xy^2 + 24yx^2$

$R_{xx}(x, y) = 8 + 24y^2$

$R_{yy}(x, y) = -30xy + 24x^2$

$R_{xy}(x, y) = -15y^2 + 48xy$

$\qquad = R_{yx}(x, y)$

25. $r(x, y) = \dfrac{6y}{x + y}$

$r_x(x, y) = \dfrac{(x + y)(0) - 6y(1)}{(x + y)^2}$

$\qquad = -6y(x + y)^{-2}$

$r_y(x, y) = \dfrac{(x + y)(6) - 6y(1)}{(x + y)^2}$

$\qquad = 6x(x + y)^{-2}$

$r_{xx}(x, y) = -6y(-2)(x + y)^{-3}(-1)$

$\qquad = \dfrac{12y}{(x + y)^3}$

$r_{yy}(x, y) = 6x(-2)(x + y)^{-3}(1)$

$\qquad = -\dfrac{12x}{(x + y)^3}$

$r_{xy}(x, y) = r_{yx}(x, y)$

$\qquad = -6y(-2)(x + y)^{-3}(1) + (x + y)^{-2}(-6)$

$\qquad = \dfrac{12y - 6(x + y)}{(x + y)^3}$

$\qquad = \dfrac{6y - 6x}{(x + y)^3}$

27. $z = 9ye^x$

$z_x = 9ye^x$

$z_y = 9e^x$

$z_{xx} = 9ye^x$

$z_{yy} = 0$

$z_{xy} = z_{yx} = 9e^x$

29. $r = \ln|x + y|$

$r_x = \dfrac{1}{x + y}$

$r_y = \dfrac{1}{x + y}$

$r_{xx} = \dfrac{-1}{(x + y)^2}$

$r_{yy} = \dfrac{-1}{(x + y)^2}$

$r_{xy} = r_{yx} = \dfrac{-1}{(x + y)^2}$

31. $z = x \ln|xy|$

$z_x = \ln|xy| + 1$

$z_y = \dfrac{x}{y}$

$z_{xx} = \dfrac{1}{x}$

$z_{yy} = -xy^{-2} = \dfrac{-x}{y^2}$

$z_{xy} = z_{yx} = \dfrac{1}{y}$

33. $f(x, y) = 6x^2 + 6y^2 + 6xy + 36x - 5$

First, $f_x = 12x + 6y + 36$ and

$f_y = 12y + 6x.$

We must solve the system

$$12x + 6y + 36 = 0$$
$$12y + 6x = 0.$$

Multiply both sides of the first equation by -2 and add.

$$
\begin{array}{r}
-24x - 12y - 72 = 0 \\
6x + 12y \qquad\quad = 0 \\
\hline
-18x \qquad\quad - 72 = 0 \\
x = -4
\end{array}
$$

Substitute into either equation to get $y = 2$.

The solution is $x = -4, y = 2$.

35. $f(x, y) = 9xy - x^3 - y^3 - 6$

First, $f_x = 9y - 3x^2$ and $f_y = 9x - 3y^2$.

We must solve the system

$$9y - 3x^2 = 0$$
$$9x - 3y^2 = 0.$$

From the first equation, $y = \frac{1}{3}x^2$.

Substitute into the second equation to get

$$9x - 3\left(\frac{1}{3}x^2\right)^2 = 0$$

$$9x - 3\left(\frac{1}{9}x^4\right) = 0$$

$$9x - \frac{1}{3}x^4 = 0.$$

Multiply by 3 to get

$$27x - x^4 = 0.$$

Now factor.

$$x(27 - x^3) = 0$$

Set each factor equal to 0.

$$x = 0 \quad \text{or} \quad 27 - x^3 = 0$$
$$x = 3$$

Substitute into $y = \frac{x^2}{3}$.

$$y = 0 \quad \text{or} \quad y = 3$$

The solutions are $x = 0, y = 0$ and $x = 3, y = 3$.

37. $f(x, y, z) = x^4 + 2yz^2 + z^4$

$f_x(x, y, z) = 4x^3$

$f_y(x, y, z) = 2z^2$

$f_z(x, y, z) = 4yz + 4z^3$

$f_{yz}(x, y, z) = 4z$

39. $f(x, y, z) = \dfrac{6x - 5y}{4z + 5}$

$f_x(x, y, z) = \dfrac{6}{4z + 5}$

$f_y(x, y, z) = \dfrac{-5}{4z + 5}$

$f_z(x, y, z) = \dfrac{-4(6x - 5y)}{(4z + 5)^2}$

$f_{yz}(x, y, z) = \dfrac{20}{(4z + 5)^2}$

41. $f(x, y, z) = \ln|x^2 - 5xz^2 + y^4|$

$f_x(x, y, z) = \dfrac{2x - 5z^2}{x^2 - 5xz^2 + y^4}$

$f_y(x, y, z) = \dfrac{4y^3}{x^2 - 5xz^2 + y^4}$

$f_z(x, y, z) = \dfrac{-10xz}{x^2 - 5xz^2 + y^4}$

$f_{yz}(x, y, z) = \dfrac{4y^3(10zx)}{(x^2 - 5xz^2 + y^4)^2}$

$\qquad\qquad = \dfrac{40xy^3z}{(x^2 - 5xz^2 + y^4)^2}$

43. $f(x, y) = \left(x + \dfrac{y}{2}\right)^{x + y/2}$

(a) $f_x(1, 2) = \lim\limits_{h \to 0} \dfrac{f(1 + h, 2) - f(1, 2)}{h}$

We will use a small value for h. Let $h = 0.00001$.

$f_x(1, 2) \approx \dfrac{f(1.00001, 2) - f(1, 2)}{0.00001}$

$\approx \dfrac{\left(1.00001 + \frac{2}{2}\right)^{1.00001 + 2/2} - \left(1 + \frac{2}{2}\right)^{1 + 2/2}}{0.00001}$

$\approx \dfrac{2.00001^{2.00001} - 2^2}{0.00001}$

≈ 6.773

(b) $f_y(1, 2) = \lim\limits_{h \to 0} \dfrac{f(1, 2 + h) - f(1, 2)}{h}$

Again, let $h = 0.00001$.

$f_y(1, 2) \approx \dfrac{f(1, 200001) - f(1, 2)}{0.00001}$

$\approx \dfrac{\left(1 + \frac{2.00001}{2}\right)^{1+2.00001/2} - \left(1 + \frac{2}{2}\right)^{1+2/2}}{0.00001}$

$\approx \dfrac{2.000005^{2.000005} - 2^2}{0.00001}$

≈ 3.386

---

45. $M(x, y) = 45x^2 + 40y^2 - 20xy + 50$

(a) $M_y(x, y) = 80y - 20x$

$M_y(4, 2) = 80(2) - 20(4) = 80$

(b) $M_x(x, y) = 90x - 20y$

$M_x(3, 6) = 90(3) - 20(6) = 150$

(c) $\dfrac{\partial M}{\partial x}(2, 5) = 90(2) - 20(5) = 80$

(d) $\dfrac{\partial M}{\partial y}(6, 7) = 80(7) - 20(6) = 440$

---

47. $f(p, i) = 99p - 0.5pi - 0.0025p^2$

(a) $f(19, 400, 8)$

$= 99(19, 400) - 0.5(19, 400)(8)$

$- 0.0025(19, 400)^2$

$= \$902, 100$

The weekly sales are $902,100.

(b) $f_p(p, i) = 99 - 0.5i - 0.005p$, which represents the rate of change in weekly sales revenue per unit change in price when the interest rate remains constant.

$f_i(p, i) = -0.5p$, which represents the rate of change in weekly sales revenue per unit change in interest rate when the list price remains constant.

(c) $p = 19, 400$ remains constant and i changes by 1 unit from 8 to 9.

$f_i(p, i) = f_i(19, 400, 8)$

$= -0.5(19, 400)$

$= -9700$

Therefore, sales revenue declines by $9700.

---

49. $f(x, y) = \left(\dfrac{1}{4}x^{-1/4} + \dfrac{3}{4}y^{-1/4}\right)^{-4}$

(a) $f(16, 81) = \left[\dfrac{1}{4}(16)^{-1/4} + \dfrac{3}{4}(81)^{-1/4}\right]^{-4}$

$= \left(\dfrac{1}{4} \cdot \dfrac{1}{2} \cdot \dfrac{3}{4} \cdot \dfrac{1}{3}\right)^{-4}$

$= \left(\dfrac{3}{8}\right)^{-4} \approx 50.56790123$

50.57 hundred units are produced.

(b)

$f_x(x, y) = -4\left(\dfrac{1}{4}x^{-1/4} + \dfrac{3}{4}y^{-1/4}\right)^{-5}\left[\dfrac{1}{4}\left(-\dfrac{1}{4}\right)x^{-5/4}\right]$

$= \dfrac{1}{4}x^{-5/4}\left(\dfrac{1}{4}x^{-1/4} + \dfrac{3}{4}y^{-1/4}\right)^{-5}$

$f_x(16, 81) = \dfrac{1}{4}(16)^{-5/4}\left[\dfrac{1}{4}(16)^{-1/4} + \dfrac{3}{4}(81)^{-1/4}\right]^{-5}$

$= \dfrac{1}{4}\left(\dfrac{1}{32}\right)\left(\dfrac{3}{8}\right)^{-5} = \dfrac{256}{243}$

≈ 1.053497942

$f_x(16, 81) = 1.053$ hundred units and is the rate at which production is changing when labor changes by one unit (from 16 to 17) and capital remains constant.

$f_y(x, y) = -4\left(\dfrac{1}{4}x^{-1/4} + \dfrac{3}{4}y^{-1/4}\right)^{-5}\left[\dfrac{3}{4}\left(-\dfrac{1}{4}\right)y^{-5/4}\right]$

$= \dfrac{3}{4}y^{-5/4}\left(\dfrac{1}{4}x^{-1/4} + \dfrac{3}{4}y^{-1/4}\right)^{-5}$

$f_y(16, 81) = \dfrac{3}{4}(81)^{-5/4}\left[\dfrac{1}{4}(16)^{-1/4} + \dfrac{3}{4}(81)^{-1/4}\right]^{-5}$

$= \dfrac{3}{4}\left(\dfrac{1}{243}\right)\left(\dfrac{3}{8}\right)^{-5} = \dfrac{8192}{19,683}$

≈ 0.4161967180

$f_y(16, 81) = 0.4162$ hundred units and is the rate at which production is changing when capital changes by one unit (from 81 to 82) and labor remains constant.

(c)   Using the value of $f_x(16, 81)$ found in (b), production would increase by approximately 105 units.

51. $z = x^{0.4}y^{0.6}$

The marginal productivity of labor is

$$\frac{\partial z}{\partial x} = 0.4x^{-0.6}y^{0.6} + x^{0.4}\cdot 0$$

$$= 0.4x^{-0.6}y^{0.6}.$$

The marginal productivity of capital is

$$\frac{\partial z}{\partial y} = x^{0.4}(0.6y^{-0.4}) + y^{0.6}\cdot 0$$

$$= 0.6x^{0.4}y^{-0.4}.$$

53. (c) $f(x,y) = \dfrac{y(1+x)e^{-x}}{x+y}$

$$f_x(x,y) = \frac{\left[\dfrac{\partial}{\partial x}\left(y(1+x)e^{-x}\right)(x+y)\right. }{(x+y)^2}$$
$$\left. - \left(y(1+x)e^{-x}\right)\dfrac{\partial}{\partial x}(x+y)\right]$$

$$= \frac{\left[\left((y)\left(e^{-x}\right) + (y(1+x))\left(-e^{-x}\right)\right)(x+y)\right.}{(x+y)^2}$$
$$\left. - \left(y(1+x)e^{-x}\right)(1)\right]$$

$$= \frac{\left((-xy)(x+y) - (y+xy)\right)\left(e^{-x}\right)}{(x+y)^2}$$

$$= \frac{y\left(-x^2 - x - xy - 1\right)\left(e^{-x}\right)}{(x+y)^2}$$

Since x and y are positive, $f_x(x,y) < 0$.

$$f_y(x,y) = \frac{\left[\dfrac{\partial}{\partial y}\left(y(1+x)e^{-x}\right)(x+y)\right.}{(x+y)^2}$$
$$\left. - \left(y(1+x)e^{-x}\right)\dfrac{\partial}{\partial y}(x+y)\right]$$

$$= \frac{\left[\left((1+x)e^{-x}\right)(x+y)\right.}{(x+y)^2}$$
$$\left. - \left(y(1+x)e^{-x}\right)(1)\right]$$

$$= \frac{\left((1+x)(x+y) - (y+xy)\right)\left(e^{-x}\right)}{(x+y)^2}$$

$$= \frac{x(1+x)e^{-x}}{(x+y)^2}$$

Since x and y are positive, $f_y(x,y) > 0$.

55. $f(w,v) = 25.92w^{0.68} + \dfrac{3.62w^{0.75}}{v}$

(a) $f(300,10) = 25.92(300)^{0.68} + \dfrac{3.62(300)^{0.75}}{10}$

$$\approx 1279.46$$

The value is about 1279 kcal/hr.

(b) $f_w(w,v) = 25.92(0.68)w^{-.32}$

$$+ \frac{3.62(0.75)w^{-0.25}}{v}$$

$$= \frac{17.6256}{w^{0.32}} + \frac{2.715}{w^{0.25}v}$$

$$f_w(300,10) = \frac{17.6256}{(300)^{0.32}} + \frac{2.715}{(300)^{0.25}(10)}$$

$$\approx 2.906$$

The value is about 2.906 kcal/hr/g. This means the instantaneous rate of change of energy usage for a 300 kg animal traveling at 10 kilometers per hour to walk or run 1 kilometer is about 2.906 kcal/hr/g.

57. $A = 0.024265h^{0.3964}m^{0.5378}$

(a) $A_m = (0.024265)(0.5378)h^{0.3964}m^{(0.5378-1)}$

$$= 0.013050h^{0.3964}m^{-0.4622}$$

When the mass m increases from 72 to 73 while the height h remains at 180 cm, the approximate change in body surface area is

$$0.013050(180)^{0.3964}(72)^{-0.4622} \approx 0.0142$$

or about 0.0142 m^2.

(b) $A_h = (0.024265)(0.3964)h^{(0.3964-1)}m^{0.5378}$

$$= 0.0096186h^{-0.6036}m^{0.5378}$$

When the height h increases from 160 to 161 while the mass m remains at 70, the approximate change in body surface area is

$$0.0096186(160)^{-0.6036}(70)^{0.5378}$$
$$\approx 0.00442$$

or about 0.00442 m^2.

59. $f(n,c) = \dfrac{1}{8}n^2 - \dfrac{1}{5}c + \dfrac{1937}{8}$

(a) $f(4,1200) = \dfrac{1}{8}(4) - \dfrac{1}{5}(1200) + \dfrac{1937}{8}$

$$= 2 - 240 + \frac{1937}{8} = 4.125$$

The client could expect to lose 4.125 lb.

(b) $\dfrac{\partial f}{\partial n} = \dfrac{1}{8}(2n) - \dfrac{1}{5}(0) + 0 = \dfrac{1}{4}n,$

which represents the rate of change of weight loss per unit change in number of workouts.

(c) $f_n(3,1100) = \dfrac{1}{4}(3) = \dfrac{3}{4}\text{lb}$

represents an additional weight loss by adding the fourth workout.

61. $ABSI = \dfrac{w}{b^{2/3}h^{1/2}}$

(a) For $w = 0.864, b = 23.1,$ and $h = 1.85,$

$$ABSI = \dfrac{0.864}{(23.1)^{2/3}(1.85)^{1/2}} \approx 0.0783.$$

(b) $\dfrac{\partial}{\partial w} ABSI = \dfrac{1}{b^{2/3}h^{1/2}}$

For the person in part (a) this is

$$\dfrac{1}{(23.1)^{2/3}(1.85)^{1/2}} = 0.0906 \text{ per m.}$$

(c) $\dfrac{\partial}{\partial h} ABSI = -\dfrac{1}{2}\dfrac{w}{b^{2/3}h^{3/2}}$

For the person in part (a) this is

$$\left(-\dfrac{1}{2}\right)\dfrac{0.864}{(23.1)^{2/3}(1.85)^{3/2}} = -0.0212 \text{ per m.}$$

63. $R(x,t) = x^2(a-x)t^2e^{-t} = (ax^2 - x^3)t^2e^{-t}$

(a) $\dfrac{\partial R}{\partial x} = (2ax - 3x^2)t^2e^{-t}$

(b) $\dfrac{\partial R}{\partial t} = x^2(a-x) \cdot [t^2 \cdot (-e^{-t}) + e^{-t} \cdot 2t]$

$$= x^2(a-x)(-t^2 + 2t)e^{-t}$$

(c) $\dfrac{\partial^2 R}{\partial x^2} = (2a - 6x)t^2e^{-t}$

(d) $\dfrac{\partial^2 R}{\partial x \partial t} = (2ax - 3x^2)(-t^2 + 2t)e^{-t}$

(e) $\dfrac{\partial R}{\partial x}$ gives the rate of change of the reaction per unit of change in the amount of drug administered.

$\dfrac{\partial R}{\partial t}$ gives the rate of change of the reaction for a 1-hour change in the time after the drug is administered.

65. $W(V,T)$

$$= 91.4 - \dfrac{(10.45 + 6.69\sqrt{V} - 0.447V)(91.4 - T)}{22}$$

(a)

$W(20,10)$

$$= 91.4 - \dfrac{(10.45 + 6.69\sqrt{20} - 0.447(20))(91.4 - 10)}{22}$$

$$\approx -24.9$$

The wind chill is $-24.9°F$ when the wind speed is 20 mph and the temperature is $10°F$.

(b) Solve

$$-25 = 91.4 - \dfrac{(10.45 + 6.69\sqrt{V} - 0.447V)(91.4 - 5)}{22}$$

for V.

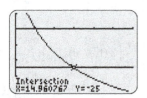

The wind speed is approximately 15 mph.

(c) $W_V = -\dfrac{1}{22}\left(\dfrac{6.69}{2\sqrt{V}} - 0.447\right)(91.4 - T)$

$W_T = -\dfrac{1}{22}(10.45 + 6.69\sqrt{V} - 0.447V)(-1)$

$$= \dfrac{1}{22}(10.45 + 6.69\sqrt{V} - 0.447V)$$

$W_V(20,10) = \dfrac{1}{22}\left(\dfrac{6.69}{2\sqrt{20}} - 0.447\right)$
$$(91.4 - 10)$$
$$\approx -1.114$$

When the temperature is held fixed at $10°\,F$, the wind chill decreases approximately 1.1 degrees when the wind velocity increases by 1 mph.

$W_T(20,10) =$

$$\dfrac{1}{22}[10.45 + 6.69\sqrt{20} - 0.447(20)] \approx 1.429$$

When the wind velocity is held fixed at 20 mph, the wind chill increases approximately $1.429°F$ when the temperature increases from $10°\,F$ to $11°\,F$.

(d) A sample table is

$T\backslash V$	5	10	15	20
30	27	16	9	4
20	16	3	−5	−11
10	6	−9	−18	−25
0	−5	−21	−32	−39

67. The rate of change in lung capacity with respect to age can be found by comparing the change in two lung capacity measurements to the difference in the respective ages when the height is held constant. So for a woman 58 inches tall, at age 20 the measured lung capacity is 1900 ml, and at age 25 the measured lung capacity is 1850 ml. So the rate of change in lung capacity with respect to age is

$$\frac{1900 - 1850}{20 - 25} = \frac{50}{-5}$$
$$= -10 \text{ ml per year.}$$

The rate of change in lung capacity with respect to height can be found by comparing the change in two lung capacity measurements to the difference in the respective heights when the age is held constant. So for a 20-year old woman the measured lung capacity for a woman 58 inches tall is 1900 ml and the measured lung capacity for a woman 60 inches tall is 2100 ml. So the rate of change in lung capacity with respect to height is

$$\frac{1900 - 2100}{58 - 60} = \frac{-200}{-2}$$
$$= 100 \text{ ml per in.}$$

The two rates of change remain constant throughout the table.

69. $F = \dfrac{mgR^2}{r^2} = mgR^2 r^{-2}$

(a) $F_m = \dfrac{gR^2}{r^2}$ is the approximate rate of change in gravitational force per unit change in mass while distance is held constant.

$F_r = \dfrac{-2mgR^2}{r^3}$ is the approximate rate of change in gravitational force per unit change in distance while mass is held constant.

(b) $F_m = \dfrac{gR^2}{r^2}$, where all quantities are positive. Therefore, $F_m > 0$.

$F_r = \dfrac{-2mgR^2}{r^3}$, where m, g, R^2, and r^3 are positive.

Therefore, $F_r < 0$.

These results are reasonable since gravitational force increases when mass increases (m is in the numerator) and gravitational force decreases when distance increases (r is in the denominator).

71. $T = (s, w) = 105 + 265 \log_2\left(\dfrac{2s}{w}\right)$

(a) $T(3, 0.5) = 105 + 265 \log_2\left[\dfrac{2(3)}{0.5}\right]$
$$= 105 + 265 \log_2 12$$
$$\approx 1055$$

(b) $T(s, w) = 105 + 265 \dfrac{\ln\left(\frac{2s}{w}\right)}{\ln 2}$
$$= 105 + \frac{265}{\ln 2}[\ln(2s) - \ln(w)]$$

$T_s(s, w) = \dfrac{265}{\ln 2}\left(\dfrac{1}{s}\right)$

$T_w(s, w) = -\dfrac{265}{\ln 2}\left(\dfrac{1}{w}\right)$

$T_s(3, 0.5) = \dfrac{265}{3\ln 2} \approx 127.4$ msec/ft

If the distance the object is being moved increases from 3 feet to 4 feet, while keeping w fixed at 0.5 foot, the time to move the object increases by approximately 127.4 msec.

$T_w(3, 0.5) = -\dfrac{265}{0.5 \ln 2}$
$$\approx -764.5 \text{ msec/ft}$$

It the width of the target area is increased by 1 foot, while keeping the distance fixed at 3 feet, the movement time decreases by approximately 764.5 msec.

9.3 Maxima and Minima

Your Turn 1

$$f(x, y) = 4x^3 + 3xy + 4y^3$$
$$f_x(x, y) = 12x^2 + 3y, \ f_y(x, y) = 3x + 12y^2$$

$$12x^2 + 3y = 0$$
$$3x + 12y^2 = 0$$

Solve this system by substitution. From the first equation,

$$y = -4x^2.$$

Substituting for y in the second equation gives

$$3x + 12\left(-4x^2\right)^2 = 0.$$

$$x + 64x^4 = 0$$
$$x(1 + 64x^3) = 0$$

$$x = 0 \quad \text{or} \quad 1 + 64x^3 = 0$$

$$x = 0 \quad \text{or} \quad x = -\frac{1}{4}$$

If $x = 0$, $y = -4(0)^2 = 0$.

If $x = -\frac{1}{4}$, $y = -4\left(-\frac{1}{4}\right)^2 = -\frac{1}{4}$.

So the critical points are

$$(0,0) \quad \text{and} \quad \left(-\frac{1}{4}, -\frac{1}{4}\right).$$

Your Turn 2

Use the information from Your Turn 1:

$$f(x, y) = 4x^3 + 3xy + 4y^3$$
$$f_x(x, y) = 12x^2 + 3y$$
$$f_y(x, y) = 3x + 12y^2$$

The critical points are $(0, 0)$ and $(-1/4, -1/4)$.

Now compute the second partial derivatives.

$$f_{xx}(x, y) = 24x, \; f_{yy}(x, y) = 24y, \; f_{xy}(x, y) = 3$$

$$D(a,b) = f_{xx}(a,b) \cdot f_{yy}(a,b) - [f_{xy}(a,b)]^2$$
$$= (24a)(24b) - 9$$
$$= 576ab - 9$$

At the critical point $(0, 0)$,

$D(0,0) = 576(0)(0) - 9 = -9 < 0$, so this critical point is a saddle point.

At the critical point $(-1/4, -1/4)$,

$$D\left(-\frac{1}{4}, -\frac{1}{4}\right) = 576\left(-\frac{1}{4}\right)\left(-\frac{1}{4}\right) - 9$$
$$= 36 - 9$$
$$= 24$$

and

$$f_{xx}\left(-\frac{1}{4}, -\frac{1}{4}\right) = 24\left(-\frac{1}{4}\right) = -\frac{1}{6}.$$

Since $D > 0$ and $f_{xx} < 0$, there is a relative maximum at $(-1/4, -1/4)$.

9.3 Warmup Exercises

W1. $f(x) = x^3 - 3x^2 - 24x + 5$
$$f'(x) = 3x^2 - 6x - 24$$
$$= 3(x^2 - 2x - 8)$$
$$f''(x) = 6x - 6$$
Find the critical points:

$$f'(x) = 0$$
$$3(x^2 - 2x - 8) = 0$$
$$(x - 4)(x + 2) = 0$$

The critical values are 4 and -2. The second derivative is positive at 4 so there is a relative minimum at $x = 4$. The second derivative is negative at -2, so there is a relative maximum at $x = -2$.
$$f(4) = -75, f(-2) = 33.$$
Thus there is a relative minimum at $(4, -75)$ and a relative maximum at $(-2, 33)$.

W2. $f(x) = x^4 - 8x^2 + 1$
$$f'(x) = 4x^3 - 16x$$
$$= 4x(x^2 - 4)$$
$$f''(x) = 12x^2 - 16$$

Find the critical points:
$$f'(x) = 0$$
$$4x(x^2 - 4) = 0$$
$$4x(x - 2)(x + 2) = 0$$
The critical numbers are $-2, 0$, and 2. The second derivative is negative at 0, so there is a relative maximum for $x = 0$. The second derivative is positive at -2 and 0, so there are relative minima at these values.
$$f(0) = 1, f(2) = f(-2) = -15.$$
Thus there is a relative maximum at $(0, 1)$ and relative minima at $(-2, -15)$ and $(2, -15)$.

9.3 Exercises

1. $f(x, y) = xy + y - 2x$

   $f_x(x, y) = y - 2, f_y(x, y) = x + 1$

   If $f_x(x, y) = 0, y = 2.$

   If $f_y(x, y) = 0, x = -1.$

   Therefore, $(-1, 2)$ is the critical point.

   $$f_{xx}(x, y) = 0$$
   $$f_{yy}(x, y) = 0$$
   $$f_{xy}(x, y) = 1$$

   For $(-1, 2)$,

   $$D = 0 \cdot 0 - 1^2 = -1 < 0.$$

   A saddle point is at $(-1, 2)$.

3. $f(x, y) = 3x^2 - 4xy + 2y^2 + 6x - 10$

   $f_x(x, y) = 6x - 4y + 6$

   $f_y(x, y) = -4x + 4y$

   Solve the system $f_x(x, y) = 0, f_y(x, y) = 0.$

   $$\begin{array}{r} 6x - 4y + 6 = 0 \\ \underline{-4x + 4y \qquad = 0} \\ 2x \qquad + 6 = 0 \\ x = -3 \\ -4(-3) + 4y = 0 \\ y = -3 \end{array}$$

   Therefore, $(-3, -3)$ is a critical point.

   $$f_{xx}(x, y) = 6$$
   $$f_{yy}(x, y) = 4$$
   $$f_{xy}(x, y) = -4$$
   $$D = 6 \cdot 4 - (-4)^2 = 8 > 0$$

   Since $f_{xx}(x, y) = 6 > 0$, there is a relative minimum at $(-3, -3)$.

5. $f(x, y) = x^2 - xy + y^2 + 2x + 2y + 6$

   $f_x(x, y) = 2x - y + 2,$

   $f_y(x, y) = -x + 2y + 2$

   Solve the system $f_x(x, y) = 0, f_y(x, y) = 0.$

   $$\begin{array}{r} 2x - y + 2 = 0 \\ \underline{-x + 2y + 2 = 0} \\ 2x - y + 2 = 0 \\ \underline{-2x + 4y + 4 = 0} \\ 3y + 6 = 0 \\ y = -2 \\ -x + 2(-2) + 2 = 0 \\ x = -2 \end{array}$$

   $(-2, -2)$ is the critical point.

   $$f_{xx}(x, y) = 2$$
   $$f_{yy}(x, y) = 2$$
   $$f_{xy}(x, y) = -1$$

   For $(-2, -2)$,

   $$D = (2)(2) - (-1)^2 = 3 > 0.$$

   Since $f_{xx}(x, y) > 0$, a relative minimum is at $(-2, -2)$.

7. $f(x, y) = x^2 + 3xy + 3y^2 - 6x + 3y$

   $f_x(x, y) = 2x + 3y - 6,$

   $f_y(x, y) = 3x + 6y + 3$

   Solve the system $f_x(x, y) = 0, f_y(x, y) = 0.$

   $$\begin{array}{r} 2x + 3y - 6 = 0 \\ 3x + 6y + 3 = 0 \\ -4x - 6y + 12 = 0 \\ \underline{3x + 6y + 3 = 0} \\ -x + 15 = 0 \\ x = 15 \\ 3(15) + 6y + 3 = 0 \\ 6y = -48 \\ y = -8 \end{array}$$

   $(15, -8)$ is the critical point.

   $$f_{xx}(x, y) = 2$$
   $$f_{yy}(x, y) = 6$$
   $$f_{xy}(x, y) = 3$$

   For $(15, -8)$,

   $$D = 2 \cdot 6 - 9 = 3 > 0.$$

   Since $f_{xx}(x, y) > 0$, a relative minimum is at $(15, -8)$.

9. $f(x, y) = 4xy - 10x^2$
$\qquad - 4y^2 + 8x + 8y + 9$
$f_x(x, y) = 4y - 20x + 8$
$f_y(x, y) = 4x - 8y + 8$

$$4y - 20x + 8 = 0$$
$$4x - 8y + 8 = 0$$

$$\begin{aligned} 4y - 20x + 8 &= 0 \\ \underline{-4y + 2x + 4 = 0} \\ -18x + 12 &= 0 \end{aligned}$$
$$x = \frac{2}{3}$$

$$4y - 20\left(\frac{2}{3}\right) + 8 = 0$$

The critical point is $\left(\frac{2}{3}, \frac{4}{3}\right)$.

$$f_{xx}(x, y) = -20$$
$$f_{yy}(x, y) = -8$$
$$f_{xy}(x, y) = 4$$

For $\left(\frac{2}{3}, \frac{4}{3}\right)$,

$$D = (-20)(-8) - 16 = 144 > 0.$$

Since $f_{xx}(x, y) < 0$, a relative maximum is at $\left(\frac{2}{3}, \frac{4}{3}\right)$.

11. $f(x, y) = x^2 + xy - 2x - 2y + 2$
$f_x(x, y) = 2x + y - 2$
$f_y(x, y) = x - 2$

$$\begin{aligned} 2x + y - 2 &= 0 \\ x \qquad\quad - 2 &= 0 \\ x &= 2 \\ 2(2) + y - 2 &= 0 \\ y &= -2 \end{aligned}$$

The critical point is $(2, -2)$.

$$f_{xx}(x, y) = 2$$
$$f_{yy}(x, y) = 0$$
$$f_{xy}(x, y) = 1$$

For $(2, -2)$,

$$D = 2 \cdot 0 - 1^2 = -1 < 0.$$

A saddle point is at $(2, -2)$.

13. $f(x, y) = 3x^2 + 2y^3 - 18xy + 42$
$f_x(x, y) = 6x - 18y$
$f_y(x, y) = 6y^2 - 18x$

If $f_x(x, y) = 0$, $6x - 18y = 0$, or $x = 3y$.
Substitute $3y$ for x in $f_y(x, y) = 0$ and solve for y.

$$6y^2 - 18(3y) = 0$$
$$6y(y - 9) = 0$$

$$y = 0 \quad \text{or} \quad y = 9$$
Then $\qquad\qquad x = 0 \quad \text{or} \quad x = 27.$
Therefore, $(0, 0)$ and $(27, 9)$ are critical points.

$$f_{xx}(x, y) = 6$$
$$f_{yy}(x, y) = 12y$$
$$f_{xy}(x, y) = -18$$

For $(0, 0)$,

$$D = 6 \cdot 12(0) - (-18)^2 = -324 < 0.$$

There is a saddle point at $(0, 0)$.
For $(27, 9)$,

$$D = 6 \cdot 12(9) - (-18)^2 = 324 > 0.$$

Since $f_{xx}(x, y) = 6 > 0$, there is a relative minimum at $(27, 9)$.

15. $f(x, y) = x^2 + 4y^3 - 6xy - 1$
$f_x(x, y) = 2x - 6y$, $f_y(x, y) = 12y^2 - 6x$
Solve $f_x(x, y) = 0$ for x.

$$2x + 6 = 0$$
$$x = 3y$$

Substitute for x in $12y^2 - 6x = 0$.

$$12y^2 - 6(3y) = 0$$
$$6y(2y - 3) = 0$$

$$y = 0 \quad \text{or} \quad y = \frac{3}{2}$$
Then $\qquad\qquad x = 0 \quad \text{or} \quad x = \frac{9}{2}.$

The critical points are $(0, 0)$ and $\left(\frac{9}{2}, \frac{3}{2}\right)$.

$$f_{xx}(x, y) = 2$$
$$f_{yy}(x, y) = 24y$$
$$f_{xy}(x, y) = -6$$

For $(0, 0)$,

$$D = 2 \cdot 24(0) - (-6)^2$$
$$= -36 < 0.$$

A saddle point is at $(0, 0)$.

For $\left(\frac{9}{2}, \frac{3}{2}\right)$,

$$D = 2 \cdot 24\left(\frac{3}{2}\right) - (-6)^2$$
$$= 36 > 0.$$

Since $f_{xx}(x, y) > 0$, a relative minimum is at $\left(\frac{9}{2}, \frac{3}{2}\right)$.

17. $f(x, y) = e^{x(y+1)}$

$$f_x(x, y) = (y + 1)e^{x(y+1)}$$
$$f_y(x, y) = xe^{x(y+1)}$$

If $f_x(x, y) = 0$
$$(y + 1)e^{x(y+1)} = 0$$
$$y + 1 = 0$$
$$y = -1.$$

If $f_y(x, y) = 0$
$$xe^{x(y+1)} = 0$$
$$x = 0.$$

Therefore, $(0, -1)$ is a critical point.

$$f_{xx}(x, y) = (y + 1)^2 e^{x(y+1)}$$
$$f_{yy}(x, y) = x^2 e^{x(y+1)}$$
$$f_{xy}(x, y) = (y + 1)e^{x(y+1)} \cdot x + e^{x(y+1)} \cdot 1$$
$$= (xy + x + 1)e^{x(y+1)}$$

For $(0, -1)$,

$$f_{xx}(0, -1) = (0)^2 e^0 = 0$$
$$f_{yy}(0, -1) = (0)^2 e^0 = 0$$
$$f_{xy}(0, -1) = (0 + 0 + 1)e^0 = 1$$
$$D = 0 \cdot 0 - 1^2 = -1 < 0$$

There is a saddle point at $(0, -1)$.

21. $z = -3xy + x^3 - y^3 + \dfrac{1}{8}$

$$f_x(x, y) = -3y + 3x^2, \quad f_y(x, y) = -3x - 3y^2$$

Solve the system $f_x = 0$, $f_y = 0$.

$$-3y + 3x^2 = 0$$
$$-3x - 3y^2 = 0$$
$$-y + x^2 = 0$$
$$-x - y^2 = 0$$

Solve the first equation for y, substitute into the second, and solve for x.

$$y = x^2$$
$$-x - x^4 = 0$$
$$x(1 + x^3) = 0$$
$$x = 0 \quad \text{or} \quad x = -1$$

Then $y = 0 \quad \text{or} \quad y = 1.$

The critical points are $(0, 0)$ and $(-1, 1)$.

$$f_{xx}(x, y) = 6x$$
$$f_{yy}(x, y) = -6y$$
$$f_{xy}(x, y) = -3$$

For $(0, 0)$,

$$D = 0 \cdot 0 - (-3)^2 = -9 < 0.$$

A saddle point is at $(0, 0)$.

For $(-1, 1)$,

$$D = -6(-6) - (-3)^2 = 27 > 0.$$

$$f_{xx}(x, y) = 6(-1) = -6 < 0.$$

$$f(-1, 1) = -3(-1)(1) + (-1)^3 - 1^3 + \frac{1}{8}$$
$$= \frac{9}{8}$$

A relative maximum of $\frac{9}{8}$ is at $(-1, 1)$.

The equation matches graph (a).

23. $z = y^4 - 2y^2 + x^2 - \dfrac{17}{16}$

$$f_x(x, y) = 2x, \quad f_y(x, y) = 4y^3 - 4y$$

Solve the system $f_x = 0$, $f_y = 0$.

$$2x = 0 \qquad (1)$$
$$4y^3 - 4y = 0 \qquad (2)$$
$$4y(y^2 - 1) = 0$$
$$4y(y + 1)(y - 1) = 0$$

Equation (1) gives $x = 0$ and equation (2) gives $y = 0$, $y = -1$, or $y = 1$.

The critical points are $(0, 0)$, $(0, -1)$ and $(0, 1)$.

$$f_{xx}(x, y) = 2,$$
$$f_{yy}(x, y) = 12y^2 - 4,$$
$$f_{xy}(x, y) = 0$$

For $(0, 0)$,

$$D = 2(12 \cdot 0^2 - 4) - 0 = -8 < 0.$$

A saddle point is at $(0, 0)$.

For $(0, -1)$,

$$D = 2[12(-1)^2 - 4] - 0 = 16 > 0.$$

$$f_{xx}(x, y) = 2 > 0$$

$$f(0, -1) = (-1)^4 - 2(-1)^2 + 0^2 - \frac{17}{16}$$
$$= -2\frac{1}{16}$$

A relative minimum of $-\dfrac{33}{16}$ is at $(0, -1)$.

For $(0, 1)$,

$$D = 2(12 \cdot 1^2 - 4) - 0 = 16 > 0$$

$$f_{xx}(x, y) = 2 > 0$$

$$f(0, 1) = 1^4 - 2 \cdot 1^2 + 0^2 - \frac{17}{16}$$
$$= -\frac{33}{16}$$

A relative minimum of $-\frac{33}{16}$ is at $(0, 1)$.

The equation matches graph (b).

25. $z = -x^4 + y^4 + 2x^2 - 2y^2 + \dfrac{1}{16}$

$f_x(x, y) = -4x^3 + 4x$, $f_y(x, y) = 4y^3 - 4y$

Solve $f_x(x, y) = 0$, $f_y(x, y) = 0$.

$$-4x^3 + 4x = 0 \quad (1)$$
$$4y^3 - 4y = 0 \quad (2)$$
$$-4x(x^2 - 1) = 0 \quad (1)$$
$$-4x(x + 1)(x - 1) = 0$$
$$4y(y^2 - 1) = 0 \quad (2)$$
$$4y(y + 1)(y - 1) = 0$$

Equation (1) gives $x = 0$, -1, or 1.

Equation (2) gives $y = 0$, -1, or 1.

Critical points are $(0, 0)$, $(0, -1)$, $(0, 1)$, $(-1, 0)$, $(-1, -1)$, $(-1, 1)$, $(1, 0)$, $(1, -1)$, $(1, 1)$.

$$f_{xx}(x, y) = -12x^2 + 4,$$
$$f_{yy}(x, y) = 12y^2 - 4$$
$$f_{xy}(x, y) = 0$$

For $(0, 0)$,

$$D = 4(-4) - 0 = -16 < 0.$$

For $(0, -1)$,

$$D = 4(8) - 0 = 32 > 0,$$

and $f_{xx}(x, y) = 4 > 0$.

$$f(0, -1) = -\frac{15}{16}$$

For $(0, 1)$,

$$D = 4(8) - 0 = 32 > 0,$$

and $f_{xx}(x, y) = 4 > 0$.

$$f(0, 1) = -\frac{15}{16}$$

For $(-1, 0)$,

$$D = -8(-4) - 0 = 32 > 0,$$

and $f_{xx}(x, y) = -8 < 0$.

$$f(-1, 0) = \frac{17}{16}$$

For $(-1, -1)$,

$$D = -8(8) - 0 = -64 < 0.$$

For $(-1, 1)$,

$$D = -8(8) - 0 = -64 < 0.$$

For $(1, 0)$,

$$D = -8(-4) = 32 > 0,$$

and $f_{xx}(x, y) = -8 < 0.$

$$f(1, 0) = 1\frac{1}{16}$$

For $(1, -1)$,

$$D = -8(8) - 0 = -64 < 0.$$

For $(1, 1)$,

$$D = -8(8) - 0 = -64 < 0.$$

Saddle points are at $(0, 0)$, $(-1, -1)$, $(-1, 1)$, $(1, -1)$, and $(1, 1)$.

Relative maximum of $\frac{17}{16}$ is at $(-1, 0)$ and $(1, 0)$.

Relative minimum of $-\frac{15}{16}$ is at $(0, -1)$ and $(0, 1)$.

The equation matches graph (e).

27. $f(x, y) = 1 - x^4 - y^4$

$f_x(x, y) = -4x^3, f_y(x, y) = -4y^3$

The system

$f_x(x, y) = -4x^3 = 0, f_y(x, y) = -4y^3 = 0$

gives the critical point $(0, 0)$.

$$f_{xx}(x, y) = -12x^2$$
$$f_{yy}(x, y) = -12y^3$$
$$f_{xy}(x, y) = 0$$

For $(0, 0)$,

$$D = 0 \cdot 0 - 0^2 = 0.$$

Therefore, the test gives no information. Examine a graph of the function drawn by using level curves.

If $f(x, y) = 1$, then $x^4 + y^4 = 0$. The level curve is the point $(0, 0, 1)$.

If $f(x, y) = 0$, then $x^4 + y^4 = 1$. The level curve is the circle with center $(0, 0, 0)$ and radius 1.

If $f(x, y) = -15$, then $x^4 + y^4 = 16$. The level curve is the curve with center $(0, 0, -15)$ and radius 2.

The xz-trace is

$$z = 1 - x^4.$$

This curve has a maximum at $(0, 0, 1)$ and opens downward.

The yz-trace is

$$z = 1 - y^4.$$

This curve also has a maximum at $(0, 0, 1)$ and opens downward.

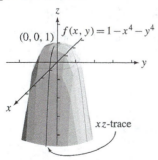

If $f(x, y) > 1$, then $x^4 + y^4 < 0$, which is impossible, so the function does not exist. Thus, the function has a relative maximum of 1 at $(0, 0)$.

31. $f(x, y) = x^2(y + 1)^2 + k(x + 1)^2 y^2$

(a) $f_x(x, y) = 2x + 2ky^2(x + 1)$

$f_y(x, y) = 2x^2(y + 1) + 2k(x + 1)^2 y$

$f_x(0, 0) = 2(0) + 2k(0)^2(0 + 1)$

$\qquad = 0$

$f_y(0, 0) = 2(0)^2(0 + 1) + 2k(0 + 1)^2(0)$

$\qquad = 0$

Thus, $(0, 0)$ is a critical point for all values of k.

(b) $f_{xx}(x, y) = 2 + 2ky^2$

$f_{yy}(x, y) = 2x^2 + 2k(x + 1)^2$

$f_{xy}(x, y) = 4ky(x + 1)$

$f_{xx}(0, 0) = 2 + 2k(0)^2 = 2$

$f_{yy}(0, 0) = 2(0)^2 = 2k(0 + 1)^2 = 2k$

$f_{xy}(0, 0) = 4k(0)(0 + 1) = 0$

$$D = 2 \cdot 2k - 0^2 = 4k$$

$(0, 0)$ is a relative minimum when $4k > 0$, hence when $k > 0$. When $k = 0$, $D = 0$ so the test for relative extrema gives no information. But if $k = 0$, $f(xy) = x^2(y + 1)^2$, which is always greater than or equal to $f(0, 0) = 0$. So $(0, 0)$ is a relative minimum for $k \geq 0$.

35. $L(x, y) = \frac{3}{2}x^2 + y^2 - 2x - 2y - 2xy + 68$,

where x is the number of skilled hours and y is the number of semiskilled hours.

$$L_x(x, y) = 3x - 2 - 2y,$$
$$L_y(x, y) = 2y - 2 - 2x$$

$$3x - 2 - 2y = 0$$
$$\underline{-2x - 2 + 2y = 0}$$
$$x - 4 \qquad\quad = 0$$
$$x = 4$$

$$-2(4) - 2 + 2y = 0$$
$$2y = 10$$
$$y = 5$$

Let $L_{xx}(x, y) = 3$, $L_{yy}(x, y) = 2$,
$L_{xy}(x, y) = -2$.

$$D = 3(2) - (-2)^2 = 2 > 0 \text{ and } L_{xx}(x, y) > 0.$$

Relative minimum at (4, 5) is

$$L(4, 5) = \frac{3}{2}(4)^2 + (5)^2 - 2(4) - 2(5)$$
$$- 2(4)(5) + 68$$
$$= 59.$$

So $59 is a minimum cost, when $x = 4$ and
$y = 5$.

37. $R(x, y) = 15 + 169x + 182y$
$$- 5x^2 - 7y^2 - 7xy$$
$R_x(x, y) = 169 - 10x - 7y$
$R_y(x, y) = 182 - 14y - 7x$

Solve the system $R_x = 0, R_y = 0$.

$$-10x - 7y + 169 = 0$$
$$-7x - 14y + 182 = 0$$

$$20x + 14y - 338 = 0$$
$$\underline{-7x - 14y + 182 = 0}$$
$$13x \qquad\quad - 156 = 0$$
$$x = 12$$

$$-10(12) - 7y + 169 = 0$$
$$-7y = -49$$
$$y = 7$$

(12, 7) is a critical point.

$$R_{xx} = -10$$
$$R_{yy} = -14$$
$$R_{xy} = -7$$

$$D = (-10)(-14) - (-7)^2 = 91 > 0$$

Since $R_{xx} = -10 < 0$, there is a relative
maximum at (12, 7).

$$R(12, 7) = 15 + 169(12) + 182(7) - 5(12)^2$$
$$- 7(7)^2 - 7(12)(7)$$
$$= 1666 \text{ (hundred dollars)}$$

12 spas and 7 solar heaters should be sold to
produce a maximum revenue of $166,600.

39. $T(x, y) = x^4 + 16y^4 - 32xy + 40$
$T_x(x, y) = 4x^3 - 32y$
$T_y(x, y) = 64y^3 - 32x$
$T_x(x, y) = 0$

$$4x^3 - 32y = 0$$
$$4x^3 = 32y$$
$$\frac{1}{8}x^3 = y$$

$$T_y(x, y) = 0$$
$$64y^3 - 32x = 0$$
$$64y^3 = 32x$$
$$2y^3 = x$$

Use substitution to solve the system of equations

$$\frac{1}{8}x^3 = y$$
$$2y^3 = x$$

$$y = \frac{1}{8}(2y^3)^3$$
$$y = \frac{1}{8}(8)y^9$$
$$y = y^9$$
$$y^9 - y = 0$$
$$y(y^8 - 1) = 0$$

$$y = 0 \quad \text{or} \quad y^8 - 1 = 0$$
$$y = 0 \quad \text{or} \qquad y^8 = 1$$
$$y = 0 \quad \text{or} \qquad y = 1$$

If $y = 0$, $x = 2(0)^3 = 0$.

If $y = 1$, $x = 2(1)^3 = 2$.

The critical points are (0, 0) and (2, 1).

$T_{xx}(x, y) = 12x^2$

$T_{yy}(x, y) = 192y^2$

$T_{xy}(x, y) = -32$

$T_{xx}(0,0) = 0$

$T_{yy}(0,0) = 0$

$T_{xy}(0,0) = -32$

$$D = 0 \cdot 0 - (-32)^2 = -1024$$

Since $D < 0$, there is a saddle point at $(0, 0)$.

$T_{xx}(2,1) = 48$

$T_{yy}(2,1) = 192$

$T_{xy}(2,1) = -32$

$$D = 48 \cdot 192 - (-32)^2 = 8192$$

Since $D > 0$ and $T_{xx} > 0$, there is a relative minimum at $(2, 1)$.

$$T(2,1) = 2^4 + 16(1)^4 - 32(2)(1) + 40$$
$$= 16 + 16 - 64 + 40$$
$$= 8$$

Spend $2000 on quality control and $1000 on consulting, for a minimum time of 8 hours.

41. Using the procedure suggested, take the natural logarithm of the number-of-transistors data.

Time in years since 1985, t	Natural log of number of transistors in millions w	
0	ln(0.275)	$\approx$ -1.291
4	ln(1.2)	$\approx$ 0.182
8	ln(3.1)	$\approx$ 1.131
12	ln(7.5)	$\approx$ 2.015
14	ln(9.5)	$\approx$ 2.251
15	ln(42)	$\approx$ 3.738
20	ln(291)	$\approx$ 5.673
22	ln(820)	$\approx$ 6.709
24	ln(1900)	$\approx$ 7.550

(a) A linear fit of the data in the table above gives

$$w = r + st$$
$$w = -1.722 + 0.3653t$$

where w is the natural logarithm of the number of transistors in millions and t is the number of years since 1985. To convert this back to an exponential fit for the original data, compute

$$a = e^r = e^{-1.722} = 0.1787$$

and

$$b = e^s = e^{0.3652} = 1.441.$$

Thus an exponential fit to the original data has the form $y = a \cdot b^t$ with $a = 0.1787$ and $b = 1.441$:

$$y = 0.1787(1.441)^t$$

(b) Same as (a).

(c) Same as (a).

9.4 Lagrange Multipliers

Your Turn 1

Find the minimum value of $f(x, y) = x^2 + 2x + 9y^2 + 3y + 6xy$ subject to the constraint $2x + 3y = 12$.

Step 1 Rewrite the constraint as
$g(x, y) = 2x + 3y - 12$.

Step 2 Form the Lagrange function.

$$F(x, y, \lambda) = f(x, y) - \lambda \cdot g(x, y)$$
$$= x^2 + 2x + 9y^2 + 3y$$
$$\quad + 6xy - \lambda(2x + 3y - 12)$$
$$= x^2 + 9y^2 + 6xy - 2x\lambda$$
$$\quad - 3y\lambda + 2x + 3y + 12\lambda$$

Step 3 Find the first partial derivatives of F.

$$F_x(x, y, \lambda) = 2x + 6y - 2\lambda + 2$$
$$F_y(x, y, \lambda) = 6x + 18y - 3\lambda + 3$$
$$F_\lambda(x, y, \lambda) = -2x - 3y + 12$$

Step 4 Form the system of equations

$F_x(x, y, \lambda) = 0,\ F_y(x, y, \lambda) = 0,\ F_\lambda(x, y, \lambda) = 0.$

$$2x + 6y - 2\lambda + 2 = 0$$
$$6x + 18y - 3\lambda + 3 = 0$$
$$-2x - 3y + 12 = 0$$

Step 5 Solve the system of equations from Step 4.

First solve the first two equations for λ and then set these results equal to each other (remove the common factors of 2 from the first equation and 3 from the second equation).

$$x + 3y + 1 = \lambda$$
$$2x + 6y + 1 = \lambda$$

$$x + 3y = 2x + 6y$$
$$x + 3y = 0$$
$$x = -3y$$

Now substitute for x in the last equation.

$$-2(-3y) - 3y + 12 = 0$$
$$6y - 3y = -12$$
$$3y = -12$$
$$y = -4$$

Since $x = -3y$ we have $x = 12$ and $y = -4$. So a candidate for a minimum value of the function f will be $f(12, -4) = 12$.

To see if this is really a minimum we can evaluate f at a few nearby points that satisfy the constraint $2x + 3y = 12$. For example, let $x = 11$. Then $2(11) + 3y = 12$, so $y = -10/3$. $f(11, -10/3) = 13$, which is larger than our candidate. We conclude that the minimum value of f subject to the given constraint is 12.

Your Turn 2

As in Example 3, let x, y, and z be the dimensions of the box. If the front and top are missing, the surface area is $2xy + xz + yz$ so the constraint is $2xy + xz + yz = 6$. As in Example 3, the volume is xyz. The problem is thus to maximize $f(x, y, z) = xyz$ subject to $2xy + xz + yz = 6$.

Step 1   Rewrite the constraint as

$$g(x, y, z) = 2xy + xz + yz - 6.$$

Step 2   Form the Lagrange function.

$$F(x, y, z, \lambda) = f(x, y, z) - \lambda \cdot g(x, y, z)$$
$$= xyz - \lambda(2xy + xz + yz - 6)$$
$$= xyz - 2xy\lambda - xz\lambda - yz\lambda + 6\lambda$$

Step 3   Find the first partial derivatives of F.

$$F_x(x, y, z, \lambda) = yz - 2y\lambda - z\lambda$$
$$F_y(x, y, z, \lambda) = xz - 2x\lambda - z\lambda$$
$$F_z(x, y, z, \lambda) = xy - x\lambda - y\lambda$$
$$F_\lambda(x, y, z, \lambda) = -2xy - xz - yz + 6$$

Step 4   Form the system of equations

$$F_x(x, y, z, \lambda) = 0,$$
$$F_y(x, y, z, \lambda) = 0,$$
$$F_z(x, y, z, \lambda) = 0,$$
$$F_\lambda(x, y, z, \lambda) = 0.$$

$$yz - 2y\lambda - z\lambda = 0$$
$$xz - 2x\lambda - z\lambda = 0$$
$$xy - x\lambda - y\lambda = 0$$
$$-2xy - xz - yz + 6 = 0$$

Step 5   Solve the system of equations from Step 4. First solve each of the first three equations for λ. This gives three expressions for λ.

$$\lambda = \frac{yz}{2y + z}, \quad \lambda = \frac{xz}{2x + z}, \quad \lambda = \frac{xy}{x + y}$$

Set the second and third expressions equal.

$$\frac{xz}{2x + z} = \frac{xy}{x + y}$$
$$x^2z + xyz = 2x^2y + xyz$$
$$x^2z = x^2(2y)$$

Since x is not zero, this shows that $z = 2y$.

Exactly the same calculation using the first and third expressions will show that $z = 2x$. Thus $x = y$. Now use the fourth equation:

$$-2xy - xz - yz + 6 = 0$$
$$-2x^2 - x(2x) - x(2x) + 6 = 0$$
$$6x^2 = 6$$
$$x = 1 \text{ (since x is positive)}$$

Now we know that $x = 1$, $y = 1$, and $z = 2$. The box should be 2 ft. wide and 1 ft. high and long.

9.4 Warmup Exercises

W1. $f(x, y) = 4x^3y^2 - 6x^4 - 3y^5$

$\qquad f_x(x, y) = 12x^2y^2 - 24x^3$

$\qquad f_y(x, y) = 8x^3y - 15y^4$

W2.

$f(x, y, z) = x^2y^3z^5 - 4x^2y^5 + 3xz^6 - 2y^8z^7$

$f_x(x, y, z) = 2xy^3z^5 - 8xy^5 + 3z^6$

$f_y(x, y, z) = 3x^2y^2z^5 - 20x^2y^4 - 16y^7z^7$

$f_z(x, y, z) = 15x^2y^2z^4 + 18xz^5 - 14y^8z^6$

9.4 Exercises

1.   Maximize $f(x, y) = 4xy$,

    subject to $x + y = 16$.

    1. $g(x, y) = x + y - 16$

2. $F(x, y, \lambda) = 4xy - \lambda(x + y - 16)$.

3. $F_x(x, y, \lambda) = 4y - \lambda$
    $F_y(x, y, \lambda) = 4x - \lambda$
    $F_\lambda(x, y, \lambda) = -(x + y - 16)$

4. $\quad 4y - \lambda = 0 \quad (1)$
    $\quad 4x - \lambda = 0 \quad (2)$
    $x + y - 16 = 0 \quad (3)$

5. Equations (1) and (2) give $\lambda = 4y$ and $\lambda = 4x$. Thus,

    $$4y = 4x$$
    $$y = x.$$

    Substituting into equation (3),

    $$x + (x) - 16 = 0$$
    $$x = 8.$$
    So $\quad\quad y = 8.$

    Maximum is $f(8, 8) = 4(8)(8) = 256$.

3. Maximize $f(x, y) = xy^2$,
    subject to $x + 2y = 15$.

    1. $g(x, y) = x + 2y - 15$

    2. $F(x, y, \lambda) = xy^2 - \lambda(x + 2y - 15)$

    3. $F_x(x, y, \lambda) = y^2 - \lambda$
        $F_y(x, y, \lambda) = 2xy - 2\lambda$
        $F_\lambda(x, y, \lambda) = -(x + 2y - 15)$

    4. $\quad\quad y^2 - \lambda = 0 \quad (1)$
        $\quad 2xy - 2\lambda = 0 \quad (2)$
        $x + 2y - 15 = 0 \quad (3)$

    5. Equations (1) and (2) give $\lambda = y^2$ and $\lambda = xy$. Thus,

        $$y^2 = xy$$
        $$y(y - x) = 0$$

        $$y = 0 \quad \text{or} \quad y = x$$

        Substituting $y = 0$ into equation (3),

        $$x + 2(0) - 15 = 0$$
        $$x = 15.$$

        Substituting $y = x$ into equation (3)

$$x + 2(x) - 15 = 0$$
$$x = 5.$$
So $\quad\quad y = x = 5.$

Thus,

$$f(15, 0) = 15(0)^2 = 0, \quad \text{and}$$
$$f(5, 5) = 5(5)^2 = 125.$$

Since $f(5, 5) > f(15, 0)$, $f(5, 5) = 125$ is a maximum.

5. Minimize $f(x, y) = x^2 + 2y^2 - xy$,
    subject to $x + y = 8$.

    1. $g(x, y) = x + y - 8$

    2. $F(x, y, \lambda)$
        $= x^2 + 2y^2 - xy - \lambda(x + y - 8)$

    3. $F_x(x, y, \lambda) = 2x - y - \lambda$
        $F_y(x, y, \lambda) = 4y - x - \lambda$
        $F_\lambda(x, y, \lambda) = -(x + y - 8)$

    4. $2x - y - \lambda = 0$
        $4y - x - \lambda = 0$
        $x + y - 8 = 0$

    5. Subtracting the second equation from the first equation to eliminate λ gives the new system of equations

        $$x + y = 8$$
        $$3x - 5y = 0.$$

        Solve this system.

        $$5x + 5y = 40$$
        $$\underline{3x - 5y = 0}$$
        $$8x = 40$$
        $$x = 5$$

        But $x + y = 8$, so $y = 3$.

        Thus, $f(5, 3) = 25 + 18 - 15 = 28$ is a minimum.

7. Maximize $f(x, y) = x^2 - 10y^2$,
    subject to $x - y = 18$.

    1. $g(x, y) = x - y - 18$

    2. $F(x, y, \lambda)$
        $= x^2 - 10y^2 - \lambda(x - y - 18)$

3. $F_x(x, y, \lambda) = 2x - \lambda$

$F_y(x, y, \lambda) = -20y - \lambda$

$F_\lambda(x, y, \lambda) = -(x - y - 18)$

4. $\quad 2x - \lambda = 0$

$\quad -20 + \lambda = 0$

$\quad x - y - 18 = 0$

5. Adding the first two equations to eliminate λ gives

$$2x - 20y = 0$$

$$x = 10y.$$

Substituting $x = 10y$ in the third equation gives

$$10y - y = 18$$

$$y = 2$$

$$x = 20.$$

Thus,

$$f(20, 2) = 20^2 - 10(2)^2$$

$$= 400 - 40 = 360.$$

$f(20, 2) = 360$ is a maximum.

9. Maximize $f(x, y, z) = xyz^2$,

subject to $x + y + z = 6$.

1. $g(x, y, z) = x + y + z - 6$

2. $F(x, y, \lambda)$

$$= xyz^2 - \lambda(x + y + z - 6)$$

3. $F_x(x, y, z, \lambda) = yz^2 - \lambda$

$F_y(x, y, z, \lambda) = xz^2 - \lambda$

$F_z(x, y, z, \lambda) = 2zxy - \lambda$

$F_\lambda(x, y, z, \lambda) = -(x + y + z - 6)$

4. Setting F_x, F_y, F_z and F_λ equal to zero yields

$$yz^2 - \lambda = 0 \quad (1)$$

$$xz^2 - \lambda = 0 \quad (2)$$

$$2xyz - \lambda = 0 \quad (3)$$

$$x + y + z - 6 = 0. \quad (4)$$

5. $\lambda = yz^2, \lambda = xz^2,$ and $\lambda = 2xyz$

$$yz^2 = xz^2$$

$$z^2(y - x) = 0$$

$$x = y \quad \text{or} \quad z = 0$$

$$yz^2 = 2xyz$$

$$2xyz - yz^2 = 0$$

$$yz(2x - z) = 0$$

$$y = 0 \text{ or } z = 0 \text{ or } z = 2x$$

In a similar way, the third equation

$$xz^2 = 2xyz$$

implies that $x = 0$ or $z = 0$ or $z = 2y$.

By the nature of the function to be maximized, $f(x, y, z) = xyz^2$, a nonzero maximum can come only from those points with nonzero coordinates.

Therefore, assume $y = x$ and $z = 2y$

$= 2x$.

If $y = x$ and $z = 2x$ are substituted into equation (4), then

$$x + x + 2x - 6 = 0$$

$$x = \frac{3}{2}.$$

Thus, $y = \frac{3}{2}$ and $z = 3$, and

$$f\left(\frac{3}{2}, \frac{3}{2}, 3\right) = \frac{3}{2} \cdot \frac{3}{2} \cdot 9$$

$$= \frac{81}{4} > 0.$$

So, $f\left(\frac{3}{2}, \frac{3}{2}, 3\right) = \frac{81}{4} = 20.25$ is a maximum.

11. The problem can be restated as

Maximize $f(x, y) = 3xy^2$,

subject to $x + y = 24, x > 0, y > 0$.

1. $g(x, y) = x + y - 24$

2. $F(x, y, \lambda) = 3xy^2 - \lambda(x + y - 24)$

3. $F_x(x, y, \lambda) = 3y^2 - \lambda$

$F_y(x, y, \lambda) = 6xy - \lambda$

$F_\lambda(x, y, \lambda) = -(x + y - 24)$

4.
$$3y^2 - \lambda = 0 \quad (1)$$
$$6xy - \lambda = 0 \quad (2)$$
$$x + y - 24 = 0 \quad (3)$$

5. Equations (1) and (2) give $\lambda = 3y^2$ and $\lambda = 6xy$. Thus,

$$3y^2 = 6xy$$
$$3y^2 - 6xy = 0$$
$$3y(y - 2x) = 0$$
$$y = 0 \quad \text{or} \quad y = 2x.$$

Substituting $y = 0$ into equation (3),

$$x + (0) - 24 = 0$$
$$x = 24.$$

Substituting $y = 2x$ into equation (3),

$$x + (2x) - 24 = 0$$
$$3x - 24 = 0$$
$$x = 8.$$

So
$$y = 2x = 16.$$

Thus,

$$f(24, 0) = 3(24)(0)^2 = 0, \text{ and}$$
$$f(8, 16) = 3(8)(16)^2 = 6144.$$

Since $f(8, 16) > f(24, 0)$, $x = 8$ and $y = 16$ will maximize $f(x, y) = 3xy^2$.

13. Let x, y, and z be three number such that

$$x + y + z = 90$$
and
$$f(x, y, z) = xyz.$$

1. $g(x, y, z) = x + y + z - 90$

2. $F(x, y, z)$
$$= xyz - \lambda(x + y + z - 90)$$

3. $F_x(x, y, z, \lambda) = yz - \lambda$
$F_y(x, y, z, \lambda) = xz - \lambda$
$F_\lambda(x, y, z, \lambda) = xy - \lambda$
$F_\lambda(x, y, z, \lambda) = -(x + y + z - 90)$

4.
$$yz - \lambda = 0 \quad (1)$$
$$xz - \lambda = 0 \quad (2)$$
$$xy - \lambda = 0 \quad (3)$$
$$x + y + z - 90 = 0 \quad (4)$$

5. $\lambda = yz$, $\lambda = xz$, and $\lambda = xy$

$$yz = xz$$
$$yz - xz = 0$$
$$(y - x)z = 0$$
$$y - x = 0 \quad \text{or} \quad z = 0$$
$$xz - xy = 0$$
$$x(z - y) = 0$$
$$x = 0 \quad \text{or} \quad z - y = 0$$

Since $x = 0$ or $z = 0$ would not maximize $f(x, y, z) = xyz$, then $y - x = 0$ and $z - y = 0$ imply that $y = x = z$. Substituting into equation (4) gives

$$x + x + x - 90 = 0$$
$$x = 30.$$

$x = y = z = 30$ will maximize $f(x, y, z) = xyz$. The numbers are 30, 30, and 30.

15. Find the maximum and minimum of $f(x, y) = x^3 + 2xy + 4y^2$ subject to $x + 2y = 12$.

1. $g(x, y) = x + 2y - 12$

2. $F(x, y) = x^3 + 2xy + 4y^2$
$$- \lambda(x + 2y - 12)$$

3. $F_x(x, y, \lambda) = 3x^2 + 2y - \lambda$
$F_y(x, y, \lambda) = 2x + 8y - 2\lambda$
$F_\lambda(x, y, \lambda) = -x - 2y + 12$

4.
$$3x^2 + 2y - \lambda = 0$$
$$2x + 8y - 2\lambda = 0$$
$$-x - 2y + 12 = 0$$

5. Solve the second equation for λ and substitute into the first equation.

$$2x + 8y - 2\lambda = 0$$
$$\lambda = x + 4y$$

The first equation is now

$$3x^2 + 2y - (x + 4y) = 0$$
or
$$3x^2 - x - 2y = 0$$

Solve the last equation for $-2y$ and substitute into this new equation.

$$-x - 2y + 12 = 0$$
$$-2y = x - 12$$
$$3x^2 - x + (x - 12) = 0$$
$$3x^2 - 12 = 0$$

Now solve for x.

$$3x^2 - 12 = 0$$
$$x^2 = 4$$
$$x = 2 \text{ or } x = -2$$

Find the corresponding y.

When $x = 2$, $-2 - 2y + 12 = 0$ so $y = 5$.

When $x = -2$, $-(-2) - 2y + 12 = 0$ so $y = 7$.

Thus our candidates for the locations of maxima or minima of f subject to the given constraint are $(2, 5)$ and $(-2, 7)$.

$f(2, 5) = 128$ and $f(-2, 7) = 160$, so probably the maximum is 160 at $(-2, 7)$ and the minimum is 128 at $(2, 5)$.

Try some nearby points that satisfy the constraints to check.

When $x = -2.2$, $y = (12 + 2.2)/2 = 7.1$; $f(-2.2, 7.1) = 159.752$.

When $x = 2.2$, $y = (12 - 2.2)/2 = 4.9$; $f(2.2, 4.9) = 128.248$.

This confirms our answer: There is a maximum value of 160 at $(-2, 7)$ and a minimum value of 128 at $(2, 5)$.

19. Consider the constraint and solve for y in terms of x.

$$3x - y = 9$$
$$y = 3x - 9$$

Then

$$f(x, y) = 8x^2 y$$
$$= 8x^2(3x - 9)$$
$$= 24x^3 - 72x^2$$

So, $f(x, y) = 24x^3 - 72x^2 = f(x)$. Notice that f is unbounded; more specifically,

$$\lim_{y \to \infty} f(x) = \infty$$

and $\lim_{y \to -\infty} f(x) = -\infty$.

Therefore f, subject to the given constraint, has neither an absolute maximum nor an absolute minimum.

21. Minimize
$$f(x, y) = x^2 + 2x + 9y^2 + 4y + 8xy \text{ subject to } x + y = 1.$$

(a) 1. $g(x, y) = x + y = 1$

2. $F(x, y) = x^2 + 2x + 9y^2 + 4y$
$$+ 8xy - \lambda(x + y - 1)$$

3. $F_x(x, y, \lambda) = 2x + 8y - \lambda + 2$
$F_y(x, y, \lambda) = 18y + 8x - \lambda + 4$
$F_\lambda(x, y, \lambda) = -x - y + 1$

4   $2x + 8y - \lambda + 2 = 0$
$18y + 8x - \lambda + 4 = 0$
$-x - y + 1 = 0$

5. Solve the system in Step 4. Solve the first two equations for λ and eliminate λ.

$$2x + 8y + 2 = 18y + 8x + 4$$
$$6x + 10y + 2 = 0$$
$$3x + 5y + 1 = 0$$

Now combine this equation in x and y with the last equation to form a system.

$$3x + 5y + 1 = 0$$
$$-x - y + 1 = 0$$

The solution of this system is $(3, -2)$.

$f(3, -2) = -5$. Test nearby points that satisfy the constraint to see if -5 is indeed a minimum.

When $x = 3.2$, $y = 1 - 3.2 = -2.2$.

When $x = 2.8$, $y = 1 - 2.8 = -1.8$.

$f(3.2, -2.2) = f(2.8, -1.8) = -4.92$, so subject to the given constraint f has a minimum of -5 at $(3, -2)$.

(d) $f_{xx}(x, y) = 2, f_{yy}(x, y) = 18,$
$f_{xy}(x, y) = 8.$

$D(3, -2) = 18 \cdot 2 - 64 = -28$, so $(3, -2)$ is a saddle point of the function f.

23. Maximize $f(x,y) = xy^2$ subject to
$x + 2y = 60$.

1. $g(x,y) = x + 2y - 60$

2. $F(x,y) = xy^2 - \lambda(x + 2y - 60)$

3. $F_x(x,y,\lambda) = y^2 - \lambda$
   $F_y(x,y,\lambda) = 2xy - 2\lambda$
   $F_\lambda(x,y,\lambda) = -x - 2y + 60$

4. $y^2 - \lambda = 0$
   $2xy - 2\lambda = 0$
   $-x - 2y + 60 = 0$

5. Solve the first two equations for λ and eliminate λ.

$$y^2 = \lambda$$
$$xy = \lambda$$
$$y^2 = xy$$

Either $y = 0$ or $x = y$. If $y = 0$, then $x = 60$; if $x = y$, then $-3x = 60$ and $x = y = 20$. So our candidates for a maximum of f are $(60, 0)$ and $(20, 20)$. But $f(60,0) = 0$ so the maximum must be $f(20,20) = 8000$.

We can confirm this by solving the constraint for x.

$$x = 60 - 2y$$
$$f(x,y) = xy^2 = (60 - 2y)y^2 = 60y^2 - 2y^3.$$

This is now a function of one variable. The second derivative with respect to y is $120 - 12y$ which is negative when $y = 20$, so the value of 8000 found above is the maximum utility, obtained by purchasing 20 units of x and 20 units of y.

25. Maximize $f(x,y) = x^4y^2$ subject to
$2x + 4y = 60$.

1. $g(x,y) = 2x + 4y - 60$

2. $F(x,y) = x^4y^2 - \lambda(2x + 4y - 60)$

3. $F_x(x,y,\lambda) = 4x^3y^2 - 2\lambda$
   $F_y(x,y,\lambda) = 2x^4y - 4\lambda$
   $F_\lambda(x,y,\lambda) = -2x - 4y + 60$

4. $4x^3y^2 - 2\lambda = 0$
   $2x^4y - 4\lambda = 0$
   $-2x - 4y + 60 = 0$

5. Solve the first two equations for 2λ and eliminate 2λ.

$$4x^3y^2 = 2\lambda$$
$$x^4y = 2\lambda$$
$$4x^3y^2 = x^4y$$

If either x or y equals 0, the utility will have a minimum value of 0. So we can assume $xy \neq 0$ and divide by x^3y to find that $4y = x$. Substitute for x in the last equation.

$$-2x - x + 60 = 0$$
$$x = 20, \ y = 20/4 = 5$$
$$f(20,5) = 4,000,000.$$

The maximum utility is 4,000,000, obtained by purchasing 20 units of x and 5 units of y.

27. Let x be the width and y be the length of a field such that the cost in dollars to enclose the field is

$$6x + 6y + 4x + 4y = 1200$$
$$10x + 10y = 1200.$$

The area is

$$f(x,y) = xy.$$

1. $g(x,y) = 10x + 10y - 1200$

2. $F(x,y) = xy - \lambda(10x + 10y - 1200)$

3.　$F_x(x, y, \lambda) = y - 10\lambda$

　　$F_y(x, y, \lambda) = x - 10\lambda$

　　$F_\lambda(x, y, \lambda) = -(10x + 10y - 1200)$

4.　　　　　$y - 10\lambda = 0$

　　　　　　$x - 10\lambda = 0$

　　　$10x + 10y - 1200 = 0$

5.　　$10\lambda = y$ and $10\lambda = x$

$$y = x$$

Substituting into the third equation gives

$$10x + 10x - 1200 = 0$$
$$20x - 1200 = 0$$
$$x = 60$$
$$y = 60.$$

These dimensions, 60 feet by 60 feet, will maximize the area.

29.　Maximize $C(x, y) = 2x^2 + 6y^2 + 4xy + 10$, subject to $x + y = 10$.

1.　$g(x, y) = x + y - 10$

2.　$F(x, y)$
　　$= 2x^2 + 6y^2 + 4xy + 10$
　　$- \lambda(x + y - 10)$

3.　$F_x(x, y, \lambda) = 4x + 4y - \lambda$
　　$F_y(x, y, \lambda) = 12y + 4x - \lambda$
　　$F_\lambda(x, y, \lambda) = -(x + y - 10)$

4.　　$4x + 4y - \lambda = 0$
　　　$12y + 4x - \lambda = 0$
　　　　　$x + y - 10 = 0$

5.　$\lambda = 4x + 4y$ and $\lambda = 12y + 4x$.

$$4x + 4y = 12y + 4x$$
$$8y = 0$$
$$y = 0$$

Since $x + y = 10$, $x = 10$.

10 large kits and no small kits will maximize the cost.

31.　Maximize $f(x, y) = 3x^{1/3}y^{2/3}$, subject to $80x + 150y = 40{,}000$.

1.　$g(x, y) = 80x + 150y - 40{,}000$

2.　$F(x, y)$
　　$= 3x^{1/3}y^{2/3} - \lambda(80x + 150y - 40{,}000)$

3,4.　$F_x(x, y, \lambda) = x^{-2/3}y^{2/3} - 80\lambda = 0$
　　　$F_y(x, y, \lambda) = 2x^{1/3}y^{-1/3} - 150\lambda = 0$
　　　$F_\lambda(x, y, \lambda) = -(80x + 150y - 40{,}000)$
　　　　　$= 0$

5.　$\dfrac{x^{-2/3}y^{2/3}}{80} = \dfrac{2x^{1/3}y^{-1/3}}{150}$

　　$\dfrac{15y}{16} = x$

Substitute into the third equation.

$$80\left(\frac{15y}{16}\right) + 150y - 40{,}000 = 0$$

$$y = 178 \,(\text{rounded})$$

$$x = \frac{15(178)}{16}$$

$$\approx 167$$

Use about 167 units of labor and 178 units of capital to maximize production.

33.　Let x and y be the dimensions of the field such that $2x + 2y = 500$, and the area is $f(x, y) = xy$.

1.　$g(x, y) = 2x + 2y - 500$

2.　$F(x, y) = xy - \lambda(2x + 2y - 500)$

3.　$F_x(x, y, \lambda) = y - 2\lambda$
　　$F_y(x, y, \lambda) = x - 2\lambda$
　　$F_\lambda(x, y, \lambda) = -(2x + 2y - 500)$

4.　　　　　$y - 2\lambda = 0$
　　　　　　$x - 2\lambda = 0$
　　　$2x + 2y - 500 = 0$

5.　$2\lambda = y$ and $2\lambda = x$, so $x = y$.

$$2x + 2x - 500 = 0$$
$$4x - 500 = 0$$
$$x = 125$$

Thus, $$y = 125.$$

Dimensions of 125 m by 125 m will maximize the area.

35.   Let x be the radius r of the circular base and y the height h of the can, such that the volume is
$$\pi x^2 y = 250\pi.$$

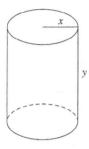

The surface area is

$$f(x, y) = 2\pi xy + 2\pi x^2.$$

1. $$g(x, y) = \pi x^2 y - 250\pi$$

2. $$F(x, y) = 2\pi xy + 2\pi x^2$$
$$- \lambda(\pi x^2 y - 250\pi)$$

3. $$F_x(x, y, \lambda) = 2\pi y + 4\pi x - \lambda(2\pi xy)$$
$$F_y(x, y, \lambda) = 2\pi x - \lambda(\pi x^2)$$
$$F_\lambda(x, y, \lambda) = -(\pi x^2 y - 250\pi)$$

4. $$2\pi y + 4\pi x - \lambda(2\pi xy) = 0$$
$$2\pi x - \lambda \pi x^2 = 0$$
$$\pi x^2 y - 250\pi = 0$$

Simplifying these equations gives

$$y + 2x - 1\lambda xy = 0$$
$$2x - 1\lambda x^2 = 0$$
$$x^2 y - 250 = 0.$$

5.   From the second equation,

$$x(2 - \lambda x) = 0$$

$$x = 0 \quad \text{or} \quad \lambda = \frac{2}{x}.$$

If $x = 0$, the volume will be 0, which is not possible.

Substituting $x = \frac{2}{\lambda}$ into the first equation gives

$$y + 2\left(\frac{2}{\lambda}\right) - \lambda\left(\frac{2}{\lambda}\right)y = 0$$

$$y + \frac{4}{\lambda} - 2y = 0$$

$$\frac{4}{\lambda} = y$$

$$\lambda = \frac{4}{y}.$$

Since $\lambda = \frac{2}{x}, y = 2x.$

Substituting into third equation gives

$$x^2(2x) - 250 = 0$$
$$2x^3 - 250 = 0$$
$$x = 5$$
$$y = 10.$$

Since $g(1, 250) = 0$ and

$$f(1, 250) = 502\pi > f(5, 10) = 150\pi,$$

a can with radius of 5 inches and height of 10 inches will have a minimum surface area.

37.   Let x, y, and z be the dimensions of the box such that the surface area is

$$xy + 2yz + 2xz = 500$$

and the volume is

$$f(x, y, z) = xyz.$$

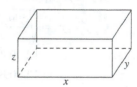

1. $$g(x, y, z) - 500 = 0$$

2. $$F(x, y, z)$$
$$= xyz - \lambda(xy + 2yz + 2xz - 500)$$

3,4.

$$F_x(x, y, z, \lambda) = yz - \lambda(y + 2z) = 0 \quad (1)$$

$$F_y(x, y, z, \lambda) = xz - \lambda(x + 2z) = 0 \quad (2)$$

$$F_z(x, y, z, \lambda) = xy - \lambda(2y + 2x) = 0 \quad (3)$$

$$F_\lambda(x, y, z, \lambda) = -(xy + 2xz + 2yz - 500)$$

$$= 0 \quad (4)$$

Multiplying equation (1) by x, equation (2) by y, and equation (3) by z gives

$$xyz - \lambda x(y + 2z) = 0$$

$$xyz - \lambda y(x + 2z) = 0$$

$$xyz - \lambda z(2y + 2z) = 0.$$

5. Subtracting the first equation from the second equation gives

$$\lambda x(y + 2z) - \lambda y(x + 2z) = 0$$

$$2\lambda xz - 2\lambda yz = 0$$

$$\lambda z(x - y) = 0,$$

so $\qquad\qquad x = y.$

Subtracting the third equation from the second equation gives

$$\lambda z(2y + 2x) - \lambda y(x + 2z) = 0$$

$$2\lambda xz - \lambda xy = 0$$

$$\lambda x(2z - y) = 0,$$

so $\qquad\qquad z = \dfrac{y}{2}.$

Substituting into the fourth equation gives

$$y^2 + 2y\left(\frac{y}{2}\right) + 2y\left(\frac{y}{2}\right) - 500 = 0$$

$$3y^2 = 500$$

$$y = \sqrt{\frac{500}{3}}$$

$$\approx 12.9099$$

$$x \approx 12.9099$$

$$z \approx \frac{12.9099}{2}$$

$$\approx 6.4549.$$

The dimensions are 12.91 m by 12.91 m by 6.455 m.

39. Let x, y, and z be the dimensions of the box. The surface area is

$$2xy + 2xz + 2yz.$$

We must minimize

$$f(x, y, z) = 2xy + 2xz + 2yz$$

subject to $xyz = 125$.

1. $g(x, y, z) = xyz - 125$

2. $F(x, y, z)$
$$= 2xy + 2xz + 2yz - \lambda(xyz - 125)$$

3. $F_x(x, y, z, \lambda) = 2y + 2z - \lambda yz$
$F_y(x, y, z, \lambda) = 2x + 2z - \lambda xz$
$F_z(x, y, z, \lambda) = 2x + 2y - \lambda xy$
$F_\lambda(x, y, z, \lambda) = -(xyz - 125)$

4. $2y + 2z - \lambda yz = 0 \quad (1)$
$2x + 2z - \lambda xz = 0 \quad (2)$
$2x + 2y - \lambda xy = 0 \quad (3)$
$\qquad xyz - 125 = 0 \quad (4)$

5. Equations (1) and (2) give

$$\frac{2y + 2z}{yz} = \lambda \quad \text{and} \quad \frac{2x + 2z}{xz} = \lambda.$$

Thus,

$$\frac{2y + 2z}{yz} = \frac{2x + 2z}{xz}$$

$$2xyz + 2xz^2 = 2xyz + 2yz^2$$

$$2xz^2 - 2yz^2 = 0$$

$2z^2 = 0 \qquad\qquad \text{or} \quad x - y = 0$
$z = 0 \quad \text{(impossible)} \quad \text{or} \qquad x = y.$

Equations (2) and (3) give

$$\frac{2x + 2z}{xz} = \lambda \quad \text{and} \quad \frac{2x + 2y}{xy} = \lambda.$$

$$\frac{2x + 2z}{xz} = \frac{2x + 2y}{xy}$$

Thus,

$$2x^2y + 2xyz = 2x^2z + 2xyz$$

$$2x^2y - 2x^2z = 0$$

$$2x^2(y - z) = 0$$

$2x^2 = 0 \qquad\qquad \text{or} \quad y - z = 0$
$x = 0 \quad \text{(impossible)} \quad \text{or} \qquad y = z.$

Therefore, $x = y = z$. Substituting into equation (4) gives

$$x^3 - 125 = 0$$

$$x^3 = 125$$

$$x = 5.$$

Thus,

$$y = 5 \text{ and } z = 5.$$

The dimensions that will minimize the surface area are 5 m by 5 m by 5 m.

41. (a) The total area of the two sides and bottom made out of free material is $2xz + xy$. This material does not add to the cost, so we would like to make at as large as possible, which corresponds to the constraint $2xz + xy = 8$. The cost of the two reinforced ends will be $(1000)(2yz)$. The number of trips will be 400 divided by the volume of the box, which is xyz. Thus the shipping cost will be $10 \cdot \dfrac{400}{xyz}$ and the total cost will be $f(x, y, z) = \dfrac{4000}{xyz} + 2000yz$.

(b) The solver gives $x = 4$ yd, $y = 1$ yd, $z = \dfrac{1}{2}$ yd, for a minimum cost of $3000.

9.5 Total Differentials and Approximations

Your Turn 1

$$f(x, y) = 3x^2y^4 + 6\sqrt{x^2 - 7y^2}$$

$$f_x(x, y) = 6xy^4 + \frac{6x}{\sqrt{x^2 - 7y^2}}$$

$$f_y(x, y) = 12x^2y^3 - \frac{42y}{\sqrt{x^2 - 7y^2}}$$

(a) $dz = f_x(x, y)\,dx + f_y(x, y)\,dy$

$$= \left(6xy^4 + \frac{6x}{\sqrt{x^2 - 7y^2}}\right)dx$$

$$+ \left(12x^2y^3 - \frac{42y}{\sqrt{x^2 - 7y^2}}\right)dy$$

(b) Evaluate dz for $x = 4$, $y = 1$, $dx = 0.02$, $dy = -0.03$.

$$dz = \left(6(4)(1)^4 + \frac{6(4)}{\sqrt{(4)^2 - 7(1)^2}}\right)(0.02)$$

$$+ \left(12(4)^2(1)^3 - \frac{42(1)}{\sqrt{(4)^2 - 7(1)^2}}\right)(-0.03)$$

$$= \left(24 + \frac{24}{3}\right)(0.02) + \left(192 - \frac{42}{3}\right)(-0.03)$$

$$= (32)(0.02) - 178(0.03)$$

$$= -4.7$$

Your Turn 2

As in Example 2, we let $f(x, y) = \sqrt{x^2 + y^2}$. As we found in Example 2,

$$dz = \left(\frac{x}{\sqrt{x^2 + y^2}}\right)dx + \left(\frac{y}{\sqrt{x^2 + y^2}}\right)dy$$

To approximate $\sqrt{5.03^2 + 11.99^2}$ we let $x = 5$, $y = 12$, $dx = 0.03$, $dy = -0.01$.

Then $f(x + dx, y + dy) \approx f(x, y) + dz$

$$\sqrt{5.03^2 + 11.99^2}$$

$$\approx \sqrt{5^2 + 12^2} + \left(\frac{5}{\sqrt{5^2 + 12^2}}\right)(0.03)$$

$$+ \left(\frac{12}{\sqrt{5^2 + 12^2}}\right)(-0.01)$$

$$= 13 + \frac{0.15}{13} - \frac{0.12}{13}$$

$$\approx 13.0023$$

Your Turn 3

As calculated in Example 3, $\dfrac{dV}{V} = 2\dfrac{dr}{r} + \dfrac{dh}{h}$. Thus the maximum percent error in the volume if the errors in the radius and length are at most 4% and 2% is $\dfrac{dV}{V} = 2(0.04) + (0.02) = 0.10$ or 10%.

9.5 Warmup Exercises

W1. $y = 2x^3 + 3x$

$$dy = y'(x)\,\Delta x$$

$$= \left(6x^2 + 3\right)\Delta x$$

When $x = 2$ and $\Delta x = 0.1$,

$$dy = \left(6x^2 + 3\right)\Delta x$$

$$= (6(2)^2 + 3)(0.1)$$

$$= 2.7.$$

W2. $y = \ln(x^2 + 1)$

$$dy = y'(x)\,\Delta x$$

$$= \left(\frac{2x}{x^2 + 1}\right)\Delta x$$

When $x = 2$ and $\Delta x = -0.02$,

$$dy = \left(\frac{2x}{x^2 + 1}\right)\Delta x$$

$$= \left(\frac{2(2)}{(2)^2 + 1}\right)(-0.02)$$

$$= -0.016.$$

W3. Use $y = \sqrt{x}$ with $x = 25$ and $\Delta x = -1$.

$$\Delta y \approx y'(5)\Delta x = \frac{1}{2\sqrt{25}}(-1) = -0.1$$

Thus $\sqrt{24} \approx \sqrt{25} - 0.1 = 4.9$.

W4. Use $y = e^x$ with $x = 0$ and $\Delta x = 0.01$.

$$\Delta y \approx y'(0)\Delta x = e^0(0.01) = 0.01$$

Thus $e^{0.01} \approx e^0 + 0.01 = 1.01$.

9.5 Exercises

1. $z = f(x, y) = 2x^2 + 4xy + y^2$

    $x = 5, dx = 0.03, y = -1, dy = -0.02$

$$f_x(x, y) = 4x + 4y$$

$$f_y(x, y) = 4x + 2y$$

$$dz = (4x + 4y)dx + (4x + 2y)dy$$

$$= [4(5) + 4(-1)](0.03)$$

$$+ [4(5) + 2(-1)](-0.02)$$

$$= 0.48 - 0.36 = 0.12$$

3. $z = \dfrac{y^2 + 3x}{y^2 - x}$, $x = 4$, $y = -4$,

    $dx = 0.01$, $dy = 0.03$

$$dz = \frac{(y^2 - x)\cdot 3 - (y^2 + 3x)\cdot(-1)}{(y^2 - x)^2}\,dx$$

$$+ \frac{(y^2 - x)\cdot 2y - (y^2 + 3x)\cdot 2y}{(y^2 - x)^2}\,dx$$

$$= \frac{4y^2}{(y^2 - x)^2}\,dx - \frac{8xy}{(y^2 - x)^2}\,dy$$

$$= \frac{4(-4)^2}{[(-4)^2 - 4]^2}(0.01) - \frac{8(4)(-4)}{[(-4)^2 - 4]^2}(0.03)$$

$$\approx 0.0311$$

5. $w = \dfrac{5x^2 + y^2}{z + 1}$

    $x = -2, y = 1, z = 1$

    $dx = 0.02, dy = -0.03, dz = 0.02$

$$f_x(x, y) = \frac{(z + 1)10x - (5x^2 + y^2)(0)}{(z + 1)^2}$$

$$= \frac{10x}{z + 1}$$

$$f_y(x, y) = \frac{(z + 1)(2y) - (5x^2 + y^2)(0)}{(z + 1)^2}$$

$$= \frac{2y}{z + 1}$$

$$f_z(x, y) = \frac{(z + 1)(0) - (5x^2 + y^2)(1)}{(z + 1)^2}$$

$$= \frac{-5x^2 - y^2}{(z + 1)^2}$$

$$dw = \frac{10x}{z + 1}\,dx + \frac{2y}{z + 1}\,dy + \frac{-5x^2 - y^2}{(z + 1)^2}\,dz$$

Substitute the given values.

$$dw = \frac{-20}{2}(0.02) + \frac{2}{2}(-0.03)$$

$$+ \frac{[-5(4) - 1](0.02)}{(2)^2}$$

$$= -0.2 - 0.03 - \frac{21}{4}(0.02)$$

$$= -0.335$$

7. Let $z = f(x, y) = \sqrt{x^2 + y^2}$.

Then

$$dz = f_x(x, y)dx + f_y(x, y)dy$$

$$= \frac{1}{2}(x^2 + y^2)^{-1/2}(2x)dx$$

$$+ \frac{1}{2}(x^2 + y^2)^{-1/2}(2y)dy$$

$$= \frac{x\,dx + y\,dy}{\sqrt{x^2 + y^2}}.$$

To approximate $\sqrt{8.05^2 + 5.97^2}$, we let $x = 8$, $dx = 0.05$, $y = 6$ and $dy = -0.03$.

$$dz = \frac{8(0.05) + 6(-0.03)}{\sqrt{8^2 + 6^2}}$$

$$= \frac{4}{5}(0.05) + \frac{3}{5}(-0.03)$$

$$= 0.04 - 0.018 = 0.022$$

$$f(8.05, 5.97) = f(8, 6) + \Delta z$$

$$\approx f(8, 6) + dz$$

$$= \sqrt{8^2 + 6^2} + 0.222$$

$$= 10.022$$

Thus, $\sqrt{8.05^2 + 5.97^2} \approx 10.022$.

Using a calculator, $\sqrt{8.05^2 + 5.97^2} \approx 10.0221$.

The absolute value of the difference of the two results is $|10.022 - 10.0221| = 0.0001$.

9. Let $z = f(x, y) = (x^2 + y^2)^{1/3}$.

Then

$$dz = f_x(x, y)dx + f_y(x, y)dy$$

$$dz = \frac{1}{3}(x^2 + y^2)^{-2/3}(2x)dx$$

$$+ \frac{1}{3}(x^2 + y^2)^{-2/3}(2y)dy$$

$$= \frac{2x}{3(x^2 + y^2)^{2/3}}dx + \frac{2y}{3(x^2 + y^2)^{2/3}}dy$$

To approximate $(1.92^2 + 2.1^2)^{1/3}$, we let $x = 2$, $dx = -0.08$, $y = 2$, and $dy = 0.1$.

$$dz = \frac{2(2)}{3[(2)^2 + (2)^2]^{2/3}}(-0.08)$$

$$+ \frac{2(2)}{3[(2)^2 + (2)^2]^{2/3}}(0.1)$$

$$= \frac{4}{12}(-0.08) + \frac{4}{12}(0.1)$$

$$= 0.00\overline{6}$$

$$f(1.92, 2.1) = f(2, 2) + \Delta z$$

$$\approx f(2, 2) + dz$$

$$= 2 + 0.00\overline{6}$$

$$f(1.92, 2.1) \approx 2.0067$$

Using a calculator, $(1.92^2 + 2.1^2)^{1/3} \approx 2.0080$.

The absolute value of the difference of the two results is $|2.0067 - 2.0080| = 0.0013$.

11. Let $z = f(x, y) = xe^y$.

Then

$$dz = f_x(x, y)dx + f_y(x, y)dy$$

$$= e^y dx + xe^y dy.$$

To approximate $1.03e^{0.04}$, we let $x = 1$, $dx = 0.03$, $y = 0$, and $dy = 0.04$.

$$dz = e^0(0.03) + 1 \cdot e^0(0.04)$$

$$= 0.07$$

$$f(1.03, 0.04) = f(1, 0) + \Delta z$$

$$\approx f(1, 0) + dz$$

$$= 1 \cdot e^0 + 0.07$$

$$= 1.07$$

Thus, $1.03e^{0.04} \approx 1.07$.

Using a calculator, $1.03e^{0.04} \approx 1.0720$.

The absolute value of the difference of the two results is $|1.07 - 1.0720| = 0.0020$.

13. Let $z = f(x, y) = x \ln y$.

Then

$$dz = f_x(x, y)dx + f_y(x, y)dy$$

$$= \ln y\, dx + \frac{x}{y}dy$$

To approximate $0.99 \ln 0.98$, we let $x = 1$, $dx = -0.01$, $y = 1$, and $dy = -0.02$.

$$dz = \ln(1) \cdot (-0.01) + \frac{1}{1}(-0.02)$$

$$= -0.02$$

$$f(0.99, 0.98) = f(1,1) + \Delta z$$
$$\approx f(1,1) + dz$$
$$= 1 \cdot \ln(1) - 0.02$$
$$\approx -0.02$$

Thus, $0.99 \ln 0.98 \approx -0.02$.

Using a calculator, $0.99 \ln 0.98 \approx -0.0200$.

The absolute value of the difference of the two results is $|-0.02 - (-0.0200)| = 0$.

15. The volume of the can is

$$V = \pi r^2 h,$$

With
$r = 2.5\,\text{cm}, h = 14\,\text{cm}, dr = 0.08, dh = 0.16.$

$$dV = 2\pi r h\, dr + \pi r^2 dh$$
$$= 2\pi(2.5)(14)(0.08) + \pi(2.5)^2(0.16)$$
$$\approx 20.73$$

Approximately 20.73 cm³ of aluminum are needed.

17. The volume of the box is
$$V = LWH$$

with $L = 10, W = 9,$ and $H = 18.$

Since 0.1 inch is applied to each side and each dimension has a side at each end,

$$dL = dW = dH = 2(0.1) = 0.2$$
$$dV = WH\, dL + LH\, dW + LW\, dH.$$

Substitute.

$$dV = (9)(18)(0.2) + (10)(18)(0.2)$$
$$+ (10)(9)(0.2)$$
$$= 86.4$$

Approximately 86.4 in³ are needed.

19. $z = x^{0.65}y^{0.35}$

$x = 50, y = 29,$
$dx = 52 - 50 = 2$
$dy = 27 - 29 = -2$

$$f_x(x, y) = y^{0.35}(0.65)(x^{-0.35})$$
$$= 0.65\left(\frac{y}{x}\right)^{0.35}$$
$$f_y(x, y) = (x^{0.65})(0.35)(y^{-0.65})$$
$$= 0.35\left(\frac{x}{y}\right)^{0.65}$$

$$dz = 0.65\left(\frac{y}{x}\right)^{0.35}dx + 0.35\left(\frac{x}{y}\right)^{0.65}dy$$

Substitute.

$$dz = 0.65\left(\frac{29}{50}\right)^{0.35}(2) + 0.35\left(\frac{50}{29}\right)^{0.65}(-2)$$
$$= 0.07694 \text{ unit}$$

21. The volume of the bone is

$$V = \pi r^2 h,$$

with $h = 7, r = 1.4, dr = 0.09, dh = 2(0.09)$
$$= 0.18$$

$$dV = 2\pi r h\, dr + \pi r^2 dh$$
$$= 2\pi(1.4)(7)(0.09) + \pi(1.4)^2(0.18)$$
$$= 6.65$$

6.65 cm³ of preservative are used.

23. $C = \dfrac{b}{a - v} = b(a - v)^{-1}$

$a = 160,$
$b = 200, v = 125$
$da = 145 - 160 = -15$
$db = 190 - 200 = -10$
$dv = 130 - 125 = 5$

$$dC = -b(a - v)^{-2}da$$
$$+ \frac{1}{a - v}db + b(a - v)^{-2}dv$$
$$= \frac{-b}{(a - v)^2}da + \frac{1}{a - v}db + \frac{b}{(a - v)^2}dv$$
$$= \frac{-200}{(160 - 125)^2}(-15) + \frac{1}{160 - 125}(-10)$$
$$+ \frac{200}{(160 - 125)^2}(5)$$
$$\approx 2.98 \text{ liters}$$

25. $L(E, P) = 23E^{0.6}P^{-0.267}$

As we found in Exercise 62 of Section 9.2,

$$\frac{\partial}{\partial E}L(E, P) = 23(0.6)E^{-0.4}P^{-0.267}$$

$$\frac{\partial}{\partial P}L(E, P) = 23(-0.267)E^{0.6}P^{-1.267}$$

The total differential is thus

$$dL = 23(0.6)E^{-0.4}P^{-0.267}dE$$
$$+ 23(-0.267)E^{0.6}P^{-1.267}dP$$

To approximate the change called for we take $E = 14{,}100$, $P = 68{,}700$, $dE = 200$, and $dP = -300$. For these inputs the value of dL is then
3.5117 ≈ 3.51 years. For comparison,

$$L(14{,}300,\ 68{,}400) - L(14{,}100,\ 68{,}700)$$
$$= 3.5077 \approx 3.51 \text{ years.}$$

27. $P(A, B, D) = \dfrac{1}{1 + e^{3.68 - 0.016A - 0.77B - 0.12D}}$

(a) Since bird pecking is present, $B = 1$.

$$P(150, 1, 20) = \frac{1}{1 + e^{3.68 - 0.016(150) - 0.77(1) - 0.12(20)}}$$

$$= \frac{1}{1 + e^{-1.89}} \approx 0.8688$$

The probability is about 87%.

(b) Since bird pecking is not present, $B = 0$.

$$P(150, 0, 20)$$

$$= \frac{1}{1 + e^{3.68 - 0.016(150) - 0.77(0) - 0.12(20)}}$$

$$= \frac{1}{1 + e^{-1.12}} \approx 0.7540$$

The probability is about 75%.

(c) Let $B = 0$. To simplify the notation, let $X = 3.68 - 0.016A - 0.12D$. Then

$$P(A, 0, D) = \frac{1}{1 + e^{3.68 - 0.016A - 0.12D}}$$

$$= \frac{1}{1 + e^X}.$$

Some other values that we will need are

$$dA = 160 - 150 = 10$$
$$dD = 25 - 20 = 4$$
$$X(150, 20) = 3.68 - 0.016(150) - 0.12(20)$$
$$= -1.12$$

$$X_A = \frac{\partial X}{\partial A} = -0.016$$

$$X_D = \frac{\partial X}{\partial D} = -0.12.$$

$$P_A(A, 0, D) = \frac{X_A e^X}{(1 + e^X)^2} = \frac{0.016 e^X}{(1 + e^X)^2}$$

$$P_D(A, 0, D) = \frac{X_D e^X}{(1 + e^X)^2} = \frac{0.12 e^X}{(1 + e^X)^2}$$

$$dP = P_A(A, 0, D)\,dA + P_D(A, 0, D)\,dD$$

$$= \frac{0.016 e^X}{(1 + e^X)^2}\,dA + \frac{0.12 e^X}{(1 + e^X)^2}\,dD$$

Substituting the given and calculated values,

$$dP = \frac{0.016 e^{-1.12}}{(1 + e^{-1.12})^2}(10) + \frac{0.12 e^{-1.12}}{(1 + e^{-1.12})^2}(5)$$

$$= (0.016 \cdot 10 + 0.12 \cdot 5)\frac{e^{-1.12}}{(1 + e^{-1.12})^2}$$

$$\approx 0.76 \cdot 0.1855 \approx 0.14.$$

Therefore,

$$P(160, 0, 25) = P(150, 0, 20) + \Delta P$$
$$\approx P(150, 0, 20) + dP$$
$$= 0.75 + 0.14 = 0.89.$$

The probability is about 89%.
Using a calculator, $P(160, 0, 25) \approx 0.8676$, or about 87%.

29. $t(x, y, p, C) = \dfrac{\sqrt{x^2 + (y - p)^2}}{331.45 + 0.6C}$

(a) $t(5, -2, 20, 20) = \dfrac{\sqrt{5^2(-2 - 20)^2}}{331.45 + 0.6(20)}$

$$= \frac{\sqrt{509}}{343.45} \approx 0.06569$$

$$t(5,-2,10,20) = \frac{\sqrt{5^2 + (-2-10)^2}}{331.45 + 0.6(20)}$$

$$= \frac{\sqrt{169}}{343.45} \approx 0.0379$$

In a close race, this difference could certainly affect the outcome.

(b) Since the starter remains stationary, $dx = dy = 0$, so $t_x(x,y,p,C)$ and $t_y(x,y,p,C)$ do not need to be computed.

$$t_p(x,y,p,C)$$

$$= \frac{-2(y-p)}{2(331.45 + 0.6C)\sqrt{x^2 + (y-p)^2}}$$

$$= \frac{p-y}{(331.45 + 0.6C)\sqrt{x^2 + (y-p)^2}}$$

$$t_C(x,y,p,C) = -\frac{0.6\sqrt{x^2 + (y-p)^2}}{(331.45 + 0.6C)^2}$$

$dx = 0, dy = 0, dp = 0.5, dC = -5$

$dt = t_x(5,-2,20,20) \cdot 0 + t_y(5,-2,20,20) \cdot 0$

$\quad + t_p(5,-2,20,20) \cdot 0.5$

$\quad + t_C(5,-2,20,20) \cdot (-5)$

$$= \frac{20 - (-2)}{(331.45 + 0.6(20))\sqrt{5^2 + (-2-20)^2}} \cdot 0.5$$

$$- \frac{0.6\sqrt{5^2 + (-2-20)^2}}{(331.45 + 0.6(20))^2} \cdot (-5)$$

≈ 0.001993 sec

This is the approximate change in the time when the swimmer stands 0.5 m farther from the starter in the y direction and the temperature decreases by 5°C.

31. The area is $A = \frac{1}{2}bh$ with $b = 15.8$ cm, $h = 37.5$ cm, $db = 1.1$ cm, and $dh = 0.8$ cm.

$$dA = \frac{1}{2}b\,dh + \frac{1}{2}h\,db$$

$$= \frac{1}{2}(15.8)(0.8) + \frac{1}{2}(37.5)(1.1)$$

$$= 26.945$$

The maximum possible error is 26.945 cm².

33. Let $z = f(L,W,H) = LWH$

Then

$$dz = f_L(L,W,H)\,dL + f_W(L,W,H)\,dW$$
$$\quad + f_H(L,W,H)\,dH$$
$$= WH\,dL + LH\,dW + LW\,dH.$$

A maximum 1% error in each measurement means that the maximum values of dL, dW, and dH are given by $dL = 0.01L$, $dW = 0.01W$, and $dH = 0.01H$. Therefore,

$$dz = WH(0.01L) + LH(0.01W) + LW(0.01H)$$
$$= 0.01LWH + 0.01LWH + 0.01LWH$$
$$= 0.03LWH.$$

Thus, an estimate of the maximum error in calculating the volume is 3%.

35. The volume of a cone is $V = \frac{\pi}{3}r^2h.$

$$\frac{dV}{V} = \frac{\frac{2\pi}{3}rh}{\frac{\pi}{3}r^2h}dr + \frac{\frac{\pi}{3}r^2}{\frac{\pi}{3}r^2h}dh$$

$$= \frac{2}{r}dr + \frac{1}{h}dh$$

When $r = 1$ and $h = 4$, a 1% change in radius changes the volume by 2%, and a 1% change in height changes the volume by $\frac{1}{4}$%. So the change produced by changing the radius is 8 times the change produced by changing the height.

9.6 Double Integrals

Your Turn 1

$$\int_1^3 (6x^2y^2 + 4xy + 8x^3 + 10y^4 + 3)\,dy$$

$$= (2x^2y^3 + 2xy^2 + 8x^3y + 2y^5 + 3y)\Big|_{y=1}^{y=3}$$

$$= (2x^2(3)^3 + 2x(3)^2 + 8x^3(3) + 2(3)^5 + 3(3))$$
$$\quad - (2x^2(1)^3 + 2x(1)^2 + 8x^3(1) + 2(1)^5 + 3(1))$$

$$= 52x^2 + 16x + 16x^3 + 484 + 6$$

$$= 16x^3 + 52x^2 + 16x + 490$$

Your Turn 2

$$\int_0^2\left[\int_1^3 (6x^2y^2 + 4xy + 8x^3 + 10y^4 + 3)\,dy\right]dx$$

Use the result from Your Turn 1 to evaluate the inner integral.

$$\int_0^2 (16x^3 + 52x^2 + 16x + 490)\,dx$$

$$= \left(4x^4 + \frac{52}{3}x^3 + 8x^2 + 490x\right)\Big|_0^2$$

$$= 4(16) + \frac{52}{3}(8) + 8(4) + 490(2)$$

$$= \frac{3644}{3}$$

Integrating in the other order:

$$\int_1^3\left[\int_0^2 (6x^2y^2 + 4xy + 8x^3 + 10y^4 + 3)\,dx\right]dy$$

$$= \int_1^3\left[(2x^3y^2 + 2x^2y + 2x^4 + 10xy^4 + 3x)\Big|_{x=0}^{x=2}\right]dy$$

$$= \int_1^3 (16y^2 + 8y + 32 + 20y^4 + 6)\,dy$$

$$= \int_1^3 (20y^4 + 16y^2 + 8y + 38)\,dy$$

$$= \left(4y^5 + \frac{16}{3}y^3 + 4y^2 + 38y\right)\Big|_1^3$$

$$\left(972 + 144 + 36 + 114\right) - \left(4 + \frac{16}{3} + 4 + 38\right)$$

$$= \frac{3644}{3}$$

Your Turn 3

Let the region R be defined by $0 \le x \le 5$ and $1 \le y \le 6$.

$$\iint_R \frac{1}{\sqrt{x + y + 3}}\,dx\,dy$$

$$= \int_1^6 \int_0^5 \frac{1}{\sqrt{x + y + 3}}\,dx\,dy$$

$$= \int_1^6 \left(2\sqrt{x + y + 3}\right)\Big|_{x=0}^{x=5}\,dy$$

$$= 2\int_1^6 \left(\sqrt{y + 8} - \sqrt{y + 3}\right)dy$$

$$= 2\frac{2}{3}\left[(y + 8)^{3/2} - (y + 3)^{3/2}\right]\Big|_1^6$$

$$= \frac{4}{3}\left[(14^{3/2} - 9^{3/2}) - (9^{3/2} - 4^{3/2})\right]$$

$$= \frac{4}{3}(14\sqrt{14} - 46)$$

$$= \frac{56\sqrt{14} - 184}{3}$$

Your Turn 4

The function $4 - x^3 - y^3$ is positive over the region $0 \le x \le 1$ and $0 \le y \le 1$, so the volume under the surface $z = 4 - x^3 - y^3$ over this region is

$$\iint_R (4 - x^3 - y^3)\,dx\,dy$$

$$= \int_0^1 \int_0^1 (4 - x^3 - y^3)\,dx\,dy.$$

$$\int_0^1 \int_0^1 (4 - x^3 - y^3)\,dx\,dy$$

$$= \int_0^1 \left(4x - \frac{x^4}{4} - xy^3\right)\Big|_{x=0}^{x=1}\,dy$$

$$= \int_0^1 \left(\frac{15}{4} - y^3\right)dy$$

$$= \left(\frac{15}{4}y - \frac{y^4}{4}\right)\Big|_0^1$$

$$= \frac{15}{4} - \frac{1}{4} = \frac{7}{2}$$

Your Turn 5

Find $\displaystyle\iint_R \left(x^3 + 4y\right)dy\,dx$ over the region bounded by

$y = 4x$ and $y = x^3$ for $0 \le x \le 2$.

The region is shown in the figure below.

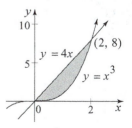

Note that throughout the region, $4x \ge x^3$.

$$\iint\limits_{R} (x^3 + 4y)dy\, dx$$

$$= \int_0^2 \int_{x^3}^{4x} (x^3 + 4y)\, dy\, dx$$

$$= \int_0^2 (x^3 y + 2y^2)\Big|_{y=x^3}^{y=4x} dx$$

$$= \int_0^2 [(4x^4 + 32x^2) - (x^6 + 2x^6)]\, dx$$

$$= \int_0^2 (4x^4 + 32x^2 - 3x^6)\, dx$$

$$= \left(\frac{4}{5}x^5 + \frac{32}{3}x^3 - \frac{3}{7}x^7 \right)\Big|_0^2$$

$$= \frac{128}{5} + \frac{256}{3} - \frac{384}{7}$$

$$= \frac{5888}{105}$$

9.6 Warmup Exercises

W1.

$$\int \left(3x^3 - 6x^2 \right) dx = \frac{3}{4}x^4 - 2x^3 + C$$

$$\int_0^2 \left(3x^3 - 6x^2 \right) dx = \left(\frac{3}{4}x^4 - 2x^3 \right)\Big|_0^2$$

$$= \left(\frac{3}{4}(16) - 2(8) \right) - (0 - 0)$$

$$= -4$$

W2.

$$\int \sqrt{x + 6}\, dx = \frac{2}{3}(x + 6)^{3/2} + C$$

$$\int_{-2}^3 \sqrt{x + 6}\, dx = \frac{2}{3}(x + 6)^{3/2}\Big|_{-2}^3$$

$$= \frac{2}{3}\left((3 + 6)^{3/2} - (-2 + 6)^{3/2} \right)$$

$$= \frac{2}{3}(27 - 8) = \frac{38}{3}$$

W3. $\displaystyle \int \frac{3}{x}\, dx = 3\ln|x| + C$

$$\int_2^4 \frac{3}{x}\, dx = \left(3\ln|x| \right)\Big|_2^4$$

$$= 3(\ln 4 - \ln 2)$$

$$= 3\ln 2 = \ln 8 \approx 2.0794$$

W4. $\displaystyle \int e^{6x} dx = \frac{e^{6x}}{6} + C$

$$\int_0^1 e^{6x} dx = \frac{e^{6x}}{6}\Big|_0^1$$

$$= \frac{e^6 - e^0}{6}$$

$$= \frac{e^6 - 1}{6} \approx 67.0715$$

W5. Use the substitution $u = x^3,\ du = 3x^2 dx$.

$$\int x^2 e^{x^3} dx = \frac{1}{3}\int e^u\, du$$

$$= \frac{1}{3}e^u + C = \frac{1}{3}e^{x^3} + C$$

$$\int_1^2 x^2 e^{x^3} dx = \frac{1}{3}e^{x^3}\Big|_1^2 = \frac{1}{3}\left(e^8 - e^1 \right)$$

$$= \frac{e^8 - e}{3} \approx 992.7466$$

W6. Use the substitution $u = x^2 + 9,\ du = 2x\, dx$.

$$\int x\sqrt{x^2 + 9}\, dx = \frac{1}{2}\int \sqrt{u}\, du$$

$$= \frac{1}{2} \cdot \frac{2}{3} \cdot u^{3/2} + C$$

$$= \frac{1}{3}\left(x^2 + 9 \right)^{3/2} + C$$

$$\int_0^4 x\sqrt{x^2 + 9}\, dx = \frac{1}{3}\left(x^2 + 9 \right)^{3/2}\Big|_0^4$$

$$= \frac{25^{3/2} - 9^{3/2}}{3}$$

$$= \frac{125 - 27}{3} = \frac{98}{3}$$

9.6 Exercises

1. $\int_0^5 (x^4y + y)dx = \left(\frac{x^5y}{5} - xy \right)\Big|_0^5$

$= (625y + 5y) - 0 = 630y$

3. $\int_4^5 x\sqrt{x^2 + 3y}\,dy$

$= \int_4^5 x(x^2 + 3y)^{1/2}\,dy$

$= \frac{2x}{9}[(x^2 + 3y)^{3/2}\Big|_4^5$

$= \frac{2x}{9}[(x^2 + 15)^{3/2} - (x^2 + 12)^{3/2}]$

5. $\int_4^9 \frac{3 + 5y}{\sqrt{x}}\,dx = (3 + 5y)\int_4^9 x^{-1/2}\,dx$

$= (3 + 5y)2x^{1/2}\Big|_4^9$

$= (3 + 5y)2\left[\sqrt{9} - \sqrt{4}\right]$

$= 6 + 10y$

7. $\int_2^6 e^{2x+3y}\,dx = \frac{1}{2}e^{2x+3y}\Big|_2^6$

$= \frac{1}{2}(e^{12+3y} - e^{4+3y})$

9. $\int_0^3 ye^{4x+y^2}\,dx$

Let $u = 4x + y^2$; then $du = 2y\,dy$.

If $y = 0$ then $u = 4x$.

If $y = 3$ then $u = 4x + 9$.

$\int_{4x}^{4x+9} e^u \cdot \frac{1}{2}\,du = \frac{1}{2}e^u\Big|_{4x}^{4x+9}$

$= \frac{1}{2}(e^{4x+9} - e^{4x})$

11. $\int_1^2 \int_0^5 (x^4y + y)\,dx\,dy$

From Exercise 1

$\int_0^5 (x^4y + y)\,dx = 630y.$

Therefore,

$\int_1^2 \left[\int_0^5 (x^4y + y)\,dx \right] dy$

$= \int_1^2 630y\,dy$

$= 315y^2\Big|_1^2$

$= 315(4 - 1) = 945.$

13. $\int_0^1 \left[\int_3^6 x\sqrt{x^2 + 3y}\,dx \right] dy$

From Exercise 4,

$\int_3^6 x\sqrt{x^2 + 3y}\,dx$

$= \frac{1}{3}[(36 + 3y)^{3/2} - (9 + 3y)^3].$

$\int_0^1 \left[\int_3^6 x\sqrt{x^2 + 3y}\,dx \right] dy$

$= \int_0^1 \frac{1}{3}[(36 + 3y)^{3/2} - (9 + 3y)^{3/2}]\,dy$

Let $u = 36 + 3y$. Then $du = 3\,dy$.

When $y = 0, u = 36$.

When $y = 1, u = 39$.

Let $z = 9 + 3y$. Then $dz = 3\,dy$.

When $y = 0, z = 9$.

When $y = 1, z = 12$.

$\frac{1}{9}\left[\int_{36}^{39} u^{3/2}\,du - \int_9^{12} z^{3/2}\,dz \right]$

$= \frac{1}{9} \cdot \frac{2}{5}[(39)^{5/2} - (36)^{5/2}$

$\quad - (12)^{5/2} + (9)^{5/2}]$

$= \frac{2}{45}[(39)^{5/2} - (12)^{5/2} - 6^5 + 3^5]$

$= \frac{2}{45}(39^{5/2} - 12^{5/2} - 7533)$

15. $\displaystyle\int_1^2\left[\int_4^9\frac{3+5y}{\sqrt{x}}\,dx\right]dy$

From Exercise 5,

$$\int_4^9\frac{3+5y}{\sqrt{x}}\,dx=6+10y.$$

$$\int_1^2\left[\int_4^9\frac{3+5y}{\sqrt{x}}\,dx\right]dy$$

$$=\int_1^2(6+10y)\,dy$$

$$=6y\,\Big|_1^2+5y^2\,\Big|_1^2$$

$$=6(2-1)+5(4-1)$$

$$=6+15$$

$$=21$$

17. $\displaystyle\int_1^3\int_1^3\frac{dy\,dx}{xy}=\int_1^3\left[\int_1^3\frac{1}{xy}\,dy\right]dx$

$$=\int_1^3\left(\frac{1}{x}\ln|y|\right)\Big|_1^3\,dx$$

$$=\int_1^3\frac{\ln 3}{x}\,dx$$

$$=(\ln 3)\ln|x|\,\big|_1^3$$

$$=(\ln 3)(\ln 3-0)$$

$$=(\ln 3)^2$$

19. $\displaystyle\int_2^4\int_3^5\left(\frac{x}{y}+\frac{y}{3}\right)dx\,dy$

$$=\int_2^4\left(\frac{x^2}{2y}+\frac{yx}{3}\right)\Big|_3^5\,dy$$

$$=\int_2^4\left[\frac{25}{2y}+\frac{5y}{3}-\left(\frac{9}{2y}+\frac{3y}{3}\right)\right]dy$$

$$=\int_2^4\left(\frac{16}{2y}+\frac{2y}{3}\right)dy$$

$$=\left(8\ln|y|+\frac{y^2}{3}\right)\Big|_2^4$$

$$=8\,(\ln 4-\ln 2)+\frac{16}{3}-\frac{4}{3}$$

$$=8\ln\frac{4}{2}+\frac{12}{3}$$

$$=8\ln 2+4$$

21. $\displaystyle\iint_R(3x^2+4y)\,dx\,dy;$

$$0\le x\le 3, 1\le y\le 4$$

$$\iint_R(3x^2+4y)\,dx\,dy$$

$$=\int_1^4\int_0^3(3x^2+4y)\,dx\,dy$$

$$=\int_1^4(x^3+4xy)\Big|_0^3\,dy$$

$$=\int_1^4(27+12y)\,dy$$

$$=(27y+6y^2)\Big|_1^4$$

$$=(108+96)-(27+6)=171$$

23. $\displaystyle\iint_R\sqrt{x+y}\,dy\,dx; 1\le x\le 3, 0\le y\le 1$

$$\iint_R\sqrt{x+y}\,dy\,dx$$

$$=\int_1^3\int_0^1(x+y)^{1/2}\,dy\,dx$$

$$=\int_1^3\left[\frac{2}{3}(x+y)^{3/2}\right]\Big|_0^1\,dx$$

$$=\int_1^3\frac{2}{3}[(x+1)^{3/2}-x^{3/2}]\,dx$$

$$=\frac{2}{3}\cdot\frac{2}{5}\left[(x+1)^{5/2}-x^{5/2}\right]\Big|_1^3$$

$$=\frac{4}{15}(4^{5/2}-3^{5/2}-2^{5/2}+1^{5/2})$$

$$=\frac{4}{15}(32-3^{5/2}-2^{5/2}+1)$$

$$=\frac{4}{15}(33-3^{5/2}-2^{5/2})$$

25. $\displaystyle\iint_R \frac{3}{(x+y)^2}\,dy\,dx;\, 2 \le x \le 4,\, 1 \le y \le 6$

$\displaystyle\iint_R \frac{3}{(x+y)^2}\,dy\,dx$

$\displaystyle = -3\int_2^4\int_1^6 (x+y)^{-2}\,dy\,dx$

$\displaystyle = -3\int_2^4 (x+y)^{-1}\Big|_1^6\,dx$

$\displaystyle = -3\int_2^4 \left(\frac{1}{x+6} - \frac{1}{x+1}\right)dx$

$\displaystyle = -3\big(\ln|x+6| - \ln|x+1|\big)\Big|_2^4$

$\displaystyle = -3\left(\ln\left|\frac{x+6}{x+1}\right|\right)\Big|_2^4$

$\displaystyle = -3\left(\ln 2 - \ln\frac{8}{3}\right)$

$\displaystyle = -3\ln\frac{2}{\frac{8}{3}}$

$\displaystyle = -3\ln\frac{3}{4}\ \text{or}\ 3\ln\frac{4}{3}$

27. $\displaystyle\iint_R ye^{(x+y^2)}\,dx\,dy;\, 2 \le x \le 3,\, 0 \le y \le 2$

$\displaystyle\iint_R ye^{(x+y^2)}\,dx\,dy$

$\displaystyle = \int_0^2\int_2^3 ye^{x+y^2}\,dx\,dy$

$\displaystyle = \int_0^2 ye^{x+y^2}\Big|_2^3\,dy$

$\displaystyle = \int_0^2 (ye^{3+y^2} - ye^{2+y^2})\,dy$

$\displaystyle = e^3\int_0^2 ye^{y^2}\,dy - e^2\int_0^2 ye^{y^2}\,dy$

$\displaystyle = \frac{e^3}{2}(e^{y^2})\Big|_0^2 - \frac{e^2}{2}(e^{y^2})\Big|_0^2$

$\displaystyle = \frac{e^3}{2}(e^4 - e^0) - \frac{e^2}{2}(e^4 - e^0)$

$\displaystyle = \frac{1}{2}(e^7 - e^6 - e^3 + e^2)$

29. $z = 8x + 4y + 10;\, -1 \le x \le 1,\, 0 \le y \le 3$

$\displaystyle V = \int_{-1}^1\int_0^3 (8x + 4y + 10)\,dy\,dx$

$\displaystyle = \int_{-1}^1 (8xy + 2y^2 + 10y)\Big|_0^3\,dx$

$\displaystyle = \int_{-1}^1 (24x + 18 + 30 - 0)\,dx$

$\displaystyle = \int_{-1}^1 (24x + 48)\,dx$

$\displaystyle = (12x^2 + 48x)\Big|_{-1}^1$

$\displaystyle = (12 + 48) - (12 - 48) = 96$

31. $z = x^2;\, 0 \le x \le 2,\, 0 \le y \le 5$

$\displaystyle V = \int_0^2\int_0^5 x^2\,dy\,dx$

$\displaystyle = \int_0^2 x^2 y\Big|_0^5\,dx$

$\displaystyle = \int_0^2 5x^2\,dx$

$\displaystyle = \frac{5}{3}x^3\Big|_0^2$

$\displaystyle = \frac{40}{3}$

33. $z = x\sqrt{x^2 + y};\, 0 \le x \le 1,\, 0 \le y \le 1$

$\displaystyle V = \int_0^1\int_0^1 x\sqrt{x^2 + y}\,dx\,dy$

Let $u = x^2 + y$. Then $du = 2x\,dx$.

When $x = 0, u = y$.

When $x = 1, u = 1 + y$.

$\displaystyle = \int_0^1\left[\int_y^{1+y} u^{1/2}\,du\right]dy$

$\displaystyle = \int_0^1 \frac{1}{2}\left(\frac{2}{3}u^{3/2}\right)\Big|_y^{1+y}\,dy$

$\displaystyle = \int_0^1 \frac{1}{3}[(1+y)^{3/2} - y^{3/2}]\,dy$

$\displaystyle = \frac{1}{3}\cdot\frac{2}{5}[(1+y)^{5/2} - y^{5/2}]\Big|_0^1$

$\displaystyle = \frac{2}{15}(2^{5/2} - 1 - 1)$

$\displaystyle = \frac{2}{15}(2^{5/2} - 2)$

35. $z = \dfrac{xy}{(x^2 + y^2)^2}; 1 \le x \le 2, 1 \le y \le 4$

$V = \displaystyle\int_1^2 \int_1^4 \frac{xy}{(x^2 + y^2)^2}\, dy\, dx$

$= \displaystyle\int_1^2 \left[\int_1^4 xy(x^2 + y^2)^{-2}\, dy \right] dx$

$= \displaystyle\int_1^2 \left[\int_1^4 \frac{1}{2}x(x^2 + y^2)^{-2}(2y)dy \right] dx$

$= \displaystyle\int_1^2 \left[-\frac{1}{2}x(x^2 + y^2)^{-1} \Big|_1^4 \right] dx$

$= \displaystyle\int_1^2 \left[-\frac{1}{2}x(x^2 + 16)^{-1} + \frac{1}{2}x(x^2 + 1)^{-1} \right] dx$

$= -\frac{1}{2}\displaystyle\int_1^2 \frac{1}{2}(x^2 + 16)^{-1}(2x)\, dx$

$\quad + \frac{1}{2}\displaystyle\int_1^2 \frac{1}{2}(x^2 + 1)^{-1}(2x)\, dx$

$= -\frac{1}{2} \cdot \frac{1}{2}\ln\left|x^2 + 16\right| \Big|_1^2$

$\quad + \frac{1}{2} \cdot \frac{1}{2}\ln\left|x^2 + 1\right| \Big|_1^2$

$= -\frac{1}{4} \cdot \ln 20 + \frac{1}{4}\ln 17$

$\quad + \frac{1}{4}\ln 5 - \frac{1}{4}\ln 2$

$= \frac{1}{4}(-\ln 20 + \ln 17 + \ln 5 - \ln 2)$

$= \frac{1}{4}\ln \frac{(17)(5)}{(20)(2)}$

$= \frac{1}{4}\ln \frac{17}{8}$

37. $\displaystyle\iint_R xe^{xy}\, dx\, dy; 0 \le x \le 2; 0 \le y \le 1$

$\displaystyle\iint_R xe^{xy}\, dx\, dy$

$= \displaystyle\int_0^2 \int_0^1 xe^{xy}\, dy\, dx$

$= \displaystyle\int_0^2 \frac{x}{x}e^{xy}\Big|_0^1 dx$

$= \displaystyle\int_0^2 (e^x - e^0)\, dx$

$= (e^x - x)\Big|_0^2$

$= e^2 - 2 - e^0 + 0$

$= e^2 - 3$

39. $\displaystyle\int_2^4 \int_2^{x^2} (x^2 + y^2)dy\, dx$

$= \displaystyle\int_2^4 \left(x^2 y + \frac{y^3}{3} \right)\Big|_2^{x^2} dx$

$= \displaystyle\int_2^4 \left(x^4 + \frac{x^6}{3} - 2x^2 - \frac{8}{3} \right) dx$

$= \left(\frac{x^5}{5} + \frac{x^7}{21} - \frac{2}{3}x^3 - \frac{8}{3}x \right)\Big|_2^4$

$= \frac{1024}{5} + \frac{16{,}384}{21} - \frac{2}{3}(64) - \frac{8}{3}(4)$

$\quad - \left(\frac{32}{5} + \frac{128}{21} - \frac{16}{3} - \frac{16}{3} \right)$

$= \frac{1024}{5} - \frac{32}{5} + \frac{16{,}384 - 128}{21}$

$\quad - \frac{128}{3} - \frac{32}{3} - \left(\frac{-32}{3} \right)$

$= \frac{992}{5} + \frac{16{,}256}{21} - \frac{128}{3}$

$= \frac{20{,}832}{105} + \frac{81{,}280}{105} - \frac{4480}{105}$

$= \frac{97{,}632}{105}$

41. $\displaystyle\int_0^4 \int_0^x \sqrt{xy}\, dy\, dx$

$\displaystyle = \int_0^4 \int_0^x (xy)^{1/2}\, dy\, dx$

$\displaystyle = \int_0^4 \left[\frac{2(xy)^{3/2}}{3x}\right]\Bigg|_0^x dx$

$\displaystyle = \frac{2}{3}\int_0^4 \left[\frac{\left(\sqrt{x^2}\right)^3}{x} - \frac{0}{x}\right] dx$

$\displaystyle = \frac{2}{3}\int_0^4 x^2\, dx = \frac{2}{3}\cdot\frac{x^3}{3}\Bigg|_0^4 = \frac{2}{9}(64)$

$\displaystyle = \frac{128}{9}$

43. $\displaystyle\int_2^6 \int_{2y}^{4y} \frac{1}{x}\, dx\, dy$

$\displaystyle = \int_2^6 (\ln|x|)\Bigg|_{2y}^{4y} dy$

$\displaystyle = \int_2^6 (\ln|4y| - \ln|2y|)\, dy$

$\displaystyle = \int_2^6 \ln\left|\frac{4y}{2y}\right| dy$

$\displaystyle = \int_2^6 \ln 2\, dy$

$\displaystyle = (\ln 2)y\Big|_2^6$

$\displaystyle = (\ln 2)(6 - 2) = 4\ln 2$

Note: We can write $4\ln 2$ as $\ln 2^4$, or $\ln 16$.

45. $\displaystyle\int_0^4 \int_1^{e^x} \frac{x}{y}\, dy\, dx$

$\displaystyle = \int_0^4 (x\ln|y|)\Bigg|_1^{e^x} dx$

$\displaystyle = \int_0^4 (x\ln e^x - x\ln 1)\, dx$

$\displaystyle = \int_0^4 x^2\, dx = \frac{x^3}{3}\Bigg|_0^4 = \frac{64}{3}$

47. $\displaystyle\iint_R (5x + 8y)\, dy\, dx;\ 1 \le x \le 3,$

$0 \le y \le x - 1$

$\displaystyle\iint_R (5x + 8y)\, dy\, dx$

$\displaystyle = \int_1^3 \int_0^{x-1} (5x + 8y)\, dy\, dx$

$\displaystyle = \int_1^3 (5xy + 4y^2)\Bigg|_0^{x-1} dx$

$\displaystyle = \int_1^3 [5x(x - 1) + 4(x - 1)^2 - 0]\, dx$

$\displaystyle = \int_1^3 (9x^2 - 13x + 4)\, dx$

$\displaystyle = \left(3x^3 - \frac{13}{2}x^2 + 4x\right)\Bigg|_1^3$

$\displaystyle = \left(81 - \frac{117}{2} + 12\right) - \left(3 - \frac{13}{2} + 4\right)$

$= 34$

49. $\displaystyle\iint_R (4 - 4x^2)\, dy\, dx;\ 0 \le x \le 1,$

$0 \le y \le 2 - 2x$

$\displaystyle\iint_R (4 - 4x^2)\, dy\, dx$

$\displaystyle = \int_0^2 \int_0^{2-2x} 4(1 - x^2)\, dy\, dx$

$\displaystyle = \int_0^1 [4(1 - x^2)y]\Bigg|_0^{2(1-x)} dx$

$\displaystyle = \int_0^1 4(1 - x^2)(2)(1 - x)\, dx$

$\displaystyle = 8\int_0^1 (1 - x - x^2 + x^3)\, dx$

$\displaystyle = 8\left(x - \frac{x^2}{2} - \frac{x^3}{3} + \frac{x^4}{4}\right)\Bigg|_0^1$

$\displaystyle = 8\left(1 - \frac{1}{2} - \frac{1}{3} + \frac{1}{4}\right)$

$\displaystyle = 8\left(\frac{1}{2} - \frac{1}{12}\right)$

$\displaystyle = 8\cdot\frac{5}{12} = \frac{10}{3}$

51. $\displaystyle\iint_R e^{x/y^2}\,dx\,dy;\, 1 \le y \le 2, 0 \le x \le y^2$

$\displaystyle\iint_R e^{x/y^2}\,dx\,dy$

$\displaystyle = \int_1^2 \int_0^{y^2} e^{x/y^2}\,dx\,dy$

$\displaystyle = \int_1^2 \left[y^2 e^{x/y^2}\right]\Big|_0^{y^2}\,dy$

$\displaystyle = \int_1^2 \left(y^2 e^{y^2/y^2} - y^2 e^0\right)\,dy$

$\displaystyle = \int_1^2 \left(ey^2 - y^2\right)\,dy$

$\displaystyle = (e-1)\frac{y^3}{3}\Big|_1^2$

$\displaystyle = (e-1)\left(\frac{8}{3} - \frac{1}{3}\right)$

$\displaystyle = \frac{7(e-1)}{3}$

53. $\displaystyle\iint_R x^3 y\,dy\,dx;\, R \text{ bounded by } y = x^2,\ y = 2x$

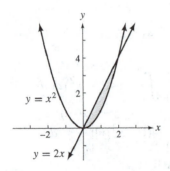

The points of intersection can be determined by solving the following system for x.

$$y = x^2$$
$$y = 2x$$

$$x^2 = 2x$$
$$x(x-2) = 0$$
$$x = 0 \quad\text{or}\quad x = 2$$

Therefore,

$\displaystyle\iint_R x^3 y\,dx\,dy$

$\displaystyle = \int_0^2 \int_{x^2}^{2x} x^3 y\,dy\,dx = \int_0^2 \left(x^3 \frac{y^2}{2}\right)\Big|_{x^2}^{2x}\,dx$

$\displaystyle = \int_0^2 \left[x^3 \frac{(4x^2)}{2} - x^3 \frac{(x^4)}{2}\right]\,dx$

$\displaystyle = \int_0^2 \left(2x^5 - \frac{x^7}{2}\right)\,dx$

$\displaystyle = \left(\frac{1}{3}x^6 - \frac{1}{16}x^8\right)\Big|_0^2$

$\displaystyle = \frac{1}{3}\cdot 2^6 - \frac{1}{16}\cdot 2^8$

$\displaystyle = \frac{64}{3} - 16$

$\displaystyle = \frac{16}{3}.$

55. $\displaystyle\iint_R \frac{dy\,dx}{y};\, R \text{ bounded by } y = x,\ y = \frac{1}{x},$

$x = 2.$

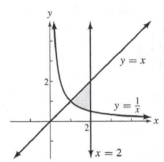

The graphs of $y = x$ and $y = \dfrac{1}{x}$ intersect at $(1, 1)$.

$\displaystyle\int_1^2 \int_{1/x}^x \frac{dy}{y}\,dx = \int_1^2 \ln y\Big|_{1/x}^x\,dx$

$\displaystyle = \int_1^2 \left(\ln x - \ln\frac{1}{x}\right)\,dx$

$\displaystyle = \int_1^2 2\ln x\,dx$

$\displaystyle = 2(x\ln x - x)\Big|_1^2$

$\displaystyle = 2[(2\ln 2 - 2) - (\ln 1 - 1)]$

$\displaystyle = 4\ln 2 - 2$

57. $\displaystyle\int_0^{\ln 2} \int_{e^y}^2 \frac{1}{\ln x}\, dx\, dy$

Changing the order of integration,

$$\int_0^{\ln 2} \int_{e^y}^2 \frac{1}{\ln x}\, dx\, dy$$

$$= \int_1^2 \int_0^{\ln x} \frac{1}{\ln x}\, dy\, dx$$

$$= \int_1^2 \left[\frac{1}{\ln x}\, y \Big|_0^{\ln x} \right] dx$$

$$= \int_1^2 (1 - 0)\, dx$$

$$= x \Big|_1^2$$

$$= 2 - 1 = 1$$

61. $f(x, y) = 6xy + 2x;\ 2 \le x \le 5, 1 \le y \le 3$

The area of region R is

$$A = (5 - 2)(3 - 1) = 6.$$

The average value of f over R is

$$\frac{1}{A} \iint_R f(x, y)\, dy\, dx$$

$$= \frac{1}{6} \int_2^5 \int_1^3 (6xy + 2x)\, dy\, dx$$

$$= \frac{1}{6} \int_2^5 \left(3xy^2 + 2xy \right) \Big|_1^3 dx$$

$$= \frac{1}{6} \int_2^5 \left[(27x + 6x) - (3x + 2x) \right] dx$$

$$= \frac{1}{6} \int_2^5 28x\, dx$$

$$= \frac{1}{6} 14x^2 \Big|_2^5$$

$$= \frac{7}{3}(25 - 4) = 49.$$

63. $f(x,y) = e^{-5y+3x}; \; 0 \le x \le 2, 0 \le y \le 2$

The area of region R is

$(2-0)(2-0) = 4.$

The average value of f over R is

$$\frac{1}{4}\int_0^2 \int_0^2 e^{-5y+3x}\, dy\, dx = \frac{1}{4}\int_0^2 -\frac{1}{5}e^{-5y+3x}\bigg|_0^2 dx = \frac{1}{4}\int_0^2 -\frac{1}{5}[e^{3x-10} - e^{3x}]\, dx$$

$$= -\frac{1}{20}\left[\frac{1}{3}e^{3x-10} - \frac{1}{3}e^{3x}\right]\bigg|_0^2 = -\frac{1}{60}[e^{-4} - e^6 - e^{-10} + 1] = \frac{e^6 + e^{-10} - e^{-4} - 1}{60}.$$

65. The plane that intersects the axes has the equation

$z = 6 - 2x - 2y.$

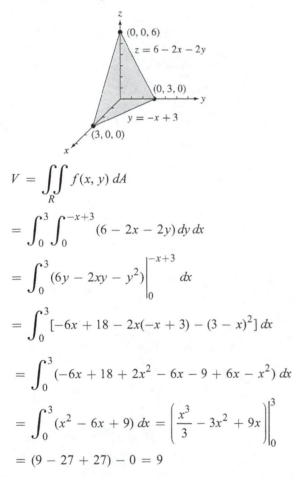

$$V = \iint_R f(x,y)\, dA$$

$$= \int_0^3 \int_0^{-x+3} (6 - 2x - 2y)\, dy\, dx$$

$$= \int_0^3 (6y - 2xy - y^2)\bigg|_0^{-x+3} dx$$

$$= \int_0^3 [-6x + 18 - 2x(-x+3) - (3-x)^2]\, dx$$

$$= \int_0^3 (-6x + 18 + 2x^2 - 6x - 9 + 6x - x^2)\, dx$$

$$= \int_0^3 (x^2 - 6x + 9)\, dx = \left(\frac{x^3}{3} - 3x^2 + 9x\right)\bigg|_0^3$$

$$= (9 - 27 + 27) - 0 = 9$$

The volume is 9 in³.

67. $P(x, y) = 500x^{0.2}y^{0.8}, 10 \leq x \leq 50,$
$20 \leq y \leq 40$

$A = 40 \cdot 20 = 800$

Average production:

$$\frac{1}{800} \int_{10}^{50} \int_{20}^{40} 500x^{0.2}y^{0.8} \, dy \, dx$$

$$= \frac{5}{8} \int_{10}^{50} \left. \frac{x^{0.2}y^{1.8}}{1.8} \right|_{20}^{40} dx = \frac{25}{72} \int_{10}^{50} x^{0.2}(40^{1.8} - 20^{1.8}) \, dx$$

$$= \frac{25(40^{1.8} - 20^{1.8})}{72} \cdot \left. \frac{x^{1.2}}{1.2} \right|_{10}^{50} = \frac{125}{432}(40^{1.8} - 20^{1.8})(50^{1.2} - 10^{1.2})$$

$$\approx 14{,}753 \text{ units}$$

69. $R = q_1p_1 + q_2p_2$ where
$q_1 = 300 - 2p_1,$
$q_2 = 500 - 1.2p_2, 25 \leq p_1 \leq 50,$
and $50 \leq p_2 \leq 75.$

$A = 25 \cdot 25 = 625$
$R = (300 - 2p_1)p_1 + (500 - 1.2p_2)p_2$
$R = 300p_1 - 2p_1^2 + 500p_2 - 1.2p_2^2$

Average Revenue:

$$\frac{1}{625} \int_{25}^{50} \int_{50}^{75} (300p_1 - 2p_1^2 + 500p_2 - 1.2p_2^2) dp_2 dp_1$$

$$= \frac{1}{625} \int_{25}^{50} \left. (300p_1p_2 - 2p_1^2 p_2 + 250p_2^2 - 0.4p_2^3) \right|_{50}^{75} dp_1$$

$$= \frac{1}{625} \int_{25}^{50} \left[\begin{array}{l} 22{,}500p_1 - 150p_1^2 + 1{,}406{,}250 - 168{,}750) \\ - (15{,}000p_1 - 100p_1^2 + 625{,}000 - 50{,}000) \end{array} \right] dp_1$$

$$= \frac{1}{625} \int_{25}^{50} (662{,}500 + 7500p_1 - 50p_1^2) dp_1$$

$$= \frac{1}{625} \left. \left(662{,}500p_1 + 3750p_1^2 - \frac{50p_1^3}{3} \right) \right|_{25}^{50}$$

$$= \frac{1}{625} \left(33{,}125{,}000 + 9{,}375{,}000 - \frac{6{,}250{,}000}{3} - 16{,}562{,}500 - 2{,}343{,}750 + \frac{781{,}250}{3} \right)$$

$$\approx \$34{,}833$$

71. $P(x, y) = 36xy - x^3 - 8y^3$

Areas $= (8 - 0)(4 - 0)$

$= 32$

The average profit is

$$\frac{1}{32} \iint_R (36xy - x^3 - 8y^3) \, dy \, dx$$

$$= \frac{1}{32} \int_0^8 \int_0^4 (36xy - x^3 - 8y^3) \, dy \, dx$$

$$= \frac{1}{32} \int_0^8 \left. \left(\frac{36xy^2}{2} - x^3 y - \frac{8y^4}{4} \right) \right|_0^4 dx$$

$$= \frac{1}{32} \int_0^8 \left[\frac{36x(4 - 0)^2}{2} - x^3(4 - 0) - \frac{8(4 - 0)^4}{4} \right] dx$$

$$= \frac{1}{32} \int_0^8 (288x - 4x^3 - 512) \, dx$$

$$= \frac{1}{32} \left. \left(\frac{288x^2}{2} - \frac{4x^4}{4} - 512x \right) \right|_0^8$$

$$= \frac{1}{32} \left[\frac{288(8 - 0)^2}{2} - \frac{4(8 - 0)^4}{4} - 512(8 - 0) \right]$$

$$= \frac{1}{32}(9216 - 4096 - 4096)$$

$$= \$32,000$$

Chapter 9   Review Exercises

1.   True

2.   True

3.   True

4.   True

5.   False: $f(x + h, y) = 3(x + h)^2$
$\qquad\qquad\qquad\quad + 2(x + h)y + y^2$

6.   False: (a, b) could be a saddle point.

7.   False: No; near a saddle point the function takes on values both larger and smaller than its value at the saddle point.

8.   True

9.   False: We need to test values of the function at nearby points that satisfy the constraints to tell if the point found represents a maximum or minimum.

10.   False: When dx and dy are interchanged, the limits on the first integral must be exchanged with the limits on the second integral.

11.   True

12.   False: The two integrals are over different regions, and neither region is a simple region of the sort that we deal with in this chapter.

17. $f(x, y) = -4x^2 + 6xy - 3$

$f(-1, 2) = -4(-1)^2 + 6(-1)(2) - 3 = -19$

$f(6, -3) = -4(6)^2 + 6(6)(-3) - 3$

$\qquad\qquad = -4(36) + (-108) - 3$

$\qquad\qquad = -255$

19. $f(x, y) = \dfrac{x - 2y}{x + 5y}$

$f(-1, 2) = \dfrac{(-1) - 2(2)}{(-1) + 5(2)} = \dfrac{-5}{9} = -\dfrac{5}{9}$

$f(6, -3) = \dfrac{(6) - 2(-3)}{(6) + 5(-3)} = \dfrac{12}{-9} = -\dfrac{4}{3}$

21.   The plane $x + y + z = 4$ intersects the axes at $(4, 0, 0)$, $(0, 4, 0)$, and $(0, 0, 4)$.

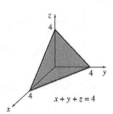

23.   The plane $5x + 2y = 10$ intersects the x- and y-axes at $(2, 0, 0)$ and $(0, 5, 0)$. Note that there is no z-intercept since $x = y = 0$ is not a solution of the equation of the plane.

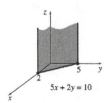

25. $x = 3$

The plane is parallel to the yz-plane. It intersects the x-axis at $(3, 0, 0)$.

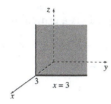

27. $z = f(x, y) = 3x^3 + 4x^2 y - 2y^2$

(a) $\dfrac{\partial z}{\partial x} = 9x^2 + 8xy$

(b) $\dfrac{\partial z}{\partial y} = 4x^2 - 4y$

$\left. \dfrac{\partial z}{\partial y} \right|(-1, 4) = 4(-1)^2 - 4(4) = -12$

(c) $\quad f_{xy}(x, y) = 8x$

$f_{xy}(x, y)(2, -1) = 8(2) = 16$

29. $f(x, y) = 6x^2 y^3 - 4y$

$f_x(x, y) = 12xy^3$

$f_y(x, y) = 18x^2 y^2 - 4$

31. $f(x, y) = \sqrt{4x^2 + y^2}$

$f_x(x, y) = \dfrac{1}{2}(4x^2 + y^2)^{-1/2}(8x)$

$= \dfrac{4x}{(4x^2 + y^2)^{-1/2}}$

$f_y(x, y) = \dfrac{1}{2}(4x^2 + y^2)^{-1/2}(2y)$

$= \dfrac{y}{(4x^2 + y^2)^{1/2}}$

33. $f(x, y) = x^3 e^{3y}$

$f_x(x, y) = 3x^2 e^{3y}$

$f_y(x, y) = 3x^3 e^{3y}$

35. $f(x, y) = \ln |2x^2 + y^2|$

$f_x(x, y) = \dfrac{1}{2x^2 + y^2} \cdot 4x$

$= \dfrac{4x}{2x^2 + y^2}$

$f_y(x, y) = \dfrac{1}{2x^2 + y^2} \cdot 2y$

$= \dfrac{2y}{2x^2 + y^2}$

37. $f(x, y) = 5x^3 y - 6xy^2$

$f_x(x, y) = 15x^2 y - 6y^2$

$f_{xx}(x, y) = 30xy$

$f_{xy}(x, y) = 15x^2 - 12y$

39. $f(x, y) = \dfrac{3x}{2x - y}$

$f_x(x, y) = \dfrac{(2x - y) \cdot 3 - 3x \cdot 2}{(2x - y)^2}$

$= \dfrac{-3y}{(2x - y)^2}$

$f_{xx}(x, y) = \dfrac{(2x - y)^2 \cdot 0 - (-3y) \cdot 2(2x - y) \cdot 2}{(2x - y)^4}$

$= \dfrac{12y}{(2x - y)^3}$

$f_{xy}(x, y) = \dfrac{\left[\begin{array}{c} (2x - y)^2 \cdot (-3) \\ - (-3y) \cdot 2(2x - y) \cdot (-1) \end{array} \right]}{(2x - y)^4}$

$= \dfrac{-6x - 3y}{(2x - y)^3}$

41. $f(x, y) = 4x^2 e^{2y}$

$f_x(x, y) = 8x e^{2y}$

$f_{xx}(x, y) = 8e^{2y}$

$f_{xy}(x, y) = 16x e^{2y}$

43. $f(x, y) = \ln |2 - x^2 y|$

$f_x(x, y) = \dfrac{1}{2 - x^2 y} \cdot (-2xy)$

$= \dfrac{2xy}{x^2 y - 2}$

$$f_{xx}(x, y) = \frac{(x^2 y - 2)2y - 2xy(2xy)}{(x^2 y - 2)^2}$$

$$= \frac{2y[(x^2 y - 2) - 2x^2 y]}{(x^2 y - 2)^2}$$

$$= \frac{2y(-x^2 y - 2)}{(x^2 y - 2)^2}$$

$$= \frac{-2x^2 y^2 - 4y}{(2 - x^2 y)^2}$$

$$f_{xy}(x, y) = \frac{2x(x^2 y - 2) - x^2(2xy)}{(x^2 y - 2)^2}$$

$$= \frac{2x[(x^2 y - 2) - x^2 y]}{(x^2 y - 2)^2}$$

$$= \frac{2x(-2)}{(x^2 y - 2)^2}$$

$$= \frac{-4x}{(2 - x^2 y)^2}$$

45. $z = 2x^2 - 3y^2 + 12y$

$$z_x(x, y) = 4x$$
$$z_y(x, y) = -6y + 12$$

If $z_x(x, y) = 0, x = 0$. If $z_y(x, y) = 0, y = 2$.

Therefore, $(0, 2)$ is a critical point.

$$z_{xx}(x, y) = 4$$
$$z_{yy}(x, y) = -6$$
$$z_{xy}(x, y) = 0$$

$$D = 4(-6) - 0^2 = -24 < 0$$

There is a saddle point at $(0, 2)$.

47. $f(x, y) = x^2 + 3xy - 7x + 5y^2 - 16y$
$f_x(x, y) = 2x + 3y - 7$
$f_y(x, y) = 3x + 10y - 16$

Solve the system $f_x(x, y) = 0, f_y(x, y) = 0$.

$$2x + 3y - 7 = 0$$
$$3x + 10y - 16 = 0$$

$$-6x - 9y + 21 = 0$$
$$\underline{6x + 20y - 32 = 0}$$
$$11y - 11 = 0$$
$$y = 1$$

$$2x + 3(1) - 7 = 0$$
$$2x = 4$$
$$x = 2$$

Therefore, $(2, 1)$ is a critical point.

$$f_{xx}(x, y) = 2$$
$$f_{yy}(x, y) = 10$$
$$f_{xy}(x, y) = 3$$
$$D = 2 \cdot 10 - 3^2 = 11 > 0$$

Since $f_{xx} = 2 > 0$, there is a relative minimum at $(2, 1)$.

49. $z = \frac{1}{2}x^2 + \frac{1}{2}y^2 + 2xy - 5x - 7y + 10$

$$z_x(x, y) = x + 2y - 5$$
$$z_y(x, y) = y + 2x - 7$$

Setting $z_x = z_y = 0$ and solving yields

$$x + 2y = 5$$
$$2x + y = 7$$

$$-2x - 4y = -10$$
$$\underline{2x + y = 7}$$
$$-3y = -3$$

$$y = 1, \ x = 3.$$

$z_{xx}(x, y) = 1, z_{yy}(x, y) = 1, z_{xy}(x, y) = 2$

For $(3, 1)$,

$$D = 1 \cdot 1 - 4 = -3 < 0.$$

Therefore, z has a saddle point at $(3, 1)$.

51. $z = x^3 + y^2 + 2xy - 4x - 3y - 2$

$$z_x(x, y) = 3x^2 + 2y - 4$$
$$z_y(x, y) = 2y + 2x - 3$$

Setting $z_x(x, y) = z_y(x, y) = 0$ yields

$$3x^2 + 2y - 4 = 0 \quad (1)$$
$$2y + 2x - 3 = 0. \quad (2)$$

Solving for $2y$ in equation (2) gives
$2y = -2x + 3.$

Substitute into equation (1).

$$3x^2 + (-2x) + 3 - 4 = 0$$
$$3x^2 - 2x - 1 = 0$$
$$(3x + 1)(x - 1) = 0$$

$$x = -\frac{1}{3} \quad \text{or} \quad x = 1$$
$$y = \frac{11}{6} \quad \text{or} \quad y = \frac{1}{2}$$

$z_{xx}(x, y) = 6x, \; z_{yy}(x, y) = 2, \; z_{xy}(x, y) = 2$

For $\left(-\frac{1}{3}, \frac{11}{6}\right)$,

$$D = 6\left(-\frac{1}{3}\right)(2) - 4$$
$$= -4 - 4 = -8 < 0,$$

so z has a saddle point at $\left(-\frac{1}{3}, \frac{11}{6}\right)$.

$$D = 6(1)(2) - 4 = 8 > 0.$$

$z_{xx}\left(1, \frac{1}{2}\right) = 6 > 0$, so z has a relative minimum at $\left(1, \frac{1}{2}\right)$.

55. $f(x, y) = x^2 + y^2; \; x = y - 6.$

1. $g(x, y) = x - y + 6$

2. $F(x, y, \lambda) = x^2 + y^2 - \lambda(x - y + 6)$

3. $F_x(x, y, \lambda) = 2x - \lambda$
   $F_y(x, y, \lambda) = 2y + \lambda$
   $F_\lambda(x, y, \lambda) = -(x - y + 6)$

4. $\quad 2x - \lambda = 0 \quad (1)$
   $\quad 2y + \lambda = 0 \quad (2)$
   $\quad x - y + 6 = 0 \quad (3)$

5. Equations (1) and (2) give $\lambda = 2x$, and $\lambda = -2y$. Thus,

$$2x = -2y$$
$$x = -y.$$

Substituting into equation (3),
$$(-y) - y + 6 = 0$$
$$y = 3.$$
So $\quad\quad x = -3.$

And $\quad f(-3, 3) = (-3)^2 + (3)^2 = 18.$

$$F_{xx}(x, y, \lambda) = 2$$
$$F_{yy}(x, y, \lambda) = 2$$
$$F_{xy}(x, y, \lambda) = 0$$

$$D = 2 \cdot 2 - 0^2 = 4 > 0$$

Since $F_{xx} > 0$, there is a relative minimum of 18 at $(-3, 3)$.

57. Maximize $f(x, y) = xy^2$, subject to $x + y = 75$.

1. $g(x, y) = x + y - 75$

2. $F(x, y, \lambda) = xy^2 - \lambda(x + y - 75)$

3. $F_x(x, y, \lambda) = y^2 - \lambda$
   $F_y(x, y, \lambda) = 2xy - \lambda$
   $F_\lambda(x, y, \lambda) = -(x + y - 75)$

4. $\quad y^2 - \lambda = 0 \quad (1)$
   $\quad 2xy - \lambda = 0 \quad (2)$
   $\quad x + y - 75 = 0 \quad (3)$

5. Equations (1) and (2) give $\lambda = y^2$ and $\lambda = 2xy$. Thus,

$$y^2 = 2xy$$
$$y(y - 2x) = 0$$
$$y = 0 \quad \text{or} \quad y = 2x$$

Substituting $y = 0$ into equation (3),
$$x + (0) - 75 = 0$$
$$x = 75.$$

Substituting $y = 2x$ into equation (3),
$$x + (2x) - 75 = 0$$
$$x = 25.$$
So $\quad\quad y = 2x = 50.$

Thus,

$$f(75, 0) = 75(0)^2 = 0, \text{ and}$$
$$f(25, 50) = 25(50)^2 = 62{,}500.$$

Since $f(25, 50) > f(75, 0), x = 25$ and $y = 50$ will maximize $f(x, y) = xy^2$.

59. $z = f(x, y) = 6x^2 - 7y^2 + 4xy$
$x = 3, y = -1, dx = 0.03, dy = 0.01$

$f_x(x,y) = 12x + 4y$

$f_y(x,y) = -14y + 4x$

$dz = (12x + 4y)\,dx + (-14y + 4x)\,dy$

$\quad = [12(3) + 4(-1)](0.03)$

$\qquad + [-14(-1) + 4(3)](0.01)$

$\quad = 0.96 + 0.26 = 1.22$

61. Let $z = f(x,y) = \sqrt{x^2 + y^2}$.

Then

$$dz = f_x(x,y)\,dx + f_y(x,y)\,dy.$$

$$dz = \frac{1}{2}(x^2 + y^2)^{-1/2}(2x)\,dx$$

$$\quad + \frac{1}{2}(x^2 + y^2)(2y)\,dx$$

$$= \frac{x}{\sqrt{x^2 + y^2}}\,dx + \frac{y}{\sqrt{x^2 + y^2}}\,dy$$

To approximate $\sqrt{5.1^2 + 12.05^2}$, we let $x = 5$, $dx = 0.1$, $y = 12$, and $dy = 0.05$.

Then,

$$dz = \frac{5}{\sqrt{5^2 + 12^2}}(0.1) + \frac{12}{\sqrt{5^2 + 12^2}}(0.05)$$

$$= \frac{5}{13}(0.1) + \frac{12}{13}(0.05) \approx 0.0846.$$

Therefore,

$$f(5.1, 12.05) = f(5,12) + \Delta z$$

$$\approx f(5,12) + dz$$

$$= \sqrt{5^2 + 12^2} + 0.0846$$

$$f(5.1, 12.05) \approx 13.0846$$

Using a calculator, $\sqrt{5.1^2 + 12.05^2} \approx 13.0848$.

The absolute value of the difference of the two results is $|13.0846 - 13.0848| = 0.0002$.

63. $\displaystyle\int_1^4 \frac{4y - 3}{\sqrt{x}}\,dx$

$$= (4y - 3)(2\sqrt{x})\Big|_1^4$$

$$= (4y - 3)(2 \cdot 2 - 2 \cdot 1)$$

$$= 8y - 6$$

65. $\displaystyle\int_0^5 \frac{6x}{\sqrt{4x^2 + 2y^2}}\,dx$

Let $u = 4x^2 + 2y^2$; then $du = 8x\,dx$.

When $x = 0$, $u = 2y^2$.

When $x = 5$, $u = 100 + 2y^2$.

$$= \frac{3}{4}\int_{2y^2}^{100 + 2y^2} u^{-1/2}\,du$$

$$= \frac{3}{4}(2u^{1/2})\Big|_{2y^2}^{100 + 2y^2}$$

$$= \frac{3}{4} \cdot 2[(100 + 2y^2)^{1/2} - (2y^2)^{1/2}]$$

$$= \frac{3}{2}[(100 + 2y^2)^{1/2} - (2y^2)^{1/2}]$$

67. $\displaystyle\int_0^2\left[\int_0^4 (x^2 y^2 + 5x)\,dx\right]dy$

$$= \int_0^2 \left(\frac{1}{3}x^3 y^2 + \frac{5}{2}x^2\right)\Big|_0^4\,dx$$

$$= \int_0^2 \left(\frac{64}{3}y^2 + 40\right)dy$$

$$= \left(\frac{64y^3}{9} + 40y\right)\Big|_0^2$$

$$= \frac{64}{9}(8) + 40(2)$$

$$= \frac{512}{9} + \frac{720}{9}$$

$$= \frac{1232}{9}$$

69. $\displaystyle\int_3^4\left[\int_2^5 \sqrt{6x + 3y}\,dx\right]dy$

$$= \int_3^4 \frac{1}{9}(6x + 3y)^{3/2}\Big|_2^5\,dx$$

$$= \int_3^4 \frac{1}{9}[(30 + 3y)^{3/2} - (12 + 3y)^{3/2}]dy$$

$$= \frac{1}{3} \cdot \frac{1}{9} \cdot \frac{2}{5} \cdot [(30 + 3y)^{5/2} - (12 + 3y)^{5/2}]\Big|_3^4$$

$$= \frac{2}{135}[(42)^{5/2} - (24)^{5/2} - (39)^{5/2} + (21)^{5/2}]$$

71. $\displaystyle\int_2^4 \int_2^4 \frac{dx\,dy}{y} = \int_2^4 \left(\frac{1}{y}x\right)\Big|_2^4 dy$

$\displaystyle = \int_2^4 \left[\frac{1}{y}(4-2)\right] dy$

$\displaystyle = 2\ln|y|\,\Big\|_2^4$

$\displaystyle = 2\ln\left|\frac{4}{2}\right|$

$= 2\ln 2 \text{ or } \ln 4$

73. $\displaystyle\iint_R (x^2 + 2y^2)\,dx\,dy;\ 0 \le x \le 5, 0 \le y \le 2$

$\displaystyle\iint_R (x^2 + 2y^2)\,dx\,dy$

$\displaystyle = \int_0^2 \int_0^5 (x^2 + 2y^2)\,dx\,dy$

$\displaystyle = \int_0^2 \left(\frac{1}{3}x^3 + 2xy^2\right)\Big|_0^5 dy$

$\displaystyle = \int_0^2 \left[\left(\frac{125}{3} + 10y^2\right) - 0\right] dy$

$\displaystyle = \int_0^2 \left(\frac{125}{3} + 10y^2\right) dy$

$\displaystyle = \left(\frac{125}{3}y + \frac{10}{3}y^3\right)\Big|_0^2$

$\displaystyle = \frac{250}{3} + \frac{80}{3} = 110$

75. $\displaystyle\iint_R \sqrt{y + x}\,dx\,dy;\ 0 \le x \le 7, 1 \le y \le 9$

$\displaystyle\iint_R \sqrt{y + x}\,dx\,dy$

$\displaystyle = \int_0^7 \int_1^9 \sqrt{y + x}\,dy\,dx$

$\displaystyle = \int_0^7 \left[\frac{2}{3}(y + x)^{3/2}\right]\Big|_1^9 dx$

$\displaystyle = \int_0^7 \frac{2}{3}[(9 + x)^{3/2} - (1 + x)^{3/2}]\,dx$

$\displaystyle = \frac{2}{3}\cdot\frac{2}{5}[(9 + x)^{5/2} - (1 + x)^{5/2}]\Big|_0^7$

$\displaystyle = \frac{4}{15}[(16)^{5/2} - (8)^{5/2} - (9)^{5/2} + (1)^{5/2}]$

$\displaystyle = \frac{4}{15}[4^5 - (2\sqrt{2})^5 - 3^5 + 1]$

$\displaystyle = \frac{4}{15}(1024 - 32(4\sqrt{2}) - 243 + 1)$

$\displaystyle = \frac{4}{15}(782 - 128\sqrt{2})$

$\displaystyle = \frac{4}{15}(782 - 8^{5/2})$

77. $z = x + 8y + 4;\ 0 \le x \le 3, 1 \le y \le 2$

$\displaystyle V = \int_0^3 \int_1^2 (x + 8y + 4)\,dy\,dx$

$\displaystyle = \int_0^3 (xy + 4y^2 + 4y)\Big|_1^2 dx$

$\displaystyle = \int_0^3 [(2x + 16 + 8) - (x + 4 + 4)]\,dx$

$\displaystyle = \int_0^3 (x + 16)\,dx$

$\displaystyle = \left(\frac{1}{2}x^2 + 16x\right)\Big|_0^3$

$\displaystyle = \left(\frac{9}{2} + 48\right) - 0 = \frac{105}{2}$

79. $\displaystyle\int_0^1\int_0^{2x} xy\,dy\,dx$

$\displaystyle = \int_0^1 \left(\frac{xy^2}{2}\right)\Big|_0^{2x} dx$

$\displaystyle = \int_0^1 \frac{x}{2}(4x^2 - 0)\,dx$

$\displaystyle = \int_0^1 2x^3\,dx$

$\displaystyle = \left(\frac{1}{2}x^4\right)\Big|_0^1 = \frac{1}{2}$

81. $\displaystyle\int_0^1\int_{x^2}^x x^3 y\,dy\,dx$

$\displaystyle \int_0^1 \left(\frac{x^3}{2}y^2\right)\Big|_{x^2}^x dx$

$\displaystyle = \int_0^1 \frac{x^3}{2}(x^2 - x^4)\,dx$

$\displaystyle = \frac{1}{2}\int_0^1 (x^5 - x^7)\,dx$

$\displaystyle = \frac{1}{2}\left(\frac{x^6}{6} - \frac{x^8}{8}\right)\Big|_0^1$

$\displaystyle = \frac{1}{2}\left(\frac{1}{6} - \frac{1}{8}\right) = \frac{1}{2}\cdot\frac{1}{24} = \frac{1}{48}$

83. $\displaystyle\int_0^2\int_{x/2}^1 \frac{1}{y^2 + 1}\,dy\,dx$

Change the order of integration.

$\displaystyle\int_0^2\int_{x/2}^1 \frac{1}{y^2 + 1}\,dy\,dx$

$\displaystyle = \int_0^1\int_0^{2y} \frac{1}{y^2 + 1}\,dx\,dy$

$\displaystyle = \int_0^1 \frac{x}{y^2 + 1}\Big|_0^{2y}\,dy$

$\displaystyle = \int_0^1 \left[\frac{1}{y^2 + 1}(2y) - \frac{1}{y^2 + 1}(0)\right]dy$

$\displaystyle = \int_0^1 \frac{2y}{y^2 + 1}\,dy$

$\displaystyle = \ln(y^2 + 1)\Big|_0^1$

$= \ln 2 - \ln 1$

$= \ln 2 - 0 = \ln 2$

85. $\displaystyle\iint_R (2x + 3y)\,dx\,dy;\ 0 \le y \le 1,$

$y \le x \le 2 - y$

$\displaystyle\int_0^1\int_y^{2-y} (2x + 3y)\,dx\,dy$

$\displaystyle = \int_0^1 (x^2 + 3xy)\Big|_y^{2-y}\,dy$

$\displaystyle = \int_0^1 [(2 - y)^2 - y^2 + 3y(2 - y - y)]\,dy$

$\displaystyle = \int_0^1 (4 - 4y + y^2 - y^2 + 6y - 6y^2)\,dy$

$\displaystyle = \int_0^1 (4 + 2y - 6y^2)\,dy$

$\displaystyle = (4y + y^2 - 2y^3)\Big|_0^1$

$= 4 + 1 - 2 = 3$

87. $C(x, y) = 4x^2 + 5y^2 - 4xy + \sqrt{x}$

(a) $C(10, 5)$

$= 4(10)^2 + 5(5)^2 - 4(10)(5) + \sqrt{10}$

$= 400 + 125 - 200 + \sqrt{10}$

$= 325 + \sqrt{10} \approx 328.16$

The cost is about \$328.16.

(b) $C(15, 10)$

$= 4(15)^2 + 5(10)^2 - 4(15)(10) + \sqrt{15}$

$= 900 + 500 - 600 + \sqrt{15}$

$= 800 + \sqrt{15} \approx 803.87$

The cost is about \$803.87.

(c) $C(20, 20)$

$= 4(20)^2 + 5(20)^2 - 4(20)(20) + \sqrt{20}$

$= 1600 + 2000 - 1600 + \sqrt{20}$

$= 2000 + \sqrt{20} \approx 2004.47$

The cost is about \$2004.47.

89. $z = x^{0.7}y^{0.3}$

(a) The marginal productivity of labor is

$\displaystyle\frac{\partial z}{\partial x} = 0.7x^{0.7-1}y^{0.3} = \frac{0.7y^{0.3}}{x^{0.3}}.$

(b) The marginal productivity of capital is

$\displaystyle\frac{\partial z}{\partial y} = 0.3x^{0.7}y^{0.3-1} = \frac{0.3x^{0.7}}{y^{0.7}}.$

91. Maximize $f(x, y) = xy^3$ subject to
$2x + 4y = 80$.

1. $g(x, y) = 2x + 4y - 80$

2. $F(x, y) = xy^3 - \lambda(2x + 4y - 80)$

3. $F_x(x, y, \lambda) = y^3 - 2\lambda$
    $F_y(x, y, \lambda) = 3xy^2 - 4\lambda$
    $F_\lambda(x, y, \lambda) = -2x - 4y + 80$

4. $y^3 - 2\lambda = 0$
    $3xy^2 - 4\lambda = 0$
    $-2x - 4y + 80 = 0$

5. Use the first and second equations to express
    4λ and eliminate 4λ.

    $$2y^3 = 4\lambda$$
    $$3xy^2 = 4\lambda$$
    $$2y^3 = 3xy^2$$

    If y equals 0, the utility will have a
    minimum value of 0. So we can assume
    $y \neq 0$ and divide by y^2 to find that
    $2y = 3x$. Substitute $6x$ for $4y$ in the last
    equation.
    $$-2x - 6x + 80 = 0$$
    $$x = 10, y = \frac{3}{2}x = 15.$$
    $$f(10, 15) = 33,750.$$

    The maximum utility is 33,750, obtained by
    purchasing 10 units of x and 15 units of y.

93. $C(x, y) = 100 \ln(x^2 + y) + e^{xy/20}$
$x = 15$, $y = 9$, $dx = 1$, $dy = -1$

$$dC = \left(\frac{200x}{x^2 + y} + \frac{y}{20} e^{xy/20} \right) dx$$
$$+ \left(\frac{100}{x^2 + y} + \frac{x}{20} e^{xy/20} \right) dy$$

$dC(15, 9)$
$$= \left(\frac{200(15)}{15^2 + 9} + \frac{9}{20} e^{(15)(9)/20} \right)(1)$$
$$+ \left(\frac{100}{15^2 + 9} + \frac{15}{20} e^{(9)(15)/20} \right)(-1)$$
$$= \frac{1450}{117} - \frac{3}{10} e^{27/4}$$
$$= -243.82$$

Costs decrease by $243.82.

95. $V = \frac{4}{3}\pi r^3, r = 2$ ft,

$dr = 1$ in $= \frac{1}{12}$ ft

$dV = 4\pi r^2 dr = 4\pi(2)^2 \left(\frac{1}{12} \right) \approx 4.19$ ft^3

97. $P(x, y) = 0.01(-x^2 + 3xy + 160x - 5y^2$
$+ 200y + 2600)$
with $x + y = 280$.

(a) $y = 280 - x$

$P(x) = 0.01[-x^2 + 3x(280 - x) + 160x$
$- 5(280 - x)^2 + 200(280 - x)$
$+ 2600]$
$= 0.01(-x^2 + 840x - 3x^2 + 160x$
$- 392,000 + 2800x - 5x^2$
$+ 56,000 - 200x + 2600)$

$P(x) = 0.01(-9x^2 + 3600x - 333,400)$
$P'(x) = 0.01(-18x + 3600)$
$0.01(-18x + 3600) = 0$
$-18x = -3600$
$x = 200$

If $x < 200, P'(x) > 0$, and if $x > 200$,
$P'(x) < 0$.

Therefore, P is maximum when $x = 200$.
If $x = 200$, $y = 80$.

$P(200, 80)$

$$= 0.01[-200^2 + 3(200)(80) + 160(200)$$
$$- 5(80)^2 + 200(80) + 2600]$$
$$= 0.01(26,600) = 266$$

Thus, \$200 spent on fertilizer and \$80 spent on seed will produce a maximum profit of \$266 per acre.

(b) $P(x, y) = 0.01(-x^2 + 3xy + 160x - 5y^2$
$$+ 200y + 2600)$$

$P_x(x, y) = 0.01(-2x + 3y + 160)$

$P_y(x, y) = 0.01(3x - 10y + 200)$

$$0.01(-2x + 3y + 160) = 0$$
$$0.01(3x - 10y + 200) = 0$$

These equations simplify to

$$-2x + 3y = -160$$
$$3x - 10y = -200.$$

Solve this system.

$$-6x + 9y = -480$$
$$\underline{6x - 20y = -400}$$
$$-11y = -880$$
$$y = 80$$

If $y = 80$,

$$3x - 10(80) = -200$$
$$3x = 600$$
$$x = 200.$$

$P_{xx}(x, y) = 0.01(-2) = -0.02$

$P_{yy}(x, y) = 0.01(-10) = -0.1$

$P_{xy}(x, y) = 0$

For $(200, 80)$, $D = (-0.02)(-0.1) - 0^2$
$= 0.002 > 0$, and $P_{xx} < 0$, so there is a relative maximum at $(200, 80)$.

$P(200, 80) = 266$, as in part (a) Thus, \$200 spent on fertilizer and \$80 spent on seed will produce a maximum profit of \$ 266 per acre.

(c) Maximize $P(x, y)$

$$= 0.01(-x^2 + 3xy + 160x - 5y^2$$
$$+ 200y + 2600)$$

subject to $x + y = 280$.

1.　$g(x, y) = x + y - 280$

2　$F(x, y, \lambda)$

$$= 0.01(-x^2 + 3xy + 160x - 5y^2$$
$$+ 200y + 2600) - \lambda(x + y - 280)$$

3.　$F_x = 0.01(-2x + 3y + 160) - \lambda$

　　$F_y = 0.01(3x - 10y + 200) - \lambda$

　　$F_\lambda = -(x + y - 280)$

4.　$0.01(-2x + 3y + 160) - \lambda = 0$　(1)

　　$0.01(3x - 10y + 200) - \lambda = 0$　(2)

　　　　　　　　$x + y - 280 = 0$　(3)

5.　Equations (1) and (2) give

$$0.01(-2x + 3y + 160)$$
$$= 0.01(3x - 10y + 200)$$
$$-2x + 3y + 160$$
$$= 3x - 10y + 200$$
$$-5x + 13y$$
$$= 40.$$

Multiplying equation (3) by 5 gives

$$5x + 5y - 1400 = 0.$$

$$-5x + 13y = 40$$
$$\underline{5x + 5y = 1400}$$
$$18y = 1440$$
$$y = 80$$

If $y = 80$,

$$5x + 5(80) = 1400$$
$$5x = 1000$$
$$x = 200.$$

Thus, $P(200, 80)$ is a maximum. As before, $P(200, 80) = 266$.

Thus, \$200 spent on fertilizer and \$80 spent on seed will produce a maximum profit of \$266 per acre.

99. The average weekly cost will be given by the integral

$$\frac{1}{2500} \iint\limits_R C(x, y) \, dx \, dy,$$

where R is the region defined the inequalities $100 \le x \le 150$ and $50 \le y \le 100$, which has area
$(50)(50) = 2500$. Compute the integral as an iterated integral:

$$\frac{1}{2500}\int_{50}^{100}\int_{100}^{150}\left(0.03x^2 + 6y + 2xy + 10\right)dx\,dy$$

First compute the inner integral:

$$\int_{100}^{150}\left(0.03x^2 + 6y + 2xy + 10\right)dx$$

$$= \left(0.01x^3 + 6xy + x^2y + 10x\right)\Big|_{100}^{150}$$

$$= 24{,}250 + 12{,}800y$$

The outer integral is then:

$$\frac{1}{2500}\int_{50}^{100}\left(24{,}250 + 12{,}800y\right)dy$$

$$= \frac{1}{2500}\left(24{,}250y + 6400y^2\right)\Big|_{50}^{100}$$

$$= \frac{49{,}212{,}500}{2500} = 19{,}685$$

The average weekly cost for the two products is $19,685.

101. $T(A,W,S) = -18.37 - 0.09A$
$\qquad\qquad\quad + 0.34W + 0.25S$

(a) $T(65,85,180) = -18.37 - 0.09(65)$
$\qquad\qquad\qquad\qquad + 0.34(85) + 0.25(180)$
$\qquad\qquad\qquad = 49.68$

The total body water is 49.68 liters.

(b) $\qquad T_A(A,M,S) = -0.09$

The approximate change in total body water if age is increased by 1 yr and mass and height are held constant is −0.09 liter.

$$T_M(A,M,S) = 0.34$$

The approximate change in total body water if mass is increased by 1 kg and height are held constant is 0.34 liter.

$$T_S(A,M,S) = 0.25$$

The approximate change in total body water if height is increased by 1 cm and age and mass are held constant is 0.25 liter.

103. (a) $f(60,1900) \approx 50$

In 1900, 50% of those born 60 years earlier are still alive.

(b) $f(70,2000) \approx 75$

In 2000, 75% of those born 70 years earlier are still alive.

(c) $f_x(60,1900) \approx -1.25$

In 1900, the percent of those born 60 years earlier who are still alive was dropping at a rate of 1.25 percent per additional year of life.

(d) $f_x(70,2000) \approx -2$

In 2000, the percent of those born 70 years earlier who are still alive was dropping at a rate of 2 percent per additional year of life.

(b) $A = \dfrac{1}{4}b\sqrt{4a^2 - b^2}$

$$dA = \frac{1}{4}b \cdot \frac{1}{2}(4a^2 - b^2)^{-1/2}(8a)da$$

$$+ \left[\frac{1}{4}b \cdot \frac{1}{2}(4a^2 - b^2)^{-1/2}(-2b)\right.$$

$$\left. + \frac{1}{4}(4a^2 - b^2)^{1/2}\right]db$$

$$dA = \frac{ab}{\sqrt{4a^2 - b^2}}da$$

$$+ \frac{1}{4}\left(\frac{-b^2}{\sqrt{4a^2 - b^2}} + \sqrt{4a^2 - b^2}\right)db$$

If $a = 3, b = 2, da = 0,$ and $db = 0.5,$

$$dA = \frac{1}{4}\left(\frac{-2^2}{\sqrt{4(3)^2 - 2^2}} + \sqrt{4(3)^2 - 2^2}\right)(0.5)$$

$dA \approx 0.6187.$

The approximate effect on the area is an increase of 0.6187 ft².

105. Let x be the length of each of the square faces of the box and y be the length of the box.

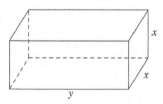

Since the volume must be 125, the constraint is $125 = x^2y.$

$f(x,y) = 2x^2 + 4xy$ is the surface area of the box.

1. $g(x) = x^2y - 125$

2. $F(x,y,\lambda) = 2x^2 + 4xy - \lambda(x^2y - 125)$

3. $F_x(x, y, \lambda) = 4x + 4y - 2xy\lambda$

     $F_y(x, y, \lambda) = 4x - x^2\lambda$

     $F_\lambda(x, y, \lambda) = -(x^2y - 125)$

4. $4x + 4y - 2xy\lambda = 0$ (1)

         $4x - x^2\lambda = 0$ (2)

          $x^2y - 125 = 0$ (3)

5.   Factoring equation (2) gives

$$x(4 - x\lambda) = 0$$

$$x = 0 \quad \text{or} \quad 4 - x\lambda = 0.$$

Since $x = 0$ is not a solution of equation (3), then

$$4 - x\lambda = 0$$

$$\lambda = \frac{4}{x}.$$

Substituting into equation (1) gives

$$4x + 4y - 2xy\left(\frac{4}{x}\right) = 0$$

or $4x + 4y - 8y = 0$

$$x = y.$$

Substituting $x = y$ into equation (3) gives

$$x^2y - 125 = 0$$

$$y^3 = 125$$

$$y = 5.$$

Therefore, $x = y = 5$. The dimensions are 5 inches by 5 inches by 5 inches.

Chapter 10

DIFFERENTIAL EQUATIONS

10.1 Solutions of Elementary and Separable Differential Equations

Your Turn 1

Find all solutions of $\dfrac{dy}{dx} = 12x^5 + \sqrt{x} + e^{5x}$.

$$y = \int (12x^5 + \sqrt{x} + e^{5x})\,dx$$

$$= 2x^6 + \frac{2}{3}x^{3/2} + \frac{1}{5}e^{5x} + C$$

Your Turn 2

Find the particular solution of $\dfrac{dy}{dx} - 12x^3 = 6x^2$, $y(2) = 60$.

First solve for dy/dx.

$$\frac{dy}{dx} = 12x^3 + 6x^2$$

$$y = 3x^4 + 2x^3 + C$$

Use the initial condition to find the value of C.

$$y(2) = 60$$
$$60 = 3(2)^4 + 2(2)^3 + C$$
$$60 = 48 + 16 + C$$
$$60 = 64 + C$$
$$-4 = C$$

The particular solution is $y = 3x^4 + 2x^3 - 4$.

Your Turn 3

Find the general solution of $\dfrac{dy}{dx} = \dfrac{x^2 + 1}{xy^2}$.

Separate the variables.

$$y^2\,dy = \frac{x^2 + 1}{x}\,dx$$

$$y^2\,dy = \left(x + \frac{1}{x}\right)dx$$

Now integrate.

$$y^2\,dy = \left(x + \frac{1}{x}\right)dx$$

$$\int y^2\,dy = \int \left(x + \frac{1}{x}\right)dx$$

$$\int y^2\,dy = \int x\,dx + \int \frac{1}{x}\,dx$$

$$\frac{1}{3}y^3 = \frac{1}{2}x^2 + \ln|x| + C$$

$$y^3 = \frac{3}{2}x^2 + 3\ln|x| + C$$

$$y = \left(\frac{3}{2}x^2 + 3\ln |x| + C\right)^{1/3}$$

Your Turn 4

Find the goat population in 5 years if the reserve can support 6000 goats, the growth rate is 15%, and there are currently 1200 goats in the area.

The general solution will be $y = N - Me^{-kt}$ as in Example 5, where N is the maximum population, k is the growth rate constant, and M is a constant to be determined using the initial population. For this problem, $N = 6000$ and $k = 20 = 15\% = 0.15$. Solve for M.

$$1200 = 6000 - Me^{(-0.15)(0)}$$
$$M = 6000 - 1200$$
$$M = 4800$$

The model is $y = 6000 - 4800e^{-0.15t}$.

Now find $y(5)$.

$$y(5) = 1200 - 4800e^{-0.15(5)}$$
$$= 6000 - 4800e^{-0.75} \approx 3733$$

The goat population in 5 years will be 3733.

10.1 Warmup Exercises

W1. Integrate the sum term by term.

$$\int \left(e^{2x} + 4x + x^{1/2} + x^{-1} \right) dx$$

$$= \frac{1}{2}e^{2x} + 2x^2 + \frac{2}{3}x^{3/2} + \ln|x| + C$$

W2. Use the substitution $u = 2x^3$, $du = 6x^2 dx$.

$$\int x^2 e^{2x^3} dx = \int \frac{1}{6} e^u du$$

$$= \frac{1}{6}e^u + C$$

$$= \frac{1}{6}e^{2x^3} + C$$

W3. Use the substitution $u = x^2 + 1$, $du = 2x\,dx$.

$$\int \frac{4x}{x^2 + 1} dx = \int \frac{2}{u} du$$

$$= 2\ln|u| + C$$

$$= 2\ln\left(x^2 + 1 \right) + C$$

W4. Use integration by parts with

$$u = 3x, du = 3\,dx;\ dv = e^{5x}dx, v = \frac{1}{5}e^{5x}.$$

$$\int 3xe^{5x} dx = \int u\,dv$$

$$= uv - \int v\,du$$

$$= \frac{3}{5}xe^{5x} - \frac{3}{5}\int e^{5x}\,dx$$

$$= \frac{3}{5}xe^{5x} - \frac{3}{25}e^{5x} + C$$

W5. Use tabular integration.

$$\int x^2 e^{-3x} dx$$

Choose x^2 as the part to be differentiated and e^{-3x} as the part to be integrated.

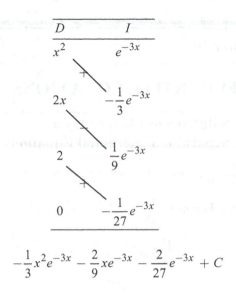

$$-\frac{1}{3}x^2 e^{-3x} - \frac{2}{9}xe^{-3x} - \frac{2}{27}e^{-3x} + C$$

10.1 Exercises

1. $\dfrac{dy}{dx} = -4x + 6x^2$

$$y = \int (-4x + 6x^2)\,dx$$

$$= -2x^2 + 2x^3 + C$$

3. $4x^3 - 2\dfrac{dy}{dx} = 0$

Solve for $\dfrac{dy}{dx}$.

$$\frac{dy}{dx} = 2x^3$$

$$y = 2\int x^3 dx$$

$$= 2\left(\frac{x^4}{4} \right) + C$$

$$= \frac{x^4}{2} + C$$

5. $y\dfrac{dy}{dx} = x^2$

Separate the variables and take antiderivatives.

$$\int y\,dy = \int x^2\,dx$$

$$\frac{y^2}{2} = \frac{x^3}{3} + K$$

$$y^2 = \frac{2}{3}x^3 + 2K$$

$$y^2 = \frac{2}{3}x^3 + C$$

7. $\dfrac{dy}{dx} = 2xy$

$$\int \dfrac{dy}{y} = \int 2x\,dx$$

$$\ln|y| = \dfrac{2x^3}{2} + C$$

$$\ln|y| = x^2 + C$$

$$e^{\ln|y|} = e^{x^2} + C$$

$$y = \pm e^{x^2} + C$$

$$y = \pm e^{x^2} \cdot e^C$$

$$y = ke^{x^2}$$

9. $\dfrac{dy}{dx} = 3x^2 y - 2xy$

$$\dfrac{dy}{dx} = y(3x^2 - 2x)$$

$$\int \dfrac{dy}{y} = \int (3x^2 - 2x)\,dx$$

$$\ln|y| = \dfrac{3x^3}{3} - \dfrac{2x^2}{2} + C$$

$$e^{\ln|y|} = e^{x^3 - x^2 + C}$$

$$y = \pm\left(e^{x^3 - x^2}\right)e^C$$

$$y = ke^{x^3 - x^2}$$

11. $\dfrac{dy}{dx} = \dfrac{y}{x},\ x > 0$

$$\int \dfrac{dy}{dx} = \int \dfrac{dx}{x}$$

$$\ln|y| = \ln x + C_1$$

$$e^{\ln|y|} = e^{\ln x + C_1}$$

$$y = \pm e^{\ln x} \cdot e^{C_1}$$

$$y = Ce^{\ln x}$$

$$y = Cx$$

13. $\dfrac{dy}{dx} = \dfrac{y^2 + 6}{2y}$

$$\dfrac{2y}{y^2 + 6}\,dy = dx$$

$$\int \dfrac{2y}{y^2 + 6}\,dy = \int dx$$

$$\ln|(y^2 + 6)| = x + C$$

Since $y^2 + 6$ is always greater than 0 we can write this as $\ln(y^2 + 6) = x + C$.

15. $\dfrac{dy}{dx} = y^2 e^{2x}$

$$\int y^{-2}\,dy = \int e^{2x}\,dx$$

$$-y^{-1} = \dfrac{1}{2}e^{2x} + C$$

$$-\dfrac{1}{y} = \dfrac{1}{2}e^{2x} + C$$

$$y = \dfrac{-1}{\frac{1}{2}e^{2x} + C}$$

17. $\dfrac{dy}{dx} + 3x^2 = 2x$

$$\dfrac{dy}{dx} = 2x = 3x^2$$

$$y = \dfrac{2x^2}{2} - \dfrac{3x^3}{3} + C$$

$$y = x^2 - x^3 + C$$

Since $y = 5$ when $x = 0$,

$$5 = 0 - 0 + C$$

$$C = 5.$$

Thus,

$$y = x^2 - x^3 + 5.$$

19. $2\dfrac{dy}{dx} = 4xe^{-x}$

$$\dfrac{dy}{dx} = 2xe^{-x}$$

Use the table of integrals or integrate by parts.

$$y = 2(-x - 1)e^{-x} + C$$

Since $y = 42$ when $x = 0$,

$$42 = 2(0 - 1)(1) + C$$

$$42 = -2 + C$$

$$C = 44$$

Thus,

$$y = -2xe^{-x} - 2e^{-x} + 44.$$

21. $\dfrac{dy}{dx} = \dfrac{x^3}{y}$; $y = 5$ when $x = 0$.

$$\int y\,dy = \int x^3\,dx$$

$$\frac{y^2}{2} = \frac{x^4}{4} + C$$

$$y^2 = \frac{1}{2}x^4 + 2C$$

$$y^2 = \frac{1}{2}x^4 + k$$

Since $y = 5$ when $x = 0$,

$$25 = 0 + k$$
$$k = 25.$$

So $y^2 = \dfrac{1}{2}x^4 + 25$.

23. $(2x + 3)y = \dfrac{dy}{dx}$; $y = 1$ when $x = 0$.

$$\int (2x + 3)dx = \int \frac{dy}{y}$$

$$\frac{2x^2}{2} + 3x + C = \ln|y|$$

$$e^{x^2+3x+C} = e^{\ln|y|}$$

$$y = (e^{x^2+3x})(\pm e^C)$$

$$y = ke^{x^2+3x}$$

Since $y = 1$ when $x = 0$.

$$1 = ke^{0+0}$$
$$k = 1.$$

So $y = e^{x^2+3x}$.

25. $\dfrac{dy}{dx} = \dfrac{2x+1}{y-3}$; $y = 4$ when $x = 0$.

$$\int (y - 3)\,dy = \int (2x + 1)dx$$

$$\frac{y^2}{2} - 3y = \frac{2x^2}{2} + x + C$$

Since $y = 4$ when $x = 0$,

$$\frac{16}{2} - 12 = 0 + 0 + C$$
$$C = -4.$$

So,

$$\frac{y^2}{2} - 3y = x^2 + x - 4.$$

27. $\dfrac{dy}{dx} = \dfrac{y^2}{x}$; $y = 3$ when $x = e$.

$$\int y^{-2}dy = \int \frac{dx}{x}$$

$$-y^{-1} = \ln|x| + C$$

$$-\frac{1}{y} = \ln|x| + C$$

$$y = \frac{-1}{\ln|x| + C}$$

Since $y = 3$ when $x = e$,

$$3 = \frac{-1}{\ln e + C}$$

$$3 = \frac{-1}{1 + C}$$

$$3 + 3C = -1$$

$$3C = -4$$

$$C = -\frac{4}{3}.$$

So $y = \dfrac{-1}{\ln|x| - \frac{4}{3}} = \dfrac{-3}{3\ln|x| - 4}$.

29. $\dfrac{dy}{dx} = (y - 1)^2 e^{x-1}$; $y = 2$ when $x = 1$.

$$\frac{dy}{(y - 1)^2} = e^{x-1}\,dx$$

$$\int (y - 1)^{-2}dy = \int e^{x-1}dx$$

$$\frac{(y - 1)^{-1}}{-1} = e^{x-1} + C$$

$$-\frac{1}{y - 1} = e^{x-1} + C$$

$$-(y - 1) = \frac{1}{e^{x-1} + C}$$

$$-y + 1 = \frac{1}{e^{x-1} + C}$$

$$1 - \frac{1}{e^{x-1} + C} = y$$

$$y = \frac{e^{x-1} + C}{e^{x-1} + C} - \frac{1}{e^{x-1} + C}$$

$$y = \frac{e^{x-1} + C - 1}{e^{x-1} + C}$$

$$y = 2, \text{ when } x = 1.$$

$$2 = \frac{e^0 + C - 1}{e^0 + C}$$

$$2 = \frac{C}{1 + C}$$

$$2 + 2C = C$$

$$C = -2$$

$$y = \frac{e^{x-1} - 3}{e^{x-1} - 2}.$$

31. $\frac{dy}{dx} = y(y^2 - 1)$

The equilibrium points are solutions of
$y(y^2 - 1) = 0$, that is,
$y = -1, y = 0,$ and $y = 1$.

Checking the sign of dy/dx in the four intervals
into which these points divide the y-axis we get
the following diagram.

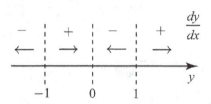

Thus -1 and 1 are unstable equilibrium points
and 0 is a stable equilibrium point.

33. $\frac{dy}{dx} = (e^y - 1)(y - 3)$

The equilibrium points are solutions of
$(e^y - 1)(y - 3) = 0$, that is,
$y = 0$ and $y = 3$.

Checking the sign of dy/dx in the three intervals
into which these points divide the y-axis we get
the following diagram.

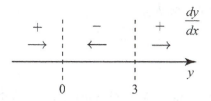

Thus 0 is a stable equilibrium point and 3 is an
unstable equilibrium point.

35. $\frac{dy}{dx} = \frac{k}{N}(N - y)y$

(a) $\frac{N \, dy}{(N - y)y} = k \, dx$

Since $\frac{1}{y} + \frac{1}{N - y} = \frac{N}{(N - y)y}$,

$$\int \frac{dy}{y} + \int \frac{dy}{N - y} = k \, dx$$

$$\ln \left| \frac{y}{N - y} \right| = kx + C$$

$$\frac{y}{N - y} = Ce^{kx}.$$

For $0 < y < N, Ce^{kx} > 0$.

For $0 < N < y, Ce^{kx} < 0$.

Solve for y.

$$y = \frac{Ce^{kx}N}{1 + Ce^{kx}} = \frac{N}{1 + C^{-1}e^{-kx}}$$

Let $b = C^{-1} > 0$ for $0 < y < N$.

$$y = \frac{N}{1 + be^{-kx}}$$

Let $-b = C^{-1} < 0$ for $0 < N < y$.

$$y = \frac{N}{1 - be^{-kx}}$$

(b) For $0 < y < N; t = 0, y = y_0$.

$$y_0 = \frac{N}{1 + be^0} = \frac{N}{1 + b}$$

Solve for b.

$$b = \frac{N - y_0}{y_0}$$

(c) For $0 < N < y; t = 0, y = y_0$.

$$y_0 = \frac{N}{1 - be^0} = \frac{N}{1 - b}$$

Solve for b.

$$b = \frac{y_0 - N}{y_0}$$

37. **(a)** $0 < y_0 < N$ implies that $y_0 > 0, N > 0$, and $N - y_0 > 0$.

Therefore,

$$b = \frac{N - y_0}{y_0} > 0.$$

Also, $e^{-kx} > 0$ for all x, which implies that $1 + be^{-kx} > 1$.

(1) $y(x) = \dfrac{N}{1 + be^{-kx}} < N$ since

$1 + be^{-kx} > 1$.

(2) $y(x) = \dfrac{N}{1 + be^{-kx}} > 0$ since $N > 0$

and $1 + be^{-kx} > 0$.

Combining statements (1) and (2), we have

$$0 < \frac{N}{1 + be^{-kx}} = y(x)$$

$$= \frac{N}{1 + be^{-kx}} < N$$

or $0 < y(x) < N$ for all x.

(b) $\displaystyle\lim_{x \to \infty} \frac{N}{1 + be^{-kx}} = \frac{N}{1 + b(0)} = N$

$\displaystyle\lim_{x \to -\infty} \frac{N}{1 + be^{-kx}} = 0$

Note that as $x \to -\infty, 1 + be^{-kx}$ becomes infinitely large.

Therefore, the horizontal asymptotes are $y = N$ and $y = 0$.

(c) $y'(x) = \dfrac{(1 + be^{-kx})(0) - N(-kbe^{-kx})}{(1 + be^{-kx})^2}$

$\quad = \dfrac{Nkbe^{-kx}}{(1 + be^{-kx})^2} > 0$ for all x.

Therefore, $y(x)$ is an increasing function.

(d) To find $y''(x)$, apply the quotient rule to find the derivation of $y'(x)$. The numerator of $y''(x)$, is

$$y''(x) = (1 + be^{-kx})^2 (-Nk^2be^{-kx})$$
$$\qquad - Nkbe^{-kx}[-2kbe^{-kx}(1 + be^{-kx})]$$
$$\qquad = -Nk^2be^{-kx}(1 - be^{-kx})(1 + be^{-kx}),$$

and the denominator is

$$[(1 + be^{-kt})^2]^2 = (1 + be^{-kx})^4.$$

Thus,

$$y''(x) = \frac{-Nk^2be^{-kx}(1 - be^{-kx})}{(1 + be^{-kx})^3}.$$

$y''(x) = 0$ when

$$k - kbe^{-kx} = 0$$
$$be^{-kx} = 1$$
$$e^{-kx} = \frac{1}{b}$$
$$-kx = \ln\left(\frac{1}{b}\right)$$
$$x = \frac{\ln\left(\frac{1}{b}\right)}{k}$$
$$= \frac{\ln\left(\frac{1}{b}\right)^{-1}}{k} = \frac{\ln b}{k}.$$

When $x = \dfrac{\ln b}{k}$,

$$y = \frac{N}{1 + be^{-k\left(\frac{\ln b}{k}\right)}} = \frac{N}{1 + be^{(-\ln b)}}$$
$$= \frac{N}{1 + be^{\ln(1/b)}} = \frac{N}{1 + b\left(\frac{1}{b}\right)} = \frac{N}{2}.$$

Therefore, $\left(\dfrac{\ln b}{k}, \dfrac{N}{2}\right)$ is a point of inflection.

(e) To locate the maximum of $\dfrac{dy}{dx}$, we must consider, from part (d),

$$\frac{d}{dx}\left(\frac{dy}{dx}\right) = \frac{-Nkbe^{-kx}(k - kbe^{-kx})}{(1 + be^{-kx})^3}.$$

Since $y''(x) > 0$ for $x < \dfrac{\ln b}{k}$ and

$$y''(x) < 0 \text{ for } x > \frac{\ln b}{k},$$

we know that $x = \dfrac{\ln b}{k}$ locates a relative maximum of $\dfrac{dy}{dx}$.

39. $\dfrac{dy}{dx} = \dfrac{100}{32 - 4x}$

$y = 100\left(-\dfrac{1}{4}\right)\ln|32 - 4x| + C$

$y = -25\ln|32 - 4x| + C$

Now, $y = 1000$ when $x = 0$.

$1000 = -25\ln|32| + C$

$C = 1000 + 25\ln 32$

$C \approx 1086.64$

Thus,

$y = -25\ln|32 - 4x| + 1086.64.$

(a) Let $x = 3$.

$y = -25\ln|32 - 12| + 1086.64$

$\approx \$1011.75$

(b) Let $x = 5$.

$y = -25\ln|32 - 20| + 1086.64$

$\approx \$1024.52$

(c) Advertising expenditures can never reach $8000. If $x = 8$, the denominator becomes zero.

41. $\dfrac{dy}{dt} = -0.05y$

See Example 5.

$\displaystyle\int \dfrac{dy}{y} = \int -0.05\, dt$

$\ln|y| = -0.05t + C$

$e^{\ln|y|} = e^{-0.05t + C}$

$e^{\ln|y|} = e^{-0.05t} \cdot e^{C}$

$|y| = e^{-0.05t} \cdot e^{C}$

$y = Me^{-0.05t}$

Let $y = 1$ when $t = 0$.

Solve for M:

$1 = Me^{0}$

$M = 1.$

So $y = e^{-0.05t}$.

If $y = 0.50$,

$0.5 = e^{-0.05t}$

$t = \dfrac{-\ln 0.5}{0.05} \approx 13.9$

It will take about 13.9 years for $1 to lose half its value.

43. $E = -\dfrac{p}{q} \cdot \dfrac{dq}{dp}$ with $p > 0$ and $q > 0$

If $E = 2$,

$2 = -\dfrac{p}{q} \cdot \dfrac{dq}{dp}$

$\dfrac{2}{p}\, dp = -\dfrac{1}{q}\, dq$

$\displaystyle\int \dfrac{2}{p}\, dp = -\int \dfrac{1}{q}\, dq$

$2\ln p = -\ln q + K$

$\ln p^2 + \ln q = K$

$\ln(p^2 q) = K$

$p^2 q = e^{K}$

$p^2 q = C$

$q = \dfrac{C}{p^2}.$

45. $\dfrac{dA}{dt} = Ai$

$\dfrac{dA}{A} = i\, dt$

$\displaystyle\int \dfrac{dA}{A} = \int i\, dt$

$\ln A = it + C$

$e^{\ln A} = e^{it + C}$

$A = Me^{it}$

When $t = 0$, $A = 5000$. Therefore, $M = 5000$.
Find i so that $A = 20,000$
when $t = 24$.

$20,000 = 5000e^{24i}$

$4 = e^{24i}$

$\ln 4 = 24i$

$i = \dfrac{\ln 4}{24}$

$= \dfrac{2\ln 2}{24}$

$= \dfrac{\ln 2}{12}$

The answer is d.

47. (a) $\dfrac{dI}{dW} = 0.088(2.4 - I)$

Separate the variables and take anti-derivatives.

$$\int \frac{dI}{2.4 - I} = \int 0.088 \, dW$$

$$-\ln|2.4 - I| = 0.088 W + k$$

Solve for I.

$$\ln|2.4 - I| = -0.088W - k$$

$$|2.4 - I| = e^{-0.088W - k} = e^{-k} e^{-0.088W}$$

$$I - 2.4 = Ce^{-0.088W}, \text{ where } C = \pm e^{-k}.$$

$$I = 2.4 + Ce^{-0.088W}$$

Since $I(0) = 1$, then

$$1 = 2.4 + Ce^{0}$$

$$C = 1 - 2.4 = -1.4.$$

Therefore, $I = 2.4 - 1.4e^{-0.088W}$.

(b) Note that as W gets larger and larger
$e^{-0.088W}$ approaches 0, so

$$\lim_{W \to \infty} I = \lim_{W \to \infty} (2.4 - 1.4e^{-0.088W})$$
$$= 2.4 - 1.4(0) = 2.4,$$

so I approaches 2.4.

49.   (a) $\dfrac{dw}{dt} = k(C - 17.5w)$

C being constant implies that the calorie
intake per day is constant.

(b)   pounds/day $= k$(calories/day)

$$\frac{\text{pounds/day}}{\text{calories/day}} = k$$

The units of k are pounds/calorie.

(c)   Since 3500 calories is equivalent to 1 pound,
$k = \frac{1}{3500}$ and

$$\frac{dw}{dt} = \frac{1}{3500}(C - 17.5w).$$

(d) $\dfrac{dw}{dt} = \dfrac{1}{3500}(C - 17.5w); \; w = w_0$
when $t = 0$.

$$\frac{3500}{C - 17.5w} \, dw = dt$$

$$\frac{3500}{-17.5} \int \frac{-17.5}{C - 17.5w} \, dw = \int dt$$

$$-200 \ln|C - 17.5w| = t + k$$

$$\ln|C - 17.5w| = 0.005t - 0.005k$$

$$|C - 17.5w| = e^{-0.005t - 0.005k}$$

$$|C - 17.5w| = e^{-0.005t} \cdot e^{-0.005k}$$

$$C - 17.5w = e^{-0.005M} e^{-0.005t}$$

$$-17.5w = -C + e^{-0.005M} e^{-0.005t}$$

$$w = \frac{C}{17.5} - \frac{e^{-0.005M}}{17.5} e^{-0.005t}$$

(e)   Since $w = w_0$ when $t = 0$,

$$w_0 = \frac{C}{17.5} - \frac{e^{-0.005M}}{17.5} \tag{1}$$

$$w_0 - \frac{C}{17.5} = -\frac{e^{-0.005M}}{17.5}$$

$$\frac{e^{-0.005M}}{17.5} = \frac{C}{17.5} - w_0.$$

Therefore,

$$w = \frac{C}{17.5} - \left(\frac{C}{17.5} - w_0\right) e^{-0.005t}$$

$$w = \frac{C}{17.5} + \left(w_0 - \frac{C}{17.5}\right) e^{-0.005t}.$$

51.   (a)

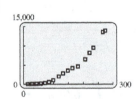

(b) $y = \dfrac{25,538}{1 + 110.28e^{-0.01819t}}$

(c)

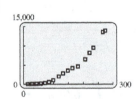

(d)   As t gets very large, the value of the function
in (b) approaches 25,538, so this is the
limiting number of deaths predicted by
the model.

53. $\dfrac{dy}{dt} = ky$

First separate the variables and integrate.

$$\frac{dy}{y} = k \, dt$$

$$\int \frac{dy}{y} = \int k \, dt$$

$$\ln|y| = kt + C.$$

Solve for y.

$$|y| = e^{kt+C_1} = e^{C_1}e^{kt}$$

$$y = Ce^{kt}, \text{ where } C = \pm e^{C_1}.$$

$$y(0) = 35.6, \text{ so } 35.6 = Ce^0 = C, \text{ and}$$

$$y = 35.6e^{kt}.$$

Since $y(50) = 102.6$, then $102.6 = 35.6e^{50k}$.

Solve for k.

$$e^{50k} = \frac{102.6}{35.6}$$

$$50k = \ln\left(\frac{102.3}{35.6}\right)$$

$$k = \frac{\ln\left(\frac{102.6}{35.6}\right)}{50} \approx 0.02117,$$

so $y = 35.6e^{0.02117t}$.

55.
$$\frac{dP}{dR} = -4\pi DRP^2$$

$$\frac{dP}{P^2} = -4\pi DR\, dR$$

$$\int \frac{1}{P^2}\, dP = -4\pi D \int R\, dR.$$

$$\frac{1}{P} = 2\pi DR^2 + C$$

$$P = \frac{1}{C + 2\pi DR^2}$$

Since $P(0) = 1$,

$$1 = \frac{1}{C + 0} \text{ and thus } C = 1.$$

$$P(R) = \frac{1}{1 + 2\pi DR^2}$$

57. $\frac{dy}{dx} = 7.5e^{-0.3y}$, $y = 0$ when $x = 0$.

$$e^{0.3y}\, dy = 7.5\, dx$$

$$\int e^{0.3y}\, dy = \int 7.5\, dx$$

$$\frac{e^{0.3y}}{0.3} = 7.5x + C$$

$$e^{0.3y} = 2.25x + C$$

$$1 = 0 + C = C$$

$$e^{0.3y} = 2.25x + 1$$

$$0.3y = \ln(2.25x + 1)$$

$$y = \frac{\ln(2.25x + 1)}{0.3}$$

When $x = 8$,

$$y = \frac{\ln[2.25(8) + 1]}{0.3} \approx 10 \text{ items}.$$

59. Let $t = 0$ be the time it started snowing. If h is the height of the snow and if the rate of snowfall is constant, $\frac{dh}{dt} = k_1$, where k_1 is a constant.

$$\frac{dh}{dt} = k_1 \text{ and } h = 0 \text{ when } t = 0.$$

$$dh = k_1\, dt$$

$$\int dh = \int k_1\, dt$$

$$h = k_1t + C_1$$

Since $h = 0$ and $t = 0$, $0 = k_1(0) + C_1$.
Thus, $C_1 = 0$ and $h = k_1t$.

Since the snowplow removes a constant volume of snow per hour and the volume is proportional to the height of the snow, the rate of travel of the snowplow is inversely proportional to the height of the snow.

$$\frac{dx}{dt} = \frac{k^2}{h}, \text{ where } k_2 \text{ is a constant.}$$

When $t = T$, $x = 0$.

When $t = T + 1$, $x = 2$.

When $t = T + 2$, $x = 3$.

Since $\frac{dy}{dt} = \frac{k^2}{h}$ and $h = k_1t$,

$$\frac{dy}{dt} = \frac{k_2}{k_1t}$$

$$\frac{dx}{dt} = \frac{k_2}{k_1} \cdot \frac{1}{t}.$$

Let $k_3 = \frac{k_2}{k_1}$. Then

$$\frac{dx}{dt} = k_3\frac{1}{t}$$

$$dx = k_3\frac{1}{t}\, dt$$

$$\int dx = \int k_3\frac{1}{t}\, dt$$

$$x = k_3 \ln t + C_2.$$

Since $x = 0$, when $t = T$,

$$0 = k_3 \ln T + C_2$$

$$C_2 = -k_3 \ln T.$$

Thus,

$$x = k_3 \ln t - k_3 \ln T$$
$$x = k_3(\ln t - \ln T)$$
$$x = k_3 \ln\left(\frac{t}{T}\right).$$

Since $x = 2$, when $t = T + 1$,

$$2 = k_3 \ln\left(\frac{T + 1}{T}\right). \qquad (1)$$

Since $x = 3$ when $t = T + 2$,

$$3 = k_3 \ln\left(\frac{T + 2}{T}\right). \qquad (2)$$

We want to solve for T, so we divide equation (1) by equation (2).

$$\frac{2}{3} = \frac{k_3 \ln\left(\frac{T+1}{T}\right)}{k_3 \ln\left(\frac{T+2}{T}\right)}$$

$$\frac{2}{3} = \frac{\ln(T + 1) - \ln T}{\ln(T + 2) - \ln T}$$

$$2\ln(T + 2) - 2\ln T = 3\ln(T + 1) - 3\ln T$$

$$\ln(T + 2)^2 - \ln T^2 - \ln(T + 1)^3 + \ln T^3 = 0$$

$$\ln \frac{(T + 2)^2 T^3}{T^2(T + 1)^3} = 0$$

$$\frac{T(T + 2)^2}{(T + 1)^3} = 1$$

$$T(T^2 + 4T + 4) = T^3 + 3T^2 + 3T + 1$$

$$T^3 + 4T^2 + 4T = T^3 + 3T^2 + 3T + 1$$

$$T^2 + T - 1 = 0$$

$$T = \frac{-1 \pm \sqrt{1 + 4}}{2}$$

$T = \frac{-1-\sqrt{5}}{2}$ is negative and is not a possible solution.

Thus, $T = \frac{-1+\sqrt{5}}{2} \approx 0.618$ hr.

0.618 hr ≈ 37 min and $5\,\text{sec}$

Now, 37 min and 5 sec before 8:00 A.M. is 7:22:55 A.M.

Thus, it started snowing at 7:22:55 A.M.

63. Use the formula from Exercise 62:

$$T = Ce^{-kt} + T_M.$$

(a) We know that $T_M = 68$, $T(0) = 98.6$ and $T(1) = 90$.

$$98.6 = (C)(1) + 68$$
$$C = 30.6$$
$$T = 30.6e^{-kt} + 68$$
$$90 = 30.6e^{-k(1)} + 68$$
$$30.6e^{-kt} = 22$$
$$e^{-k} = \frac{22}{30.6}$$
$$-k = \ln\left(\frac{22}{30.6}\right)$$
$$k = -\ln\left(\frac{22}{30.6}\right)$$
$$\approx 0.33$$

Therefore, $T = 30.6e^{-0.33t} + 68$.

When $t = 2$, $T = 30.6e^{-0.33(2)} + 68$
$$= 83.8.$$

After two hours the temperature of the body will be 83.8°F.

(b) $$75 = 30.6e^{-0.33t} + 68$$
$$7 = 30.6e^{-0.33t}$$
$$e^{-0.33t} = \frac{7}{30.6}$$
$$-0.33t = \ln\left(\frac{7}{30.6}\right)$$
$$t = \frac{\ln\left(\frac{7}{30.6}\right)}{-0.33}$$
$$\approx 4.5$$

The temperature of the body will be 75°F in approximately 4.5 hours.

(c) $$68.01 = 30.6e^{-0.33t} + 68$$
$$0.01 = 30.6e^{-0.33t}$$
$$e^{-0.33t} = \frac{0.01}{30.6}$$
$$-0.33t = \ln\left(\frac{0.01}{30.6}\right)$$
$$t = \frac{\ln\left(\frac{0.01}{30.6}\right)}{-0.33}$$
$$\approx 24.3$$

The temperature of the body will be within 0.01° of the surrounding air in approximately 24.3 hours.

10.2 Linear First-Order Differential Equations

Your Turn 1

Give the general solution of

$$x\frac{dy}{dx} - y - x^2 e^x = 0, \; x > 0.$$

Step 1: Put the equation in the required from

$$\frac{dy}{dx} + P(x)y = Q(x).$$

$$\frac{dy}{dx} + \left(-\frac{1}{x}\right)y = xe^x$$

Step 2: Find the integrating factor $I(x)$.

$$I(x) = e^{\int\left(-\frac{1}{x}\right)dx}$$

$$= e^{-\ln x}$$

$$= \frac{1}{x}$$

Step 3: Multiply each term in the equation by the integrating factor x^{-1}.

$$(x^{-1})\frac{dy}{dx} - (x^{-2})y = e^x$$

Step 4: Write the left side as the derivative of a product with respect to x.

$$D_x(x^{-1}y) = e^x$$

Step 5: Integrate on both sides.

$$\int D_x(x^{-1}y)\,dx = \int e^x dx$$

$$x^{-1}y = e^x + C$$

Step 6: Multiply both sides by x to solve for y.

$$y = xe^x + Cx$$

Your Turn 2

Solve the initial value problem

$$\frac{dy}{dx} + 2xy - xe^{-x^2} = 0, \; y(0) = 3.$$

$$\frac{dy}{dx} + (2x)y = xe^{-x^2}$$

The integrating factor is

$$I(x) = e^{\int 2x\,dx}$$

$$= e^{x^2}$$

Multiply all terms by the integrating factor, express the left side as the derivative of a product, and integrate on both sides.

$$e^{x^2}\frac{dy}{dx} + 2xe^{x^2}y = x$$

$$D_x\left(e^{x^2}y\right) = x$$

$$e^{x^2}y = \int x\,dx$$

$$e^{x^2}y = \frac{1}{2}x^2 + C$$

$$y = \frac{\frac{1}{2}x^2 + C}{e^{x^2}}$$

$$= \frac{x^2 + C}{2e^{x^2}}$$

To find the particular solution, substitute 0 for x and 3 for y.

$$3 = \frac{(0)^2 + C}{2(1)}$$

$$C = 6$$

The particular solution is $y = \dfrac{x^2 + 6}{2e^{x^2}}$.

10.2   Warmup Exercises

W1. $\displaystyle\int \frac{4}{x}\,dx = 4\ln|x| + C$

W2. $\displaystyle\int e^{-0.1t}\,dt = -10e^{-0.1t} + C$

W3. Use the substitution $u = 3x^2$, $du = 6x\,dx$.

$$\int 3xe^{3x^2}\,dx = \frac{1}{2}\int e^u\,du$$

$$= \frac{1}{2}e^u + C$$

$$= \frac{1}{2}e^{3x^2} + C$$

W4.   Use the substitution

$$u = t^3 - 3t, du = (3t^2 - 3)\, dt.$$

$$\int (t^2 - 1)e^{t^3 - 3t}\, dt$$

$$= \frac{1}{3}\int e^u\, du = \frac{1}{3}e^u + C$$

$$= \frac{1}{3}e^{t^3 - 3t} + C$$

10.2   Exercises

1. $\dfrac{dy}{dx} + 3y = 6$

$$I(x) = e^{3\int dx} = e^{3x}$$

Multiply each term by e^{3x}.

$$e^{3x}\frac{dy}{dx} + 3e^{3x}y = 6e^{3x}$$

$$D_x(e^{3x}y) = 6e^{3x}$$

Integrate both sides.

$$e^{3x}y = \int 6e^{3x}\, dx$$

$$= 2e^{3x} + C$$

$$y = 2 + Ce^{-3x}$$

3. $\dfrac{dy}{dx} + 2xy = 4x$

$$I(x) = e^{\int 2x\, dx} = e^{x^2}$$

$$e^{x^2}\frac{dy}{dx} + 2xe^{x^2}y = 4xe^{x^2}$$

$$D_x\left(e^{x^2}y\right) = 4xe^{x^2}$$

$$e^{x^2}y = \int 4xe^{x^2}\, dx$$

$$= 2e^{x^2} + C$$

$$y = 2 + Ce^{-x^2}$$

5. $x\dfrac{dy}{dx} - y - x = 0; \ x > 0$

$$\frac{dy}{dx} - \frac{1}{x}y = 1$$

$$I(x) = e^{-\int 1/x\, dx}$$

$$= e^{-\ln x} = \frac{1}{x}$$

$$\frac{1}{x}\frac{dy}{dx} - \frac{1}{x^2}y = \frac{1}{x}$$

$$D_x\left(\frac{1}{x}y\right) = \frac{1}{x}$$

$$\frac{y}{x} = \int \frac{1}{x}\, dx$$

$$\frac{y}{x} = \ln x + C$$

$$y = x\ln x + Cx$$

7. $2\dfrac{dy}{dx} - 2xy - x = 0$

$$\frac{dy}{dx} - xy = \frac{x}{2}$$

$$I(x) = e^{-\int x\, dx} = e^{-x^2/2}$$

$$e^{-x^2/2}\frac{dy}{dx} - xe^{-x^2/2}y = \frac{x}{2}e^{-x^2/2}$$

$$D_x(e^{-x^2/2}y) = \frac{x}{2}e^{-x^2/2}$$

$$e^{-x^2/2}y = \int \frac{x}{2}e^{-x^2/2}\, dx$$

$$= \frac{-1}{2}e^{-x^2/2} + C$$

$$y = -\frac{1}{2} + Ce^{x^2/2}$$

9. $x\dfrac{dy}{dx} + 2y = x^2 + 6x; \ x > 0$

$$\frac{dy}{dx} + \frac{2}{x}y = x + 6$$

$$I(x) = e^{\int 2/x\, dx} = e^{2\ln x} = x^2$$

$$x^2\frac{dy}{dx} + 2xy = x^3 + 6x^2$$

$$D_x(x^2 y) = x^3 + 6x^2$$

$$x^2 y = \int (x^3 + 6x^2)\, dx$$

$$= \frac{x^4}{4} + 2x^3 + C$$

$$y = \frac{x^2}{4} + 2x + \frac{C}{x^2}$$

11. $y - x\dfrac{dy}{dx} = x^3; \ x > 0$

$$\frac{dy}{dx} - \frac{y}{x} = -x^2$$

$$I(x) = e^{-\int 1/x\, dx} = e^{-\ln x} = x^{-1}$$

$$\frac{1}{x}\frac{dy}{dx} - \frac{y}{x^2} = -x$$

$$D_x\left(\frac{1}{x}y\right) = -x$$

$$\frac{y}{x} = \int -x\,dx$$

$$= \frac{-x^2}{2} + C$$

$$y = \frac{-x^3}{2} + Cx$$

Since $y = 20$ when $x = 1$,

$$20 = -2 + Ce^1$$

$$22 = Ce$$

$$C = \frac{22}{e}.$$

Therefore,

$$y = -2 + \frac{22}{e}\left(e^{x^2}\right)$$

$$= -2 + 22e^{(x^2-1)}.$$

13. $\dfrac{dy}{dx} + y = 4e^x$; $y = 50$ when $x = 0$.

$$I(x) = e^{\int dx} = e^x$$

$$e^x\frac{dy}{dx} + ye^x = 4e^{2x}$$

$$D_x(e^x y) = 4e^{2x}$$

$$e^x y = \int 4e^{2x}\,dx$$

$$= 2e^{2x} + C$$

$$y = 2e^x + Ce^{-x}$$

Since $y = 50$ when $x = 0$,

$$50 = 2e^0 + Ce^0$$

$$50 = 2 + C$$

$$C = 48.$$

Therefore,

$$y = 2e^x + 48e^{-x}.$$

15. $\dfrac{dy}{dx} - 2xy - 4x = 0$; $y = 20$ when $x = 1$.

$$\frac{dy}{dx} - 2xy - 4x = 0$$

$$\frac{dy}{dx} - 2xy = 4x$$

$$I(x) = e^{-\int 2x\,dx} = e^{-x^2}$$

$$e^{-x^2}\frac{dy}{dx} - 2xe^{-x^2}y = 4xe^{-x^2}$$

$$D_x(e^{-x^2}y) = 4xe^{-x^2}$$

$$e^{-x^2}y = \int 4xe^{-x^2}\,dx$$

$$e^{-x^2}y = -2e^{-x^2} + C$$

$$y = -2 + Ce^{x^2}$$

17. $x\dfrac{dy}{dx} + 5y = x^2$; $y = 12$ when $x = 2$.

$$x\frac{dy}{dx} + 5y = x^2$$

$$\frac{dy}{dx} + \frac{5}{x}y = x$$

$$I(x) = e^{\int 5/x\,dx} = e^{5\ln x} = x^5$$

$$x^5\frac{dy}{dx} + 5x^4 y = x^6$$

$$D_x(x^5 y) = x^6$$

$$x^5 y = \int x^6\,dx$$

$$x^5 y = \frac{x^7}{7} + C$$

$$y = \frac{x^2}{7} + \frac{C}{x^5}$$

Since $y = 12$, when $x = 2$,

$$12 = \frac{4}{7} + \frac{C}{32}$$

$$\frac{80}{7} = \frac{C}{32}$$

$$C = \frac{2560}{7}.$$

Therefore,

$$y = \frac{x^2}{7} + \frac{2560}{7x^5}.$$

19. $x\dfrac{dy}{dx} + (1 + x)y = 3$; $y = 50$ when $x = 4$

$$\frac{dy}{dx} + \left(\frac{1+x}{x}\right)y = \frac{3}{x}$$

$$I(x) = e^{\int (1+x)\,dx/x}$$

$$= e^{\int (1/x)\,dx + dx}$$

$$= e^{(\ln x) + x}$$

$$= e^{\ln x} \cdot e^x$$

$$= xe^x$$

$$xe^x \frac{dy}{dx} + (1+x)e^x y = 3e^x$$

$$D_x(xe^x y) = 3e^x$$

$$xe^x y = \int 3e^x\,dx$$

$$xe^x y = 3e^x + C$$

$$y = \frac{3}{x} + \frac{C}{xe^x}$$

Since $y = 50$ when $x = 4$,

$$50 = \frac{3}{4} + \frac{C}{4e^4}$$

$$\frac{197}{4} = \frac{C}{4e^4}$$

$$C = 197e^4.$$

Therefore,

$$y = \frac{3}{x} + \frac{197e^4}{xe^x}$$

$$= \frac{3}{x} + \frac{197}{x}e^{4-x}$$

$$= \frac{3 + 197e^{4-x}}{x}.$$

21. (a) $\dfrac{dA}{dt} = 0.05A - 50$

(b) Rearrange in standard linear form:

$$\frac{dA}{dt} + (-0.05)A(t) = -50$$

The integrating factor will be

$$e^{\int -0.05\,dt} = e^{-0.05t}.$$

Multiply both sides of the differential equation by the integrating factor:

$$e^{-0.05t}\frac{dA}{dt} + (-0.05)A(t)e^{-0.05t} = -50e^{-0.05t}$$

The left side is now the derivative of $e^{-0.05t}A(t)$, so integrating both sides with respect to t yields

$$e^{-0.05t}A(t) = 1000e^{-0.05t} + C \text{ or, multiplying}$$

through by $e^{0.05t}$, $A(t) = 1000 + Ce^{0.05t}$.
Since $A(0) = 2000$, $C = 1000$ and

$$A(t) = 1000 + 1000e^{0.05t}.$$

(c) $A(1) = 2051.27$ or \$2051.27
$A(5) = 2284.03$ or \$2284.03
$A(10) = 2648.72$ or \$2648.72

(d) The effective yield after n years is
$\dfrac{A(n) - 2000}{2000 \cdot n}$.

For 1 year: $\dfrac{51.27}{2000} = 2.56\%$

For 5 years: $\dfrac{284.03}{2000 \cdot 5} = 2.84\%$

For 10 years: $\dfrac{648.72}{2000 \cdot 10} = 3.24\%$

(e) If Carrie invests \$1000, then the value of C will be 0 and $A(t) = 1000$ for all t.

23. $\dfrac{dy}{dx} = cy - py^2$

(a) Let $y = \frac{1}{z}$ and $\frac{dy}{dx} = -\frac{z'}{z^2}$.

$$-\frac{z'}{z^2} = c\left(\frac{1}{z}\right) - p\left(\frac{1}{z^2}\right)$$

$$z' = -cz + p$$

$$z' + cz = p$$

$$I(x) = e^{\int c\,dx} = e^{cx}$$

$$D_x(e^{cx} \cdot z) = \int pe^{cx}\,dx$$

$$e^{cx} \cdot z = \frac{p}{c}e^{cx} + K$$

$$z = \frac{p}{c} + Ke^{-cx}$$

$$= \frac{p + Kce^{-cx}}{c}$$

Therefore,

$$y = \frac{c}{p + Kce^{-cx}}.$$

(b) Let $z(0) = \frac{1}{y_0}$.

$$\frac{1}{y_0} = \frac{p + Kce^0}{c} = \frac{p + Kc}{c}$$

$$\frac{c}{y_0} = p + Kc$$

$$Kc = \frac{c}{y_0} - p = \frac{c - py_0}{y_0}$$

$$K = \frac{c - py_0}{cy_0}$$

$$y = \frac{c}{p + \left(\frac{c - py_0}{cy_0}\right)ce^{-cx}}$$

From part (a)

$$= \frac{cy_0}{py_0 + (c - py_0)e^{-cx}}$$

(c) $\displaystyle \lim_{x \to \infty} y = \lim_{x \to \infty}\left(\frac{cy_0}{py_0 + (c - py_0)e^{-cx}}\right)$

$$= \frac{cy_0}{py_0 - 0}$$

$$= \frac{c}{p}$$

25. **(a)** $\dfrac{dC}{dt} = -kC + D(t)$

$$\frac{dC}{dt} + kC = D(t)$$

$$I = e^{\int k\, dt} = e^{kt}$$

$$e^{kt}\frac{dC}{dt} + e^{kt}kC = e^{kt}D(t)$$

$$Dt(e^{kt}C) = e^{kt}D(t)$$

$$e^{kt}C = \int e^{kt}D(t)d(t)$$

$$= \int_0^t e^{kt}D(y)dy$$

$$C(t) = e^{-kt}\int_0^t e^{ky}D(y)dy + C_2$$

If $C(0) = 0$,

$$C(0) = e^0 \int_0^0 e^{ky}D(y)dy + C_2$$

$$0 = 0 + C_2$$

Therefore,

$$C(t) = e^{-kt}\int_0^t e^{ky}D(y)dy.$$

(b) Let $D(y) = D$, a constant.

$$C(t) = e^{-kt}\int_0^t e^{ky}D\, dy$$

$$= De^{-kt}\int_0^t e^{ky}dy$$

$$= De^{-kt}\left(\frac{1}{k}\right)\left(e^{kt} - e^{k(0)}\right)$$

$$C(t) = \frac{D(1 - e^{-kt})}{k}$$

27. **(a)** The differential equation is

$$\frac{dy}{dt} = \alpha(1 - y) - \beta y.$$

Rearrange in standard linear form:

$$\frac{dy}{dt} + (\alpha + \beta)y = \alpha$$

The integrating factor is

$$e^{\int (\alpha + \beta)\, dt} = e^{(\alpha + \beta)t}.$$

Multiply through by the integrating factor:

$$\frac{dy}{dt}e^{(\alpha+\beta)t} + (\alpha + \beta)ye^{(\alpha+\beta)t} = \alpha e^{(\alpha+\beta)t}$$

The left side is now the derivative of $ye^{(\alpha+\beta)t}$.

Integrate both sides with respect to t.

$$ye^{(\alpha+\beta)t} = \frac{\alpha}{\alpha + \beta}e^{(\alpha+\beta)t} + C$$

$$y = \frac{\alpha}{\alpha + \beta} + C \cdot e^{-(\alpha+\beta)t}$$

Since $y(0) = y_0$,

$$y_0 = \frac{\alpha}{\alpha + \beta} + C$$

$$C = y_0 - \frac{\alpha}{\alpha + \beta}$$

$$y = \frac{\alpha}{\alpha + \beta} + \left(y_0 - \frac{\alpha}{\alpha + \beta}\right)e^{-(\alpha+\beta)t}$$

(b) As $t \to \infty$ the second term in the expression for y above goes to 0 because the exponential factor has limit 0, so the limiting value of y is $\alpha/(\alpha + \beta)$.

29. $\dfrac{dy}{dt} = 0.02y + e^t; y = 10,000$ when $t = 0$.

$$\frac{dy}{dt} - 0.02y = e^t$$

$$I(t) = e^{\int -0.02dt} = e^{-0.02t}$$

$$e^{-0.02t}\frac{dy}{dt} - 0.02e^{-0.02t}y = e^{-0.02t}\cdot e^t$$

$$D_t(e^{-0.02t}y) = e^{0.98t}$$

$$e^{-0.02t}y = \int e^{0.98t}\,dt$$

$$= \frac{e^{0.98t}}{0.98} + C$$

$$y = \frac{e^t}{0.98} + Ce^{0.02t}$$

$$10{,}000 = \frac{1}{0.98} + C$$

$$C \approx 9999$$

$$y \approx \frac{e^t}{0.98} + 9999e^{0.02t}$$

$$= 1.02e^t + 9999e^{0.02t}$$

31. $\dfrac{dy}{dt} = 0.02y - t;\ y = 10{,}000$ when $t = 0.$

$$\frac{dy}{dt} - 0.02y = -t$$

$$I(t) = e^{\int -0.02} = e^{-0.02t}$$

$$e^{-0.02t}\frac{dy}{dt} - 0.02e^{-0.02t}y = -te^{-0.02t}$$

$$D_t(e^{-0.02t}y) = -te^{-0.02t}$$

$$e^{-0.02t}y = \int -te^{-0.02t}\,dt$$

Integration by parts:

Let $\ u = -t \qquad dv = e^{-0.02t}\,dt$

$$du = -dt \qquad v = \frac{e^{-0.02t}}{-0.02}$$

$$e^{-0.02t} = \frac{te^{-0.02t}}{0.02} - \int \frac{e^{-0.02t}}{-0.02}\,dt$$

$$e^{-0.02t}y = \frac{te^{-0.02t}}{0.02} + \frac{e^{-0.02t}}{0.0004} + C$$

$$y = 50t + 2500 + Ce^{0.02t}$$

$$10{,}000 = 2500 + C$$

$$C = 7500$$

$$y = 50t + 2500 + 7500e^{0.02t}$$

10.3 Euler's Method

Your Turn 1

Use Euler's method to approximate the solution of $dy/dx - x^2y^2 = 1,\ y(0) = 2,$ for the interval $[0, 1]$ with $h = 0.2.$

First rewrite the equation with the derivative on the left.

$$\frac{dy}{dx} = 1 + x^2y^2$$

$$g(x, y) = 1 + x^2y^2$$

The following table shows the calculations, including the value of $g(x_i, y_i)$ used in computing $y_{i+1}.$

i	x_i	y_i	$g(x_i, y_i)$
0	0	2	1
1	0.2	2.2	1.1936
2	0.4	2.43872	1.95157684
3	0.6	2.82903537	3.88123880
4	0.8	3.60528313	9.31876252
5	1.0	5.46903563	

10.3 Exercises

Note: In each step of the calculation shown in this section, all digits should be kept in your calculator as you proceed through Euler's method. Do not round intermediate results.

1. $\dfrac{dy}{dx} = x^2 + y^2;\ y(0) = 2, h = 0.1.$ Find $y(0.5).$

$$g(x, y) = x^2 + y^2$$

$$x_0 = 0;\ y_0 = 2$$

$$g(x_0, y_0) = 0 + 4 = 4$$

$$x_1 = 0.1;\ y_1 = 2 + 4(0.1)$$

$$= 2.4$$

$$g(x_1, y_1) = (0.1)^2 + (2.4)^2$$

$$= 5.77$$

$$x_2 = 0.2;\ y_2 = 2.4 + 5.77(0.1)$$

$$= 2.977$$

$$g(x_2, y_2) = (0.2)^2 + (2.977)^2$$

$$\approx 8.903$$

$$x_3 = 0.3;\ y_3 = 2.977 + 8.903(0.1)$$

$$\approx 3.867$$

$$g(x_3, y_3) = (0.3)^2 + (3.867)^2$$

$$\approx 15.046$$

$x_4 = 0.4; y_4 = 3.867 + 15.046(0.1)$

≈ 5.372

$g(x_4, y_4) = (0.4)^2 + (5.372)^2$

≈ 29.016

$x_5 = 0.5; y_5 = 5.372 + 29.016(0.1)$

≈ 8.273

These results are tabulated as follows.

x_i	y_i
0	2
0.1	2.4
0.2	2.977
0.3	3.867
0.4	5.372
0.5	8.273

$y(0.5) \approx 8.273$.

Use Euler's method as outlined as in the solutions for Exercises 1 and 2 in the following exercises. The results are tabulated.

3. $\dfrac{dy}{dx} = 1 + y; y(0) = 2, h = 0.1$; find y(0.6).

x_i	y_i
0	2
0.1	2.3
0.2	2.63
0.3	2.993
0.4	3.3923
0.5	3.8315
0.6	4.31465

$y(0.6) \approx 4.315$

5. $\dfrac{dy}{dx} = x + \sqrt{y}; y(0) = 1, h = 0.1$; find $y(0.4)$.

x_i	y_i
0	1
0.1	1.1
0.2	1.215
0.3	1.345
0.4	1.491

$y(0.4) \approx 1.491$

7. $\dfrac{dy}{dx} = 2x\sqrt{1 + y^2}; y(1) = 0, h = 0.1$;

find $y(1.5)$.

x_i	y_i
1	2
1.1	2.447
1.2	3.029
1.3	3.794
1.4	4.815
1.5	6.191

$y(1.5) \approx 6.191$.

9. $\dfrac{dy}{dx} = -4 + x; y(0) = 1, h = 0.1$, find $y(0.4)$.

x_i	y_i
0	1
0.1	0.6
0.2	0.21
0.3	−0.17
0.4	−0.540

$y(0.4) \approx -0.540$

Exact solution:

$$\frac{dy}{dx} = -4 + x$$

$$y = -4x + \frac{x^2}{2} + C$$

At $y(0) = 1$,

$$1 = -4(0) + \frac{0}{2} + C$$

$$C = 1$$

Therefore,

$$y = -4x + \frac{x^2}{2} + 1$$

$$y(0.4) = -4(0.4) + \frac{(0.4)^2}{2} + 1$$

$$= -0.520.$$

11. $\dfrac{dy}{dx} = x^3; y(0) = 4, h = 0.1$, find $y(0.4)$.

x_i	y_i
0	4
0.1	4
0.2	4.0001
0.3	4.0009
0.4	4.0036
0.5	4.010

$y(0.5) \approx 4.010$

Exact solution:

$$\frac{dy}{dx} = x^3$$

$$y = \frac{x^4}{4} + C$$

At $y(0) = 4$,

$$4 = \frac{0}{4} + C$$

$$C = 4.$$

Therefore,

$$y = \frac{x^4}{4} + 4$$

$$y(0.5) = \frac{(0.5)^4}{4} + 4 \approx 4.016.$$

13. $\frac{dy}{dx} = 2xy$; $y(1) = 1$, $h = 0.1$, find $y(1.6)$.

$$g(x, y) = 2xy$$

x_i	y_i
1	1
1.1	1.2
1.2	1.464
1.3	1.815
1.4	2.287
1.5	2.927
1.6	3.806

$y(1.6) \approx 3.806$

Exact solution:

$$\frac{dy}{y} = 2x\,dx$$

$$\int \frac{dy}{y} = \int 2x\,dx$$

$$\ln|y| = x^2 + C$$

$$|y| = e^{x^2} + C$$

$$y = ke^{x^2}$$

At $y(1) = 1$,

$$1 = ke^1 = ke$$

$$k = \frac{1}{e}.$$

Therefore,

$$y = \frac{1}{e}\left(e^{x^2}\right)$$

$$= e^{x^2 - 1}$$

$$y(1.6) = e^{(1.6)^2 - 1}$$

$$= 4.759.$$

15. $\frac{dy}{dx} = ye^x$; $y(0) = 2$, $h = 0.1$, find $y(0.4)$.

x_i	y_i
0	2
0.1	2.2
0.2	2.443
0.3	2.742
0.4	3.112

So, $y(0.4) \approx 3.112$.

Exact solution:

$$\frac{dy}{y} = e^x\,dx$$

$$\int \frac{dy}{y} = \int e^x + c$$

$$\ln|y| = e^x + c$$

$$|y| = e^{e^x + c} = e^c e^{e^x}$$

$$y = ke^x, \text{ where } k = \pm e^c.$$

At $y(0) = 2$, $2 = ke^{e^0} = ke$, so $k = \dfrac{2}{e}$.

Therefore,

$$y = \frac{2}{e}e^{e^x} = 2e^{e^x - 1},$$

so

$$y(0.4) = 2e^{e^{0.4} - 1} \approx 3.271.$$

17. $\frac{dy}{dx} + y = 2e^x$; $y(0) = 100$, $h = 0.1$.

Find $y(0.3)$.

x_i	y_i
0	100
0.1	90.2
0.2	81.401
0.3	73.505

$y(0.3) \approx 73.505$.

Exact solution:

$$I(x) = e^{\int dx} = e^x$$

$$e^x \frac{dy}{dx} + e^x y = 2e^x e^x$$

$$D_x(e^x y) = 2e^{2x}$$

$$e^x y = \int 2e^{2x} dx + C$$

$$= e^{2x} + C$$

$$y = e^x + Ce^{-x}$$

$$100 = 1 + C$$

$$C = 99$$

$$y = e^x + 99e^{-x}$$

$$y(0.3) = e^{0.3} + 99e^{-0.3} \approx 74.691$$

19. $\frac{dy}{dx} = ye^x$; $y(0) = 2$, $h = 0.05$, find $y(0.4)$.

x_i	y_i
0	2
0.05	2.1
0.1	2.21
0.15	2.333
0.2	2.468
0.25	2.619
0.3	2.787
0.35	2.975
0.4	3.186

So, $y(0.4) \approx 3.186$.

Exact solution:

$$y = \frac{2}{e}e^{e^x} = 2e^{e^x - 1},$$

so

$$y(0.4) = 2e^{e^{0.4} - 1} \approx 3.271.$$

21. $\frac{dy}{dx} = \sqrt[3]{x}$, $y(0) = 0$

Using the program for Euler's method in the Graphing Calculator Manual, the following values are obtained:

x_i	y_i	$y(x_i)$	$y_i - y(x_i)$
0	0	0	0
0.2	0	0.08772053	−0.08772053
0.4	0.11696071	0.22104189	−0.10408118
0.6	0.26432197	0.37954470	−0.11522273
0.8	0.43300850	0.55699066	−0.12398216
1.0	0.61867206	0.75000000	−0.13132794

23. $\frac{dy}{dx} = 4 - y$, $y(0) = 0$

Using the program for Euler's method in the Graphing Calculator Manual, the following values are obtained:

x_i	y_i	$y(x_i)$	$y_i - y(x_i)$
0	0	0	0
0.2	0.8	0.725077	0.07492
0.4	1.44	1.3187198	0.12128
0.6	1.952	1.8047535	0.14725
0.8	2.3616	2.2026841	0.15892
1.0	2.68928	2.5284822	0.16080

25. $\frac{dy}{dx} = \sqrt[3]{x}$; $y(0) = 0$

$y = \frac{3}{4}x^{4/3}$ See Exercise 21.

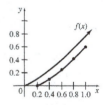

27. $\dfrac{dy}{dx} = 4 - y;\ y(0) = 0$

$y = 4(1 - e^{-x})$ See Exercise 23.

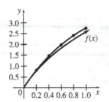

29. $\dfrac{dy}{dx} = y^2;\ y(0) = 1$

(a)

x_i	y_i
0	1
0.2	1.2
0.4	1.488
0.6	1.9308288
0.8	2.676448771
1.0	4.109124376

Thus, $y(1.0) \approx 4.109$.

(b) $\dfrac{dy}{dx} = y^2;\ y = 1$ when $x = 0$

$$\frac{dy}{dx} = y^2$$

$$\frac{1}{y^2}\,dy = dx$$

$$\int \frac{1}{y^2}\,dy = \int dx$$

$$-\frac{1}{y} = x + C$$

When $x = 0$, $y = 1$.

$$-\frac{1}{1} = 0 + C$$

$$C = -1$$

$$-\frac{1}{y} = x - 1$$

$$-1 = (x - 1)y$$

$$y = \frac{-1}{x - 1}$$

$$y = \frac{1}{1 - x}$$

As x approaches 1 from the left, y approaches ∞.

31. Let $y =$ the number of algae (in thousands) at time.

$$y \le 500;\ y(0) = 5$$

(a) $\dfrac{dy}{dt} = 0.01y(500 - y)$

$$= 5y - 0.01y^2$$

(b) Find $y(2);\ h = 0.5$.

x_i	y_i
0	5
0.5	17.375
1	59.303
1.5	189.976
2	484.462

Therefore, $y(2) \approx 484$, so about 484,000 algae are present when $t = 2$.

33. $\dfrac{dy}{dt} = 0.05y - 0.1y^{1/2};\ y(0) = 60, h = 1;$ find $y(6)$.

x_i	y_i
0	60
1	62.22541
2	64.54785
3	66.97182
4	69.50205
5	72.14347
6	74.90127

Therefore, $y(6) \approx 75$, so about 75 insects are present after 6 weeks.

35. $\dfrac{dW}{dt} = -0.01189W + 0.92389W^{0.016}$

$W(0) = 3.65;\ h = 1;$ find $W(5)$

Using the program for Euler's method in *The Graphing Calculator Manual*, the following values are obtained:

t_4	W_i
0	3.6500000
1	4.5498301
2	5.4422927
3	6.3268604
4	7.2032009
5	8.0710987

The weight of a goat at 5 weeks is about 8.07 kg.

37.

$$\frac{dN}{dt} = 0.02(500 - N)N^{1/2}; \quad N(0) = 2, \, h = 0.5;$$

Find $N(3)$.

t_i	N_i
0	2
0.5	9.043
1	23.806
1.5	47.041
2	78.108
2.5	115.394
3	156.709

Therefore, $N(3) \approx 157$, so about 157 people have heard the rumor.

10.4 Applications of Differential Equations

Your Turn 1

For this problem, $r = 0.05$ and $D = 1200$ so the differential equation is

$$\frac{dA}{dt} = 0.05A + 1200$$

Separate variables and integrate on both sides.

$$\frac{dA}{0.05A + 1200} = dt$$

$$\frac{1}{0.05}\ln(0.05A + 1200) = t + C$$

$$\ln(0.05A + 1200) = 0.05t + K$$

$$0.05A + 1200 = e^{0.05t+K}$$

$$0.05A = -1200 + Me^{0.05t}$$

$$A = -24,000 + \frac{M}{0.05}e^{0.05t}$$

Use the fact that $A(0) = 6000$.

$$6000 = -24,000 + \frac{M}{0.05}e^{0.05(0)}$$

$$30,000(0.05) = M$$

$$1500 = M$$

$$A = -24,000 + \frac{1500}{0.05}e^{0.05t}$$

$$A = -24,000 + 30,000e^{0.05t}$$

Find A when $t = 21$.

$$A = -24,000 + 30,000e^{0.05(21)}$$

$$A \approx 61,729.53$$

When Michael is 21 the account will be worth \$61,729.53.

Your Turn 2

Solve the equation given in Example 2 with $p = 4$, $q = 1, r = 3$, and $s = 5$, given that there was a time when $x = 1$ and $y = 1$.

$$\frac{dy}{dx} = \frac{y(4 - x)}{x(5y - 3)}$$

$$\frac{5y - 3}{y}dy = \frac{4 - x}{x}dx$$

$$\int \frac{5y - 3}{y}dy = \int \frac{4 - x}{x}dx$$

$$5y - 3\ln y = 4\ln x - x + C$$

Use the condition $y = 1$ when $x = 1$ to find C.

$$5(1) - 3\ln(1) = 4\ln(1) - 1 + C$$

$$6 = C$$

The equation relating x and y is

$$x + 5y - 4\ln x - 3\ln y = 6.$$

Your Turn 3

Using the notation of Example 3,

$$b = \frac{N - y_0}{y_0}$$

$$= \frac{50,000 - 80}{80}$$

$$= 624.$$

Thus the formula for y is

$$y = \frac{50,000}{1 + 624e^{-kt}}.$$

Use the fact that $y(15) = 640$ to find k.

$$640 = \frac{50,000}{1 + 624e^{-k(15)}}$$

$$640 + 399,360e^{-15k} = 50,000$$

$$e^{-15k} = \frac{49,360}{399,360}$$

$$k = \frac{1}{15}\ln\left(\frac{399,360}{49,360}\right)$$

$$k \approx 0.13938$$

Therefore, $y = \dfrac{50,000}{1 + 624e^{-0.13938t}}.$

When $t = 25$,

$$y = \frac{50{,}000}{1 + 624e^{-0.13938(25)}}$$

$$= 2483.$$

About 2483 people are infected 25 days into the epidemic.

Your Turn 4

Suppose that a tank initially contains 500 liters of a solution of water and 5 kg of salt, and that pure water flows in at a rate of 6 L/min and solution flows out at a rate of 4 L/min. These last two numbers indicate that the volume of solution increases at the rate of 2 L/min, so at time t the volume $V(t) = 500 + 2t$. How many kilograms of salt remain after 20 minutes? We will assume that the tank is large enough so that it does not overflow during the 20 minutes of interest. Following the argument in Example 4 we can write the differential equation for this process as

$$\frac{dy}{dt} = -\frac{4y}{V(t)}$$

$$\frac{dy}{dt} = -\frac{4y}{500 + 2t}$$

with initial condition $y(0) = 5$, where y is the amount of salt in the tank in kilograms.

Separate the variables and integrate on both sides.

$$\frac{dy}{dt} = -\frac{4y}{500 + 2t} dy$$

$$\frac{1}{y} dy = -\frac{4}{500 + 2t} dt$$

$$\int \frac{1}{y} dy = -\int \frac{4}{500 + 2t} dt$$

$$\int \frac{1}{y} dy = -2 \int \frac{1}{500 + 2t} 2dt$$

$$\ln y = -2\ln(500 + 2t) + C$$

Use the initial condition $y(0) = 5$.

$$\ln 5 = -2\ln[500 + 2(0)] + C$$
$$C = \ln 5 + 2\ln 500$$
$$C = \ln(5 \cdot 500^2)$$
$$\ln y = -2\ln(500 + 2t) + \ln(5 \cdot 500^2)$$
$$y = \frac{5 \cdot 500^2}{(500 + 2t)^2}$$

When $t = 20$,

$$y = \frac{5 \cdot 500^2}{(500 + 40)^2}$$

$$\approx 4.29.$$

After twenty minutes, the tank contains 540 L of solution and 4.29 kg of salt.

10.4 Warmup Exercises

W1. $\dfrac{dy}{dx} = \dfrac{x^2}{y^2}$

Separate the variables:

$$y^2 dy = x^2 dx$$

Integrate on both sides:

$$\frac{y^3}{3} = \frac{x^3}{3} + C \text{ so}$$
$$y^3 = x^3 + D$$
$$y = \sqrt[3]{x^3 + D}$$

Since $y(0) = 2$, $D = 8$ and $y = \sqrt[3]{x^3 + 8}$.

W2. $x^2 \dfrac{dy}{dx} = y$

Separate the variables:

$$\frac{dy}{y} = \frac{dx}{x^2}$$

Integrate on both sides:

$$\ln|y| = -\frac{1}{x} + C$$
$$y = De^{-1/x} \text{ with } D \text{ positive or negative}$$

Since $y(1) = e^{-2}$, $D = e^{-1}$ and $y = e^{-(1/x)-1}$.

W3. $x\dfrac{dy}{dx} + xy - x^2 = 0$

The equation will hold for $x = 0$ as long as the derivative of y is defined there, so assume $x \neq 0$.

Then

$$\frac{dy}{dx} + y = x$$

The integrating factor is $e^{\int 1\, dx} = e^x$.

$$e^x \frac{dy}{dx} + y \cdot e^x = x \cdot e^x$$

Integrate on both sides, integrating by parts on the right.

$$ye^x = (x - 1)e^x + C$$

$$y = x - 1 + Ce^{-x}$$

Since $y(0) = 3, D = 4$ and $y = x - 1 + 4e^{-x}$.

(Note that y and its derivative are continuous at $x = 0$, so our solution is well-behaved there.)

W4. $2\dfrac{dy}{dx} - 4xy = 10x$

Separate the variables by rewriting the equation:

$$\frac{dy}{dx} = (2y + x) \cdot x$$

$$\frac{dy}{2y + 5} = \frac{dx}{x}$$

Integrate on both sides:

$$\frac{1}{2} \ln|2y + 5| = \ln|x| + C$$

$$2y + 5 = e^{x^2} \cdot D \quad (D \text{ positive or negative})$$

$$y = -\frac{5}{2} + e^{x^2} \cdot E \quad (E \text{ positive or negative})$$

Since $y(0) = \dfrac{1}{2}$, $E = 8$ and $y = -\dfrac{5}{2} + 3e^{x^2}$.

10.4 Exercises

1. $\dfrac{dA}{dt} = rA + D; r = 0.06, D = \$5000$

$$\frac{dA}{dt} = 0.06A + 3000$$

$$\int \frac{dA}{0.06A + 3000} = \int dt$$

$$\ln|0.06A + 3000| = 0.06t + C$$

$$|0.06A + 3000| = e^{0.06t + C}$$

$$0.06A + 3000 = Me^{0.06t}$$

$$A = \frac{M}{0.06}e^{0.06t} - \frac{3000}{0.06}$$

$$A = 16.66667Me^{0.06t} - 50,000$$

$$t = 0, A = 5000$$

$$5000 = 16.66667Me^0 - 50,000$$

$$M = 3300$$

$$A = 16.66667(3300)e^{0.06t} - 50,000$$

$$= 55,000e^{0.06t} - 50,000$$

When $t = 10$ yr,

$$A = 55,000e^{0.06(10)} - 50,000$$

$$= \$50,216.53.$$

3. $\dfrac{dA}{dt} = rA + D; r = 0.01; D = \$50,000;$

$t = 0, A = 0$

$$\frac{dA}{dt} = 0.1A + 50,000$$

$$\frac{dA}{0.1A + 50,000} = dt$$

$$\frac{1}{0.1} \ln|0.1A + 50,000| = t + C$$

$$\ln|0.1A + 50,000| = 0.1t + C$$

$$0.1A + 50,000 = e^{0.1t + C}$$

$$= Me^{0.1t}$$

$$A = \frac{M}{0.1}e^{0.1t} - \frac{50,000}{0.1}$$

When $t = 0, A = 0$.

$$0 = \frac{M}{0.1}(1) - \frac{50,000}{0.1}$$

$$M = 50,000$$

$$A = \frac{50,000}{0.1}e^{0.1t} - \frac{50,000}{0.1}$$

$$= 500,000(e^{0.1t} - 1)$$

Find t when $A = \$500,000$.

$$500,000 = 500,000(e^{0.1t} - 1)$$

$$e^{0.1t} = 2$$

$$t = \frac{1}{0.1} \ln 2$$

$$\approx 6.9 \text{ yr}$$

It will take about 6.9 years to accumulate $500,000.

5. (a) $\dfrac{dA}{dt} = rA + D; r = 0.06; D = \$1200;$

$A(0) = 8000$

$$\frac{dA}{dt} = 0.06A - 1200$$

(b) To solve for A, first separate the variables.

$$\frac{dA}{0.06A - 1200} = dt$$

Integrate.

$$\int \frac{dA}{0.06A - 1200} = \int dt$$

$$\frac{1}{0.06} \ln|0.06A - 1200| = t + k$$

Solve for A. $\ln|0.06A - 1200| = 0.06t + C$, where $C = 0.06k$.

$$|0.06A - 1200| = e^{0.06t+C} = e^C e^{0.06t}$$

$$0.06A - 1200 = Me^{0.06t}, \text{where } M = \pm e^C$$

$$0.06A = 1200 + Me^{0.06t}, \text{ so}$$

$$A = 20{,}000 + C_1 e^{0.06t},$$

Where $\quad C_1 = \dfrac{M}{0.06}$.

Solve for C_1. Since $A(0) = 8000$, then

$$8000 = 20{,}000 + C_1 e^0, \text{ so}$$
$$C_1 = -12{,}000. \text{ Therefore,}$$
$$A = 20{,}000 - 12{,}000 e^{0.06t}$$

Since $A(2) = 20{,}000 - 12{,}000 e^{0.06(2)}$ ≈ 6470.04, the account will have $\$6470.04$ left after two years.

(c) If $A = 0$, then

$$20{,}000 - 12{,}000 e^{0.06t} = 0.$$

$$12{,}000 e^{0.06t} = 20{,}000$$

$$e^{0.06t} = \frac{5}{3}$$

$$0.06t = \ln\left(\frac{5}{3}\right), \text{ so}$$

$$t = \frac{1}{0.06} \ln\left(\frac{5}{3}\right) \approx 8.51.$$

Therefore, the account will be completely depleted in 8.51 years.

7. (a)

$$\frac{dy}{dt} = 4y - 2xy$$

$$= y(4 - 2x)$$

$$\frac{dx}{dt} = -3x + 2xy$$

$$= x(-3 + 2y)$$

$$\frac{dy}{dx} = \frac{\frac{dy}{dt}}{\frac{dx}{dt}} = \frac{y(4 - 2x)}{x(-3 + 2y)}$$

$$\int \left(\frac{2y - 3}{y}\right) dy = \int \left(\frac{4 - 2x}{x}\right) dx$$

$$\int \left(2 - \frac{3}{y}\right) dy = \int \left(\frac{4}{x} - 2\right) dx$$

$$2y - 3\ln y = 4\ln x - 2x + C$$

When $x = 1, y = 1,$

$$2 - 3\ln 1 = 4\ln 1 - 2 + C$$

$$4 = C.$$

Therefore, $2y - 3\ln y - 4\ln x + 2x = 4$ is an equation relating x and y in this case.

(b) $0 = 4y - 2xy = -y(-4 + 2x)$
$\quad\;\; 0 = -3x + 2xy = x(-3 + 2y)$

$$0 = -y(-4 + 2x) \text{ and} \qquad 0 = x(-3 + 2y)$$
$$y = 0 \qquad\qquad \text{and} \qquad\qquad x = 0 \text{ or}$$
$$-4 + 2x = 0 \qquad\qquad\qquad -3 + 2y = 0$$
$$x = 2 \qquad\qquad \text{and} \qquad\qquad y = \frac{3}{2}$$

9. (a) Let $y =$ the number of individuals infected. The differential equation is

$$\frac{dy}{dt} = k\left(1 - \frac{y}{N}\right)y.$$

The solution is Equation 7 in Example 3:

$$y = \frac{N}{1 + (N - 1)e^{-kt}}.$$

The number of individuals uninfected at time t is

$$y = N - \frac{N}{1 + (N - 1)e^{-kt}}$$

$$= \frac{N + N(N - 1)e^{-kt} - N}{1 + (N - 1)e^{-kt}}$$

$$= \frac{N(N - 1)}{N - 1 + e^{kt}}.$$

Now substitute $k = 0.25$.

$$y = \frac{5000(5000 - 1)}{5000 - 1 + e^{0.25t}}$$

$$= \frac{24,995,000}{4999 + e^{0.25t}}$$

(b) $t = 30$

$$y = \frac{24,995,000}{4999 + e^{0.25(30)}} = 3672$$

(c) $t = 50$

(d) From Example 3,

$$t_m = \frac{\ln(N - 1)}{k}$$

$$= \frac{\ln(5000 - 1)}{0.25}$$

$$= 34.$$

The maximum infection rate will occur on the 34th day.

11. (a) The differential equation is

$$\frac{dy}{dt} = a(N - y)y.$$

$y_0 = 100; \ y = 400$ when $t = 10$, $N = 20,000$.

The solution is

$$y = \frac{N}{1 + be^{-kt}},$$

where $b = \frac{N - y_0}{y_0}$ and $k = aN$.

Since $y_0 = 100$ and $N = 20,000$,

$$b = \frac{20,000 - 100}{100} = 199; \ k = 20,000a.$$

Therefore,

$$y = \frac{20,000}{1 + 199e^{-20,000\,at}}.$$

$y = 400$ when $t = 10$.

$$400 = \frac{20,000}{1 + 199e^{-20,000(10)a}}$$

$$400 + 400(199)e^{-200,000a} = 20,000$$

$$e^{-200,000a} = \frac{19,600}{400(199)}$$

$$= 0.2462312$$

$$a = \frac{\ln(0.2462312)}{-200,000}$$

$$= 7 \times 10^{-6}$$

$$k = 20,000a = 20,000(7 \times 10^{-6}) = 0.14$$

Therefore,

$$y = \frac{20,000}{1 + 199e^{-0.14t}} \quad \text{or} \quad \frac{20,000^{0.14t}}{e^{0.14t} + 199}.$$

(b) Half the community is $y = 10,000$. Find t for $y = 10,000$.

$$10,000 = \frac{20,000}{1 + 19e^{-0.14t}}$$

$$10,000 + 10,000(199)e^{-0.14t} = 20,000$$

$$e^{-0.14t} = \frac{10,000}{10,000(199)} = 0.005$$

$$t = \frac{\ln(0.005)}{-0.14}$$

$$= 37.77$$

Half the community will be infected in about 38 days.

13. (a) $\dfrac{dy}{dt} = -ay + b(f - y)Y$

$a = 1, b = 1, f = 0.5, Y = 0.01;$
$y = 0.02$ when $t = 0$.

$$\frac{dy}{dt} = -y + 1(0.5 - y)(0.01)$$

$$= -1.010y + 0.005$$

$$\int \frac{dy}{-1.010y + 0.005} = \int dt$$

$$\frac{1}{-1.010}\ln|-1.010y + 0.005| = t + C_2$$

$$\ln|-1.010y + 0.005| = -1.010t + C_1$$

$$|-1.010y + 0.005| = e^{-1.010t + C_1}$$

$$= e^{C_1}e^{-1.010t}$$

$$-1.010y + 0.005 = Ce^{-1.010t}$$

$$y = 0.005 - 0.990Ce^{-1.010t}$$

Since $y = 0.02$ when $t = 0$,

$$0.02 = 0.005 - 0.990Ce^0$$

$$-0.990C = 0.015.$$

Therefore,

$$y = 0.005 + 0.015e^{-1.010t}.$$

(b) $\dfrac{dY}{dt} = -AY + B(F - Y)y$

$A = 1, B = 1, y = 0.1, F = 0.03$;

$Y = 0.01$ when $t = 0$.

$$\frac{dY}{dt} = -Y + 1(0.03 - Y)(0.1)$$

$$= -1.1Y + 0.003$$

$$\frac{dY}{-1.1Y - 0.003} = dt$$

$$-\frac{1}{1.1}\ln|-1.1Y + 0.003| = t + C_2$$

$$\ln|-1.1Y + 0.003| = -1.1t + C_1$$

$$|-1.1Y + 0.003| = e^{-1.1t+C_1}$$

$$= e^{C_1}e^{-1.1t}$$

$$-1.1Y + 0.003 = Ce^{-1.1t}$$

$$Y = \frac{C}{-1.1}e^{-1.1t} - \frac{0.003}{-1.1}$$

$$= 0.909Ce^{-1.1t} + 0.00273$$

Since $Y = 0.01$ when $t = 0$,

$$0.01 = -0.909Ce^0 + 0.00273$$

$$-0.909C = 0.00727.$$

Therefore,

$$Y = 0.00727e^{-1.1t} + 0.00273.$$

15. (a) $\dfrac{dy}{dt} = a(N - y)y$

$y_0 = 3; y = 12$ when $t = 3; N = 45$

From equations (4) and (5) in Section 1 of this chapter, the solution is

$$y = \frac{N}{1 + be^{-kt}} \text{ where } b = \frac{N - y_0}{y_0} \text{ and }$$

$$k = aN.$$

Since $y_0 = 3$ and $N = 45$,

$$b = \frac{45 - 3}{3} = 14; k = 45a.$$

$$y = \frac{45}{1 + 14e^{-45at}}.$$

$y = 12$ when $t = 3$, so

$$12 = \frac{45}{1 + 14e^{-135a}}$$

$$12 + 168e^{-135a} = 45$$

$$e^{-135a} = \frac{33}{168} = \frac{11}{56}$$

$$-135a = \ln\frac{11}{56} = -1.627$$

$$-45a = -0.542.$$

Therefore, $\quad y \approx \dfrac{45}{1 + 14e^{-0.54t}}.$

(b) When $y = 30$,

$$30 = \frac{45}{1 + 14e^{-0.54t}}$$

$$30 + 450e^{-0.54t} = 45$$

$$e^{-0.54t} = \frac{15}{450} = \frac{1}{30}$$

$$t = -\frac{1}{0.54}\ln\frac{1}{30} = 6.30.$$

In about 6 days, 30 employees have heard the rumor.

17. (a) $\dfrac{dy}{dt} = kye^{-at}; a = 0.1; y = 5$ when $t = 0$;

$y = 15$ when $t = 3$.

$$\int \frac{dy}{y} = k\int e^{-0.1t}\,dt$$

$$\ln|y| = -10ke^{-0.1t} + C_1$$

$$|y| = e^{-10ke^{-0.1t} + C_1}$$

$$= e^{C_1}e^{-10ke^{-0.1t}}$$

$$y = Ce^{-10ke^{-0.1t}}$$

Since $y = 5$ when $t = 0$,

$$5 = Ce^{-10k}$$

$$C = 5e^{10k}.$$

Since $y = 15$ when $t = 3$,

$$15 = Ce^{-10ke^{-0.3}} = Ce^{-7.41k}$$

$$C = 15e^{7.41k}.$$

Solve the system

$$C = 5e^{10k}$$
$$C = 15e^{7.41k}.$$
$$5e^{10k} = 15e^{7.41k}$$
$$e^{10k} = 3e^{7.41k}$$

Take natural logarithms on both sides.

$$10k \ln e = \ln 3 + 7.41k$$
$$2.59k = \ln 3$$
$$k = \frac{1}{2.59} \ln 3 = 0.424$$
$$C = 5e^{10(0.424)} = 347$$

Therefore, $y = 347e^{-10(0.424)e^{-0.1t}}$

$$= 347e^{-4.24e^{-0.1t}}.$$

(b) If $y = 30$,

$$30 = 347e^{-4.24e^{-0.1t}}$$
$$e^{-4.24e^{-0.1t}} = \frac{30}{347} = 0.0865$$
$$-4.24e^{-0.1t} \ln e = \ln 0.0865$$
$$e^{-0.1t} = -\frac{1}{4.24} \ln 0.0865$$
$$= 0.5773$$
$$-0.1t \ln e = \ln 0.5773$$
$$t = -10 \ln 0.5773$$
$$= 5.493.$$

30 employees have heard the rumor in about 5.5 days.

19. Let $y =$ the amount of salt present at time t.

(a) $\dfrac{dy}{dt} = $ (rate of salt in)

$- $ (rate of salt out)

rate of salt in $= $ (3 gal/min)(2 lb/gal)

$= 6$ lb/min

rate of salt out $= \left(\dfrac{y}{V} \text{ lb/gal}\right)(2 \text{ gal/min})$

$= \left(\dfrac{2y}{V} \text{ lb/min}\right)$

$$\frac{dy}{dt} = 6 - \frac{2y}{V}; \quad y(0) = 20 \text{ lb}$$

$$\frac{dV}{dt} = \text{(rate of liquid in)}$$

$- \text{(rate of liquid out)}$

$= 3 \text{ gal/min} - 2 \text{ gal/min}$

$= 1 \text{ gal/min}$

$$\frac{dV}{dt} = 1$$

$$V = t + C_1$$

When $t = 0$, $V = 100$. Thus,

$$C_1 = 100.$$
$$V = t + 100.$$

Therefore,

$$\frac{dy}{dt} = 6 - \frac{2y}{t + 100}$$

$$\frac{dy}{dt} + \frac{2}{t + 100} y = 6.$$

$$I(t) = e^{\int 2\,dt/(t+100)}$$
$$= e^{2 \ln|t + 100|}$$
$$= (t + 100)^2$$

$$\frac{dy}{dt}(t + 100)^2 + 2y(t + 100) = 6(t + 100)^2$$

$$D_t[y(t + 100)^2] = 6(t + 100)^2$$

$$y(t + 100)^2 = 6\int (t + 100)^2\,dt$$

$$y(t + 100)^2 = 2(t + 100)^3 + C$$

$$y = 2(t + 100) + \frac{C}{(t + 100)^2}$$

Since $t = 0$ when $y = 20$,

$$20 = 2(100) + \frac{C}{100^2}$$

$$C = -1{,}800{,}000.$$

$$y = 2(t + 100) - \frac{1{,}800{,}000}{(t + 100)^2}$$

$$= \frac{2(t + 100)^3 - 1{,}800{,}000}{(t + 100)^2}.$$

(b) $t = 1\,\text{hr} = 60$ min

$$y = \frac{2(160)^3 - 1{,}800{,}000}{(160)^2} = 249.69$$

After 1 hr, about 250 lb of salt are present.

(c) As time increases, salt concentration continues to increase.

21. Let $y =$ the amount of salt present at time t minutes.

(a) $\dfrac{dy}{dt} =$ (rate of salt in) $-$ (rate of salt out)

rate of salt in $= 0$

rate of salt out $= \left(\dfrac{y}{V} \text{lb/gal}\right)(2 \text{ gal/min})$

$= \dfrac{2y}{V} \text{lb/min}$

$\dfrac{dV}{dt} = -\dfrac{2y}{V}; y(0) = 20$

$\dfrac{dV}{dt} =$ (rate of liquid in)

$-$ (rate of liquid out)

$= 2 \text{ gal/min} - 2 \text{ gal/min} = 0$

$\dfrac{dV}{dt} = 0$

$V = C_1$

When $t = 0, V = 100$, so $C_1 = 100$.

Therefore,

$\dfrac{dy}{dt} = -\dfrac{2y}{100} = -0.02y$

$\dfrac{dy}{y} = -0.02 \, dt$

$\ln |y| = -0.02t + C_1$

$|y| = e^{-0.02t+C_1} = e^{C_1}e^{-0.02t}$

$= Ce^{-0.02t}$

Since $t = 0$ when $y = 20$,

$20 = Ce^0$

$C = 20.$

$y = 20e^{-0.02t}$

(b) $t = 1 \text{ hr} = 60 \text{ min}$

$y = 20e^{-0.02(60)} = 6.024$

After 1 hr, about 6 lb of salt are present.

(c) As time increases, salt concentration continues to decrease.

23. Let $y =$ amount of the chemical at time t.

(a) $\dfrac{dy}{dt} =$ (rate of chemical in)

$-$ (rate of chemical out)

rate of chemical in

$= (2 \text{ liters/min})(0.25 \text{ g/liter})$

$= 0.5 \text{ g/min}$

rate of chemical out

$= \left(\dfrac{y}{V} \text{g/liter}\right)(1 \text{ liter/min})$

$= \dfrac{y}{V} \text{g/liter}$

$\dfrac{dy}{dt} = 0.5 - \dfrac{y}{V}; y(0) = 5$

$\dfrac{dV}{dt} =$ (rate of liquid in)

$-$ (rate of liquid out)

$= 2 \text{ liter/min} - 1 \text{ liter/min}$

$= 1 \text{ liter/min}$

$\dfrac{dV}{dt} = 1$

$V = t + C_1$

When $t = 0, V = 100$, so $C_1 = 100$.

$V = t + 100$

Therefore,

$\dfrac{dy}{dt} = 0.5 - \dfrac{y}{t + 100}$

$\dfrac{dy}{dt} + \dfrac{1}{t + 100} \cdot y = 0.5$

$I(t) = e^{\int dt/(t+100)} = e^{\ln|t+100|}$

$= t + 100$

$\dfrac{dy}{dt}(t + 100) + y = 0.5(t + 100)$

$D_x(t + 100)y = 0.5(t + 100)$

$(t + 100)y = \int 0.5(t + 100) \, dt$

$(t + 100)y = 0.25(t + 100)^2 + C$

$y = 0.25(t + 100) + \dfrac{C}{t + 100}$

$t = 0, \; y = 5$

$5 = 0.25(100) + \dfrac{C}{100}$

$500 = 2500 + C$

$C = -2000$

Therefore,

$y = 0.25(t + 100) + \dfrac{-2000}{t + 100}$

$= \dfrac{0.25(t + 100)^2 - 2000}{t + 100}.$

(b) When $t = 30$ min,

$$y = \frac{0.25(130)^2 - 2000}{130}$$

$$= 17.115,$$

After 30 min, about 17.1 g of chemical are present.

Chapter 10 Review Exercises

1. True

2. False: No; $y = e^{2x}$ satisfies the given differential equation.

3. True

4. False: No; the term $y(dy/dx)$ makes the equation nonlinear.

5. False: Many (most) differential equations are neither separable nor linear.

6. True

7. False: There is no way to separate the variables in this equation.

8. True

9. False: The integrating factor is x^5.

10. False: Euler's method finds a numerical approximation to a particular solution over some interval.

11. True

12. True

17. $y\dfrac{dy}{dx} = 2x + y$

Since you cannot separate the variables, that is, rewrite the equation in the form $g(y)dy = f(x)dx$ where g is a function of y alone and f is a function of x alone, then the equation is not separable. Since you cannot rewrite the equation in the form $\dfrac{dy}{dx} + P(x)y = Q(x)$, then the equation is not a linear first-order differential equation. Therefore, the equation is neither linear nor separable.

19. $\sqrt{x}\,\dfrac{dy}{dx} = \dfrac{1 + \ln x}{y}$

Since you can rewrite the equation in the form $y\,dy = \dfrac{1 + \ln x^x}{\sqrt{x}}\,dx$, then the equation is separable, but since it cannot be rewritten in the form $\dfrac{dy}{dx} + P(x)y = Q(x)$, then the equation is not linear.

21. $\dfrac{dy}{dx} + x = xy$

Since you can rewrite the equation in the form $\dfrac{dy}{dx} + (-x)y = -x$, then the equation is linear. Since the equation can be rewritten in the form $\dfrac{dy}{y-1} = x\,dx$, then it is also separable. Therefore, it is both linear and separable.

23. $x\dfrac{dy}{dx} + y = e^x(1 + y)$

Since the equation can be rewritten in the form

$$\frac{dy}{dx} + y\left(\frac{1 - e^x}{x}\right) = \frac{e^x}{x},$$

then the equation is linear. Since the equation cannot be rewritten in the form $g(y)dy = f(x)dx$, then the equation is not separable.

25. $\dfrac{dy}{dx} = 3x^2 + 6x$

$$dy = (3x^2 + 6x)dx$$

$$y = x^3 + 3x^2 + C$$

27. $\dfrac{dy}{dx} = 4e^{2x}$

$$dy = 4e^{2x}dx$$

$$y = 2e^{2x} + C$$

29. $\dfrac{dy}{dx} = \dfrac{3x + 1}{y}$

$$y\,dy = (3x + 1)\,dx$$

$$\frac{y^2}{2} = \frac{3x^2}{2} + x + C_1$$

$$y^2 = 3x^2 + 2x + C$$

31.
$$\frac{dy}{dx} = \frac{2y+1}{x}$$

$$\frac{dy}{2y+1} = \frac{dy}{x}$$

$$\frac{1}{2}\left(\frac{2\,dy}{2y+1}\right) = \frac{dx}{x}$$

$$\frac{1}{2}\ln|2y+1| = \ln|x| + C_1$$

$$\ln|2y+1|^{1/2} = \ln|x| + \ln k$$

$$\text{Let } \ln k = C_1$$

$$\ln|2y+1|^{1/2} = \ln k|x|$$

$$|2y+1|^{1/2} = k|x|$$

$$2y+1 = k^2 x^2$$

$$2y+1 = Cx^2$$

$$2y = Cx^2 - 1$$

$$y = \frac{Cx^2 - 1}{2}$$

33.
$$\frac{dy}{dx} + y = x$$

$$I(x) = e^{\int dx} = e^x$$

$$e^x \frac{dy}{dx} + e^x y = xe^x$$

$$D_x(e^x y) = xe^x$$

Integrate both sides, integrating the right side by parts.

$$e^x y = xe^x - e^x + C$$

$$y = x - 1 + Ce^{-x}$$

35.
$$x\ln x \frac{dy}{dx} + y = 2x^2$$

$$\frac{dy}{dx} + \frac{1}{x\ln x}y = \frac{2x}{\ln x}$$

$$I(x) = e^{\int \frac{1}{x\ln x}dx}$$

$$= e^{\int \frac{1}{\ln x}\left(\frac{1}{x}dx\right)}$$

$$= e^{\ln(\ln x)}$$

$$= \ln x$$

Multiply each term by the integrating factor, express the left side as the derivative of a product and integrate on both sides.

$$\ln x \frac{dy}{dx} + \frac{1}{x}y = 2x$$

$$D_x[(\ln x)y] = 2x$$

$$(\ln x)y = x^2 + C$$

$$y = \frac{x^2 + C}{\ln x}$$

37.
$$\frac{dy}{dx} = x^2 - 6x; \ y = 3 \text{ when } x = 0.$$

$$dy = (x^2 - 6x)dx$$

$$y = \frac{x^3}{3} - 3x^2 + C$$

When $x = 0$, $y = 3$.

$$3 = 0 - 0 + C$$

$$C = 3$$

$$y = \frac{x^3}{3} - 3x^2 + 3$$

39.
$$\frac{dy}{dx} = (x+2)^3 e^y; \ y = 0 \text{ when } x = 0.$$

$$e^{-y}dy = (x+2)^3 dx$$

$$-e^{-y} = \frac{1}{4}(x+2)^4 + C$$

$$-1 = \frac{1}{4}(2)^4 + C$$

$$C = -5$$

$$e^{-y} = 5 - \frac{1}{4}(x+2)^4$$

$$y = -\ln\left[5 - \frac{1}{4}(x+2)^4\right]$$

Notice that $x < \sqrt[4]{20} - 2$.

41.
$$\frac{dy}{dx} = \frac{1-2x}{y+3}; \ y = 16 \text{ when } x = 0.$$

$$(y+3)\,dy = (1-2x)\,dx$$

$$\frac{y^2}{2} + 3y = x - x^2 + C$$

$$\frac{16^2}{2} + 3(16) = 0 + C$$

$$176 = C$$

$$\frac{y^2}{2} + 3y = x - x^2 + 176$$

$$y^2 + 6y = 2x - 2x^2 + 352$$

43. $e^x \dfrac{dy}{dx} - e^x y = x^2 - 1; \; y = 42$ when $x = 0$.

$$\frac{dy}{dx} - y = (x^2 - 1)e^{-x}$$

$$I(x) = e^{\int -1\, dx} = e^{-x}$$

$$e^{-x} y = \int (x^2 - 1)e^{-x} \cdot e^{-x} dx$$

$$= \int (x^2 - 1)e^{-2x}\, dx$$

Integration by parts:

Let $u = x^2 - 1 \qquad\qquad dv = e^{-2x} dx$

$\qquad du = 2x\, dx \qquad\qquad v = -\dfrac{1}{2}e^{-2x}.$

$$e^{-x} y = \frac{(x^2 - 1)}{2} e^{-2x} + \int x e^{-2x} dx$$

Let $u = x \qquad\qquad dv = e^{-2x} dx$

$\qquad du = dx \qquad\qquad v = -\dfrac{1}{2}e^{-2x}.$

$$e^{-x} y = -\frac{(x^2 - 1)}{2} e^{-2x} - \frac{x}{2} e^{-2x} + \int \frac{1}{2} e^{-2x}\, dx$$

$$= -\frac{(x^2 - 1)}{2} e^{-2x} - \frac{x}{2} e^{-2x} - \frac{1}{4} e^{-2x} + C$$

$$y = -\frac{(x^2 - 1)}{2} e^{-x} - \frac{x}{2} e^{-x} - \frac{1}{4} e^{-x} + C e^{x}$$

$$42 = \frac{1}{2} - 0 - \frac{1}{4} + C$$

$$C = 41.75$$

$$y = -\frac{(x^2 - 1)}{2} e^{-x} - \frac{x}{2} e^{-x} - \frac{1}{4} e^{-x} + 41.75 e^{x}$$

$$= e^{-x}\left[-\frac{x^2}{2} - \frac{x}{2} + \frac{1}{4} \right] + 41.75 e^{x}$$

$$= \frac{-x^2 e^{-x}}{2} - \frac{x e^{-x}}{2} + \frac{e^{-x}}{4} + 41.75 e^{x}$$

45. $x \dfrac{dy}{dx} - 2x^2 y + 3x^2 = 0$

$y = 15$ when $x = 0$.

$$\frac{dy}{dx} - 2xy = -3x$$

$$I(x) = e^{\int -2x\, dx} = e^{-x^2}$$

$$e^{-x^2} y - 2x e^{-x^2} y = -3x e^{-x^2}$$

$$D_x(e^{-x^2} y) = -3x e^{-x^2}$$

$$e^{-x^2} y = \int -3x e^{-x^2}\, dx$$

$$= \frac{3}{2} e^{-x^2} + C$$

$$y = \frac{3}{2} + C e^{x^2}$$

Since $x = 0$ when $y = 15$,

$$15 = \frac{3}{2} + C e^{0}$$

$$C = \frac{27}{2}.$$

Therefore,

$$y = \frac{3}{2} + \frac{27 e^{x^2}}{2}.$$

47. $\dfrac{dy}{dx} = (1 - e^y)(y - 2)$

The equilibrium points are solutions of
$(1 - e^y)(y - 2) = 0$, that is, $y = 8$ and $y = 2$.

Checking the sign of dy/dx in the three intervals into which these points divide the y-axis we get the following diagram.

Thus 0 is an unstable equilibrium point and 2 is a stable equilibrium point.

51. $\dfrac{dy}{dx} = e^x + y; y(0) = 1, h = 0.2$; find $y(0.6)$.

$g(x, y) = e^x + y$

$x_0 = 0; y_0 = 1$

$g(x_0, y_0) = e^0 + 1 = 2$

$x_1 = 0.2; y_1 = y_0 + g(x_0, y_0)h$
$\qquad = 1 + 2(0.2) = 1.4$

$g(x_1, y_1) = e^2 + 1.4 \approx 2.6214$

$x_2 = 0.4; y_2 = y_1 + g(x_1, y_1)h$
$\qquad = 1.4 + 2.6214(0.2)$
$\qquad \approx 1.9243$

$g(x_2, y_2) = e^{0.4} + 1.9243 \approx 3.4161$

$x_3 = 0.6; y_3 = y_2 + g(x_2, y_2)h$
$\qquad = 1.9243 + 3.4161(0.2)$
$\qquad \approx 2.6075$

x_i	y_i
0	1
0.2	1.4
0.4	1.9243
0.6	2.608

So, $y(0.6) \approx 2.608$.

53. $\dfrac{dy}{dx} = 3 + \sqrt{y},\ y(0) = 0, h = 0.2$, find $y(1)$.

x_i	y_i
0	0
0.2	0.6
0.4	1.354919
0.6	2.187722
0.8	3.083541
1	4.034741

Therefore, $y(1) \approx 4.035$.

55. **(a)** $\dfrac{dy}{dx} = 6e^{0.3x}$

$dy = 6e^{0.3x}dx$

$y = 20e^{0.3x} + C$

When $x = 0,\ y = 0$

$0 = 20e^0 + C$

$C = -20$

$y = 20e^{0.3x} - 20.$

When $x = 6$,

$y = 20e^{1.8} - 20$
$\qquad \approx 100.99.$

Sales are $10,099.

(b) When $x = 12$,

$y = 20e^{3.6} - 20$
$\qquad \approx 711.96.$

Sales are $71,196.

57. $A = 300,000$ when $t = 0; r = 0.05$,
$D = -20,000$

(a) $\dfrac{dA}{dt} = 0.05A - 20,000$

(b) $\dfrac{1}{0.05A - 20,000}\,dA = dt$

$\dfrac{1}{0.05}\ln|0.05A - 20,000| = t + C$

$\ln|0.05A - 20,000| = 0.05t + k$

$\ln|0.05(300,000) - 20,000| = k$

$k = \ln|-5000| = \ln 5000$

$\ln|0.05A - 20,000| = 0.05t + \ln 5000$

$|0.05A - 20,000| = 5000e^{0.05t}$

Since $0.05A < 20,000$,

$$|0.05A - 20,000| = 20,000 - 0.05A$$

$$20,000 - 0.05A = 5000e^{0.05t}$$

$$A = \frac{1}{0.05 = 5}(20,000 - 5000e^{0.05t})$$

$$= 100,000(4 - e^{0.05t}).$$

When $t = 10$,

$$A = 100,000\left(4 - e^{0.05(10)}\right)$$

$$\approx \$235,127.87.$$

59. $\dfrac{dy}{dt} = \dfrac{-10}{1 + 5t}$; $y = 50$ when $t = 0$.

$$y = -2\ln(1 + 5t) + C$$

$$50 = -2\ln 1 + C$$

$$= C$$

$$y = 50 - 2\ln(1 + 5t)$$

(a) If $t = 24$,

$$y = 50 - 2\ln[1 + 5t(24)]$$

$$\approx 40 \text{ insects.}$$

(b) If $y = 0$,

$$50 = 2\ln(1 + 5t)$$

$$1 + 5t = e^{25}$$

$$t = \frac{e^{25} - 1}{5}$$

$$\approx 1.44 \times 10^{10} \text{ hours}$$

$$\approx 6 \times 10^{8} \text{ days}$$

$$\approx 1.6 \text{ million years.}$$

61.

$$\frac{dx}{dt} = 0.2x - 0.5xy$$

$$\frac{dy}{dt} = -0.3y + 0.4xy$$

$$\frac{dy}{dt} = \frac{\frac{dy}{dt}}{\frac{dx}{dt}}$$

$$= \frac{-0.3y + 0.4xy}{0.2x - 0.5xy}$$

$$= \frac{y(-0.3 + 0.4x)}{x(0.2 - 0.5y)}$$

$$\frac{0.2 - 0.5y}{y}dy = \frac{-0.3 + 0.4x}{x}dx$$

$$\left(\frac{0.2}{y} - 0.5\right)dy = \left(\frac{-0.3}{x} + 0.4\right)dx$$

$$0.2\ln y - 0.5y = -0.3\ln x + 0.4x + C$$

$$0.3\ln x + 0.2\ln y - 0.4x - 0.5y = C$$

Both growth rates are 0 if

$0.2x - 0.5xy = 0$ and $-0.3y + 0.4xy = 0$.

If $x \neq 0$ and $y \neq 0$, we have

$0.2 - 0.5y = 0$ and $-0.3 + 0.4x = 0$, so

$x = \frac{3}{4}$ units and $y = \frac{2}{5}$ units.

63. Let y = the amount in parts per million (ppm) of smoke at time t.

When $t = 0$, $y = 20$ ppm, $V = 15,000$ ft^3.

rate of smoke in $= 5$ ppm,

rate of smoke out

$$= (1200 \text{ ft}^3/\text{min})\left(\frac{y}{V} \text{ ppm/ft}^3\right)$$

Rate of air in = rate of air out,

so $V = 15,000$ft^3 for all t.

$$\frac{dy}{dt} = 5 - \frac{1200y}{15,000} = 5 - \frac{2y}{25}$$

$$= \frac{125 - 2y}{25}$$

$$\frac{1}{125 - 2y}dy = \frac{dt}{25}$$

$$-\frac{1}{2}\ln(125 - 2y) = \frac{t}{25} + C$$

$$-\frac{1}{2}\ln(125 - 2(20)) = C$$

$$C = -\frac{1}{2}\ln 85$$

If $y = 10$,

$$-\frac{1}{2}\ln[125 - 2(10)] = \frac{t}{25} - \frac{1}{2}\ln 85$$

$$\ln 105 = \ln 85 - \frac{2t}{25}$$

$$t = \frac{25}{2}[\ln 85 - \ln 105],$$

Which is negative.

It is impossible to reduce y to 10 ppm.

65. $y = \dfrac{N}{1 + be^{-kt}}$; $y = y_i$ when $x = x_i$,

$i = 1, 2, 3$.

t_1, t_2, t_3 are equally spaced:

$t_3 = 2t_2 - t_1$, so $t_1 + t_3 = 2t_2$,

or $t_2 = \dfrac{t_1 + t_3}{2}$.

Show $N = \dfrac{\dfrac{1}{y_1} + \dfrac{1}{y_3} - \dfrac{2}{y_2}}{\dfrac{1}{y_1 y_3} - \dfrac{1}{y_2^2}}$.

Let

$$A = \frac{1}{y_1} + \frac{1}{y_3} - \frac{2}{y_2}$$

$$= \frac{1 + be^{-kt_1}}{N} + \frac{1 + be^{-kt_3}}{N} - \frac{2\left(1 + be^{-kt_2}\right)}{N}$$

$$= \frac{1}{N}\left(1 + be^{-kt_1} + 1 + be^{-kt_3} - 2 - 2be^{-kt_2}\right)$$

$$= \frac{b}{N}\left[e^{-kt_1} + e^{-kt_3} - 2e^{-kt_2}\right]$$

Let

$$B = \frac{1}{y_1 y_3} - \frac{1}{y_2^2}$$

$$= \frac{\left(1 + be^{-kt_1}\right)\left(1 + be^{-kt_3}\right)}{N^2}$$

$$- \frac{\left(1 + be^{-kt_2}\right)^2}{N^2}$$

$$= \frac{1}{N^2}\left[1 + be^{-kt_1} + e^{-kt_3} + b^2 e^{-k(t_1 + t_3)}\right.$$

$$\left. -1 - 2be^{-kt_2} - b^2 e^{-2kt_2}\right]$$

$$= \frac{b}{N^2}\left[e^{-kt_1} + be^{-kt_3} + be^{-k(2t_2)}\right.$$

$$\left. -2e^{-kt_2} - be^{-2kt_2}\right]$$

$$= \frac{b}{N^2}\left[e^{-kt_1} + e^{-kt_3} - 2e^{-kt_2}\right]$$

Clearly, $\frac{A}{B} = N$.

Hence,

$$N = \frac{\dfrac{1}{y_1} + \dfrac{1}{y_3} - \dfrac{2}{y_2}}{\dfrac{1}{y_1 y_3} - \dfrac{1}{y_2^2}}.$$

67. From Exercise 66,

$$N = \frac{\dfrac{1}{39.8} + \dfrac{1}{203.3} - \dfrac{2}{105.7}}{\dfrac{1}{(39.8)(203.3)} - \dfrac{1}{(105.7)^2}} = 326.347 \approx 326.$$

(a) $y_0 = 39.8$,

$$b = \frac{N - y_0}{y_0} = \frac{326.347 - 39.8}{39.8}$$

$$\approx 7.20.$$

If 1920 corresponds to $t = 5$ decades, then

$$105.7 = \frac{326.3}{1 + 7.23e^{-5k}}$$

$$1 + 7.20e^{-5k} = \frac{326.3}{105.7}$$

$$e^{-5k} = \frac{1}{7.20}\left(\frac{326.3}{105.7} - 1\right)$$

so $k = -\dfrac{1}{5}\ln\left[\dfrac{1}{7.20}\left(\dfrac{326.3}{105.7} - 1\right)\right]$

$$\approx 0.248.$$

(b) $y = \dfrac{326}{1 + 7.20e^{-0.248t}}$

In 2010, $t = 14$. If $t = 14$,

$$y = \frac{326}{1 + 7.20e^{-0.248(14)}} \approx 266 \text{ million}$$

The predicated population is 266 million which is less than the table value of 308.7 million.

(c) In 2030, $t = 16$, so

$$y = \frac{326}{1 + 7.20e^{-0.248(16)}} \approx 287 \text{ million}.$$

In 2050, $t = 18$, so

$$y = \frac{326}{1 + 7.20e^{-0.248(18)}} \approx 301 \text{ million}.$$

69. The models are

Exponential: $y = Me^{kt}$

Limited growth: $y = N - (N - y_0)e^{-kt}$

Logistic: $y = \dfrac{N}{1 + be^{-kt}}$

$k > 0$, $N > y_0$, $b > 0$

(a) All three models increase for all t, as one can see by looking at the behavior of the exponential (either e^{kt} or e^{-kt}).

(b) The second derivatives are:

Exponential: $y'' = Mk^2 e^{kt}$

Limited growth: $y'' = -(N - y_0)k^2 e^{-kt}$

Logistic: $y'' = \dfrac{Nbk^2 e^{-kt}\left(be^{-kt} - 1\right)}{(1 + be^{-kt})^3}$

For the exponential, y'' is always positive and the model is concave upward everywhere. For the limited growth, y'' is always negative, so the model is concave downward everywhere. The logistic second derivative is positive for

$$be^{-kt} > 1, \text{ or } \ln b - kt > 0, \text{ or } t < \frac{\ln b}{k},$$

so the model is concave upward for $t < (\ln b)/k$.

For $t > (\ln b)/k$ the second derivative is negative and the model is concave downward. There is an inflection point at $t = (\ln b)/k$.

71. (a) $\dfrac{dx}{dt} = 1 - kx$

Separate the variables and integrate.

$$\int \frac{dx}{1 - kx} = \int dt$$

$$-\frac{1}{k}\ln|1 - kx| = t + C_1$$

Solve for x.

$$\ln|1 - kx| = -kt - kC_1$$

$$|1 - kx| = e^{-kC_1}e^{-kt}$$

$$1 - kx = Me^{-kt}, \text{ where } M = \pm e^{-kC_1}.$$

$$x = -\frac{1}{k}(Me^{-kt} - 1) = \frac{1}{k} + Ce^{-kt},$$

where $C = -\dfrac{M}{k}$.

(b) Write the linear first-order differential equation in the linear form

$$\frac{dx}{dt} + kx = 1.$$

The integrating factor is $I(t) = e^{\int k\, dt} = e^{kt}$.

Multiply both sides of the differential equation by $I(t)$.

$$\frac{dx}{dt}e^{kt} + kxe^{kt} = e^{kt}$$

Replace the left side of this equation by

$$D_t(xe^{kt}) = \frac{dx}{dt}e^{kt} + xke^{kt}.$$

$$D_t(xe^{kt}) = e^{kt}$$

Integrate both sides with respect to t.

$$xe^{kt} = \int e^{kt}\, dt = \frac{1}{k}e^{kt} + C$$

Solve for x.

$$x = \frac{1}{k} + Ce^{-kt}.$$

(c) Since $k > 0$, then as t gets larger and larger, Ce^{-kt} approaches 0,

so $\displaystyle\lim_{x \to \infty}\left(\frac{1}{k} + Ce^{-kt}\right) = \frac{1}{k}$.

73. $\dfrac{dT}{dt} = k(T - T_F);\ T_F = 300;\ T = 40$ when $t = 0$.

$T = 150$ when $t = 1$.

From Exercise 61 in Section 1 of this chapter, the solution to the differential equation is

$$T = Ce^{kt} + T_M$$

where C is a constant ($-k$ has been replaced by k in this exercise.)

Here, $T = Ce^{kt} + 300$.

$$40 = C + 300$$

$$C = -260$$

$$T = 300 - 260e^{kt}$$

$$150 = 300 - 260e^{k}$$

$$e^k = \frac{15}{26}$$

$$k = \ln\left(\frac{15}{26}\right) \approx -0.55$$

$$T = 300 - 260e^{-0.55t}$$

$$250 = 300 - 260e^{-0.55t}$$

$$e^{-0.55t} = \frac{5}{26}$$

$$t = -\frac{1}{0.55}\ln\left(\frac{5}{26}\right) \approx 3 \text{ hr}$$

PROBABILITY AND CALCULUS

11.1 Continuous Probability Models

Your Turn 1

$$f(x) = \frac{2}{x^2} \text{ on } [1, 2]$$

Condition 1 holds because $f(x) \geq 0$ on $[1, 2]$.

Condition 2 holds because $\displaystyle\int_1^2 f(x)\,dx = \int_1^2 \frac{2}{x^2}\,dx$

$$= -\frac{2}{x}\Big|_1^2$$

$$= -1 - (-2)$$

$$= 1.$$

Thus $f(x)$ is a probability density function for the interval $[1, 2]$.

$$P(3/2 \leq X \leq 2) = \int_{3/2}^2 \frac{2}{x^2}\,dx$$

$$= -\frac{2}{x}\Big|_{3/2}^2$$

$$= -1 - \left(-\frac{2}{3/2}\right)$$

$$= -1 + \frac{4}{3}$$

$$= \frac{1}{3}$$

Your Turn 2

The function $f(x) = kx^3$ will be nonnegative on the interval $[0, 4]$ for any positive k.

$$\int_0^4 kx^3\,dx = \frac{k}{4}x^4\Big|_0^4$$

$$= \frac{k}{4}4^4$$

$$= 64k$$

Thus $f(x)$ will be a probability density function on $[0, 4]$ when $64k = 1$, or $k = 1/64$.

Your Turn 3

$$P(0 \leq X \leq 1) = \int_0^1 2xe^{-x^2}\,dx$$

$$= \left(-e^{-x^2}\right)\Big|_0^1$$

$$= -\frac{1}{e} - (-1)$$

$$= 1 - \frac{1}{e} \approx 0.6321$$

Your Turn 4

The cumulative distribution function from Example 5 is $F(x) = -e^{-x^2} + 1$. The probability that there is a bird's nest within 1 km of the given point is

$$F(1) = -\frac{1}{e} + 1 \approx 0.6321.$$

11.1 Warmup Exercises

W1.

$$\int_0^4 \left(12x^2 + 9\sqrt{x}\right)dx$$

$$= \left(4x^3 + 6x^{3/2}\right)\Big|_0^4$$

$$= 4 \cdot 4^3 + 6 \cdot 4^{3/2} = 304$$

W2.

$$\int_1^2 \left(\frac{4}{x} - \frac{6}{x^4}\right)dx$$

$$= \left(4\ln x + 2x^{-3}\right)\Big|_1^2$$

$$= \left(4\ln 2 + \frac{1}{4}\right) - (0 + 2)$$

$$= 4\ln 2 - \frac{7}{4} \approx 1.023$$

W3.

$$\int_0^1 20e^{5x}dx$$

$$= 4e^{5x}\Big|_0^1$$

$$= 4e^5 - 4 \approx 589.7$$

11.1   Exercises

1. $f(x) = \dfrac{1}{9}x - \dfrac{1}{18}$; $[2, 5]$

Show that condition 1 holds.

Since $2 \le x \le 5$,

$$\frac{2}{9} \le \frac{1}{9}x \le \frac{5}{9}$$

$$\frac{1}{6} \le \frac{1}{9}x - \frac{1}{18} \le \frac{1}{2}.$$

Hence, $f(x) \ge 0$ on $[2, 5]$.

Show that condition 2 holds.

$$\int_2^5 \left(\frac{1}{9}x - \frac{1}{18} \right) dx = \frac{1}{9} \int_2^5 \left(x - \frac{1}{2} \right) dx$$

$$= \frac{1}{9} \left(\frac{x^2}{2} - \frac{1}{2}x \right) \Bigg|_2^5$$

$$= \frac{1}{9} \left(\frac{25}{2} - \frac{5}{2} - \frac{4}{2} + 1 \right)$$

$$= \frac{1}{9}(8 + 1)$$

$$= 1$$

Yes, $f(x)$ is a probability density function.

3. $f(x) = \dfrac{1}{21}x^2$; $[1, 4]$

Since $x^2 \ge 0$, $f(x) \ge 0$ on $[1, 4]$.

$$\frac{1}{21} \int_1^4 x^2 \, dx = \frac{1}{21} \left(\frac{x^3}{3} \right) \Bigg|_1^4$$

$$= \frac{1}{21} \left(\frac{64}{3} - \frac{1}{3} \right) = 1$$

Yes, $f(x)$ is a probability density function.

5. $f(x) = 4x^3$; $[0, 3]$

$$4 \int_0^3 x^3 \, dx = 4 \left(\frac{x^4}{4} \right) \Bigg|_0^3$$

$$= 4 \left(\frac{81}{4} - 0 \right)$$

$$= 81 \ne 1$$

No, $f(x)$ is not a probability density function.

7. $f(x) = \dfrac{x^2}{16}$; $[-2, 2]$

$$\frac{1}{16} \int_{-2}^2 x^2 \, dx = \frac{1}{16} \left(\frac{x^3}{3} \right) \Bigg|_{-2}^2$$

$$= \frac{1}{16} \left(\frac{8}{3} + \frac{8}{3} \right)$$

$$= \frac{1}{3} \ne 1$$

No, $f(x)$ is not a probability density function.

9. $f(x) = \dfrac{5}{3}x^2 - \dfrac{5}{90}$; $[-1, 1]$

Let $x = 0$. Then $f(x) = f(0) = -\dfrac{5}{90} < 0$.

So $f(x) < 0$ for at least one x-value in $[-1, 1]$.

No, $f(x)$ is not a probability density function.

11. $f(x) = kx^{1/2}$; $[1, 4]$

$$\int_1^4 kx^{1/2} \, dx = \frac{2}{3} kx^{3/2} \Bigg|_1^4$$

$$= \frac{2}{3} k(8 - 1)$$

$$= \frac{14}{3} k$$

If $\frac{14}{3}k = 1$,

$$k = \frac{3}{14}.$$

Notice that $f(x) = \frac{3}{4}x^{1/2} \ge 0$ for all x in $[1, 4]$.

13. $f(x) = kx^2$; $[0, 5]$

$$\int_0^5 kx^2 \, dx = k \frac{x^3}{3} \Bigg|_0^5$$

$$= k \left(\frac{125}{3} - 0 \right) = k \left(\frac{125}{3} \right)$$

If $k \left(\frac{124}{3} \right) = 1$,

$$k = \frac{3}{125}.$$

Notice that $f(x) = \frac{3}{125}x^2 \geq 0$ for all x in $[0, 5]$.

15. $f(x) = kx;\ [0, 3]$

$$\int_0^3 kx\,dx = k\frac{x^2}{2}\Big|_0^3$$

$$= k\left(\frac{9}{2} - 0\right)$$

$$= \frac{9}{2}k$$

If $\frac{9}{2}k = 1$,

$$k = \frac{2}{9}.$$

Notice that $f(x) = \frac{2}{9}x \geq 0$ for all x in $[0, 3]$.

17. $f(x) = kx;\ [1, 5]$

$$\int_1^5 kx\,dx = k\frac{x^2}{2}\Big|_1^5$$

$$= k\left(\frac{25}{2} - \frac{1}{2}\right)$$

$$= 12k$$

If $12k = 1$,

$$k = \frac{1}{12}.$$

Notice that $f(x) = \frac{1}{12}x \geq 0$ for all x in $[1, 5]$.

19.   For the probability density function
$f(x) = \frac{1}{9}x - \frac{1}{18}$ on $[2, 5]$, the cumulative
distribution function is

$$F(x) = \int_a^x f(t)\,dt$$

$$= \int_a^x \left(\frac{1}{9}t - \frac{1}{18}\right)dt$$

$$= \left(\frac{1}{18}t^2 - \frac{1}{18}t\right)\Big|_2^x$$

$$= \frac{1}{18}[(x^2 - x) - (4 - 2)]$$

$$= \frac{1}{18}(x^2 - x - 2),\ 2 \leq x \leq 5.$$

21.   For the probability density function $f(x) = \frac{x^2}{21}$
on $[1, 4]$, the cumulative distribution function is

$$F(x) = \int_1^x \frac{t^2}{21}\,dt$$

$$= \frac{t^3}{63}\Big|_1^x$$

$$= \frac{1}{63}(x^3 - 1),\ 1 \leq x \leq 4.$$

23.   The value of k was found to be $\frac{3}{14}$. For the
probability density function $f(x) = \frac{3}{14}x^{1/2}$ on
$[1, 4]$, the cumulative distribution function is

$$F(x) = \int_1^x \frac{3}{14}t^{1/2}\,dt$$

$$= \frac{3}{14} \cdot \frac{2}{3}t^{3/2}\Big|_1^x$$

$$= \frac{1}{7}(x^{3/2} - 1),\ 1 \leq x \leq 4.$$

25.   The total area under the graph of a probability
density function always equals 1.

29. $f(x) = \frac{1}{2}(1 + x)^{-3/2};\ [0, \infty)$

$$\frac{1}{2}\int_0^\infty (1 + x)^{-3/2}\,dx$$

$$= \lim_{a \to \infty} \frac{1}{2}\int_0^a (1 + x)^{-3/2}\,dx$$

$$= \lim_{a \to \infty} \frac{1}{2}(1 + x)^{-1/2}\left(\frac{-2}{1}\right)\Big|_0^a$$

$$= \lim_{a \to \infty} [-(1 + a)^{-1/2} + 1]$$

$$= \lim_{a \to \infty}\left(\frac{-1}{\sqrt{1 + a}} + 1\right)$$

$$= 0 + 1 = 1$$

Since $x \geq 0,\ f(x) \geq 0$.

$f(x)$ is a probability density function.

(a) $P(0 \le X \le 2)$

$$= \frac{1}{2} \int_0^2 (1 + x)^{-3/2} \, dx$$

$$= -(1 + x)^{-1/2} \Big|_0^2$$

$$= -3^{-1/2} + 1$$

$$\approx 0.4226$$

(b) $P(1 \le X \le 3)$

$$= \frac{1}{2} \int_1^3 (1 + x)^{-3/2} \, dx$$

$$= -(1 + x)^{-1/2} \Big|_1^3$$

$$= -4^{-1/2} + 2^{-1/2}$$

$$\approx 0.2071$$

(c) $P(X \ge 5)$

$$= \frac{1}{2} \int_5^\infty (1 + x)^{-3/2} \, dx$$

$$= \lim_{a \to \infty} \frac{1}{2} \int_5^a (1 + x)^{-3/2} \, dx$$

$$= \lim_{a \to \infty} [-(1 + x)^{-1/2}] \Big|_5^a$$

$$= \lim_{a \to \infty} [-(1 + a)^{-1/2} + 6^{-1/2}]$$

$$= \lim_{a \to \infty} \left(\frac{-1}{\sqrt{1 + a}} + 6^{-1/2} \right)$$

$$\approx 0 + 0.4082$$

$$= 0.4082$$

31. $f(x) = \frac{1}{2} e^{-x/2}; \ [0, \infty)$

$$\frac{1}{2} \int_0^\infty e^{-x/2} \, dx$$

$$= \lim_{a \to \infty} \frac{1}{2} \int_0^a e^{-x/2} \, dx$$

$$= \lim_{a \to \infty} \frac{1}{2} \left(\frac{-2}{1} e^{-x/2} \right) \Big|_0^a$$

$$= \lim_{a \to \infty} -e^{-x/2} \Big|_0^a$$

$$= \lim_{a \to \infty} \left(\frac{-1}{e^{a/2}} + 1 \right)$$

$$= 0 + 1$$

$$= 1$$

$f(x) > 0$ for all x.

$f(x)$ is a probability density function.

(a) $P(0 \le X \le 1) = \frac{1}{2} \int_0^1 e^{-x/2} \, dx$

$$= -e^{-x/2} \Big|_0^1$$

$$= \frac{-1}{e^{x/2}} + 1$$

$$\approx 0.3935$$

(b) $P(1 \le X \le 3) = \frac{1}{2} \int_1^3 e^{-x/2} \, dx$

$$= -e^{-x/2} \Big|_1^3$$

$$= \frac{-1}{e^{3/2}} + \frac{1}{e^{1/2}}$$

$$\approx 0.3834$$

(c) $P(X \ge 2) = \frac{1}{2} \int_2^\infty e^{-x/2} \, dx$

$$= \lim_{a \to \infty} \frac{1}{2} \int_2^a e^{-x/2} \, dx$$

$$= \lim_{a \to \infty} (-e^{-x/2}) \Big|_2^a$$

$$= \lim_{a \to \infty} \left(\frac{-1}{e^{a/2}} + \frac{1}{e} \right)$$

$$\approx 0.3679$$

33.
$$f(x) = \begin{cases} \dfrac{x^3}{12} & \text{if } 0 \le x \le 2 \\ \dfrac{16}{3x^3} & \text{if } x > 2 \end{cases}$$

First, note that $f(x) > 0$ for $x > 0$. Next,

$$\int_0^\infty f(x) \, dx$$

$$= \int_0^2 \frac{x^3}{12} \, dx + \lim_{a \to \infty} \int_2^a \frac{16}{3x^3} \, dx$$

$$= \left(\frac{x^4}{48} \right) \Big|_0^2 + \lim_{a \to \infty} \left(-\frac{8}{3x^2} \right) \Big|_2^a$$

$$= \left(\frac{1}{3} - 0 \right) + \left[\lim_{a \to \infty} \left(-\frac{8}{3a^2} \right) - \left(-\frac{8}{12} \right) \right]$$

$$= \frac{1}{3} + \frac{2}{3}$$

$$= 1.$$

Therefore, $f(x)$ is a probability density function.

(a) $P(0 \le X \le 2) = \displaystyle\int_0^2 f(x)\,dx$

$$= \left(\frac{x^4}{48}\right)\Big|_0^2$$

$$= \frac{1}{3}$$

(b) $P(X \ge 2) = P(X > 2)$

$$= \int_2^\infty \frac{16}{3x^3}\,dx$$

$$= \lim_{a \to \infty} \int_2^a \frac{16}{3x^3}\,dx$$

$$= \lim_{a \to \infty} \left(-\frac{8}{3x^2}\right)\Big|_2^a$$

$$= \lim_{a \to \infty} \left(-\frac{8}{3a^2}\right) - \left(-\frac{8}{3 \cdot 2^2}\right)$$

$$= 0 - \left(-\frac{2}{3}\right)$$

$$= \frac{2}{3}$$

(c) $P(1 \le X \le 3)$

$$= \int_1^2 \frac{x^3}{12}\,dx + \int_2^3 \frac{16}{3x^3}\,dx$$

$$= \left(\frac{x^4}{48}\right)\Big|_1^2 + \left(-\frac{8}{3x^2}\right)\Big|_2^3$$

$$= \left(\frac{1}{3} - \frac{1}{48}\right) + \left(-\frac{8}{27} + \frac{2}{3}\right)$$

$$= \frac{295}{432}$$

35. $f(t) = \dfrac{1}{2}e^{-t/2}; \ [0, \infty)$

(a) $P(0 \le T \le 12) = \dfrac{1}{2}\displaystyle\int_0^{12} e^{-t/2}\,dt$

$$= -e^{-t/2}\Big|_0^{12}$$

$$= \frac{-1}{e^6} + 1$$

$$\approx 0.9975$$

(b) $P(12 \le T \le 20) = \dfrac{1}{2}\displaystyle\int_{12}^{20} e^{-t/2}\,dt$

$$= -e^{-t/2}\Big|_{12}^{20}$$

$$= \frac{-1}{e^{10}} + \frac{1}{e^6}$$

$$\approx 0.0024$$

(c) $F(t) = \displaystyle\int_0^t \frac{1}{2}e^{-s/2}\,ds$

$$= \frac{1}{2}(-2)e^{-s/2}\Big|_0^t$$

$$= -(e^{-t/2} - 1)$$

$$= 1 - e^{-t/2}, \ t \ge 0.$$

(d) $F(6) = 1 - e^{-6/2}$

$$= 1 - e^{-3}$$

$$\approx 0.9502$$

The probability is 0.9502.

The probability is 0.8155.

37. If $f(x)$ is proportional to $(10 + x)^{-2}$, then, for some value of k, $f(x) = k(10 + x)^{-2}$ on $[0, 40]$. Find k. We know the total probability must equal 1.

$$\int_0^{10} k(10 + x)^{-2}\,dx = -k(10 + x)^{-1}\Big|_0^{40}$$

$$= -k(50^{-1} - 10^{-1})^{-1}$$

$$= -k\left(\frac{1}{50} - \frac{1}{10}\right)$$

$$= \frac{2}{25}x$$

If $\frac{2}{25}k = 1$, then $k = \frac{25}{2}$. Therefore

$$f(x) = \frac{25}{2}(10 + x)^{-2}, 0 \le x \le 40$$

So the probability distribution function is

$$F(x) = \int_0^x \frac{25}{2}(10 + t)^{-2}\, dt$$

$$= -\frac{25}{2}(10 + t)^{-1}\Big|_0^x$$

$$= -\frac{25}{2}[(10 + x)^{-1} - 10^{-1}]$$

$$= -\frac{25}{2}\left(\frac{1}{10 + x} - \frac{1}{10}\right)$$

$$= \frac{25}{2}\left(\frac{1}{10} - \frac{1}{10 + x}\right)$$

$$F(6) = \frac{25}{2}\left(\frac{1}{10} - \frac{1}{106}\right)$$

$$\approx 0.47$$

The correct answer choice is **c**.

39. $f(x) = \dfrac{1}{2\sqrt{x}}$; [1, 4]

(a) $P(3 \le X \le 4) = \displaystyle\int_3^4 \left(\frac{1}{2\sqrt{x}}\right) dx$

$$= \frac{1}{2}\int_3^4 x^{-1/2}\, dx$$

$$= \frac{1}{2}(2)x^{1/2}\Big|_3^4$$

$$= 2 - 3^{1/2} \approx 0.2679$$

(b) $P(1 \le X \le 2) = \displaystyle\int_1^2 \left(\frac{1}{2\sqrt{x}}\right) dx$

$$= \frac{1}{2}(2)x^{1/2}\Big|_1^2$$

$$= 2^{1/2} - 1 = 0.4142$$

(c) $P(2 \le X \le 3) = \displaystyle\int_2^3 \left(\frac{1}{2\sqrt{x}}\right) dx$

$$= \frac{1}{2}(2)x^{1/2}\Big|_2^3$$

$$= 3^{1/2} - 2^{1/2} = 0.3178$$

41. $f(x) = 1.185 \cdot 10^{-9} x^{4.5222} - 0.049846x$

(a) $P(0 \le X \le 150)$

$$= \int_0^{150} 1.185 \cdot 10^{-9} x^{4.5222} e^{-0.049846x}\, dx$$

$$\approx 0.8131$$

(b) $P(100 \le x \le 200)$

$$= \int_{100}^{200} 1.185 \cdot 10^{-9} x^{4.5222} e^{-0.049846x}\, dx$$

$$\approx 0.4901$$

43. **(a)**

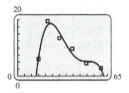

A polynomial function could fit the data.

(b) $N(t) = -0.00007445t^4 + 0.01243t^3$
$$- 0.7419t^2 + 18.18t - 137.5$$

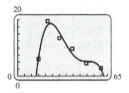

(c) $\displaystyle\int_{13.4}^{62.0} N(t)\, dt \approx 466.26$, as found using a

calculator. Thus the density function corresponding to the quartic fit is

$$S(t) = \frac{1}{466.26} N(t)$$

$$= \frac{1}{466.26}(-0.00007445t^4$$

$$+ 0.01243t^3 - 0.7419t^2$$

$$+ 18.18t - 137.5).$$

(d) Use a calculator to compute the integrals needed in (d).

$$P(35 \le \text{age} < 45) = \int_{35}^{45} S(t)\, dt$$

$$\approx 0.1688;$$

actual relative frequency

$$\frac{9.701}{56.065} \approx 0.1730$$

$$P(18 \le \text{age} < 35) = \int_{18}^{35} S(t)\, dt$$

$$\approx 0.5896;$$

actual relative frequency

$$\frac{19.461 + 13.423}{56.065} \approx 0.5865$$

$$P(45 \leq age) = \int_{45}^{62} S(t)\, dt$$

$$\approx 0.1610;$$

actual relative frequency

$$= \frac{4.582 + 2.849}{56.065} \approx 0.1325$$

45. $f(x) = \dfrac{5.5 - x}{15};\ [0, 5]$

(a) $P(3 \leq X \leq 5) = \displaystyle\int_3^5 \frac{5.5 - x}{15}\, dx$

$$= \left(\frac{5.5}{15}x - \frac{1}{15} \cdot \frac{x^2}{2} \right)\Big|_3^5$$

$$= \left(\frac{5.5}{15} \cdot 5 - \frac{1}{15} \cdot \frac{5^2}{2} \right)$$

$$- \left(\frac{5.5}{15} \cdot 3 - \frac{1}{15} \cdot \frac{3^2}{2} \right)$$

$$= 0.2$$

(b) $P(0 \leq X \leq 2) = \displaystyle\int_0^2 \frac{5.5 - x}{15}\, dx$

$$= \left(\frac{5.5}{15}x - \frac{1}{15} \cdot \frac{x^2}{2} \right)\Big|_0^2$$

$$= -\left(\frac{5.5}{15} \cdot 2 - \frac{1}{15} \cdot \frac{2^2}{2} \right)$$

$$- \left(\frac{5.5}{15} \cdot 0 - \frac{1}{15} \cdot \frac{0^2}{2} \right)$$

$$= 0.6$$

(c) $P(1 \leq X \leq 4) = \displaystyle\int_1^4 \frac{5.5 - x}{15}\, dx$

$$= \left(\frac{5.5}{15}x - \frac{1}{15} \cdot \frac{x^2}{2} \right)\Big|_1^4$$

$$= \left(\frac{5.5}{15} \cdot 4 - \frac{1}{15} \cdot \frac{4^2}{2} \right)$$

$$- \left(\frac{5.5}{15} \cdot 1 - \frac{1}{15} \cdot \frac{1^2}{2} \right)$$

$$= 0.6$$

47. $f(t) = \dfrac{1}{3650.1} e^{-t/3650.1}$

(a) $P(365 < T < 1095)$

$$= \int_{365}^{1095} \frac{1}{3650.1} e^{-t/3650.1}\, dt$$

$$= (-e^{-t/3650.1})\Big|_{365}^{1095}$$

$$= -e^{-1095/3650.1} + e^{-365/3650.1}$$

$$\approx 0.1640$$

(b) $P(T > 7300)$

$$= \int_{7300}^{\infty} \frac{1}{3650.1} e^{-t/3650.1}\, dt$$

$$= \lim_{b \to \infty} \int_{7300}^{b} \frac{1}{3650.1} e^{-t/3650.1}\, dt$$

$$= \lim_{b \to \infty} (-e^{-t/3650.1})\Big|_{7300}^{b}$$

$$= \lim_{b \to \infty} (-e^{-b/3650.1} + e^{-7300/3650.1})$$

$$= 0 + e^{-7300/3650.1} \approx 0.1353$$

49. $f(t) = 0.06049 e^{-0.03211t};\ [16, 84]$

(a) $P(16 \leq T \leq 25)$

$$= \int_{16}^{25} f(t)\, dt$$

$$= \int_{16}^{25} 0.06049 e^{-0.03211t}\, dt$$

$$= \frac{0.06049}{-0.03211} (e^{-0.03211t})\Big|_{16}^{25}$$

$$\approx -1.88384 (e^{-0.03211t})\Big|_{16}^{25}$$

$$= -1.88384 (e^{-0.03211 \cdot 25} - e^{-0.03211 \cdot 16})$$

$$\approx 0.2829$$

(b) $P(35 \leq T \leq 84)$

$$= \int_{35}^{84} 0.06049 e^{-0.03211t}\, dt$$

$$\approx -1.88384 (e^{-0.03211t})\Big|_{35}^{84}$$

$$\approx 0.4853$$

(c) $P(21 \le T \le 30)$

$$= \int_{21}^{30} 0.06049 e^{-0.03211t}\, dt$$

$$\approx -1.88384 \left(e^{-0.03211t} \right) \Big|_{21}^{30}$$

$$\approx 0.2409$$

(d) $F(t) = \int_{16}^{t} 0.06049 e^{-0.03211s}\, ds$

$$= 0.06049 \cdot \frac{1}{-0.03211} e^{-0.03211s} \Big|_{16}^{t}$$

$$= -1.8838 \left(e^{-0.03211t} - e^{-0.03211 \cdot 16} \right)$$

$$= 1.8838 \left(0.5982 - e^{-0.03211t} \right),$$
$$16 \le t \le 84$$

(e) $F(21) = 1.8838 \left(0.5982 - e^{-0.03211 \cdot 21} \right)$

$$= 1.8838(0.0887)$$

$$= 0.1671$$

The probability is 0.1671.

51. **(a)**

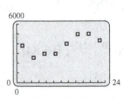

A polynomial function could fit the data.

(b) $T(t) = -2.416t^3 + 90.91t^2$
$$\qquad\qquad - 846.8t + 4880$$

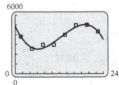

(c) $\int_{0}^{24} T(t)\, dt \approx 91{,}762$, as found using a

calculator. Thus the density function corresponding to the cubic fit is

$$S(t) = \frac{1}{91{,}762} T(t)$$

$$= \frac{1}{91{,}762} (-2.416t^3 + 90.91t^2$$
$$\qquad\qquad - 846.8t + 4880).$$

(d) Use a calculator to compute the integrals needed in (d).

$P(12\ \text{am} \le \text{time} \le 2\ \text{am})$

$$= \int_{0}^{2} S(t)\, dt$$

$$\approx 0.09044$$

$P(4\ \text{pm} \le \text{time} \le 5{:}30\ \text{pm})$

$$= \int_{16}^{17.5} S(t)\, dt$$

$$\approx 0.07916$$

11.2 Expected Value and Variance of Continuous Random Variables

Your Turn 1

Find the expected value and the variance of the random variable X with probability density function
$f(x) = \frac{8}{3x^3}$ on $[1, 2]$.

$$\mu = \int_{1}^{2} x f(x)\, dx$$

$$= \int_{1}^{2} x \frac{8}{3x^3}\, dx$$

$$= \int_{1}^{2} \frac{8}{3} x^{-2}\, dx$$

$$= -\frac{8}{3} x^{-1} \Big|_{1}^{2}$$

$$= -\frac{8}{6} - \left(-\frac{8}{3} \right)$$

$$= \frac{4}{3}$$

$$\text{Var}(x) = \int_{1}^{2} \left(x - \frac{4}{3} \right)^2 \frac{8}{3x^3}\, dx$$

$$= \int_{1}^{2} \left(x^2 - \frac{8}{3} x + \frac{16}{9} \right) \frac{8}{3x^3}\, dx$$

$$= \int_{1}^{2} \left(\frac{8}{3} x^{-1} - \frac{64}{9} x^{-2} + \frac{128}{27} x^{-3} \right) dx$$

$$= \left(\frac{8}{3} \ln(x) + \frac{64}{9x} - \frac{64}{27x^2} \right) \Big|_{1}^{2}$$

$$= \left(\frac{8}{3} \ln 2 + \frac{32}{9} - \frac{16}{27} \right) - \left(\frac{64}{9} - \frac{64}{27} \right)$$

$$= \frac{8}{3} \ln 2 - \frac{16}{9} \approx 0.0706$$

$$\sigma \approx \sqrt{0.0706} \approx 0.2657$$

Your Turn 2

Use the alternative formula to find the variance of the random variable X with probability density function $f(x) = \frac{4}{x^5}$ for $x \geq 1$. First find the mean μ.

$$\mu = \int_1^\infty x\left(\frac{4}{x^5}\right) dx$$

$$= \int_1^\infty 4x^{-4}\, dx$$

$$= \lim_{b \to \infty} \int_1^b 4x^{-4}\, dx$$

$$= \lim_{b \to \infty} \left(-\frac{4}{3}x^{-3}\Big|_1^b\right)$$

$$= \lim_{b \to \infty} \left(-\frac{4}{3b^3} - \left(-\frac{4}{3}\right)\right)$$

$$= \frac{4}{3}$$

$$\text{Var}(X) = \int_1^\infty x^2 f(x)\, dx - \mu^2$$

$$= \int_1^\infty x^2 \frac{4}{x^5}\, dx - \left(\frac{4}{3}\right)^2$$

$$= 2\int_1^\infty 2x^{-3}\, dx - \frac{16}{9}$$

$$= 2 \lim_{b \to \infty} \left(-x^{-2}\Big|_1^b\right) - \frac{16}{9}$$

$$= 2\left(1 - \lim_{b \to \infty} b^{-2}\right) - \frac{16}{9}$$

$$= 2 - \frac{16}{9} = \frac{2}{9}$$

Your Turn 3

Find the median m for the probability density function $f(x) = \frac{4}{x^5}$ for $x \geq 1$.

$$\int_1^m \frac{4}{x^5}\, dx = \frac{1}{2}$$

$$\left(-\frac{1}{x^4}\right)\Big|_1^m = \frac{1}{2}$$

$$-\frac{1}{m^4} - (-1) = \frac{1}{2}$$

$$m^4 = 2$$

$$m = \sqrt[4]{2} \approx 1.1892$$

11.2 Warmup Exercises

W1.
$$P(1 \leq X \leq 2) = \int_1^2 f(x)\, dx$$

$$= \int_1^2 \frac{2}{x^3}\, dx = \left(-\frac{1}{x^2}\right)\Big|_1^2$$

$$= -\frac{1}{4} + 1 = \frac{3}{4}$$

W2.
$$P(1 \leq X \leq 2) = \int_1^2 f(x)\, dx$$

$$= \int_1^2 \frac{5}{62} x^{3/2}\, dx = \left(\frac{1}{31} x^{5/2}\right)\Big|_1^2$$

$$= \frac{1}{31}\left(2^{5/2} - 1\right) = \frac{4\sqrt{2} - 1}{31} \approx 0.1502$$

11.2 Exercises

1. $f(x) = \frac{1}{4}$; $[3, 7]$

$$E(X) = \mu = \int_3^7 \frac{1}{4}x\, dx = \frac{1}{4}\left(\frac{x^2}{2}\right)\Big|_3^7$$

$$= \frac{49}{8} - \frac{9}{8}$$

$$= 5$$

$$\text{Var}(X) = \int_3^7 (x - 5)^2 \left(\frac{1}{4}\right) dx$$

$$= \frac{1}{4} \cdot \frac{(x-5)^3}{3}\Big|_3^7$$

$$= \frac{8}{12} + \frac{8}{12}$$

$$= \frac{4}{3} \approx 1.33$$

$$\sigma \approx \sqrt{\text{Var}(X)}$$

$$= \sqrt{\frac{4}{3}}$$

$$\approx 1.15$$

3. $f(x) = \dfrac{x}{8} - \dfrac{1}{4}$; $[2, 6]$

$$\mu = \int_2^6 x\left(\frac{x}{8} - \frac{1}{4}\right)dx$$

$$= \int_2^6 \left(\frac{x^2}{8} - \frac{x}{4}\right)dx$$

$$= \left(\frac{x^3}{24} - \frac{x^2}{8}\right)\Bigg|_2^6$$

$$= \left(\frac{216}{24} - \frac{36}{8}\right) - \left(\frac{8}{24} - \frac{4}{8}\right)$$

$$= \frac{208}{24} - 4$$

$$= \frac{26}{3} - 4$$

$$= \frac{14}{3} \approx 4.67$$

Use the alternative formula to find

$$\text{Var}(X) = \int_2^6 x^2\left(\frac{x}{8} - \frac{1}{4}\right)dx - \left(\frac{14}{3}\right)^2$$

$$= \int_2^6 \left(\frac{x^3}{8} - \frac{x^2}{4}\right)dx - \frac{196}{9}$$

$$= \left(\frac{x^4}{32} - \frac{x^3}{12}\right)\Bigg|_2^6 - \frac{196}{9}$$

$$= \left(\frac{1296}{32} - \frac{216}{12}\right)$$

$$- \left(\frac{16}{32} - \frac{8}{12}\right) - \frac{196}{9}$$

$$\approx 0.89.$$

$$\sigma = \sqrt{\text{Var}(X)} \approx \sqrt{0.89} \approx 0.94$$

5. $f(x) = 1 - \dfrac{1}{\sqrt{x}}$; $[1, 4]$

$$\mu = \int_1^4 x(1 - x^{-1/2})dx$$

$$= \int_1^4 (x - x^{1/2})dx$$

$$= \left(\frac{x^2}{2} - \frac{2x^{3/2}}{3}\right)\Bigg|_1^4$$

$$= \frac{16}{2} - \frac{16}{3} - \frac{1}{2} + \frac{2}{3}$$

$$= \frac{17}{6} \approx 2.83$$

$$\text{Var}(X) = \int_1^4 x^2(1 - x^{-1/2})dx - \left(\frac{17}{6}\right)^2$$

$$= \int_1^4 (x^2 - x^{3/2})dx - \frac{289}{36}$$

$$= \left(\frac{x^3}{3} - \frac{2x^{5/2}}{5}\right)\Bigg|_1^4 - \frac{289}{36}$$

$$= \frac{64}{3} - \frac{64}{5} - \frac{1}{3} + \frac{2}{5} - \frac{289}{36}$$

$$\approx 0.57$$

$$\sigma \approx \sqrt{\text{Var}(X)} \approx 0.76$$

7. $f(x) = 4x^{-5}$; $[1, \infty)$

$$\mu = \int_1^\infty x(4x^{-5})dx$$

$$= \lim_{a \to \infty} \int_1^a 4x^{-4}dx$$

$$= \lim_{a \to \infty} \left(\frac{4x^{-3}}{-3}\right)\Bigg|_1^a$$

$$= \lim_{a \to \infty} \left(\frac{-4}{3a^3} + \frac{4}{3}\right)$$

$$= \frac{4}{3} \approx 1.33$$

$$\text{Var}(X) = \int_1^\infty x^2(4x^{-5})dx - \left(\frac{4}{3}\right)^2$$

$$= \lim_{a \to \infty} \int_1^a 4x^{-3}\,dx - \frac{16}{9}$$

$$= \lim_{a \to \infty} \left(\frac{4x^{-2}}{-2}\right)\Bigg|_1^a - \frac{16}{9}$$

$$= \lim_{a \to \infty} \left(\frac{-2}{a^2} + 2\right) - \frac{16}{9}$$

$$= 2 - \frac{16}{9} = \frac{2}{9} \approx 0.22$$

$$\sigma = \sqrt{\text{Var}(X)} = \sqrt{\frac{2}{9}} \approx 0.47$$

11. $f(x) = \dfrac{\sqrt{x}}{18}; \ [0, 9]$

(a) $E(X) = \mu = \displaystyle\int_0^9 \frac{x\sqrt{x}}{18}dx$

$$= \int_0^9 \frac{x^{3/2}}{18}dx$$

$$= \frac{2x^{5/2}}{90}\Bigg|_0^9 = \frac{x^{5/2}}{45}\Bigg|_0^9$$

$$= \frac{243}{45} = \frac{27}{5} = 5.40$$

(b) $\text{Var}(X) = \displaystyle\int_0^9 \frac{x^2\sqrt{x}}{18}\,dx - \left(\frac{27}{5}\right)^2$

$$= \int_0^9 \frac{x^{5/2}}{18}\,dx - \left(\frac{27}{5}\right)^2$$

$$= \frac{x^{7/2}}{63}\Bigg|_0^9 - \left(\frac{27}{5}\right)^2$$

$$= \frac{2187}{63} - \left(\frac{27}{5}\right)^2 \approx 5.55$$

(c) $\sigma = \sqrt{\text{Var}(X)} \approx 2.36$

(d) $P(5.40 < X \le 9)$

$$= \int_{5.4}^9 \frac{x^{1/2}}{18}\,dx$$

$$= \frac{x^{3/2}}{27}\Bigg|_{5.4}^9$$

$$= \frac{27}{27} - \frac{(5.4)^{1.5}}{27}$$

$$\approx 0.5352$$

(e) $P(5.40 - 2.36 \le X \le 5.40 + 2.36)$

$$= \int_{3.04}^{7.76} \frac{x^{1/2}}{18}\,dx$$

$$= \frac{x^{3/2}}{27}\Bigg|_{3.04}^{7.76}$$

$$= \frac{7.76^{3/2}}{27} - \frac{3.04^{3/2}}{27}$$

$$\approx 0.6043$$

13. $f(x) = \dfrac{1}{4}x^3; \ [0, 2]$

(a) $E(X) = \mu = \displaystyle\int_0^2 \frac{1}{4}x^4\,dx = \frac{x^5}{20}\Bigg|_0^2$

$$= \frac{32}{20} = \frac{8}{5} = 1.6$$

(b) $\text{Var}(X) = \displaystyle\int_0^2 \frac{1}{4}x^5 dx - \frac{64}{25}$

$$= \frac{x^6}{24}\Bigg|_0^2 - \frac{64}{25}$$

$$= \frac{8}{3} - \frac{64}{25}$$

$$= \frac{8}{75} \approx 0.11$$

(c) $\sigma = \sqrt{\text{Var}(X)}$

$$= \sqrt{\frac{8}{75}}$$

$$\approx 0.3266$$

$$\approx 0.33$$

(d) $P\left(8/5 < X \leq 2\right) = \displaystyle\int_{8/5}^{2} \dfrac{x^3}{4}\,dx$

$= \dfrac{x^4}{16}\Big|_{8/5}^{2}$

$= 1 - \dfrac{256}{625}$

$= \dfrac{369}{625} \approx 0.5904$

(e) Use a four-place value for the standard deviation.

$P\left(1.6 - 0.3266 \leq X \leq 1.6 + 0.3266\right)$

$= \displaystyle\int_{1.2734}^{1.9266} \dfrac{x^3}{4}\,dx = \dfrac{x^4}{16}\Big|_{1.2734}^{1.9266}$

$= \dfrac{1.9266^4}{16} - \dfrac{0.1.2734^4}{16}$

≈ 0.6967

15. $f(x) = \dfrac{1}{4};\ [3, 7]$

(a) m = median: $\displaystyle\int_{3}^{m} \dfrac{1}{4}\,dx = \dfrac{1}{2}$

$\dfrac{1}{4}x\Big|_{3}^{m} = \dfrac{1}{2}$

$\dfrac{m}{4} - \dfrac{3}{4} = \dfrac{1}{2}$

$m - 3 = 2$

$m = 5$

(b) $E(X) = \mu = 5$ (from Exercise 1)

$P(X = 5) = \displaystyle\int_{5}^{5} \dfrac{1}{4}\,dx = 0$

17. $f(x) = \dfrac{x}{8} - \dfrac{1}{4};\ [2, 6]$

(a) m = median:

$\displaystyle\int_{2}^{m}\left(\dfrac{x}{8} - \dfrac{1}{4}\right)dx = \dfrac{1}{2}$

$\left(\dfrac{x^2}{16} - \dfrac{x}{4}\right)\Big|_{2}^{m} = \dfrac{1}{2}$

$\dfrac{m^2}{16} - \dfrac{m}{4} - \dfrac{1}{4} + \dfrac{1}{2} = \dfrac{1}{2}$

$m^2 - 4m - 4 + 8 = 8$

$m^2 - 4m - 4 = 0$

$m = \dfrac{4 \pm \sqrt{16 + 16(1)}}{2}$

Reject $\dfrac{4-\sqrt{32}}{2}$ since it is not in $[2, 6]$.

$m = \dfrac{4 + \sqrt{32}}{2}$

$= 2 + 2\sqrt{2} \approx 4.8284$

(b) $E(X) = \mu = \dfrac{14}{3}$ (from Exercise 3)

$P\left(\dfrac{14}{3} \leq X \leq 2 + 2\sqrt{2}\right)$

$= \displaystyle\int_{14/3}^{2+2\sqrt{2}}\left(\dfrac{x}{8} - \dfrac{1}{4}\right)dx$

$= \left(\dfrac{x^2}{16} - \dfrac{x}{4}\right)\Big|_{14/3}^{2+2\sqrt{2}}$

$= \dfrac{\left(2 + 2\sqrt{2}\right)^2}{16} - \dfrac{2 + 2\sqrt{2}}{4}$

$- \dfrac{(14/3)^2}{16} + \dfrac{14/3}{4}$

$= \dfrac{1}{18} \approx 0.0556$

If you do the integration on a calculator using rounded values for the limits you may get a slightly different answer, such as 0.0553.

19. $f(x) = 4x^{-5}$; $[1, \infty)$

(a) $m = $ median:

$$\int_1^m 4x^{-5}\, dx = \frac{1}{2}$$

$$\frac{4x^{-4}}{-4}\Big|_1^m = \frac{1}{2}$$

$$-m^{-4} + 1 = \frac{1}{2}$$

$$1 - \frac{1}{m^4} = \frac{1}{2}$$

$$2m^4 - 2 = m^4$$

$$m^4 = 2$$

$$m = \sqrt[4]{2} \approx 1.189$$

(b) $E(X) = \mu = \frac{4}{3}$ (from Exercise 7)

$$P\left(1.19 \le X \le \tfrac{4}{3}\right) \approx \int_{1.189}^{1.333} 4x^{-5}\, dx$$

$$\approx -x^{-4}\Big|_{1.189}^{1.333}$$

$$\approx -\frac{1}{1.333^4} + \frac{1}{1.189^4}$$

$$\approx 0.1836$$

21.
$$f(x) = \begin{cases} \dfrac{x^3}{12} & \text{if } 0 \le x \le 2 \\[2mm] \dfrac{16}{3x^3} & \text{if } x > 2 \end{cases}$$

Expected value:

$$E(X) = \mu = \int_0^\infty x f(x)\, dx$$

$$= \int_0^2 x\left(\frac{x^3}{12}\right) dx + \lim_{a\to\infty} \int_2^a x\left(\frac{16}{3x^3}\right) dx$$

$$= \int_0^2 \frac{x^4}{12}\, dx + \lim_{a\to\infty} \int_2^a \frac{16}{3x^2}\, dx$$

$$= \left(\frac{x^5}{60}\right)\Big|_0^2 + \lim_{a\to\infty}\left(-\frac{16}{3x}\right)\Big|_2^a$$

$$= \left(\frac{8}{15} - 0\right) + \left[\lim_{a\to\infty}\left(-\frac{16}{3a}\right) - \left(-\frac{16}{6}\right)\right]$$

$$= \frac{16}{5}$$

Variance:

$$\text{Var}(X) = \int_0^\infty x^2 f(x)\, dx - \mu^2$$

$$= \int_0^2 x^2\left(\frac{x^3}{12}\right) dx + \int_2^\infty \left(\frac{16}{3x^3}\right) dx - \left(\frac{16}{5}\right)^2$$

Examine the second integral.

$$\int_2^\infty x^2\left(\frac{16}{3x^3}\right) dx$$

$$= \lim_{a\to\infty} \int_2^a x^2\left(\frac{16}{3x^3}\right) dx$$

$$= \lim_{a\to\infty} \int_2^a \frac{16}{3x}\, dx$$

$$= \lim_{a\to\infty} \frac{16}{3}\ln|a| - \frac{16}{3}\ln|2|$$

Since the limit diverges, neither the variance nor the standard deviation exists.

23. $f(x) = \begin{cases} \dfrac{|x|}{10} & \text{for } -2 \le x \le 4 \\[2mm] 0 & \text{otherwise} \end{cases}$

First, note that

$$|x| = \begin{cases} -x \text{ for } -2 \le x \le 0 \\ x \text{ for } 0 \le x \le 4 \end{cases}$$

The expected value is

$$E(X) = \mu = \int_{-2}^4 x \cdot \frac{|x|}{10}\, dx$$

$$= \int_{-2}^0 x \cdot \frac{-x}{10}\, dx + \int_0^4 x \cdot \frac{x}{10}\, dx$$

$$= \int_{-2}^0 -\frac{x^2}{10}\, dx + \int_0^4 \frac{x^2}{10}\, dx$$

$$= -\frac{x^3}{30}\Big|_{-2}^0 + \frac{x^3}{30}\Big|_0^4$$

$$= -\left(0 - \frac{-8}{30}\right) + \left(\frac{64}{30} - 0\right)$$

$$= \frac{56}{30} = \frac{28}{15}$$

The correct answer choice is **d.**

25. $f(t) = \dfrac{1}{11}\left(1 + \dfrac{3}{\sqrt{t}}\right); [4, 9]$

(a) From Exercise 6, $\mu = \dfrac{141}{22}$ yr

≈ 6.409 yr.

(b) $\sigma = 1.447$ yr.

(c) $P\left(T > \dfrac{141}{22}\right)$

$= \displaystyle\int_{141/22}^{9} \dfrac{1}{11}(1 + 3t^{-1/2})\,dt$

$= \dfrac{1}{11}(t + 6t^{1/2})\Big|_{141/22}^{9}$

$= \dfrac{1}{11}\left[9 + 18 - \dfrac{141}{22} - 6\left(\dfrac{141}{22}\right)^{1/2}\right]$

≈ 0.4910

27. Using the hint, we have

$$\text{loss not paid} = \begin{cases} x \text{ for } 0.6 < x < 2 \\ 2 \text{ for } x > 2 \end{cases}$$

Therefore, the mean of the manufacturer's annual losses not paid will be

$\mu = \displaystyle\int_{0.6}^{0} x \cdot f(x)\,dx + \int_{2}^{\infty} 2 \cdot f(x)\,dx$

$= \displaystyle\int_{0.6}^{2} x \dfrac{2.5(0.6)^{2.5}}{x^{3.5}}\,dx$

$\quad + \displaystyle\int_{2}^{\infty} 2\dfrac{2.5(0.6)^{2.5}}{x^{3.5}}\,dx$

$= 2.5(0.6)^{2.5} \displaystyle\int_{0.6}^{2} \dfrac{1}{x^{2.5}}\,dx$

$\quad + 5(0.6)^{2.5} \displaystyle\int_{2}^{\infty} \dfrac{1}{x^{3.5}}\,dx$

$= 2.5(0.6)^{2.5}\left(\dfrac{1}{-1.5}\right)\dfrac{1}{x^{1.5}}\Big|_{0.6}^{2}$

$\quad + 5(0.6)^{2.5}\left(\dfrac{1}{-2.5}\right)\dfrac{1}{x^{2.5}}\Big|_{2}^{\infty}$

$= -\dfrac{5}{3}(0.6)^{2.5}\left(\dfrac{1}{2^{1.5}} - \dfrac{1}{0.6^{1.5}}\right)$

$\quad - 2(0.6)^{2.5}\left(0 - \dfrac{1}{2^{2.5}}\right)$

$\approx 0.8357 + 0.0986 \approx 0.93$

The correct answer choice is **c**.

29. Since the probability density function is proportional to $(1 + x)^{-4}$, we have $f(x) = k(1 + x)^{-4}$, $0 < x < \infty$. To determine k, solve the equation $\int_{0}^{\infty} k f(x)\,dx = 1$.

$$\int_{0}^{\infty} k(1 + x)^{-4}\,dx = 1$$

$$k\left(-\dfrac{1}{3}\right)(1 + x)^{-3}\Big|_{0}^{\infty} = 1$$

$$-\dfrac{k}{3}(0 - 1) = 1$$

$$\dfrac{k}{3} = 1$$

$$k = 3$$

Thus, $f(x) = 3(1 + x)^{-4}, 0 < x < \infty$.

The expected monthly claims are

$$\int_{0}^{\infty} x \cdot 3(1 + x)^{-4}\,dx = 3\int_{0}^{\infty} \dfrac{x}{(1 + x)^4}\,dx$$

The antiderivative can be found using the substitution $u = 1 - x$.

$$\int \dfrac{x}{(1 + x)^4}\,dx = \int \dfrac{u - 1}{u^4}\,du$$

$$= \int\left(\dfrac{1}{u^3} - \dfrac{1}{u^4}\right)du$$

$$= -\dfrac{1}{2u^2} + \dfrac{1}{3u^3}$$

Resubstitute $u = 1 + x$.

$$3\int_{0}^{\infty} \dfrac{x}{(1 + x)^4}\,dx$$

$$= 3\left(-\dfrac{1}{2(1 + x)^2} + \dfrac{1}{3(1 + x)^3}\right)\Big|_{0}^{\infty}$$

$$= 3\left[0 - \left(-\dfrac{1}{2} + \dfrac{1}{3}\right)\right] = 3\left(\dfrac{1}{6}\right) = \dfrac{1}{2}$$

The correct answer choice is **c**.

31. $f(t) = \dfrac{1}{(\ln 20)t}; [1, 20]$

(a) $\mu = \displaystyle\int_{1}^{20} t \cdot \dfrac{1}{(\ln 20)t}\,dt$

$$= \int_1^{20} \frac{1}{\ln 20} dt$$

$$= \frac{t}{\ln 20} \bigg|_1^{20}$$

$$= \frac{19}{\ln 20} \approx 6.3424 \approx 6.342 \text{ seconds}$$

(b) $\text{Var}(T) = \int_1^{20} t^2 \cdot \frac{1}{(\ln 20)t} dt - \mu^2$

$$= \int_1^{20} \frac{t}{\ln 20} dt - \mu^2$$

$$= \frac{t^2}{2 \ln 20} \bigg|_1^{20} - (6.3424)^2$$

$$= \frac{399}{2 \ln 20} - (6.3424)^2$$

$$\approx 26.3687$$

$$\sigma \approx \sqrt{26.3687}$$

$$\approx 5.135 \text{ sec}$$

(c) $P(6.3424 - 5.1350 < T < 6.3424 + 5.1350)$

$$= P(1.2074 < T < 11.4774)$$

$$= \int_{1.2074}^{11.4774} \frac{1}{(\ln 20)t} dt$$

$$= \frac{\ln t}{\ln 20} \bigg|_{1.2074}^{11.4774}$$

$$= \frac{1}{\ln 20} (\ln 11.4774 - \ln 1.2074)$$

$$\approx 0.7517$$

If you do the integration on a calculator using differently rounded values for the limits you may get a slightly different answer, such as 0.7518.

(d) The median clotting time is the value of m such that $\int_a^m f(t)dt = \frac{1}{2}$.

$$\int_1^m \frac{1}{(\ln 20)t} dt = \frac{1}{2}$$

$$\frac{1}{\ln 20} \ln t \bigg|_1^m = \frac{1}{2}$$

$$\frac{1}{\ln 20} (\ln m - 0) = \frac{1}{2}$$

$$\ln m = \frac{\ln 20}{2}$$

$$m = e^{\ln 20/2}$$

$$\approx 4.472$$

33.

$$f(x) = \frac{1}{2\sqrt{x}}; \ [1, 4]$$

(a) $\mu = \int_1^4 x \cdot \frac{1}{2\sqrt{x}} dx$

$$= \int_1^4 \frac{x^{1/2}}{2} dx = \frac{x^{3/2}}{3} \bigg|_1^4$$

$$= \frac{1}{3}(8 - 1)$$

$$= \frac{7}{3} \approx 2.333 \text{ cm}$$

(b) $\text{Var}(X) = \int_1^4 x^2 \cdot \frac{1}{2\sqrt{x}} dx - \left(\frac{7}{3}\right)^2$

$$= \int_1^4 \frac{x^{3/2}}{2} dx - \frac{49}{9}$$

$$= \frac{x^{5/2}}{5} \bigg|_1^4 - \frac{49}{9}$$

$$= \frac{1}{5}(32 - 1) - \frac{49}{9}$$

$$\approx 0.7556$$

$$\sigma = \sqrt{\text{Var}(X)}$$

$$\approx 0.8692 \text{ cm}$$

(c) $P(X > 2.33 + 2(0.87))$

$$= P(X > 4.07)$$

$$= 0$$

The probability is 0 since two standard deviations falls out of the given interval $[1, 4]$.

(d) The median petal length is the value of m such that $\int_a^m f(x)dx = \frac{1}{2}$.

$$\int_1^m \frac{1}{2\sqrt{x}}\,dx = \frac{1}{2}$$

$$\sqrt{x}\Big|_1^m = \frac{1}{2}$$

$$\sqrt{m} - 1 = \frac{1}{2}$$

$$\sqrt{m} = \frac{3}{2}$$

$$m = \frac{9}{4} = 2.25$$

The median petal length is 2.25 cm.

35. $f(x) = 1.185 \cdot 10^{-9} x^{4.5222} e^{-0.049846x}$

$$E(X) = \int_1^{1000} x\, f(x)\,dx$$

Using the integration function on our calculator.

$$E(X) \approx 110.80$$

The expected size is about 111.

37. $S(t) = \dfrac{1}{466.26}(-0.00007445t^4 + 0.01243t^3$

$$- \ 0.7419t^2 + 18.18t - 137.5)$$

for t in the interval $[13.4, 62.0]$.
Evaluate the required integrals with a calculator.

$$\mu = \int_{13.4}^{62.0} t\, S(t)\,dt$$

$$\approx 31.75 \text{ years}$$

$$\text{Var}(X) = \int_{13.4}^{62.0} t^2 S(t)\,dt - (31.75)^2$$

$$\approx 133.44$$

$$\sigma \approx \sqrt{133.44}$$

$$\approx 11.55 \text{ years}$$

39. $f(x) = \dfrac{5.5 - x}{15}; \ [0, 5]$

(a) $\mu = \displaystyle\int_0^5 x\left(\frac{5.5 - x}{15}\right)dx$

$$= \int_0^5 \left(\frac{5.5}{15}x - \frac{1}{15}x^2\right)dx$$

$$= \frac{5.5}{30}x^2 - \frac{1}{45}x^3 \Big|_0^5$$

$$= \left(\frac{5.5}{30} \cdot 25 - \frac{1}{45} \cdot 125\right) - 0$$

$$\approx 1.806$$

(b) $\text{Var}(X) = \displaystyle\int_0^5 x^2\left(\frac{5.5 - x}{15}\right)dx - \mu^2$

$$= \int_0^5 \left(\frac{5.5}{15}x^2 - \frac{1}{15}x^3\right)dx - \mu^2$$

$$= \left(\frac{5.5}{45}x^3 - \frac{1}{60}x^4\right)\Big|_0^5 - \mu^2$$

$$= \frac{5.5}{45} \cdot 125 - \frac{1}{60} \cdot 625 - 0 - \mu^2$$

$$\approx 1.60108$$

$$\sigma = \sqrt{\text{Var}(X)} \approx 1.265$$

(c) $P(X \leq \mu - \sigma)$

$$= P(X \leq 1.806 - 1.265)$$

$$= P(X \leq 0.541)$$

$$= \int_0^{0.541} \frac{5.5 - x}{15}\,dx$$

$$= \left(\frac{5.5}{15}x - \frac{1}{30}x^2\right)\Big|_0^{0.541}$$

$$= \left(\frac{5.5}{15}(0.541) - \frac{1}{30}(0.541)^2 - 0\right)$$

$$\approx 0.1886$$

41. $f(t) = 0.06049e^{-0.03211t}$; $[16, 84]$

(a) Expected value:

$$E(T) = \mu$$

$$= \int_{16}^{84} t(0.06049e^{-0.03211t})\,dt$$

Use integration by parts.

$$\int_{16}^{84} t(0.06049e^{-0.03211t})\,dt$$

$$\approx -1.88384(t + 31.14295)\,e^{-0.03211t}\Big|_{16}^{84}$$

$$= -1.88384[(84 + 31.14295)e^{-0.03211\cdot 84}$$

$$\qquad - (16 + 31.14295)e^{-0.03211\cdot 16}$$

$$\approx 38.512$$

$$\approx 38.51$$

The expected value is 38.51 years.

(b) Var(T)

$$= \int_{16}^{84} t^2(0.06049e^{-0.03211t})\,dt - \mu^2$$

Use integration by parts twice.

$$\int_{16}^{84} t^2(0.06049e^{-0.03211t})\,dt - \mu^2$$

$$\approx -1.88384(t^2 + 62.28589t$$

$$\qquad + 1939.76619)\,e^{-0.03211t}\Big|_{16}^{84} - 38.512^2$$

$$\approx 308.305$$

$$\sigma = \sqrt{\text{Var}(T)} = \sqrt{308.290} \approx 17.558$$

$$s \approx 17.56 \text{ years}$$

(c) $P(16 \le T \le 38.512 - 17.558)$

$$= \int_{16}^{20.954} 0.06049e^{-0.03211t}\,dt$$

$$= -1.88384\,e^{-0.03211t}\Big|_{16}^{20.954}$$

$$\approx 0.1657$$

If you do the integration on a calculator using differently rounded values for the limits you may get a slightly different answer, such as 0.1656.

(d) To find the median age, find the value of m such that $\int_a^m f(t)\,dt = \frac{1}{2}$.

$$\int_{16}^{m} 0.06049e^{-0.03211t}\,dt = \frac{1}{2}$$

$$\frac{0.06049}{-0.03211}e^{-0.03211t}\Big|_{16}^{m} = \frac{1}{2}$$

$$\frac{0.06049}{-0.03211}\left(e^{-0.03211m} - e^{-0.03211(16)}\right) = \frac{1}{2}$$

$$e^{-0.03211m} - e^{-0.51376} = -\frac{0.03211}{2\cdot 0.06049}$$

$$e^{-0.03211m} = e^{-0.51376} - \frac{0.03211}{2\cdot 0.06049}$$

$$-0.03211m = \ln\left(e^{-0.51376} - \frac{0.03211}{2\cdot 0.06049}\right)$$

$$m = \frac{1}{-0.03211}\ln\left(e^{-0.51376} - \frac{0.03211}{2\cdot 0.06049}\right)$$

$$m \approx 34.26$$

The median age is 34.26 years.

43.
$$S(t) = \frac{1}{91,762}(-2.416t^3 + 90.91t^2$$

$$\qquad - 846.8t + 4880) \text{ for } t \text{ in } [0, 24]$$

$$\mu = \int_0^{24} tS(t)\,dt$$

$$= \frac{1}{91,762}\int_0^{24}(-2.416t^4 + 90.91t^3$$

$$\qquad - 846.8t^2 + 4880t)\,dt$$

Use a calculator to evaluate the integral.

$$\mu \approx 13.037$$

The expected time of day at which a fatal accident will occur is about 1 P.M.

11.3 Special Probability Density Functions

Your Turn 1

$42 - 27 = 15$, so the uniform distribution for the maximum daily temperature T is $f(t) = \frac{1}{15}$ for t in $[27, 42]$.

$$\mu = \frac{27 + 42}{2}$$

$$= \frac{69}{2}$$

$$= 34.5$$

The expected maximum daily temperature is 34.5°C.

$$\sigma = \frac{1}{\sqrt{12}}(42 - 27)$$

$$= \frac{15}{\sqrt{12}}$$

$$= \frac{5\sqrt{3}}{2} \approx 4.33$$

A temperature one standard deviation below the mean is
$34.5 - \frac{5\sqrt{3}}{2} \approx 30.16987$. A temperature one standard

deviation above the mean is $34.5 + \frac{5\sqrt{3}}{2} \approx 38.83013$.

$$P(\mu - \sigma \le T \le \mu + \sigma) = \int_{30.16987}^{38.83013} \frac{1}{15}\,dx$$

This integral is 1/15 times the difference of the limits, but this difference is just twice the standard deviation, so the probability is

$$\frac{5\sqrt{3}}{2} \cdot \frac{2}{15} = \frac{\sqrt{3}}{3} \approx 0.5774$$

Your Turn 2

$$f(t) = \frac{1}{25}e^{-t/25} \text{ for } t \ge 0$$

(a) The probability that a randomly selected battery has a useful life less than 100 hours is

$$P(T \le 100) = \int_0^{100} \frac{1}{25}e^{-t/25}\,dt$$

$$= \frac{1}{25}(-25e^{-t/25})\Big|_0^{100}$$

$$= -(e^{-100/25} - e^0)$$

$$= 1 - e^{-4}$$

$$\approx 0.9817.$$

(b) $\mu = \frac{1}{1/25} = 25$

$$\sigma = \frac{1}{1/25} = 25$$

(c) $P(T > 40) = \int_{40}^{\infty} \frac{1}{25}e^{-t/25}\,dt$

$$= \lim_{b \to \infty} \frac{1}{25}(-25e^{-t/25})\Big|_{40}^{b}$$

$$= \lim_{b \to \infty} (-e^{-b/25} + e^{-40/25})$$

$$= e^{-8/5}$$

$$\approx 0.2019$$

Your Turn 3

(a) The z-score for an age of 79 is

$$z = \frac{79 - \mu}{\sigma}$$

$$= \frac{79 - 75}{16}$$

$$= 0.25.$$

$$P(X > 79) = P(z > 0.25)$$

$$= 1 - P(z \le 0.25)$$

$$= 1 - 0.5987$$

$$= 0.4013$$

(We find the value 0.5987 in the row for 0.2 and the column for 0.05 in the table giving the area under the normal curve.)

(b) Compute the z-scores for 67 and 83.

$$z = \frac{67 - 75}{16} = -0.5$$

$$z = \frac{83 - 75}{16} = 0.5$$

$$P(67 \le X \le 83) = P(-0.5 \le z \le 0.5)$$

$$= 0.6915 - 0.3085$$

$$= 0.3830$$

(We find the value 0.6915 in the row for 0.5 and the column for 0.00 in the normal table, and the value 0.3085 in the row for −0.5 and the column for 0.00.)

11.3   Warmup Exercises

W1. $\int_0^{\infty} xe^{-2x}\,dx$

Using integration by parts with
$dv = e^{-2x}$ and $u = x$, so that

$$v = -\frac{1}{2}e^{-2x} \text{ and } du = dx,$$

$$\int xe^{-2x}\,dx$$

$$= -\frac{x}{2}e^{-2x} + \frac{1}{2}\int e^{-2x}\,dx$$

$$= -\frac{x}{2}e^{-2x} - \frac{1}{4}e^{-2x} = \left(-\frac{x}{2} - \frac{1}{4}\right)e^{-2x} + C$$

$$\int_0^\infty xe^{-2x}\,dx = \lim_{b\to\infty}\int_0^b xe^{-2x}\,dx$$

$$= \lim_{b\to\infty}\left[\left(-\frac{x}{2} - \frac{1}{4}\right)e^{-2x}\right]_0^b$$

$$= \frac{1}{4} + \lim_{b\to\infty}\left[\left(-\frac{b}{2} - \frac{1}{4}\right)e^{-2b}\right]$$

$$= \frac{1}{4}$$

W2. $\displaystyle\int_0^\infty x^2 e^{-3x}\,dx$

Use tabular integration.

D		I
x^2	$+$	e^{-3x}
$2x$	$-$	$-\dfrac{1}{3}e^{-3x}$
2	$+$	$\dfrac{1}{9}e^{-3x}$
0		$-\dfrac{1}{27}e^{-3x}$

$$\int x^2 e^{-3x}\,dx$$

$$= -\frac{1}{27}\left(9x^2 + 6x + 2\right)e^{-3x} + C$$

$$\int_0^\infty x^2 e^{-3x}\,dx = \lim_{b\to\infty}\int_0^b x^2 e^{-3x}\,dx$$

$$= \lim_{b\to\infty}\left[-\frac{1}{27}\left(9x^2 + 6x + 2\right)e^{-3x}\right]_0^b$$

$$= \frac{2}{27}$$

11.3 Exercises

1. $f(x) = \dfrac{5}{7}$ for x in $[3, 4.4]$

This is a uniform distribution: $a = 3, b = 4.4$.

(a) $\mu = \dfrac{1}{2}(4.4 + 3) = \dfrac{1}{2}(7.4)$

$\qquad = 3.7$ cm

(b) $\sigma = \dfrac{1}{\sqrt{12}}(4.4 - 3)$

$\qquad = \dfrac{1}{\sqrt{12}}(1.4)$

$\qquad \approx 0.4041$ cm

(c) $P(3.7 < X < 3.7 + 0.4041)$

$\qquad = P(3.7 < X < 4.1041)$

$\qquad = \displaystyle\int_{3.7}^{4.1041}\frac{5}{7}\,dx$

$\qquad = \dfrac{5}{7}x\Big|_{3.7}^{4.1041}$

$\qquad \approx 0.2886$

3. $f(t) = 4e^{-4t}$ for t in $[0, \infty)$

This is an exponential distribution: $a = 4$.

(a) $\mu = \dfrac{1}{4} = 0.25$ year

(b) $\sigma = \dfrac{1}{4} = 0.25$ year

(c) $P(0.25 < T < 0.25 + 0.25)$

$\qquad = P(0.25 < T < 0.5)$

$\qquad = \displaystyle\int_{0.25}^{0.5} 4e^{-4t}\,dt$

$\qquad = -e^{-4t}\Big|_{0.25}^{0.5}$

$\qquad = -\dfrac{1}{e^{-2}} + \dfrac{1}{e^{-1}}$

$\qquad \approx 0.2325$

5. $f(t) = \dfrac{e^{-t/3}}{3}$ for t in $[0, \infty)$

This is an exponential distribution: $a = \frac{1}{3}$.

(a) $\mu = \dfrac{1}{\frac{1}{3}} = 3$ days

(b) $\sigma = \dfrac{1}{\frac{1}{3}} = 3$ days

(c) $P(3 < T < 3 + 3) = P(3 < T < 6)$

$$= \int_3^6 \frac{e^{-t/3}}{3} dt$$

$$= e^{-t/3} \Big|_3^6$$

$$= -\frac{1}{e^{-2}} + \frac{1}{e^{-1}}$$

$$\approx 0.2325$$

7. $z = 3.50$

Area to the left of $z = 3.50$ is 0.9998. Given mean $\mu = z - 0$, so area to left of μ is 0.5.

Area between μ and z is

$$0.9998 - 0.5 = 0.4998.$$

Therefore, this area represents 49.98% of total area under normal curve.

9.   Between $z = 1.28$ and $z = 2.05$

Area to left of $z = 2.05$ is 0.9798 and area to left of $z = 1.28$ is 0.8997.

$$0.9798 - 0.8997 = 0.0801$$

Percent of total area $= 8.01\%$

11.   Since $10\% = 0.10$, the z-score that corresponds to the area of 0.10 to the left of z is -1.28.

13. 18% of the total area to the right of z means $1 - 0.18$ of the total area is to the left of z.

$$1 - 0.18 = 0.82$$

The closest z-score that corresponds to the area of 0.82 is 0.92

19.   Let m be the median of the exponential distribution $f(x) = ae^{-ax}$ for $[0, \infty)$.

$$\int_0^m ae^{-ax} dx = 0.5$$

$$-e^{-ax} \Big|_0^m = 0.5$$

$$-e^{-am} + 1 = 0.5$$

$$0.5 = e^{-am}$$

$$-am = \ln 0.5$$

$$m = -\frac{\ln 0.5}{a}$$

or $-am = \ln \frac{1}{2}$

$$-am = -\ln 2$$

$$m = \frac{\ln 2}{a}$$

21.   The area that is to the left of x is

$$A = \int_{-\infty}^x \frac{1}{\sigma\sqrt{2\pi}} e^{-\frac{(t-\mu)^2}{2\sigma^2}} dt.$$

Let $u = \frac{(t-\mu)}{\sigma}$. Then $du = \frac{1}{\sigma} dt$ and

$dt = \sigma du.$

If $t = x$,

$$u = \frac{x - \mu}{\sigma} = z.$$

As $t \to -\infty$, $u \to -\infty$.

Therefore,

$$A = \int_{-\infty}^z \frac{1}{\sigma\sqrt{2\pi}} e^{(-1/2)u^2} \sigma \, du$$

$$= \frac{\sigma}{\sigma} \int_{-\infty}^z \frac{1}{\sqrt{2\pi}} e^{-u^2/2} du$$

$$= \int_{-\infty}^z \frac{1}{\sqrt{2\pi}} e^{-u^2/2} du.$$

This is the area to the left of z for the standard normal curve.

23.   (a) $\int_0^{35} 0.5e^{-0.5x} dx \approx 1.00000$

(b) $\int_0^{35} 0.5xe^{-0.5x} dx \approx 1.99999$

(c) $\int_0^{35} 0.5x^2 e^{-0.5x} dx = 7.999998$

25.   Use Simpson's rule with $n = 40$ and limits of -6 and 6 to approximate the mean and standard deviation of a normal probability distribution.

(a) $\int_{-\infty}^{\infty} \frac{x}{\sqrt{2\pi}} e^{-x^2/2} dx$

$$\approx \int_{-6}^6 \frac{x}{\sqrt{2\pi}} e^{-x^2/2} dx$$

$\mu = 0$

(For the integral of an odd function over an interval symmetric to 0, Simpson's rule will give 0; there is no need to do any calculation.)

(b) $\displaystyle\int_{-\infty}^{\infty} \frac{x^2}{\sqrt{2\pi}} e^{-x^2/2}\, dx$

$$\approx \int_{-6}^{6} \frac{x^2}{\sqrt{2\pi}} e^{-x^2/2}\, dx$$

$\sigma \approx 0.9999999224$ (Simpson's rule)

$\sigma \approx 0.9999999251$ (calculator)

27. The probability density function for the uniform distribution is $f(x) = \frac{1}{b-a}$ for x in $[a, b]$.

The cumulative distribution function for f is

$$F(x) = P(X \le x)$$

$$= \int_{a}^{x} f(t)\, dt$$

$$= \int_{a}^{x} \frac{1}{b-a}\, dt$$

$$= \frac{1}{b-a} t \Big|_{a}^{x}$$

$$= \frac{1}{b-a}(x-a)$$

$$= \frac{x-a}{b-a}, \quad a \le x \le b.$$

29. For a uniform distribution,

$$f(x) = \frac{1}{b-a} \quad \text{for } [a, b].$$

Thus, we have

$$f(x) = \frac{1}{85-10} = \frac{1}{75}$$

for $[10, 85]$.

(a) $\mu = \dfrac{1}{2}(10 + 85) = \dfrac{1}{2}(95)$

$\qquad = 47.5$ thousands

Therefore, the agent sells $\$47,500$ in insurance.

(b) $P(50 < X < 85) = \displaystyle\int_{50}^{85} \frac{1}{75}\, dx$

$$= \frac{x}{75} \Big|_{50}^{85}$$

$$= \frac{85}{75} - \frac{50}{75}$$

$$= \frac{35}{75} = 0.4667$$

31. (a) Since we have exponential distribution with $\mu = 4.25$,

$$\mu = \frac{1}{a} = 4.25$$

$$a = 0.235.$$

Therefore, $f(x) = 0.235e^{-0.235x}$ on $[0, \infty)$.

(b) $P(X > 10)$

$$= \int_{10}^{\infty} 0.235e^{-0.235x}\, dx$$

$$= \lim_{a \to \infty} \int_{10}^{a} 0.235e^{-0.235x}\, dx$$

$$= \lim_{a \to \infty} (-e^{-0.235x}) \Big|_{10}^{a}$$

$$= \lim_{a \to \infty} \left(-\frac{1}{e^{0.235a}} + \frac{1}{e^{2.35}}\right)$$

$$= \frac{1}{e^{2.35}} = 0.0954$$

33. (a) $\mu = 2.5,\ \sigma = 0.2,\ x = 2.7$

$$z = \frac{2.7 - 2.5}{0.2} = 1$$

Area to the right of $z = 1$ is

$1 - 0.8413 = 0.1587.$

Probability $= 0.1587$

(b) Within 1.2 standard deviations of the mean is the area between $z = -1.2$ and $z = 1.2$.

Area to left of $z = 1.2 = 0.8849$

Area to the left of $z = -1.2 = 0.1151$

$0.8849 - 0.1151 = 0.7698$

Probability $= 0.7698$

Using the TI-84 Plus C command normalcdf$(-1.2, 1.2)$ you will get an answer of 0.7699.

35. If X has a uniform distribution on $[0, 1000]$, then its density function is $f(x) = \frac{1}{1000}$ for x in $[0, 1000]$. The expected payment with no deductible is

$$E(X) = \int_0^{1000} x \cdot \frac{1}{1000} \, dx$$

$$= \frac{1}{1000} \cdot \frac{1}{2} x^2 \Big|_0^{1000}$$

$$= \frac{1}{2000} \cdot (1000^2 - 0)$$

$$= 500.$$

Now, let the deductible be D. According to the hint,

$$\text{payment} = \begin{cases} 0 & \text{for } x \leq D \\ x - D & \text{for } x > D. \end{cases}$$

The expected payment with the deductible is therefore

$$E(X) = \int_0^D 0 \cdot \frac{1}{1000} \, dx + \int_D^{1000} (x - D) \cdot \frac{1}{1000} \, dx$$

$$= 0 + \frac{1}{1000} \cdot \left(\frac{1}{2} x^2 - Dx \right) \Big|_D^{1000}$$

$$= \frac{1}{1000} \cdot \left[\left(\frac{1}{2} 1000^2 - 1000D \right) - \left(\frac{1}{2} D^2 - D^2 \right) \right]$$

$$= 500 - D + \frac{1}{2000} D^2.$$

For this amount to be 25% of the amount with no deductible, we must have

$$500 - D + \frac{1}{2000} \cdot D^2 = 0.25 \cdot 500$$

$$\frac{1}{2000} D^2 - D + 500 = 125$$

$$\frac{1}{2000} D^2 - D + 375 = 0$$

$$D^2 - 2000D + 750,000 = 0$$

$$(D - 1500)(D - 500) = 0$$

$$D = 1500 \text{ or } D = 500$$

We reject $D = 1500$ since it is not in $[0, 1000]$.

Therefore, $D = 500$. The correct answer choice is **c**.

37. Let the random variable X be the lifetime of the printer in a years. Then it has exponential distribution $f(x) = ae^{-ax}$ for $x \geq 0$.

If the mean is 2 years, then $\frac{1}{a} = 2$, or $a = \frac{1}{2}$ and the function is $f(x) = \frac{1}{2} e^{-x/2}$ for $x \geq 0$.

We wish to find $P(0 \leq X \leq 1)$ and $P(1 \leq X \leq 2)$.

$$P(0 \leq X \leq 1) = \int_0^1 \frac{1}{2} e^{-x/2} \, dx$$

$$= -e^{-x/2} \Big|_0^1$$

$$= -e^{-1/2} + 1$$

$$P(1 \leq X \leq 2) = \int_1^2 \frac{1}{2} e^{-x/2} \, dx$$

$$= -e^{-x/2} \Big|_1^2$$

$$= -e^{-1} + e^{-1/2}$$

If 100 printers are sold, then $(1 - e^{-1/2} + 1)(100)$ will fail in the first year, and $(e^{-1/2} - e^{-1/2})(100)$ will fail in the second year.

The manufacturer pays a full refund on those failing first year and one-half refund on those failing during the second year.

$$\begin{aligned} \text{Refunds} &= (1 - e^{-1/2} + 1)(100)(\$200) \\ &\quad + (e^{-1/2} - e^{-1})(100)(\$100) \\ &\approx \$10,255.90 \end{aligned}$$

The correct answer choice is **d**.

39. For a uniform distribution,

$$f(x) = \frac{1}{b - a} \text{ for } x \text{ in } [a, b].$$

$$f(x) = \frac{1}{36 - 20} = \frac{1}{16} \text{ for } x \text{ in } [20, 36]$$

(a) $\mu = \frac{1}{2}(20 + 36) = \frac{1}{2}(56)$

$= 28$ days

(b) $P(30 < X \leq 36)$

$$= \int_{30}^{36} \frac{1}{16} \, dx = \frac{1}{16} x \Big|_{30}^{36}$$

$$= \frac{1}{16}(36 - 30)$$

$$= 0.375$$

41. We have an exponential distribution, with $a = 1$.

$$f(t) = e^{-t}, [0, \infty)$$

(a) $\mu = \dfrac{1}{1} = 1$ hr

(b) $P(T < 30 \text{ min})$

$$= \int_0^{0.5} e^{-t} \, dt$$

$$= -e^{-t} \Big|_0^{0.5}$$

$$1 - e^{-0.5} \approx 0.3935$$

43. $f(x) = ae^{-ax}$ for $[0, \infty]$

Since $\mu = 25$ and $\mu = \dfrac{1}{a}$,

$$a = \frac{1}{25} = 0.04.$$

This, $f(x) = 0.04e^{-0.04x}$.

(a) We must find t such that $P(X \le t) = 0.90$.

$$\int_0^t 0.04e^{-0.04x} \, dx = 0.90$$

$$-e^{-0.04x} \Big|_0^t = 0.90$$

$$-e^{-0.04t} + 1 = 0.90$$

$$0.10 = -e^{-0.04t}$$

$$-0.04t = \ln 0.10$$

$$t = \frac{\ln 0.10}{-0.04}$$

$$t \approx 57.56$$

The longest time within which the predator will be 90% certain of finding a prey is approximately 58 min.

(b) $P(X \ge 60)$

$$= \int_{60}^{\infty} 0.04e^{-0.04x} \, dx$$

$$= \lim_{b \to \infty} \int_{60}^{b} 0.04e^{-0.04x} \, dx$$

$$= \lim_{b \to \infty} (e^{-0.04x}) \Big|_{60}^{b}$$

$$= \lim_{b \to \infty} \left[-e^{-0.04b} + e^{-0.04(60)} \right]$$

$$= 0 + e^{-2.4}$$

$$\approx 0.0907$$

The probability that the predator will have to spend more than one hour looking for a prey is approximately 0.0907.

45. For an exponential distribution,

$$f(x) = ae^{-ax} \text{ for } x \text{ in } [0, \infty).$$

Since $\mu = \dfrac{1}{a} = 12.3$, $a = \dfrac{1}{12.3}$.

(a) $P(X \ge 20) = \displaystyle\int_{20}^{\infty} \dfrac{1}{12.3} e^{-x/12.3} \, dx$

$$= \lim_{b \to \infty} \int_{20}^{b} \frac{1}{12.3} e^{-x/12.3} \, dx$$

$$= \lim_{b \to \infty} \left(\frac{1}{12.3} e^{-x/12.3} \Big|_{20}^{b} \right)$$

$$= \lim_{b \to \infty} (-e^{-b/12.3} + e^{-20/12.3})$$

$$= e^{-20/12.3}$$

$$\approx 0.1967$$

(b) $P(10 \le X \le 20) = \displaystyle\int_{10}^{20} \dfrac{1}{12.3} e^{-x/12.3} \, dx$

$$= (-e^{-x/12.3}) \Big|_0^{20}$$

$$= -e^{-20/12.3} + e^{-10/12.3}$$

$$\approx 0.2468$$

47. Use the TI-84 Plus C command invNorm.

invNorm$(0.85, 52, 8) = 60.291$.

The 85th percentile is a speed of 69.29 mph.

49. We have an exponential distribution, with $a = 0.229$.

So $f(t) = 0.229e^{-0.229t}$, for $[0, \infty)$.

(a) The life expectancy is

$$\mu = \frac{1}{a} = \frac{1}{0.229} \approx 4.37 \text{ millennia.}$$

The standard deviation is

$$\sigma = \frac{1}{a} = \frac{1}{0.229} \approx 4.37 \text{ millennia.}$$

(b) $P(T \geq 2) = \int_2^\infty 0.229e^{-0.229t}\,dt$

$$= 1 - \int_0^2 0.229e^{-0.229t}\,dt$$

$$= 1 + \left[e^{-0.229t} \Big|_0^2 \right]$$

$$= 1 + \left[e^{-0.229(2)} - 1 \right]$$

$$= e^{-0.458} \approx 0.6325$$

51. For an exponential distribution,

$$f(x) = ae^{-ax} \text{ for } [0, \infty).$$

Since $\mu = \dfrac{1}{a} = 8, a = \dfrac{1}{8}$.

(a) $P(X \geq 10) = \int_{10}^\infty \dfrac{1}{8} e^{-x/8}\,dx$

$$= 1 - \int_0^{10} \dfrac{1}{8} e^{-x/8}\,dx$$

$$= 1 + \left[e^{-x/8} \Big|_0^{10} \right]$$

$$= 1 + [e^{-10/8} - 1]$$

$$= e^{-10/8} \approx 0.2865$$

(b) $P(X < 2) = \int_0^2 \dfrac{1}{8} e^{-x/8}\,dx$

$$= -e^{-x/8} \Big|_0^2$$

$$= -e^{-2/8} + 1$$

$$\approx 0.2212$$

53. We have an exponential distribution $f(x)$
$= ae^{-ax}$ for $x \geq 0$. Since $a = \frac{1}{90}, f(x) =$
$\frac{1}{90} e^{-x/90}$ for $x \geq 0$.

(a) The probability that the time for a goal is no more than 71 minutes is

$$P(0 < X < 71) = \int_0^{71} \dfrac{1}{90} e^{-x/90}\,dx$$

$$= -e^{-x/90} \Big|_0^{71}$$

$$= -e^{-71/90} + 1$$

$$\approx 0.5457.$$

(b) The probability that the time for a goal is 499 minutes or more is

$$P(X \geq 499) = \int_{499}^\infty \dfrac{1}{90} e^{-x/90}\,dx$$

$$= e^{-x/90} \Big|_{499}^\infty$$

$$= 0 + e^{-499/90}$$

$$\approx 0.0039.$$

Chapter 11 Review Exercises

1. True

2. True

3. True

4. False: A density function is always nonnegative.

5. False: If the random variable takes on negative values the expectation may also be negative.

6. True

7. True

8. True

9. False: The normal distribution is symmetrical; the exponential distribution has a long tail to the right.

10. False: The expected value is 0 and the standard deviation is 1.

11. In a probability function, the y-values (or function values) represent probabilities.

13. A probability density function f for $[a, b]$ must satisfy the following two conditions:

(1) $f(x) \geq 0$ for all x in the interval $[a, b]$;

(2) $\int_a^b f(x)\,dx = 1.$

15. $f(x) = \sqrt{x}; [4, 9]$

$$\int_4^9 x^{1/2}\,dx = \frac{2}{3}x^{3/2}\Big|_4^9$$

$$= \frac{2}{3}(27 - 8)$$

$$= \frac{38}{3} \neq 1$$

$f(x)$ is not a probability density function.

17. $f(x) = 0.7e^{-0.7x}; [0, \infty)$

$$\int_0^\infty 0.7e^{-0.7x}\,dx = -e^{-0.7x}\Big|_0^\infty$$

$$= \lim_{b \to \infty}(-e^{-0.7b}) + e^0$$

$$= \lim_{b \to \infty}\left(-\frac{1}{e^{0.7b}}\right) + 1$$

$$= 0 + 1 = 1$$

$f(x) \geq 0$ for all x in $[0, \infty)$.

Therefore, $f(x)$ is a probability density function.

19. $f(x) = kx^2; [1, 4]$

$$\int_1^4 kx^2\,dx = \frac{kx^3}{3}\Big|_1^4$$

$$= 21k$$

Since $f(x)$ is a probability density function,

$$21k = 1$$

$$k = \frac{1}{21}.$$

21. $f(x) = \frac{1}{10}$ for $[10, 20]$

(a) $P(10 \leq X \leq 12)$

$$= \int_{10}^{12} \frac{1}{10}\,dx$$

$$= \frac{x}{10}\Big|_{10}^{12}$$

$$= \frac{1}{5} = 0.2$$

(b) $P\left(\frac{31}{2} \leq X \leq 20\right)$

$$= \int_{31/2}^{20} \frac{1}{10}\,dx$$

$$= \frac{x}{10}\Big|_{31/2}^{20}$$

$$= 2 - \frac{31}{20}$$

$$= \frac{9}{20} = 0.45$$

(c) $P(10.8 \leq X \leq 16.2)$

$$= \int_{10.8}^{16.2} \frac{1}{10}\,dx$$

$$= \frac{x}{10}\Big|_{10.8}^{16.2} = 0.54$$

23.   If we consider the probabilities as weights, the expected value or mean of a probability distribution represents the point at which the distribution balances.

25. $f(x) = \frac{2}{9}(x - 2); [2, 5]$

(a) $\mu = \int_2^5 \frac{2x}{9}(x - 2)\,dx$

$$= \int_2^5 \frac{2}{9}(x^2 - 2x)\,dx$$

$$= \frac{2}{9}\left(\frac{x^3}{3} - x^2\right)\Big|_2^5$$

$$= \frac{2}{9}\left(\frac{125}{3} - 25 - \frac{8}{3} + 4\right) = 4$$

(b) $\text{Var}(X) = \int_2^5 \frac{2x^2}{9}(x - 2)\,dx - (4)^2$

$$= \int_2^5 \frac{2}{9}(x^3 - 2x^2)\,dx - 16$$

$$= \frac{2}{9}\left(\frac{x^4}{4} - \frac{2x^3}{3}\right)\Big|_2^5 - 16$$

$$= \frac{2}{9}\left(\frac{625}{4} - \frac{250}{3} - 4 + \frac{16}{3}\right) - 16$$

$$= 0.5$$

(c) $\sigma = \sqrt{0.5} \approx 0.7071$

(d) $\int_2^m \frac{2}{9}(x-2)dx = \frac{1}{2}$

$$\frac{1}{9}(m-2)^2 \Big|_2^m = \frac{1}{2}$$

$$\frac{1}{9}[(m-2)^2 - 0] = \frac{1}{2}$$

$$m^2 - 4m + 4 = \frac{9}{2}$$

$$m^2 - 4m - \frac{1}{2} = 0$$

$$m = \frac{4 \pm 3\sqrt{2}}{2}$$

$$\approx -0.121, 4.121$$

We reject -0.121 since it is not in $[2, 5]$. So, $m = 4.121$

(e) $\int_2^x \frac{2}{9}(t-2)dt = \frac{1}{9}(t-2)^2 \Big|_2^x$

$$= \frac{1}{9}[(x-2)^2 - 0]$$

$$= \frac{(x-2)^2}{9}, 2 \le x \le 5$$

27. $f(x) = 5x^{-6}; [1, \infty)$

(a) $\mu = \int_1^\infty x \cdot 5x^{-6} \, dx = \int_1^\infty 5x^{-5} \, dx$

$$= \lim_{b \to \infty} \int_1^b 5x^{-5} \, dx = \lim_{b \to \infty} \frac{5x^{-4}}{-4} \Big|_1^b$$

$$= \lim_{b \to \infty} \frac{5}{4}\left(1 - \frac{1}{b^4}\right) = \frac{5}{4}$$

(b) $\text{Var}(X) = \int_1^\infty x^2 \cdot 5x^{-6} \, dx - \left(\frac{5}{4}\right)^2$

$$= \lim_{b \to \infty} \int_1^b 5x^{-4} \, dx - \frac{25}{16}$$

$$= \lim_{b \to \infty} \frac{5x^{-3}}{-3} \Big|_1^b - \frac{25}{16}$$

$$= \lim_{b \to \infty} \frac{5}{3}\left(1 - \frac{1}{b^3}\right) - \frac{25}{16}$$

$$= \frac{5}{3} - \frac{25}{16} = \frac{5}{48} \approx 0.1042$$

(c) $\sigma \approx \sqrt{\text{Var}(X)} \approx 0.3227$

(d) $\int_1^m 5x^{-6}dx = \frac{1}{2}$

$$-x^{-5} \Big|_1^m = \frac{1}{2}$$

$$-m^{-5} + 1 = \frac{1}{2}$$

$$m^{-5} = \frac{1}{2}$$

$$m^5 = 2$$

$$m = \sqrt[5]{2} \approx 1.149$$

(e) $\int_1^x 5t^{-6}dt = -t^{-5} \Big|_1^x$

$$= -x^{-5} + 1$$

$$= 1 - \frac{1}{x^5}, x \ge 1$$

29. $f(x) = 4x - 3x^2; [0, 1]$

(a) $\mu = \int_0^1 x(4x - 3x^2)dx$

$$= \int_0^1 (4x^2 - 3x^3)dx$$

$$= \left(\frac{4x^3}{3} - \frac{3x^4}{4}\right) \Big|_0^1$$

$$= \frac{4}{3} - \frac{3}{4} = \frac{7}{12}$$

$$\approx 0.5833$$

(b) $\text{Var}(X) = \int_0^1 x^2(4x - 3x^2)dx - \left(\frac{7}{12}\right)^2$

$$= \int_0^1 (4x^3 - 3x^4)dx - \left(\frac{7}{12}\right)^2$$

$$= \left(x^4 - \frac{3x^5}{5}\right) \Big|_0^1 - \left(\frac{7}{12}\right)^2$$

$$= 1 - \frac{3}{5} - \left(\frac{7}{12}\right)^2$$

$$\approx 0.0597$$

$$\sigma \approx \sqrt{\text{Var}(X)}$$

$$\approx 0.2444$$

(c) $P\left(0 \le X \le \frac{7}{12}\right)$

$$= \int_0^{7/12} (4x - 3x^3)dx$$

$$= (2x^2 - x^3)\Big|_0^{7/12}$$

$$= 2\left(\frac{7}{12}\right)^2 - \left(\frac{7}{12}\right)^3$$

$$\approx 0.4821$$

(d) $P(\mu - \sigma \le X \le \mu + \sigma)$

$$\approx P(0.3389 \le x \le 0.8277)$$

$$= \int_{0.3389}^{0.8277} (4x - 3x^2)dx$$

$$= (2x^2 - x^3)\Big|_{0.3389}^{0.8277}$$

$$= 2(0.8277)^2 - (0.8277)^3$$

$$\quad - 2(0.3389)^2 + (0.3389)^3$$

$$\approx 0.6123$$

31. $f(x) = 0.01e^{-0.01x}$ for $[0, \infty)$ is an exponential distribution.

(a) $\mu = \dfrac{1}{0.01} = 100$

(b) $\sigma = \dfrac{1}{0.01} = 100$

(c) $P(100 - 100 < X < 100 + 100)$

$$= P(0 < X < 200)$$

$$= \int_0^{200} 0.01e^{-0.01x}\, dx$$

$$= -e^{-0.01x}\Big|_0^{200}$$

$$= 1 - e^{-2} \approx 0.8647$$

For Exercises 33–40, use the table in the Appendix for the areas under the normal curve.

33. Area to the left of $z = -0.43$ is 0.3336.

Percent of area is 33.36%.

35. Area between $z = -1.17$ and $z = -0.09$ is

$$0.4641 - 0.1210 = 0.3431.$$

Percent of area is 34.31%.

37. The region up to 1.2 standard deviations below the mean is the region to the left of $z = -1.2$. The area is 0.1151, so the percent of area is 11.51%.

39. 52% of area is to the right implies that 48% is to the left.

$$P(z < a) = 0.48 \text{ for } a = -0.05$$

Thus, 52% of the area lies to the right of $z = -0.05$.

41. $f(x) = 0.05$ for $[10, 30]$

(a) This is a uniform distribution.

(b) The domain of f is $[10, 30]$.

The range of f is $\{0.05\}$.

(c)

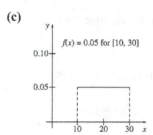

(d) For a uniform distribution,

$$\mu = \frac{1}{2}(b + a) \text{ and}$$

$$\text{Var}(X) = \frac{b^2 - 2ab + a^2}{12}.$$

Thus,

$$\mu = \frac{1}{2}(30 + 10) = \frac{1}{2}(40) = 20$$

$$\text{Var}(X) = \frac{30^2 - 2(10)(30) + 10^2}{12}$$

$$= \frac{400}{12}.$$

$$\sigma = \sqrt{\frac{400}{12}} \approx 5.77$$

(e) $P(\mu - \sigma \le X \le \mu + \sigma)$

$$= P(20 - 5.77 \le X \le 20 + 5.77)$$

$$= P(14.23 \le X \le 25.77)$$

$$= \int_{14.23}^{25.77} 0.05\, dx$$

$$= 0.05x\Big|_{14.23}^{25.77}$$

$$= 0.05(25.77 - 14.23)$$

$$\approx 0.577$$

43. $f(x) = \dfrac{e^{-x^2}}{\sqrt{\pi}}$ for $(-\infty, \infty)$

(a) Since the exponent of e in $f(x)$ may be written

$$-x^2 = \dfrac{-(x-0)^2}{2\left(\frac{1}{\sqrt{2}}\right)^2},$$

and

$$\dfrac{1}{\sqrt{\pi}} = \dfrac{1}{\frac{1}{\sqrt{2}}\sqrt{2\pi}},$$

$f(x)$ is a normal distribution with $\mu = 0$ and $\sigma = \frac{1}{\sqrt{2}}$.

(b) The domain of f is $(-\infty, \infty)$.

The range of f is $\left(0, \frac{1}{\sqrt{\pi}}\right]$.

(c)

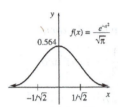

(d) For this normal distribution, $\mu = 0$ and $\sigma = \frac{1}{\sqrt{2}}$.

(e) $P(\mu - \sigma \le X \le \mu + \sigma)$
$= 2P(0 \le X \le \mu + \sigma)$
$= 2P\left(0 \le X \le \frac{1}{\sqrt{2}}\right)$

If $x = \frac{1}{\sqrt{2}}$, $z = \dfrac{\frac{1}{\sqrt{2}} - 0}{\frac{1}{\sqrt{2}}} = 1.00$.

Thus,

$$P(\mu - \sigma \le X \le \mu + \sigma)$$
$$= 2P(0 \le z \le 1.00)$$
$$= 2(0.3413)$$
$$\approx 0.6826$$

45. $f(x) = \dfrac{x^{-1/2}e^{-x/2}}{\sqrt{2\pi}}$ for x in $(0, \infty)$

(a) Using integration by parts:

Let $u = e^{-x/2}$ and $dv = x^{-1/2}dx$

Then $du = -\frac{1}{2}e^{-x/2}$ and $v = 2x^{1/2}$.

$$\frac{1}{\sqrt{2\pi}}\int x^{-1/2}e^{-x/2}dx$$

$$= \frac{1}{\sqrt{2\pi}}\left[2x^{1/2}e^{-x/2} - \int 2x^{1/2}\left(-\frac{1}{2}\right)e^{-x/2}dx\right]$$

$$= \frac{1}{\sqrt{2\pi}}\left[2x^{1/2}e^{-x/2} + \int x^{1/2}e^{-x/2}dx\right]$$

Thus,

$P(0 < X \le b)$

$$= \frac{1}{\sqrt{2\pi}}\int_0^b x^{-1/2}e^{-x/2}dx$$

$$= \frac{1}{\sqrt{2\pi}}\left[2x^{1/2}e^{-x/2}\Big|_0^b + \int_0^b x^{1/2}e^{-x/2}dx\right].$$

(b) $P(0 < X \le 1)$

$$= \frac{1}{\sqrt{2\pi}}\left[2x^{1/2}e^{-x/2}\Big|_0^1\right.$$

$$\left. + \int_0^1 x^{1/2}e^{-x/2}dx\right]$$

Notice that

$$2x^{1/2}e^{-x/2}\Big|_0^1$$

$$= 2e^{-1/2} - 0$$

$$\approx 1.2131.$$

Using Simpson's rule with $n = 12$ to evaluate the improper integral, we have

$$\int_0^1 x^{1/2}e^{-x/2}dx \approx 0.4962$$

Therefore,

$$P(0 \le X \le 1)$$

$$= \frac{1}{\sqrt{2\pi}}\int_0^1 x^{-1/2}e^{-x/2}dx$$

$$\approx \frac{1}{\sqrt{2\pi}}(1.2131 + 0.4962)$$

$$\approx 0.6819.$$

(c) $P(0 < X \le 10)$

$$= \frac{1}{\sqrt{2\pi}}\left[2x^{1/2}e^{-x/2}\Big|_0^{10}\right.$$

$$\left. + \int_0^{10} x^{1/2}e^{-x/2}dx\right]$$

First,

$$2x^{1/2}e^{-x/2}\Big|_0^{10}$$

$$= 2\sqrt{10}e^{-5} - 0$$

$$\approx 0.0426.$$

Using Simpson's rule with $n = 12$ to evaluate the improper integral, we have

$$\int_0^{10} x^{1/2}e^{-x/2}\,dx \approx 2.3928$$

Therefore,

$$P(0 \le X \le 1)$$

$$= \frac{1}{\sqrt{2\pi}}\int_0^{10} x^{-1/2}e^{-x/2}\,dx$$

$$\approx \frac{1}{\sqrt{2\pi}}(0.0426 + 2.3928)$$

$$\approx 0.9716.$$

(d) Since $f(x)$ is a probability density function, the limit as b $\to \infty$ should be 1. The previous results do support this conclusion.

47. (a) $f(t) = \dfrac{5}{112}(1 - t^{-3/2})$; $[1, 25]$

P(No repairs in years 1-3)

$= P$(First repair needed in years 4-25)

$$= \int_4^{25} \frac{5}{112}(1 - t^{-3/2})\,dt$$

$$= \frac{5}{112}(t + 2t^{-1/2})\Big|_4^{25}$$

$$= \frac{5}{112}\left[25 + \frac{2}{5} - 4 - 1\right]$$

$$= \frac{51}{56} \approx 0.9107$$

(b) $\mu = \dfrac{5}{112}\displaystyle\int_1^{25} t(1 - t^{-3/2})\,dt$

$$= \frac{5}{112}\int_1^{25}(t - t^{-1/2})\,dt$$

$$= \frac{5}{112}\left(\frac{t^2}{2} - 2t^{1/2}\right)\Big|_1^{25}$$

$$= \frac{5}{112}\left(\frac{625}{2} - 10 - \frac{1}{2} + 2\right)$$

$$= \frac{95}{7}$$

$$\approx 13.57$$

The expected value for the number of years before the machine requires repairs is 13.57 years.

(c)

$$\text{Var} = \frac{5}{112}\int_1^{25} t^2(1 - t^{-3/2})\,dt - \left(\frac{95}{7}\right)^2$$

$$= \frac{5}{112}\int_1^{25}(t^2 - t^{1/2})\,dt - \left(\frac{95}{7}\right)^2$$

$$= \frac{5}{112}\left(\frac{t^3}{3} - \frac{2}{3}t^{3/2}\right)\Big|_1^{25} - \left(\frac{95}{7}\right)^2$$

$$= \frac{5}{112}\left(\frac{15{,}625}{3} - \frac{250}{3} - \frac{1}{3} + \frac{2}{3}\right) - \frac{9025}{49}$$

$$= \frac{6560}{147}$$

$$\sigma = \sqrt{\frac{6560}{147}}$$

$$\approx 6.68$$

The standard deviation of the number of years before repairs are required is 6.68 years.

49. (a) $\mu = 8$

$$\frac{1}{a} = 8$$

$$a = \frac{1}{8}$$

$$f(x) = \frac{1}{8}e^{-x/8} \text{ for } [0, \infty)$$

(b) Expected number $= \mu = 8$

(c) $\sigma = \mu = 8$

(d) $P(5 \le X \le 10) = \displaystyle\int_5^{10} \frac{1}{8}e^{-x/8}\,dx$

$$= -e^{-x/8}\Big|_5^{10}$$

$$= -e^{-10/8} + e^{-5/8}$$

$$\approx 0.2488$$

51. Let the random variable X be the number of printers. We have an exponential distribution $f(t) = ae^{-at}$ for t in $[0, \infty)$. Since $\mu = 10, a = \frac{1}{\mu} = 0.1$ so that

$$f(t) = 0.1e^{-0.1t}, t \ge 0.$$

We need to determine the portion of the printers sold in the first year and in the second and third year. In other words, we need to calculate $P(0 \le X \le 1)$ and $P(1 \le X \le 3)$.

$$P(0 \le X \le 1) = \int_0^1 0.1e^{-0.1t}\,dt$$

$$= -e^{-0.1t}\Big|_0^1$$

$$= -e^{-0.1} + e^0$$

$$= 1 - e^{-0.1}$$

$$P(1 \le X \le 3) = \int_1^3 0.1e^{-0.1t}\,dt$$

$$= -e^{-0.1t}\Big|_1^3$$

$$= -e^{-0.3} + e^{-0.1}$$

$$= e^{-0.1} - e^{-0.3}$$

Since

$$\text{payment} = \begin{cases} x & \text{for } 0 \le x \le 1 \\ 0.5x & \text{for } 1 \le x \le 3 \\ 0 & \text{for } x > 3 \end{cases}$$

the expected payment will be

$$E(X) = (1 - e^{-0.1})(x)$$
$$+ (e^{-0.1} - e^{-0.3})(0.5x) + 0$$
$$= x - 1.5xe^{-0.1} - 0.5xe^{-0.3}.$$

To determine the level x must be set for this to be 1000, solve

$$E(X) = x - 1.5xe^{-0.1} - 0.5xe^{-0.3}$$
$$= 1000.$$

Using our calculators, we find $x \approx 5644$. The correct answer choice is **d**.

53. $f(x) = 0.01e^{-0.01x}$ for $[0, \infty)$ is an exponential distribution.

$$P(0 \le X \le 100)$$

$$= \int_0^{100} 0.01e^{-0.01x}\,dx$$

$$= -e^{-0.01x}\Big|_0^{100}$$

$$= 1 - \frac{1}{e}$$

$$\approx 0.6321$$

55. $f(x) = \dfrac{3}{19,696}(x^2 + x)$ for x in $[38, 42]$

(a) $\mu = \dfrac{3}{19,696}\displaystyle\int_{38}^{42} x(x^2 + x)\,dx$

$$= \frac{3}{19,696}\int_{38}^{42} (x^3 + x^2)\,dx$$

$$= \frac{3}{19,696}\left(\frac{x^4}{4} + \frac{x^3}{3}\right)\Bigg|_{38}^{42}$$

$$= \frac{3}{19,696}\left(\frac{(42)^4}{4} + \frac{(42)^3}{3}\right.$$
$$\left. - \frac{(38)^4}{4} - \frac{(38)^3}{3}\right)$$

$$\approx 40.07$$

The expected body temperature of the species is 40.07°C.

(b) $P(X \le \mu)$

$$= \frac{3}{19,696}\int_{38}^{40.07} (x^2 + x)\,dx$$

$$= \frac{3}{19,696}\left(\frac{x^3}{3} + \frac{x^2}{2}\right)\Bigg|_{38}^{40.07}$$

$$= \frac{3}{19,696}\left(\frac{(40.07)^3}{3} + \frac{(40.07)^2}{2}\right.$$
$$\left. - \frac{(38)^3}{3} - \frac{(38)^2}{2}\right)$$

$$\approx 0.4928$$

The probability of a body temperature below the mean is 0.4928.

57.   Normal distribution,
$\mu = 2.2$ g, $\sigma = 0.4$ g, $X = $ tension

$$P(X < 1.9)$$

$$= P\left(\frac{x - 2.2}{0.4} < \frac{1.9 - 2.2}{0.4}\right)$$

$$= P(z < -0.75)$$

$$\approx 0.2266.$$

59. For an exponential distribution,

$$f(x) = ae^{-ax} \text{ for } x \text{ in } [0, \infty).$$

Since $\mu = \dfrac{1}{a} = 32.5$, $a = \dfrac{1}{32.5}$.

(a) $P(X \geq 40) = \displaystyle\int_{40}^{\infty} \dfrac{1}{32.5} e^{-x/32.5} \, dx$

$$= \lim_{b \to \infty} \int_{40}^{b} \dfrac{1}{32.5} e^{-x/32.5} \, dx$$

$$= \lim_{b \to \infty} \left(\dfrac{1}{32.5} e^{-x/32.5} \Big|_{40}^{b} \right)$$

$$= \lim_{b \to \infty} \left(-e^{-b/32.5} + e^{-40/32.5} \right)$$

$$= e^{-40/32.5}$$

$$\approx 0.2921$$

(b) $P(30 \leq X < 50)$

$$= \int_{30}^{50} \dfrac{1}{32.5} e^{-x/32.5} \, dx$$

$$= \left(-e^{-x/32.5} \right) \Big|_{30}^{50}$$

$$= -e^{-50/32.5} + e^{-30/32.5}$$

$$\approx 0.1826$$

61. $f(t) = \dfrac{1}{3650.1} e^{-t/3650.1}$

This is an exponential distribution with

$$a = \dfrac{1}{3650.1}.$$

So the expected value is $\mu = \dfrac{1}{a} = 3650.1$ days.

The standard deviation is $\sigma = \dfrac{1}{a} = 3650.1$ days.

SEQUENCES AND SERIES

12.1 Geometric Sequences

Your Turn 1

If $= 1$, $a_n = 3(1) - 6 = 3 - 6 = -3$.

If $= 2$, $a_n = 3(2) - 6 = 6 - 6 = 0$.

If $= 3$, $a_n = 3(3) - 6 = 9 - 6 = 3$.

If $= 4$, $a_n = 3(4) - 6 = 12 - 6 = 6$.

Your Turn 2

Verify that $r = -3$.

$$-\frac{6}{2} = \frac{18}{-6} = \frac{-54}{18} = -3$$

$$a_7 = 2(-3)^{7-1}$$
$$= 2(-3)^6$$
$$= 2(729)$$
$$= 1458$$

Your Turn 3

Since the machine loses 10% of its value each year, at the end of each year its value is 0.9 times its value at the end of the year before. Thus its values form a geometric sequence with $r = 0.9$ and $a_1 = 10,000$. Its value at the end of 10 years is its value at $n = 11$.

$$a_{11} = a_0 r^{11-1}$$
$$= 10,000(0.9)^{10}$$
$$= 3486.78$$

The value of the machine at the end of its tenth year is $3486.78.

Your Turn 4

Since $-8/2 = -4$, this is a geometric sequence with $a_1 = 2$ and $r = -4$.

$$S_7 = \frac{2[(-4)^7 - 1]}{-4 - 1}$$
$$= \frac{2(-16,384 - 1)}{-5}$$
$$= 6554$$

Your Turn 5

Here $a = 81$ and $r = 1/3$. The index of summation i runs from 0 to 4, so $n = 5$.

$$\sum_{i=0}^{4} 81\left(\frac{1}{3}\right)^i = S_5$$

$$= \frac{81\left[\left(\frac{1}{3}\right)^5 - 1\right]}{\frac{1}{3} - 1}$$

$$= \frac{81\left(-\frac{242}{243}\right)}{-\frac{2}{3}}$$

$$= \frac{-\frac{242}{243}}{-\frac{2}{3}} = 121$$

12.1 Exercises

1. $a_1 = 2$, $r = 3$, $n = 4$

Since $n = 4$, we must find a_1, a_2, a_3, and a_4 with $a = a_1 = 2$.

$$a_1 = 2$$
$$a_2 = 2(3)^{2-1} = 2(3)^1 = 2(3) = 6$$
$$a_3 = 2(3)^{3-1} = 2(3)^2 = 2(9) = 18$$
$$a_4 = 2(3)^{4-1} = 2(3)^3 = 2(27) = 54$$

The first four terms of this geometric sequence are 2, 6, 18, and 54.

3. $a_1 = \frac{1}{2}$, $r = 4$, $n = 4$

Since $n = 4$, we must find a_1, a_2, a_3, and a_4, with $a = a_1 = \frac{1}{2}$.

$$a_1 = \frac{1}{2}$$
$$a_2 = \frac{1}{2}(4)^{2-1} = \frac{1}{2}(4)^1 = \frac{1}{2}(4) = 2$$
$$a_3 = \frac{1}{2}(4)^{3-1} = \frac{1}{2}(4)^2 = \frac{1}{2}(16) = 8$$
$$a_4 = \frac{1}{2}(4)^{4-1} = \frac{1}{2}(4)^3 = \frac{1}{2}(64) = 32$$

The first four terms of this geometric sequence are $\frac{1}{2}$, 2, 8, and 32.

5. $a_3 = 6$, $a_4 = 12$, $n = 5$

Since $n = 5$, we must find a_1, a_2, a_3, a_4 and a_5

with $r = \frac{a_4}{a_3} = \frac{12}{6} = 2$. To find a, use $a_3 = 6$,

$r = 2$, and $n = 3$ in the formula.

$$6 = a(2)^{3-1}$$
$$6 = a(2)^2$$
$$6 = 4a$$
$$\frac{3}{2} = \frac{3}{2} = a$$

$a_1 = \dfrac{3}{2}$

$a_2 = \dfrac{3}{2}(2)^{2-1} = \dfrac{3}{2}(2)^1 = \dfrac{3}{2}(2) = 3$

$a_3 = \dfrac{3}{2}(2)^{3-1} = \dfrac{3}{2}(2)^2 = \dfrac{3}{2}(4) = 6$

$a_4 = \dfrac{3}{2}(2)^{4-1} = \dfrac{3}{2}(2)^3 = \dfrac{3}{2}(8) = 12$

$a_5 = \dfrac{3}{2}(2)^{5-1} = \dfrac{3}{2}(2)^4 = \dfrac{3}{2}(16) = 24$

The first five terms of this geometric sequence are $\frac{3}{2}$, 3, 6, 12, and 24.

7. $a_1 = 4$, $r = 3$

Since we want a_5, use $n = 5$ in the formula with $a = a_1 = 4$ and $r = 3$.

$$a_5 = 4(3)^{5-1} = 4(3)^4 = 4(81) = 324$$
$$a_n = 4(3)^{n-1}$$

9. $a_1 = -3$, $r = -5$

Since we want a_5, use $n = 5$ in the formula with $a = a_1 = -3$ and $r = -5$.

$$a_5 = -3(-5)^{5-1} = -3(-5)^4$$
$$= -3(625) = -1875$$
$$a_n = -3(-5)^{n-1}$$

11. $a_2 = 12$, $r = \dfrac{1}{2}$

To find a, use $a_2 = 12$, $r = \frac{1}{2}$, and $n = 2$ in the formula.

$$12 = a\left(\frac{1}{2}\right)^{2-1} = \frac{a}{2}$$
$$a = 24$$

Since we want a_5, use $n = 5$ in the formula with $a = 24$ and $r = \frac{1}{2}$.

$$a_5 = 24\left(\frac{1}{2}\right)^{5-1} = 24\left(\frac{1}{2}\right)^4 = 24\left(\frac{1}{16}\right) = \frac{3}{2}$$
$$a_n = 24\left(\frac{1}{2}\right)^{n-1} \quad \text{or} \quad \frac{24}{2^{n-1}}$$

13. $a_4 = 64$, $r = -4$

To find a, use $a_4 = 64$, $r = -4$, and $n = 4$ in the formula.

$$64 = a(-4)^{4-1}$$
$$64 = a(-4)^3$$
$$64 = a(-64)$$
$$-1 = a$$

Since we want a_5, use $n = 5$ in the formula with $a = -1$ and $r = -4$.

$$a_5 = -1(-4)^{5-1} = -1(-4)^4$$
$$= -1(256) = -256$$
$$a_n = -(-4)^{n-1}$$

15. $6, 12, 24, 48, \dots$

$$r = \frac{12}{6} = \frac{24}{12} = \frac{48}{24} = 2$$

Since $r = 2$ and $a = a_1 = 6$, $a_n = 6(2)^{n-1}$.

17. $\dfrac{3}{4}, \dfrac{3}{2}, 3, 6, 12, \dots$

$$r = \frac{\frac{3}{2}}{\frac{3}{4}} = \frac{3}{\frac{3}{2}} = \frac{6}{3} = \frac{12}{6} = 2$$

Since $r = 2$ and $a = a_1 = \frac{3}{4}$, $a_n = \frac{3}{4}(2)^{n-1}$.

19. $4, 8, -16, 32, 64, -128, \dots$

Since $\frac{8}{4} = 2$ and $\frac{-16}{8} = -2$, the ratio is not constant, so the sequence is not geometric.

21. $-\dfrac{5}{8}, \dfrac{5}{12}, -\dfrac{5}{18}, \dfrac{5}{27}, \dots$

$$r = \dfrac{\frac{5}{12}}{-\frac{5}{8}} = \dfrac{-\frac{5}{18}}{\frac{5}{12}} = \dfrac{\frac{5}{27}}{-\frac{5}{18}} = -\dfrac{2}{3}$$

Since $r = -\dfrac{2}{3}$ and $a = a_1 = -\dfrac{5}{8}$,

$$a_n = -\dfrac{5}{8}\left(-\dfrac{2}{3}\right)^{n-1}.$$

23. $3, 6, 12, 24, \dots$

Since $a = a_1 = 3$ and $r = \dfrac{6}{3} = 2$,

$$S_5 = \dfrac{3(2^5 - 1)}{2 - 1}$$
$$= \dfrac{3(32 - 1)}{1}$$
$$= 93.$$

The sum of the first five terms of this geometric sequence is 93.

25. $12, -6, 3, -\dfrac{3}{2}, \dots$

Since $a = a_1 = 12$ and $r = \dfrac{-6}{12} = -\dfrac{1}{2}$,

$$S_5 = \dfrac{12\left[\left(-\frac{1}{2}\right)^5 - 1\right]}{\left(-\frac{1}{2}\right) - 1}$$
$$= \dfrac{12\left[-\frac{1}{32} - 1\right]}{-\frac{3}{2}}$$
$$= \dfrac{33}{4}.$$

The sum of the first five terms of this geometric sequence is $\dfrac{33}{4}$.

27. $a_1 = 3, r = -2$

Since $a = a_1 = 3$,

$$S_5 = \dfrac{3[(-2)^5 - 1]}{-2 - 1} = \dfrac{3(-32 - 1)}{-3}.$$

The sum of the first five terms of this geometric sequence is 33.

29. $a_1 = 6.324, r = 2.598$

Since $a = a_1 = 6.324$,

$$S_5 = \dfrac{6.324(2.598^5 - 1)}{2.598 - 1}$$
$$\approx \dfrac{6.324(118.3575 - 1)}{1.598}$$
$$\approx 464.4.$$

The sum of the first five terms of this sequence is about 464.4.

31. For $\displaystyle\sum_{i=0}^{7} 8(2)^i$, use the formula with $a = 8$, $r = 2$, and $n = 8$.

$$S_8 = \dfrac{8(2^8 - 1)}{2 - 1}$$
$$= \dfrac{8(256 - 1)}{1}$$
$$= 2040$$

33. For $\displaystyle\sum_{i=0}^{8} \dfrac{3}{2}(4)^i$, use the formula with $a = \dfrac{3}{2}$, $r = 4$, and $n = 9$.

$$S_9 = \dfrac{\frac{3}{2}(4^9 - 1)}{4 - 1}$$
$$= \dfrac{\frac{3}{2}(262{,}144 - 1)}{3}$$
$$= \dfrac{262{,}143}{2}$$

35. For $\displaystyle\sum_{i=0}^{4} \dfrac{3}{4}(-3)^i$, use the formula with $a = \dfrac{3}{4}$, $r = -3$, and $n = 5$.

$$S_5 = \dfrac{\frac{3}{4}[(-3)^5 - 1]}{-3 - 1}$$
$$= \dfrac{\frac{3}{4}(-243 - 1)}{-4}$$
$$= \dfrac{183}{4}$$

37. For $\displaystyle\sum_{i=0}^{8} 64\left(\dfrac{1}{2}\right)^i$, use the formula with $n = 9$, $r = \dfrac{1}{2}$, and $a = 64$.

$$S_9 = \frac{64\left[\left(\frac{1}{2}\right)^9 - 1\right]}{\frac{1}{2} - 1}$$

$$= \frac{64\left(\frac{1}{512} - 1\right)}{-\frac{1}{2}}$$

$$= \frac{511}{4}$$

Therefore, $\displaystyle\sum_{i=0}^{8} 64\left(\frac{1}{2}\right)^i = \frac{511}{4}$.

39. (a) A machine that loses 20% of its value maintains 80% of its value. If the initial value of the machine is \$12,000, then its value at the end of the first year will be 80% of \$12,000, and its value at the end of each subsequent year will be 80% of its value at the end of the previous year. This means that the end-of-year values form a geometric sequence with $r = 0.80$. If we let $a_1 = 12,000$, then a_2 will represent the value at the end of the first year, a_3 will represent the value at the end of the second year, and so on. Thus, to find the value of the machine at the end of the fifth year, we are looking for a_6 in the geometric sequence with $n = 6$, $r = 0.80$ and $a = a_1 = 12,000$.

$$a_6 = 12,000(0.80)^{6-1} = 12,000(0.80)^5$$
$$= 12,000(0.32768) = 3932.16$$

Therefore, the value of the machine at the end of the fifth year is about \$3932.

(b) To find the value of the machine at the end of the eighth year, we are looking for a_9 in the geometric sequence.

$$a_9 = 12,000(0.80)^{9-1} = 12,000(0.80)^8$$
$$= 12,000(0.16777216) \approx 2013.27$$

Therefore, the value of the machine at the end of the eighth year is about \$2013.

41. The amounts saved form a geometric sequence with $a_1 = 1$ and $r = 2$. To determine the amount saved on January 31, we want to use the formula to find a_n with $n = 31$, $r = 2$, and $a = a_1 = 1$.

$$a_{31} = 1(2)^{31-1} = 2^{30}$$
$$= 1,073,741,824$$

To determine the total amount saved during January, we want to use the formula to find S_n with $n = 31$, $r = 2$, and $a = a_1 = 1$.

$$S_{31} = \frac{1(2^{31} - 1)}{2 - 1} = \frac{2,147,483,648 - 1}{1}$$
$$= 2,147,483,647$$

The savings on January 31 would be \$2^{30}$ or \$1,073,741,824 and for the month of January would be \$2^{31} - \$1$ or \$2,147,483,647.

43. If we let a_1 represent the initial bacteria population, then the population at the end of the first hour will be $a_1 + 5\%(a_1)$, or $105\%(a_1)$. The populations form a geometric sequence with $r = 105\% = 1.05$. To determine the percent increase in the population after 7 hours, we want to use the formula to find a_n with $n = 8$, $r = 1.05$, and $a = a_1$.

$$a_6 = a(1.05)^{8-1} \approx 1.40710a$$

Since $a = a_1$, $a_8 \approx 1.40710a_1$. This means that the population at the end of 7 hours is about 141% of the initial population.

After five hours, the population has increased by about 41%.

45. The number of rotations form a geometric sequence with $r = \frac{3}{4}$ and $a_1 = 400$. If we let $a_1 = 400$, then a_2 represents the number of rotations at the end of the first minute, a_3 is the number of rotations at the end of the second minute, and so on. To determine the number of rotations in the fifth minute, we want a_6. Use the formula to find a_n with $n = 6$, $r = \frac{3}{4}$, and $a = a_1 = 400$.

$$a_6 = 400\left(\frac{3}{4}\right)^{6-1} = 400\left(\frac{3}{4}\right)^5$$
$$= 400\left(\frac{243}{1024}\right) = 94.921875$$

In the fifth minute after the rider's feet are removed from the pedals, the wheel will rotate about 95 times.

47. (a) Initially, there are 64 teams that play $a_1 = 32$ games in round one. After each round, the number of teams (and the number of games) decreases by $\frac{1}{2}$. At the end of the first round, there are 32 teams left to play $a_2 = 16$ games in round two. At the end of the second round, there will be 16 teams left to play $a_3 = 8$ games in round 3, and so on.

Note that $a_1 = 2^5$, $a_2 = 2^4$, $a_3 = 2^3$, so that by the sixth round there will be 2 teams left playing $a_6 = 2^0 = 1$ game to decide the champion. Therefore, to determine the total number of games, we need to sum the terms of the geometric sequence

$a_1 = 2^5$, $a_2 = 2^4$, $a_3 = 2^3$, $a^4 = 2^2$,
$a_5 = 2^1$, $a_6 = 1$ or $1 + 2 + 2^2 + 2^3$
$+ 2^4 + 2^5$.

(b) To find the sum from part (a), use the formula to find S_n with $n = 6, r = \frac{1}{2}$, and $a = a_1 = 2^5$.

$$S_6 = \frac{2^5\left[\left(\frac{1}{2}\right)^6 - 1\right]}{\frac{1}{2} - 1}$$

$$= \frac{\left(\frac{1}{2} - 32\right)}{-\frac{1}{2}} = 63,$$

so 63 games must be played to produce the champion.

(c) To generalize the methods of (a) and (b) for 2^n teams rather than $64 = 2^6$ teams, first note that 2^n teams play 2^{n-1} games in the initial round. In part (a), note that it required 6 rounds to determine a champion when 2^6 teams are initially involved. Similarly, for 2^n teams it will require n rounds. So, the total number of games is given by

$$S_n = \frac{a(r^n - 1)}{r - 1}$$

with $a = a_1 = 2^{n-1}$, and $r = \frac{1}{2}$, which is the desired sum

$$2^{n-1} + 2^{n-2} + \cdots + 2^2 + 2^1 + 2^0.$$

$$S_n = \frac{2^{n-1}\left[\left(\frac{1}{2}\right)^n - 1\right]}{\frac{1}{2} - 1} = \frac{\frac{1}{2} - 2^{n-1}}{-\frac{1}{2}}$$

$$= -2\left(\frac{1}{2} - 2^{n-1}\right)$$

$$= -1 + 2^n$$

$$= 2^n - 1.$$

12.2 Annuities: An Application of Sequences

Your Turn 1

This is an ordinary annuity with $R = 22,000$, $n = 7$, and $i = 0.05$.

$$S = 22,000\left[\frac{(1.05)^7 - 1}{0.05}\right]$$

$$= (22,000)(8.142008453125)$$

$$\approx 179,124.19$$

Erin will have $179,124.19 on deposit.

Your Turn 2

Since the interest is compounded monthly for 5 years, the number of periods n is $(5)(12) = 60$. The annual interest is 2%, so the interest per monthly compounding period is $0.02/12$. The initial deposit is $125.

$$S = 125\left[\frac{\left(1 + \frac{0.02}{12}\right)^{60} - 1}{\frac{0.02}{12}}\right]$$

$$\approx (125)(63.047356)$$

$$\approx 7880.92$$

The amount of the annuity is $7880.92.

Your Turn 3

The only difference from Example 3 is that the interest rate per period is now $0.025/4$. There are still $(4)(3) = 12$ compounding periods.

$$2400 = R\left[\frac{\left(1 + \frac{0.025}{4}\right)^{12} - 1}{\frac{0.025}{4}}\right]$$

$$2400 \approx R(12.421246)$$

$$R \approx \frac{2400}{12.421246} \approx 193.22$$

Melissa must make twelve deposits of $193.22.

Your Turn 4

We need to find the present value of an annuity of payments of $2500 made at the end of each year for 12 years, with interest at 4.5% compounded annually.

$$P = R\left[\frac{1 - (1 + i)^{-n}}{i}\right]$$

$$= 2500\left[\frac{1 - (1.045)^{-12}}{0.045}\right]$$

$$\approx (2500)(9.118581)$$

$$\approx 22{,}796.45$$

The amount accumulated at the end of 12 years will be $22,796.45.

Your Turn 5

The present value is $11,000, the interest rate is 0.06/12 per period, or 0.005, and there will be 48 payments.

$$P = R\left[\frac{1 - (1 + i)^{-n}}{i}\right]$$

$$11{,}000 = R\left[\frac{1 - (1 + 0.005)^{-48}}{0.005}\right]$$

$$11{,}000 \approx R(42.580318)$$

$$R \approx \frac{11{,}000}{42.580318} \approx 258.34$$

Each payment will be $258.34.

Your Turn 6

The present value is $256{,}000 - \$32{,}000 = \$224{,}000$. The monthly interest rate is $0.049/12$ and there are $(12)(30) = 360$ periods.

$$P = R\left[\frac{1 - (1 + i)^{-n}}{i}\right]$$

$$224{,}000 = R\left[\frac{1 - \left(1 + \frac{0.049}{12}\right)^{-360}}{\frac{0.049}{12}}\right]$$

$$224{,}000 \approx R(188.420888)$$

$$R \approx \frac{224{,}000}{188.420888}$$

$$\approx 1188.83$$

Each monthly payment will be $1188.83.

12.2 Exercises

1. $R = 120, i = 0.05, n = 10$

$$S = R \cdot s_{\overline{n}|i}$$

$$= 120 \cdot s_{\overline{10}|0.05}$$

$$= 120\left[\frac{(1.05)^{10} - 1}{0.05}\right]$$

$$\approx 120(12.57789254)$$

$$\approx 1509.35$$

The amount is $1509.35.

3. $R = 9000, i = 0.06, n = 18$

$$S = 9000 \cdot s_{\overline{18}|0.06}$$

$$= 9000\left[\frac{(1.06)^{18} - 1}{0.06}\right]$$

$$\approx 9000(30.90565255)$$

$$\approx 278{,}150.87$$

The amount is $278,150.87.

5. $R = 11{,}500, i = 0.055, n = 30$

$$S = 11{,}500 \cdot s_{\overline{30}|0.055}$$

$$= 11{,}500\left[\frac{(1.055)^{30} - 1}{0.055}\right]$$

$$\approx 11{,}500(72.43547797)$$

$$\approx 833{,}008.00.$$

The amount is $833,008.00

7. $R = 10{,}500$

Interest is earned semiannually for 7 years, so $i = \frac{0.10}{2} = 0.05$ and $n = 2 \cdot 7 = 14$ periods.

$$S = 10{,}500\left[\frac{(1.05)^{14} - 1}{0.05}\right]$$

$$\approx 10{,}500(19.59863199)$$

$$\approx 205{,}785.6359$$

The amount is $205,785.64.

9. $R = 1800$

Interest is earned quarterly for 12 years, so $i = \frac{0.08}{4} = 0.02$ and $n = 4 \cdot 12 = 48$ periods.

$$S = 1800\left[\frac{(1.02)^{48} - 1}{0.02}\right]$$

$$\approx 1800(79.35351927)$$

$$\approx 142,836.33$$

The amount is $142,836.33.

11. This describes an ordinary annuity with
$S = 10,000$, $i = 0.08$, and $n = 12$ periods.

$$10,000 = R \cdot s_{\overline{12}|0.08}$$

$$10,000 = R(18.97712646)$$

$$R \approx 526.95$$

The periodic payment should be $526.95.

13. This describes an ordinary annuity with
$S = 50,000$, $i = 0.03\left(= \frac{12\%}{4}\right)$, and

$n = 8 \cdot 4 = 32$ periods.

$$50,000 = R \cdot s_{\overline{32}|0.03}$$

$$50,000 \approx R(52.50275852)$$

$$R \approx 952.33$$

The periodic payment should be $952.33.

15. $R = 5000$, $i = 0.06$, and $n = 11$ payments.

$$a_{\overline{11}|0.06} = \left[\frac{1 - (1.06)^{-11}}{0.06}\right] = 7.886874577,$$

so

$$P \approx 5000(7.886874577) \approx 39,434.37,$$

or $39,434.37.

17. $R = 1400$, $i = \frac{0.06}{2} = 0.03$, and

$n = 2 \cdot 8 = 16$ periods.

$$P = 1400\left[\frac{1 - (1.03)^{-16}}{0.03}\right]$$

$$\approx 1400(12.56110203)$$

$$\approx 17,585.54$$

The present value is $17,585.54.

19. $R = 50,000$, $i = \frac{0.08}{4} = 0.02$, and

$n = 10 \cdot 4 = 40$ payments.

$$a_{\overline{40}|0.02} = \left[\frac{1 - (1.02)^{-40}}{0.02}\right]$$

$$\approx 27.35547924,$$

so

$$P \approx 50,000(27.35547924)$$

$$\approx 1,367,773.96,$$

or $1,367,773.96.

21. $a_{\overline{15}|0.04} = \left[\frac{1 - (1.04)^{-15}}{0.04}\right] \approx 11.11838743$, so

$P \approx 10,000(11.11838743) \approx 111,183.87$, or
$111,183.87. A lump sum deposit of $111,183.87
today at 4% compounded annually will yield the
same total after 15 years as deposits of $10,000
at the end of each year for 15 years at 4%
compounded annually.

23. $a_{\overline{15}|0.06} = \left[\frac{1 - (1.06)^{-15}}{0.06}\right] \approx 9.712248988$, so

$P \approx 10,000(9.712248988) \approx 97,122.49$, or
$97,122.49. A lump sum deposit of $97,122.49
today at 6% compounded annually will yield the
same total after 15 years as deposits of $10,000
at the end of each year for 15 years at 6%
compounded annually.

25. $2500 is the present value of this annuity of R
dollars, with 6 periods, and $i = \frac{16\%}{4} = 4\%$

$= 0.04$ per period.

$$P = R \cdot a_{\overline{n}|i}$$

$$2500 = R \cdot a_{\overline{6}|0.04}$$

$$R = \frac{2500}{a_{\overline{6}|0.04}}$$

$$\approx \frac{2500}{5.242136857}$$

$$\approx 476.90$$

Each payment is $476.90.

27. $90,000 is the present value of this annuity of R
dollars, with 12 periods, and $i = 8\% = 0.08$
per period.

$$P = R \cdot a_{\overline{n}|i}$$

$$90,000 = R \cdot a_{\overline{12}|0.08}$$

$$R = \frac{90,000}{a_{\overline{12}|0.08}}$$

$$\approx \frac{90,000}{7.536078017}$$

$$\approx 11,942.55$$

Each payment is $11,942.55.

29. $55,000 is the present value of this annuity of R
dollars, with $i = \frac{0.06}{12} = 0.005$, and $n = 36$
periods.

$$P = R \cdot a_{\overline{n}|i}$$

$$R = \frac{55,000}{a_{\overline{36}|0.005}}$$

$$\approx \frac{55,000}{32.87101624}$$

$$\approx 1673.21$$

Each payment is \$1673.21.

31. **(a)** Sarah's payments form an ordinary annuity with $R = 12,000$, $n = 9$, and $i = 0.05$. The amount of this annuity is

$$S = 12,000s_{\overline{9}|0.05}$$

$$= 12,000\left[\frac{(1.05)^9 - 1}{0.05}\right]$$

$$\approx (12,000)(11.026564)$$

$$\approx 132,318.77,$$

or \$132,318.77.

(b) Using her brother-in-law's bank gives her an annuity with $R = 12,000$, $n = 9$, and $i = 0.03$. The amount of this annuity is

$$S = 12,000s_{\overline{9}|0.03}$$

$$S = 12,000\left[\frac{(1.03)^9 - 1}{0.03}\right]$$

$$\approx (12,000)(10.159106)$$

$$\approx 121,909.27,$$

or \$121,909.27.

(c) The amount Sarah would lose using the second option is $132,318.77 - 121,909.27 = 10,409.50$, or \$10,409.50.

33. **(a)** Steve's payments will form an ordinary annuity with final amount 20,000, R unknown, $n = (4)(8) = 32$, and $i = 0.06/4 = 0.0015$.

$$20,000 = Rs_{\overline{32}|0.015}$$

$$= R\left[\frac{(1.05)^{32} - 1}{0.015}\right]$$

$$\approx R(40.688288)$$

$$R \approx \frac{20,000}{40.688288}$$

$$\approx 491.54$$

The required payments at the end of each quarter are \$491.54.

(b) In this scenario, $i = 0.04/4 = 0.01$.

$$20,000 = Rs_{\overline{32}|0.01}$$

$$= R\left[\frac{(1.01)^{32} - 1}{0.01}\right]$$

$$\approx R(37.494068)$$

$$R \approx \frac{20,000}{37.494068} \approx 533.42$$

The required payments at the end of each quarter are \$533.42.

35. Harv's payments will form an ordinary annuity with final amount 12,000, R unknown, $n = (2)(4) = 8$, and $i = 0.04/2 = 0.02$.

$$12,000 = Rs_{\overline{8}|0.02}$$

$$= R\left[\frac{(1.02)^8 - 1}{0.02}\right]$$

$$\approx R(8.582969)$$

$$R \approx \frac{12,000}{8.582969} \approx 1398.12$$

The required payments at the end of each quarter are \$1398.12.

37. Interest of $\frac{6\%}{2} = 3\%$ is earned semiannually. In $65 - 40 = 25$ years, there are $25 \cdot 2 = 50$ semiannual periods. Since

$$s_{\overline{50}|0.03} = \left[\frac{(1.03)^{50} - 1}{0.03}\right]$$

$$\approx 112.7968673,$$

the \$1000 semiannual deposits will produce a total of

$$S = 1000(112.7968673)$$

$$\approx 112,796.87,$$

or \$112,796.87.

39. Interest of $\frac{10\%}{2} = 5\%$ is earned semiannually. In $65 - 40 = 25$ years, there are $25 \cdot 2 = 50$ semiannual periods. Since

$$s_{\overline{50}|0.05} = \left[\frac{(1.05)^{50} - 1}{0.05}\right] \approx 209.3479957,$$

the \$1000 semiannual deposits will produce a total of $S \approx 1000(209.3479957) \approx 209,348.00$, or \$209,348.00.

41. **(a)** The total amount of interest paid is
$(60,000)(0.08)(7) = 33,600$, or $33,600.
This total is divided into $7 \cdot 4 = 28$ equal
quarterly interest payments. Since
$\frac{33,600}{28} = 1200$, each quarterly interest
payment will be $1200.

(b) This ordinary annuity will amount to
$60.000 in 7 years at 6% compounded
semiannually. Thus, $S = 60,000$,

$n = 7 \cdot 2 = 14$, and $i = \frac{6\%}{2} = 3\%$

$= 0.03$, so

$60,000 = R \cdot s_{\overline{14}|0.03}$

$R = \dfrac{60,000}{s_{\overline{14}|0.03}}$ or $3511.58.

$\approx \dfrac{60,000}{17.08632416} \approx 3511.58,$

(c)

Payment Number	Amount of Deposit	Interest Earned	Total in Account
1	$3511.58	$0	$3511.58
2	$3511.58	$(3511.58)(0.03) = \$105.35$	$3511.58 + 3511.58 + 105.35 = \$7128.51$
3	$3511.58	$(7128.51)(0.03) = \$213.86$	$7128.51 + 3511.58 + 213.86 = \$10,853.95$
4	$3511.58	$(10,853.95)(0.03) = \$325.62$	$10,853.95 + 3511.58 + 325.62 = \$14.691.15$
5	$3511.58	$(14,691.15)(0.03) = \$440.73$	$14,691.15 + 3511.58 + 440.73 = \$18,643.46$
6	$3511.58	$(18,643.46)(0.03) = \$559.30$	$18,643.46 + 3511.58 + 559.30 = \$22,714.34$
7	$3511.58	$(22,714.34)(0.03) = \$681.43$	$22,714.34 + 3511.58 + 681.43 = \$26,907.35$
8	$3511.58	$(26,907.35)(0.03) = \$807.22$	$26,907.35 + 3511.58 + 807.22 = \$31,226.15$
9	$3511.58	$(31,226.15)(0.03) = \$936.78$	$31,226.15 + 3511.58 + 936.78 = \$35,674.51$
10	$3511.58	$(35,674.51)(0.030) = \$1070.24$	$35,674.51 + 3511.58 + 1070.24 = \$40,256.33$
11	$3511.58	$(40,256.33)(0.03) = \$1207.69$	$40,256.33 + 3511.58 + 1207.69 = \$44,975.60$
12	$3511.58	$(44,975.60)(0.03) = \$1349.27$	$44,975.60 + 3511.58 + 1349.27 = \$49,836.45$
13	$3511.58	$(49,836.45)(0.03) = \$1495.09$	$49,836.45 + 3511.58 + 1495.09 = \$54,843.12$
14	$3511.58	$(54,843.12)(0.03) = \$1645.29$	$54,843.12 + 3511.58 + 1645.29 = \$59,999.99$

So that the final total in the account is $60,000, add $0.01 to the last amount of deposit.
Thus, line 14 of the table will be:

14    $3511.59 $(54,843.12)(0.03) = \$1645.29$ $54,843.12 + 3511.59 + 1645.29 = \$60,000.00$

43. $150,000 is the future value of an annuity over 79 yr compounded quarterly. So, there are $79(4) = 316$ payment periods.

(a) The interest per quarter is $\frac{5.25\%}{4} = 1.3125\%$.

Thus, $S = 150,000$, $n = 316$, $i = 0.013125$, and we must find the quarterly payment R in the formula

$$S = R\left[\frac{(1+i)^n - 1}{i}\right]$$

$$150,000 = R\left[\frac{(1.013125)^{316} - 1}{0.013125}\right]$$

$$R \approx 32.4923796$$

She would have to put $32.49 into her savings at the end of every three months.

(b) For a 2% interest rate, the interest per quarter is $\frac{2\%}{4} = 0.5\%$. Thus, $S = 150,000$, $n = 316$, $i = 0.005$, and we must find the quarterly payment R in the formula

$$S = R\left[\frac{(1+i)^n - 1}{i}\right]$$

$$150,000 = R\left[\frac{(1.005)^{316} - 1}{0.005}\right]$$

$$R \approx 195.5222794$$

She would have to put $195.52 into her savings at the end of every three months. For a 7% interest rate, the interest per quarter is $\frac{7\%}{4} = 1.75\%$. Thus, $S = 150,000$, $n = 316$, $i = 0.0175$, and we must find the quarterly payment R in the formula

$$S = R\left[\frac{(1+i)^n - 1}{i}\right]$$

$$150,000 = R\left[\frac{(1.0175)^{316} - 1}{0.0175}\right]$$

$$R \approx 10.9663932$$

She would have to put $10.97 into her savings at the end of every three months.

45. We want to find the present value of an annuity of $2000 per year for 9 years at 8% compounded annually.

$$a_{\overline{9}|0.08} = \left[\frac{1 - (1.08)^{-9}}{0.08}\right] \approx 6.246887911, \text{ so}$$

$$P \approx 2000\,(6.246887911) \approx 12,493.78,$$

or $12,493.78. A lump sum deposit of $12,493.78 today at 8% compounded annually will yield the same total after 9 years deposits of $2000 at the end of each year for 9 years at 8% compounded annually.

47. For parts (a) and (b), if $1 million is divided into 20 equal payments, each payment is $50,000.

(a) $i = 0.05, n = 20$

$$P = R\left[\frac{1 - (1+i)^{-n}}{i}\right]$$

$$= 50,000\left[\frac{1 - (1+0.05)^{-20}}{0.05}\right]$$

$$\approx 623,110.52$$

The present value is $623,110.52.

(b) $i = 0.09, n = 20$

$$P = R\left[\frac{1 - (1+i)^{-n}}{i}\right]$$

$$= 50,000\left[\frac{1 - (1+0.09)^{-20}}{0.09}\right]$$

$$\approx 456,427.28$$

The present value is $456,427.28.
For parts (c) and (d), if $1 million is divided into 25 equal payments, each payment is $40,000.

(c) $i = 0.05, n = 25$

$$P = R\left[\frac{1 - (1+i)^{-n}}{i}\right]$$

$$= 40,000\left[\frac{1 - (1+0.05)^{-25}}{0.05}\right]$$

$$\approx 563,757.78$$

The present value is $563,7573.78.

(d) $i = 0.09, n = 25$

$$P = R\left[\frac{1 - (1+i)^{-n}}{i}\right]$$

$$= 40,000\left[\frac{1 - (1+0.09)^{-25}}{0.09}\right]$$

$$\approx 392,903.18$$

The present value is $392,903.18.

49. The present value, P, is 249,560, $i = \frac{0.0775}{12} \approx 0.0064583333$, and $n = 12 \cdot 25 = 300$.

$$249{,}560 = R \cdot a_{\overline{300}|0.0064583333} = R\left[\frac{1 - (1 + 0.0064583333)^{-300}}{0.0064583333}\right] \approx R\left[\frac{1 - 0.1449639356}{0.0064583333}\right] \approx R\left[\frac{0.8550360644}{0.0064583333}\right]$$

or $R \approx 1885.00$.

Monthly payments of $1885.00 will be required to amortize the loan.

Use the formula for the unpaid balance of a loan with $R = 1885$, $i \approx 0.0064583333$, $n = 12 \cdot 15 = 300$, and $x = 12 \cdot 5 = 60$.

$$y = R\left[\frac{1 - (1 + i)^{-(n-x)}}{i}\right] \approx 1885\left[\frac{1 - (1 + 0.0064583333)^{-240}}{0.0064583333}\right] \approx 229{,}612.44$$

The unpaid balance after 5 years is $229,612.44.

51. The present value, P, is 353,700, $i = \frac{0.0795}{12} = 0.006625$, and $n = 12 \cdot 30 = 360$ periods.

$$P = R \cdot a_{\overline{n}|i}$$

$$R = \frac{353{,}700}{a_{\overline{360}|0.006625}} = 353{,}700\left[\frac{1 - (1.006625)^{-360}}{0.006625}\right] \approx \frac{353{,}700}{136.9334083} \approx 2583.01$$

Monthly payments of $2583.01 will be required to amortize the loan.

To find the unpaid balance, use the formula for the unpaid balance of a loan with $R = 2583.01$, $i = 0.006625$, $n = 360$, and $x = 12 \cdot 5 = 60$.

$$y = R \cdot \left[\frac{1 - (1 + i)^{-(n-x)}}{i}\right] = 2583.01\left[\frac{1 - (1 + 0.006625)^{-300}}{0.006625}\right] \approx 2583.01(130.1224488) \approx 336{,}107.59$$

The unpaid balance after 5 years is $336,107.59.

53. (a) $25,000 is the present value of this annuity of 8 years with interest of 6% per year.

$a_{\overline{8}|0.06} = 6.209793811$, so

$25{,}000 = R(6.209793811)$

$R \approx 4025.90$

Annual withdrawals of $4025.90 each will be needed.

(b) $25,000 is the present value of this annuity of 12 years with interest of 6% per year.

$a_{\overline{12}|0.06} = 8.38384394$, so

$25{,}000 = R(8.38384394)$

$R \approx 2981.93$

Annual withdrawals of $2981.93 each will be needed.

55.   First, find the amount of each payment. $4000 is the present value of an annuity of R dollars, with 4 periods, and
$i = 0.08$ per period.

$$P = R \cdot a_{\overline{n}|i}$$

$$4000 = R \cdot a_{\overline{4}|0.08}$$

$$R = \frac{4000}{a_{\overline{4}|0.08}} \approx \frac{4000}{3.31212684} \approx 1207.68$$

Each payment is $1207.68.

Payment Number	Amount of Payment	Interest for Period	Portion to Principal	Principal at End of Period
0	—	—	—	$4000.00
1	$1207.68	$(4000)(0.08)(1) = \$320.00$	$1207.68 - 320.00 = \$887.68$	$4000.00 - 887.68 = \$3112.32$
2	$1207.68	$(3112.32)(0.08)(1) = \$248.99$	$1207.68 - 248.99 = \$958.69$	$3112.32 - 958.69 = \$2153.63$
3	$1207.68	$(2153.63)(0.08)(1) = \$172.29$	$1207.68 - 172.29 = \$1035.39$	$2153.63 - 1035.39 = \$1118.24$
4	$1207.68	$(1118.24)(0.08)(1) = \$89.46$	$1207.68 - 89.46 = \$1118.22$	$1118.24 - 1118.22 = \$0.02$

So that the final principal is zero, add $0.02 to the last payment. Thus, line 4 of the table will be:

| 4 | $1207.70 | $(1118.24)(0.08)(1) = \$89.46$ | $1207.70 - 89.46 = \$1118.24$ | $1118.24 - 1118.24 = \$0$ |

57.   The firm's total cost is $8(1048) = \$8384.00$. After making a down payment of $1200, a balance of
$8384 - 1200 = \$7184$ is owed. To amortize this balance, first find the amount of each payment. $7184 is the
present value of an annuity of R dollars, with $4 \cdot 12 = 48$ periods, and $i = \frac{10.5\%}{12} = .875\% = 0.00875$
per period.

$$P = R \cdot a_{\overline{n}|i}$$

$$7184 = R \cdot a_{\overline{48}|0.00875}$$

$$R = \frac{7184}{a_{\overline{48}|0.00875}} \approx \frac{7184}{39.05734359} \approx 183.93$$

Each payment is $183.93.

Payment Number	Amount of Payment	Interest for Period	Portion to Principal	Principal at End of Period
0	—	—	—	$7184.00
1	$183.93	$(7184.00)(0.105)\left(\frac{1}{12}\right) = \$62.86$	$183.93 - 62.86 = \$121.07$	$7184.00 - 121.07 = \$7062.93$
2	$183.93	$(7062.93)(0.105)\left(\frac{1}{12}\right) = \$61.80$	$183.93 - 61.80 = \$122.13$	$7062.93 - 122.13 = \$6940.80$
3	$183.93	$(6940.80)(0.105)\left(\frac{1}{12}\right) = \$60.73$	$183.93 - 60.73 = \$123.20$	$6940.80 - 123.20 = \$6817.60$
4	$183.93	$(6817.60)(0.105)\left(\frac{1}{12}\right) = \$59.65$	$183.93 - 59.65 = \$124.28$	$6817.60 - 124.28 = \$6693.32$
5	$183.93	$(6693.32)(0.105)\left(\frac{1}{12}\right) = \$58.57$	$183.93 - 58.57 = \$125.36$	$6693.32 - 125.36 = \$6567.96$
6	$183.93	$(6567.96)(0.105)\left(\frac{1}{12}\right) = \$57.47$	$183.93 - 57.47 = \$126.46$	$6567.96 - 126.46 = \$6441.50$

12.3   Taylor Polynomials at 0

Your Turn 1

$$P_5(-0.15) = 1 + \frac{1}{1!}(-0.15) + \frac{1}{2!}(-0.15)^2 + \frac{1}{3!}(-0.15)^3 + \frac{1}{4!}(-0.15)^4 + \frac{1}{5!}(-0.15)^5$$
$$\approx 0.860708$$

Your Turn 2

The derivatives of $f(x) = \sqrt{x + 4}$ are exactly as shown in Example 2 except that $x + 1$ is replaced everywhere by $x + 4$.

Derivative	Value at 0
$f(x) = (x + 4)^{1/2}$	$f(0) = 2$
$f^{(1)}(x) = \dfrac{1}{2(x + 4)^{1/2}}$	$f^{(1)}(0) = \dfrac{1}{4}$
$f^{(2)}(x) = \dfrac{-1}{4(x + 4)^{3/2}}$	$f^{(2)}(0) = -\dfrac{1}{32}$
$f^{(3)}(x) = \dfrac{3}{8(x + 4)^{5/2}}$	$f^{(3)}(0) = \dfrac{3}{256}$
$f^{(4)}(x) = \dfrac{-15}{16(x + 4)^{7/2}}$	$f^{(4)}(0) = \dfrac{-15}{2048}$

$$P_4(x) = f(0) + \frac{f^{(1)}(0)}{1!} + \frac{f^{(2)}(0)}{2!} + \frac{f^{(3)}(0)}{3!} + \frac{f^{(4)}(0)}{4!}$$
$$= 2 + \frac{1/4}{1!} + \frac{-1/32}{2!} + \frac{3/256}{3!} + \frac{-15/2048}{4!}$$
$$= 2 + \frac{1}{4}x - \frac{1}{64}x^2 + \frac{1}{512}x^3 - \frac{5}{16,384}x^4$$

Your Turn 3

To approximate $\sqrt{4.05}$ we evaluate $f(0.05)$ with $f(x) = \sqrt{x + 4}$. Using the approximation from Your Turn 2,

$$\sqrt{4.05} = f(0.05)$$
$$\approx 2 + \frac{1}{4}(0.05) - \frac{1}{64}(0.05)^2 + \frac{1}{512}(0.05)^3 - \frac{5}{16,384}(0.05)^4$$
$$\approx 2.0124612.$$

12.3 Warmup Exercises

W1. $f(x) = \sqrt{2x + 5} = (2x + 5)^{1/2}$

$f'(x) = \dfrac{1}{2}(2x + 5)^{-1/2}(2) = (2x + 5)^{-1/2}$

$\qquad = \dfrac{1}{\sqrt{2x + 5}}$

$f''(x) = \left(-\dfrac{1}{2}\right)(2x + 5)^{-3/2}(2)$

$\qquad = -(2x + 5)^{-3/2}$

$f'''(x) = \left(-\dfrac{3}{2}\right)\left[-(2x + 5)^{-5/2}\right](2)$

$\qquad = 3(2x + 5)^{-5/2}$

W2. $f(x) = \dfrac{1}{x + 2} = (x + 2)^{-1}$

$f'(x) = -(x + 2)^{-2}$

$f''(x) = (-2)\left[-(x + 2)^{-3}\right] = 2(x + 2)^{-3}$

$f'''(x) = (-3)\left[2(x + 2)^{-4}\right] = -6(x + 2)^{-4}$

W3. $f(x) = e^{3x}$

$f'(x) = 3e^{3x}$

$f''(x) = 3\left(3^{3x}\right) = 9e^{3x}$

$f'''(x) = 3\left(9e^{3x}\right) = 27e^{3x}$

W4. $f(x) = \ln(1 + 2x)$

$f'(x) = (2)\dfrac{1}{1 + 2x} = \dfrac{2}{1 + 2x} = 2(1 + 2x)^{-1}$

$f''(x) = (-1)\left[2(1 + 2x)^{-2}\right](2)$

$\qquad = -4(1 + 2x)^{-2}$

$f'''(x) = (-2)\left[-4(1 + 2x)^{-3}\right](2)$

$\qquad = 16(1 + 2x)^{-3}$

12.3 Exercises

1.

Derivative	Value at 0
$f(x) = e^{-2x}$	$f(0) = 1$
$f^{(1)}(x) = -2e^{-2x}$	$f^{(1)}(0) = -2$
$f^{(2)}(x) = 4e^{-2x}$	$f^{(2)}(0) = 4$
$f^{(3)}(x) = -8e^{-2x}$	$f^{(3)}(0) = -8$
$f^{(4)}(x) = 16e^{-2x}$	$f^{(4)}(0) = 16$

$$P_4(x) = f(0) + \frac{f^{(1)}(0)}{1!}x + \frac{f^{(2)}(0)}{2!}x^2 + \frac{f^{(3)}(0)}{3!}x^3 + \frac{f^{(4)}(0)}{4!}x^4$$

$$= 1 + \frac{-2}{1!}x + \frac{4}{2!}x^2 + \frac{-8}{3!}x^3 + \frac{16}{4!}x^4$$

$$= 1 - 2x + 2x^2 - \frac{4}{3}x^3 + \frac{2}{3}x^4$$

3.

Derivative	Value at 0
$f(x) = e^{x+1}$	$f(0) = e$
$f^{(1)}(x) = e^{x+1}$	$f^{(1)}(0) = e$
$f^{(2)}(x) = e^{x+1}$	$f^{(2)}(0) = e$
$f^{(3)}(x) = e^{x+1}$	$f^{(3)}(0) = e$
$f^{(4)}(x) = e^{x+1}$	$f^{(4)}(0) = e$

$$P_4(x) = f(0) + \frac{f^{(1)}(0)}{1!}x + \frac{f^{(2)}(0)}{2!}x^2 + \frac{f^{(3)}(0)}{3!}x^3 + \frac{f^{(4)}(0)}{4!}x^4$$

$$= e + \frac{e}{1!}x + \frac{e}{2!}x^2 + \frac{e}{3!}x^3 + \frac{e}{4!}x^4$$

$$= e + ex + \frac{e}{2}x^2 + \frac{e}{6}x^3 + \frac{e}{24}x^4$$

5.

Derivative	Value at 0
$f(x) = \sqrt{x+9} = (x+9)^{1/2}$	$f(0) = 3$
$f^{(1)}(x) = \frac{1}{2}(x+9)^{-1/2} = \frac{1}{2(x+9)^{1/2}}$	$f^{(1)}(0) = \frac{1}{6}$
$f^{(2)}(x) = -\frac{1}{4}(x+9)^{-3/2} = -\frac{1}{4(x+9)^{3/2}}$	$f^{(2)}(0) = -\frac{1}{108}$
$f^{(3)}(x) = \frac{3}{8}(x+9)^{-5/2} = \frac{3}{8(x+9)^{5/2}}$	$f^{(3)}(0) = \frac{1}{648}$
$f^{(4)}(x) = -\frac{15}{16}(x+9)^{-7/2} = -\frac{15}{16(x+9)^{7/2}}$	$f^{(4)}(0) = -\frac{5}{11,664}$

$$P_4(x) = f(0) + \frac{f^{(1)}(0)}{1!}x + \frac{f^{(2)}(0)}{2!}x^2 + \frac{f^{(3)}(0)}{3!}x^3 + \frac{f^{(4)}(0)}{4!}x^4$$

$$= 3 + \frac{\frac{1}{6}}{1!}x + \frac{-\frac{1}{108}}{2!}x^2 + \frac{\frac{1}{648}}{3!}x^3 + \frac{-\frac{5}{11,664}}{4!}x^4$$

$$= 3 + \frac{1}{6}x - \frac{1}{216}x^2 + \frac{1}{3888}x^3 - \frac{5}{279,936}x^4$$

7.

Derivative	Value at 0
$f(x) = \sqrt[3]{x-1} = (x-1)^{1/3}$	$f(0) = -1$
$f^{(1)}(x) = \dfrac{1}{3}(x-1)^{-2/3} = \dfrac{1}{3(x-1)^{3/2}}$	$f^{(1)}(0) = \dfrac{1}{3}$
$f^{(2)}(x) = -\dfrac{2}{9}(x-1)^{-5/3} = -\dfrac{2}{9(x-1)^{5/3}}$	$f^{(2)}(0) = \dfrac{2}{9}$
$f^{(3)}(x) = \dfrac{10}{27}(x-1)^{-8/3} = \dfrac{10}{27(x-1)^{8/3}}$	$f^{(3)}(0) = \dfrac{10}{27}$
$f^{(4)}(x) = -\dfrac{80}{81}(x-1)^{-11/3} = -\dfrac{80}{81(x-1)^{11/3}}$	$f^{(4)}(0) = \dfrac{80}{81}$

$$P_4(x) = f(0) + \frac{f^{(1)}(0)}{1!}x + \frac{f^{(2)}(0)}{2!}x^2 + \frac{f^{(3)}(0)}{3!}x^3 + \frac{f^{(4)}(0)}{4!}x^4$$

$$= -1 + \frac{\frac{1}{3}}{1!}x + \frac{\frac{2}{9}}{2!}x^2 + \frac{\frac{10}{27}}{3!}x^3 + \frac{\frac{80}{81}}{4!}x^4$$

$$= -1 + \frac{1}{3}x + \frac{1}{9}x^2 + \frac{5}{81}x^3 + \frac{10}{243}x^4$$

9.

Derivative	Value at 0
$f(x) = \sqrt[4]{x+1} = (x+1)^{1/4}$	$f(0) = 1$
$f^{(1)}(x) = \dfrac{1}{4}(x+1)^{-3/4} = \dfrac{1}{4(x+1)^{3/4}}$	$f^{(1)}(0) = \dfrac{1}{4}$
$f^{(2)}(x) = -\dfrac{3}{16}(x+1)^{-7/4} = -\dfrac{3}{16(x+1)^{7/4}}$	$f^{(2)}(0) = -\dfrac{3}{16}$
$f^{(3)}(x) = \dfrac{21}{64}(x+1)^{-11/4} = \dfrac{21}{64(x+1)^{11/4}}$	$f^{(3)}(0) = \dfrac{21}{64}$
$f^{(4)}(x) = -\dfrac{231}{256}(x+1)^{-15/4} = \dfrac{231}{256(x+1)^{15/4}}$	$f^{(4)}(0) = -\dfrac{231}{256}$

$$P_4(x) = f(0) + \frac{f^{(1)}(0)}{1!}x + \frac{f^{(2)}(0)}{2!}x^2 + \frac{f^{(3)}(0)}{3!}x^3 + \frac{f^{(4)}(0)}{4!}x^4$$

$$= 1 + \frac{\frac{1}{4}}{1!}x + \frac{-\frac{3}{16}}{2!}x^2 + \frac{\frac{21}{64}}{3!}x^3 + \frac{-\frac{231}{256}}{4!}x^4$$

$$= 1 + \frac{1}{4}x - \frac{3}{32}x^2 + \frac{7}{128}x^3 - \frac{77}{2048}x^4$$

11.

Derivative	Value at 0
$f(x) = \ln(1 - x)$	$f(0) = 0$
$f^{(1)}(x) = -\dfrac{1}{1-x} = \dfrac{1}{x-1} = (x-1)^{-1}$	$f^{(1)}(0) = -1$
$f^{(2)}(x) = -(x-1)^{-2} = -\dfrac{1}{(x-1)^2}$	$f^{(2)}(0) = -1$
$f^{(3)}(x) = 2(x-1)^{-3} = \dfrac{2}{(x-1)^3}$	$f^{(3)}(0) = -2$
$f^{(4)}(x) = -6(x-1)^{-4} = -\dfrac{6}{(x-1)^4}$	$f^{(4)}(0) = -6$

$$P_4(x) = f(0) + \frac{f^{(1)}(0)}{1!}x + \frac{f^{(2)}(0)}{2!}x^2 + \frac{f^{(3)}(0)}{3!}x^3 + \frac{f^{(4)}(0)}{4!}x^4$$

$$= 0 + \frac{-1}{1!}x + \frac{-1}{2!}x^2 + \frac{-2}{3!}x^3 + \frac{-6}{4!}x^4$$

$$= -x - \frac{1}{2}x^2 - \frac{1}{3}x^3 - \frac{1}{4}x^4$$

13.

Derivative	Value at 0
$f(x) = \ln(1 + 2x^2)$	$f(0) = 0$
$f^{(1)}(x) = \dfrac{4x}{1+2x^2}$	$f^{(1)}(0) = 0$
$f^{(2)}(x) = \dfrac{4 - 8x^2}{(1+2x^2)^2}$	$f^{(2)}(0) = 4$
$f^{(3)}(x) = \dfrac{32x^3 - 48x}{(1+2x^2)^3}$	$f^{(3)}(0) = 0$
$f^{(4)}(x) = \dfrac{-192x^4 + 576x^2 - 48}{(1+2x^2)^4}$	$f^{(4)}(0) = -48$

$$P_4(x) = f(0) + \frac{f^{(1)}(0)}{1!}x + \frac{f^{(2)}(0)}{2!}x^2 + \frac{f^{(3)}(0)}{3!}x^3 + \frac{f^{(4)}(0)}{4!}x^4$$

$$= 0 + \frac{0}{1!}x + \frac{4}{2!}x^2 + \frac{0}{3!}x^3 + \frac{-48}{4!}x^4$$

$$= 2x^2 - 2x^4$$

15.

Derivative	Value at 0
$f(x) = xe^{-x}$	$f(0) = 0$
$f^{(1)}(x) = -xe^{-x} + e^{-x}$	$f^{(1)}(0) = 1$
$f^{(2)}(x) = xe^{-x} - 2e^{-x}$	$f^{(2)}(0) = -2$
$f^{(3)}(x) = -xe^{-x} + 3e^{-x}$	$f^{(3)}(0) = 3$
$f^{(4)}(x) = xe^{-x} - 4e^{-x}$	$f^{(4)}(0) = -4$

$$P_4(x) = f(0) + \frac{f^{(1)}(0)}{1!}x + \frac{f^{(2)}(0)}{2!}x^2 + \frac{f^{(3)}(0)}{3!}x^3 + \frac{f^{(4)}(0)}{4!}x^4$$

$$= 0 + \frac{1}{1!}x + \frac{-2}{2!}x^2 + \frac{3}{3!}x^3 + \frac{-4}{4!}x^4$$

$$= x - x^2 + \frac{1}{2}x^3 - \frac{1}{6}x^4$$

17.

Derivative	Value at 0
$f(x) = (9 - x)^{3/2}$	$f(0) = 27$
$f^{(1)}(x) = -\frac{3}{2}(9 - x)^{1/2}$	$f^{(1)}(0) = -\frac{9}{2}$
$f^{(2)}(x) = \frac{3}{4}(9 - x)^{-1/2} = \frac{3}{4(9 - x)^{1/2}}$	$f^{(2)}(0) = \frac{1}{4}$
$f^{(3)}(x) = \frac{3}{8}(9 - x)^{-3/2} = \frac{3}{8(9 - x)^{3/2}}$	$f^{(3)}(0) = \frac{1}{72}$
$f^{(4)}(x) = \frac{9}{16}(9 - x)^{-5/2} = \frac{9}{16(9 - x)^{5/2}}$	$f^{(4)}(0) = \frac{1}{432}$

$$P_4(x) = f(0) + \frac{f^{(1)}(0)}{1!}x + \frac{f^{(2)}(0)}{2!}x^2 + \frac{f^{(3)}(0)}{3!}x^3 + \frac{f^{(4)}(0)}{4!}x^4$$

$$= 27 + \frac{-\frac{9}{2}}{1!}x + \frac{\frac{1}{4}}{2!}x^2 + \frac{\frac{1}{72}}{3!}x^3 + \frac{\frac{1}{432}}{4!}x^4$$

$$= 27 - \frac{9}{2}x + \frac{1}{8}x^2 + \frac{1}{432}x^3 + \frac{1}{10,368}x^4$$

19.

Derivative	Value at 0
$f(x) = \frac{1}{1 + x} = (1 + x)^{-1}$	$f(0) = 1$
$f^{(1)}(x) = -(1 + x)^{-2} = -\frac{1}{(1 + x)^2}$	$f^{(1)}(0) = -1$
$f^{(2)}(x) = 2(1 + x)^{-3} = \frac{2}{(1 + x)^3}$	$f^{(2)}(0) = 2$
$f^{(3)}(x) = -6(1 + x)^{-4} = -\frac{6}{(1 + x)^4}$	$f^{(3)}(0) = -6$
$f^{(4)}(x) = 24(1 + x)^{-5} = \frac{24}{(1 + x)^5}$	$f^{(4)}(0) = 24$

$$P_4(x) = f(0) + \frac{f^{(1)}(0)}{1!}x + \frac{f^{(2)}(0)}{2!}x^2 + \frac{f^{(3)}(0)}{3!}x^3 + \frac{f^{(4)}(0)}{4!}x^4$$

$$= 1 + \frac{-1}{1!}x + \frac{2}{2!}x^2 + \frac{-6}{3!}x^3 + \frac{24}{4!}x^4$$

$$= 1 - x + x^2 - x^3 + x^4$$

For Exercises 21–34 each approximation can be determined using the Taylor polynomials from Exercises 1–14, respectively.

21. Using the result of Exercise 1, with $f(x) = e^{-2x}$ and $P_4(x) = 1 - 2x + 2x^2 - \frac{4}{3}x^3 + \frac{2}{3}x^4$, we can

approximate $e^{-0.04}$ by evaluating $f(0.01) = e^{-2(0.02)} = e^{-0.04}$. Using $P_4(x)$ from Exercise 1 with $x = 0.02$ gives

$$P_4(0.02) = 1 - 2(0.02) + 2(0.02)^2 - \frac{4}{3}(0.02)^3 + \frac{2}{3}(0.02)^4$$

$$\approx 0.096078944.$$

To four decimal places, $P_4(0.02)$ approximates the value of $e^{-0.04}$ as 0.9608.

23. Using the result of Exercise 3, with $f(x) = e^{x+1}$ and $P_4(x) = e + ex + \frac{e}{2}x^2 + \frac{e}{6}x^3 + \frac{e}{24}x^4$, we can

approximate $e^{1.02}$ by evaluating $f(0.02) = e^{0.02+1} = e^{1.02}$. Using $P_4(x)$ from Exercise 3 with $x = 0.02$ gives

$$P_4(0.02) = e + e(0.02) + \frac{e}{2}(0.02)^2 + \frac{e}{6}(0.02)^3 + \frac{e}{24}(0.02)^4$$

$$\approx 2.773194764.$$

To four decimal places, $P_4(0.02)$ approximates the value of $e^{1.02}$ as 2.7732.

25. Using the result of Exercise 5, with $f(x) = \sqrt{x+9}$ and $P_4(x) = 3 + \frac{1}{6}x - \frac{1}{216}x^2 + \frac{1}{3888}x^3 - \frac{5}{279,936}x^4$, we

can approximate $\sqrt{8.92}$ by evaluating $f(-0.08) = \sqrt{-0.08+9} = \sqrt{8.92}$. Using $P_4(x)$ from Exercise 5 with $x = -0.08$ gives

$$P_4(-0.08) = 3 + \frac{1}{6}(-0.08) - \frac{1}{216}(-0.08)^2 + \frac{1}{3888}(-0.08)^3 - \frac{5}{279,936}(-0.08)^4$$

$$\approx 2.986636905.$$

To four decimal places, $P_4(-0.08)$ approximates the value of $\sqrt{8.92}$ as 2.9866.

27. Using the result of Exercise 7, with $f(x) = \sqrt[3]{x-1}$ and $P_4(x) = -1 + \frac{1}{3}x + \frac{1}{9}x^2 + \frac{5}{81}x^3 + \frac{10}{243}x^4$, we can

approximate $\sqrt[3]{-1.05}$ by evaluating $f(-0.05) = \sqrt[3]{-0.05-1} = \sqrt[3]{-1.05}$. Using $P_4(x)$ from Exercise 7 with $x = -0.05$ gives

$$P_4(-0.05) = -1 + \frac{1}{3}(-0.05) + \frac{1}{9}(-0.05)^2 + \frac{5}{81}(-0.05)^3 + \frac{10}{243}(-0.05)^4$$

$$\approx -1.016396348.$$

To four decimal places, $P_4(-0.05)$ approximates the value of $\sqrt[3]{-1.05}$ as -1.0164.

29. Using the result of Exercise 9, with $f(x) = \sqrt[4]{x+1}$ and $P_4(x) = 1 + \frac{1}{4}x - \frac{3}{32}x^2 + \frac{7}{128}x^3 - \frac{77}{2048}x^4$,

we can approximate $\sqrt[4]{1.06}$ by evaluating $f(0.06) = \sqrt[4]{0.06+1} = \sqrt[4]{1.06}$. Using $P_4(x)$ from Exercise 9 with $x = 0.06$ gives

$$P_4(0.06) = 1 + \frac{1}{4}(0.06) - \frac{3}{32}(0.06)^2 + \frac{7}{128}(0.06)^3 - \frac{77}{2048}(0.06)^4$$

$$\approx 1.014673825.$$

To four decimal places, $P_4(0.06)$ approximates the value of $\sqrt[4]{1.06}$ as 1.0147.

31.   Using the result of Exercise 11, with $f(x) = \ln(1 - x)$ and $P_4(x) = -x - \frac{1}{2}x^2 - \frac{1}{3}x^3 - \frac{1}{4}x^4$, we can approximate $\ln 0.97$ by evaluating $f(0.03) = \ln(1 - 0.03) = \ln 0.97$. Using $P_4(x)$ from Exercise 11 with $x = 0.03$ gives

$$P_4(0.03) = -(0.03) - \frac{1}{2}(0.03)^2 - \frac{1}{3}(0.03)^3 - \frac{1}{4}(0.03)^4$$
$$\approx -0.0304592025.$$

To four decimal places, $P_4(0.03)$ approximates the value of $\ln 0.97$ as -0.0305.

33.   Using the result of Exercise 13, with $f(x) = \ln(1 + 2x^2)$ and $P_4(x) = 2x^2 - 2x^4$, we can compute $\ln 1.008$ by

evaluating $f\left(\frac{\sqrt{10}}{50}\right) = \ln\left[1 + 2\left(\frac{\sqrt{10}}{50}\right)^2\right] = \ln[1 + 2(0.004)] = \ln 1.008$. Using $P_4(x)$ with $x = \frac{\sqrt{10}}{50}$ gives

$$P_4\left(\frac{\sqrt{10}}{50}\right) = 2\left(\frac{\sqrt{10}}{50}\right)^2 - 4\left(\frac{\sqrt{10}}{50}\right)^4$$
$$= 0.007968$$

To four decimal places, $P_4\left(\frac{\sqrt{10}}{50}\right)$ approximates $\ln 1.008$ as 0.0080.

35. $P_3(x) = f(0) + \dfrac{f^{(1)}(0)}{1!}x + \dfrac{f^{(2)}(0)}{2!}x^2 + \dfrac{f^{(3)}(0)}{3!}x^3 = 3 + \dfrac{6}{1}x + \dfrac{12}{2}x^2 + \dfrac{24}{6}x^3 = 3 + 6x + 6x^2 + 4x^3$

37.   **(a)**

Derivative	Value at 0
$f(x) = (a + x)^{\frac{1}{n}}$	$f(0) = a^{\frac{1}{n}}$
$f^{(1)}(x) = \dfrac{1}{n}(a + x)^{\frac{1}{n}-1}$	$f^{(1)}(0) = \dfrac{1}{n}a^{\frac{1}{n}-1} = \dfrac{a^{\frac{1}{n}}}{na}$
$f^{(2)}(x) = \dfrac{1}{n}\left(\dfrac{1}{n} - 1\right)(a + x)^{\frac{1}{n}-2}$	$f^{(2)}(0) = \dfrac{1}{n}\left(\dfrac{1 - n}{n}\right)a^{\frac{1}{n}-2} = \dfrac{(1 - n)a^{\frac{1}{n}}}{n^2 a^2}$
$f^{(3)}(x) = \dfrac{1}{n}\left(\dfrac{1}{n} - 1\right)\left(\dfrac{1}{n} - 2\right)(a + x)^{\frac{1}{n}-3}$	$f^{(3)}(0) = \dfrac{1}{n}\left(\dfrac{1 - n}{n}\right)\left(\dfrac{1 - 2n}{n}\right)a^{\frac{1}{n}-3} = \dfrac{(1 - n)(1 - 2n)a^{\frac{1}{n}}}{n^3 a^3}$
	$f^{(4)}(0) = \dfrac{1}{n}\left(\dfrac{1 - n}{n}\right)\left(\dfrac{1 - 2n}{n}\right)\left(\dfrac{1 - 3n}{n}\right)a^{\frac{1}{n}-4}$
$f^{(4)}(x) = \dfrac{1}{n}\left(\dfrac{1}{n} - 1\right)\left(\dfrac{1}{n} - 2\right)\left(\dfrac{1}{n} - 3\right)(a + x)^{\frac{1}{n}-4}$	$= \dfrac{(1 - n)(1 - 2n)(1 - 3n)\,a^{\frac{1}{n}}}{n^4 a^4}$

$$P_4(x) = a^{1/n} + \frac{\frac{a^{1/n}}{na}}{1!}x + \frac{\frac{(1-n)a^{1/n}}{n^2 a^2}}{2!}x^2 + \frac{\frac{(1-n)(1-2n)a^{1/n}}{n^3 a^3}}{3!}x^3 + \frac{\frac{(1-n)(1-2n)(1-3n)a^{1/n}}{n^4 a^4}}{4!}x^4$$

$$= a^{1/n} + \frac{xa^{1/n}}{na} + \frac{x^2 a^{1/n}(1 - n)}{2!\,n^2 a^2} + \frac{x^3 a^{1/n}(1 - n)(1 - 2n)}{3!\,n^3 a^3} + \frac{x^4 a^{1/n}(1 - n)(1 - 2n)(1 - 3n)}{4!\,n^4 a^4}$$

$$= a^{1/n} + \frac{xa^{1/n}}{na}\left[1 + \frac{x(1 - n)}{2!\,na} + \frac{x^2(1 - n)(1 - 2n)}{3!\,n^2 a^2} + \frac{x^3(1 - n)(1 - 2n)(1 - 3n)}{4!\,n^3 a^3}\right]$$

For values of x that are small compared with a, the terms after 1 in the brackets get closer and closer to zero. For example, consider the term $\frac{x(1-n)}{2!na}$ where $x = \frac{1}{10}a$. Then $\frac{x(1-n)}{2!na} = \frac{1-n}{20n}$. For even the smallest value of n, $n = 2$, the value of this term is $-\frac{1}{40}$. This term, and the terms which follow will have little impact on our approximation. Thus the value of the expression is approximately 1, so $P_4(x) \approx a^{1/n} + \frac{xa^{1/n}}{na}$. Thus, if x is small compared with a, the Taylor polynomial $P_4(x)$ approximates $(a + x)^{1/n}$ as $a^{1/n} + \frac{xa^{1/n}}{na}$.

(b) Using the result of part (a), we can approximate $\sqrt[3]{66}$ by evaluating $f(2)$ given $f(x) = (64 + x)^{1/3}$ since $f(2) = (64 + 2)^{1/3} = 66^{1/3} = \sqrt[3]{66}$. Using $P_4(x)$ from part (a) with $x = 2$, $a = 64$, and $n = 3$ gives

$$(64 + 2)^{1/3} = 64^{1/3} + \frac{2 \cdot 64^{1/3}}{3 \cdot 64}$$

$$= 4 + \frac{8}{192}$$

$$= 4.041\overline{6}$$

To five decimal places, $P_4(x) \approx (64 + x)^{1/3}$ approximates the value of $\sqrt[3]{66}$ as 4.04167. To five decimal places, a calculator gives an approximation of 4.04124.

39. (a)

Derivative	Value at 0
$f(N) = e^{\lambda N}$	$f(0) = 1$
$f^{(1)}(N) = \lambda e^{\lambda N}$	$f^{(1)}(0) = \lambda$
$f^{(2)}(N) = \lambda^2 e^{\lambda N}$	$f^{(2)}(0) = \lambda^2$

$$P_2(N) = f(0) + \frac{f^{(1)}(0)}{1!}N + \frac{f^{(2)}(0)}{2!}N^2$$

$$= 1 + \frac{\lambda}{1!}N + \frac{\lambda^2}{2!}N^2 = 1 + \lambda N + \frac{\lambda^2}{2}N^2$$

The Taylor polynomial of degree 2 at $N = 0$ for $e^{\lambda N}$ is $P_2(N) = 1 + \lambda N + \frac{\lambda^2}{2}N^2$.

(b) Substituting $P_2(N) = 1 + \lambda N + \frac{\lambda^2}{2}N^2$ for $e^{\lambda N}$ in the original equation gives:

$$\frac{1 + \lambda N + \frac{\lambda^2}{2}N^2}{\lambda} - N = \frac{1}{\lambda} + k$$

$$\left(1 + \lambda N + \frac{\lambda^2}{2}N^2\right) - \lambda N = 1 + \lambda k$$

$$\frac{\lambda^2}{2}N^2 = \lambda k$$

$$N^2 = \frac{2k}{\lambda}$$

$$N = \pm\sqrt{\frac{2k}{\lambda}}$$

Since N is a positive quantity, use the positive square root. So, $N = \sqrt{\frac{2k}{\lambda}}$.

41.

Derivative	Value at 0
$P(x) = \ln(100 + 3x)$	$P(0) = 4.605$
$P^{(1)}(x) = \dfrac{3}{100 + 3x}$	$P^{(1)}(0) = 0.03$
$P^{(2)}(x) = -\dfrac{9}{(100 + 3x)^2}$	$P^{(2)}(0) = -0.0009$

$$P_2(x) = P(0) + \frac{P^{(1)}(0)}{1!}x + \frac{P^{(2)}(0)}{2!}x^2$$

$$= 4.605 + \frac{0.03}{1}x + \frac{-0.0009}{2}x^2$$

$$= 4.605 + 0.03x - 0.00045x^2$$

$$P_2(0.6) = 4.605 + 0.03(0.6) - 0.00045(0.6)^2$$

$$= 4.622838$$

The Taylor polynomial $P_2(0.6)$ approximates $P(0.6)$ as 4.622838. The approximate profit is 4.623 thousand dollars, or $4623.

Finding $P(0.6)$ by direct substitution gives:

$$P(0.6) = \ln 100 + 3(0.6)$$

$$= \ln 101.8$$

$$= 4.623010104$$

Direct substitution also estimates the profit as 4.623 thousand dollars, or $4623.

43.

Derivative	Value at 0
$R(x) = 500 \ln\left(4 + \dfrac{x}{50}\right)$	$R(0) = 500(1.3863) = 693.15$
$R^{(1)}(x) = \dfrac{500}{200 + x}$	$R^{(1)}(0) = 2.5$
$R^{(2)}(x) = -\dfrac{500}{(200 + x)^2}$	$R^{(2)}(0) = -0.0125$

$$P_2(x) = R(0) + \frac{R^{(1)}(0)}{1!}x + \frac{R^{(2)}(0)}{2!}x^2 = 693.15 + \frac{2.5}{1}x + \frac{-0.0125}{2}x^2 = 693.15 + 2.5x - 0.00625x^2$$

$$P_2(10) = 693.15 + 2.5(10) - 0.00625(10)^2 = 717.525$$

The Taylor polynomial $P_2(10)$ approximates $R(10)$ as 717.525. Thus, the approximate revenue is $718.

Finding $R(10)$ by direct substitution gives:

$$R(10) = 500 \ln\left(4 + \frac{10}{50}\right) \approx 717.5422626$$

Direct substitution estimates the revenue as $718.

45.

Derivative	Value at 0
$P(k) = 1 - e^{-2k}$	$P(0) = 1 - e^0 = 0$
$P^{(1)}(k) = 2e^{-2k}$	$P^{(1)}(0) = 2$

$$P_1(k) = P(0) + \frac{P^{(1)}(0)}{1!}k = 0 + \frac{2}{1!}k = 2k$$

So, if k is small, that is, close to 0, then

$$P(k) \approx P_1(k) = 2k.$$

12.4 Infinite Series

Your Turn 1

The sequence is $1, \frac{1}{4}, \frac{1}{9}, \frac{1}{16}, \frac{1}{25}, \ldots$.

$$S_1 = 1$$

$$S_2 = 1 + \frac{1}{4} = \frac{5}{4}$$

$$S_3 = 1 + \frac{1}{4} + \frac{1}{9} = \frac{49}{36}$$

$$S_4 = 1 + \frac{1}{4} + \frac{1}{9} + \frac{1}{16} = \frac{205}{144}$$

$$S_5 = 1 + \frac{1}{4} + \frac{1}{9} + \frac{1}{16} + \frac{1}{25} = \frac{5269}{3600}$$

Your Turn 2

(a) This is a geometric series with $a = a_1 = 2$ and $r = 1/3$. Since r is in $(-1, 1)$, the series converges and has sum

$$\frac{a}{1 - r} = \frac{2}{1 - (1/3)}$$

$$= \frac{2}{2/3}$$

$$= 3.$$

3. $2 + 6 + 18 + 54 + \cdots$ is geometric series with $a = a_1 = 2$ and $r = 3$.

(b) This is a geometric series with $a = a_1 = 4$ and $r = -1/4$. Since r is in $(-1, 1)$, the series converges and has sum

$$\frac{a}{1 - r} = \frac{5}{1 - (-1/4)}$$

$$= \frac{5}{5/4}$$

$$= 4.$$

(c) This is a geometric series with $a = a_1 = 2$ and $r = 1.01$. Since r is not in $(-1, 1)$, the series diverges.

Your Turn 3

(a) Solve using algebra.
Let t be the number of hours since 2:00 pm.
Between 8:00 am and 2:00 pm, Turtle has traveled $(6)(60)(15) = 5400$ feet.

Turtle's distance from the starting line at any number of hours t after 2:00 pm is
$$d_1 = 5400 + (t)(60)(15) = 5400 + 900t \text{ feet.}$$

Rabbit's distance from the starting line at time t is $d_2 = (t)(60)(45) = 2700t$ feet.

$$d_1 = d_2$$
$$5400 + 900t = 2700t$$
$$5400 = 1800t$$
$$t = 3$$

Rabbit catches up with Turtle 3 hours after 2 P.M., that is, at 5 P.M..

(b) Solve using a geometric series.
At 2 P.M. Turtle has traveled 5400 ft. Traveling at 45 feet per minute, it will take Rabbit $5400/45 = 120$ minutes to make up this distance. During this time, Turtle will have traveled a further $(120)(15) = 1800$ feet, and it will take Rabbit $1800/45 = 40$ minutes to make up this distance. The total number of minutes it takes Rabbit to make up all these distances is $120 + 40 + 40/3 + \cdots$. This is a geometric series with $a = a_1 = 120$ and $r = 1/3$. Since r is in $(-1, 1)$, the series converges and has sum

$$\frac{a}{1 - r} = \frac{120}{1 - (1/3)}$$

$$= \frac{120}{2/3}$$

$$= 180,$$

that is, 180 minutes or 3 hours after 2 P.M., or 5 P.M., as found in (a).

12.4 Exercises

1. $20 + 10 + 5 + \frac{5}{2} + \cdots$ is a geometric series with $a = a_1 = 20$ and $r = \frac{1}{2}$. Since r is in $(-1, 1)$, the series converges and has sum

$$\frac{a}{1 - r} = \frac{20}{1 - \frac{1}{2}} = \frac{20}{\frac{1}{2}} = 40.$$

3. $2 + 6 + 18 + 54 + \cdots$ is geometric series with $a = a_1 = 2$ and $r = 3$. Since $r > 1$, the series diverges.

5. $27 + 9 + 3 + 1 + \cdots$ is a geometric series with
$a = a_1 = 27$ and $r = \frac{1}{3}$. Since r is in $(-1, 1)$,
the series converges and has sum

$$\frac{a}{1 - r} = \frac{27}{1 - \frac{1}{3}} = \frac{27}{\frac{2}{3}} = \frac{81}{2}.$$

7. $100 + 10 + 1 + \cdots$ is a geometric series
with $a = a_1 = 100$ and $r = \frac{1}{10}$. Since r
is in $(-1, 1)$, the series converges and has sum

$$\frac{a}{1 - r} = \frac{100}{1 - \frac{1}{10}} = \frac{100}{\frac{9}{10}} = \frac{1000}{9}.$$

9. $\frac{5}{4} + \frac{5}{8} + \frac{5}{16} + \cdots$ is a geometric series with
$a = a_1 = \frac{5}{4}$ and $r = \frac{1}{2}$. Since r is in $(-1, 1)$,
the series converges and has sum

$$\frac{a}{1 - r} = \frac{\frac{5}{4}}{1 - \frac{1}{2}} = \frac{\frac{5}{4}}{\frac{1}{2}} = \frac{5}{2}.$$

11. $\frac{1}{3} - \frac{2}{9} + \frac{4}{27} - \frac{8}{81} + \cdots$ is a geometric series
with $a = a_1 = \frac{1}{3}$ and $r = -\frac{2}{3}$. Since r is in
$(-1, 1)$, the series converges and has sum

$$\frac{a}{1 - r} = \frac{\frac{1}{3}}{1 - \left(-\frac{2}{3}\right)} = \frac{\frac{1}{3}}{1 + \frac{2}{3}} = \frac{\frac{1}{3}}{\frac{5}{3}} = \frac{1}{5}.$$

13. $e - 1 + \frac{1}{e} - \frac{1}{e^2} + \cdots$ is a geometric series with
$a = a_1 = e$ and $r = -\frac{1}{e}$. Since r is in $(-1, 1)$,
the series converges and the sum

$$\frac{a}{1 - r} = \frac{e}{1 - \left(-\frac{1}{e}\right)} = \frac{e}{1 + \frac{1}{e}}$$

$$= \frac{e}{\frac{e+1}{e}} = \frac{e^2}{e + 1}.$$

15. $S_1 = a_1 = \dfrac{1}{1} = 1$

$S_2 = a_1 + a_2 = 1 + \dfrac{1}{2} = \dfrac{3}{2}$

$S_3 = a_1 + a_2 + a_3 = 1 + \dfrac{1}{2} + \dfrac{1}{3} = \dfrac{11}{6}$

$S_4 = a_1 + a_2 + a_3 + a_4 = 1 + \dfrac{1}{2} + \dfrac{1}{3} + \dfrac{1}{4} = \dfrac{25}{12}$

$S_5 = a_1 + a_2 + a_3 + a_4 + a_5 = 1 + \dfrac{1}{2} + \dfrac{1}{3} + \dfrac{1}{4} + \dfrac{1}{5} = \dfrac{137}{60}$

17. $S_1 = a_1 = \dfrac{1}{2(1) + 5} = \dfrac{1}{7}$

$S_2 = a_1 + a_2 = \dfrac{1}{2(1) + 5} + \dfrac{1}{2(2) + 5} = \dfrac{1}{7} + \dfrac{1}{9} = \dfrac{16}{63}$

$S_3 = a_1 + a_2 + a_3 = \dfrac{1}{2(1) + 5} + \dfrac{1}{2(2) + 5} + \dfrac{1}{2(3) + 5} = \dfrac{1}{7} + \dfrac{1}{9} + \dfrac{1}{11} = \dfrac{239}{693}$

$S_4 = a_1 + a_2 + a_3 + a_4$

$= \dfrac{1}{2(1) + 5} + \dfrac{1}{2(2) + 5} + \dfrac{1}{2(3) + 5} + \dfrac{1}{2(4) + 5} = \dfrac{1}{7} + \dfrac{1}{9} + \dfrac{1}{11} + \dfrac{1}{13} = \dfrac{3800}{9009}$

$S_5 = a_1 + a_2 + a_3 + a_4 + a_5$

$= \dfrac{1}{2(1) + 5} + \dfrac{1}{2(2) + 5} + \dfrac{1}{2(3) + 5} + \dfrac{1}{2(4) + 5} + \dfrac{1}{2(5) + 5}$

$= \dfrac{1}{7} + \dfrac{1}{9} + \dfrac{1}{11} + \dfrac{1}{13} + \dfrac{1}{15} = \dfrac{22,003}{45,045}$

19. $S_1 = a_1 = \dfrac{1}{(1 + 1)(1 + 2)} = \dfrac{1}{6}$

$S_2 = a_1 + a_2 = \dfrac{1}{(1 + 1)(1 + 2)} + \dfrac{1}{(2 + 1)(2 + 2)} = \dfrac{1}{6} + \dfrac{1}{12} = \dfrac{1}{4}$

$S_3 = a_1 + a_2 + a_3$

$= \dfrac{1}{(1 + 1)(1 + 2)} + \dfrac{1}{(2 + 1)(2 + 2)} + \dfrac{1}{(3 + 1)(3 + 2)} = \dfrac{1}{6} + \dfrac{1}{12} + \dfrac{1}{20} = \dfrac{3}{10}$

$S_4 = a_1 + a_2 + a_3 + a_4 = \dfrac{1}{(1 + 1)(1 + 2)} + \dfrac{1}{(2 + 1)(2 + 2)} + \dfrac{1}{(3 + 1)(3 + 2)} + \dfrac{1}{(4 + 1)(4 + 2)}$

$= \dfrac{1}{6} + \dfrac{1}{12} + \dfrac{1}{20} + \dfrac{1}{30} = \dfrac{1}{3}$

$S_5 = a_1 + a_2 + a_3 + a_4 + a_5$

$= \dfrac{1}{(1 + 1)(1 + 2)} + \dfrac{1}{(2 + 1)(2 + 2)} + \dfrac{1}{(3 + 1)(3 + 2)} + \dfrac{1}{(4 + 1)(4 + 2)} + \dfrac{1}{(5 + 1)(5 + 2)}$

$= \dfrac{1}{6} + \dfrac{1}{12} + \dfrac{1}{20} + \dfrac{1}{30} + \dfrac{1}{42} = \dfrac{5}{14}$

21. The infinite geometric series

$$0.2 + 0.2\left(\dfrac{1}{10}\right) + 0.2\left(\dfrac{1}{10}\right)^2 + 0.2\left(\dfrac{1}{10}\right)^3 + \cdots$$

is a geometric series with $a = a_1 = 0.2$ and $r = 1/10$. Since r is in $(-1,1)$, the series converges and has sum

$$\frac{a}{1-r} = \frac{0.2}{1-(1/10)}$$
$$= \frac{0.2}{0.9} = 2/9.$$

23.   **(a)**   It follows from Viète's formula that

$$\frac{2}{\pi} \approx \frac{\sqrt{2}}{2} \cdot \frac{\sqrt{2+\sqrt{2}}}{2} \cdot \frac{\sqrt{2+\sqrt{2+\sqrt{2}}}}{2}$$
$$= \frac{\sqrt{2}\left(\sqrt{2+\sqrt{2}}\right)\left(\sqrt{2+\sqrt{2+\sqrt{2}}}\right)}{8}$$
$$\approx 0.641$$

Thus, $\pi \approx \frac{2}{0.641} \approx 3.12$.

It follows from Liebniz's formula that

$$\frac{\pi}{4} \approx 1 - \frac{1}{3} + \frac{1}{5} - \frac{1}{7} \approx 0.724.$$

Thus, $\pi \approx 4\,(0.724) \approx 2.90$.

Therefore, approximating π by multiplying the first three terms of Viète's formula together is more accurate than approximating π by adding the first four terms of Liebniz's formula together.

   (b)   Using the graphing calculator with $Y_1 = 4 * \text{sum}\left(\text{seq}\left(\frac{(-1)^{N-1}}{2N-1}, N, 1, X\right)\right)$ and the table function gives the

following values for Y_1 for $35 \le X \le 41$.

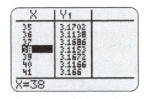

Thus, 38 terms of the second formula must be added together to produce the same accuracy as the product of the first three terms of the first formula.

25.   **(a)**   The total expenditure is the sum of an infinite geometric series with $a = a_1 = 200$ and $r = 0.9$. The sum of this convergent series is

$$\frac{a}{1-r} = \frac{200}{1-0.9}$$
$$= \frac{200}{0.1}$$
$$= 2000.$$

The total expenditure, including the government's original $200 payout, will be $2000.

   (b)   Since the original payout was $200, the multiplier is $2000/$200 = 10.

27. Let S_n be the total medical malpractice payments the company pays after n years. If each year's payments are 20% less than those of the previous year, then they are 80% of the previous year's payments. Thus,

$$S_0 = 60$$
$$S_1 = 60 + 60(0.8)$$
$$S_2 = 60 + 60(0.8) + 60(0.8)^2$$
$$\vdots$$
$$S_n = 60 + 60(0.8) + \cdots + 60(0.8)^n.$$

The total medical malpractice payments that the company pays in all years is, therefore,

$$S = 60 + 60(0.8) + \cdots + 60(0.8)^n + \cdots$$

This is a geometric series with $a = 60$ and $r = 0.8$.

$$S = \frac{a}{1-r} = \frac{60}{1-0.8} = \frac{60}{0.2} = 300$$

The correct answer choice is **d**.

29. The height that the ball returns to after the first bounce is given by $10\left(\frac{3}{4}\right)$, after the second bounce $10\left(\frac{3}{4}\right)^2$, after the third bounce $10\left(\frac{3}{4}\right)^3$, and so on. Thus, the distance that the ball travels before it comes to rest is given by

$$10 + 2(10)\left(\frac{3}{4}\right) + 2(10)\left(\frac{3}{4}\right)^2 + 2(10)\left(\frac{3}{4}\right)^3 + \cdots.$$

$2(10)\left(\frac{3}{4}\right) + 2(10)\left(\frac{3}{4}\right)^2 + 2(10)\left(\frac{3}{4}\right)^3 + \ldots$ is an infinite geometric series with $a_1 = a = 2(10)\left(\frac{3}{4}\right) = 15$ and $r = \frac{3}{4}$. This series converges to

$$\frac{a}{1-r} = \frac{15}{1-\frac{3}{4}} = \frac{15}{\frac{1}{4}} = 60.$$

So, the total distance traveled by the ball is $10 + 60 = 70$ meters.

31. On its second swing, the pendulum bob will swing through an arc $40(0.8)$ centimeters long, on its third swing $40(0.8)^2$ centimeters long, on its fourth swing $40(0.8)^3$ centimeters long, and so on. Thus, the distance it will swing altogether before coming to a complete stop is given by

$$40 + 40(0.8) + 40(0.8)^2 + 40(0.8)^3 + \cdots.$$

This is an infinite geometric series with $a_1 = a = 40$ and $r = 0.8$. This series converges to

$$\frac{a}{1-r} = \frac{40}{1-0.8} = \frac{40}{0.2} = 200.$$

So, the total distance altogether before coming to a complete stop is 200 centimeters.

So, the total perimeter of all the triangles is 12 meters.

33. The first triangle has sides 2 meters in length, the second triangle has sides 1 meter in length, the third triangle has sides $\frac{1}{2}$ meter in length, the fourth triangle has sides $\frac{1}{4}$ meter in length, and so on. Thus, the height of the first triangle is $\sqrt{3}$ meters, the height of the second triangle is $\frac{\sqrt{3}}{2}$ meter, the height of the third triangle is $\frac{\sqrt{3}}{4}$ meter, the height of the fourth triangle is $\frac{\sqrt{3}}{8}$ meter, and so on. The total area of the triangles, disregarding the overlaps, is given by

$$\frac{1}{2}(2)(\sqrt{3}) + \frac{1}{2}(1)\left(\frac{\sqrt{3}}{2}\right) + \frac{1}{2}\left(\frac{1}{2}\right)\left(\frac{\sqrt{3}}{4}\right) + \frac{1}{2}\left(\frac{1}{4}\right)\left(\frac{\sqrt{3}}{8}\right) + \cdots = \sqrt{3} + \frac{\sqrt{3}}{4} + \frac{\sqrt{3}}{16} + \frac{\sqrt{3}}{64} + \cdots.$$

This is an infinite geometric series with $a = a_1 = \sqrt{3}$ and $r = \frac{1}{4}$.

This series converges to $\dfrac{a}{1-r} = \dfrac{\sqrt{3}}{1 - \frac{1}{4}} = \dfrac{\sqrt{3}}{\frac{3}{4}} = \dfrac{4\sqrt{3}}{3}$.

The total area of the triangles, disregarding the overlaps, is given by $\frac{4\sqrt{3}}{3}$ square meters.

35. **(a)** The tortoise's starting point is 10 meters. Since Achilles runs 10 meters per second, it takes him 1 second to reach the tortoise's starting point. The tortoise has traveled 1 meter in this 1 second since the tortoise's rate is 1 meter per second. After Achilles reaches the tortoise's starting point, the tortoise has traveled 1 meter. Since Achilles travels 10 meters per second, it takes Achilles $\frac{1}{10}$ second to reach this new point. The tortoise has traveled $\frac{1}{10}$ meter in this $\frac{1}{10}$ second since the tortoise's rate is 1 meter per second. The time that it takes Achilles to reach the tortoise is given by

$$1 + \frac{1}{10} + \frac{1}{100} + \cdots.$$

This is an infinite geometric series with $a = a_1 = 1$ and $r = \frac{1}{10}$. This series converges to $\dfrac{a}{1-r} = \dfrac{1}{1 - \frac{1}{10}} = \dfrac{10}{9}$.

It takes Achilles $\frac{10}{9}$ sec to catch the tortoise

(b) Using distance $=$ rate $\times$ time, Achilles' distance is given by $10t$ meters in t seconds and the tortoise's distance is given by $t + 10$ meters in t seconds. The distances are equal when

$$10t = t + 10$$
$$t = \frac{10}{9}$$

It takes Achilles $\frac{10}{9}$ seconds to catch the tortoise.

(c) The error is the assumption that an infinite series must have an infinite sum.

37. (a) To win the next point, A has to win on B's serve and then on her own serve. This can happen immediately with probability x^2, after one trade of serve with probability $(x)(1 - x)x^2$, after two trades of serve, probability $[x(1 - x)]x^2$, and so on. Note that each trade of serve happens when A wins on B's serve (probability x) and then loses on her own serve (probability $1 - x$). So the probability that A wins the next point is

$$x^2 \sum_{n=0}^{\infty} [x(1 - x)]^n.$$

(b) Since $0 < x(1 - x) < 1$, $\displaystyle\sum_{n=0}^{\infty} [x(1 - x)]^n$ is a

convergent geometric series with sum $\dfrac{1}{1 - [x(1 - x)]}$.

Thus the probability in (a) simplifies to $\dfrac{x^2}{1 - x + x^2}$.

(d) The only difference from the computation in (a) is that A already has the serve, so the factor of x can be omitted from each of the summands in (a), giving a probability of $\dfrac{x}{1 - x + x^2}$.

12.5   Taylor Series

Your Turn 1

(a) Use the Taylor series for $\ln(1 + x)$ with property (2).

$$\ln(1 + x) = x - \frac{x^2}{2} + \frac{x^3}{3} - \frac{x^4}{4} + \cdots + \frac{(-1)^{n+1}x^n}{n} + \cdots$$

$$-7\ln(1 + x) = -7x + \frac{7x^2}{2} - \frac{7x^3}{3} + \frac{7x^4}{4} + \cdots + \frac{7(-1)^n x^n}{n} + \cdots$$

for all x in $(-1, 1]$

(b) Use the Taylor series for $\frac{1}{1-x}$ with property (3).

$$\frac{1}{1 - x} = 1 + x + x^2 + x^3 + \cdots + x^n + \cdots$$

$$\frac{x^2}{1 - x} = x^2 + x^3 + x^4 + x^5 + \cdots + x^n + \cdots$$

for all x in $(-1, 1)$

Your Turn 2

(a) Use the Taylor series for $\ln(1 + x)$ together with composition.

$$\ln(1 + x) = x - \frac{x^2}{2} + \frac{x^3}{3} - \frac{x^4}{4} + \cdots + \frac{(-1)^{n+1}x^n}{n} + \cdots$$

$$\ln(1 + 2x^2) = (2x^2) - \frac{(2x^2)^2}{2} + \frac{(2x^2)^3}{3} - \frac{(2x^2)^4}{4} + \cdots + \frac{(-1)^{n+1}(2x^2)^n}{n} + \cdots$$

$$= 2x^2 - 2x^4 + \frac{8x^6}{3} - 4x^8 + \cdots + \frac{(-1)^{n+1}2^n x^{2n}}{n} + \cdots$$

The interval of convergence of the original series is $-1 < x \le -1$.

If $-\frac{1}{\sqrt{2}} \le x \le \frac{1}{\sqrt{2}}$, then $-1 < 0 \le 2x^2 \le 1$, so the interval of convergence is now $\left[-\frac{1}{\sqrt{2}}, \frac{1}{\sqrt{2}} \right]$.

(b) Use the series for $\frac{1}{1-x}$ together with composition.

$$g(x) = \frac{3}{4 - x^2} = \frac{3/4}{1 - \left(\frac{x}{2}\right)^2}$$

$$= \frac{3}{4} f\left(\frac{x}{2}\right) \text{ where } f(x) = \frac{1}{1 - x}$$

$$\frac{1}{1 - x} = 1 + x + x^2 + x^3 + \cdots + x^n + \cdots$$

$$\frac{3/4}{1 - \left(\frac{x}{2}\right)^2} = \frac{3}{4} + \frac{3}{4}\left(\frac{x}{2}\right)^2 + \frac{3}{4}\left[\left(\frac{x}{2}\right)^2\right]^2 + \frac{3}{4}\left[\left(\frac{x}{2}\right)^2\right]^3 + \cdots + \frac{3}{4}\left[\left(\frac{x}{2}\right)^2\right]^n + \cdots$$

$$= \frac{3}{4} + \frac{3x^2}{16} + \frac{3x^4}{64} + \frac{3x^6}{256} + \cdots + \frac{3x^{2n}}{4^{n+1}} + \cdots$$

The interval of convergence of the original series is $-1 < x < -1$.

When $-2 < x < 2$, $-1 < 0 \le \left(\frac{x}{2}\right)^2 < 1$ so the interval of convergence of the new series is $(-2, 2)$.

Your Turn 3

For an interest rate of 3.5%, the doubling time is $\frac{\ln 2}{\ln(1+0.035)} \approx 20.149$. Since this interest rate is less than 5%, we can use the

rule of 70 to estimate this doubling time; the estimate is $\frac{70}{100(0.035)} = 20$ years.

For an interest rate of 8%, the doubling time is $\frac{\ln 2}{\ln(1+0.08)} \approx 9.006$. Since this interest rate is more than 5%, we can use the rule

of 72 to estimate this doubling time; the estimate is $\frac{72}{100(0.08)} = 9$ years.

12.5   Exercises

1. Use the Taylor series for $\frac{1}{1-x}$ with property (2).

$$\frac{1}{1-x} = 1 + x + x^2 + x^3 + \cdots + x^n + \cdots$$

$$\frac{6}{1-x} = 6 + 6x + 6x^2 + 6x^3 + \cdots + 6x^n + \cdots$$

for all x in $(-1, 1)$

3. Use the Taylor series for e^x with property (3).

$$e^x = 1 + \frac{x}{1!} + \frac{x^2}{2!} + \frac{x^3}{3!} + \cdots + \frac{x^n}{n!} + \cdots$$

$$x^2 e^x = x^2 + \frac{x^3}{1!} + \frac{x^4}{2!} + \frac{x^5}{3!} + \cdots + \frac{x^{n+2}}{n!} + \cdots$$

for x in $(-\infty, \infty)$

5. This function most nearly matches $\frac{1}{1-x}$. To get 1 in the denominator, instead of 2, divide the numerator and denominator by 2.

$$\frac{5}{2-x} = \frac{\frac{5}{2}}{1 - \frac{x}{2}}$$

Thus, we can find the Taylor series for $\frac{\frac{5}{2}}{1 - \frac{x}{2}}$ by starting with the Taylor series for $\frac{1}{1-x}$, multiplying each term by $\frac{5}{2}$, and replacing each x with $\frac{x}{2}$.

$$\frac{5}{2-x} = \frac{\frac{5}{2}}{1 - \frac{x}{2}}$$

$$= \frac{5}{2} \cdot 1 + \frac{5}{2}\left(\frac{x}{2}\right) + \frac{5}{2}\left(\frac{x}{2}\right)^2 + \frac{5}{2}\left(\frac{x}{2}\right)^3 + \cdots + \frac{5}{2}\left(\frac{x}{2}\right)^n + \cdots$$

$$= \frac{5}{2} + \frac{5x}{4} + \frac{5x^2}{8} + \frac{5x^3}{16} + \cdots + \frac{5x^n}{2^{n+1}} + \cdots$$

The Taylor series for $\frac{1}{1-x}$ is valid when $-1 < x < 1$. Replacing x with $\frac{x}{2}$ gives

$$-1 < \frac{x}{2} < 1 \text{ or } -2 < x < 2.$$

The interval of convergence of the new series is $(-2, 2)$.

7. $\dfrac{8x}{1+3x} = x \cdot \dfrac{8}{1-(-3x)}$

Use the Taylor series for $\frac{1}{1-x}$, multiply each term by 8, and replace x with $-3x$. Also, use property (3) with $k = 1$.

$$\frac{8x}{1+3x} = x \cdot \frac{8}{1-(-3x)}$$

$$= x \cdot 8 \cdot 1 + x \cdot 8(-3x) + x \cdot 8(-3x)^2 + x \cdot 8(-3x)^3 + \cdots + x \cdot 8(-3x)^n + \cdots$$

$$= 8x - 24x^2 + 72x^3 - 216x^4 + \cdots + (-1)^n \cdot 8 \cdot 3^n x^{n+1} + \cdots$$

The Taylor series for $\frac{1}{1-x}$ is valid when $-1 < x < 1$. Replacing x with $-3x$ gives

$$-1 < -3x < 1 \quad \text{or} \quad \frac{1}{3} > x > -\frac{1}{3}.$$

The interval of convergence of the new series is $\left(-\frac{1}{3}, \frac{1}{3}\right)$.

9. $\dfrac{x^2}{4-x} = x^2 \cdot \dfrac{\frac{1}{4}}{1-\frac{x}{4}}$

Use the Taylor series for $\frac{1}{1-x}$, multiply each term by $\frac{1}{4}$, and replace x with $\frac{x}{4}$. Also, use property (3) with $k = 2$.

$$\frac{x^2}{4-x} = x^2 \cdot \frac{\frac{1}{4}}{1-\frac{x}{4}}$$

$$= x^2 \cdot \frac{1}{4} \cdot 1 + x^2 \cdot \frac{1}{4}\left(\frac{x}{4}\right) + x^2 \cdot \frac{1}{4}\left(\frac{x}{4}\right)^2 + x^2 \cdot \frac{1}{4}\left(\frac{x}{4}\right)^3 + \cdots + x^2 \cdot \frac{1}{4}\left(\frac{x}{4}\right)^n + \cdots$$

$$= \frac{x^2}{4} + \frac{x^3}{16} + \frac{x^4}{64} + \frac{x^5}{256} + \cdots + \frac{x^{n+2}}{4^{n+1}} + \cdots$$

The Taylor series for $\frac{1}{1-x}$ is valid when $-1 < x < 1$. Replacing x with $\frac{x}{4}$ gives

$$-1 < \frac{x}{4} < 1 \text{ or } -4 < x < 4 \cdot$$

The interval of convergence of the new series is $(-4, 4)$.

11. We find the Taylor series of $\ln(1 + 4x)$ by starting with the Taylor series for $\ln(1 + x)$ and replacing each x with $4x$.

$$\ln(1 + 4x) = 4x - \frac{(4x)^2}{2} + \frac{(4x)^3}{3} - \frac{(4x)^4}{4} + \cdots + \frac{(-1)^n(4x)^{n+1}}{n+1} + \cdots$$

$$= 4x - 8x^2 + \frac{64}{3}x^3 - 64x^4 + \cdots + \frac{(-1)^n 4^{n+1} x^{n+1}}{n+1} + \cdots$$

The Taylor series for $\ln(1 + x)$ is valid when $-1 < x \leq 1$. Replacing x with $4x$ gives

$$-1 < 4x \leq 1 \text{ or } -\frac{1}{4} < x \leq \frac{1}{4}.$$

The interval of convergence of the new series is $\left(-\frac{1}{4}, \frac{1}{4}\right]$.

13. We find the Taylor series for e^{4x^2} by starting with the Taylor series for e^x and replacing each x with $4x^2$.

$$e^{4x^2} = 1 + (4x^2) + \frac{1}{2!}(4x^2)^2 + \frac{1}{3!}(4x^2)^3 + \cdots + \frac{1}{n!}(4x^2)^n + \cdots$$

$$= 1 + 4x^2 + 8x^4 + \frac{32}{3}x^6 + \cdots + \frac{4^n x^{2n}}{n!} + \cdots$$

The Taylor series for e^{4x^2} has the same interval of convergence, $(-\infty, \infty)$, as the Taylor series for e^x.

15. Use the Taylor series for e^x and replace x with $-x$. Also, use property (3) with $k = 3$.

$$x^3 e^{-x} = x^3 \cdot 1 + x^3(-x) + x^3 \cdot \frac{1}{2!}(-x)^2 + x^3 \cdot \frac{1}{3!}(-x)^3 + \cdots + x^3 \cdot \frac{1}{n!}(-x)^n + \cdots$$

$$= x^3 - x^4 + \frac{1}{2}x^5 - \frac{1}{6}x^6 + \cdots + \frac{(-1)^n x^{n+3}}{n!} + \cdots$$

The Taylor series for $x^3 e^{-x}$ has the same interval of convergence, $(-\infty, \infty)$, as the Taylor series for e^x.

17. $\dfrac{2}{1+x^2} = \dfrac{2}{1-(-x^2)}$

Use the Taylor series for $\frac{1}{1-x}$, multiply each term by 2, and replace x with $-x^2$.

$$\dfrac{2}{1+x^2} = \dfrac{2}{1-(-x^2)}$$

$$= 2\cdot 1 + 2(-x^2) + 2(-x^2)^2 + 2(-x^2)^3 + \cdots + 2(-x^2)^n + \cdots$$

$$= 2 - 2x^2 + 2x^4 - 2x^6 + \cdots + (-1)^n \cdot 2 \cdot x^{2n} + \cdots$$

The Taylor series for $\frac{1}{1-x}$ is valid when $-1 < x < 1$. Replacing x with $-x^2$ gives

$$-1 < -x^2 < 1 \text{ or } 1 > x^2 > -1.$$

The inequality is satisfied by any x in the interval $(-1, 1)$.

19. $\dfrac{e^x + e^{-x}}{2} = \dfrac{1}{2}e^x + \dfrac{1}{2}e^{-x}$

We find the Taylor series for $\frac{1}{2}e^x$ by starting with the Taylor series for e^x and multiplying each term by $\frac{1}{2}$.

$$\frac{1}{2}e^x = \frac{1}{2}\cdot 1 + \frac{1}{2}\cdot x + \frac{1}{2}\cdot\frac{1}{2!}x^2 + \frac{1}{2}\cdot\frac{1}{3!}x^3 + \cdots + \frac{1}{2}\cdot\frac{1}{n!}x^n + \cdots$$

$$= \frac{1}{2} + \frac{1}{2}x + \frac{1}{4}x^2 + \frac{1}{12}x^3 + \cdots + \frac{1}{2n!}x^n + \cdots$$

We find the Taylor series for $\frac{1}{2}e^{-x}$ by starting with the Taylor series for e^x, multiplying each term by $\frac{1}{2}$, and replacing x with $-x$.

$$\frac{1}{2}e^{-x} = \frac{1}{2}\cdot 1 + \frac{1}{2}(-x) + \frac{1}{2}\cdot\frac{1}{2!}(-x)^2 + \frac{1}{2}\cdot\frac{1}{3!}(-x)^3 + \cdots + \frac{1}{2}\cdot\frac{1}{n!}(-x)^n + \cdots$$

$$= \frac{1}{2} - \frac{1}{2}x + \frac{1}{4}x^2 - \frac{1}{12}x^3 + \cdots + \frac{(-1)^n}{2n!}x^n + \cdots$$

Use property (1) with $f(x) = \frac{1}{2}e^x$ and $g(x) = \frac{1}{2}e^{-x}$.

$$\frac{e^x + e^{-x}}{2}$$

$$= \frac{1}{2}e^x + \frac{1}{2}e^{-x}$$

$$= \left(\frac{1}{2} + \frac{1}{2}\right) + \left(\frac{1}{2} - \frac{1}{2}\right)x + \left(\frac{1}{4} + \frac{1}{4}\right)x^2 + \left(\frac{1}{12} - \frac{1}{12}\right)x^3 + \cdots + \left(\frac{1}{2n!} + \frac{(-1)^n}{2n!}\right)x^n + \cdots$$

$$= 1 + \frac{1}{2}x^2 + \frac{1}{24}x^4 + \frac{1}{720}x^6 + \cdots + \frac{1 + (-1)^n}{2n!}x^n + \cdots$$

$$= 1 + \frac{1}{2}x^2 + \frac{1}{24}x^4 + \frac{1}{720}x^6 + \cdots + \frac{1}{(2n)!}x^{2n} + \cdots$$

The new series has the same interval of convergence, $(-\infty, \infty)$, as the Taylor series for e^x.

21. Use the Taylor series for $\ln(1 + x)$ and replace x with $2x^4$.

$$\ln(1 + 2x^4) = 2x^4 - \frac{(2x^4)^2}{2} + \frac{(2x^4)^3}{3} - \frac{(2x^4)^4}{4} + \cdots + \frac{(-1)^n(2x^4)^{n+1}}{n+1} + \cdots$$

$$= 2x^4 - 2x^8 + \frac{8}{3}x^{12} - 4x^{16} + \cdots + \frac{(-1)^n \cdot 2^{n+1}x^{4n+4}}{n+1} + \cdots$$

The Taylor series for $\ln(1 + x)$ is valid when $-1 < x \le 1$. Replacing x with $2x^4$ gives

$$-1 < 2x^4 \le 1 \text{ or } -\frac{1}{2} < x^4 \le \frac{1}{2}.$$

The inequality is satisfied by any x in the interval $\left[-\frac{1}{\sqrt[4]{2}}, \frac{1}{\sqrt[4]{2}}\right]$.

23. $\dfrac{1 + x}{1 - x} = \dfrac{1}{1 - x} + \dfrac{x}{1 - x}$

The Taylor series for $\frac{1}{1-x}$ is given as

$$\frac{1}{1 - x} = 1 + x + x^2 + x^3 + \cdots + x^n + \cdots$$

We find the Taylor series for $\frac{x}{1-x}$ by starting with the Taylor series for $\frac{1}{1-x}$ and using property (3) with $k = 1$.

$$\frac{x}{1 - x} = x \cdot \frac{1}{1 - x}$$
$$= x \cdot 1 + x \cdot x + x \cdot x^2 + x \cdot x^3 + \cdots + x \cdot x^n + \cdots$$
$$= x + x^2 + x^3 + x^4 + \cdots + x^{n+1} + \cdots$$

Use property (1) with $f(x) = \frac{1}{1-x}$ and $g(x) = \frac{x}{1-x}$.

$$\frac{1 + x}{1 - x} = \frac{1}{1 - x} + \frac{x}{1 - x}$$
$$= 1 + (1 + 1)x + (1 + 1)x^2 + (1 + 1)x^3 + \cdots + (1 + 1)x^n + \cdots$$
$$= 1 + 2x + 2x^2 + 2x^3 + \cdots + 2x^n + \cdots$$

25.   The Taylor series for e^x is given as

$$e^x = 1 + x + \frac{1}{2!}x^2 + \frac{1}{3!}x^3 + \frac{1}{4!}x^4 + \frac{1}{5!}x^5 + \cdots + \frac{1}{n!}x^n + \cdots$$
$$= 1 + x + \frac{x^2}{2} + \frac{x^3}{6} + \frac{x^4}{24} + \frac{x^5}{120} + \cdots + \frac{x^n}{n!} + \cdots$$
$$= 1 + x + \frac{x^2}{2}\left[1 + \frac{x}{3} + \frac{x^2}{12} + \frac{x^3}{60} + \cdots + \frac{x^{n-2}}{\frac{n!}{2}} + \cdots\right]$$

Compare the series $\frac{x}{3} + \frac{x^2}{12} + \frac{x^3}{60} + \cdots + \frac{x^{n-2}}{\frac{n!}{2}} + \cdots$, for $n \ge 3$, with the series

$\frac{x}{3} + \frac{x^2}{12} + \frac{x^3}{48} + \cdots + \frac{x^{n-2}}{3 \cdot 4^{n-3}} + \cdots$, for $n \ge 3$. The second series is an infinite geometric series that is larger than

or equal to the first series term by term. The geometric series has $a = \frac{x}{3}$ and $r = \frac{x}{4}$, and sums to $\frac{4x}{12-3x}$.
Thus, for $x > 0$,

$$1 + x + \frac{x^2}{2} < 1 + x + \frac{x^2}{2}\left[1 + \frac{x}{3} + \frac{x^2}{12} + \frac{x^3}{60} + \cdots + \frac{x^{n-2}}{\frac{n!}{2}} + \cdots\right] < 1 + x + \frac{x^2}{2}\left[1 + \frac{4x}{12 - 3x}\right],$$

or $$1 + x + \frac{x^2}{2} < e^x < 1 + x + \frac{x^2}{2}\left[1 + \frac{4x}{12 - 3x}\right].$$

For values of x sufficiently close to 0, $\frac{4x}{12-3x}$ approaches 0, and $1 + x + \frac{x^2}{2}\left[1 + \frac{4x}{12-3x}\right] \approx 1 + x + \frac{x^2}{2}$.

Thus, $e^x \approx 1 + x + \frac{x^2}{2}$.

For $x < 0$, a similar argument can be made.

27. The Taylor series for e^x is given as

$$e^x = 1 + x + \frac{1}{2!}x^2 + \frac{1}{3!}x^3 + \frac{1}{4!}x^4 + \frac{1}{5!}x^5 + \cdots + \frac{1}{n!}x^n + \cdots$$

For $x > 0, \frac{1}{n!}x^n > 0$ when $n \geq 2$. Thus,

$$\frac{1}{2!}x^2 + \frac{1}{3!}x^3 + \frac{1}{4!}x^4 + \frac{1}{5!}x^5 + \cdots + \frac{1}{n!}x^n + \cdots > 0$$

since each term on the left-hand side is positive. Adding $1 + x$ to both sides we have

$$1 + x + \frac{1}{2!}x^2 + \frac{1}{3!}x^3 + \frac{1}{4!}x^4 + \frac{1}{5!}x^5 + \cdots + \frac{1}{n!}x^n + \cdots > 1 + x$$

or $e^x > 1 + x$.

For $x = 0, e^x = e^0 = 1 = 1 + 0 = 1 + x$.

For $-1 \leq x < 0$, when b is an even positive integer, $\frac{x^b}{b} > 0$. Also, when $a \geq 3, \frac{|x|}{a} < 1$, or $1 + \frac{x}{a} > 0$. Thus,

$\frac{x^b}{b}\left(1 + \frac{x}{a}\right) > 0$ for b an even positive integer and $a \geq 3$, since both factors are positive. Thus

$$\frac{x^2}{2!}\left(1 + \frac{x}{3}\right) + \frac{x^4}{4!}\left(1 + \frac{x}{5}\right) + \frac{x^6}{6!}\left(1 + \frac{x}{7}\right) + \cdots + \frac{x^{2n}}{(2n)!}\left(1 + \frac{x}{2n+1}\right) + \cdots > 0,$$

since each term has been shown to be positive. Consider the Taylor series for e^x,

$$e^x = 1 + x + \frac{1}{2!}x^2 + \frac{1}{3!}x^3 + \frac{1}{4!}x^4 + \frac{1}{5!}x^5 + \cdots + \frac{1}{n!}x^n + \cdots$$

$$= 1 + x + \frac{x^2}{2!}\left(1 + \frac{x}{3}\right) + \frac{x^4}{4!}\left(1 + \frac{x}{5}\right) + \cdots + \frac{x^{2n}}{(2n)!}\left(1 + \frac{x}{2n+1}\right) + \cdots$$

We have already shown the series to be positive from the third term on, so $e^x > 1 + x$.

For $x < -1, 1 + x < 0$ and $e^x > 0$. Thus $e^x > 1 + x$.

29. The area is given by

$$\int_0^{\frac{1}{3}} e^{x^2}\, dx.$$

Find the Taylor series for e^{x^2} by using the Taylor series for e^x and replacing x with x^2.

$$e^{x^2} = 1 + (x^2) + \frac{1}{2!}(x^2)^2 + \frac{1}{3!}(x^2)^3 + \frac{1}{4!}(x^2)^4 + \cdots + \frac{1}{n!}(x^2)^n + \cdots$$

Using the first five terms of this series gives

$$\int_0^{\frac{1}{3}} e^{x^2}\, dx \approx \int_0^{\frac{1}{3}}\left[1 + x^2 + \frac{1}{2!}(x^2)^2\, dx + \frac{1}{3!}(x^2)^3 + \frac{1}{4!}(x^2)^4\right] dx$$

$$= \int_0^{\frac{1}{3}}\left[1 + x^2 + \frac{1}{2}x^4 + \frac{1}{6}x^6 + \frac{1}{24}x^8\right] dx$$

$$= \left[x + \frac{1}{3}x^3 + \frac{1}{10}x^5 + \frac{1}{42}x^7 + \frac{1}{216}x^9\right]\Big|_0^{\frac{1}{3}}$$

$$= \frac{1}{3} + \frac{1}{81} + \frac{1}{2430} + \frac{1}{91,854} + \frac{1}{4,251,528} - 0$$

$$\approx 0.3461.$$

31.   The area is given by

$$\int_{\frac{1}{4}}^{\frac{1}{3}} \frac{1}{1 - \sqrt{x}} \, dx.$$

Find the Taylor series for $\frac{1}{1-\sqrt{x}}$ by using the Taylor series for $\frac{1}{1-x}$ and replacing x with $\sqrt{x}$.

$$\frac{1}{1 - \sqrt{x}} = 1 + \sqrt{x} + \sqrt{x}^2 + \sqrt{x}^3 + \cdots + \sqrt{x}^n + \cdots$$

Using the first five terms of this series gives

$$\int_{\frac{1}{4}}^{\frac{1}{3}} \frac{1}{1 - \sqrt{x}} \, dx \approx \int_{\frac{1}{4}}^{\frac{1}{3}} (1 + \sqrt{x} + x + x^{3/2} + x^2) \, dx = \left(x + \frac{2}{3} x^{3/2} + \frac{x^2}{2} + \frac{2}{5} x^{5/2} + \frac{x^3}{3} \right) \Bigg|_{\frac{1}{4}}^{\frac{1}{3}}$$

$$= \left(\frac{1}{3} + \frac{2}{3}\left(\frac{1}{3}\right)^{3/2} + \frac{1}{18} + \frac{2}{5}\left(\frac{1}{3}\right)^{5/2} + \frac{1}{81} \right) - \left(\frac{1}{4} + \frac{1}{12} + \frac{1}{32} + \frac{1}{80} + \frac{1}{192} \right)$$

$$\approx 0.1729.$$

33.   The area is given by

$$\int_{0}^{0.4} \frac{1}{\sqrt{2\pi}} e^{-x^2/2} \, dx = \frac{1}{\sqrt{2\pi}} \int_{0}^{0.4} e^{-x^2/2} \, dx.$$

Find the Taylor series for $e^{-x^2/2}$ by using the Taylor series for e^x and replacing x with $-\frac{x^2}{2}$.

$$e^{-x^2/2} = 1 - \frac{1}{2} x^2 + \frac{1}{2!2^2} x^4 - \frac{1}{3!2^3} x^6 + \frac{1}{4!2^4} x^8 + \cdots + \frac{(-1)^n}{n!2n} x^{2n} + \cdots$$

Using the first five terms of this series gives

$$\frac{1}{\sqrt{2\pi}} \int_{0}^{0.4} e^{-x^2/2} \, dx \approx \frac{1}{\sqrt{2\pi}} \int_{0}^{0.4} \left(1 - \frac{1}{2} x^2 + \frac{1}{2!2^2} x^4 - \frac{1}{3!2^3} x^6 + \frac{1}{4!2^4} x^8 \right) dx$$

$$= \frac{1}{\sqrt{2\pi}} \int_{0}^{0.4} \left(1 - \frac{1}{2} x^2 + \frac{1}{8} x^4 - \frac{1}{48} x^6 + \frac{1}{384} x^8 \right) dx$$

$$= \frac{1}{\sqrt{2\pi}} \left(x - \frac{1}{6} x^3 + \frac{1}{40} x^5 - \frac{1}{336} x^7 + \frac{1}{3456} x^9 \right) \Bigg|_{0}^{0.4}$$

$$= \frac{1}{\sqrt{2\pi}} \left(0.4 - \frac{1}{6}(0.4)^3 + \frac{1}{40}(0.4)^5 - \frac{1}{336}(0.4)^7 + \frac{1}{3456}(0.4)^9 - 0 \right)$$

$$\approx \frac{1}{\sqrt{2(3.1416)}} (0.389585)$$

$$\approx 0.1554.$$

35.   The doubling time n for a quantity that increases at an annual rate r is given by

$$n = \frac{\ln 2}{\ln(1 + r)} = \frac{\ln 2}{\ln 0.0475} \approx 14.94.$$

It will take about 14.94 years.
According to the Rule of 70, the doubling time is given by

$$\text{Doubling time} \approx \frac{70}{100r} = \frac{70}{100(0.0475)}$$

$$= 14.74.$$

It will take about 14.74 years, a difference of 0.2 years, or about 10 weeks.

37. **(a)** $\displaystyle\sum_{x=0}^{\infty} f(x) = \sum_{x=0}^{\infty} \frac{\lambda^x e^{-\lambda}}{x!}$

$$= \frac{\lambda^0 e^{-\lambda}}{0!} + \frac{\lambda^1 e^{-\lambda}}{1!} + \frac{\lambda^2 e^{-\lambda}}{2!} + \frac{\lambda^3 e^{-\lambda}}{3!} + \cdots + \frac{\lambda^n e^{-\lambda}}{n!} + \cdots$$

$$= e^{-\lambda} + \lambda e^{-\lambda} + \frac{\lambda^2 e^{-\lambda}}{2!} + \frac{\lambda^3 e^{-\lambda}}{3!} + \cdots + \frac{\lambda^n e^{-\lambda}}{n!} + \cdots$$

$$= e^{-\lambda}\left(1 + \lambda + \frac{\lambda^2}{2!} + \frac{\lambda^3}{3!} + \cdots + \frac{\lambda^n}{n!} + \cdots\right)$$

$$= e^{-\lambda} \cdot e^{\lambda}$$

$$= e^0$$

$$= 1$$

(b) $\displaystyle\sum_{x=0}^{\infty} x f(x) = \sum_{x=0}^{\infty} x \frac{\lambda^x e^{-\lambda}}{x!}$

$$= 0\frac{\lambda^0 e^{-\lambda}}{0!} + 1\frac{\lambda^1 e^{-\lambda}}{1!} + 2\frac{\lambda^2 e^{-\lambda}}{2!} + 3\frac{\lambda^3 e^{-\lambda}}{3!} + \cdots + n\frac{\lambda^n e^{-\lambda}}{n!} + \cdots$$

$$= 0 + \lambda e^{-\lambda} + \lambda^2 e^{-\lambda} + \frac{\lambda^3 e^{-\lambda}}{2!} + \cdots + \frac{\lambda^n e^{-\lambda}}{(n-1)!} + \cdots$$

$$= \lambda e^{-\lambda}\left(1 + \lambda + \frac{\lambda^2}{2!} + \cdots + \frac{\lambda^{n-1}}{(n-1)!} + \cdots\right)$$

$$= \lambda e^{-\lambda} \cdot e^{\lambda}$$

$$= \lambda e^0$$

$$= \lambda$$

(c) Since 6.17 is the expected value from part b we have $\lambda = 6.17$.

$$P(x < 4) = P(x = 0) + P(x = 1) + P(x = 2) + P(x = 3)$$

$$= \frac{6.17^0 e^{-6.17}}{0!} + \frac{6.17^1 e^{-6.17}}{1!} + \frac{6.17^2 e^{-6.17}}{2!} + \frac{6.17^3 e^{-6.17}}{3!}$$

$$= e^{-6.17}\left(1 + 6.17 + \frac{38.0689}{2} + \frac{234.885113}{6}\right)$$

$$\approx 0.1367$$

39. **(a)** The probability, p, of popping a 6 is $p = \frac{1}{6}$. From Exercise 38b, the expected value is given by

$$\sum_{x=1}^{\infty} x f(x) = \frac{1}{p} = \frac{1}{\frac{1}{6}} = 6.$$

(b) For $x \geq 4$, find

$$\sum_{x=4}^{\infty} f(x) = 1 - \sum_{x=1}^{3} f(x)$$

$$= 1 - \sum_{x=1}^{3} (1-p)^{x-1} p$$

$$= 1 - \sum_{x=1}^{3} \left(1 - \frac{1}{6}\right)^{x-1} \left(\frac{1}{6}\right)$$

$$= 1 - \left[\left(1 - \frac{1}{6}\right)^{1-1} \left(\frac{1}{6}\right) + \left(1 - \frac{1}{6}\right)^{2-1} \left(\frac{1}{6}\right) + \left(1 - \frac{1}{6}\right)^{3-1} \left(\frac{1}{6}\right)\right]$$

$$\approx 1 - 0.4213$$

$$= 0.5787.$$

12.6   Newton's Method

Your Turn 1

$$f(x) = 2x^3 - 5x^2 + 6x - 10$$

First note that $f(1) = -7$ and $f(3) = 17$, so f has a zero on $[1, 3]$. Use $c_1 = 2$ as a first guess.

For Newton's method we need to know f', which is $f'(x) = 6x^2 - 10x + 6$.

$$c_2 = c_1 - \frac{f(c_1)}{f'(c_1)}$$

$$= 2 - \frac{-2}{10}$$

$$= 2.2$$

$$c_3 = c_2 - \frac{f(c_2)}{f'(c_2)}$$

$$= 2.2 - \frac{0.296}{13.04}$$

$$\approx 2.177301$$

$$c_4 = c_3 - \frac{f(c_3)}{f'(c_3)}$$

$$\approx 2.177301 - \frac{0.004207}{12.670828}$$

$$\approx 2.176969$$

To three decimal places the root is 2.177. In fact c_4 is correct to six decimal places.

Your Turn 2

We find $\sqrt[3]{15}$ by finding a zero of the function

$f(x) = x^3 - 15.$ $f'(x) = 3x^2.$ Use 2.5 as the initial guess c_1.

$$c_2 = c_1 - \frac{f(c_1)}{f'(c_1)}$$

$$= 2.5 - \frac{0.625}{18.75}$$

$$\approx 2.466667$$

$$c_3 = c_2 - \frac{f(c_2)}{f'(c_2)}$$

$$= 2.466667 - \frac{0.008302}{18.25334}$$

$$\approx 2.466212$$

$\sqrt[3]{15} \approx 2.466.$ In fact, c_3 is correct to six decimal places.

12.6   Warmup Exercises

W1. $f(x) = 6x^{1/3} + 3\ln x$

$$f'(x) = \left(\frac{1}{3}\right)\left(6x^{-2/3}\right) + \frac{3}{x}$$

$$= 2x^{-2/3} + \frac{3}{x}$$

W2. Use the product rule.

$$f(x) = x^2 e^{-3x}$$

$$f'(x) = x^2\left(-3e^{-3x}\right) + (2x)\left(e^{-3x}\right)$$

$$= \left(-3x^2 + 2x\right)e^{-3x}$$

12.6   Exercises

1. $f(x) = 5x^2 - 3x - 3$

$f'(x) = 10x - 3$

$f(1) = -1 < 0$ and $f(2) = 11 > 0$ so a solution exists in $(1, 2)$.

Let $c_1 = 1$.

$$c_2 = c_1 - \frac{f(c_1)}{f'(c_1)} = 1 - \frac{-1}{7} = 1.1429$$

$$c_3 = c_2 - \frac{f(c_2)}{f'(c_2)} = 1.1429 - \frac{0.1020}{8.4286} = 1.1308$$

$$c_4 = c_3 - \frac{f(c_3)}{f'c_3} = 1.1308 - \frac{0.0007}{8.3075} = 1.1307$$

Subsequent approximations will agree with c_3 and c_4 to the nearest hundredth. Thus, $x = 1.13$.

3. $f(x) = 2x^3 - 6x^2 - x + 2$

$f'(x) = 6x^2 - 12x - 1$

$f(3) = -1 < 0$ and $f(4) = 30 > 0$ so a solution exists in $(3, 4)$.

Let $c_1 = 3$.

$$c_2 = c_1 - \frac{f(c_1)}{f'(c_1)} = 3 - \frac{-1}{17} = 3.0588$$

$$c_3 = c_2 - \frac{f(c_2)}{f'(c_2)} = 3.0588 - \frac{0.0419}{18.4325} = 3.0565$$

Subsequent approximations will agree with c_2 and c_3 to the nearest hundredth. Thus, $x = 3.06$.

5. $f(x) = -3x^3 + 5x^2 + 3x + 2$

$f'(x) = -9x^2 + 10x + 3$

$f(2) = 4 > 0$ and $f(3) = -25 < 0$ so a solution exists in $(2, 3)$.

Let $c_1 = 2$.

$$c_2 = c_1 - \frac{f(c_1)}{f'(c_1)} = 2 - \frac{4}{-13} = 2.3077$$

$$c_3 = c_2 - \frac{f(c_2)}{f'(c_2)} = 2.3077 - \frac{-1.3182}{-21.8521} = 2.2474$$

$$c_4 = c_3 - \frac{f(c_3)}{f'(c_3)} = 2.2474 - \frac{-0.0567}{-19.9824} = 2.2445$$

$$c_5 = c_4 - \frac{f(c_4)}{f'(c_4)} = -2.2445 - \frac{-0.0001}{-19.8960} = 2.2445$$

Subsequent approximations will agree with c_4 and c_5 to the nearest hundredth. Thus, $x = 2.24$.

7. $f(x) = 2x^4 - 2x^3 - 3x^2 - 5x - 8$

$f'(x) = 8x^3 - 6x^2 - 6x - 5$

In the interval $[-2, -1]$: $f(-2) = 38 > 0$ and $f(-1) = -2 < 0$ so a solution exists in $(-2, -1)$.

Let $c_1 = -2$.

$$c_2 = c_1 - \frac{f(c_1)}{f'(c_1)} = -2 - \frac{38}{-81} = -1.5309$$

$$c_3 = c_2 - \frac{f(c_2)}{f'(c_2)} = -1.5309 - \frac{10.7834}{-38.5773} = -1.2513$$

$$c_4 = c_3 - \frac{f(c_3)}{f'(c_3)} = -1.2513 - \frac{2.3817}{-22.5622} = -1.1458$$

$$c_5 = c_4 - \frac{f(c_4)}{f'(c_4)} = -1.1458 - \frac{0.2457}{-18.0356} = -1.1322$$

$$c_6 = c_5 - \frac{f(c_5)}{f'(c_5)} = -1.1322 - \frac{0.0036}{-17.5069} = -1.1319$$

Subsequent approximations will agree with c_5 and c_6 to the nearest hundredth. Thus, $x = -1.13$ is a solution in $[-2, -1]$.

In the interval $[2, 3]$: $f(2) = -14 < 0$ and $f(3) = 58 > 0$ so a solution exists in $(2, 3)$.

Let $c_1 = 2$.

$$c_2 = c_1 - \frac{f(c_1)}{f'(c_1)} = 2 - \frac{-14}{23} = 2.6087$$

$$c_3 = c_2 - \frac{f(c_2)}{f'(c_2)} = 2.6087 - \frac{15.6588}{80.5396} = 2.4143$$

$$c_4 = c_3 - \frac{f(c_3)}{f'(c_3)} = 2.4143 - \frac{2.2460}{58.1188} = 2.3756$$

$$c_5 = c_4 - \frac{f(c_4)}{f'(c_4)} = 2.3756 - \frac{0.0774}{54.1413} = 2.3742$$

$$c_6 = c_5 - \frac{f(c_5)}{f'(c_5)} = 2.3742 - \frac{0.0001}{53.9972} = 2.3742$$

Subsequent approximations will agree with c_5 and c_6 to the nearest hundredth. Thus, $x = 2.37$ is a solution in $[2, 3]$.

9. $f(x) = 4x^{1/3} - 2x^2 + 4$

$$f'(x) = \frac{4}{3}x^{-2/3} - 4x$$

$f(-3) = -19.7690 < 0$ and $f(0) = 4 > 0$ so a solution exists in $(-3, 0)$.

Let $c_1 = -3$.

$$c_2 = c_1 - \frac{f(c_1)}{f'(c_1)} = -3 - \frac{-19.7690}{12.6410} = -1.4361$$

$$c_3 = c_2 - \frac{f(c_2)}{f'(c_2)} = -1.4361 - \frac{-4.6378}{6.7920} = -0.7533$$

$$c_4 = c_3 - \frac{f(c_3)}{f'(c_3)} = -0.7533 - \frac{-0.7744}{4.6236} = -0.5858$$

$$c_5 = c_4 - \frac{f(c_4)}{f'(c_4)} = -0.5858 - \frac{-0.0332}{4.2477} = -0.5780$$

$$c_6 = c_5 - \frac{f(c_5)}{f'(c_5)} = -0.5780 - \frac{-0.0001}{4.2335} - 0.5780$$

Subsequent approximations will agree with c_5 and c_6 to the nearest hundredth. Thus, $x = -0.58$.

11. $f(x) = e^x + x - 2$

$$f'(x) = e^x + 1$$

$f(0) = -1 < 0$ and $f(3) = 21.085537$ so a solution exists in $(0, 3)$.

Let $c_1 = 0$.

$$c_2 = c_1 - \frac{f(c_1)}{f'(c_1)} = 0 - \frac{-1}{2} = 0.5$$

$$c_3 = c_2 - \frac{f(c_2)}{f'(c_2)} = 0.5 - \frac{0.14872}{2.6487} = 0.4439$$

$$c_4 = c_3 - \frac{f(c_3)}{f'(c_3)} = 0.4439 - \frac{0.00267}{2.5588} = 0.4429$$

Subsequent approximations will agree with c_3 and c_4 to the nearest hundredth. Thus, $x = 0.44$.

13. $f(x) = x^2 e^{-x} + x^2 - 2$

$$f'(x) = -x^2 e^{-x} + 2xe^{-x} + 2x$$

$f(0) = -2$ and $f(3) = 7.4480836$ so a solution exists in $(0, 3)$.

Let $c_1 = 0$.

$$c_2 = c_1 - \frac{f(c_1)}{f'(c_1)} = 0 - \frac{-2}{0}$$

Since c_2 is undefined, let $c_1 = 1$.

$$c_2 = c_1 - \frac{f(c_1)}{f'(c_1)} = 1 - \frac{-0.6321}{2.3679} = 1.2670$$

$$c_3 = c_2 - \frac{f(c_2)}{f'(c_2)} = 1.2670 - \frac{0.05746}{2.7956} = 1.2464$$

$$c_4 = c_4 - \frac{f(c_3)}{f'(c_3)} = 1.2464 - \frac{2.1 \cdot 10^{-4}}{2.7629} = 1.2463$$

Subsequent approximations will agree with c_3 and c_4 to the nearest hundredth. Thus, $x = 1.25$.

15. $f(x) = \ln x + x - 2$

$$f'(x) = \frac{1}{x} + 1$$

$f(1) = -1 < 0$ and $f(4) = 3.3862944 > 0$ so a solution exists in $(1, 4)$.

Let $c_1 = 1$.

$$c_2 = c_1 - \frac{f(c_1)}{f'(c_1)} = 1 - \frac{-1}{2} = 1.5$$

$$c_3 = c_2 - \frac{f(c_2)}{f'(c_2)} = 1.5 - \frac{-0.0945}{1.6667} = 1.5567$$

$$c_4 = c_3 - \frac{f(c_3)}{f'(c_3)} = 1.5567 - \frac{-7 \cdot 10^{-4}}{1.6424} = 1.5571$$

Subsequent approximations will agree with c_3 and c_4 to the nearest hundredth. Thus, $x = 1.56$.

17. $\sqrt{2}$ is a solution of $x^2 - 2 = 0$.

$$f(x) = x^2 - 2$$
$$f'(x) = 2x$$

Since $1 < \sqrt{2} < 2$, let $c_1 = 1$.

$$c_2 = 1 - \frac{-1}{2} = 1.5$$

$$c_3 = 1.5 - \frac{0.25}{3} = 1.417$$

$$c_4 = 1.417 - \frac{0.00789}{2.834} = 1.414$$

$$c_5 = 1.414 - \frac{-6 \cdot 10^{-4}}{2.828} = 1.414$$

Since $c_4 = c_5 = 1.414$, to the nearest thousandth, $\sqrt{2} = 1.414$.

19. $\sqrt{11}$ is a solution of $x^2 - 11 = 0$.

$$f(x) = x^2 - 11$$
$$f'(x) = 2x$$

Since $3 < \sqrt{11} < 4$, let $c_1 = 3$.

$$c_2 = 3 - \frac{-2}{6} = 3.333$$

$$c_3 = 3.333 - \frac{0.10889}{6.666} = 3.317$$

$$c_4 = 3.317 - \frac{0.00249}{6.634} = 3.317$$

Since $c_3 = c_4 = 3.317$, to the nearest thousandth, $\sqrt{11} = 3.317$.

21. $\sqrt{250}$ is a solution of $x^2 - 250 = 0$.

$$f(x) = x^2 - 250$$
$$f'(x) = 2x$$

Since $15 < \sqrt{250} < 16$, let $c_1 = 15$.

$$c_2 = 15 - \frac{-25}{30} = 15.833$$

$$c_3 = 15.833 - \frac{0.68389}{31.666} = 15.811$$

$$c_4 = 15.811 - \frac{-0.0123}{31.622} = 15.811$$

Since $c_3 = c_4 = 15.811$, to the nearest thousandth, $\sqrt{250} = 15.811$.

23. $\sqrt[3]{9}$ is a solution of $x^3 - 9 = 0$.

$$f(x) = x^3 - 9$$
$$f'(x) = 3x^2$$

Since $2 < \sqrt[3]{9} < 3$, let $c_1 = 2$.

$$c_2 = 2 - \frac{-1}{12} = 2.083$$

$$c_3 = 2.083 - \frac{0.03791}{13.017} = 2.080$$

$$c_4 = 2.080 - \frac{-0.0011}{12.979} = 2.080$$

Since $c_3 = c_4 = 2.080$, to the nearest thousandth, $\sqrt[3]{9} = 2.080$.

25. $\sqrt[3]{100}$ is a solution of $x^3 - 100 = 0$.

$$f(x) = x^3 - 100$$
$$f'(x) = 3x^2$$

Since $4 < \sqrt[3]{100} < 5$, let $c_1 = 4$.

$$c_2 = 4 - \frac{-36}{48} = 4.75$$

$$c_3 = 4.75 - \frac{7.1719}{67.688} = 4.644$$

$$c_4 = 4.644 - \frac{0.15592}{64.7} = 4.642$$

$$c_5 = 4.642 - \frac{0.02658}{64.644} = 4.642$$

Since $c_4 = c_5 = 4.642$, to the nearest thousandth, $\sqrt[3]{100} = 4.642$.

27. $f(x) = x^3 - 3x^2 - 18x + 4$

$f'(x) = 3x^2 - 6x - 18$

To find critical points, solve
$f'(x) = 3x^2 - 6x - 18 = 0$.

$$f''(x) = 6x - 6$$

$f'(-2) = 6 > 0$ and $f'(-1) = -9 < 0$ so a solution exists in $(-2, -1)$.

Let $c_1 = -2$.

$$c_2 = -2 - \frac{6}{-18} = -1.67$$

$$c_3 = -1.67 - \frac{0.3867}{-16.02} = -1.65$$

$$c_4 = -1.65 - \frac{0.0675}{-15.9} = -1.65$$

Subsequent approximations will agree with c_3 and c_4 to the nearest hundredth. Thus, $x = -1.65$. Since $f''(-1.65) = -15.9 < 0$, the graph has a relative maximum at $x = -1.65$.

$f'(3) = -9 < 0$ and $f'(4) = 6 > 0$ so a solution exists in $(3, 4)$.

Let $c_1 = 3$.

$$c_2 = 3 - \frac{-9}{12} = 3.75$$

$$c_3 = 3.75 - \frac{1.6875}{16.5} = 3.65$$

$$c_4 = 3.65 - \frac{0.0675}{15.9} = 3.65$$

Subsequent approximations will agree with c_3 and c_4 to the nearest hundredth. Thus, $x = 3.65$. Since $f''(3.65) = 15.9 > 0$, the graph has a relative minimum at $x = 3.65$.

29. $f(x) = x^4 - 3x^3 + 6x - 1$

$f'(x) = 4x^3 - 9x^2 + 6$

To find critical points, solve
$f'(x) = 4x^3 - 9x^2 + 6 = 0$.

$$f''(x) = 12x^2 - 18x$$

$f'(-1) = -7 < 0$ and $f'(0) = 6$ so a solution exists in $(-1, 0)$.

Let $c_1 = -1$.

$$c_2 = -1 - \frac{-7}{30} = -0.77$$

$$c_3 = -0.77 - \frac{-1.162}{20.975} = -0.71$$

$$c_4 = -0.71 - \frac{0.03146}{18.829} = -0.71$$

Subsequent approximations will agree with c_3 and c_4 to the nearest hundredth. Thus, $x = -0.71$. Since $f''(-0.71) = 18.829 > 0$, the graph has a relative minimum at $x = -0.71$.

$f'(1) = 1 > 0$ and $f'(1.5) = -0.75 < 0$ so a solution exists in $(1, 1.5)$.

Let $c_1 = 1$.

$$c_2 = 1 - \frac{1}{-6} = 1.17$$

$$c_3 = 1.17 - \frac{0.08635}{-4.633} = 1.19$$

$$c_4 = 1.19 - \frac{-0.00043}{-4.427} = 1.19$$

Subsequent approximations will agree with c_3 and c_4 to the nearest hundredth. Thus, $x = 1.19$. Since $f''(1.19) = -4.427 < 0$, the graph has a relative maximum at $x = 1.19$.

$f'(1.5) = -0.75 < 0$ and $f'(2) = 2 > 0$ so a solution exists in $(1.5, 2)$.

Let $c_1 = 1.5$.

$$c_2 = 1.5 - \frac{-0.75}{0}$$

Since c_2 is undefined, let $c_1 = 1.6$.

$$c_2 = 1.6 - \frac{-0.656}{1.92} = 1.94$$

$$c_3 = 1.94 - \frac{1.3331}{10.243} = 1.81$$

$$c_4 = 1.81 - \frac{0.23406}{6.7332} = 1.78$$

$$c_5 = 1.78 - \frac{0.04341}{5.9808} = 1.77$$

$$c_6 = 1.77 - \frac{-0.0152}{5.7348} = 1.77$$

Subsequent approximations will agree with c_5 and c_6 to the nearest hundredth. Thus, $x = 1.77$. Since $f''(1.77) = 5.7348 > 0$, the graph has a relative minimum at $x = 1.77$.

31. $f(x) = (x - 1)^{1/3}$

$$f'(x) = \frac{1}{3}(x - 1)^{-2/3}$$

$f(0) = -1 < 0$ and $f(2) = 1 > 0$ so a solution exists in $(0, 2)$.

Let $c_1 = 0$.

$$c_2 = c_1 - \frac{f(c_1)}{f'(c_1)} = 0 - \frac{-1}{0.3333} = 3$$

$$c_3 = c_2 - \frac{f(c_2)}{f'(c_2)} = 3 - \frac{1.2599}{0.2100} = -3$$

$$c_4 = c_3 - \frac{f(c_3)}{f'(c_3)} = -3 - \frac{-1.5874}{0.1323} = 9$$

$$c_5 = c_4 - \frac{f(c_4)}{f'(c_4)} = 9 - \frac{2}{0.0833} = -15$$

Not only are successive approximations alternating in sign but they are moving further and further apart. Thus, the approximations are not approaching any specific value. The method fails in this case because the derivative, $f'(x) = \frac{1}{3}(x - 1)^{-2/3}$, is undefined at $x = 1$; the function has a vertical tangent line there.

33.   The process should be used at least until the savings produced is equal to the increased costs incurred, costs incurred, or when $S(x) = C(x)$. Therefore, solve $S(x) - C(x)$.

$$
\begin{aligned}
f(x) &= S(x) - C(x) \\
&= (x^3 + 5x^2 + 9) - (x^2 + 40x + 20) \\
&= x^3 + 4x^2 - 40x - 11
\end{aligned}
$$

$$f'(x) = 3x^2 + 8x - 40$$

$$f'(x) = 3x^2 + 8x - 40$$

$f(4) = -43 < 0$ and $f(5) = 14 > 0$ so a solution exists in $(4, 5)$.

Let $c_1 = 4$.

$$c_2 = 4 - \frac{-43}{40} = 5.08$$

$$c_3 = 5.08 - \frac{20.122}{78.059} = 4.82$$

$$c_4 = 4.82 - \frac{1.1098}{68.257} = 4.80$$

$$c_5 = 4.80 - \frac{-0.248}{67.52} = 4.80$$

Subsequent approximations will agree with c_4 and c_5 to the nearest hundredth. Thus, $x = 4.80$. The process should be used for at least 4.80 years.

35. $i_2 = 0.02 - \dfrac{57(0.02) - 57(0.02)(1 + 0.02)^{-12} - 600(0.02)^2}{57[-1 + (12)(0.02)(1 + 0.02)]^{-12-1} + (1 + 0.02)^{-12}]}$

$ = 0.02 - \dfrac{0.0011177798}{-1.480804049}$

$ = 0.02075485$

$i_3 = 0.02075485 - \dfrac{57(0.02075485) - 57(0.02075485)(1 + 0.02075485)^{-12} - 600(0.02075485)^2}{57[-1 + (12)(0.02075485)(1 + 0.02075485)]^{-12-1} + (1 + 0.02075485)^{-12}]}$

$ = 0.02075485 - \dfrac{4.0638916 \cdot 10^{-6}}{-1.583924743}$

$ = 0.02075742$

12.7 L'Hospital's Rule

Your Turn 1

Find $\lim\limits_{x \to 4} \dfrac{3x - 12}{\sqrt[3]{x + 4} - 2}$.

First note that $\lim\limits_{x \to 4} 3x - 12 = 0$ and
$\lim\limits_{x \to 4} \sqrt[3]{x + 4} - 2 = 0$, so the conditions of l'Hospital's
rule are satisfied. Differentiate the numerator and
denominator.

For $f(x) = 3x - 12$, $f'(x) = 3$.

For $g(x) = \sqrt[3]{x + 4} - 2$, $g'(x) = \dfrac{1}{3}(x + 4)^{-2/3}$.

$\lim\limits_{x \to 4} \dfrac{f'(x)}{g'(x)} = \lim\limits_{x \to 4} \dfrac{3}{(1/3)(x + 4)^{-2/3}}$

$\phantom{\lim\limits_{x \to 4} \dfrac{f'(x)}{g'(x)}} = \dfrac{3}{(1/3)(8)^{-2/3}}$

$\phantom{\lim\limits_{x \to 4} \dfrac{f'(x)}{g'(x)}} = 36$

By l'Hospital's rule, $\lim\limits_{x \to 4} \dfrac{3x - 12}{\sqrt[3]{x + 4} - 2} = 36$.

Your Turn 2

Find $\lim\limits_{x \to 1} \dfrac{e^{x-1} - 1}{x^2 - 2x + 1}$.

First note that $\lim\limits_{x \to 1} e^{x-1} - 1 = 0$ and $\lim\limits_{x \to 1} x^2 - 2x + 1$
$= 0$, so the conditions of l'Hospital's rule are satisfied.
Differentiate the numerator and denominator.

For $f(x) = e^{x-1} - 1$, $f'(x) = e^{x-1}$.

For $g(x) = x^2 - 2x + 1$, $g'(x) = 2x - 2$.

$\lim\limits_{x \to 1} \dfrac{f'(x)}{g'(x)} = \lim\limits_{x \to 1} \dfrac{e^{x-1}}{2x - 2}$

The numerator has limit 1 but the denominator has limit 0,
so the limit of the quotient does not exist.

By l'Hospital's rule, $\lim\limits_{x \to 1} \dfrac{e^{x-1} - 1}{x^2 - 2x + 1}$ does not exist.

Your Turn 3

Find $\lim\limits_{x \to 0} \dfrac{x^3}{\ln(x + 1)}$.

First note that $\lim\limits_{x \to 0} x^3 = 0$ and $\lim\limits_{x \to 0} \ln(x + 1) = 0$, so
the conditions of l'Hospital's rule are satisfied.
Differentiate the numerator and denominator.

For $f(x) = x^3$, $f'(x) = 3x^2$.

For $g(x) = \ln(x + 1)$, $g'(x) = \dfrac{1}{x + 1}$.

$\lim\limits_{x \to 0} \dfrac{f'(x)}{g'(x)} = \lim\limits_{x \to 0} \dfrac{3x^2}{\frac{1}{x+1}}$

$\phantom{\lim\limits_{x \to 0} \dfrac{f'(x)}{g'(x)}} = \lim\limits_{x \to 0} [3x^2(x + 1)]$

$\phantom{\lim\limits_{x \to 0} \dfrac{f'(x)}{g'(x)}} = 0$

By l'Hospital's rule, $\lim\limits_{x \to 0} \dfrac{x^3}{\ln(x+1)} = 0$.

Your Turn 4

Find $\lim\limits_{x \to 0} \dfrac{e^{3x} - \frac{9}{2}x^2 - 3x - 1}{x^3}$.

First note that $\lim\limits_{x \to 0} e^{3x} - \frac{9}{2}x^2 - 3x - 1 = 1 - 1 = 0$,

and $\lim\limits_{x \to 0} x^3 = 0$, so the conditions of l'Hospital's rule are satisfied. Differentiate the numerator and denominator.

For $f(x) = e^{3x} - \dfrac{9}{2}x^2 - 3x - 1$,

$$f'(x) = 3e^{3x} - 9x - 3.$$

For $g(x) = x^3$, $g'(x) = 3x^2$.

$$\lim\limits_{x \to 0} \frac{f'(x)}{g'(x)} = \lim\limits_{x \to 0} \frac{3e^{3x} - 9x - 3}{3x^2}$$

Both the numerator and denominator have limit equal to 0, so we apply l'Hospital's rule again with

$$f(x) = 3e^{3x} - 9x - 3 \quad \text{and} \quad g(x) = 3x^2.$$
$$f'(x) = 9e^{3x} - 9$$
$$g'(x) = 6x$$

We see that both f' and g' still have limit 0 as $x \to 0$, so we need one more application of l'Hospital's rule, now with $f(x) = 9e^{3x} - 9$ and $g(x) = 6x$.

$$f'(x) = 27e^{3x}$$
$$g'(x) = 6$$

Thus $\lim\limits_{x \to 0} \dfrac{e^{3x} - \frac{9}{2}x^2 - 3x - 1}{x^3} = \lim\limits_{x \to 0} \dfrac{27e^{3x}}{6}$

$$= \frac{27}{6}$$
$$= \frac{9}{2}.$$

Your Turn 5

Find $\lim\limits_{x \to 0^+} x^2 \ln(3x)$.

Write this as

$$\lim\limits_{x \to 0^+} \frac{\ln(3x)}{\frac{1}{x^2}}.$$

Now both numerator and denominator become infinite in magnitude as $x \to 0$, so we can apply l'Hospital's rule, differentiating the numerator and the denominator.

$$\lim\limits_{x \to 0^+} \frac{\ln(3x)}{\frac{1}{x^2}} = \lim\limits_{x \to 0^+} \frac{3/x}{-2/x^3}$$

$$= \lim\limits_{x \to 0^+} -\frac{3}{2}x^2 = 0$$

Thus $\lim\limits_{x \to 0^+} x^2 \ln(3x) = 0$.

Your Turn 6

Find $\lim\limits_{x \to \infty} \dfrac{\ln x}{e^x}$ and $\lim\limits_{x \to \infty} \dfrac{\ln x}{x^2}$.

Each expression is a quotient in which both the numerator and denominator tend to infinity as $x \to \infty$, so we can apply l'Hospital's rule.

$$\lim\limits_{x \to \infty} \frac{\ln x}{e^x} = \lim\limits_{x \to \infty} \frac{1/x}{e^x} = \lim\limits_{x \to \infty} \frac{1}{xe^x} = 0.$$

$$\lim\limits_{x \to \infty} \frac{\ln x}{x^2} = \lim\limits_{x \to \infty} \frac{1/x}{2x} = \lim\limits_{x \to \infty} \frac{1}{2x^2} = 0.$$

12.7   Warmup Exercises

W1. $f(x) = \sqrt{3x^2 + 4x + 5}$

$$f'(x) = (6x + 4)\left(\frac{1}{2}\left(3x^2 + 4x + 5\right)^{-1/2}\right)$$

$$= \frac{3x + 2}{\sqrt{3x^2 + 4x + 5}}$$

W2. $f(x) = \left(\ln(5x + 3)\right)^2$

$$f'(x) = (2)\left(\ln(5x + 3)\right)\frac{d}{dx}\ln(5x + 3)$$

$$= (2)\left(\ln(5x + 3)\right)\frac{5}{5x + 3}$$

$$= \frac{10\ln(5x + 3)}{5x + 3}$$

12.7   Exercises

1. The limit in the numerator is 0, as is the limit in the denominator, so that l'Hospital's rule applies. Taking derivatives separately in the numerator and denominator gives

$$\lim_{x \to 1} \frac{3x^2 + 2x - 1}{2x - 1} = \frac{3(1)^2 + 2(1) - 1}{2(1) - 1} = 4.$$

By l'Hospital's rule,

$$\lim_{x \to 1} \frac{x^3 + x^2 - x - 1}{x^2 - x} = 4.$$

3. The limit in the numerator is 0, as is the limit in the denominator, so that l'Hospital's rule applies. Taking derivatives separately in the numerator and denominator gives

$$\lim_{x \to 0} \frac{5x^4 - 6x^2 + 8x}{40x^4 - 4x + 5} = \frac{5(0)^4 - 6(0)^2 + 8(0)}{40(0)^4 - 4(0) + 5} = 0.$$

By l'Hospital's rule,

$$\lim_{x \to 0} \frac{x^5 - 2x^3 + 4x^2}{8x^5 - 2x^2 + 5x} = 0.$$

5. The limit in the numerator is 0, as is the limit in the denominator, so that l'Hospital's rule applies. Taking derivatives separately in the numerator and denominator gives

$$\lim_{x \to 2} \frac{\frac{1}{x-1}}{1} = \frac{1}{2 - 1} = 1.$$

By l'Hospital's rule,

$$\lim_{x \to 2} \frac{\ln(x - 1)}{x - 2} = 1.$$

7. The limit in the numerator is 0, as is the limit in the denominator, so that l'Hospital's rule applies. Taking derivatives separately in the numerator and denominator gives

$$\lim_{x \to 0} \frac{e^x}{4x^3} \text{ which does not exist.}$$

By l'Hospital's rule,

$$\lim_{x \to 0} \frac{e^x - 1}{x^4} \text{ does not exist.}$$

9. The limit in the numerator is 0, as is the limit in the denominator, so that l'Hospital's rule applies. Taking derivatives separately in the numerator and denominator gives

$$\lim_{x \to 0} \frac{e^x + xe^x}{e^x} = \frac{e^0 + 0 \cdot e^0}{e^0} = 1.$$

By l'Hospital's rule,

$$\lim_{x \to 0} \frac{xe^x}{e^x - 1} = 1.$$

11. $\lim_{x \to 0} e^x = 1$ and l'Hospital's rule does not apply. However,

$$\lim_{x \to 0} \frac{e^x}{2x^3 + 9x^2 - 11x} \text{ does not exist.}$$

13. The limit in the numerator is 0, as is the limit in the denominator, so that l'Hospital's rule applies. Taking derivatives separately in the numerator and denominator gives

$$\lim_{x \to 0} \frac{\frac{1}{2(2+x)^{1/2}}}{1} = \frac{1}{2(2)^{1/2}} = \frac{1}{2\sqrt{2}} \text{ or } \frac{\sqrt{2}}{4}.$$

By l'Hospital's rule,

$$\lim_{x \to 0} \frac{\sqrt{2 + x} - \sqrt{2}}{x} = \frac{1}{2\sqrt{2}} \text{ or } \frac{\sqrt{2}}{4}.$$

15. The limit in the numerator is 0, as is the limit in the denominator, so that l'Hospital's rule applies. Taking derivatives separately in the numerator and denominator gives

$$\lim_{x \to 4} \frac{\frac{1}{2x^{1/2}}}{1} = \frac{1}{2 \cdot 4^{1/2}} = \frac{1}{4}.$$

By l'Hospital's rule,

$$\lim_{x \to 4} \frac{\sqrt{x} - 2}{x - 4} = \frac{1}{4}.$$

17. The limit in the numerator is 0, as is the limit in the denominator, so that l'Hospital's rule applies. Taking derivatives separately in the numerator and denominator gives

$$\lim_{x \to 8} \frac{\frac{1}{3x^{2/3}}}{1} = \frac{1}{3 \cdot 8^{2/3}} = \frac{1}{12}.$$

By l'Hospital's rule,

$$\lim_{x \to 8} \frac{\sqrt[3]{x} - 2}{x - 8} = \frac{1}{12}.$$

19. The limit in the numerator is 0, as is the limit in the denominator, so that l'Hospital's rule applies. Taking derivatives separately in the numerator and denominator gives

$$\lim_{x \to 1} \frac{9x^8 + 24x^7 + 20x^4}{1} = \frac{9 + 24 + 20}{1} = 53.$$

By l'Hospital's rule,

$$\lim_{x \to 1} \frac{x^9 + 3x^8 + 4x^5 - 8}{x - 1} = 53.$$

21. The limit in the numerator is 0, as is the limit in the denominator, so that l'Hospital's rule applies. Taking derivatives separately in the numerator and denominator gives

$$\lim_{x \to 0} \frac{e^x - e^{-x}}{1} = e^0 - e^0 = 0.$$

By l'Hospital's rule,

$$\lim_{x \to 0} \frac{e^x + e^{-x} - 2}{x} = 0.$$

23. The limit in the numerator is 0, as is the limit in the denominator, so that l'Hospital's rule applies. Taking derivatives separately in the numerator and denominator gives

$$\lim_{x \to 3} \frac{\frac{x}{(x^2+7)^{1/2}}}{2x} = \lim_{x \to 3} \frac{1}{2(x^2 + 7)^{1/2}}$$

$$= \frac{1}{2(3^2 + 7)^{1/2}} = \frac{1}{8}.$$

By l'Hospital's rule,

$$\lim_{x \to 3} \frac{\sqrt{x^2 + 7} - 4}{x^2 - 9} = \frac{1}{8}.$$

25. The limit in the numerator is 0, as is the limit in the denominator, so that l'Hospital's rule applies. Taking derivatives separately in the numerator and denominator gives

$$\lim_{x \to 0} \frac{\frac{1}{3} - \frac{1}{3(1+x)^{2/3}}}{2x} = \frac{\frac{1}{3} - \frac{1}{3(1+0)^{2/3}}}{2 \cdot 0} = \frac{0}{0}.$$

Using l'Hospital's rule a second time, gives

$$\lim_{x \to 0} \frac{\frac{2}{9(1+x)^{5/3}}}{2} = \lim_{x \to 0} \frac{1}{(1 + x)^{5/3}}$$

$$= \frac{1}{9(1 + 0)^{5/3}} = \frac{1}{9}.$$

By l'Hospital's rule,

$$\lim_{x \to 0} \frac{1 + \frac{1}{3}x - (1 + x)^{1/3}}{x^2} = \frac{1}{9}.$$

27. The limit in the numerator is 0, as is the limit in the denominator, so that l'Hospital's rule applies. Taking derivatives separately in the numerator and denominator gives

$$\lim_{x \to 0} \frac{\frac{1}{2(1+x)^{1/2}} + \frac{1}{2(1-x)^{1/2}}}{1}$$

$$= \lim_{x \to 0} \left(\frac{1}{2(1 + x)^{1/2}} + \frac{1}{2(1 - x)^{1/2}} \right)$$

$$= \frac{1}{2(1 + 0)^{1/2}} + \frac{1}{2(1 - 0)^{1/2}} = 1.$$

By l'Hospital's rule,

$$\lim_{x \to 0} \frac{\sqrt{1 + x} - \sqrt{1 - x}}{x} = 1.$$

29. $\lim\limits_{x \to 0} \sqrt{x^2 - 5x + 4} = 2$ and l'Hospital's rule does not apply. However,

$$\lim_{x \to 0} \frac{\sqrt{x^2 - 5x + 4}}{x} \text{ does not exist.}$$

31. The limit in the numerator is 0, as is the limit in the denominator, so that l'Hospital's rule applies. Taking derivatives separately in the numerator and denominator gives

$$\lim_{x \to 0} \frac{\ln(x + 1) + \frac{5+x}{x+1}}{e^x}$$

$$= \lim_{x \to 0} \frac{(x + 1)\ln(x + 1) + 5 + x}{e^x(x + 1)}$$

$$= \frac{(0 + 1)\ln(0 + 1) + 5 + 0}{e^0(0 + 1)} = 5.$$

By l'Hospital's rule,

$$\lim_{x \to 0} \frac{(5 + x)\ln(x + 1)}{e^x - 1} = 5.$$

33. We have a limit of the form $0 \times \infty$. Rewrite the expression.

$$x^2(\ln x)^2 = \frac{(\ln x)^2}{\frac{1}{x^2}}$$

Now both the numerator and denominator become infinite and l'Hospital's rule applies to the limit of the form ∞/∞. Differentiate the numerator and denominator.

$$\lim_{x \to 0^+} \frac{(\ln x)^2}{\frac{1}{x^2}} = \lim_{x \to 0^+} \frac{\frac{2(\ln x)}{x}}{\frac{-2}{x^3}}$$

$$= \lim_{x \to 0^+} -x^2 \ln x$$

This problem is similar to what we started with so we handle it in the same manner.

$$\lim_{x \to 0^+} -x^2 \ln x = \lim_{x \to 0^+} \frac{-\ln x}{\frac{1}{x^2}}$$

$$= \lim_{x \to 0^+} \frac{\frac{-1}{x}}{\frac{-2}{x^3}}$$

$$= \lim_{x \to 0^+} \frac{x^2}{2} = 0$$

Therefore, by l'Hospital's rule,

$$\lim_{x \to 0^+} x^2 (\ln x)^2 = 0.$$

35. We have a limit of the form $0 \times \infty$. Rewrite the expression.

$$x \ln(e^x - 1) = \frac{(\ln e^x - 1)}{\frac{1}{x}}$$

Now both the numerator and denominator become infinite and l'Hospital's rule applies to the limit of the form ∞/∞. Differentiate the numerator and denominator.

$$\lim_{x \to 0^+} \frac{\ln(e^x - 1)}{\frac{1}{x}} = \lim_{x \to 0^+} \frac{\frac{e^x}{(e^x - 1)}}{\frac{-1}{x^2}} = \lim_{x \to 0^+} \frac{-x^2 e^x}{e^x - 1}$$

Notice that $\frac{e^x}{e^x - 1} = 1 + \frac{1}{e^x - 1}$.

$$\lim_{x \to 0^+} \frac{-x^2 e^x}{e^x - 1}$$

$$= \lim_{x \to 0^+} -x^2 \cdot \left(1 + \frac{1}{e^x - 1}\right)$$

$$= 0 \cdot 1 = 0$$

Therefore, by l'Hospital's rule,

$$\lim_{x \to 0^+} x \ln(e^x - 1) = 0.$$

37. The limit in the numerator is ∞, as in the limit in the denominator, so l'Hospital's rule applies. Taking derivatives separately in the numerator and the denominator gives

$$\lim_{x \to \infty} \frac{\sqrt{x}}{\ln(\ln x)} = \lim_{x \to \infty} \frac{\frac{1}{(2\sqrt{x})}}{\frac{1}{(x \ln x)}}$$

$$= \lim_{x \to \infty} \frac{\sqrt{x} \ln x}{2} = \infty.$$

By l'Hospital's rule,

$$\lim_{x \to \infty} \frac{\sqrt{x}}{\ln(\ln x)} = \infty \text{ (does not exist).}$$

39. The limit in the numerator is ∞, as is the limit in the denominator, so l'Hospital's rule applies. Taking derivatives separately in the numerator and the denominator gives

$$\lim_{x \to 0^+} \frac{\ln(e^x + 1)}{5x} = \lim_{x \to 0^+} \frac{\frac{e^x}{(e^x + 1)}}{5}$$

$$= \lim_{x \to 0^+} \frac{e^x}{5(e^x + 1)}$$

Notice that $\frac{e^x}{e^x + 1} = 1 - \frac{1}{e^x + 1}$.

$$\lim_{x \to 0^+} \frac{e^x}{5(e^x + 1)} = \frac{1}{5} \cdot \lim_{x \to 0^+} \frac{e^x}{e^x + 1}$$

$$= \frac{1}{5} \cdot \lim_{x \to 0^+} \left(1 - \frac{1}{e^x + 1}\right)$$

$$= \frac{1}{5} \cdot 1 = \frac{1}{5}$$

Therefore, by l'Hospital's rule,

$$\lim_{x \to 0^+} \frac{\ln(e^x + 1)}{5x} = \frac{1}{5}.$$

41. We have a limit of the form $\infty \cdot 0$. Rewrite the expression as a quotient.

$$x^5 e^{-0.001x} = \frac{x^5}{e^{0.001x}}$$

The limit in the numerator is ∞, as is the limit in the denominator, so l'Hospital's rule applies. Taking derivatives separately in the numerator and the denominator gives

$$\lim_{x \to \infty} \frac{x^5}{e^{0.001x}} = \lim_{x \to \infty} \frac{5x^4}{0.001 e^{0.001x}}$$

This problem is similar to what we started with so we handle it in the same manner. In fact, continue the process four more times.

$$\lim_{x \to \infty} \frac{5x^4}{0.001e^{0.001x}} = \lim_{x \to \infty} \frac{20x^3}{(0.001)^2 e^{0.001x}}$$

$$= \lim_{x \to \infty} \frac{60x^2}{(0.001)^3 e^{0.001x}}$$

$$= \lim_{x \to \infty} \frac{120x}{(0.001)^4 e^{0.001x}}$$

$$= \lim_{x \to \infty} \frac{120}{(0.001)^5 e^{0.001x}}$$

$$= 0$$

Therefore, by l'Hospital's rule,

$$\lim_{x \to \infty} x^5 e^{-0.001x} = 0.$$

43. Find $\lim_{x \to 0} \left(\dfrac{e^x}{x^2} - \dfrac{1}{x^2} - \dfrac{1}{x} \right)$.

Rewrite over a common denominator.

$$\lim_{x \to 0} \left(\frac{e^x - 1 - x}{x^2} \right)$$

The numerator and denominator have limit 0, so l'Hospital's rule applies. Differentiate the numerator and denominator.

$$\lim_{x \to 0} \left(\frac{e^x - 1 - x}{x^2} \right) = \lim_{x \to 0} \left(\frac{e^x - 1}{2x} \right)$$

The numerator and denominator still have limit 0, so differentiate the numerator and denominator again.

$$\lim_{x \to 0} \left(\frac{e^x - 1}{2x} \right) = \lim_{x \to 0} \left(\frac{e^x}{2} \right) = \frac{1}{2}$$

Thus

$$\lim_{x \to 0} \left(\frac{e^x}{x^2} - \frac{1}{x^2} - \frac{1}{x} \right) = \frac{1}{2}.$$

45.   Find $\displaystyle\lim_{x\to 1}\left(\frac{x}{x-1}-\frac{1}{\ln x}\right)$.

Rewrite over a common denominator: $\displaystyle\lim_{x\to 1}\left(\frac{x\ln x-x+1}{(x-1)\ln(x)}\right)$.

The numerator and denominator have limit 0, so l'Hospital's rule applies. Differentiate the numerator and denominator.

$$\lim_{x\to 1}\left(\frac{x\ln x-x+1}{(x-1)\ln(x)}\right)=\lim_{x\to 1}\left(\frac{\ln x}{\dfrac{x-1}{x}+\ln x}\right)=\lim_{x\to 1}\left(\frac{x\ln x}{x-1+x\ln x}\right)$$

The numerator and denominator still have limit 0, so differentiate the numerator and denominator again.

$$\lim_{x\to 1}\left(\frac{x\ln x}{x-1+x\ln x}\right)=\lim_{x\to 1}\left(\frac{1+\ln x}{1+1+\ln x}\right)=\frac{1}{2}$$

Thus

$$\lim_{x\to 1}\left(\frac{x}{x-1}-\frac{1}{\ln x}\right)=\frac{1}{2}.$$

47. $\displaystyle\lim_{x\to 0}x^2+3\neq 0$, so l'Hospital's rule does not apply.

49. (b)   Since $s(b)$ and $1-b$ are both 0 at $b=1$, we can apply

l'Hospital's rule and evaluate $\displaystyle\lim_{b\to 1}\frac{s(b)}{1-b}$ as

$$\frac{\dfrac{d}{db}s(b)}{\dfrac{d}{db}(1-b)}=\frac{s'(1)}{-1}=-s'(1).$$

Chapter 12 Review Exercises

1. True

2. False: The amounts are a constant arithmetic sequence.

3. True

4. True

5. False: In general the fifth derivatives will be different at 0.

6. True

7. False: It converges as long as $-1 < r < 1$.

8. True

9. True

10. False: It converges only for x in $(-1, 1]$.

11. False: Newton's method may fail to converge by alternating between two values, neither of which is a zero.

12. False: The rule applies to limits of quotients, not derivatives of quotients.

13. $a_4 = a_1 \cdot r^3$

$\quad = 5(-2)^3$

$\quad = -40$

$a_n = 5(-2)^{n-1}$

$S_5 = \dfrac{a_1(r^5 - 1)}{r - 1}$

$\quad = \dfrac{5[(-2)^5 - 1]}{(-2) - 1}$

$\quad = \dfrac{5(-33)}{-3}$

$\quad = 55$

15. $a_4 = a_1 \cdot r^3$

$\quad = 27\left(\dfrac{1}{3}\right)^3$

$\quad = 1$

$a_n = 27\left(\dfrac{1}{3}\right)^{n-1}$

$S_5 = \dfrac{a_1(r^5 - 1)}{r - 1}$

$\quad = \dfrac{27[(1/3)^5 - 1]}{(1/3) - 1}$

$\quad = \dfrac{27\left(-\dfrac{242}{243}\right)}{-\dfrac{2}{3}}$

$\quad = \dfrac{242}{6}$

$\quad = \dfrac{121}{3}$

17.

Derivative	Value at 0
$f(x) = e^{2-x}$	$f(0) = e^2$
$f^{(1)}(x) = -e^{2-x}$	$f^{(1)}(0) = -e^2$
$f^{(2)}(x) = e^{2-x}$	$f^{(2)}(0) = e^2$
$f^{(3)}(x) = -e^{2-x}$	$f^{(3)}(0) = -e^2$
$f^{(4)}(x) = e^{2-x}$	$f^{(4)}(0) = e^2$

$$P_4(x) = f(0) + \frac{f^{(1)}(0)}{1!}x + \frac{f^{(2)}(0)}{2!}x^2 + \frac{f^{(3)}(0)}{3!}x^3 + \frac{f^{(4)}(0)}{4!}x^4$$

$$= e^2 + \frac{-e^2}{1!}x + \frac{e^2}{2!}x^2 + \frac{-e^2}{3!}x^3 + \frac{e^2}{4!}x^4$$

$$= e^2 - e^2 x + \frac{e^2}{2}x^2 - \frac{e^2}{6}x^3 + \frac{e^2}{24}x^4$$

19.

Derivative	Value at 0
$f(x) = \sqrt{x+1} = (x+1)^{1/2}$	$f(0) = 1$
$f^{(1)}(x) = \frac{1}{2}(x+1)^{-1/2} = \frac{1}{2(x+1)^{1/2}}$	$f^{(1)}(0) = \frac{1}{2}$
$f^{(2)}(x) = -\frac{1}{4}(x+1)^{-3/2} = -\frac{1}{4(x+1)^{3/2}}$	$f^{(2)}(0) = -\frac{1}{4}$
$f^{(3)}(x) = \frac{3}{8}(x+1)^{-5/2} = \frac{3}{8(x+1)^{5/2}}$	$f^{(3)}(0) = \frac{3}{8}$
$f^{(4)}(x) = -\frac{15}{16}(x+1)^{-7/2} = -\frac{15}{16(x+1)7/2}$	$f^{(4)}(0) = -\frac{15}{16}$

$$P_4(x) = f(0) + \frac{f^{(1)}(0)}{1!}x + \frac{f^{(2)}(0)}{2!}x^2 + \frac{f^{(3)}(0)}{3!}x^3 + \frac{f^{(4)}(0)}{4!}x^4$$

$$= 1 + \frac{\frac{1}{2}}{1}x + \frac{-\frac{1}{4}}{2}x^2 + \frac{\frac{3}{8}}{6}x^3 + \frac{-\frac{15}{16}}{24}x^4$$

$$= 1 + \frac{1}{2}x - \frac{1}{8}x^2 + \frac{1}{16}x^3 - \frac{5}{128}x^4$$

21.

Derivative	Value at 0
$f(x) = \ln(2 - x)$	$f(0) = \ln 2$
$f^{(1)}(x) = -\dfrac{1}{2 - x} = -(2 - x)^{-1}$	$f^{(1)}(0) = -\dfrac{1}{2}$
$f^{(2)}(x) = -(2 - x)^{-2} = -\dfrac{1}{(2 - x)^2}$	$f^{(2)}(0) = -\dfrac{1}{4}$
$f^{(3)}(x) = -2(2 - x)^{-3} = -\dfrac{2}{(2 - x)^3}$	$f^{(3)}(0) = -\dfrac{1}{4}$
$f^{(4)}(x) = -6(2 - x)^{-4} = -\dfrac{6}{(2 - x)^4}$	$f^{(4)}(0) = -\dfrac{3}{8}$

$$P_4(x) = f(0) + \frac{f^{(1)}(0)}{1!}x + \frac{f^{(2)}(0)}{2!}x^2 + \frac{f^{(3)}(0)}{3!}x^3 + \frac{f^{(4)}(0)}{4!}x^4$$

$$= \ln 2 + \frac{-\frac{1}{2}}{1}x + \frac{-\frac{1}{4}}{2}x^2 + \frac{-\frac{1}{4}}{6}x^3 + \frac{-\frac{3}{8}}{24}x^4$$

$$= \ln 2 - \frac{1}{2}x - \frac{1}{8}x^2 - \frac{1}{24}x^3 - \frac{1}{64}x^4$$

23.

Derivative	Value at 0
$f(x) = (1 + x)^{2/3}$	$f(0) = 1$
$f^{(1)}(x) = \dfrac{2}{3}(1 + x)^{-1/3} = \dfrac{2}{3(1 + x)^{1/3}}$	$f^{(1)}(0) = \dfrac{2}{3}$
$f^{(2)}(x) = -\dfrac{2}{9}(1 + x)^{-4/3} = -\dfrac{2}{9(1 + x)^{4/3}}$	$f^{(2)}(0) = -\dfrac{2}{9}$
$f^{(3)}(x) = \dfrac{8}{27}(1 + x)^{-7/3} = \dfrac{8}{27(1 + x)^{7/3}}$	$f^{(3)}(0) = \dfrac{8}{27}$
$f^{(4)}(x) = -\dfrac{56}{81}(1 + x)^{-10/3} = -\dfrac{56}{81(1 + x)^{10/3}}$	$f^{(4)}(0) = -\dfrac{56}{81}$

$$P_4(x) = f(0) + \frac{f^{(1)}(0)}{1!}x + \frac{f^{(2)}(0)}{2!}x^2 + \frac{f^{(3)}(0)}{3!}x^3 + \frac{f^{(4)}(0)}{4!}x^4$$

$$= 1 + \frac{\frac{2}{3}}{1}x + \frac{-\frac{2}{9}}{2}x^2 + \frac{\frac{8}{27}}{6}x^3 + \frac{-\frac{56}{81}}{24}x^4$$

$$= 1 + \frac{2}{3}x - \frac{1}{9}x^2 + \frac{4}{81}x^3 - \frac{7}{243}x^4$$

25. Using the result of Exercise 17, with $f(x) = e^{2-x}$ and $P_4(x) = e^2 - e^2 x + \frac{e^2}{2}x^2 - \frac{e^2}{6}x^3 + \frac{e^2}{24}x^4$, we can approximate $e^{1.93}$ by evaluating $f(0.07) = e^{2-0.07} = e^{2-0.07} = e^{1.93}$. Using $P_4(x)$ from Exercise 17 with $x = 0.07$ gives

$$P_4(0.07) = e^2 - e^2(0.07) + \frac{e^2}{2}(0.07)^2 - \frac{e^2}{6}(0.07)^3 + \frac{e^2}{24}(0.07)^4$$

$$\approx 6.88951034388.$$

To four decimal places, $P_4(0.07)$ approximates the value of $e^{1.93}$ as 6.8895.

27. Using the result of Exercise 19, with $f(x) = \sqrt{x+1}$ and $P_4(x) = 1 + \frac{1}{2}x - \frac{1}{8}x^2 + \frac{1}{16}x^3 - \frac{5}{128}x^4$, we can approximate $\sqrt{1.03}$ by evaluating $f(0.03) = \sqrt{0.03+1} = \sqrt{1.03}$. Using $P_4(x)$ from Exercise 19 with $x = 0.03$ gives

$$P_4(0.03) = 1 + \frac{1}{2}(0.03) - \frac{1}{8}(0.03)^2 + \frac{1}{16}(0.03)^3 - \frac{5}{128}(0.03)^4$$
$$\approx 1.01488915586.$$

To four decimal places, $P_4(0.03)$ approximates the value of $\sqrt{1.03}$ as 1.0149.

29. Using the result of Exercise 21, with $f(x) = \ln(2-x)$ and $P_4(x) = \ln 2 - \frac{1}{2}x - \frac{1}{8}x^2 - \frac{1}{24}x^3 - \frac{1}{64}x^4$, we can approximate $\ln 2.05$ by evaluating $f(-0.05) = \ln(2 - (-0.05)) = \ln 2.05$. Using $P_4(x)$ from Exercise 21 with $x = -0.05$ gives

$$P_4(-0.05) = \ln 2 - \frac{1}{2}(-0.05) - \frac{1}{8}(-0.05)^2 - \frac{1}{24}(-0.05)^3 - \frac{1}{64}(-0.05)^4$$
$$\approx 0.717842610677 \text{ (using } \ln 2 = 0.69315).$$

To four decimal places, $P_4(-0.05)$ approximates the value of $\ln 2.05$ as 0.7178.

31. Using the result of Exercise 23, with $f(x) = (1+x)^{2/3}$ and $P_4(x) = 1 + \frac{2}{3}x - \frac{1}{9}x^2 + \frac{4}{81}x^3 - \frac{7}{243}x^4$, we can approximate $(0.92)^{2/3}$ by evaluating $f(-0.08) = (1 + (-0.08))^{2/3} = (0.92)^{2/3}$. Using $P_4(x)$ from Exercise 23 with $x = -0.08$ gives

$$P_4(-0.08) = 1 + \frac{2}{3}(-0.08) - \frac{1}{9}(-0.08)^2 + \frac{4}{81}(-0.08)^3 - \frac{7}{243}(-0.08)^4$$
$$\approx 0.945929091687.$$

To four decimal places, $P_4(-0.08)$ approximates the value of $(0.92)^{2/3}$ as 0.9459.

33. $9 - 6 + 4 - \frac{8}{3} + \cdots$ is a geometric series with $a = a_1 = 9$ and $r = -\frac{2}{3}$. Since r is in $(-1, 1)$, the series converges and has sum

$$\frac{a}{1-r} = \frac{9}{1 - (-\frac{2}{3})} = \frac{9}{1 + \frac{2}{3}} = \frac{9}{\frac{5}{3}} = \frac{27}{5}.$$

35. $3 + 9 + 27 + 81 + \cdots$ is a geometric series with $a = a_1 = 3$ and $r = 3$. Since $r > 1$, the series diverges.

37. $\frac{2}{5} - \frac{2}{25} + \frac{2}{125} - \frac{2}{625} + \cdots$ is a geometric series with $a = a_1 = \frac{2}{5}$ and $r = -\frac{1}{5}$. Since r is in $(-1, 1)$, the series converges and has sum

$$\frac{a}{1-r} = \frac{\frac{2}{5}}{1 - (-\frac{1}{5})} = \frac{\frac{2}{5}}{1 + \frac{1}{5}} = \frac{\frac{2}{5}}{\frac{6}{5}} = \frac{1}{3}.$$

39.
$$S_1 = a_1 = \frac{1}{2(1) - 1} = 1$$

$$S_2 = a_1 + a_2 = \frac{1}{2(1) - 1} + \frac{1}{2(2) - 1} = 1 + \frac{1}{3} = \frac{4}{3}$$

$$S_3 = a_1 + a_2 + a_3 = \frac{1}{2(1) - 1} + \frac{1}{2(2) - 1} + \frac{1}{2(3) - 1} = 1 + \frac{1}{3} + \frac{1}{5} = \frac{23}{15}$$

$$S_4 = a_1 + a_2 + a_3 + a_4 = \frac{1}{2(1) - 1} + \frac{1}{2(2) - 1} + \frac{1}{2(3) - 1} + \frac{1}{2(4) - 1}$$

$$= 1 + \frac{1}{3} + \frac{1}{5} + \frac{1}{7} = \frac{176}{105}$$

$$S_5 = a_1 + a_2 + a_3 + a_4 + a_5 = \frac{1}{2(1) - 1} + \frac{1}{2(2) - 1} + \frac{1}{2(3) - 1} + \frac{1}{2(4) - 1} + \frac{1}{2(5) - 1}$$

$$1 + \frac{1}{3} + \frac{1}{5} + \frac{1}{7} + \frac{1}{9} = \frac{563}{315}$$

41. This function most nearly matches $\frac{1}{1-x}$. To get 1 in the denominator, instead of 3, divide the numerator and denominator by 3.

$$\frac{4}{3 - x} = \frac{\frac{4}{3}}{1 - \frac{x}{3}}$$

Thus, we can find the Taylor series for $\frac{\frac{4}{3}}{1 - \frac{x}{3}}$ by starting with the Taylor series for $\frac{1}{1-x}$, multiplying each term by $\frac{4}{3}$, and replacing x with $\frac{x}{3}$.

$$\frac{4}{3 - x} = \frac{\frac{4}{3}}{1 - \frac{x}{3}}$$

$$= \frac{4}{3} \cdot 1 + \frac{4}{3}\left(\frac{x}{3}\right) + \frac{4}{3}\left(\frac{x}{3}\right)^2 + \frac{4}{3}\left(\frac{x}{3}\right)^3 + \cdots + \frac{4}{3}\left(\frac{x}{3}\right)^n + \cdots$$

$$= \frac{4}{3} + \frac{4x}{9} + \frac{4x^2}{27} + \frac{4x^3}{81} + \cdots + \frac{4x^n}{3^{n+1}} + \cdots$$

The Taylor series for $\frac{1}{1-x}$ is valid when $-1 < x < 1$. Replacing x with $\frac{x}{3}$ gives

$$-1 < \frac{x}{3} < 1 \quad \text{or} \quad -3 < x < 3.$$

The interval of convergence of the new series is $(-3, 3)$.

43.
$$\frac{x^2}{x + 1} = x^2 \cdot \frac{1}{1 - (-x)}$$

Use the Taylor series for $\frac{1}{1-x}$ and replace x with $-x$. Also, use property (3) with $k = 2$.

$$\frac{x^2}{x + 1} = x^2 \cdot \frac{1}{1 - (-x)}$$

$$= x^2 \cdot 1 + x^2(-x) + x^2(-x)^2 + x^2(-x)^3 + \cdots + x^2(-x)^n + \cdots$$

$$= x^2 - x^3 + x^4 - x^5 + \cdots + (-1)^n x^{n+2} + \cdots$$

The Taylor series for $\frac{1}{1-x}$ is valid when $-1 < x < 1$. Replacing x with $-x$ gives

$$-1 < -x < 1 \quad \text{or} \quad 1 > x > -1.$$

The interval of convergence of the new series is $(-1, 1)$.

45. We find the Taylor series for $\ln(1 - 2x)$ by starting with the Taylor series for $\ln(1 + x)$ and replacing each x with $-2x$.

$$\ln(1 - 2x) = -2x - \frac{(-2x)^2}{2} + \frac{(-2x)^3}{3} - \frac{(-2x)^4}{4} + \cdots + \frac{(-1)^n(-2x)^{n+1}}{n+1} + \cdots$$

$$= -2x - 2x^2 - \frac{8}{3}x^3 - 4x^4 - \cdots - \frac{2^{n+1}x^{n+1}}{n+1} - \cdots$$

The Taylor series for $\ln(1 + x)$ is valid when $-1 < x \leq 1$. Replacing x with $-2x$ gives

$$-1 < -2x \leq 1 \quad \text{or} \quad \frac{1}{2} > x \geq -\frac{1}{2}.$$

The interval of convergence of the new series is $\left[-\frac{1}{2}, \frac{1}{2}\right)$.

47. We find the Taylor series for e^{-2x^2} by starting with the Taylor series for e^x and replacing each x with $-2x^2$.

$$e^{-2x^2} = 1 + (-2x^2) + \frac{1}{2!}(-2x^2)^2 + \frac{1}{3!}(-2x^2)^3 + \cdots + \frac{1}{n!}(-2x^2)^n + \cdots$$

$$= 1 - 2x^2 + 2x^4 - \frac{4}{3}x^6 + \cdots + \frac{(-1)^n \cdot 2^n x^{2n}}{n!} + \cdots$$

The Taylor series for e^{-2x^2} has the same interval of convergence, $(-\infty, \infty)$, as the Taylor series for e^x.

49. Use the Taylor series for e^x, multiply each term by 2, and replace x with $-3x$. Also, use property (3) with $k = 3$.

$$2x^3 e^{-3x} = 2x^3 \cdot 1 + 2x^3(-3x) + 2x^3 \cdot \frac{1}{2!}(-3x)^2 + 2x^3 \cdot \frac{1}{3!}(-3x)^3 + \cdots + 2x^3 \cdot \frac{1}{n!}(-3x)^n + \cdots$$

$$= 2x^3 - 6x^4 + 9x^5 - 9x^6 + \cdots + \frac{(-1)^n \cdot 2 \cdot 3^n x^{n+3}}{n!} + \cdots$$

The Taylor series for $2x^3 e^{-3x}$ has the same interval of convergence, $(-\infty, \infty)$, as the Taylor series for e^x.

51. The limit in the numerator is 0, as is the limit in the denominator, so that l'Hospital's rule applies. Taking derivatives separately in the numerator and denominator gives,

$$\lim_{x \to 2} \frac{3x^2 - 2x - 1}{2x} = \frac{3(2)^2 - 2(2) - 1}{2(2)} = \frac{7}{4}.$$

By l'Hospital's rule,

$$\lim_{x \to 2} \frac{x^3 - x^2 - x - 2}{x^2 - 4} = \frac{7}{4}.$$

53. $\lim\limits_{x\to-5} x^3 - 3x^2 + 4x - 1 = -221$ and l'Hospital's rule does not apply. However,

$$\lim_{x\to-5} \frac{x^3 - 3x^2 + 4x - 1}{x^2 - 25} \text{ does not exist.}$$

55. The limit in the numerator is 0, as is the limit in the denominator, so that l'Hospital's rule applies. Taking derivatives separately in the numerator and denominator gives,

$$\lim_{x\to0} \frac{5e^x}{3x^2 - 16x + 7} = \frac{5e^0}{3(0)^2 - 16(0) + 7} = \frac{5}{7}.$$

By l'Hospital's rule,

$$\lim_{x\to0} \frac{5e^x - 5}{x^3 - 8x^2 + 7x} = \frac{5}{7}.$$

57. The limit in the numerator is 0, as is the limit in the denominator, so that l'Hospital's rule applies. Taking derivatives separately in the numerator and denominator gives,

$$\lim_{x\to0} \frac{-e^{2x} - 2xe^{2x}}{2e^{2x}} = \frac{-e^{2(0)} - 2(0)e^{2(0)}}{2e^{2(0)}}$$
$$= -\frac{1}{2}.$$

By l'Hospital's rule,

$$\lim_{x\to0} \frac{-xe^{2x}}{e^{2x} - 1} = -\frac{1}{2}.$$

59. The limit in the numerator is 0, as is the limit in the denominator, so that l'Hospital's rule applies. Taking derivatives separately in the numerator and denominator gives,

$$\lim_{x\to0} \frac{2 - \frac{1}{2}(1 + x)^{-1/2}}{3x^2} = \lim_{x\to0} \frac{4(1 + x)^{1/2} - 1}{3x^2(2(1 + x)^{1/2})}$$
$$= \frac{3}{0} \text{ which does not exist.}$$

By l'Hospital's rule,

$$\lim_{x\to0} \frac{1 + 2x - (1 + x)^{1/2}}{x^3} \text{ does not exist.}$$

61. We have a limit of the form $\infty \cdot 0$. Rewrite the expression.

$$x^2 e^{-\sqrt{x}} = \frac{x^2}{e^{\sqrt{x}}}$$

Now both the numerator and denominator become infinite and l'Hospital's rule applies to the limit of the form ∞/∞. Differentiate the numerator and the denominator.

$$\lim_{x\to\infty} \frac{x^2}{e^{\sqrt{x}}} = \lim_{x\to\infty} \frac{2x}{\frac{e^{\sqrt{x}}}{(2\sqrt{x})}} = \lim_{x\to\infty} \frac{4x^{3/2}}{e^{\sqrt{x}}}$$

This problem is similar to what we started with so we handle it in the same manner. In fact, continue the process three more times.

$$\lim_{x\to\infty}\frac{4x^{3/2}}{e^{\sqrt{x}}} = \lim_{x\to\infty}\frac{6x^{1/2}}{\frac{e^{\sqrt{x}}}{(2\sqrt{x})}} = \lim_{x\to\infty}\frac{12x}{e^{\sqrt{x}}} = \lim_{x\to\infty}\frac{12}{\frac{e^{\sqrt{x}}}{(2\sqrt{x})}} = \lim_{x\to\infty}\frac{24\sqrt{x}}{e^{\sqrt{x}}} = \lim_{x\to\infty}\frac{\frac{12}{\sqrt{x}}}{\frac{e^{\sqrt{x}}}{(2\sqrt{x})}} = \lim_{x\to\infty}\frac{24}{e^{\sqrt{x}}} = 0$$

Therefore, by l'Hospital's rule,

$$\lim_{x\to\infty} x^2 e^{-\sqrt{x}} = 0$$

63. $\lim\limits_{x\to 0}\left(\dfrac{e^{3x}}{x^2} - \dfrac{1}{x^2} - \dfrac{3}{x}\right).$

Rewrite over a common denominator.

$$\lim_{x\to 0}\left(\frac{e^{3x}-1-3x}{x^2}\right)$$

The numerator and denominator have limit 0, so l'Hospital's rule applies. Differentiate the numerator and denominator.

$$\lim_{x\to 0}\left(\frac{e^{3x}-1-3x}{x^2}\right) = \lim_{x\to 0}\left(\frac{3e^{3x}-3}{2x}\right)$$

The numerator and denominator still have limit 0, so differentiate the numerator and denominator again.

$$\lim_{x\to 0}\left(\frac{3e^{3x}-3}{2x}\right) = \lim_{x\to 0}\left(\frac{9e^{3x}}{2}\right) = \frac{9}{2}$$

Thus

$$\lim_{x\to 0}\left(\frac{e^{3x}}{x^2} - \frac{1}{x^2} - \frac{3}{x}\right) = \frac{9}{2}.$$

65.   Find $\lim\limits_{x\to 0}\left(\dfrac{\ln(1-4x)}{x^2} + \dfrac{4}{x}\right).$

Rewrite over a common denominator.

$$\lim_{x\to 0}\left(\frac{\ln(1-4x)+4x}{x^2}\right)$$

The numerator and denominator have limit 0, so l'Hospital's rule applies. Differentiate the numerator and denominator.

$$\lim_{x\to 0}\left(\frac{\ln(1-4x)+4x}{x^2}\right) = \lim_{x\to 0}\left(\frac{\frac{-4}{1-4x}+4}{2x}\right)$$

$$= \lim_{x\to 0}\left(\frac{-16x}{(1-4x)(2x)}\right)$$

$$= \lim_{x\to 0}\left(\frac{-8}{1-4x}\right) = -8$$

Thus

$$\lim_{x\to 0}\left(\frac{\ln(1-4x)}{x^2} + \frac{4}{x}\right) = -8.$$

67. $f(x) = x^3 - 8x^2 + 18x - 12$

$f'(x) = 3x^2 - 16x + 18$

$f(4) = -4 < 0$ and $f(5) = 3 > 0$ so a solution exists in $(4, 5)$.

Let $c_1 = 4$.

$c_2 = c_1 - \dfrac{f(c_1)}{f'(c_1)} = 4 - \dfrac{-4}{2} = 6$

$c_3 = c_2 - \dfrac{f(c_2)}{f'(c_2)} = 6 - \dfrac{24}{30} = 5.2$

$c_4 = c_3 - \dfrac{f(c_3)}{f'(c_3)} = 5.2 - \dfrac{5.888}{15.92} = 4.8302$

$c_5 = c_4 - \dfrac{f(c_4)}{f'(c_4)} = 4.8302 - \dfrac{0.98953}{10.709} = 4.7378$

$c_6 = c_5 - \dfrac{f(c_5)}{f'(c_5)} = 4.7378 - \dfrac{0.05462}{9.5354} = 4.7321$

$c_7 = c_6 - \dfrac{f(c_6)}{f'(c_6)} = 4.7321 - \dfrac{4.7 \cdot 10^{-4}}{9.4647} = 4.7321$

Subsequent approximations will agree with c_6 and c_7 to the nearest hundredth. Thus, $x = 4.73$.

69. $f(x) = x^4 + 3x^3 - 4x^2 - 21x - 21$

$f'(x) = 4x^3 + 9x^2 - 8x - 21$

$f(2) = -39 < 0$ and $f(3) = 42 > 0$ so a solution exists in $(2, 3)$.

Let $c_1 = 2$.

$c_2 = c_1 - \dfrac{f(c_1)}{f'(c_1)} = 2 - \dfrac{-39}{31} = 3.2581$

$c_3 = c_2 - \dfrac{f(c_2)}{f'(c_2)} = 3.2581 - \dfrac{84.558}{186.81} = 2.8055$

$c_4 = c_3 - \dfrac{f(c_3)}{f'(c_3)} = 2.8055 - \dfrac{16.796}{115.72} = 2.6604$

$c_5 = c_4 - \dfrac{f(c_4)}{f'(c_4)} = 2.6604 - \dfrac{1.4037}{96.735} = 2.6459$

$c_6 = c_5 - \dfrac{f(c_5)}{f'(c_5)} = 2.6459 - \dfrac{0.01411}{94.933} = 2.6458$

Subsequent approximations will agree with c_5 and c_6 to the nearest hundredth. Thus, $x = 2.65$.

71. $\sqrt{37.6}$ is a solution of $x^2 - 37.6 = 0$.

$f(x) = x^2 - 37.6$

$f'(x) = 2x$

Since $6 < \sqrt{37.6} < 7$, let $c_1 = 6$.

$c_2 = c_1 - \dfrac{f(c_1)}{f'(c_1)} = 6 - \dfrac{-1.6}{12} = 6.1333$

$c_3 = c_2 - \dfrac{f(c_2)}{f'(c_2)} = 6.1333 - \dfrac{0.01737}{12.2666} = 6.1319$

$c_4 = c_3 - \dfrac{f(c_3)}{f'(c_3)} = 6.1319 - \dfrac{0.0002}{12.2638} = 6.1319$

Since $c_4 = c_3 = 6.1319$, to the nearest thousandth, $\sqrt{37.6} = 6.132$.

73. $\sqrt[3]{94.7}$ is a solution of $x^3 - 94.7 = 0$.

$f(x) = x^3 - 94.7$

$f'(x) = 3x^2$

Since $4 < \sqrt[3]{94.7} < 5$, let $c_1 = 4$

$c_2 = c_1 - \dfrac{f(c_1)}{f'(c_1)} = 4 - \dfrac{-30.7}{48} = 4.6396$

$c_3 = c_2 - \dfrac{f(c_2)}{f'(c_2)} = 4.6396 - \dfrac{5.1715}{64.5777} = 4.5595$

$c_4 = c_3 - \dfrac{f(c_3)}{f'(c_3)} = 4.5595 - \dfrac{0.08763}{62.3671} = 4.5581$

$c_5 = c_4 - \dfrac{f(c_4)}{f'(c_4)} = 4.5581 - \dfrac{0.0003}{62.3288} = 4.5581$

Since $c_3 = c_4 = 4.5581$, to the nearest thousandth, $\sqrt[3]{94.7} = 4.558$.

75.   The yearly incomes produced by the mine form a geometric sequence with $r = 118\% = 1.18$ and $a_1 = 750{,}000$. To determine the total amount produced in 8 years, use the formula to find S_n with $n = 8, r = 1.18$, and $a = a_1 = 750{,}000$.

$S_8 = \dfrac{750{,}000\left[(1.18)^8 - 1\right]}{1.18 - 1}$

$= \dfrac{750{,}000\,(3.75886 - 1)}{0.18} \approx 11{,}495{,}247$

The total amount of income produced by the mine in five years is $11,495,247.

77. The payments form an ordinary annuity with $R = 491, n = 9 \cdot 4 = 36,$ and

$i = \frac{9.4\%}{4} = 2.35\% = 0.0235.$ The amount of this annuity is

$$S = 491\left[\frac{(1.0235)^{36} - 1}{0.0235}\right].$$

The number is brackets, $s\,\overline{{}_{36}|_{0.0235}}$, is 55.64299673, so that

$$S = 491(55.64299673) \approx 27,320.71.$$

or $27,320.71

79. $20,000 is the present value of an annuity of R dollars, with 9 periods, and $i = 8.9\% = 0.089$ per period.

$$P = R \cdot a\,\overline{{}_{n}|_{i}}$$

$$20,000 = R \cdot a\,\overline{{}_{9}|_{0.089}}$$

$$R = \frac{20,000}{a\,\overline{{}_{9}|_{0.089}}} = \frac{20,000}{6.019696915}$$

$$\approx 3322.43$$

Each payment is $3322.43.

81. The present value, P, is 156,890, $i = \frac{0.0774}{12} = 0.00645,$ and $n = 12 \cdot 25 = 300.$

$$156,890 = R \cdot a\,\overline{{}_{300}|_{0.00645}}$$

$$= R\left[\frac{1 - (1 + 0.00645)^{-300}}{0.00645}\right]$$

$$= R\left[\frac{1 - 0.1453244675}{0.00645}\right]$$

$$= R\left[\frac{0.8546755325}{0.00645}\right]$$

$$R \approx 1184.01$$

Monthly payments of $1184.01 will be required to amortize the loan.

83. The doubling time n for a quantity that increases at an annual rate r is given by

$$n = \frac{\ln 2}{\ln(1 + r)} = \frac{\ln 2}{\ln 1.0325} \approx 21.67.$$

It will take about 21.67 years.

According to the Rule of 70, the doubling time is given by

$$\text{Doubling time} \approx \frac{70}{100r}$$

$$= \frac{70}{100(0.0325)}$$

$$= 21.54.$$

It will take about 21.54 years, a difference of 0.13 year, or about 7 weeks.

85. 2 hours amounts to six doubling times, so after two hours the number of bacteria will be $1000(2^6)$ $= 64,000$ bacteria.

THE TRIGONOMETRIC FUNCTIONS

13.1 Definitions of the Trigonometric Functions

Your Turn 1

(a) Since $1° = \pi/180$ radians,

$$210° = 210\left(\frac{\pi}{180}\right) \text{ radians}$$

$$= \frac{7\pi}{6} \text{ radians.}$$

(b) Since 1 radian $= 180°/\pi$,

$$\frac{3\pi}{4} \text{ radians} = \frac{3\pi}{4}\left(\frac{180°}{\pi}\right)$$

$$= 3(45°)$$

$$= 135°$$

Your Turn 2

Sketch the triangle formed by joining the origin, the point $(9, 40)$, and the point $(9, 0)$ on the x-axis.

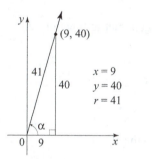

$$x = 9$$
$$y = 40$$
$$r = 41$$

To find r, use $r = \sqrt{x^2 + y^2}$. Here $x = 9$, and $y = 40$, so

$$r = \sqrt{9^2 + 40^2}$$

$$= \sqrt{1681}$$

$$= 41.$$

$$\sin\alpha = \frac{y}{r} = \frac{40}{41} \qquad \tan\alpha = \frac{y}{x} = \frac{40}{9}$$

$$\sec\alpha = \frac{r}{x} = \frac{41}{9} \qquad \cos\alpha = \frac{x}{r} = \frac{9}{41}$$

$$\cot\alpha = \frac{x}{y} = \frac{9}{40} \qquad \csc\alpha = \frac{r}{y} = \frac{41}{40}$$

Your Turn 3

Sketch the angle and label the sides of an appropriate right triangle.

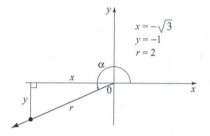

$$x = -\sqrt{3}$$
$$y = -1$$
$$r = 2$$

The triangle formed in the third quadrant is a 30°-60°-90° triangle, so we label the sides as in Example 4, except that here in the third quadrant both x and y are negative.

$$\sin\frac{7\pi}{6} = \frac{-1}{2} = -\frac{1}{2} \qquad \tan\frac{7\pi}{6} = \frac{-1}{-\sqrt{3}} = \frac{\sqrt{3}}{3}$$

$$\sec\frac{7\pi}{6} = \frac{2}{-\sqrt{3}} = -\frac{2\sqrt{3}}{3} \qquad \cos\frac{7\pi}{6} = \frac{-\sqrt{3}}{2} = -\frac{\sqrt{3}}{2}$$

$$\cot\frac{7\pi}{6} = \frac{-\sqrt{3}}{-1} = \sqrt{3} \qquad \csc\frac{7\pi}{6} = \frac{2}{-1} = -2$$

Your Turn 4

(a) $\cos 6° \approx 0.9945$

(b) $\sec 4 \approx -1.5299$

Your Turn 5

The cosine function is negative in quadrants II and III. In quadrant II we draw a triangle with an angle whose cosine is $-\sqrt{2}/2$, which we recognize as a 45°-45°-90° degree triangle (Figure (a)). Thus the angle corresponding to the terminal side has measure $\pi - (\pi/4) = 3\pi/4$. Drawing the same triangle in quadrant III (Figure (b)) we see that the angle corresponding to the terminal side is $\pi + (\pi/4) = 5\pi/4$.

Thus the equation $\cos\theta = -\sqrt{2}/2$ has two solutions between 0 and 2π, namely $3\pi/4$ and $5\pi/4$.

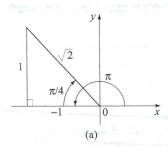

(a)

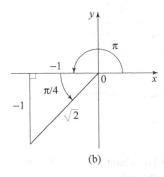

(b)

13.1   Exercises

1. $60° = 60\left(\dfrac{\pi}{180}\right) = \dfrac{\pi}{3}$

3. $150° = 150\left(\dfrac{\pi}{180}\right) = \dfrac{5\pi}{6}$

5. $270° = 270\left(\dfrac{\pi}{180}\right) = \dfrac{3\pi}{2}$

7. $495° = 495\left(\dfrac{\pi}{180}\right) = \dfrac{11\pi}{4}$

9. $\dfrac{5\pi}{4} = \dfrac{5\pi}{4}\left(\dfrac{180°}{\pi}\right) = 225°$

11. $-\dfrac{13\pi}{6} = -\dfrac{13\pi}{6}\left(\dfrac{180°}{\pi}\right) = -390°$

13. $\dfrac{8\pi}{5} = \dfrac{8\pi}{5}\left(\dfrac{180°}{\pi}\right) = 288°$

15. $\dfrac{7\pi}{12} = \dfrac{7\pi}{12}\left(\dfrac{180°}{\pi}\right) = 105°$

17. Let α = the angle with terminal side through $(-3, 4)$. Then $x = -3$, $y = 4$, and

$$r = \sqrt{x^2 + y^2} = \sqrt{(-3)^2 + (4)^2}$$
$$= \sqrt{25} = 5.$$

$\sin\alpha = \dfrac{y}{r} = \dfrac{4}{5}$ $\cot\alpha = \dfrac{x}{y} = -\dfrac{3}{4}$

$\cos\alpha = \dfrac{x}{r} = -\dfrac{3}{5}$ $\sec\alpha = \dfrac{r}{x} = -\dfrac{5}{3}$

$\tan\alpha = \dfrac{y}{x} = -\dfrac{4}{3}$ $\csc\alpha = \dfrac{r}{y} = \dfrac{5}{4}$

19. Let α = the angle with terminal side through $(7, -24)$. Then $x = 7$, $y = -24$, and

$$r = \sqrt{x^2 + y^2} = \sqrt{49 + 576}$$
$$= \sqrt{625} = 25.$$

$\sin\alpha = \dfrac{y}{r} = -\dfrac{24}{25}$ $\cot\alpha = \dfrac{x}{y} = -\dfrac{7}{24}$

$\cos\alpha = \dfrac{x}{r} = \dfrac{7}{25}$ $\sec\alpha = \dfrac{r}{x} = \dfrac{25}{7}$

$\tan\alpha = \dfrac{y}{x} = -\dfrac{24}{7}$ $\csc\alpha = \dfrac{r}{y} = -\dfrac{25}{24}$

21. In quadrant I, all six trigonometric functions are positive, so their sign is $+$.

23. In quadrant III, $x < 0$ and $y < 0$.

Furthermore, $r > 0$.

$\sin\theta = \dfrac{y}{r} < 0$, so the sign is $-$.

$\cos\theta = \dfrac{x}{r} < 0$, so the sign is $-$.

$\tan\theta = \dfrac{y}{x} > 0$, so the sign is $+$.

$\cot\theta = \dfrac{x}{y} > 0$, so the sign is $+$.

$\sec\theta = \dfrac{r}{x} < 0$, so the sign is $-$.

$\csc\theta = \dfrac{r}{y} < 0$, so the sign is $-$.

25. When an angle θ of 30° is drawn in standard position, one choice of a point on its terminal side is $(x, y) = (\sqrt{3}, 1)$. Then

$$r = \sqrt{x^2 + y^2} = \sqrt{3 + 1} = 2.$$

$$\tan \theta = \frac{y}{x} = \frac{1}{\sqrt{3}} = \frac{\sqrt{3}}{3}$$

$$\cot \theta = \frac{x}{y} = \sqrt{3}$$

$$\csc \theta = \frac{r}{y} = 2$$

27. When an angle θ of 60° is drawn in standard position, one choice of a point on its terminal side is $(x, y) = (1, \sqrt{3})$. Then

$$r = \sqrt{x^2 + y^2} = \sqrt{1 + 3} = 2.$$

$$\sin \theta = \frac{y}{r} = \frac{\sqrt{3}}{2}$$

$$\cot \theta = \frac{x}{y} = \frac{1}{\sqrt{3}} = \frac{\sqrt{3}}{3}$$

$$\csc \theta = \frac{r}{y} = \frac{2}{\sqrt{3}} = \frac{2\sqrt{3}}{3}$$

29. When an angle θ of 135° is drawn in standard position, one choice of a point on its terminal side is $(x, y) = (-1, 1)$. Then

$$r = \sqrt{x^2 + y^2} = \sqrt{1 + 1} = \sqrt{2}.$$

$$\tan \theta = \frac{y}{x} = -1$$

$$\cot \theta = \frac{x}{y} = -1$$

31. When an angle θ of 210° is drawn in standard position, one choice of a point on its terminal side is $(x, y) = (-\sqrt{3}, -1)$. Then

$$r = \sqrt{x^2 + y^2} = \sqrt{3 + 1} = 2.$$

$$\cos \theta = \frac{x}{r} = -\frac{\sqrt{3}}{2}$$

$$\sec \theta = \frac{r}{x} = \frac{2}{-\sqrt{3}} = -\frac{2\sqrt{3}}{3}$$

33. When an angle of $\frac{\pi}{3}$ is drawn in standard position, one choice of a point on its terminal side is $(x, y) = (1, \sqrt{3})$. Then

$$r = \sqrt{x^2 + y^2} = \sqrt{1 + 3} = 2.$$

$$\sin \frac{\pi}{3} = \frac{y}{r} = \frac{\sqrt{3}}{2}$$

35. When an angle of $\frac{\pi}{4}$ is drawn in standard position, one choice of a point on its terminal side is $(x, y) = (1, 1)$.

$$\tan \frac{\pi}{4} = \frac{y}{x} = 1$$

37. When an angle of $\frac{\pi}{6}$ is drawn in standard position, one choice of a point on its terminal side is $(x, y) = (\sqrt{3}, 1)$. Then

$$r = \sqrt{x^2 + y^2} = \sqrt{3 + 1} = 2.$$

$$\csc \frac{\pi}{6} = \frac{r}{y} = \frac{2}{1} = 2$$

39. When an angle of 3π is drawn in standard position, one choice of a point on its terminal side is $(x, y) = (-1, 0)$. Then

$$r = \sqrt{x^2 + y^2} = \sqrt{1} = 1.$$

$$\cos 3\pi = \frac{x}{r} = -1$$

41. When an angle of $\frac{7\pi}{4}$ is drawn in standard position, one choice of a point on its terminal side is $(x, y) = (1, -1)$. Then

$$r = \sqrt{x^2 + y^2} = \sqrt{1 + 1} = \sqrt{2}.$$

$$\sin \frac{7\pi}{4} = \frac{y}{r} = \frac{-1}{\sqrt{2}} = -\frac{\sqrt{2}}{2}$$

43. When an angle of $\frac{5\pi}{4}$ is drawn in standard position, one choice of a point on its terminal side is $(x, y) = (-1, -1)$. Then

$$r = \sqrt{x^2 + y^2} = \sqrt{1 + 1} = \sqrt{2}.$$

$$\sec \frac{5\pi}{4} = \frac{r}{x} = \frac{\sqrt{2}}{-1} = -\sqrt{2}$$

45. When an angle of $-\frac{3\pi}{4}$ is drawn in standard position, one choice of a point on its terminal side is $(x, y) = (-1, -1)$. Then

$$\cot\left(-\frac{3\pi}{4}\right) = \frac{y}{x} = \frac{-1}{-1} = 1$$

47. When an angle of $-\frac{7\pi}{6}$ is drawn in standard position, one choice of a point on its terminal side is $(x, y) = (-\sqrt{3}, 1)$. Then

$$r = \sqrt{x^2 + y^2} = \sqrt{3 + 1} = 2.$$

$$\sin\left(-\frac{7\pi}{6}\right) = \frac{y}{r} = \frac{1}{2}$$

49. The cosine function is positive in quadrants I and IV. We know that $\cos(\pi/3) = 1/2$, so the solution in quadrant I is $\pi/3$. The solution in quadrant IV is $2\pi - (\pi/3) = 5\pi/3$. The two solutions of $\cos\theta = 1/2$ between 0 and 2π are $\pi/3$ and $5\pi/3$.

51. The tangent function is negative in quadrants II and IV. We know that $\tan(\pi/4) = 1$. The solution in quadrant II is $\pi - (\pi/4) = 3\pi/4$. The solution in quadrant IV is $2\pi - (\pi/4) = 7\pi/4$. The two solutions of $\tan\theta = -1$ between 0 and 2π are $3\pi/4$ and $7\pi/4$.

53. The secant function is negative in quadrants II and III. We know that $\sec(\pi/6) = 2/\sqrt{3}$, so the solution in quadrant II is $\pi - (\pi/6) = 5\pi/6$. The solution in quadrant III is $\pi + (\pi/6) = 7\pi/6$. The two solutions of $\sec\theta = -2/\sqrt{3}$ between 0 and 2π are $5\pi/6$ and $7\pi/6$.

55. $\sin 39° \approx 0.6293$

57. $\tan 123° \approx -1.5399$

59. $\sin 0.3638 \approx 0.3558$

61. $\cos 1.2353 \approx 0.3292$

63. $f(x) = \cos(3x)$ is of the form $f(x) = a\cos(bx)$ where $a = 1$ and $b = 3$. Thus, $|a| = 1$ and $T = \frac{2\pi}{b} = \frac{2\pi}{3}$.

65. $g(t) = -2\sin\left(\frac{\pi}{4}t + 2\right)$ is of the form $g(t) = a\sin(bt + c)$ where $a = -2$, $b = \frac{\pi}{4}$, and $c = 2$. Thus, $|a| = 2$ and $T = \frac{2\pi}{b} = \frac{2\pi}{\frac{\pi}{4}} = 8$.

67. The graph of $y = 2\cos x$ is similar to the graph of $y = \cos x$ except that it has twice the amplitude. (That is, its height is twice as great.)

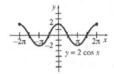

69. The graph of $y = -\frac{1}{2}\cos x$ is similar to the graph of $y = \cos x$ except that it has half the amplitude and is reflected about the x-axis.

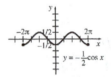

71. $y = 4\sin\left(\frac{1}{2}x + \pi\right) + 2$ has amplitude $a = 4$, period $T = \frac{2\pi}{b} = \frac{2\pi}{\frac{1}{2}} = 4\pi$, phase shift $\frac{c}{b} = \frac{\pi}{\frac{1}{2}} = 2\pi$, and vertical shift $d = 2$. Thus, the graph of $y = 4\sin\left(\frac{1}{2}x + \pi\right) + 2$ is similar to the graph of $f(x) = \sin x$ except that it has 4 times the amplitude, twice the period, and is shifted up 2 units vertically. Also, $y = \sin\left(\frac{1}{2}x + \pi\right) + 2$ is shifted 2π units to the left relative to the graph of $g(x) = \sin\left(\frac{1}{2}x\right)$.

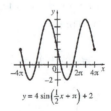

$y = 4\sin\left(\frac{1}{2}x + \pi\right) + 2$

73. The graph of $y = -3\tan x$ is similar to the graph of $y = \tan x$ except that it is reflected about the x-axis and each ordinate value is three times larger in absolute value. Note that the points $\left(-\frac{\pi}{4}, 3\right)$ and $\left(\frac{\pi}{4}, -3\right)$ lie on the graph.

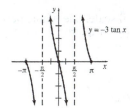

75. (a) Since the three angles θ are equal and their sum is 180°, each angle θ is 60°.

(b) The base angle on the left is still 60°. The bisector is perpendicular to the base, so the other base angle is 90°. The angle formed by bisecting the original vertex angle θ is 30°.

(c) Two sides of the triangle on the left are given in the diagram: the hypotenuse is 2, and the base is half of the original base of 2, or 1. The Pythagorean Theorem gives the length of the remaining side (the vertical bisector) as $\sqrt{2^2 - 1^2} = \sqrt{3}$.

77. $S(t) = 500 + 500\cos\frac{\pi}{6}t$

(a) November corresponds to $t = 0$.
Therefore,

$$S(0) = 500 + 500\left[\cos\left(\frac{\pi}{6}\right)(0)\right]$$
$$= 500 + 500\cos 0$$
$$= 1000 \text{ snowblowers.}$$

(b) January corresponds to $t = 2$.
Therefore,

$$S(2) = 500 + 500\left[\cos\left(\frac{\pi}{6}\right)(2)\right]$$
$$= 500 + 500\cos\frac{\pi}{3}$$
$$= 500 + 500\left(\frac{1}{2}\right)$$
$$= 750 \text{ snowblowers.}$$

(c) February corresponds to $t = 3$.
Therefore,

$$S(3) = 500 + 500\left[\cos\left(\frac{\pi}{6}\right)(3)\right]$$
$$= 500 + 500\cos\frac{\pi}{2}$$
$$= 500 + 500(0)$$
$$= 500 \text{ snowblowers.}$$

(d) May corresponds to $t = 6$.
Therefore,

$$S(6) = 500 + 500\left[\cos\left(\frac{\pi}{6}\right)(6)\right]$$
$$= 500 + 500\cos\pi$$
$$= 500 + 500(-1)$$
$$= 0 \text{ snowblowers.}$$

(e) August corresponds to $t = 9$.
Therefore,

$$S(9) = 500 + 500\left[\cos\left(\frac{\pi}{6}\right)(9)\right]$$
$$= 500 + 500\cos\frac{3\pi}{2}$$
$$= 500 + 500(0)$$
$$= 500 \text{ snowblowers.}$$

(f) Use the ordered pairs obtained in parts (a)-(e) to plot the graph.

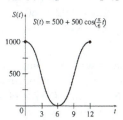

79. (a) The period is $\dfrac{2\pi}{\left(\frac{\pi}{14.77}\right)} = 29.54$

There is a lunar cycle every 29.54 days.

(b) $y = 100 + 1.8\cos\left[\dfrac{(t-6)\pi}{14.77}\right]$ reaches a maximum value when $\cos\left[\dfrac{(t-6)\pi}{14.77}\right] = 1$ which occurs when

$$t - 6 = 0$$
$$t = 6$$

Six days from October 8, 2014, is October 14, 2014.

$$y = 100 + 1.8 \cos \left[\frac{(6-6)\pi}{14.77} \right]$$

$$= 101.8$$

There is a percent increase of 1.8 percent.

(c) On October 25, $t = 17$.

$$y = 100 + 1.8 \cos \left[\frac{(17-6)\pi}{14.77} \right]$$

$$\approx 98.75$$

The formula predicts that the number of consultations was 98.75% of the daily mean.

81. (a)

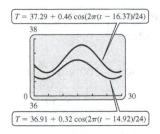

$T = 37.29 + 0.46 \cos(2\pi(t - 16.37)/24)$

38

0

36

30

$T = 36.91 + 0.32 \cos(2\pi(t - 14.92)/24)$

(b) The cosine function in the expression for the body temperature T of patients will have a maximum where $k = t$. For the patients without Alzheimer's this will be when $t = 14.92$. Since $0.92 \times 60 \approx 55$, this time is 2:55 P.M.

(c) For the patients with Alzheimer's the maximum temperature will occur when $t = k = 16.37$. Since $0.37 \times 60 \approx 22$, this time is 4:22 P.M.

83.

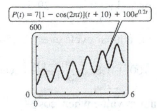

$P(t) = 7[1 - \cos(2\pi t)](t + 10) + 100 e^{0.2t}$

600

0

0   6

85. Solving $\dfrac{c_1}{c_2} = \dfrac{\sin \theta_1}{\sin \theta_2}$ for c_2 gives

$$c_2 = \frac{c_1 \sin \theta_2}{\sin \theta_1}.$$

$c_1 = 3 \cdot 10^8$, $\theta_1 = 46°$, and $\theta_2 = 31°$ so

$$c_2 = \frac{3 \cdot 10^8 (\sin 31°)}{\sin 46°}$$

$$= 214{,}796{,}150$$

$$\approx 2.1 \times 10^8 \text{ m/sec.}$$

87. On the horizontal scale, one whole period clearly spans four square, so $4 \cdot 30° = 120°$ is the period.

89. $T(t) = 60 - 30 \cos \left(\dfrac{t}{2} \right)$

(a) $t = 1$ represents February, so the maximum afternoon temperature in February is

$$T(0) = 60 - 30 \cos \frac{1}{2} \approx 34°\text{F.}$$

(b) $t = 3$ represents April, so the maximum afternoon temperature in April is

$$T(3) = 60 - 30 \cos \frac{3}{2} \approx 58°\text{F.}$$

(c) $t = 8$ represents September, so the maximum afternoon temperature in September is

$$T(8) = 60 - 30 \cos 4 \approx 80°\text{F.}$$

(d) $t = 6$ represents July, so the maximum afternoon temperature in July is

$$T(6) = 60 - 30 \cos 3 \approx 90°\text{F.}$$

(e) $t = 11$ represents December, so the maximum afternoon temperature in December is

$$T(11) = 60 - 30 \cos \frac{11}{2} \approx 39°\text{F.}$$

91. (a)

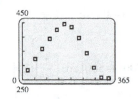

450

0

250

365

Yes; because of the cyclical nature of the days of the year, it is reasonable to assume that the times of the sunset are periodic.

(b) The function $s(t)$, derived by a TI-84 Plus using the sine regression function under the STATCALC menu, is given by

$$s(t) = 94.0872 \sin(0.0166t - 1.2213)$$
$$+ 347.4158.$$

(c) $s(60) = 94.0872\sin[0.0166(60) - 1.2213]$

$+ 347.4158$

$= 326$ minutes

$= 5{:}26$ P.M.

$s(120) = 94.0872\sin[0.0166(120) - 1.2213]$

$+ 347.4158$

$= 413$ minutes $+ 60$ minutes

(daylight savings)

$= 7{:}53$ P.M.

$s(240) = 94.0872\sin[0.0166(240) - 1.2213]$

$+ 347.4158$

$= 382$ minutes $+ 60$ minutes

(daylight savings)

$= 7{:}22$ P.M.

(d) The following graph shows $s(t)$ and $y = 360$ (corresponding to a sunset at 6:00 P.M.). These graphs first intersect on day 82. However because of daylight savings time, to find the second value we find where the graphs of $s(t)$ and $y = 360 - 60 = 300$ intersect. These graphs intersect on day 295. Thus, the sun sets at approximately 6:00 P.M. on the 82nd and 295th days of the year.

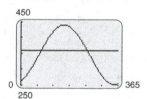

93. Let $h = $ the height of the building.

$$\tan 42.8° = \frac{h}{65}$$

$$h = 65\tan 42.8° \approx 60.2$$

The height of the building is approximately 60.2 meters.

95. Let $\theta = $ the average angle with the horizontal.

$$\tan \theta = \frac{26}{5280}$$

Using the TAN^{-1} key on the calculator,

$$\theta = \text{TAN}^{-1}\left(\frac{26}{5280}\right) \approx 0.28°.$$

97. We need to find the values of t for which

$$3.5 \le h(t) \le 4.$$

The following graphs show where $h(t) = \sin\left(\frac{t}{\pi} - 2\right) + 4$ intersects the horizontal lines $y = 3.5$ and $y = 4$.

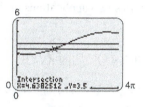

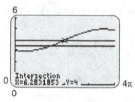

Thus, $3.5 \le h(t) \le 4$ when t is in the interval $[4.6, 6.3]$, to the nearest tenth.

13.2   Derivatives of Trigonometric Functions

Your Turn 1

Find the derivative of $y = 5\sin(3x^4)$.

By the chain rule,

$$\frac{dy}{dx} = 5\cos(3x^4) \cdot D_x(3x^4)$$

$$= 60x^3\cos(3x^4).$$

Your Turn 2

Find the derivative of $y = 2\sin^3(\sqrt{x})$.

This derivative requires two applications of the chain rule.

$$\frac{dy}{dx} = 6\sin^2(\sqrt{x}) \cdot D_x[\sin(\sqrt{x})]$$

$$= 6\sin^2(\sqrt{x}) \cdot \cos(\sqrt{x}) \cdot D_x(\sqrt{x})$$

$$= 6\sin^2(\sqrt{x}) \cdot \cos(\sqrt{x}) \cdot \frac{1}{2\sqrt{x}}$$

$$= \frac{3\sin^2(\sqrt{x}) \cdot \cos(\sqrt{x})}{\sqrt{x}}$$

Your Turn 3

Find the derivative of $y = x\cos(x^2)$.

By the product rule and two applications of the chain rule,

$$\frac{dy}{dx} = 1 \cdot \cos(x^2) + x \cdot D_x[\cos(x^2)]$$

$$= \cos(x^2) + x \cdot [-\sin(x^2)] \cdot D_x(x^2)$$

$$= \cos(x^2) + x \cdot [-\sin(x^2)] \cdot (2x)$$

$$= -2x^2\sin(x^2) + \cos(x^2).$$

Your Turn 4

Find the derivative of $y = x\tan^2(x)$.

By the product rule and two applications of the chain rule,

$$\frac{dy}{dx} = 1 \cdot \tan^2 x + x \cdot D_x(\tan^2 x)$$

$$= 1 \cdot \tan^2 x + x \cdot (2\tan x) \cdot D_x(\tan x)$$

$$= 1 \cdot \tan^2 x + x \cdot (2\tan x) \cdot \sec^2 x$$

$$= 2x\tan x\sec^2 x + \tan^2 x.$$

Your Turn 5

Find the derivative of $y = x\sec^2(\sqrt{x})$.

By two applications of the chain rule,

$$\frac{dy}{dx} = [2\sec(\sqrt{x})] \cdot D_x[\sec(\sqrt{x})]$$

$$= [2\sec(\sqrt{x})] \cdot [\sec(\sqrt{x})\tan(\sqrt{x})] \cdot D_x(\sqrt{x})$$

$$= [2\sec(\sqrt{x})] \cdot [\sec(\sqrt{x})\tan(\sqrt{x})] \cdot \frac{1}{2\sqrt{x}}$$

$$= \frac{\sec^2(\sqrt{x})\tan(\sqrt{x})}{\sqrt{x}}.$$

Your Turn 6

Find the derivative of $f(x) = \sin(\cos x)$ when $x = \pi/2$.

Use the chain rule.

$$f'(x) = \cos(\cos x) \cdot (-\sin x)$$

$$f'\left(\frac{\pi}{2}\right) = \cos\left(\cos\frac{\pi}{2}\right) \cdot \left(-\sin\frac{\pi}{2}\right)$$

$$= (\cos 0) \cdot (-1)$$

$$= (1)(-1)$$

$$= -1$$

13.2   Warmup Exercises

W1. $y = x\ln 2x$ Use the product rule.

$$\frac{dy}{dx} = x\frac{d}{dx}(\ln 2x) + (\ln 2x)\frac{d}{dx}x$$

$$= (x)\left(\frac{2}{2x}\right) + (\ln 2x)(1)$$

$$= 1 + \ln 2x$$

W2. $y = \dfrac{e^x}{x^2}$ Use the quotient rule.

$$\frac{dy}{dx} = \frac{\left(e^x\right)\left(x^2\right) - \left(e^x\right)(2x)}{\left(x^2\right)^2}$$

$$= \frac{e^x(1 - 2x)}{x^3}$$

W3. $y = (1 + 2x)^5$ Use the chain rule.

$$\frac{dy}{dx} = 5(1 + 2x)^4 \frac{d}{dx}(1 + 2x)$$

$$= 10(1 + 2x)^4$$

W4. $y = e^{5x^2 - 3x}$ Use the chain rule.

$$\frac{dy}{dx} = e^{5x^2 - 3x} \frac{d}{dx}\left(5x^2 - 3x\right)$$

$$= (10x - 3)e^{5x^2 - 3x}$$

W5. $y = \ln\left(x^2 + 1\right)$ Use the chain rule.

$$\frac{dy}{dx} = \frac{1}{\left|x^2 + 1\right|} \frac{d}{dx}\left(x^2 + 1\right)$$

$$= \frac{2x}{x^2 + 1}$$

13.2   Exercises

1. $y = \dfrac{1}{2}\sin 8x$

$$\frac{dy}{dx} = \frac{1}{2}(\cos 8x) \cdot D_x(8x)$$

$$= \frac{1}{2}(\cos 8x) \cdot 8$$

$$= 4\cos 8x$$

3. $y = 12\tan(9x + 1)$

$$\frac{dy}{dx} = [12\sec^2(9x + 1) \cdot D_x(9x + 1)$$

$$= [12\sec(9x + 1)] \cdot 9$$

$$= 108\sec^2(9x + 1)$$

5. $y = \cos^4 x$

$$\frac{dy}{dx} = [4(\cos x)^3]\, D_x(\cos x)$$

$$= (4\cos^3 x)(-\sin x)$$

$$= -4\sin x \cos^3 x$$

7. $y = \tan^8 x$

$$\frac{dy}{dx} = 8(\tan x)^7 \cdot D_x(\tan x)$$

$$= 8\tan^7 x \sec^2 x$$

9. $y = -6x \cdot \sin 2x$

$$\frac{dy}{dx} = -6x \cdot D_x(\sin 2x) + \sin 2x \cdot D_x(-6x)$$

$$= -6x(\cos 2x) \cdot D_x(2x) + (\sin 2x)(-6)$$

$$= -6x(\cos 2x) \cdot 2 - 6\sin 2x$$

$$= -12x\cos 2x - 6\sin 2x$$

11. $y = \dfrac{\csc x}{x}$

$$\frac{dy}{dx} = \frac{x \cdot D_x(\csc x) - (\csc x) \cdot D_x x}{x^2}$$

$$= \frac{-x\csc x \cot x - \csc x}{x^2}$$

$$= \frac{-(x\csc x \cot x + \csc x)}{x^2}$$

13. $y = \sin e^{4x}$

$$\frac{dy}{dx} = \cos e^{4x} \cdot D_x(e^{4x})$$

$$= (\cos e^{4x}) \cdot e^{4x} \cdot D_x(4x)$$

$$= (\cos e^{4x}) \cdot e^{4x} \cdot 4$$

$$= 4e^{4x}\cos e^{4x}$$

15. $y = e^{\cos x}$

$$\frac{dy}{dx} = e^{\cos x} \cdot D_x(\cos x)$$

$$= e^{\cos x} \cdot (-\sin x)$$

$$= (-\sin x)e^{\cos x}$$

17. $y = \sin(\ln 3x^4)$

$$\frac{dy}{dx} = [\cos(\ln 3x^4)] \cdot D_x(\ln 3x^4)$$

$$= \cos(\ln 3x^4) \cdot \frac{D_x(3x^4)}{3x^4}$$

$$= \cos(\ln 3x^4)\frac{12x^3}{3x^4}$$

$$= \cos(\ln 3x^4) \cdot \frac{4}{x}$$

$$= \frac{4}{x}\cos(\ln 3x^4)$$

19. $y = \ln|\sin x^2|$

$$\frac{dy}{dx} = \frac{D_x(\sin x^2)}{\sin x^2} = \frac{(\cos x^2) \cdot D_x(x^2)}{\sin x^2}$$

$$= \frac{(\cos x^2) \cdot 2x}{\sin x^2} = \frac{2x \cos x^2}{\sin x^2}$$

or $\quad 2x \cot x^2$

21.

$$y = \frac{2\sin x}{3 - 2\sin x}$$

$$\frac{dy}{dx} = \frac{(3 - 2\sin x)D_x(2\sin x) - (2\sin x) \cdot D_x(3 - 2\sin x)}{(3 - 2\sin x)^2}$$

$$= \frac{(3 - 2\sin x) \cdot 2D_x(\sin x) - (2\sin x) \cdot [-2D_x(\sin x)]}{(3 - 2\sin x)^2}$$

$$= \frac{6\cos x - 4\sin x \cos x + 4\sin x \cos x}{(3 - 2\sin x)^2}$$

$$= \frac{6\cos x}{(3 - 2\sin x)^2}$$

23. $y = \sqrt{\dfrac{\sin x}{\sin 3x}} = \left(\dfrac{\sin x}{\sin 3x}\right)^{1/2}$

$$\frac{dy}{dx} = \frac{1}{2}\left(\frac{\sin x}{\sin 3x}\right)^{1/2} \cdot D_x\left(\frac{\sin x}{\sin 3x}\right)$$

$$= \frac{1}{2}\left(\frac{\sin 3x}{\sin x}\right)^{-1/2} \cdot \left[\frac{(\sin 3x) \cdot D_x(\sin x)}{\frac{-(\sin x) \cdot D_x(\sin 3x)}{(\sin 3x)^2}}\right]$$

$$= \frac{1}{2}\left(\frac{\sin 3x}{\sin x}\right)^{-1/2} \cdot \left[\frac{(\sin 3x)(\cos x)}{\frac{-(\sin x)(\cos 3x) \cdot D_x(3x)}{\sin^2 3x}}\right]$$

$$= \frac{1(\sin 3x)^{1/2}}{2(\sin x)^{1/2}} \cdot \frac{\sin 3x \cos x - 3\sin x \cos 3x}{\sin^2 3x}$$

$$= \frac{(\sin 3x)^{1/2}(\sin 3x \cos x - 3\sin x \cos 3x)}{2(\sin x)^{1/2}(\sin^2 3x)}$$

$$= \frac{\sqrt{\sin 3x}\,[\sin 3x \cos x - 3\sin x \cos 3x]}{2\sqrt{\sin x}\,(\sin^2 3x)}$$

25. $y = 3\tan\left(\dfrac{1}{4}x\right) + 4\cot 2x - 5\csc x + e^{-2x}$

$$\frac{dy}{dx} = 3\tan\left(\frac{1}{4}x\right) + 4\cot 2x - 5\csc x + e^{-2x}$$

$$= 3\sec^2\left(\frac{1}{4}x\right) \cdot D_x\left(\frac{1}{4}x\right)$$

$$\quad + 4(-\csc^2 2x) \cdot D_x(2x) - 5(-\csc x \cot x)$$

$$\quad + e^{-2x} \cdot D_x(-2x)$$

$$= 3\sec^2\left(\frac{1}{4}x\right) \cdot \frac{1}{4} - (4\csc^2 2x) \cdot 2$$

$$\quad + 5\csc x \cot x + e^{-2x} \cdot (-2)$$

$$= \frac{3}{4}\sec^2\left(\frac{1}{4}x\right) - 8\csc^2 2x$$

$$\quad + 5\csc x \cot x - 2e^{-2x}$$

27. $y = \sin x; x = 0$

Let $f(x) = \sin x$.

Then $f'(x) = \cos x$, so

$$f'(0) = \cos 0 = 1.$$

The slope of the tangent line to the graph of $y = \sin x$ at $x = 0$ is 1.

29. $y = \cos x; x = -\dfrac{5\pi}{6}$

Let $f(x) = \cos x$.

Then $f'(x) = -\sin x$, so

$$f'\left(-\frac{5\pi}{6}\right) = -\sin\left(-\frac{5\pi}{6}\right) = \frac{1}{2}.$$

The slope of the tangent line to the graph of $y = \cos x$ at $x = -\frac{5\pi}{6}$ is $\frac{1}{2}$.

31. $y = \tan x; x = 0$

Let $f(x) = \tan x$.

Then $f'(x) = \sec^2 x$, so

$$f'(0) = \sec^2 0 = \frac{1}{\cos^2 0} = 1.$$

The slope of the tangent line to the graph of $y = \tan x$ at $x = 0$ is 1.

33. Since $\cot x = \dfrac{\cos x}{\sin x}$, by using the quotient rule,

$$D_x(\cos x) = D_x\left(\frac{\cos x}{\sin x}\right)$$

$$= \frac{(\sin x)(-\sin x) - (\cos x)(\cos x)}{\sin^2 x}$$

$$= \frac{-\sin^2 x - \cos^2 x}{\sin^2 x}$$

$$= -\frac{\sin^2 x + \cos^2 x}{\sin^2 x}$$

$$= -\frac{1}{\sin^2 x}$$

$$= -\csc^2 x.$$

35. Since $\csc x = \dfrac{1}{\sin x} = (\sin x)^{-1}$,

$$D_x(\csc x) = D_x\left(\frac{1}{\sin x}\right) = D_x(\sin x)^{-1}$$

$$= -1(\sin x)^{-2}\cos x$$

$$= -\frac{\cos x}{\sin^2 x}$$

$$= -\frac{1}{\sin x}\cdot\frac{\cos x}{\sin x}$$

$$= -\csc x \cot x.$$

37. $y = \sin x \qquad \dfrac{dy}{dx} = \cos x$

    (a) The critical numbers are the zeros of $\cos x$, that is, $n\pi/2$ where n is an odd integer.

    (b) The sine function is increasing where the cosine is positive, that is, on the intervals $\left(n\pi,(n+1/2)\pi\right)\cup\left((n+3/2)\pi,(n+2)\pi\right)$ where n is an even integer.

    (c) The sine function is decreasing where the cosine is negative, that is, on the intervals $\left((n+1/2)\pi,(n+3/2)\right)$ where n is an even integer.

39. The relative maxima of $y = \sin \pi x$ have value 1 and occur at the values of x for which $\pi x = (n+1/2)\pi$ where n is an even integer, that is, at $x = n + 1/2$ for n an even integer. The relative minima of $y = \sin \pi x$ have value -1 and occur at the values of x for which $\pi x = (n-1/2)\pi$ where n is an even integer, that is, at $x = n - 1/2$ where n is an even integer.

41. Use the chain rule and the product rule.

$$f(x) = \cos(x^3)$$

$$f'(x) = -\sin(x^3)\left(3x^2\right)$$

$$f''(x) = -\sin(x^3)(6x) + (-\cos(x^3))(3x^2)(3x^2)$$

$$= -9x^4\cos(x^3) - 6x\sin(x^3)$$

$$f''(0) = 0$$

$$f''(2) = -144\cos 8 - 12\sin 8$$

43. $f(x) = \sin x$

$$f'(x) = \cos x$$

$$f''(x) = -\sin x$$

$$f'''(x) = -\cos x$$

$$f^{(4)}(x) = \sin x$$

$$f^{(4n)}(x) = \sin x$$

45. $f(x) = \sin 2x$

$$f'(x) = 2\cos 2x \text{ and } f''(x) = -4\sin 2x$$

The graph of f will be concave upward where $\sin 2x < 0$, that is, for $(2n - 1)\pi < 2x < 2n\pi$, where n is an integer; x will then be in one of the intervals $\left((n - 1/2)\pi,\ n\pi\right)$ with n an integer. The graph will be concave downward on the intervals $\left(n\pi,\ (n + 1/2)\pi\right)$ with n an integer. The inflection points will occur when x is a point separating the intervals of opposite concavity, that is, for $x = n\pi/2$. The inflection points are therefore $(n\pi/2, 0)$ for integer n.

47. $f(x) = x + \cos x$

$f'(x) = 1 - \sin x$

$f''(x) = -\cos x$

Since $f'(x)$ is positive except at the isolated points $x = (2n + 1/2)\pi$ for integer n, the function f is increasing everywhere. The graph has a single y-intercept at $(0, 1)$ and a single x-intercept at approximately $(-0.739, 0)$, found by solving $x + \cos x = 0$ with a calculator. The concavity is given by the sign of $-\cos x$; the graph will be concave downward on the intervals $\left((2n - 1/2)\pi,\ (2n + 1/2)\pi\right)$ for integer n and concave upward on the intervals $\left((2n + 1/2)\pi,\ (2n + 3/2)\pi\right)$ for integer n.

Inflection points occur at the x-values separating regions of opposite concavity, that is, where $\cos x = 0$, for example at $x = -\pi/2, x = \pi/2,$ and $x = 3\pi/2$. The corresponding y-values are $-\pi/2, \pi/2,$ and $3\pi/2$.

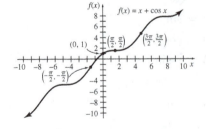

49. $f(x) = \dfrac{x}{2} - \sin x$ on $[0, \pi]$

$f'(x) = \dfrac{1}{2} - \cos x$

The only critical number in $[0, \pi]$ is $\pi/3$, since $f'(\pi/3) = 0$ and f' is negative in the second quadrant. Thus extrema could only be located at $x = 0, x = \pi/3,$ and $x = \pi.$

$f(0) = 0$

$f(\pi/3) = \dfrac{\pi - 3\sqrt{3}}{6} \approx -0.342$

$f(\pi) = \pi/2$

So there is an absolute maximum of $\pi/2$ at $x = \pi$ and an absolute minimum of $\left(\pi - 3\sqrt{3}\right)/6$ at $x = \pi/3.$

51. $\sin(xy) = x$

Differentiate both sides with respect to x.

$\dfrac{d}{dx}\sin(xy) = \dfrac{d}{dx}x$

$\cos(xy)\dfrac{d}{dx}xy = 1$

$\cos(xy)\left(x\dfrac{dy}{dx} + y\right) = 1$

Solve for dy/dx.

$\cos(xy)\left(x\dfrac{dy}{dx} + y\right) = 1$

$\dfrac{dy}{dx} = \dfrac{1}{x}\left(\dfrac{1}{\cos(xy)} - y\right)$

$= \dfrac{\sec xy}{x} - \dfrac{y}{x}$

$\cos(xy)\left(x\dfrac{dy}{dx} + y\right) = 1$

53. $\cos(\pi xy) + 2x + y^2 = 2$

Differentiate with respect to t.

$-\sin(\pi xy)(\pi)\left(x\dfrac{dy}{dt} + y\dfrac{dx}{dt}\right) + 2\dfrac{dx}{dt} + 2y\dfrac{dy}{dt} = 0$

Substitute the given values for $dx/dt, x$ and y.

$$-\sin\left[\pi(1/2)(1)\right](\pi)\left(\frac{1}{2}\frac{dy}{dt} + (1)\left(-\frac{2}{\pi}\right)\right)$$

$$+ 2\left(-\frac{2}{\pi}\right) + 2(1)\frac{dy}{dt} = 0$$

$$-\frac{\pi}{2}\frac{dy}{dt} + 2 - \frac{4}{\pi} + 2\frac{dy}{dt} = 0$$

$$\left(2 - \frac{\pi}{2}\right)\frac{dy}{dt} = \frac{4}{\pi} - 2$$

$$\frac{dy}{dt} = \frac{\frac{4}{\pi} - 2}{2 - \frac{\pi}{2}} = \frac{8 - 4\pi}{\pi(4 - \pi)}$$

55. $\dfrac{d}{dx}\sin x = \cos x$

$\sin(0 + h) \approx \sin 0 + h\left(\cos 0\right)$

$\sin(0.03) \approx 0 + (0.03)(1) = 0.03$

To four places, $\sin(0.03) = 0.0300$.

The absolute difference to four places is 0.

57. $R(t) = 120\cos 2\pi t + 150$

(a) $R'(t) = 120 \cdot (-\sin 2\pi t)(2\pi) + 0$

$= -240\pi \sin 2\pi t$

(b) Replace t with $\frac{1}{12}$ (for $\frac{1}{12}$ of a year).

$$R'\left(\frac{1}{12}\right) = -240\pi \sin 2\pi\left(\frac{1}{12}\right)$$

$$= -240\pi \sin\frac{\pi}{6}$$

$$= -240\pi\left(\frac{1}{2}\right)$$

$$= -120\pi$$

$R'(t)$ for August 1 is $-\$120\pi$ per year.

(c) January 1 is 6 months, or $\frac{6}{12} = \frac{1}{2}$ of a year

for July 1. Replace t with $\frac{1}{2}$.

$$R'\left(\frac{1}{2}\right) = -240\pi \sin 2\pi\left(\frac{1}{2}\right)$$

$$= -240\pi \sin\pi$$

$$= -240\pi(0)$$

$$= 0$$

$R'(t)$ for January 1 is $\$0$ per year.

(d) June 1, is $\frac{11}{12}$ of a year from July 1. Replace t with $\frac{11}{12}$.

$$R'\left(\frac{11}{12}\right) = -240\pi \sin 2\pi\left(\frac{11}{12}\right)$$

$$= -240\pi \sin\frac{11\pi}{6}$$

$$= -240\pi\left(-\frac{1}{2}\right)$$

$$= 120\pi$$

$R'(t)$ for June 1 is $\$120\pi$ per year.

59. $y = \dfrac{\pi}{8}\cos 3\pi\left(t - \dfrac{1}{3}\right)$

(a) The graph should resemble the graph of $y = \cos x$ with following difference: The maximum and minimum values of y are $\frac{\pi}{8}$ and $-\frac{\pi}{8}$. The period of the graph will be $\frac{2\pi}{3\pi} = \frac{2}{3}$ units. The graph will be shifted horizontally $\frac{1}{3}$ units to the right.

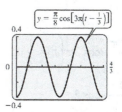

(b) velocity $= \dfrac{dy}{dt}$

$$= D_t\left[\frac{\pi}{8}\cos 3\pi\left(t - \frac{1}{3}\right)\right]$$

$$= \frac{\pi}{8}D_t\left[\cos 3\pi\left(t - \frac{1}{3}\right)\right]$$

$$= \frac{\pi}{8}\left[-\sin 3\pi\left(t - \frac{1}{3}\right)\right]D_t\left[3\pi\left(t - \frac{1}{3}\right)\right]$$

$$= \frac{\pi}{8}\left[-\sin 3\pi\left(t - \frac{1}{3}\right)\right] \cdot 3\pi$$

$$= -\frac{3\pi^2}{8}\sin\left[3\pi\left(t - \frac{1}{3}\right)\right]$$

acceleration $= \dfrac{d^2 y}{dt^2}$

$$= D_t \left[\frac{-3\pi^2 \sin 3\pi \left(t - \frac{1}{3} \right)}{8} \right]$$

$$= \frac{-3\pi^2}{8} D_t \left[\sin 3\pi \left(t - \frac{1}{3} \right) \right]$$

$$= \frac{-3\pi^2}{8} \left[\cos 3\pi \left(t - \frac{1}{3} \right) \right] D_t \left[3\pi \left(t - \frac{1}{3} \right) \right]$$

$$= \frac{-3\pi^2}{8} \left[\cos 3\pi \left(t - \frac{1}{3} \right) \right] \cdot 3\pi$$

$$= -\frac{9\pi^3}{8} \cos \left[3\pi \left(t - \frac{1}{3} \right) \right]$$

(c) $\dfrac{d^2 y}{dt^2} + 9\pi^2 y$

$$= -\frac{9\pi^3}{8} \cos 3\pi \left(t - \frac{1}{3} \right)$$

$$+ 9\pi^2 \left[\frac{\pi}{8} \cos 3\pi \left(t - \frac{1}{3} \right) \right]$$

$$= -\frac{9\pi^3}{8} \cos 3\pi \left(t - \frac{1}{3} \right)$$

$$+ \frac{9\pi^3}{8} \cos 3\pi \left(t - \frac{1}{3} \right)$$

$$= 0$$

(d) $a(1) = -\dfrac{9\pi^3}{8} \cos 3\pi \left(t - \dfrac{1}{3} \right)$

$$= -\frac{9\pi^3}{8} \cos 2\pi$$

$$= -\frac{9\pi^3}{8} \cdot 1$$

$$= -\frac{9\pi^3}{8} < 0$$

$$y(1) = \frac{\pi}{8} \cos 3\pi \left(t - \frac{1}{3} \right)$$

$$= \frac{\pi}{8} \cos 2\pi$$

$$= \frac{\pi}{8} \cdot 1 = \frac{\pi}{8}$$

Therefore, at $t = 1$ second, the force is clockwise and the arm makes an angle of $\frac{\pi}{8}$ radians forward from the vertical. The arm is moving clockwise.

$$a\left(\frac{4}{3} \right) = -\frac{9\pi^3}{8} \cos 3\pi \left(\frac{4}{3} - \frac{1}{3} \right)$$

$$= -\frac{9\pi^3}{8} \cos (3\pi)$$

$$= -\frac{9\pi^3}{8} (-1) = \frac{9\pi^3}{8} > 0$$

$$y\left(\frac{4}{3} \right) = \frac{\pi}{8} \cos 3\pi \left(\frac{4}{3} - \frac{1}{3} \right)$$

$$= \frac{\pi}{8} \cos (3\pi)$$

$$= \frac{\pi}{8} (-1) = -\frac{\pi}{8}$$

Therefore, at $t = \frac{4}{3}$ seconds, the force is counter clockwise and the arm makes an angle of $-\frac{\pi}{8}$ radians from the vertical. The arm is moving counterclockwise.

$$a\left(\frac{5}{3} \right) = -\frac{9\pi^3}{8} \cos 3\pi \left(\frac{5}{3} - \frac{1}{3} \right)$$

$$= -\frac{9\pi^3}{8} \cos (4\pi)$$

$$= -\frac{9\pi^3}{8} \cdot 1$$

$$= -\frac{9\pi^3}{8} < 0$$

$$y\left(\frac{5}{3} \right) = \frac{\pi}{8} \cos 3\pi \left(\frac{5}{3} - \frac{1}{3} \right)$$

$$= \frac{\pi}{8} \cos (4\pi)$$

$$= \frac{\pi}{8} \cdot 1$$

$$= \frac{\pi}{8}$$

Therefore, at $t = \frac{5}{3}$ seconds, the answer corresponds to $t = 1$ second. So the arm is moving clockwise and makes an angle of $\frac{\pi}{8}$ from the vertical.

61. $L(t) = 0.022t^2 + 0.55t + 316 + 3.5 \sin(2\pi t)$

(a)

(b) $L(25) = 343.5$ parts per million

$\quad\quad L(35.5) = 363.25$ parts per million

$\quad\quad L(50.2) = 402.38$ parts per million

(c)

$L'(t) = 0.044t + 0.55 + 7.0\pi \cos(2\pi t)$

$L'(50.2) = 9.55$ parts per million per year

At the beginning of 2010, the level of carbon dioxide was increasing at 9.55 parts per million per year.

63. (a) $\quad f(t) = 1000e^{2\sin(t)}$

$\quad\quad f(0.2) = 1000e^{2\sin(0.2)}$

$\quad\quad\quad \approx 1488$

(b) $\quad f(t) = 1000e^{2\sin(t)}$

$\quad\quad f(1) = 1000e^{2\sin(1)}$

$\quad\quad\quad \approx 5381$

(c) Since $f'(t) = 2000\cos(t)e^{2\sin(t)}$, $f'(0)$ is given by

$$f'(0) = 2000\cos(0)e^{2\sin(0)}$$

$$= 2000.$$

(d) Since $f'(t) = 2000\cos(t)e^{2\sin(t)}$, $f'(0.2)$ is given by

$$f'(0.2) = 2000\cos(0.2)e^{2\sin(0.2)}$$

$$\approx 2916.$$

(e)

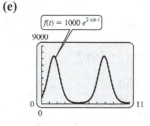

(f) Set $f'(t) = 2000\cos(t)e^{2\sin(t)}$ equal to zero to find critical numbers.

$$2000\cos(t)e^{2\sin(t)} = 0$$

$$\cos(t)e^{2\sin(t)} = 0$$

$\cos(t) = 0 \quad\quad$ or $\quad e^{2\sin(t)} = 0$

$t = \dfrac{\pi}{2} + n\pi \quad\quad e^{2\sin(t)} \neq 0,$

for n any interger $\quad\quad$ for all t

We will use the second derivative test.

$f''(t) = -2000\sin(t)e^{2\sin(t)}$

$\quad\quad + 4000\cos^2(t)e^{2\sin(t)}$

$\quad\quad = 2000e^{2\sin t}(-\sin t + 2\cos^2 t)$

Since $2000e^{2\sin t} > 0$ for all t, $f''(t) > 0$ when $-\sin t + 2\cos^2 t > 0$, when $\sin t < 2\cos^2 t$. Also, $f''(t) < 0$ when $-\sin t + 2\cos^2 t < 0$, when $\sin t > 2\cos^2 t$. Since

$\sin\left(\dfrac{\pi}{2} + n\pi\right) < 0 = 2\cos^2\left(\dfrac{\pi}{2} + n\pi\right)$ for n any odd integer, and

$\sin\left(\dfrac{\pi}{2} + n\pi\right) > 0 = 2\cos^2\left(\dfrac{\pi}{2} + n\pi\right)$

for n any even integer, f has a maximum at $t = \dfrac{\pi}{2} + n\pi$ for n any even integer and f has a minimum at $t = \dfrac{\pi}{2} + 2\pi n$ for n any odd integer. This is equivalent to f having a maximum for $t = \dfrac{\pi}{2} + 2\pi n$ for n any integer and f having a minimum for $t = \dfrac{3\pi}{2} + 2\pi n$ for n any integer. To find the maximum and minimum values we will find $f\left(\dfrac{\pi}{2}\right)$ and $f\left(\dfrac{3\pi}{2}\right)$ respectively.

$$f\left(\dfrac{\pi}{2}\right) = 1000e^{2\sin\left(\frac{\pi}{2}\right)} \approx 7389$$

$$\left(\dfrac{3\pi}{2}\right) = 1000e^{2\sin\left(\frac{3\pi}{2}\right)} \approx 135$$

65. (a) Using the graphing calculator to graph

$$P(t) = 0.003\sin(220\pi t)$$

$$+ \dfrac{0.003}{3}\sin(660\pi t)$$

$$+ \dfrac{0.003}{5}\sin(1100\pi t)$$

$$+ \dfrac{0.003}{7}\sin(1540\pi t)$$

in a $[0, 0.01]$ by $[-0.004, 0.004]$ viewing window, gives the following graph.

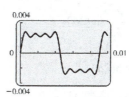

(b) Using the graphing calculator to find $P'(x)$ and evaluate $P'(0.002)$ gives .151. However, the slope of the graph of $P(x)$ at $x = 0.002$ appears to be approximately -1. Thus, we calculate $P'(x)$ by hand and use the calculator to evaluate $P'(0.002)$.

Since $P'(t) = 0.66\pi \cos(220\pi t) +$
$0.66\pi \cos(660\pi t) + 0.66\pi \cos(1100\pi t) +$
$0.66\pi \cos(1540\pi t)$, $P'(0.002)$ is given by

$P'(0.002) = 0.66\pi \cos[220\pi(0.002)] +$
$0.66\pi \cos[660\pi(0.002)] +$
$0.66\pi \cos[1100\pi(0.002)] +$
$0.66\pi \cos[1540\pi(0.002)] \approx -1.05$ pounds

per square foot per second. At 0.002 seconds the pressure is decreasing at a rate of about 1.05 pounds per square foot per second.

67.   **(a)** $y = x \tan\alpha - \dfrac{16x^2}{V^2} \sec^2\alpha$

$$= 40\tan\left(\frac{\pi}{4}\right) - \frac{16(40)^2}{44^2}\sec^2\left(\frac{\pi}{4}\right)$$

$$= 40(1) - \frac{16(40)^2}{44^2}(2) \approx 13.55 \text{ feet}$$

(b) $y = x\tan\alpha - \dfrac{16x^2}{V^2}\sec^2\alpha$

$$0 = x\tan\alpha - \frac{16x^2}{V^2}\sec^2\alpha$$

$$0 = x\left(\tan\alpha - \frac{16x}{V^2}\sec^2\alpha\right)$$

$$0 = \tan\alpha - \frac{16x}{V^2}\sec^2\alpha \quad (\text{for } x \neq 0)$$

$$\frac{16x}{V^2}\sec^2\alpha = \tan\alpha$$

$$x = \frac{V^2}{16}\cdot\frac{\tan\alpha}{\sec^2\alpha}$$

$$= \frac{V^2}{16}\cdot \sin\alpha\cos\alpha$$

$$= \frac{V^2}{32}\cdot 2\sin\alpha\cos\alpha$$

$$= \frac{V^2}{32}\sin(2\alpha)$$

(c) $x = \dfrac{V^2}{32}\sin(2\alpha)$

$$= \frac{44^2}{32}\sin\left[2\left(\frac{\pi}{3}\right)\right]$$

$$\approx 52.39 \text{ feet}$$

(d) $\dfrac{dx}{d\alpha} = \dfrac{V^2}{32}\cos(2\alpha)\cdot D_\alpha(2\alpha)$

$$= \frac{V^2}{16}\cos(2\alpha)$$

Find critical values.

$$\frac{V^2}{16}\cos(2\alpha) = 0$$

$$\cos(2\alpha) = 0$$

$$2\alpha = \frac{\pi}{2} + n\pi, \text{ for } n \text{ any integer}$$

$$\alpha = \frac{\pi}{4} + \frac{n\pi}{2}, \text{ for } n \text{ any integer}$$

Since $0 < \alpha < \frac{\pi}{2}, \frac{dx}{d\alpha} = 0$ for $\alpha = \frac{\pi}{4}$.

Furthermore, $\frac{d^2x}{d\alpha^2} = -\frac{V^2}{8}\sin(2\alpha)$ which is less than zero at $\alpha = \frac{\pi}{4}$. Therefore, x is maximized when $\alpha = \frac{\pi}{4}$.

(e)   Since 60 miles per hour is 88 feet per second, evaluate

$$x = \frac{V^2}{32}\sin(2\alpha) \text{ when } V$$

$$= 88 \text{ and } \alpha = \frac{\pi}{4}.$$

$$= \frac{88^2}{32}\sin\left[2\left(\frac{\pi}{4}\right)\right] = 242 \text{ feet}$$

69. $s(t) = \sin t + 2\cos t$

$$v(t) = s'(t) = \cos t - 2\sin t$$

(a) $v(0) = 1 - 2(0) = 1$

(b) $v\left(\dfrac{\pi}{4}\right) = \dfrac{\sqrt{2}}{2} - 2\left(\dfrac{\sqrt{2}}{2}\right) = -\dfrac{\sqrt{2}}{2}$

$$\approx -0.7071$$

(c) $v\left(\dfrac{3\pi}{2}\right) = 0 - 2(-1) = 2$

$$a(t) = v'(t) = -\sin t - 2\cos t$$

(d) $a(0) = 0 - 2(1) = -2$

(e) $a\left(\dfrac{\pi}{4}\right) = -\dfrac{\sqrt{2}}{2} - 2\left(\dfrac{\sqrt{2}}{2}\right) = -\dfrac{3\sqrt{2}}{2}$

$$\approx -2.1213$$

(f) $a(\pi) = -0 - (-2) = 2$

71. From the figure, we see that

$$\tan \theta = \frac{x}{60}$$

$$60 \tan \theta = x.$$

Differentiating with respect to time, t, gives

$$60 \sec^2 \theta \cdot \frac{d\theta}{dt} = \frac{dx}{dt}$$

$$\frac{d\theta}{dt} = \frac{\frac{dx}{dt}}{60 \sec^2 \theta}.$$

Since

$$\frac{dx}{dt} = 600,$$

$$\frac{d\theta}{dt} = \frac{600}{60 \sec^2 \theta}$$

$$= \frac{10}{\sec^2 \theta}.$$

(a) When the car is at the point on the road closest to the camera $\theta = 0$. Thus,

$$\frac{d\theta}{dt} = \frac{10}{(\sec 0)^2}$$

$$= \frac{10}{1^2} = 10 \text{ radians/min}$$

$$\frac{d\theta}{dt} = \frac{10 \text{ radians}}{\min} \cdot \frac{1 \text{ rev}}{2\pi \text{ radians}}$$

$$= \frac{5}{\pi} \text{ rev/min}$$

(b) Six seconds later is $\frac{1}{10}$ of a minute later, and the car has traveled 60 feet. Thus,

$\tan \theta = \frac{60}{60}$ and $\theta = \frac{\pi}{4}$.

$$\frac{d\theta}{dt} = \frac{10}{\left(\sec \frac{\pi}{4}\right)^2} = \frac{10}{(\sqrt{2})^2}$$

$$= 5 \text{ radians/min}$$

This is one-half the previous value, so

$$\frac{d\theta}{dt} = \frac{1}{2} \cdot \frac{5}{\pi} \text{ rev/min}$$

$$= \frac{5}{2\pi} \text{ rev/min}.$$

73. Let x represent the length of the ladder.

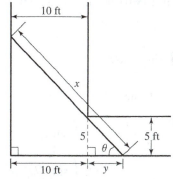

$$\tan \theta = \frac{5}{y} \quad \text{and} \quad \cos \theta = \frac{10 + y}{x}$$

$$y = 5 \cot \theta \quad \text{and} \quad x = \frac{10 + y}{\cos \theta}$$

Thus,

$$x = \frac{10 + 5 \cot \theta}{\cos \theta} = 10 \sec \theta + 5 \csc \theta$$

$$\frac{dx}{d\theta} = 10 \sec \theta \tan \theta - 5 \csc \theta \cot \theta$$

$$= 10 \left(\frac{1}{\cos \theta}\right)\left(\frac{\sin \theta}{\cos \theta}\right) - 5 \left(\frac{1}{\sin \theta}\right)\left(\frac{\cos \theta}{\sin \theta}\right)$$

$$= \frac{10 \sin \theta}{\cos^2 \theta} - \frac{5 \cos \theta}{\sin^2 \theta}$$

If $\frac{dx}{d\theta} = 0$, then

$$\frac{10 \sin \theta}{\cos^2 \theta} = \frac{5 \cos \theta}{\sin^2 \theta}$$

$$10 \sin^3 \theta = 5 \cos^3 \theta$$

$$\frac{\sin^3 \theta}{\cos^3 \theta} = \frac{1}{2}$$

$$\tan^3 \theta = \frac{1}{2}$$

$$\tan \theta = \sqrt[3]{\frac{1}{2}}$$

$$\theta \approx 0.6709 \text{ radian}.$$

If $\theta < 0.6709$ radian, $\frac{dx}{d\theta} < 0$.

If $\theta > 0.6709$ radian, $\frac{dx}{d\theta} > 0$.

Thus, x will be minimum when $\theta = 0.6709$. If $\theta = 0.6709$, then

$$x = \frac{10 + 5 \cot 0.6709}{\cos 0.6709} \approx 20.81.$$

The length of the longest possible ladder is approximately 20.81 feet.

13.3   Integrals of Trigonometric Functions

Your Turn 1

(a)   Find $\int \sin\left(\dfrac{x}{2}\right) dx$.

If $u = x/2$, then $du = (1/2)\, dx$.

$$\int \sin\left(\frac{x}{2}\right) dx = 2 \int \sin x \left(\frac{1}{2}\, dx\right)$$

$$= 2 \int \sin u \, du$$

$$= 2(-\cos u) + C$$

$$= -2\cos\left(\frac{x}{2}\right) + C$$

(b)   Find $\int 6(\sec^2 x)\sqrt{\tan x}\ dx$.

If $u = \tan x$, then $du = \sec^2 x \, dx$.

$$\int 6(\sec^2 x)\sqrt{\tan x}\ dx = 6 \int \sqrt{\tan x}\ (\sec^2 x \, dx)$$

$$= 6 \int \sqrt{u}\ du$$

$$= 6\left(\frac{2}{3} u^{3/2}\right) + C$$

$$= 4u^{3/2} + C$$

$$= 4(\tan x)^{3/2} + C$$

Your Turn 2

Find $\int \dfrac{\tan(\sqrt{x})}{\sqrt{x}}\ dx$.

If $u = \sqrt{x}$, then $du = \dfrac{1}{2\sqrt{x}}\, dx$.

$$\int \frac{\tan(\sqrt{x})}{\sqrt{x}}\ dx = 2 \int \tan(\sqrt{x})\left(\frac{1}{2\sqrt{x}}\, dx\right)$$

$$= 2 \int \tan u \, du$$

$$= -2\ln |\cos u| + C$$

$$= -2\ln |\cos(\sqrt{x})| + C$$

Your Turn 3

Find $\int x\cos(3x)\ dx$.

Let $u = x$ and $dv = \cos(3x)\, dx$.

Then $du = dx$ and $v = \dfrac{1}{3}\sin 3x$.

$$\int x\cos(3x)\ dx = \int u \, dv$$

$$= uv - \int v \, du$$

$$= \frac{x}{3}\sin(3x) - \frac{1}{3}\int \sin(3x)\ dx$$

$$= \frac{x}{3}\sin(3x) - \frac{1}{3}\left(-\frac{1}{3}\cos(3x)\right) + C$$

$$= \frac{x}{3}\sin(3x) + \frac{1}{9}\cos(3x) + C$$

Your Turn 4

Find the area under the curve $y = \sec^2(x/3)$ between $x = -\pi$ and $x = \pi$.

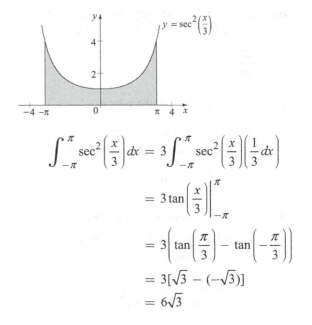

$$\int_{-\pi}^{\pi} \sec^2\left(\frac{x}{3}\right) dx = 3 \int_{-\pi}^{\pi} \sec^2\left(\frac{x}{3}\right)\left(\frac{1}{3}\, dx\right)$$

$$= 3 \tan\left(\frac{x}{3}\right)\Big|_{-\pi}^{\pi}$$

$$= 3\left(\tan\left(\frac{\pi}{3}\right) - \tan\left(-\frac{\pi}{3}\right)\right)$$

$$= 3[\sqrt{3} - (-\sqrt{3})]$$

$$= 6\sqrt{3}$$

13.3   Warmup Exercises

W1.   Let $u = x^2$, then $du = 2x\ dx$.

$$\int xe^{x^2}\ dx = \frac{1}{2}\int e^u \, du$$

$$= \frac{1}{2} e^u + C$$

$$= \frac{1}{2} e^{x^2} + C$$

W2. Let $u = x^2 + 6$, then $du = 2x\,dx$.

$$\int \frac{x}{x^2 + 6}\,dx = \frac{1}{2}\int \frac{1}{u}\,du$$

$$= \frac{1}{2}\ln|u| + C$$

$$= \frac{1}{2}\ln(x^2 + 6) + C$$

W3. Let $dv = e^x\,dx$ and $u = x + 1$.

Then $v = e^x$ and $du = dx$.

$$\int (x + 1)e^x\,dx = \int u\,dv$$

$$= uv - \int v\,du$$

$$= (x + 1)e^x - \int e^x dx$$

$$= (x + 1)e^x - e^x + C$$

$$= xe^x + C$$

W4. Let $dv = 8x\,dx$ and $u = \ln x$.

Then $v = 4x^2$ and $du = \frac{1}{x}\,dx$.

$$\int 8\ln x\,dx = \int u\,dv$$

$$= uv - \int v\,du$$

$$= 4x^2 \ln x - \int (4x^2)(1/x)\,dx$$

$$= 4x^2 \ln x - 2x^2 + C$$

13.3   Exercises

1. $\displaystyle\int \cos 3x\,dx$

Let $u = 3x$, so $du = 3\,dx$ or $\frac{1}{3}du = dx$.

$$\int \cos 3x\,dx = \int \cos u \cdot \frac{1}{3}\,du$$

$$= \frac{1}{3}\int \cos u\,du$$

$$= \frac{1}{3}\sin u + C$$

$$= \frac{1}{3}\sin 3x + C$$

3. $\displaystyle\int (3\cos x - 4\sin x)\,dx$

$$= \int 3\cos x\,dx - \int 4\sin x\,dx$$

$$= 3\int \cos x\,dx - 4\int \sin x\,dx$$

$$= 3\sin x + 4\cos x + C$$

5. $\displaystyle\int x \sin x^2\,dx$

Let $u = x^2$.

Then $du = 2x\,dx$

$$\frac{1}{2} = du = x\,dx.$$

$$\int x \sin x^2\,dx = \int \sin u \cdot \frac{1}{2}\,du$$

$$= \frac{1}{2}\int \sin u\,du$$

$$= -\frac{1}{2}\cos u + C$$

$$= -\frac{1}{2}\cos x^2 + C$$

7. $\displaystyle -\int 3\sec^2 3x\,dx$

Let $u = 3x$, so $du = 3\,dx$ or $\frac{1}{3}du = dx$.

$$-\int 3\sec^2 3x\,dx = -\int 3\sec^2 u \cdot \frac{1}{3}\,du$$

$$= -\int \sec^2 u\,du$$

$$= -\tan u + C$$

$$= -\tan 3x + C$$

9. $\displaystyle\int \sin^7 x \cos x\,dx$

Let $u = \sin x$.

Then $du = \cos x\,dx$.

$$\int \sin^7 x(\cos x)\,dx = \int u^7\,du$$

$$= \frac{1}{8}u^8 + C$$

$$= \frac{1}{8}\sin^8 x + C$$

11. $\displaystyle\int 3\sqrt{\cos x}\,(\sin x)\,dx$

Let $u = \cos x$, so $du = -\sin x\,dx$ or $-du = \sin x\,dx$.

$$\int 3\sqrt{\cos x}\,(\sin x)\,dx$$

$$= \int 3\sqrt{u}\,(-du)$$

$$= -3\int u^{1/2}\,du$$

$$= -3\left(\frac{2}{3}u^{3/2}\right) + C$$

$$= -2u^{3/2} + C$$

$$= -2(\cos x)^{3/2} + C$$

13. $\displaystyle\int \frac{\sin x}{1 + \cos x}\,dx$

Let $u = 1 + \cos x$.

Then $du = -\sin x$

$\qquad -du = -\sin x\,dx$.

$$\int \frac{\sin x}{1 + \cos x}\,dx = \int \frac{1}{u}(-du) = -\int \frac{1}{u}\,du$$

$$= -\ln|u| + C$$

$$= -\ln|1 + \cos x| + C$$

15. $\displaystyle\int 2x^7 \cos x^8\,dx$

Let $u = x^8$, so $du = 8x^7\,dx$ or $\frac{1}{8}du = x^7\,dx$.

$$\int 2x^7 \cos x^8\,dx = \int 2(\cos u)\cdot\frac{1}{8}\,du$$

$$= \frac{1}{4}\int \cos u\,du$$

$$= \frac{1}{4}\sin u + C$$

$$= \frac{1}{4}\sin x^8 + C$$

17. $\displaystyle\int \tan\frac{1}{3}x\,dx$

Let $u = \frac{1}{3}x$, so $du = \frac{1}{3}dx$ or $3\,du = dx$.

$$\int \tan\frac{1}{3}x\,dx = \int (\tan u)\cdot 3\,du$$

$$= 3\int \tan u\,du$$

$$= -3\ln|\cos u| + C$$

$$= -3\ln\left|\cos\frac{1}{3}x\right| + C$$

19. $\displaystyle\int x^5 \cot x^6\,dx$

Let $u = x^6$, so $du = 6x^5\,dx$ or $\frac{1}{6}du = x^5\,dx$.

$$\int x^5 \cot x^6\,dx = \int (\cot u)\cdot\frac{1}{6}\,du$$

$$= \frac{1}{6}\int \cot u\,du$$

$$= \frac{1}{6}\ln|\sin u| + C$$

$$= \frac{1}{6}\ln|\sin x^6| + C$$

21. $\displaystyle\int e^x \sin e^x\,dx$

Let $u = e^x$.

Then $du = e^x\,dx$.

$$\int e^x \sin e^x\,dx = \int \sin u\,du$$

$$= -\cos u + C$$

$$= -\cos e^x + C$$

23. $\displaystyle\int e^x \csc e^x \cot e^x\,dx$

Let $u = e^x$, so that $du = e^x\,dx$.

$$\int e^x \csc e^x \cot e^x\,dx = \int \csc e^x \cot e^x (e^x\,dx)$$

$$= \int \csc u \cot u\,du$$

$$= -\csc u + C$$

$$= -\csc e^x + C$$

25. $\displaystyle\int -6x \cos 5x\, dx$

Let $u = -6x$ and $dv = \cos 5x\, dx$.

Then $du = -6\, dx$ and $v = \frac{1}{5}\sin 5x$.

$\displaystyle\int -6x \cos 5x\, dx$

$\displaystyle = (-6x)\left(\frac{1}{5}\sin 5x\right) - \int \left(\frac{1}{5}\sin 5x\right)(-6\, dx)$

$\displaystyle = -\frac{6}{5}\sin 5x + \frac{6}{5}\int \sin 5x\, dx$

$\displaystyle = -\frac{6}{5}x\sin 5x + \frac{6}{5}\cdot\frac{1}{5}(-\cos 5x) + C$

$\displaystyle = -\frac{6}{5}x\sin 5x - \frac{6}{25}\cos 5x + C$

27. $\displaystyle\int 4x \sin x\, dx$

Let $u = 4x$ and $dv = \sin x\, dx$.

Then $du = 4\, dx$ and $v = -\cos x$.

$\displaystyle\int 4x \sin x\, dx$

$\displaystyle = 4x(-\cos x) - \int (-\cos x)\cdot 4\, dx$

$\displaystyle = -4x\cos x + 4\int \cos x\, du$

$\displaystyle = -4x\cos x + 4\sin x + C$

29. $\displaystyle\int -6x^2 \cos 8x\, dx$

Let $u = -6x^2$ and $dv = \cos 8x\, dx$.

Then $du = -12x\, dx$ and $v = \frac{1}{8}\sin 8x$.

$\displaystyle\int -6x^2 \cos 8x\, dx$

$\displaystyle = (-6x^2)\left(\frac{1}{8}\sin 8x\right)$

$\displaystyle \quad - \int \left(\frac{1}{8}\sin 8x\right)(-12x\, dx)$

$\displaystyle = -\frac{3}{4}x^2 \sin 8x + \frac{3}{2}\int x\sin 8x\, dx$

In $\displaystyle\int x\sin 8x\, dx$, let

$\qquad u = x \quad$ and $\quad dv = \sin 8x\, dx$.

Then $\quad du = dx \quad$ and $\quad v = -\frac{1}{8}\cos 8x$.

$\displaystyle\int -6x^2 \cos 8x\, dx$

$\displaystyle = -\frac{3}{4}x^2 \sin 8x$

$\displaystyle \quad + \frac{3}{2}\left[-\frac{1}{8}x\cos 8x - \int\left(-\frac{1}{8}\cos 8x\right)dx\right]$

$\displaystyle = -\frac{3}{4}x^2 \sin 8x - \frac{3}{16}x\cos 8x$

$\displaystyle \quad + \frac{3}{16}\int \cos 8x\, dx$

$\displaystyle = -\frac{3}{4}x^2 \sin 8x - \frac{3}{16}x\cos 8x$

$\displaystyle \quad + \frac{3}{16}\cdot\frac{1}{8}\sin 8x + C$

$\displaystyle = -\frac{3}{4}x^2 \sin 8x - \frac{3}{16}x\cos 8x$

$\displaystyle \quad + \frac{3}{128}\sin 8x + C$

31. $\displaystyle\int_0^{\pi/4} \sin x\, dx = -\cos x\Big|_0^{\pi/4}$

$\displaystyle \qquad = -\cos\frac{\pi}{4} - (-\cos 0)$

$\displaystyle \qquad = -\frac{\sqrt{2}}{2} + 1$

$\displaystyle \qquad = 1 - \frac{\sqrt{2}}{2}$

33. $\displaystyle\int_0^{\pi/6} \tan x\, dx = -\ln|\cos x|\Big|_0^{\pi/6}$

$\displaystyle \qquad = -\ln\left|\cos\frac{\pi}{6}\right| - (-\ln|\cos 0|)$

$\displaystyle \qquad = -\ln\frac{\sqrt{3}}{2} + \ln 1$

$\displaystyle \qquad = -\ln\frac{\sqrt{3}}{2} + 0$

$\displaystyle \qquad = -\ln\frac{\sqrt{3}}{2}$

35. $\displaystyle\int_{\pi/2}^{2\pi/3} \cos x\, dx = \sin x\Big|_{\pi/2}^{2\pi/3}$

$\displaystyle \qquad = \sin\frac{2\pi}{3} - \sin\frac{\pi}{2}$

$\displaystyle \qquad = \frac{\sqrt{3}}{2} - 1$

37. Use the fnInt function on the graphing calculator to enter fnInt $(e \wedge (-x) \sin(x), \ x, 0, b)$, for successively larger values of b, which returns a value of 0.5 for sufficiently large enough b.

Thus, an estimate of $\int_0^\infty e^{-x} \sin x \, dx$ is 0.5.

39. $f(x) = \sin x$ on $[0, \ 3\pi/2]$

f is positive to the left of $x = \pi$ and negative to the right of this point, so we compute the area as the sum of two integrals:

$$\text{Area} = \int_0^\pi \sin x \, dx + \left(\int_\pi^{3\pi/2} -\sin x \, dx \right)$$

$$= \left(-\cos x \right) \Big|_0^\pi - \left(-\cos x \right) \Big|_\pi^{3\pi/2}$$

$$= \left(1 - (-1) \right) - \left(0 - 1 \right)$$

$$= 3$$

41. Sketch the graphs:

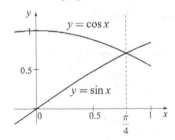

$$\text{Area} = \int_0^{\pi/4} (\cos x - \sin x) \, dx$$

$$= \left(\sin x + \cos x \right) \Big|_0^{\pi/4}$$

$$= \left(\frac{\sqrt{2}}{2} + \frac{\sqrt{2}}{2} \right) - (0 + 1)$$

$$= \sqrt{2} - 1$$

43. Sketch the graphs:

$$\text{Area} = \int_0^{\pi/4} (\tan x - \sin x) \, dx$$

$$= \left(\ln|\sec x| + \cos x \right) \Big|_0^{\pi/4}$$

$$= \left(\ln \sqrt{2} + \frac{\sqrt{2}}{2} \right) - (\ln 1 + 1)$$

$$= \frac{1}{2} \ln 2 + \frac{\sqrt{2}}{2} - 1$$

45. $S(t) = 500 + 500 \cos \left(\frac{\pi}{6} t \right)$

Total sales over a year's time are approximated by the area under the graph of S during any twelve-month period due to periodicity of S.

Total sales

$$\approx \int_0^{12} S(t) \, dt$$

$$= \int_0^{12} \left[500 + 500 \cos \left(\frac{\pi}{6} t \right) \right] dt$$

$$= \int_0^{12} 500 \, dt + 500 \int_0^{12} \cos \left(\frac{\pi}{6} t \right) dt$$

$$= 500 t \Big|_0^{12} + 500 \left[\frac{6}{\pi} \sin \left(\frac{\pi}{6} t \right) \right] \Big|_0^{12}$$

$$= 6000 + \frac{3000}{\pi} (\sin 2\pi - \sin 0)$$

$$= 6000 + \frac{3000}{\pi} (0 - 0) = 6000$$

47. $T(t) = 50 + 50 \cos \left(\frac{\pi}{6} t \right)$

Since T is periodic, the number of animals passing the checkpoint is equal to the area under the curve for any 12-month period. Let t vary from 0 to 12.

$$\text{Total} = \int_0^{12} \left[50 + 50 \cos\left(\frac{\pi}{6}t\right) \right] dt$$

$$= \int_0^{12} 50 \, dt$$

$$+ \frac{6}{\pi} \int_0^{12} 50 \cos\left(\frac{\pi}{6}t\right)\left(\frac{\pi}{6} \, dt\right)$$

$$= 50t \Big|_0^{12} + \frac{300}{\pi} \sin\left(\frac{\pi}{6}t\right)\Big|_0^{12}$$

$$= (600 - 0) + \frac{300}{\pi}(0 - 0)$$

$$= 600 \text{ (in hundreds)}$$

The total number of animals is 60,000.

49. The total amount of daylight is given by

$$\int_0^{365} N(t)\,dt = \int_0^{365} [183.549 \sin(0.0172t$$

$$- 1.329) + 728.124] dt$$

$$= \left[-\frac{183.549}{0.0172} \cos(0.0172t \right.$$

$$\left. - 1.329) + 728.124 \right]\Big|_0^{365}$$

$$\approx 265{,}819.0192 \text{ minutes}$$

$$\approx 4430 \text{ hours.}$$

The result is relatively close to the actual value.

51. The amount of water in the tank is

$$\int_0^k \sec^2 t \, dt = \tan t \Big|_0^k$$

$$= \tan k - \tan 0$$

$$= \tan k \text{ (“tank”).}$$

Chapter 13   Review Exercises

1. False; The period of cosine is 2π.

2. True

3. False; There's no reason to suppose that the Dow is periodic.

4. False; $\cos(a + b) = \cos a \cos b - \sin a \sin b$

5. True

6. False; $D_x \tan(x^2) = 2x \sec^2(x^2)$

7. True

8. False; The secant function has no absolute maximum or minimum

9. False; Since the sine function is negative on half of this interval, the true area is $\int_0^{2\pi} |\sin x|\, dx$.

10. False; Use substitution, with $u = 5 + \sin x$ and $du = \cos x \, dx$.

15. $90° = 90\left(\frac{\pi}{180}\right) = \frac{90\pi}{180} = \frac{\pi}{2}$

17. $225° = 225\left(\frac{\pi}{180}\right) = \frac{5\pi}{4}$

19. $5\pi = 5\pi\left(\frac{180°}{\pi}\right) = 900°$

21. $\frac{9\pi}{20} = \frac{9\pi}{20}\left(\frac{180°}{\pi}\right) = 81°$

23. When an angle $60°$ is drawn in standard position, one choice of a point on its terminal side is $(x, y) = (1, \sqrt{3})$. Then

$$r = \sqrt{x^2 + y^2} = \sqrt{1 + 3} = 2,$$

so

$$\sin 60° = \frac{y}{r} = \frac{\sqrt{3}}{2}.$$

25. When an angle of $-45°$ is drawn in standard position, one choice of a point on its terminal side is $(x, y) = (1, -1)$. Then

$$r = \sqrt{x^2 + y^2} = \sqrt{1 + 1} = \sqrt{2},$$

so

$$\cos(-45°) = \frac{x}{r} = \frac{1}{\sqrt{2}} = \frac{\sqrt{2}}{2}.$$

27. When an angle of $120°$ is drawn in standard position, one choice of a point on its terminal side is $(x, y) = (-1, \sqrt{3})$. Then

$$r = \sqrt{x^2 + y^2} = \sqrt{1 + 3} = 2,$$

so

$$\csc 120° = \frac{r}{y} = \frac{2}{\sqrt{3}} = \frac{2\sqrt{3}}{3}.$$

29. When an angle of $\frac{\pi}{6}$ is drawn in standard position, one choice of a point on its terminal side is $(x, y) = (\sqrt{3}, 1)$. Then

$$r = \sqrt{x^2 + y^2} = \sqrt{3 + 1} = 2,$$

so

$$\sin \frac{\pi}{6} = \frac{y}{r} = \frac{1}{2}.$$

31. When an angle of $\frac{5\pi}{3}$ is drawn in standard position, one choice of a point on its terminal side is $(x, y) = (1, -\sqrt{3})$. Then

$$r = \sqrt{x^2 + y^2} = \sqrt{1 + 3} = 2,$$

so

$$\sec \frac{5\pi}{3} = \frac{r}{x} = \frac{2}{1} = 2.$$

33. $\tan 115° \approx -2.1445$

35. $\sin 2.3581 \approx 0.7058$

37. The graph of $y = \cos x$ appears in Figure 17 in Section 1 of this chapter. To get $y = 4\cos x$, each value of y in $y = \cos x$ must be multiplied by 4. This gives a graph going through $(0, 4)$, $(\pi, -4)$ and $(2\pi, 4)$.

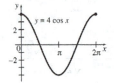

39. The graph of $y = \tan x$ appears in Figure 18 in Section 1 in this chapter. The difference between the graph of $y = \tan x$ and $y = -\tan x$ is that the y-values of points on the graph of $y = -\tan x$ are the opposites of the y-values of the corresponding points on the graph of $y = \tan x$.

A sample calculation:

When $x = \frac{\pi}{4}$,

$$y = -\tan \frac{\pi}{4} = -1.$$

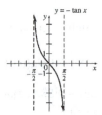

41. Because the derivative of $y = \sin x$ is $y' = \cos x$, the slope of $y = \sin x$ varies from -1 to 1.

43. $y = 2\tan 5x$

$$\frac{dy}{dx} = 2\sec^2 5x \cdot D_x(5x)$$

$$= 10\sec^2 5x$$

45. $y = \cot(6 - 3x^2)$

$$\frac{dy}{dx} = [-\csc^2(6 - 3x^2)] \cdot D_x(6 - 3x^2)$$

$$= 6x\csc^2(6 - 3x^2)$$

47. $y = 2\sin^4(4x^2)$

$$\frac{dy}{dx} = [8\sin^3(4x^2)] \cdot D_x[\sin(4x^2)]$$

$$= 8\sin^3(4x^2) \cdot \cos(4x^2) \cdot D_x(4x^2)$$

$$= 64x\sin^3(4x^2)\cos(4x^2)$$

49. $y = \cos(1 + x^2)$

$$\frac{dy}{dx} = [-\sin(1 + x^2)] \cdot D_x(1 + x^2)$$

$$= -2x\sin(1 + x^2)$$

51. $y = e^{-2x} \sin x$

$$\frac{dy}{dx} = e^{-2x} \cdot D_x(\sin x) + \sin x \cdot D_x(e^{-2x})$$

$$= e^{-2x}(\cos x)$$

$$+ (\sin x)(e^{-2x}) \cdot D_x(-2x)$$

$$= e^{-2x}(\cos x) + (\sin x)(e^{-2x})(-2)$$

$$= e^{-2x}(\cos x - 2 \sin x)$$

53.

$$y = \frac{\cos^2 x}{1 - \cos x}$$

$$\frac{dy}{dx} = \frac{(1 - \cos x)(-2 \cos x \sin x) - (\cos^2 x)(\sin x)}{(1 - \cos x)^2}$$

$$= \frac{-2 \cos x \sin x + \cos^2 x \sin x}{(1 - \cos x)^2}$$

55. $y = \dfrac{\tan x}{1 + x}$

$$\frac{dy}{dx} = \frac{(1 + x)(\sec^2 x) - (\tan x)(1)}{(1 + x)^2}$$

$$= \frac{\sec^2 x + x \sec^2 x - \tan x}{(1 + x)^2}$$

57. $y = \ln|5 \sin x|$

$$\frac{dy}{dx} = \frac{1}{5 \sin x} \cdot D_x(5 \sin x)$$

$$= \frac{\cos x}{\sin x}$$

   or $\cot x$

59. $f(x) = \tan(2x)$ is decreasing on its whole domain, so there are no intervals where it is increasing. It is decreasing wherever defined, that is, on intervals of the form

$$\left(-\frac{\pi}{4} + \frac{n\pi}{2}, \frac{\pi}{4} + \frac{n\pi}{2}\right) \text{ for integer } n.$$

61.   The relative maxima of $f(x) = 2\cos(\pi x)$ have value 2 and occur when $\pi x = 2n\pi$, that is, when $x = 2n$ for integer n. The relative minima have value -2 and occur when $\pi x = (2n + 1)\pi$, that is, when $x = 2n + 1$ for integer n.

63. $f(x) = \tan 7x$

$$f'(x) = 7 \sec^2 7x$$

$$f''(x) = (7)(14)(\sec 7x)(\sec 7x \tan 7x)$$

$$= 98 \sec^2 7x \tan 7x$$

$$f''(1) = 98 \sec^2 7 \tan 7$$

$$f''(-3) = 98 \sec^2(-21) \tan(-21)$$

$$= -98 \sec^2 21 \tan 21$$

65. $f(x) = x - \sin x$

$$f'(x) = 1 - \cos x$$

$$f''(x) = \sin x$$

Since $f'(x)$ is positive except at the isolated points $x = 2n\pi$ for integer n, the function f is increasing everywhere. The graph has a single intercept at $(0, 0)$, which is both an x-intercept and a y-intercept. The concavity is given by the sign of $\sin x$; the graph will be concave upward on the intervals $\left(2n\pi, (2n + 1)\pi\right)$ for integer n and concave downward on the intervals $\left((2n + 1)\pi, (2n + 2)\pi\right)$ for integer n.

Inflection points occur at the x-values separating regions of opposite concavity, that is, where $\sin x = 0$, for example at $x = -\pi, x = 0$, $x = \pi$, and $x = 2\pi$. The y-values at these inflection points are equal to the corresponding x-values.

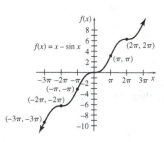

67. $f(x) = \cos x + x$ on $[0, \pi]$

$$f'(x) = -\sin x + 1$$

The only value in $[0, \pi]$ for which $f'(x) = 0$ is $x = \pi/2$, so this is the only critical number, and extrema could only be located at 0, $\pi/2$, and π.

$$f(0) = 1$$

$$f(\pi/2) = \pi/2$$

$$f(\pi) = \pi - 1$$

Thus there is an absolute maximum of $\pi - 1$ at $x = \pi$ and an absolute minimum of 1 at $x = 0$.

69. $x + \cos(x + y) = y^2$

Differentiate with respect to x.

$$1 - \sin(x + y)\left(1 + \frac{dy}{dx}\right) = 2y\frac{dy}{dx}$$

$$1 - \sin(x + y) = \left(2y + \sin(x + y)\right)\frac{dy}{dx}$$

$$\frac{dy}{dx} = \frac{1 - \sin(x + y)}{2y + \sin(x + y)}$$

71. $y = \sin x$

Differentiate with respect to t.

$$\frac{dy}{dt} = (\cos x)\left(\frac{dx}{dt}\right)$$

Substitute the given values.

$$\frac{dy}{dt} = \left(\cos\left(\frac{\pi}{3}\right)\right)(-1) = -\frac{1}{2}$$

73. $\displaystyle\int \cos 5x\, dx$

Let $u = 5x$, so $du = 5dx$ or $\frac{1}{5}du = dx$.

$$\int \cos 5x\, dx = \int \cos u \cdot \frac{1}{5}\, du$$

$$= \frac{1}{5}\int \cos u\, du$$

$$= \frac{1}{5}\sin u + C$$

$$= \frac{1}{5}\sin 5x + C$$

75. $\displaystyle\int \sec^2 5x\, dx$

Let $u = 5x$, so that $du = 5\, dx$.

$$\int \sec^2 5x\, dx = \frac{1}{5}\int (\sec^2 5x)(5\, dx)$$

$$= \frac{1}{5}\int \sec^2 u\, du$$

$$= \frac{1}{5}\tan u + C$$

$$= \frac{1}{5}\tan 5x + C$$

77. $\displaystyle\int 4\csc^2 x\, dx = -4\int -\csc^2 x\, dx$

$$= -4\cot x + C$$

79. $\displaystyle\int 5x \sec 2x^2 \tan 2x^2\, dx$

Let $u = 2x^2$, so that $du = 4x\, dx$.

$$\int 5x \sec 2x^2 \tan 2x^2\, dx$$

$$= \frac{5}{4}\int \sec 2x^2 \tan 2x^2 (4x\, dx)$$

$$= \frac{5}{4}\int \sec u \tan u\, du$$

$$= \frac{5}{4}\sec u + C$$

$$= \frac{5}{4}\sec 2x^2 + C$$

81. $\displaystyle\int \cos^8 x \sin x\, dx$

Let $u = \cos x$, so $du = -\sin x\, dx$.

$$\int \cos^8 x \sin x\, dx = -\int u^8\, du$$

$$= -\frac{1}{9}u^9 + C$$

$$= -\frac{1}{9}\cos^9 x + C$$

83. $\displaystyle\int x^2 \cot 8x^3\, dx$

Let $u = 8x^3$, so that $du = 24x^3\, dx$.

$$\int x^2 \cot 8x^3\, dx$$

$$= \frac{1}{24}\int (\cot 8x^3)(24x^2)\, dx$$

$$= \frac{1}{24}\int \cot u\, du$$

$$= \frac{1}{24}\ln|\sin u| + C$$

$$= \frac{1}{24}\ln|\sin 8x^3| + C$$

85. $\displaystyle\int (\cos x)^{-4/3} \sin x\, dx$

Let, $u = \cos x$ so that $du = -\sin x\, dx$.

$$\int (\cos x)^{-4/3} \sin x \, dx$$

$$= -\int (\cos x)^{-4/3} (\sin x) \, dx$$

$$= -\int u^{-4/3} \, du$$

$$= 3u^{-1/3} + C$$

$$= 3(\cos x)^{-1/3} + C$$

87. $\displaystyle\int_0^{\pi/2} \cos x \, dx = \sin x \Big|_0^{\pi/2}$

$$= \sin \frac{\pi}{2} - \sin 0$$

$$= 1 - 0 = 1$$

89. $\displaystyle\int_0^{2\pi} (10 + 10 \cos x) \, dx$

$$= \int_0^{2\pi} 10 \, dx + \int_0^{2\pi} 10 \cos x \, dx$$

$$= 10(2\pi) + 10 \int_0^{2\pi} \cos x \, dx$$

$$= 20\pi + 10(\sin x) \Big|_0^{2\pi}$$

$$= 20\pi + 10(\sin 2\pi - \sin 0)$$

$$= 20\pi + 10(0 - 0)$$

$$= 20\pi$$

91. Let $dv = \sin x \, dx$ and $u = x + 2$.
Then $v = -\cos x$ and $du = dx$.

$$\int (x + 2) \sin x \, dx = \int u \, dv$$

$$= uv - \int v \, du$$

$$= -(x + 2) \cos x + \int \cos x \, dx$$

$$= -(x + 2) \cos x + \sin x + C$$

93. (a) Residential usage of natural gas is probably periodic.

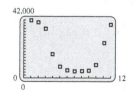

(b) $C(t) = 22,288 \sin(0.45733t + 1.3713)$
$+ 22,299.$

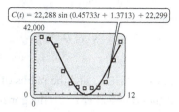

(c) Use a calculator to take the numerical derivative of C at $t = 11$.
$C'(11) \approx 10,121.$ In November the consumption is increasing at 10,121 million cubic feet per month.

(d) Integrating $C(t)$ from 0 to 12 using a calculator gives the estimate 236,377 million cubic feet; the sum of the values in the table is 232,127 million cubic feet.

(e) The function in the model has a period of

$$T = \frac{2\pi}{b}$$

$$= \frac{2\pi}{0.45733}$$

$$\approx 13.7 \text{ months.}$$

95. (a) The sketch shows θ in standard position.

From the diagram, we see that

$$\sin \theta = \frac{s}{L_2}.$$

(b) $\sin \theta = \dfrac{s}{L_2}$

$$L_2 \sin \theta = s$$

$$L_2 = \frac{s}{\sin \theta}$$

(c) Using the sketch and the definition of cotangent, we see that

$$\cot \theta = \frac{L_0 - L_1}{s}.$$

(d)
$$\cot \theta = \frac{L_0 - L_1}{s}$$
$$s \cot \theta = L_0 - L_1$$
$$L_1 + s \cot \theta = L_0$$
$$L_1 = L_0 - s \cot \theta$$

(e) The length of AD is L_1, and the radius of that section of blood vessel is r_1, so the general equation

$$k = \frac{L}{r^4}$$

is similar to

$$R_1 = k \frac{L_1}{r_1^{\,4}}$$

for that particular segment of the blood vessel.

(f) $R_2 = k \cdot \dfrac{L_2}{r_2^{\,4}}$ where R_2 is the resistance along DC.

(g) $R = R_1 + R_2$
$$= k \frac{L_1}{r_1^{\,4}} + k \frac{L_2}{r_2^{\,4}}$$
$$= k \left(\frac{L_1}{r_1^{\,4}} + \frac{L_2}{r_2^{\,4}} \right)$$

(h) $R = k \left(\dfrac{L_1}{r_1^{\,4}} + \dfrac{L_2}{r_2^{\,4}} \right)$
$$= k \left(\frac{L_0 - s \cot \theta}{r_1^{\,4}} + \frac{s}{(\sin \theta) r_2^{\,4}} \right)$$
$$= \frac{k(L_0 - s \cot \theta)}{r_1^{\,4}} + \frac{ks}{r_2^{\,4} \sin \theta}$$

(i)

$R' = D_\theta R$
$$= D_\theta \left(k \frac{L_0}{r_1^{\,4}} - \frac{sk}{r_1^{\,4}} \cot \theta + \frac{s}{r_2^{\,4}} \cdot \frac{k}{\sin \theta} \right)$$
$$= k D_\theta \left(\frac{L_0}{r_1^{\,4}} - \frac{s}{r_1^{\,4}} \cot \theta + \frac{s}{r_2^{\,4}} \cdot \frac{1}{\sin \theta} \right)$$
$$= k \left[D_\theta \left(\frac{L_0}{r_1^{\,4}} \right) - D_\theta \left(\frac{s}{r_1^{\,4}} \cot \theta \right) \right.$$
$$\left. + D_\theta \left(\frac{s}{r_2^{\,4}} \cdot \frac{1}{\sin \theta} \right) \right]$$
$$= k \left[0 - \frac{s}{r_1^{\,4}} D_\theta (\cot \theta) + \frac{s}{r_2^{\,4}} D_\theta \left(\frac{1}{\sin \theta} \right) \right]$$
$$= k \left[-\frac{s}{r_1^{\,4}} (-\csc^2 \theta) + \frac{s}{r_2^{\,4}} \left(\frac{-\cos \theta}{\sin^2 \theta} \right) \right]$$
$$= k \left(\frac{s}{r_1^{\,4}} \frac{1}{\sin^2 \theta} - \frac{s}{r_2^{\,4}} \cdot \frac{\cos \theta}{\sin^2 \theta} \right)$$
$$= \frac{ks}{\sin^2 \theta} \left(\frac{1}{r_1^{\,4}} - \frac{\cos \theta}{r_2^{\,4}} \right)$$
$$= \frac{ks \csc^2 \theta}{r_1^{\,4}} - \frac{ks \cos \theta}{r_2^{\,4} \sin^2 \theta}$$

(j) Using part (h) gives
$$\frac{ks \csc^2 \theta}{r_1^{\,4}} - \frac{ks \cos \theta}{r_2^{\,4} \sin^2 \theta} = 0.$$

(k) If the left side of the equation in the solution to part (h) is multiplied by $\frac{\sin^2 \theta}{s}$, we get

$$\frac{\sin^2 \theta}{s} \cdot \frac{ks}{\sin^2 \theta} \left(\frac{1}{r_1^{\,4}} - \frac{\cos \theta}{r_2^{\,4}} \right)$$
$$= k \left(\frac{1}{r_1^{\,4}} - \frac{\cos \theta}{r_2^{\,4}} \right)$$

This gives the equation

$$\frac{k}{r_1^{\,4}} - \frac{k \cos \theta}{r_2^{\,4}} = 0.$$

(l) Using part (k) gives

$$k\left(\frac{1}{r_1^{\,4}} - \frac{\cos\theta}{r_2^{\,4}}\right) = 0$$

$$\frac{1}{r_1^{\,4}} - \frac{\cos\theta}{r_2^{\,4}} = 0 \quad k \neq 0$$

$$\frac{1}{r_1^{\,4}} = \frac{\cos\theta}{r_2^{\,4}}$$

$$\cos\theta = \frac{r_2^{\,4}}{r_1^{\,4}}$$

(n) If $r_1 = 1$ and $r_2 = \frac{1}{4}$, then

$$\cos\theta = \left(\frac{\frac{1}{4}}{1}\right)^4 = \left(\frac{1}{4}\right)^4$$

$$= \frac{1}{256} \approx 0.0039,$$

from which we get

$$\theta \approx 90°.$$

(o) If $r_1 = 1.4$ and $r_2 = 0.8$, then

$$\cos\theta = \frac{(0.8)^4}{(1.4)^4} \approx 0.1066.$$

Thus,

$$\theta \approx 84°.$$

97. **(a)** $y = x\tan\alpha - \dfrac{16x^2}{V^2}\sec^2\alpha + h$

$$= 39\tan\frac{\pi}{24} - \frac{16(39)^2}{73^2}\sec^2\frac{\pi}{24} + 9$$

$$\approx 9.5$$

Yes, the ball will make it over the net since the height of the ball is about 9.5 feet when x is 39 feet.

(b) Entering

$$Y_1 = 39\tan x - \frac{16(39)^2}{44^2}\sec^2 x + 9 \text{ and}$$

$$Y_2 = \frac{\left[\begin{array}{l}44^2\sin x\cos x \\[4pt] + 44^2\cos^2 x\sqrt{\tan^2 x + \frac{576}{44^2}\sec^2 x}\end{array}\right]}{32}$$

into the graphing calculator and using the table function, indicates that the tennis ball will clear the net and travel between 39 and 60 feet for $0.18 \le x \le 0.41$ or $0.18 \le \alpha \le 0.41$ in radians. In degrees,

$$10.3 \le \alpha \le 23.5.$$

X	Y1	Y2
.17	2.754	44.141
.18	3.1103	44.827
.22	4.5223	47.571
.34	8.6526	55.486
.38	10.002	57.9
.41	11.006	59.602
	11.339	60.146

X=.42

(c) Using Y_2 from part (b) and the graphing calculator, we get

nDeriv(Y2,X,π/8)

57.01184054

Note that $57.01\,\frac{\text{feet}}{\text{radian}} \approx 0.995\,\frac{\text{feet}}{\text{degree}}$.

The distance the tennis ball travels will increase by approximately 1 foot by increasing the angle of the tennis racket by one degree.

99. $s(t) = A\cos(Bt + C)$

$$s'(t) = -A\sin(Bt + C) \cdot D_t(Bt + C)$$

$$= -A\sin(Bt + C) \cdot B$$

$$= -AB\sin(Bt + C)$$

$$s''(t) = -AB\cos(Bt + C) \cdot D_t(Bt + C)$$

$$= -AB\cos(Bt + C) \cdot B$$

$$= -B^2 A\cos(Bt + C)$$

$$= -B^2 s(t)$$

101. **(a)** $\dfrac{d}{dx}\ln|\sec x + \tan x|$

$$= \frac{\sec x\tan x + \sec^2 x}{\sec x + \tan x}$$

$$= \frac{\sec x(\tan x + \sec x)}{(\sec x + \tan x)}$$

$$= \sec x$$

(b) $\dfrac{d}{dx}(-\ln|\sec x - \tan x|)$

$$= -\dfrac{\sec x \tan x - \sec^2 x}{\sec x - \tan x}$$

$$= -\dfrac{\sec x(\tan x - \sec x)}{(\sec x - \tan x)}$$

$$= -(-\sec x)$$

$$= \sec x$$

(d) $D(\theta) = k \displaystyle\int_0^\theta \sec x \, dx$

$$= k \ln|\sec x + \tan x|\Big|_0^\theta$$

$$= k \ln|\sec \theta + \tan \theta|$$
$$\qquad - k \ln|\sec 0 + \tan 0|$$
$$= k \ln|\sec \theta + \tan \theta|$$
$$\qquad - k \ln|1 + 0|$$
$$= k \ln|\sec \theta + \tan \theta| - k \ln 1$$
$$= k \ln|\sec \theta + \tan \theta| - 0$$
$$= k \ln|\sec \theta + \tan \theta|$$

Since $D(34°03') = 7$,

$$7 = k \ln|\sec 34°03' + \tan 34°03'|$$

$$k = \dfrac{7}{\ln|\sec 34°03' + \tan 34°03'|}$$

$$\approx 11.0635.$$

$D(40°45') \approx 11.0635 \ln\left|\sec 40°45' + \tan 40°45'\right|$

$$\approx 8.63$$

New York City should be placed approximately 8.63 inches from the equator.

(e) $D(25°46')$

$$\approx 11.0635 \ln|\sec 25°46' + \tan 25°46'|$$

$$\approx 5.15$$

Miami should be placed approximately 5.15 inches from the equator.